Lecture Notes in Computer Scier

T0238687

Commenced Publication in 1973
Founding and Former Series Editors:
Gerhard Goos, Juris Hartmanis, and Jan van Leeuwen

Beniamino Murgante   Sanjay Misra
Maurizio Carlini   Carmelo M. Torre
Hong-Quang Nguyen   David Taniar
Bernady O. Apduhan   Osvaldo Gervasi (Eds.)

Computational Science and Its Applications – ICCSA 2013

13th International Conference
Ho Chi Minh City, Vietnam, June 24-27, 2013
Proceedings, Part I

 Springer

Volume Editors

Beniamino Murgante, Università degli Studi della Basilicata, Potenza, Italy
E-mail: beniamino.murgante@unibas.it

Sanjay Misra, Covenant University, Canaanland OTA, Nigeria
E-mail: sanjay.misra@covenantuniversity.edu.ng

Maurizio Carlini, Università degli Studi della Tuscia, Viterbo, Italy
E-mail: maurizio.carlini@unitus.it

Carmelo M. Torre, Politecnico di Bari, Italy
E-mail: torre@poliba.it

Hong-Quang Nguyen, Int. University VNU-HCM, Ho Chi Minh City, Vietnam
E-mail: htphong@hcmiu.edu.vn

David Taniar, Monash University, Clayton, VIC, Australia
E-mail: david.taniar@infotech.monash.edu.au

Bernady O. Apduhan, Kyushu Sangyo University, Fukuoka, Japan
E-mail: bob@is.kyusan-u.ac.jp

Osvaldo Gervasi, University of Perugia, Italy
E-mail: osvaldo@unipg.it

ISSN 0302-9743   e-ISSN 1611-3349
ISBN 978-3-642-39636-6   e-ISBN 978-3-642-39637-3
DOI 10.1007/978-3-642-39637-3
Springer Heidelberg Dordrecht London New York

Library of Congress Control Number: 2013942720

CR Subject Classification (1998): C.2.4, C.2, H.4, F.2, H.3, D.2, F.1, H.5, H.2.8, K.6.5, I.3

LNCS Sublibrary: SL 1 – Theoretical Computer Science and General Issues

Typesetting: Camera-ready by author, data conversion by Scientific Publishing Services, Chennai, India

Printed on acid-free paper

Springer is part of Springer Science+Business Media (www.springer.com)

Preface

These multiple volumes (LNCS volumes 7971, 7972, 7973, 7974, and 7975) consist of the peer-reviewed papers from the 2013 International Conference on Computational Science and Its Applications (ICCSA2013) held in Ho Chi Minh City, Vietnam, during June 24–27, 2013.

ICCSA 2013 was a successful event in the International Conferences on Computational Science and Its Applications (ICCSA) conference series, previously held in Salvador, Brazil (2012), Santander, Spain (2011), Fukuoka, Japan (2010), Suwon, South Korea (2009), Perugia, Italy (2008), Kuala Lumpur, Malaysia (2007), Glasgow, UK (2006), Singapore (2005), Assisi, Italy (2004), Montreal, Canada (2003), (as ICCS) Amsterdam, The Netherlands (2002), and San Francisco, USA (2001).

Computational science is a main pillar of most of the present research, industrial, and commercial activities and plays a unique role in exploiting ICT innovative technologies; the ICCSA conference series have been providing a venue to researchers and industry practitioners to discuss new ideas, to share complex problems and their solutions, and to shape new trends in computational science.

Apart from the general track, ICCSA 2013 also included 33 special sessions and workshops, in various areas of computational sciences, ranging from computational science technologies, to specific areas of computational sciences, such as computer graphics and virtual reality. We accepted 46 papers for the general track, and 202 in special sessions and workshops, with an acceptance rate of 29.8%. We would like to express our appreciation to the Workshops and Special Sessions Chairs and Co-chairs.

The success of the ICCSA conference series, in general, and ICCSA 2013, in particular, is due to the support of many people: authors, presenters, participants, keynote speakers, Workshop Chairs, Organizing Committee members, student volunteers, Program Committee members, International Liaison Chairs, and people in other various roles. We would like to thank them all. We would also like to thank Springer for their continuous support in publishing ICCSA conference proceedings.

May 2013

David Taniar
Beniamino Murgante
Hong-Quang Nguyen

Message from the General Chairs

On behalf of the ICCSA Organizing Committee it is our great pleasure to welcome you to the proceedings of the 13th International Conference on Computational Science and Its Applications (ICCSA 2013), held June 24–27, 2013, in Ho Chi Minh City, Vietnam.

ICCSA is one of the most successful international conferences in the field of computational sciences, and ICCSA 2013 was the 13th conference of this series previously held in Salvador da Bahia, Brazil (2012), in Santander, Spain (2011), Fukuoka, Japan (2010), Suwon, Korea (2009), Perugia, Italy (2008), Kuala Lumpur, Malaysia (2007), Glasgow, UK (2006), Singapore (2005), Assisi, Italy (2004), Montreal, Canada (2003), (as ICCS) Amsterdam, The Netherlands (2002), and San Francisco, USA (2001).

The computational science community has enthusiastically embraced the successive editions of ICCSA, thus contributing to making ICCSA a focal meeting point for those interested in innovative, cutting-edge research about the latest and most exciting developments in the field. It provides a major forum for researchers and scientists from academia, industry and government to share their views on many challenging research problems, and to present and discuss their novel ideas, research results, new applications and experience on all aspects of computational science and its applications. We are grateful to all those who have contributed to the ICCSA conference series.

For the successful organization of ICCSA 2013, an international conference of this size and diversity, we counted on the great support of many people and organizations.

We would like to thank all the workshop organizers for their diligent work, which further enhanced the conference level and all reviewers for their expertise and generous effort, which led to a very high quality event with excellent papers and presentations.

We especially recognize the contribution of the Program Committee and local Organizing Committee members for their tremendous support, the faculty members of the School of Computer Science and Engineering and authorities of the International University (HCM-VNU), Vietnam, for allowing us to use the venue and facilities to realize this highly successful event. Further, we would like to express our gratitude to the Office of the Naval Research, US Navy, and other institutions/organizations that supported our efforts to bring the conference to fruition.

We would like to sincerely thank our keynote speakers who willingly accepted our invitation and shared their expertise.

We also thank our publisher, Springer-Verlag, for accepting to publish the proceedings and for their kind assistance and cooperation during the editing process.

Finally, we thank all authors for their submissions and all conference attendees for making ICCSA 2013 truly an excellent forum on computational science, facilitating an exchange of ideas, fostering new collaborations and shaping the future of this exciting field.

We thank you all for participating in ICCSA 2013, and hope that you find the proceedings stimulating and interesting for your research and professional activities.

Osvaldo Gervasi

Bernady O. Apduhan

Duc Cuong Nguyen

Organization

ICCSA 2013 was organized by The Ho Chi Minh City International University (Vietnam), University of Perugia (Italy), University of Basilicata (Italy), Monash University (Australia), and Kyushu Sangyo University (Japan).

Honorary General Chairs

Phong Thanh Ho	International University (VNU-HCM), Vietnam
Antonio Laganà	University of Perugia, Italy
Norio Shiratori	Tohoku University, Japan
Kenneth C.J. Tan	Qontix, UK

General Chairs

Osvaldo Gervasi	University of Perugia, Italy
Bernady O. Apduhan	Kyushu Sangyo University, Japan
Duc Cuong Nguyen	International University (VNU-HCM), Vietnam

Program Committee Chairs

David Taniar	Monash University, Australia
Beniamino Murgante	University of Basilicata, Italy
Hong-Quang Nguyen	International University (VNU-HCM), Vietnam

Workshop and Session Organizing Chair

Beniamino Murgante	University of Basilicata, Italy

Local Organizing Committee

Hong Quang Nguyen	International University (VNU-HCM), Vietnam (Chair)
Bao Ngoc Phan	International University (VNU-HCM), Vietnam

| Van Hoang | International University (VNU-HCM), Vietnam |
| Ly Le | International University (VNU-HCM), Vietnam |

International Liaison Chairs

Jemal Abawajy	Deakin University, Australia
Ana Carla P. Bitencourt	Universidade Federal do Reconcavo da Bahia, Brazil
Claudia Bauzer Medeiros	University of Campinas, Brazil
Alfredo Cuzzocrea	ICAR-CNR and University of Calabria, Italy
Marina L. Gavrilova	University of Calgary, Canada
Robert C.H. Hsu	Chung Hua University, Taiwan
Andrés Iglesias	University of Cantabria, Spain
Tai-Hoon Kim	Hannam University, Korea
Sanjay Misra	University of Minna, Nigeria
Takashi Naka	Kyushu Sangyo University, Japan
Ana Maria A.C. Rocha	University of Minho, Portugal
Rafael D.C. Santos	National Institute for Space Research, Brazil

Workshop Organizers

Advances in Web-Based Learning (AWBL 2013)

| Mustafa Murat Inceoglu | Ege University, Turkey |

Big Data: Management, Analysis, and Applications (Big-Data 2013)

| Wenny Rahayu | La Trobe University, Australia |

Bio-inspired Computing and Applications (BIOCA 2013)

| Nadia Nedjah | State University of Rio de Janeiro, Brazil |
| Luiza de Macedo Mourell | State University of Rio de Janeiro, Brazil |

Computational and Applied Mathematics (CAM 2013)

| Ana Maria Rocha | University of Minho, Portugal |
| Maria Irene Falcao | University of Minho, Portugal |

Computer-Aided Modeling, Simulation, and Analysis (CAMSA 2013)

Jie Shen	University of Michigan, USA
Yanhui Wang	Beijing Jiaotong University, China
Hao Chen	Shanghai University of Engineering Science, China

Computer Algebra Systems and Their Applications (CASA 2013)

Andres Iglesias   University of Cantabria, Spain
Akemi Galvez   University of Cantabria, Spain

Computational Geometry and Applications (CGA 2013)

Marina L. Gavrilova   University of Calgary, Canada
Han Ming Huang   Guangxi Normal University, China

Chemistry and Materials Sciences and Technologies (CMST 2013)

Antonio Laganà   University of Perugia, Italy

Cities, Technologies and Planning (CTP 2013)

Giuseppe Borruso   University of Trieste, Italy
Beniamino Murgante   University of Basilicata, Italy

Computational Tools and Techniques for Citizen Science and Scientific Outreach (CTTCS 2013)

Rafael Santos   National Institute for Space Research, Brazil
Jordan Raddickand   Johns Hopkins University, USA
Ani Thakar   Johns Hopkins University, USA

Econometrics and Multidimensional Evaluation in the Urban Environment (EMEUE 2013)

Carmelo M. Torre   Polytechnic of Bari, Italy
Maria Cerreta   Università Federico II of Naples, Italy
Paola Perchinunno   University of Bari, Italy

Energy and Environment - Scientific, Engineering and Computational Aspects of Renewable Energy Sources, Energy Saving and Recycling of Waste Materials (ENEENV 2013)

Maurizio Carlini   University of Viterbo, Italy
Carlo Cattani   University of Salerno, Italy

Future Computing Systems, Technologies, and Applications (FISTA 2013)

Bernady O. Apduhan   Kyushu Sangyo University, Japan
Rafael Santos   National Institute for Space Research, Brazil
Jianhua Ma   Hosei University, Japan
Qun Jin   Waseda University, Japan

Geographical Analysis, Urban Modeling, Spatial Statistics (GEOG-AN-MOD 2013)

Giuseppe Borruso   University of Trieste, Italy
Beniamino Murgante   University of Basilicata, Italy
Hartmut Asche   University of Potsdam, Germany

International Workshop on Biomathematics, Bioinformatics and Biostatistics (IBBB 2013)

Unal Ufuktepe   Izmir University of Economics, Turkey
Andres Iglesias   University of Cantabria, Spain

International Workshop on Agricultural and Environmental Information and Decision Support Systems (IAEIDSS 2013)

Sandro Bimonte   IRSTEA, France
Andr Miralles   IRSTEA, France
Franois Pinet   IRSTEA, France
Frederic Flouvat   University of New Caledonia, New Caledonia

International Workshop on Collective Evolutionary Systems (IWCES 2013)

Alfredo Milani   University of Perugia, Italy
Clement Leung   Hong Kong Baptist University, Hong Kong

Mobile Communications (MC 2013)

Hyunseung Choo   Sungkyunkwan University, Korea

Mobile Computing, Sensing, and Actuation for Cyber Physical Systems (MSA4CPS 2013)

Moonseong Kim   Korean Intellectual Property Office, Korea
Saad Qaisar   NUST School of Electrical Engineering and
                          Computer Science, Pakistan

Mining Social Media (MSM 2013)

Robert M. Patton   Oak Ridge National Laboratory, USA
Chad A. Steed   Oak Ridge National Laboratory, USA
David R. Resseguie   Oak Ridge National Laboratory, USA
Robert M. Patton   Oak Ridge National Laboratory, USA

Parallel and Mobile Computing in Future Networks (PMCFUN 2013)

Al-Sakib Khan Pathan   International Islamic University Malaysia,
                                  Malaysia

Quantum Mechanics: Computational Strategies and Applications (QMCSA 2013)

Mirco Ragni   Universidad Federal de Bahia, Brazil
Vincenzo Aquilanti   University of Perugia, Italy
Ana Carla Peixoto Bitencourt   Universidade Federal do Reconcavo da Bahia,
                                  Brazil
Roger Anderson   University of California, USA
Frederico Vasconcellos
      Prudente   Universidad Federal de Bahia, Brazil

Remote Sensing Data Analysis, Modeling, Interpretation and Applications: From a Global View to a Local Analysis (RS 2013)

Rosa Lasaponara   Institute of Methodologies for Environmental
                                  Analysis - National Research Council, Italy
Nicola Masini   Archaeological and Monumental Heritage
                                  Institute - National Research Council, Italy

Soft Computing for Knowledge Discovery in Databases (SCKDD 2013)

Tutut Herawan   Universitas Ahmad Dahlan, Indonesia

Software Engineering Processes and Applications (SEPA 2013)

Sanjay Misra   Covenant University, Nigeria

Spatial Data Structures and Algorithms for Geoinformatics (SDSAG 2013)

Farid Karimipour   University of Tehran, Iran and
                                  Vienna University of Technology, Austria

Software Quality (SQ 2013)

Sanjay Misra   Covenant University, Nigeria

Security and Privacy in Computational Sciences (SPCS 2013)

Arijit Ukil   Tata Consultancy Services, India

Technical Session on Computer Graphics and Geometric Modeling (TSCG 2013)

Andres Iglesias   University of Cantabria, Spain

Tools and Techniques in Software Development Processes (TTSDP 2013)

Sanjay Misra   Covenant University, Nigeria

Virtual Reality and Its Applications (VRA 2013)

Osvaldo Gervasi   University of Perugia, Italy
Lucio Depaolis   University of Salento, Italy

Wireless and Ad-Hoc Networking (WADNet 2013)

Jongchan Lee   Kunsan National University, Korea
Sangjoon Park   Kunsan National University, Korea

Warehousing and OLAPing Complex, Spatial and Spatio-Temporal Data (WOCD 2013)

Alfredo Cuzzocrea   Istituto di Calcolo e Reti ad Alte Prestazioni -
                                  National Research Council, Italy and
                                  University of Calabria, Italy

Program Committee

Jemal Abawajy   Deakin University, Australia
Kenny Adamson   University of Ulster, UK
Filipe Alvelos   University of Minho, Portugal
Hartmut Asche   University of Potsdam, Germany
Md. Abul Kalam Azad   University of Minho, Portugal
Assis Azevedo   University of Minho, Portugal
Michela Bertolotto   University College Dublin, Ireland
Sandro Bimonte   CEMAGREF, TSCF, France
Rod Blais   University of Calgary, Canada
Ivan Blecic   University of Sassari, Italy
Giuseppe Borruso   University of Trieste, Italy
Yves Caniou   Lyon University, France
José A. Cardoso e Cunha   Universidade Nova de Lisboa, Portugal
Carlo Cattani   University of Salerno, Italy
Mete Celik   Erciyes University, Turkey
Alexander Chemeris   National Technical University of Ukraine
                                  "KPI", Ukraine
Min Young Chung   Sungkyunkwan University, Korea
Gilberto Corso Pereira   Federal University of Bahia, Brazil
M. Fernanda Costa   University of Minho, Portugal

Frank Devai	London South Bank University, UK
Rodolphe Devillers	Memorial University of Newfoundland, Canada
Prabu Dorairaj	NetApp, India/USA
M. Irene Falcao	University of Minho, Portugal
Cherry Liu Fang	U.S. DOE Ames Laboratory, USA
Edite M.G.P. Fernandes	University of Minho, Portugal
Jose-Jesus Fernandez	National Centre for Biotechnology, CSIS, Spain
Rosário Fernandes	University of Minho, Portugal
Maria Celia Furtado Rocha	PRODEBPósCultura/UFBA, Brazil
Akemi Galvez	University of Cantabria, Spain
Marina Gavrilova	University of Calgary, Canada
Jerome Gensel	LSR-IMAG, France
Maria Giaoutzi	National Technical University, Athens, Greece
Alex Hagen-Zanker	University of Cambridge, UK
Malgorzata Hanzl	Technical University of Lodz, Poland
Shanmugasundaram Hariharan	B.S. Abdur Rahman University, India
Fermin Huarte	University of Barcelona, Spain
Andres Iglesias	University of Cantabria, Spain
Farid Karimipour	Vienna University of Technology, Austria
Antonio Laganà	University of Perugia, Italy
Rosa Lasaponara	National Research Council, Italy
Jongchan Lee	Kunsan National University, Korea
Gang Li	Deakin University, Australia
Fang Liu	AMES Laboratories, USA
Xin Liu	University of Calgary, Canada
Savino Longo	University of Bari, Italy
Helmuth Malonek	University of Aveiro, Portugal
Ernesto Marcheggiani	Katholieke Universiteit Leuven, Belgium
Antonino Marvuglia	Research Centre Henri Tudor, Luxembourg
Nicola Masini	National Research Council, Italy
Alfredo Milani	University of Perugia, Italy
Fernando Miranda	University of Minho, Portugal
Sanjay Misra	Federal University of Technology Minna, Nigeria
Giuseppe Modica	University of Reggio Calabria, Italy
José Luis Montaña	University of Cantabria, Spain
Belen Palop	Universidad de Valladolid, Spain
Eric Pardede	La Trobe University, Australia
Kwangjin Park	Wonkwang University, Korea
Ana Isabel Pereira	Polytechnic Institute of Bragança, Portugal
Maurizio Pollino	Italian National Agency for New Technologies, Energy and Sustainable Economic Development, Italy
Alenka Poplin	University of Hamburg, Germany
David C. Prosperi	Florida Atlantic University, USA

Wenny Rahayu	La Trobe University, Australia
Jerzy Respondek	Silesian University of Technology, Poland
Ana Maria A.C. Rocha	University of Minho, Portugal
Humberto Rocha	INESC-Coimbra, Portugal
Alexey Rodionov	Institute of Computational Mathematics and Mathematical Geophysics, Russia
Cristina S. Rodrigues	University of Minho, Portugal
Haiduke Sarafian	The Pennsylvania State University, USA
Ricardo Severino	University of Minho, Portugal
Jie Shen	University of Michigan, USA
Qi Shi	Liverpool John Moores University, UK
Dale Shires	U.S. Army Research Laboratory, USA
Ana Paula Teixeira	University of Tras-os-Montes and Alto Douro, Portugal
Senhorinha Teixeira	University of Minho, Portugal
Graça Tomaz	University of Aveiro, Portugal
Carmelo Torre	Polytechnic of Bari, Italy
Javier Martinez Torres	Centro Universitario de la Defensa Zaragoza, Spain
Giuseppe A. Trunfio	University of Sassari, Italy
Unal Ufuktepe	Izmir University of Economics, Turkey
Mario Valle	Swiss National Supercomputing Centre, Switzerland
Pablo Vanegas	University of Cuenca, Equador
Paulo Vasconcelos	University of Porto, Portugal
Piero Giorgio Verdini	INFN Pisa and CERN, Italy
Marco Vizzari	University of Perugia, Italy
Krzysztof Walkowiak	Wroclaw University of Technology, Poland
Robert Weibel	University of Zurich, Switzerland
Roland Wismüller	Universität Siegen, Germany
Xin-She Yang	National Physical Laboratory, UK
Haifeng Zhao	University of California, Davis, USA
Kewen Zhao	University of Qiongzhou, China

Additional Reviewers

Antonio Aguilar	Universitat de Barcelona, Spain
José Alfonso Aguilar Caldern	Universidad Autnoma de Sinaloa, Mexico
Vladimir Alarcon	Geosystems Research Institute, Mississippi State University, USA
Margarita Alberti	Universitat de Barcelona, Spain
Vincenzo Aquilanti	University of Perugia, Italy
Takefusa Atsuko	National Institute of Advanced Industrial Science and Technology, Japan
Raffaele Attardi	University of Napoli Federico II, Italy

Sansanee Auephanwiriyakul	Chiang Mai University, Thailand
Assis Azevedo	University of Minho, Portugal
Thierry Badard	Université Laval, Canada
Marco Baioletti	University of Perugia, Italy
Daniele Bartoli	University of Perugia, Italy
Paola Belanzoni	University of Perugia, Italy
Massimiliano Bencardino	University of Salerno, Italy
Priyadarshi Bhattacharya	University of Calgari, Canada
Massimo Bilancia	University of Bari, Italy
Gabriele Bitelli	University of Bologna, Italy
Letizia Bollini	University of Milano Bicocca, Italy
Alessandro Bonifazi	University of Bari, Italy
Atila Bostam	Atilim University, Turkey
Maria Bostenaru Dan	University of Bucharest, Romania
Thang H. Bui	Ho Chi Minh City University of Technology, Vietnam
Michele Campagna	University of Cagliari, Italy
Francesco Campobasso	University of Bari, Italy
Maurizio Carlini	University of Tuscia, Italy
Simone Caschili	University College of London, UK
Sonia Castellucci	University of Tuscia, Italy
Filippo Celata	University of Rome La Sapienza, Italy
Claudia Ceppi	Polytechnic of Bari, Italy
Ivan Cernusak	Comenius University of Bratislava, Slovakia
Maria Cerreta	University of Naples Federico II, Italy
Aline Chiabai	Basque Centre for Climate Change, Spain
Andrea Chiancone	University of Perugia, Italy
Eliseo Clementini	University of L'Aquila, Italy
Anibal Zaldivar Colado	Universidad Autonoma de Sinaloa, Mexico
Marco Crasso	Universidad Nacional del Centro de la provincia de Buenos Aires, Argentina
Ezio Crestaz	Saipem, Italy
Maria Danese	IBAM National Research Council, Italy
Olawande Daramola	Covenant University, Nigeria
Marcelo de Alemida Maia	Universidade Federal de Uberlândia, Brazil
Roberto De Lotto	University of Pavia, Italy
Lucio T. De Paolis	University of Salento, Italy
Pasquale De Toro	University of Naples Federico II, Italy
Hendrik Decker	Universidad Politécnica de Valencia, Spain
Margherita Di Leo	Joint Research Centre, Belgium
Andrea Di Carlo	University of Rome La Sapienza, Italy
Arta Dilo	University of Twente, The Netherlands
Alberto Dimeglio	CERN, Switzerland
Young Ik Eom	Sungkyunkwan University, South Korea
Rogelio Estrada	Universidad Autonoma de Sinaloa, Mexico
Stavros C. Farantos	University of Crete, Greece

Chih-Hsiao Tsai	Takming University of Science and Technology, Taiwan
Devis Tuia	Laboratory of Geographic Information Systems, Switzerland
Arijit Ukil	Tata Consultancy Services, India
Paulo Vasconcelos	University of Porto, Portugal
Flavio Vella	University of Perugia, Italy
Mauro Villarini	University of Tuscia, Italy
Christine Voiron-Canicio	Université Nice Sophia Antipolis, France
Kira Vyatkina	Saint Petersburg State University, Russia
Jian-Da Wu	National Changhua University of Education, Taiwan
Toshihiro Yamauchi	Okayama University, Japan
Iwan Tri Riyadi Yanto	Universitas Ahmad Dahlan, Indonesia
Syed Shan-e-Hyder Zaidi	Sungkyunkwan University, South Korea
Vyacheslav Zalyubouskiy	Sungkyunkwan University, South Korea
Alejandro Zunino	National University of the Center of the Buenos Aires Province, Argentina

Sponsoring Organizations

ICCSA 2013 would not have been possible without tremendous support of many organizations and institutions, for which all organizers and participants of ICCSA 2013 express their sincere gratitude:

Ho CHi Minh City International University, Vietnam
(http://www.hcmiu.edu.vn/HomePage.aspx)

University of Perugia, Italy
(http://www.unipg.it)

 Monash University, Australia
(http://monash.edu)

 Kyushu Sangyo University, Japan
(www.kyusan-u.ac.jp)

University of Basilicata, Italy (http://www.unibas.it)

The Office of Naval Research, USA
(http://www.onr.navy.mil/Science-technology/onr-global.aspx)

ICCSA 2013 Invited Speakers

Dharma Agrawal
University of Cincinnati, USA

Manfred M. Fisher
Vienna University of Economics and Business, Austria

Wenny Rahayu
La Trobe University, Australia

Selecting LTE and Wireless Mesh Networks for Indoor/Outdoor Applications

Dharma Agrawal*

School of Computing Sciences and Informatics, University of Cincinnati, USA
dharmaagrawal@gmail.com

Abstract. The smart phone usage and multimedia devices have been increasing yearly and predictions indicate drastic increase in the upcoming years. Recently, various wireless technologies have been introduced to add flexibility to these gadgets. As data plans offered by the network service providers are expensive, users are inclined to utilize freely accessible and commonly available Wi-Fi networks indoors.

LTE (Long Term Evolution) has been a topic of discussion in providing high data rates outdoors and various service providers are planning to roll out LTE networks all over the world. The objective of this presentation is to compare usefulness of these two leading wireless schemes based on LTE and Wireless Mesh Networks (WMN) and bring forward their advantages for indoor and outdoor environments. We also investigate to see if a hybrid LTE-WMN network may be feasible. Both these networks are heterogeneous in nature, employ cognitive approach and support multi hop communication. The main motivation behind this work is to utilize similarities in these networks, explore their capability of offering high data rates and generally have large coverage areas.

In this work, we compare both these networks in terms of their data rates, range, cost, throughput, and power consumption. We also compare 802.11n based WMN with Femto cell in an indoor coverage scenario, while for outdoors; 802.16 based WMN is compared with LTE. The main objective is to help users select a network that could provide enhanced performance in a cost effective manner.

---

* More information can be found at http://www.iccsa.org/invited-speakers

Neoclassical Growth Theory, Regions and Spatial Externalities

Manfred M. Fisher*

Vienna University of Economics and Business, Austria
manfred.fischer@wu.ac.at

Abstract. The presentation considers the standard neoclassical growth model in a Mankiw-Romer-Weil world with externalities across regions.

The reduced form of this theoretical model and its associated empirical model lead to a spatial Durbin model, and this model provides very rich own- and cross-partial derivatives that quantify the magnitude of direct and indirect (spillover or externalities) effects that arise from changes in regions characteristics (human and physical capital investment or population growth rates) at the outset in the theoretical model.

A logical consequence of the simple dependence on a small number of nearby regions in the initial theoretical specification leads to a final-form model outcome where changes in a single region can potentially impact all other regions. This is perhaps surprising, but of course we must temper this result by noting that there is a decay of influence as we move to more distant or less connected regions.

Using the scalar summary impact measures introduced by LeSage and Pace (2009) we can quantify and summarize the complicated set of non-linear impacts that fall on all regions as a result of changes in the physical and human capital in any region. We can decompose these impacts into direct and indirect (or externality) effects. Data for a system of 198 regions across 22 European countries over the period 1995 to 2004 are used to test the predictions of the model and to draw inferences regarding the magnitude of regional output responses to changes in physical and human capital endowments.

The results reveal that technological interdependence among regions works through physical capital externalities crossing regional borders.

---

* More information can be found at http://www.iccsa.org/invited-speakers

Global Spatial-Temporal Data Integration to Support Collaborative Decision Making

Wenny Rahayu*

La Trobe University, Australia
W.Rahayu@latrobe.edu.au

Abstract. There has been a huge effort in the recent years to establish a standard vocabulary and data representation for the areas where a collaborative decision support is required. The development of global standards for data interchange in time critical domains such as air traffic control, transportation systems, and medical informatics, have enabled the general industry in these areas to move into a more data-centric operations and services. The main aim of the standards is to support integration and collaborative decision support systems that are operationally driven by the underlying data.

The problem that impedes rapid and correct decision-making is that information is often segregated in many different formats and domains, and integrating them has been recognised as one of the major problems. For example, in the aviation industry, weather data given to flight en-route has different formats and standards from those of the airport notification messages. The fact that messages are exchanged using different standards has been an inherent problem in data integration in many spatial-temporal domains. The solution is to provide seamless data integration so that a sequence of information can be analysed on the fly.

Our aim is to develop an integration method for data that comes from different domains that operationally need to interact together. We especially focus on those domains that have temporal and spatial characteristics as their main properties. For example, in a flight plan from Melbourne to Ho Chi Minh City which comprises of multiple international airspace segments, a pilot can get an integrated view of the flight route with the weather forecast and airport notifications at each segment. This is only achievable if flight route, airport notifications, and weather forecast at each segment are integrated in a spatial temporal system.

In this talk, our recent efforts in large data integration, filtering, and visualisation will be presented. These integration efforts are often required to support real-time decision making processes in emergency situations, flight delays, and severe weather conditions.

---

* More information can be found at http://www.iccsa.org/invited-speakers

Table of Contents – Part I

Workshop on Mobile Communications (MC 2013)

Workshop on Agricultural and Environmental Information and Decision Support Systems (IAEIDSS 2013)

Workshop on Computational and Applied Mathematics (CAM 2013)

Workshop on Mobile-Computing, Sensing, and Actuation - Cyber Physical Systems (MSA4CPS 2013)

Workshop on Soft Computing for Knowledge Discovery in Databases (SCKDD 2013)

Workshop on Bio-inspired Computing and Applications (BIOCA 2013)

Workshop on Spatial Data Structures and Algorithms for Geoinformatics (SDSAG 2013)

Workshop on Big Data: Management, Analysis, and Applications (BigData 2013) International Workshop on Biomathematics, Bioinformatics and Biostatistics (IBBB 2013)

Technical Session on Computer Graphics (TSCG 2013)

Workshop on Virtual Reality and Applications (VRA 2013)

Water $(H_2O)_m$ or Benzene $(C_6H_6)_n$ Aggregates to Solvate the K^+?

Noelia Faginas Lago[1,*], Margarita Albertí[2], Antonio Laganà[1], and Andrea Lombardi[1]

[1] Dipartimento di Chimica, Università di Perugia, Italy
noelia@dyn.unipg.it
[2] IQTCUB, Departament de Química Física, Universitat de Barcelona, Barcelona, Spain

Abstract. The main goal of this research is the rationalization of the structure and the energetics of K^+-$(C_6H_6)_n$-$(H_2O)_m$ (n=1-4; m=1-6) aggregates for which the full intermolecular potential, V, is given as a combination of few leading effective interaction components. Despite the fact that the K^+-$(C_6H_6)_n$-$(H_2O)_m$ systems are better considered as aggregates rather than solvated species, we find that the dynamics of the molecules surrounding the ion can be rationalized in terms of "a first and a second solvation shell" centered on K^+. The substitution of one C_6H_6 by two or more molecules of H_2O in the first solvation shell, has also been investigated in order to understand the role played by them in stabilizing certain structures. The interplay between molecules of the first and the second solvation shell has also been analyzed.

Keywords: Molecular Dynamics, Empirical potential energy surface, benzene-water aggregates, DLPOLY software.

1 Introduction

The rationalization of a number of important chemical and biochemical processes involve the recognition and selection of specific molecules [1]. A typical example of this is the rationalization of several enzyme and other proteins processes which use properties such as size, charge, shape, or polarity as the basis for molecular recognition. As these interactions typically involve electrostatic, hydrogen-bond, van der Waals, or dispersion interactions, they are arguably weaker than conventional chemical bonds and, therefore, are more difficult to quantify using either experimental or theoretical investigations. However, these forces play an important role in determining reaction rates [2–4], transport properties, adsorption mechanisms of ion channels size-selectivity, enzyme-substrate binding and protein structures in biological systems making of paramount importance the characterization of the different types of interactions [5, 6]. For this reason the study of weak intermolecular interactions is an ubiquitous subject and a key

---

* Corresponding author.

B. Murgante et al. (Eds.): ICCSA 2013, Part I, LNCS 7971, pp. 1–15, 2013.
© Springer-Verlag Berlin Heidelberg 2013

issue in many fields, from alignment and orientation effects in molecular collisions, to the stereodynamics of chiral molecules (see e.g. [7–13]). Recently, there
has been growing interest in investigating a specific noncovalent interaction, the
cation–π one, and its contribution to molecular recognition in biological systems. The cation–π interaction describes the interplay between the π–electron ·
cloud of an aromatic π (or any π electron system such as alkenes π) and an ion.
The extraction of alkali metal ions from aqueous solutions is fundamental to
countless molecular processes throughout chemistry and biology. Numerous ion
channels and other proteins in cellular membranes extract and transport Na^+
and K^+ from the surrounding aqueous medium. In order to be extracted out of
these systems the ion must be at least partially desolvated. At the molecular
level, this means that the ionophore (or other macromolecules) must compete
with and overcome the ion-water interaction and disrupt the hydrogen bond
network surrounding the ion. It is, therefore, important to understand how the
competition between different noncovalent interactions involving the ion, the
ionophore (or the macromolecule) and the water molecules takes place. In general, the competitive solvation of ions by different partners [14–16] (other basic
phenomena characterising organic aggregates, such as the formation of weak hydrogen bonds [14, 17], molecular recognition and selection processes [1, 18, 19])
are controlled by noncovalent interactions [20–26]. They stem from a delicate
balance between weak competing and cooperative effects of either attractive or
repulsive nature. An accurate modelling of such interactions is key to the study
of the aggregates from a theoretical point of view.

Indeed, the K^+-$(Bz)_n$-$(H_2O)_m$ (with $Bz = C_6H_6$) study presented here was
motivated by the importance that the alkali metal ions extraction from aqueous solutions plays among molecular processes of relevance for chemistry and
biology [27]. As the interplay of ion-molecule and molecule-molecule interactions
can be investigated by modelling the microscopic environment of the ion [28],
the competitive solvation of K^+ by benzene and water was studied using Molecular Dynamics (MD) simulations by considering a maximum of 4 and 6 rigid
molecules of C_6H_6 and H_2O, respectively.

The involved intermolecular interaction has been formulated by decomposing
the molecular polarizability [29] in effective components associated with atoms,
bonds (or groups of atoms) of the molecule [2, 3]. Such modelling of the intermolecular energy was applied in the past to investigate several neutral [30]
and ionic [31, 32] systems often involving weak interactions [30, 33] difficult to
calculate. The adequacy of a potential function to describe several intermolecular systems was proved by comparing ab initio energy and geometry estimates
for several configurations. In particular, the study of the alkali ion-C_6H_6 systems [34, 35] showed that chemical bonding plays a small or negligible role in
determining the formation of aggregates. Moreover, the halide-C_6H_6 calculations, pointed out that the current model is able to reproduce remarkably well
the main features of the potential energy surface for the heavier systems [36,37].
The paper is organized as follows: in section 2, we outline the construction of
the potential energy function. The details of the MD simulations are given in

section 3. Results are presented in section 4 and concluding remarks are given in section 5.

2   Potential Energy Surfaces

The potential model used in the calculations combines the contributions of some leading effective interaction components, i.e. includes the effect of other less important terms (see below), which are formulated as suitable potential functions (like those of ref. [38]). The model assumes separability of electrostatic (V_{el}) and non electrostatic (V_{nel}) interactions. Accordingly, in the global intermolecular potential V, that is represented as $V = V_{el} + V_{nel}$, the electrostatic contribution, V_{el}, is described as a sum of coulombic terms calculated from the charge distributions on the molecular frames. In particular, as K^+ and H_2O have low polarizability (α) values with respect to those of C_6H_6, α_{K^+} and α_{H_2O} the polarizability was not decomposed. This means that K^+ and H_2O are considered as single interaction centers (placed on K^+ and on O, respectively), with assigned values of the polarizability equal to α_{K^+} and α_{H_2O}. On the contrary, the polarizability of C_6H_6 was decomposed in 12 bond contributions (α_{CC} and α_{CH}) assigned to 12 interaction centers placed on the midpoint of the CC and CH bonds [39]. Accordingly, the K^+-C_6H_6 non electrostatic interaction [31] is described through 12 ion-bond contributions. Such representation, while allowing an additive formulation of the intermolecular potential V, includes, in a natural way, higher-order effects [39]. By bearing in mind that the interaction can be decompose and that a given atom of the molecule can take part in more than one bond, the CC and CH bond polarizabilities were further decomposed in atomic effective polarizabilities (different from those of the individual atoms). The use of effective atomic polarizabilities allows an indirect account of many body effects. In particular, the value assigned to each atom in the molecule is different (often smaller than that of the isolated gas phase atom) due to the many body effects arising from the formation of chemical bonds.

More in detail, the intermolecular interaction energy, V, is formulated as,

$$V = \sum_{i=1}^{n} V_{K^+-(C_6H_6)_i} + \sum_{k=1}^{m} V_{K^+-(H_2O)_k} \qquad (1)$$
$$+ \sum_{i=1}^{n} \sum_{k=1}^{m} V_{(C_6H_6)_i-(H_2O)_k} +$$
$$\sum_{i=1}^{n-1} \sum_{j>i}^{n} V_{(C_6H_6)_i-(C_6H_6)_j}$$
$$+ \sum_{k=1}^{m-1} \sum_{j>k}^{m} V_{(H_2O)_k-(H_2O)_j}$$

with all the intermolecular terms of Eq. 1 including both electrostatic and non electrostatic contributions.

Interactions between any pair of centers placed on different molecules (independently of the molecular polarizability decomposition) are described by means of the Improved Lennard Jones function V_{ILJ} [39, 40] as follows

$$V_{ILJ} = \varepsilon \left[\frac{m}{n(r) - m} \left(\frac{r_0}{r} \right)^{n(r)} - \frac{n(r)}{n(r) - m} \left(\frac{r_0}{r} \right)^m \right] \tag{2}$$

where r is the distance between the two interaction centers and ε and r_0 represent the minimum of the interaction and the associated value of r (as in the Lennard Jones (LJ) potential). In Eq. 2 the positive term defines the contribution to the repulsion while the negative term defines the attraction of each interaction pair and $n(r) = \beta + 4.0 \left(\frac{r}{r_0} \right)^2$. This expression in Eq.(2) greatly improves bothe the sizxe repulsion and the long range dispersion attraction, over the usual Lennard-Jones representation [41]. When effective quantities are considered, ε and r_0 are calculated directly from the characteristics of the interaction centers (for instance, for the K^+-H_2O interaction, with the interaction centers placed on the cation and on the oxigen atom, ε and r_0 are calculated from the average polarizability of the water molecule and both the polarizability and charge of the potassium ion) [42]. On the contrary, when the interaction is described by means of ion-bond terms, ε and r_0 are calculated from their perpendicular and parallel components ($\varepsilon_\perp$, $\varepsilon_\parallel$, $r_{0\perp}$ and $r_{0\parallel}$ derived from the polarizability and the charge of the ion) and the corresponding components of the bond polarizability (see for instance, Ref. [43]). It is worth to mention that, when groups or bonds are involved, The ILJ formula can be given additional flexibility by introducing in ϵ and r_0 the dependence on the set of angles that defines the relative orientation of the two interacting centers. One can then expand ϵ and r_0 in terms of appropriate angula functions, whose coefficient in the expansion can be estimated by selecting a small number of leading configurations of the two centers. These procedure has been previously adopted in a number of cases [7,44–47].

The parameter β, though being not a directly transferable parameter, due to its modulation when passing from an environment to another, can indirectly account for the selectivity and the strength of additional interaction components, as, for instance, perturbations due to induction and charge transfer in the formation of hydrogen bonds.

In our model, the parameters of the ILJ function derived from the involved effective polarizabilities are shown in Table 1. In particular they have been determined following the guidelines given in refs. [29,43,48], using the effective polarizabilities of C and H atoms in Bz (in agreement with data reported in refs. [49,50]) and the polarizability of H_2O [42,51].

For the H_2O monomer [52], the OH bond length was set at 0.9578 Å and the amplitude of the HOH angle was set at 104.4 degrees. On that geometry, charges of -0.65848 a.u and 0.32924 a.u were located respectively on the O and the H atoms in order to reproduce the dipole moment of the molecule. For C_6H_6, the CC and CH distances have been taken equal to 1.397 Å and 1.084 Å, respectively, and the CCC and CCH angles have been set equal to 120 degrees [52]. As in our previous studies of systems containing benzene, two negative charges of -0.04623 a.u have been placed on each C atom (above and below the aromatic plane with a separation of 1.905 Å) and one positive charge of 0.09246 a.u has been placed on each H atom [53,54].

Table 1. Perpendicular and parallel components of the well depth ($\varepsilon_\perp$, $\varepsilon_\parallel$) and of the equilibrium distances ($r_{0\perp}$, $r_{0\parallel}$) for the K$^+$-C$_6$H$_6$. Well depth (ε) and equilibrium distance (r_0) defined from K$^+$-H$_2$O, C$_6$H$_6$-H$_2$O, C$_6$H$_6$-C$_6$H$_6$ and H$_2$O-H$_2$O and β parameter for all the non electrostatic energy contributions.

ion-bond	$\varepsilon_\perp$ / meV	$\varepsilon_\parallel$/ meV	$r_{0\perp}$/ Å	$r_{0\parallel}$/ Å	β
K$^+$-CC	22.95	75.77	3.266	3.547	8.5
K$^+$-CH	39.97	42.70	3.044	3.240	8.5

atom-molecule	ε / meV	r_0 / Å	β
K$^+$-(H$_2$O)	102.1	3.161	7.0
C-(H$_2$O)	5.312	3.920	8.5
H-(H$_2$O)	3.306	3.482	8.5

atom-atom	ε / meV	r_0 / Å	β
C-C	3.400	4.073	9.0
H-H	1.610	3.099	9.0
C-H	1.720	3.733	9.0

molecule-molecule	ε / meV	r_0 / Å	β
(H$_2$O)-(H$_2$O)	9.060	3.730	6.6

3   Molecular Dynamics Calculations

In our MD calculations the molecules were, as already mentioned, treated as rigid and all simulations were carried out using a time step of 1 fs that is short enough to ensure a relative fluctuation of the total energy smaller than 10^{-5}. The dynamical evolution of the K$^+$-(C$_6$H$_6$)$_{1-4}$-(H$_2$O)$_{1-6}$ system was followed for 15 ns. This time is sufficient to observe isomerization processes and their fragmentation to release either H$_2$O or C$_6$H$_6$. A microcanonical ensemble (NVE) of particles, where the number of particles, N, the volume, V, and the total energy, E, are all conserved, was considered. Most of the calculations were performed at total energy values corresponding to a 10 K - 100 K temperature interval (related experiments were estimated to be carried out at these temperatures which are consistent with the evaporation process (see [55, 56]).

However, for some aggregates, when isomerization processes were found to be likely to occur, additional calculations were performed at higher temperatures. The total energy, E_{tot}, is evaluated as a sum of potential and kinetic energies. The kinetic energy at each step i, $E_{kin,i}$, allows determination of the instantaneous temperature, T_i, whose mean values are given at the end of the simulation (E_{kin} and T, respectively). All simulations were performed using the DL_POLY program [57]. To handle the geometry constraints and to integrate the equations of motion, use was made of the SHAKE method and the Verlet algorithm, respectively. The van der Waals interactions were calculated considering a cutoff radius of half the length of the box and the electrostatic energy was evaluated using the Ewald sum [58].

In order to figure out the dependence of the process on initial the following arrangements were considered: a) all C_6H_6 molecules were placed in the first shell around K^+ and all the H_2O ones in the second shell, b) all the H_2O molecules were placed in the first shell around K^+ and all the C_6H_6 ones in the second shell and c) C_6H_6 and H_2O molecules were distributed on both the first and the second shell around K^+. To better understand features of the binding of K^+ with (C_6H_6) or (H_2O) clusters of molecules also systems made of K^+ and either benzene or water, were investigated

4   Results

Before undertaking the analysis of the structure and energetics of K^+-$(C_6H_6)_n$-$(H_2O)_m$ (n=1-4; m=1-6) aggregates we have to emphasize that the full intermolecular potential, V, is given as a combination of few leading effective interaction components and that, although the interaction is strong enough to let us consider such aggregate stable in the investigated range of temperatures, we use the concept of "first and second solvation shell" (that is typical of solutions) beacuse our focus is on the dynamical evolution of the spatial distribution of the molecules around K^+ (in analogy with what is done by other authors when investigating similar systems [14]).

4.1   Structural Properties for K^+-$(C_6H_6)_n$-$(H_2O)_m$ (n=1-4; m=1-6) Systems: a Temperature Dependence

The competing role of the interaction components involved in some M^+-C_6H_6 [59, 60], and M^+-C_6F_6 [61] systems surrounded by Ar atoms was analyzed in the past to model the behaviour of water solvated systems (Ar atoms have been often used in the literature as a popular surrogate of water molecules). However, due to the fact that in rare gases the electrostatic component does not contribute appreciably to V, some competitive effects went undetected in those studies. For this reason, in order to evaluate in detail the role played by the different interaction components, we performed MD simulations of the K^+-$(C_6H_6)_n$-$(H_2O)_m$ (n=1,4; m=1,6) aggregates at several values of T. All pure aggregates (made of either C_6H_6 or H_2O molecules placed only in the first solvation shell) were found to be highly stable and to have all the molecules placed in the most favorable positions around K^+.

Mixed aggregates containing only one C_6H_6 molecule, showed equilibrium structures with all the molecules placed in the first solvation shell independently of the number of the H_2O molecules. However, it has been observed that, for these aggregate, several isomers with similar stability can be formed and frequent interconversions between such isomers are likely to form at all temperature values.

In Fig. 1 a two-dimensional map of the water molecules probability density obtained from MD simulations performed at 100 K for the K^+-C_6H_6-$(H_2O)_6$ aggregate, is plotted. As is apparent from the figure, four of the six H_2O molecules tend to be placed close to the cation, while the two remaining ones are placed at larger distances from the C$_6$H$_6$ center of mass and K$^+$ (blue color represents high probability density). This suggest that the π-hydrogen bond interaction (where an electron π system acts as proton acceptor) [62] and/or the hydrogen bonds (H$_2$O-H$_2$O) formation contribute to the stabilizaton of the H$_2$O molecules placed at larger distances. Nevertheless, a detailed inspection of the trajectories shows that the molecules interchange their positions quite easily. In spite of these frequent interconversions, all the K$^+$-C$_6$H$_6$-(H$_2$O)$_m$ aggregates turn out very stable and this indicates that the π-hydrogen bond interaction and the formation of hydrogen bonds can play an important role (together with both ion-quadrupole and the ion-dipole interactions) in stabilizing such aggregates.

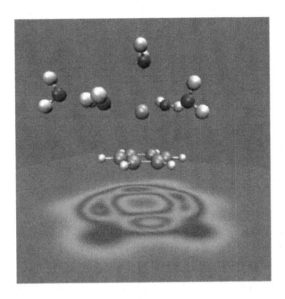

Fig. 1. A two dimensional map of the probability density of the water molecules of K$^+$-C$_6$H$_6$-(H$_2$O)$_6$ drawn on a plane parallel to that of the aromatic ring

By further increasing T, the fragmentation probability of K$^+$-C$_6$H$_6$-(H$_2$O)$_m$ to form K$^+$-C$_6$H$_6$-(H$_2$O)$_{m-1}$ +H$_2$O increases whereas that of C$_6$H$_6$ has not been observed. The effect of adding water molecules on the energy of the aggregates has been investigated. The variation of E_{cfg} and of its components, E_{el} and E_{nel}, as a function of the number of water molecules m for the K$^+$-C$_6$H$_6$-(H$_2$O)$_m$ aggregates whose geometry is close to the equilibrium is shown in Fig. 2. As expected, the addition of water molecules contributes to the decrease (more negative values) of both, electrostatic and non electrostatic energies. By focusing on the central panel of the figure, it can be seen that, while E_{el} becomes more negative almost linearly with m in the interval $n=1{:}4$, it varies very little by an addition of further H$_2$O molecules [3,63]. This can be understood in terms of the almost completeness of the first solvation shell. At low temperature, none of the molecules feels a ionic interaction sufficient to displace it in the second solvation shell and the addition of a new H_2O molecule causes a quick reorganization of the others to let it get into the first solvation shell. Accordingly, the radius of the solvation shell increases while small changes in E_{el} are observed. Indeed, the expected increase associated with the addition of a new water molecule to the first solvation shell is, partly, compensated by the decrease in stability of the set of molecules sitting around K^+.

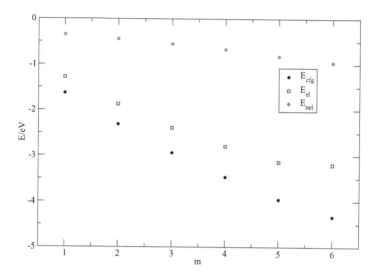

Fig. 2. Variation of the configuration energy (solid black circle) and of its electrostatic (empty red square) and non electrostatic (empty green diamonds) components for the K^+-(C_6H_6)-$(H_2O)_m$ aggregates as a function of the number of H_2O molecules

MD simulations, performed at increasing values of T, show that at low and moderate temperature, an increase of T originates only small and gradual changes of E_{cfg} and of its components E_{nel} and E_{el} for the three aggregates. This clearly indicates that the substitution of one or two C_6H_6 by one or two H_2O molecules, forming aggregates with different structures, does not modify the strength of the interaction with K^+.

4.2   The H_2O and C_6H_6 Solvation Competition

The competition between C_6H_6 and H_2O in solvating K^+ was investigated by paying special attention to aggregates containing three C_6H_6 molecules (the K^+-$(C_6H_6)_3$-$(H_2O)_{1-3}$ systems) as done for experimental observations [14]. It has been observed that the addition of one H_2O molecule to the K^+-$(C_6H_6)_3$ aggregate causes a reorganization of the three C_6H_6 molecules around K^+, allowing H_2O to get into the first solvation shell together with the three C_6H_6 molecules regardless of the initial configuration of the system. Moreover, no dissociation of the aggregate into K^+-$(C_6H_6)_3$+H_2O has been observed at temperatures lower than 250 K (a water molecule is considered to have dissociated when its distances from the other molecules and from the cation are larger than 10 Å). MD calculations for K^+-$(C_6H_6)_3$-$(H_2O)_2$ at moderate T values indicate that the mean value of E_{cfg} remains almost constant when increasing the temperature. Simulations performed at T close to 200 K show that the first solvation shell is formed by three molecules of C_6H_6 (approximately equidistant from K^+) and one molecule of H_2O. Moreover, the distance of the center of mass of the aromatic molecules from K^+ averaged over the whole trajectory, $R(C_6H_6$-$K^+)$, is equal to 2.83 Å. Such value is close to the corresponding distance of the O atom of H_2O from K^+, $R(O$-$K^+)$, that is equal to 2.86 Å. On the contrary, for the second water molecule, $R(O$-$K^+)$ is about 5 Å, which means that it is placed on the second solvation shell and is mainly stabilized by the hydrogen bonds formed by the two molecules of water and by interactions of the π-hydrogen bond type (for this aggregate, having an additional water molecule with respect to that of K^+-$(C_6H_6)_3$-H_2O, the number of hydrogen bonding sites available is double and allows the formation of a H_2O-H_2O hydrogen bond). In the top panel of Fig. 3 the evolution in time of the distance of K^+ from the center of mass of the C_6H_6 molecules and from the O atom of water obtained from MD simulations performed at T=125 K. The figure clearly shows that the dynamical evolution of the three aromatic molecules around the cation along the trajectory is similar, while that of water is different. As expected, the oscillations of the distances of water from the cation are larger for the molecule placed in the second solvation shell (the most labile molecule). However, long lasting trajectories (20 ns) do not provide any evidence for the dissociation of the aggregate. By further increasing T, the C_6H_6 molecules can reach larger distances from K^+, allowing the water molecules to get into the first solvation shell. This behavior is illustrated in the lower panel of Fig. 3.

As is for H_2O, at low temperature there is no difference in the behavior of the three C_6H_6 molecules along the trajectory. Moreover, at a temperature of about 200 K, the behaviour of the two H_2O molecules is also similar (see lower panel of Fig. 3), indicating that they are placed in the same solvation shell. As a matter of fact, the similar values of the mean $R(C_6H_6$-$K^+)$ and $R(O$-$K^+)$ distances (about 3-3.5 Å), indicates that, along the trajectory, the five molecules are preferentially placed in the first solvation shell. This means that when energy is large enough, the addition of a new water molecule may disrupt the stable structure of the aggregate formed by the three C_6H_6 and one H_2O allowing a reorganization of the first solvation shell to form the K^+-$(C_6H_6)_3$-$(H_2O)_2$ aggregate with all molecules placed close to K^+. Such behavior is independent of the initial configuration of the aggregate and it has been observed for temperatures as high as 300 K. By further increasing T, the fragmentation of the aggregate becomes more likely. An analysis of the outcomes of twenty simulations carried out at temperature values around 300 K shows that the flying away molecule is always water (the loss of a C_6H_6 molecule has never been observed). This result is in agreement

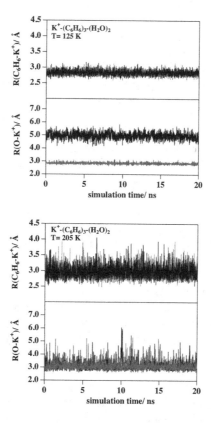

Fig. 3. Evolution of the distance of K^+ from the center of mass of C_6H_6 and from the O atom of H_2O for K^+-$(C_6H_6)_3$-$(H_2O)_2$ at two different values of T (different molecules are represented by different colors)

with the experimental data for K^+-$(C_6H_6)_{1-3}$-H_2O [14], indicating that for up to three C_6H_6 molecules, the only unimolecular loss process is that of water.

4.3   The Population of the Second Solvatation Shell

When a further molecule of water is added to K^+-$(C_6H_6)_3$-$(H_2O)_3$ the six surrounding molecules can not be all placed in the first solvation shell. Therefore, they have to compete to get into the shell closer to K^+.

The study of K^+-$(C_6H_6)_3$-$(H_2O)_3$ using different initial configurations shows that while C_6H_6 is able to displace one H_2O molecule from the first to the second solvation shell, the contrary (H_2O displacing C_6H_6) does not occur (see Fig. 4). As a matter of fact, when the water molecules are initially placed in the second solvation shell, after equilibration, the first solvated shell is formed by two molecules of C_6H_6 and three of H_2O. The third C_6H_6 molecule is left, first, in the second solvation shell. However, after 15 ns, the third C_6H_6 molecule is also able

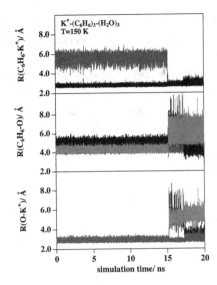

Fig. 4. Evolution of the distance to K^+ from the centre of mass of C_6H_6 (top panel), of the O atom of water from the centre of mass of C_6H_6 (middle panel) and of K^+ from the O atom of the H_2O molecules for K^+-$(C_6H_6)_3$-$(H_2O)_3$ at T= 150 K (different molecules are represented by different colours)

to get into the first solvation shell by displacing one of H_2O molecule into the second solvation shell (where as shown by longer simulations it remains). By recalling that in the K^+-$(C_6H_6)_3$-$(H_2O)_2$ aggregate the five surrounding molecules are indeed firmly placed in the first solvation shell (even at high temperature) one has to conclude that the process of adding C_6H_6 and displacing a water molecule is able to disrupt the stabling of the K^+-$(C_6H_6)_3$-$(H_2O)_2$ configuration and confines a water molecule into the second solvation shell. By further increasing T, K^+-$(C_6H_6)_3$-$(H_2O)_3$ tends to dissociate in K^+-$(C_6H_6)_3$-$(H_2O)_2$ plus a H_2O molecule. This confirms the tendency of K^+ to dehydrate in the presence of aromatic rings found in the experiment [14, 15].

The study of aggregates containing four C_6H_6 molecules singles out properties differing from those of smaller aggregates. As a matter of fact, while only the fragmentation of H_2O has been observed for the aggregates containing less than four benzene molecules, the fragmentation giving a C_6H_6 has been observed for K^+-$(C_6H_6)_4$-$(H_2O)_{1-6}$. The preference for fragmentations leading to the flying away of a C_6H_6 molecule is mainly observed for K^+-$(C_6H_6)_4$-H_2O and K^+-$(C_6H_6)_4$-$(H_2O)_2$ (see Fig. 5).

As found also before, this can be understood in terms of a stabilization of the system having four C_6H_6 molecules, by a π-hydrogen bond interaction in the second solvation shell. However, due the high stability of the K^+-$(C_6H_6)_3$-H_2O aggregate, the aromatic molecule is unable to displace water from the first solvation shell and, in agreement with experimental findings [14], it becomes the

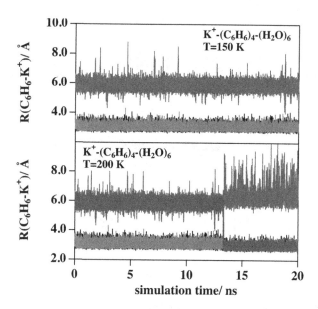

Fig. 5. Evolution of the distance of K^+ from the centre of mass of the C_6H_6 molecules at 150 K (top panel) and 200 K (lower panel) for K^+-$(C_6H_6)_4$-$(H_2O)_6$ (different molecules are represented by different colours)

most labile molecule. On the contrary, no interconversion between the benzene molecules is observed. However, by increasing T one of the C_6H_6 molecules placed in the second solvation shell can displace water molecules from the first solvation solvation shell. No interconversions are observed at 150 K, while at 200 K, after about 13 ns of simulation, one C_6H_6 placed in the second solvation shell, get into the first one, while the other C_6H_6 molecule, placed in the second solvation shell tends to dissociate.

5   Concluding Remarks

The paper shows that the competition between C_6H_6 and H_2O in solvating K^+ in mixed-solvent K^+-$(C_6H_6)_n$-$(H_2O)_m$ ionic aggregates can be investigated at molecular level and that the involved non electrostatic contributions to the total intermolecular interaction can be described by means of Improved Lennard Jones functions. The electrostatic contribution is calculated from charge distributions according to the quadrupole moment of C_6H_6 and the dipole moment of H_2O. The preference of K^+ to bind C_6H_6 better than H_2O has been investigated by running MD simulations of K^+-$(C_6H_6)_3$-H_2O and K^+-$(C_6H_6)_2$-$(H_2O)_2$ at several temperatures. At high temperatures the fragmentation of H_2O is observed, while this is not true for the fragmentation of C_6H_6. We found also that due to the different size of C_6H_6 and H_2O, the first solvation shell around K^+ can accept different numbers of solvent molecules. While only four C_6H_6 molecules can be placed in the first solvation shell, in the the mixed K^+-$(C_6H_6)_n$-$(H_2O)_m$ aggregates such number increases, due to the fact that one C_6H_6 can be substituted by more than one H_2O molecules. Accordingly, we find that an increase of the number of molecules in the first solvation shell and the change in its dimension can compensate for the different strength of the K^+-C_6H_6 and K^+-H_2O interactions we find, as well, that different aggregates can have similar configuration energies. This has been observed for aggregates formed by the cation and four molecules, such as K^+-$(C_6H_6)_4$, K^+-$(C_6H_6)_3$-H_2O and K^+-$(C_6H_6)_2$-$(H_2O)_2$, with interaction energies of -3.40, -3.47 and -3.41 eV, respectively. Larger differences can be observed when the aggregates contain four C_6H_6 molecules, for which a fragmentation of C_6H_6 is possible. In particular, our MD simulations allowed us also to find out that for K^+-$(C_6H_6)_4$-H_2O, C_6H_6 becomes more labile than H_2O because of the preference of K^+ to bind with C_6H_6 rather than H_2O and the different fragmentation of the aggregates depending on the number of C_6H_6 molecules, predicted by experiments. The validity of the used force field was found to be supported by its ability to reproduce the experimental findings observed in gas-phase infrared vibrational spectroscopy.

Acknowledgement. M. Albertí acknowledges financial support from the Ministerio de Educación y Ciencia (Spain, Projects CTQ2010-16709) and to the Comissionat per a Universitats i Recerca del DIUE (Generalitat de Catalunya, Project 2009-SGR 17). Also thanks are due to the SUR and the Departament d'Economia i Coneixement de la Generalitat de Catalunya (Project 2011 BE1-00063). The Centre de Serveis Científics i Acadèmics de Catalunya CESCA and Fundació Catalana per a la Recerca are also acknowledged for the allocated supercomputing time. Noelia Faginas Lago acknowledges financial support from MIUR PRIN 2008 (contract 2008KJX4SN 003), Phys4entry FP72007-2013 (contract 242311) and EGI Inspire. A. Lombardi also acknowledges financial support to MIUR-PRIN 2010-2011 (contract 2010ERFKXL 002). Thanks are also due to INSTM, CINECA, IGI and the COMPCHEM virtual organization for the allocation of computing time.

References

[1] Ma, J.C., Dougherty, D.A.: Chem. Rev. 97, 1303 (1997)

[2] Bartolomei, M., Pirani, F., Laganà, A., Lombardi, A.: J. Comp. Chem. 33, 1806 (2012)

[3] Lombardi, A., Faginas Lago, N., Laganà, A., Pirani, F., Falcinelli, S.: A bond-bond portable approach to intermolecular interactions: Simulations for N-methylacetamide and carbon dioxide dimers. In: Murgante, B., Gervasi, O., Misra, S., Nedjah, N., Rocha, A.M.A.C., Taniar, D., Apduhan, B.O. (eds.) ICCSA 2012, Part I. LNCS, vol. 7333, pp. 387–400. Springer, Heidelberg (2012)

[4] Lombardi, A., Laganà, A., Pirani, F., Faginas Lago, N., Palazzetti, F.: Carbon oxides in gas flows and earth and planetary atmospheres: State-to-state simulations of energy transfer and dissociation reactions. In: Murgante, B., Misra, S., Carlini, M., Torre, C.M., Quang, N.H., Taniar, D., Apduhan, B.O., Gervasi, O. (eds.) ICCSA 2013, Part II. LNCS, vol. 7972, pp. 17–31. Springer, Heidelberg (2013)

[5] Basch, H., Stevens, W.J.: Theochem. 338, 1303 (1995)

[6] Lee, J.Y., Lee, S.J., Choi, H.S., Cho, S.J., Kim, K.S., Ha, T.K.: Chem. Phys. Lett. 232, 67 (1995)

[7] Barreto, P.R.P., Albernaz, A.F., Caspobianco, A., Palazzetti, F., Lombardi, A., Grossi, G., Aquilanti, V.: Comput. Theor. Chem. 990, 56 (2012)

[8] Aquilanti, V., Grossi, G., Lombardi, A., Maciel, G.S., Palazzetti, F.: Rendiconti Lincei 22, 125 (2011)

[9] Lombardi, A., Palazzetti, F., Maciel, G.S., Aquilanti, V., Sevryuki, M.B.: Int. J. Quantum Chem. 111, 1651 (2011)

[10] Lombardi, A., Maciel, G.S., Palazzetti, F., Grossi, G., Aquilanti, V.: J. Vacuum Soc. Japan 53, 645 (2010)

[11] Aquilanti, V., Grossi, G., Lombardi, A., Maciel, G.S., Palazzetti, F.: Phys. Scripta 78, 58119 (2008)

[12] Falcinelli, S., Rosi, M., Candori, P., Vecchiocattivi, F., Bartocci, A., Lombardi, A., Faginas Lago, N., Pirani, F.: Modeling the intermolecular interactions and characterization of the dynamics of collisional autoionization processes. In: Murgante, B., Misra, S., Carlini, M.T., Nguyen, H.-Q., Taniar, D., Apduhan, B.O., Gervasi, O. (eds.) ICCSA 2013. LNCS, vol. 7971, pp. 69–83. Springer, Heidelberg (2013)

[13] Barreto, P.R.P., Palazzetti, F., Grossi, G., Lombardi, A., Maciel, G.S., Vilela, A.F.A.: Int. J. Quantum Chem. 110, 777 (2010)

[14] Cabarcos, O.M., Weinheimer, C.J., Lisy, J.M.: J. Chem. Phys. 110, 5151 (1998)

[15] Cabarcos, O.M., Weinheimer, C.J., Lisy, J.M.: J. Chem. Phys. 110, 8429 (1999)

[16] Morais-Cabral, J., Zhou, Y., MacKinnon, R.: Nature 414, 37 (2001)

[17] Alonso, J.L., Antolínez, S., Blanco, S., Lesarri, A., Lopez, J.C., Caminati, W.: J. Am. Chem. Soc. 126, 3244 (2004)

[18] Kumpf, R., Dougherty, D.: Science 261, 1708 (1993)

[19] Dougherty, D.A.: Science 271, 163 (1996)

[20] Tsuzuki, S., Yoshida, M., Uchimaru, T., Mikami, M.: J. Phys. Chem. A 105, 769 (2001)

[21] Felder, C., Jiang, H.L., Zhu, W.L., Chen, K.X., Silman, I., Botti, S.A., Sussman, J.L.: J. Phys. Chem. A 105, 1326 (2001)

[22] Mecozzi, S., West, A.P., Dougherty, D.: J. Am. Chem. Soc. 118, 2307 (1996)

[23] Caldwell, J.W., Kollman, P.A.: J. Am. Chem. Soc. 117, 4177 (1995)

[24] Nicholas, J.B., Hay, B.P., Dixon, D.A.: J. Phys. Chem. A 103, 1394 (1999)

[25] Quiñonero, D., Garau, C., Frontera, A., Ballester, P., Costa, A., Deyà, P.: J. Phys. Chem. A 109, 4632 (2005)

[26] Albout, A.F., Adamowicz, L.: J. Chem. Phys. 116, 9672 (2002)

[27] Vaden, T., Lisy, J.: J. Phys. Chem. A 109, 3880 (2005)

[28] Weinheimer, C.J., Lisy, J.M.: Int. J. of Mass Spectr. Ion Processes 159, 197 (1996)

[29] Pirani, F., Cappelletti, D., Liuti, G.: Chem. Phys. Lett. 350, 286 (2001)

[30] Albertí, M., Castro, A., Laganà, A., Pirani, F., Porrini, M., Cappelletti, D.: Chem. Phys. Lett. 392, 514 (2004)

[31] Albertí, M., Castro, A., Laganà, A., Moix, M., Pirani, F., Cappelletti, D., Liuti, G.: J. Phys. Chem. A 109, 2906 (2005)

[32] Albertí, M., Aguilar, A., Lucas, J.M., Pirani, F.: Theor. Chem. Acc. 123, 21 (2009)

[33] Albertí, M.: J. Phys. Chem. A 114, 2266 (2010)
[34] Albertí, M., Aguilar, A., Lucas, J.M., Pirani, F., Cappelletti, D., Coletti, C., Re, N.: J. Phys. Chem. A 110, 9002 (2006)
[35] Coletti, C., Re, N.: J. Phys. Chem. A 110, 6563 (2006)
[36] Coletti, C., Re, N.: J. Phys. Chem. A 113, 1578 (2009)
[37] Albertí, M., Aguilar, A., Lucas, J.M., Pirani, F., Coletti, C., Re, N.: J. Phys. Chem. A 113, 14606 (2009)
[38] Albertí, M., Castro, A., Laganà, A., Moix, M., Pirani, F., Cappelletti, D., Liuti, G.: J. Phys. Chem. A 110, 9002 (2005)
[39] Pirani, F., Albertí, M., Castro, A., Moix, M., Cappelletti, D.: Chem. Phys. Lett. 394, 37 (2004)
[40] Pirani, P., Brizi, S., Roncaratti, L., Casavecchia, P., Cappelletti, D., Vecchiocattivi, F.: Phys. Chem. Chem. Phys. 10, 5489 (2008)
[41] Lombardi, A., Palazzetti, F.: Journal of Molecular Structure: THEOCHEM 22, 852 (2008)
[42] Albertí, M., Aguilar, A., Cappelletti, D., Laganà, A., Pirani, F.: Int. J. of Mass. Spectr. 280, 50 (2009)
[43] Cambi, R., Cappelletti, D., Liuti, G., Pirani, F.: J. Chem. Phys. 95, 1852 (1991)
[44] Barreto, P.R.P., Albernaz, A.F., Palazzetti, F., Lombardi, A., Grossi, G., Aquilanti, V.: Phys. Scripta 84, 028111 (2011)
[45] Palazzetti, F., Munusamy, E., Lombardi, A., Grossi, G., Aquilanti, V.: Int. J. Quantum Chem. 111, 318 (2011)
[46] Maciel, G.S., Barreto, P.R.P., Palazzetti, F., Lombardi, A., Aquilanti, V.: J. Chem. Phys. 129, 164302 (2008)
[47] Barreto, P.R.P., Vilela, A.F.A., Lombardi, A., Maciel, G.S., Palazzetti, F., Aquilanti, V.: Phys. Chem. A 111, 12754 (2007)
[48] Capitelli, M., Cappelletti, D., Colonna, G., Gorse, C., Laricchiuta, A., Liuti, G., Longo, S., Pirani, F.: Chem. Phys. 338, 62 (2007)
[49] Ewig, C.S., Waldman, M., Maple, J.R.: J. Phys. Chem. 106, 326 (2002)
[50] Gavezzotti, A.: J. Phys. Chem. 107, 2344 (2003)
[51] Albertí, M., Aguilar, A., Bartolomei, M., Cappelletti, D., Laganà, A., Lucas, J., Pirani, F.: Phys. Script. 78, 058108 (2008)
[52] Computational Chemistry Comparison and Benchmark DataBase, http://cccbdb.nist.gov/
[53] Paolantoni, M., Faginas Lago, N., Albertí, M., Laganà, A.: J. Phys. Chem. A 113(52), 15100 (2009)
[54] Albertí, M., Faginas Lago, N., Pirani, F.: Chem. Phys. 399, 232 (2012)
[55] Vaden, T., Weinheimer, C., Lisy, J.: J. Chem. Phys. 121, 3102 (2004)
[56] Vaden, T., Lisy, J.: J. Chem. Phys. 124, 214315 (2006)
[57] Laboratory, S.D., http://www.cse.clrc.ac.uk/ccg/software/DL_POLY/index.shtml
[58] Allen, M.P., Tildesley, D.J.: Computer Simulation of Liquids. Clarendon Press, Oxford (1998)
[59] Albertí, M., Aguilar, A., Lucas, J.M., Laganà, A., Pirani, F.: J. Phys. Chem. A 111, 1780 (2007)
[60] Huarte-Larragaña, F., Aguilar, A., Lucas, J.M., Albertí, M.: J. Phys. Chem. A 111, 8072 (2007)
[61] Albertí, M., Faginas Lago, N.: J. Phys. Chem. A 116, 3094 (2012)
[62] Suzuki, S., Green, P., Bumgarner, R., Dasgupta, S., Godard III, W., Blake, G.: Science 257, 942 (1992)
[63] Albertí, M., Faginas Lago, N., Pirani, F.: The Journal of Physical Chemistry A 115, 10871–10879 (2011)

Using Vectorization and Parallelization to Improve the Application of the APH Hamiltonian in Reactive Scattering

Jeff Crawford, Zachary Eldredge, and Gregory A. Parker

University of Oklahoma

Abstract. In time-dependent quantum reactive scattering calculations, it is important to quickly and accurately apply the Hamiltonian, as this must be performed at every time-step. We present a method of separating the Hamiltonian for adiabatically-adjusting principal axes hyperspherical (APH) coordinates into its constituent parts, vectorizing these to reduce computational steps required in matrix multiplications, and finally parallelizing the application to reduce the time required to perform the calculation. This reduction in time is achieved with no modification of results for triatomic systems of either lithium or hydrogen.

1 Background

1.1 Quantum Reactive Scattering

In reactive scattering, a collision of an atom and a diatomic molecule carries the possibility of an atomic exchange. Since our system is triatomic, it has three separate arrangement channels, corresponding to each of the three atoms which can be separated from the others. In addition, each arrangement channel has a number of different possible quantum states. The reaction takes the form:

$$A + BC \rightleftharpoons \begin{cases} AB + C \\ AC + B \\ A + B + C. \end{cases} \tag{1}$$

The problem is: given that the system starts in a particular arrangement channel, in a particular initial quantum state, what is the likelihood that it ends up in a given final arrangement channel and state? Our method produces these state-to-state probability amplitudes. To do this, we must solve a partial differential equation in seven variables–six spatial coordinates for three atoms (once center-of-mass motion is discarded) and a time coordinate.

This probability amplitude is given in the scattering matrix, or S-matrix. Time-independent methods calculate the entire S-matrix for all potential incoming and outgoing states at a particular energy [1, 2, 3, 4]. In time-dependent reactive scattering, we propagate the wavefunction in time on a spatial grid [6]. At each time step, the wavepacket is projected onto an analysis line outside the interaction region [7], and these accumulated time-dependent elements

B. Murgante et al. (Eds.): ICCSA 2013, Part I, LNCS 7971, pp. 16–30, 2013.
© Springer-Verlag Berlin Heidelberg 2013

are analyzed by Fourier transforming them to energy-dependent ones and then matching to the appropriate boundary conditions [5]. In this method, we start in a given initial state, but we recover information for all open final states and a range of energies, which allows our method to scale better for systems with a large number of basis functions.

1.2 Propagation

In order to do this, we must solve the time-dependent Schrodinger equation:

$$i\hbar \frac{\partial}{\partial t} \varphi(t) = \mathcal{H} \varphi(t), \tag{2}$$

and, since the Hamiltonian $\mathcal{H}$ is time-independent, we can solve this by defining a time evolution operator:

$$\varphi(t) = e^{-i\mathcal{H}(t-t_0)/\hbar} \varphi(t_0). \tag{3}$$

We can do this most efficiently by expanding the exponential time-evolution operator as a Chebychev polynomial[8]. This paper presents an efficient scheme for applying the Hamiltionian operator to the propagating wavepacket, which allows us to greatly reduce the computational steps required to propagate the wavepacket and obtain working results.

2 APH Coordinate Systems

Throughout our propagation we make use of adiabatically-adjusing principal axes hyperspherical (APH) coordinates to describe the arrangement of the tri-atomic system [4]. The internal coordinates of the APH system are the hypperra-dius ρ and two angles, χ and θ. The remaining coordinates are Euler angles α_Q, β_Q, γ_Q which orient the triatomic plane with respect to a set of space-fixed axes. The hyperradius ρ describes the overall size of the system. θ varies from 0 to π, "bending" the system from an equilateral triangle to a colinear arrangement respectively. The other hyperangle χ describes–in a colinear arrangement–the ratio of $\mathbf{s}_\tau$ to $\mathbf{S}_\tau$, with:

$$\lim_{\chi_\tau \to 0} \frac{\mathbf{s}_\tau}{\mathbf{S}_\tau} = 0 \tag{4}$$

Here, τ represents some arrangement channel labeled as A, B, or C depending on which atom is being treated as "separate." $\mathbf{S}_\tau$ and $\mathbf{s}_\tau$ are mass-scaled Ja-cobi coordinates in that arrangement channel, where $\mathbf{s}_\tau$ is the separation of the diatomic fragment while $\mathbf{S}_\tau$ points from the center-of-mass of that fragment to the atom τ. Rotations in χ can correspond to changes in arrangement channel if large enough, which is to say that the channels are separated by differences in χ.

These internal coordinates are defined by :

$$\rho = \left(q^2 + Q^2\right)^{1/2}, \tag{5a}$$

$$\theta = \pi/2 - 2\tan^{-1}(q/Q), \tag{5b}$$

where $0 \leq \rho \leq \infty$, $0 \leq \theta \leq \pi/2$, and q and Q are vectors (related to $\mathbf{s}_\tau$ and $\mathbf{S}_\tau$ by kinematic rotation) pointing along the smallest principal moments of inertia, making this an instantaneous principal-axes system. These coordinates treat all arrangement channels equivalently. Another important aspect of the APH coordinate system is that the potential energy surface (PES) is symmetric in χ. The PES belongs to the C_{6v}, C_{2v}, or C_2 point groups depending on whether it consists of one, two, or three different types of atoms, respectively. This has two benefits. First, it allows us to reduce the size of the coordinate space needed to represent the wavefunction, allowing us to use only a fraction of the grid we would ordinarily need. Second, we can propagate each irreducible representation of the wavepacket separately, using symmetry-adapted wavefunctions. The size of the coordinate space required to represent the symmetry-adapted χ_i is $0 \leq \chi_i \leq 2\pi l_\Gamma/h$, where h is the order of the group and l_Γ is the dimension of irreducible representation Γ [9, 10].

Finally, the kinetic energy operator in APH coordinates, used later to construct the Hamiltonian, is given by [4]:

$$
\begin{aligned}
T = {} & -\frac{\hbar^2}{2\mu\rho^{5/2}} \frac{\partial^2}{\partial\rho^2} \rho^{5/2} + \frac{15\hbar^2}{8\mu\rho^2} \\
& -\frac{\hbar^2}{2\mu\rho^2} \left[\frac{4}{\sin 2\theta} \frac{\partial}{\partial\theta} \sin 2\theta \frac{\partial}{\partial\theta} + \frac{1}{\sin^2\theta} \frac{\partial^2}{\partial\chi_i^2} \right] \\
& + \frac{1}{\mu\rho^2} \left[\frac{\mathcal{A}_\theta + \mathcal{B}_\theta}{2} J^2 + \left(\mathcal{C}_\theta - \frac{\mathcal{A}_\theta - \mathcal{B}_\theta}{2} \right) J_z^2 \right], \\
& -\frac{1}{2\mu\rho^2} \left[\frac{\mathcal{A}_\theta - \mathcal{B}_\theta}{2} \left(J_+^2 + J_-^2 \right) + \hbar\mathcal{D}_\theta \left(J_- - J_+ \right) \frac{\partial}{\partial\chi_i} \right]
\end{aligned}
\tag{6}
$$

where J is the total angular momentum operator, a differential operator in the Euler angles α_Q, β_Q, γ_Q,

$$J_\pm = J_x \mp iJ_y, \tag{7}$$

are the raising and lowering operators, J_i are the components of the total angular momentum with respect to the a set of body-fixed axes, and

$$\mathcal{A}_\theta = \frac{1}{1 + \sin\theta}, \tag{8a}$$

$$\mathcal{B}_\theta = \frac{1}{2\sin^2\theta}, \tag{8b}$$

$$\mathcal{C}_\theta = \frac{1}{1 - \sin\theta}, \tag{8c}$$

$$\mathcal{D}_\theta = \frac{\cos\theta}{\sin^2\theta}. \tag{8d}$$

3   Vectorization

3.1   Motivation

Kronecker products arise in the process of applying the APH Hamiltonian. These require a large number of scalar multiplications with a corresponding increase in machine time, especially as the grid becomes large in the number of points used to represent the APH coordinates ρ, θ, and χ. Our calculations proceed more quickly when we express the action of the Hamiltonian on a wavefunction in terms of ordinary matrix multiplications. By removing the Kronecker products from our calculations we can greatly reduce the number of multiplications we need to perform and improve the computational efficiency of the Hamiltonian's application.

3.2   Notation and Background

In this section, vectors and matrices will appear in bold, as in $\mathbf{A}$. Scalar quantities will appear in plain (italicized) font, as in A. Because we will use subscripts to identify matrices, we will not use the normal notation for particular elements of a matrix. The element in the ith row and the jth column will be represented not as A_{ij} but as $A[i, j]$. Rank 3 matrices will also appear, and an element of these will be denoted as $A[i, j, k]$.

To represent an entire row or column of a matrix, the "unused" coordinate will be replaced with a colon. For instance, the vector formed from the first column of $\mathbf{A}$ (a rank 2 matrix) would be denoted as $\mathbf{A}[:, 1]$ as it would contain elements from all rows but only the first column. More explicitly, if $\mathbf{A}$ is an $n_A \times m_A$ matrix, then:

$$\mathbf{A}[i, :] = (\ A[i, 1]\ \ A[i, 2]\ \ \cdots\ \ A[i, m_A]\ ) \tag{9}$$

and

$$\mathbf{A}[:, j] = \begin{pmatrix} A[1, j] \\ A[2, j] \\ \vdots \\ A[n_A, j] \end{pmatrix}. \tag{10}$$

We can extend this easily to rank 3 matrices. Use of one colon in the coordinates describes a vector. If $\mathbf{B}$ is a $n_B \times m_B \times p_B$ matrix, then:

$$\mathbf{B}[:, j, k] = \begin{pmatrix} B[1, j, k] \\ B[2, j, k] \\ \vdots \\ B[n_B, j, k] \end{pmatrix} \tag{11}$$

and similarly for the other two coordinates. The use of two colons gives us a "slice" of the rank 3 matrix, yielding a rank 2 matrix:

$$\mathbf{B}[:,:,k] = \begin{pmatrix} B[1,1,k] & B[1,2,k] & \cdots & B[1,m_B,k] \\ B[2,1,k] & B[2,2,k] & \cdots & B[2,m_B,k] \\ \vdots & \vdots & \ddots & \vdots \\ B[n_B,1,k] & B[n_B,2,k] & \cdots & B[n_B,m_B,k] \end{pmatrix} \tag{12}$$

$$\mathbf{B}[:,j,:] = \begin{pmatrix} B[1,j,1] & B[1,j,2] & \cdots & B[1,j,p_B] \\ B[2,j,1] & B[2,j,2] & \cdots & B[2,j,p_B] \\ \vdots & \vdots & \ddots & \vdots \\ B[n_B,j,1] & B[n_B,j,2] & \cdots & B[n_B,j,p_B] \end{pmatrix} \tag{13}$$

$$\mathbf{B}[i,:,:] = \begin{pmatrix} B[i,1,1] & B[i,1,2] & \cdots & B[i,1,p_B] \\ B[i,2,1] & B[i,2,2] & \cdots & B[i,2,p_B] \\ \vdots & \vdots & \ddots & \vdots \\ B[i,m_B,1] & B[i,m_B,2] & \cdots & B[i,m_B,p_B] \end{pmatrix}. \tag{14}$$

The Kronecker product of two matrices $\mathbf{A} \in \Re^{n_A \times m_A}$ and $\mathbf{B} \in \Re^{n_B \times m_B}$ is a $n_A n_B \times m_A m_B$ matrix given by:

$$\mathbf{A} \otimes \mathbf{B} = \begin{pmatrix} A_{1,1}\mathbf{B} & A_{1,2}\mathbf{B} & \cdots & A_{1,n_A}\mathbf{B} \\ A_{2,1}\mathbf{B} & A_{2,2}\mathbf{B} & \cdots & A_{2,n_A}\mathbf{B} \\ \vdots & \vdots & \ddots & \vdots \\ A_{n_A,1}\mathbf{B} & A_{n_A,2}\mathbf{B} & \cdots & A_{n_A,n_A}\mathbf{B} \end{pmatrix} \tag{15}$$

Next we wish to define the vec operator. The vec operator acts on matrices and outputs vectors. In a two-dimensional case we consider a matrix $\mathbf{X} \in \Re^{n_X \times m_X}$. Applying the vec operator to $\mathbf{X}$ stacks the columns onto each other, giving us a vector of length $n_X m_X$.

$$\text{vec}(\mathbf{X}) = \begin{pmatrix} X[1,1] \\ \vdots \\ X[n_X,1] \\ X[1,2] \\ \vdots \\ X[n_X,2] \\ \vdots \\ \vdots \\ X[1,m_X] \\ \vdots \\ X[n_X,m_X] \end{pmatrix} = \begin{pmatrix} \mathbf{X}[:,1] \\ \mathbf{X}[:,2] \\ \vdots \\ \mathbf{X}[:,m_X] \end{pmatrix} \tag{16}$$

In a three-dimensional case, let $\mathbf{X} \in \Re^{n_X \times m_X \times p_X}$. Then applying the vec operator yields a vector of length $n_X m_X p_X$:

$$
\text{vec}(\mathbf{X}) = \begin{pmatrix} \mathbf{X}[:,1,1] \\ \vdots \\ \mathbf{X}[:,1,p_X] \\ \mathbf{X}[:,2,1] \\ \vdots \\ \mathbf{X}[:,2,p_X] \\ \vdots \\ \vdots \\ \mathbf{X}[:,m_X,1] \\ \vdots \\ \mathbf{X}[:,m_X,p_X] \end{pmatrix}
\tag{17}
$$

Other useful ways to write (17), which will be used later, are

$$
\text{vec}(\mathbf{X}) = \begin{pmatrix} \text{vec}(\mathbf{X}[:,1,:]) \\ \text{vec}(\mathbf{X}[:,2,:]) \\ \vdots \\ \text{vec}(\mathbf{X}[:,n_B,:]) \end{pmatrix}
\tag{18}
$$

and

$$
\text{vec}(\mathbf{X}) = \text{vec} \begin{pmatrix} \mathbf{X}[:,:,1] \\ \mathbf{X}[:,:,2] \\ \vdots \\ \mathbf{X}[:,:,n_C] \end{pmatrix}
\tag{19}
$$

We will deal with discretized operations of the form:

$$
\mathbf{Cx} = (\mathbf{B} \otimes \mathbf{A})\text{vec}(\mathbf{X}),
\tag{20}
$$

where $\mathbf{B} \in \Re^{n_B \times n_B}$, $\mathbf{A} \in \Re^{n_A \times n_A}$, $\mathbf{X} \in \Re^{n_A \times n_B}$, $\mathbf{C} \in \Re^{n_A n_B \times n_A n_B}$, $\mathbf{x} \in \Re^{n_A n_B \times 1}$, and $\mathbf{x} = \text{vec}(\mathbf{X})$. The operation of $\mathbf{C}$ on $\mathbf{x}$ requires $2n_A^2 n_B^2$ multiplications. The number of multiplications can be reduced by rewriting (20) as

$$
(\mathbf{B} \otimes \mathbf{A})\text{vec}(\mathbf{X}) = \text{vec}(\mathbf{AXB}^{\mathrm{T}}),
\tag{21}
$$

which requires $n_A^2 n_B + n_A n_B^2$ multiplications[11].

3.3 Vectorization of APH Hamiltonian

Referring to the APH kinetic energy operator we saw in (6), we now specialize to cases with zero total angular momentum ($J = 0$). The kinetic energy is now of the form:

$$T = -\frac{\hbar^2}{2\mu\rho^{5/2}}\frac{\partial^2}{\partial\rho^2}\rho^{5/2} + \frac{15\hbar^2}{8\mu\rho^2}$$
$$-\frac{\hbar^2}{2\mu\rho^2}\left[\frac{4}{\sin 2\theta}\frac{\partial}{\partial\theta}\sin 2\theta\frac{\partial}{\partial\theta} + \frac{1}{\sin^2\theta}\frac{\partial^2}{\partial\chi_i^2}\right]$$

(22)

Now adding in the potential, we can form the Hamiltonian:

$$\mathcal{H} = h_\rho + f_\rho h_\theta + f_\rho f_\theta h_\chi + \tilde{V}$$

(23)

where

$$h_\rho = -\frac{\hbar^2}{2\mu\rho^{5/2}}\frac{\partial^2}{\partial\rho^2}\rho^{5/2}$$

(24)

$$h_\theta = -\frac{4}{\sin 2\theta}\frac{\partial}{\partial\theta}\sin 2\theta\frac{\partial}{\partial\theta}$$

(25)

$$h_\chi = -\frac{\partial^2}{\partial\chi^2}$$

(26)

$$f_\rho = \frac{\hbar^2}{2\mu\rho^2}$$

(27)

$$f_\theta = \frac{1}{\sin^2\theta}$$

(28)

$$\tilde{V} = V(\rho,\theta,\chi) + \frac{15\hbar^2}{8\mu\rho^2}.$$

(29)

Discretizing the application of the Hamiltonian on the wave function gives the following expression:

$$\mathcal{H}\psi_t = (\mathbf{f}_\rho \otimes \mathbf{f}_\theta \otimes \mathbf{h}_\chi + \mathbf{f}_\rho \otimes \mathbf{h}_\theta \otimes \mathbf{I}_\chi + \mathbf{h}_\rho \otimes \mathbf{I}_\theta \otimes \mathbf{I}_\chi + \tilde{\mathbf{V}})\text{vec}(\boldsymbol{\Psi}_t),$$

(30)

with $\{\mathbf{f}_\rho, \mathbf{h}_\rho\} \in \Re^{n_\rho \times n_\rho}$, $\{\mathbf{f}_\theta, \mathbf{h}_\theta\} \in \Re^{n_\theta \times n_\theta}$, $\mathbf{h}_\chi \in \Re^{n_\chi \times n_\chi}$, $\tilde{\mathbf{V}} \in \Re^{n_\rho n_\theta n_\chi \times n_\rho n_\theta n_\chi}$, $\boldsymbol{\Psi}_t \in \Re^{n_\chi \times n_\rho \times n_\theta}$, and $\psi_t \in \Re^{n_\chi n_\rho n_\theta \times 1}$. Also, $\mathbf{I}_\theta$ and $\mathbf{I}_\chi$ are $(n_\theta \times n_\theta)$ and $(n_\chi \times n_\chi)$ identity matrices, respectively. Note that $\tilde{\mathbf{V}}$, $\mathbf{f}_\rho$, and $\mathbf{f}_\theta$ are all diagonal matrices. This kronecker product form requires $2n_\rho n_\theta n_\chi^2 + 2n_\rho n_\theta^2 n_\chi + 2n_\rho^2 n_\theta n_\chi + n_\rho n_\theta n_\chi$ multiplications, not counting multiplications by zero.

We would like to reduce the number of scalar multiplications required to compute $\mathcal{H}\psi_t$ by expressing the right hand side of (30) as:

$$(\mathbf{f}_\rho \otimes \mathbf{f}_\theta \otimes \mathbf{h}_\chi + \mathbf{f}_\rho \otimes \mathbf{h}_\theta \otimes \mathbf{I}_\chi + \mathbf{h}_\rho \otimes \mathbf{I}_\theta \otimes \mathbf{I}_\chi + \tilde{\mathbf{V}})\text{vec}(\boldsymbol{\Psi}_t) = \text{vec}(\mathbf{Y}) + \tilde{\mathbf{V}}\text{vec}(\boldsymbol{\Psi}_t)$$

(31)

where $\mathbf{Y} \in \Re^{n_\chi \times n_\rho \times n_\theta}$ is obtained by regular matrix multiplication, rather than by kronecker product, and can be written as

$$\mathbf{Y} = \mathbf{Y}_\chi + \mathbf{Y}_\theta + \mathbf{Y}_\rho$$

(32)

$$\text{vec}(\mathbf{Y}) = \text{vec}(\mathbf{Y}_\chi) + \text{vec}(\mathbf{Y}_\theta) + \text{vec}(\mathbf{Y}_\rho).$$

(33)

The χ, θ, and ρ components of $\mathbf{Y}$ satisfy the equations

$$(\mathbf{f}_\rho \otimes \mathbf{f}_\theta \otimes \mathbf{h}_\chi)\mathrm{vec}(\mathbf{\Psi}_t) = \mathrm{vec}(\mathbf{Y}_\chi) \tag{34}$$

$$(\mathbf{f}_\rho \otimes \mathbf{h}_\theta \otimes \mathbf{I}_\chi)\mathrm{vec}(\mathbf{\Psi}_t) = \mathrm{vec}(\mathbf{Y}_\theta) \tag{35}$$

$$(\mathbf{h}_\rho \otimes \mathbf{I}_\theta \otimes \mathbf{I}_\chi)\mathrm{vec}(\mathbf{\Psi}_t) = \mathrm{vec}(\mathbf{Y}_\rho). \tag{36}$$

$\mathbf{Y}_\chi$: First, we will find $\mathbf{Y}_\chi$ using (34):

$$(\mathbf{f}_\rho \otimes \mathbf{f}_\theta \otimes \mathbf{h}_\chi) = \begin{pmatrix} f_\rho[1,1](\mathbf{f}_\theta \otimes \mathbf{h}_\chi) & 0 & \cdots & 0 \\ 0 & f_\rho[2,2](\mathbf{f}_\theta \otimes \mathbf{h}_\chi) & \cdots & 0 \\ \vdots & \vdots & \ddots & \vdots \\ 0 & 0 & \cdots & f_\rho[n_\rho,n_\rho](\mathbf{f}_\theta \otimes \mathbf{h}_\chi) \end{pmatrix} \tag{37a}$$

$$\mathrm{vec}(\mathbf{\Psi}_t) = \begin{pmatrix} \mathrm{vec}(\mathbf{\Psi}_t[:,1,:]) \\ \mathrm{vec}(\mathbf{\Psi}_t[:,2,:]) \\ \vdots \\ \mathrm{vec}(\mathbf{\Psi}_t[:,n_\rho,:]) \end{pmatrix}$$

$$(\mathbf{f}_\rho \otimes \mathbf{f}_\theta \otimes \mathbf{h}_\chi)\mathrm{vec}(\mathbf{\Psi}_t) = \begin{pmatrix} f_\rho[1,1](\mathbf{f}_\theta \otimes \mathbf{h}_\chi)\mathrm{vec}(\mathbf{\Psi}_t[:,1,:]) \\ f_\rho[2,2](\mathbf{f}_\theta \otimes \mathbf{h}_\chi)\mathrm{vec}(\mathbf{\Psi}_t[:,2,:]) \\ \vdots \\ f_\rho[n_\rho,n_\rho](\mathbf{f}_\theta \otimes \mathbf{h}_\chi)\mathrm{vec}(\mathbf{\Psi}_t[:,n_\rho,:]) \end{pmatrix} \tag{37b}$$

Using the 2D relation from (21) provides

$$(\mathbf{f}_\theta \otimes \mathbf{h}_\chi)\mathrm{vec}(\mathbf{\Psi}_t[:,j,:]) = \mathrm{vec}(\mathbf{h}_\chi \mathbf{\Psi}_t[:,j,:]\mathbf{f}_\theta) \tag{38}$$

giving

$$(\mathbf{f}_\rho \otimes \mathbf{f}_\theta \otimes \mathbf{h}_\chi)\mathrm{vec}(\mathbf{\Psi}_t) = \begin{pmatrix} \mathrm{vec}\,(f_\rho[1,1]\mathbf{h}_\chi\mathbf{\Psi}_t[:,1,:]\mathbf{f}_\theta) \\ \mathrm{vec}\,(f_\rho[2,2]\mathbf{h}_\chi\mathbf{\Psi}_t[:,2,:]\mathbf{f}_\theta) \\ \vdots \\ \mathrm{vec}\,(f_\rho[n_\rho,n_\rho]\mathbf{h}_\chi\mathbf{\Psi}_t[:,n_\rho,:]\mathbf{f}_\theta) \end{pmatrix} = \mathrm{vec}(\mathbf{Y}_\chi). \tag{39}$$

Then, $\mathbf{Y}_\chi$ is given by

$$\mathbf{Y}_\chi[:,j,:] = f_\rho[j,j]\mathbf{h}_\chi\mathbf{\Psi}_t[:,j,:]\mathbf{f}_\theta. \tag{40}$$

Obtaining $\mathbf{Y}_\chi$ using (40) requires $n_\rho(n_\theta n_\chi^2 + n_\theta n_\chi + 1)$ multiplications, but this can be made more efficient. If we compute the χ vectors of $\mathbf{Y}_\chi$ for each set of ρ and θ values according to

$$\mathbf{Y}_\chi[:,j,k] = f_\rho[j,j]\mathbf{h}_\chi\mathbf{\Psi}_t[:,j,:]\mathbf{f}_\theta[:,k] \tag{41}$$

$$= f_\rho[j,j]\mathbf{h}_\chi\mathbf{\Psi}_t[:,j,k]f_\theta[k,k] \tag{42}$$

where (42) arises since $\mathbf{f}_\theta$ is diagonal. This requires $n_\rho n_\theta(n_\chi^2 + 2)$ multiplications to obtain the full $\mathbf{Y}_\chi$, which is $n_\rho(n_\theta n_\chi - 2n_\theta + 1)$ less than for (40).

$\mathbf{Y}_\theta$: Next, we will find $\mathbf{Y}_\theta$ using (35). Just as in (37:

$$(\mathbf{f}_\rho \otimes \mathbf{h}_\theta \otimes \mathbf{I}_\chi)\mathrm{vec}(\mathbf{\Psi}_t) = \begin{pmatrix} f_\rho[1,1](\mathbf{h}_\theta \otimes \mathbf{I}_\chi)\mathrm{vec}(\mathbf{\Psi}_t[:,1,:]) \\ f_\rho[2,2](\mathbf{h}_\theta \otimes \mathbf{I}_\chi)\mathrm{vec}(\mathbf{\Psi}_t[:,2,:]) \\ \vdots \\ f_\rho[n_\rho,n_\rho](\mathbf{h}_\theta \otimes \mathbf{I}_\chi)\mathrm{vec}(\mathbf{\Psi}_t[:,n_\rho,:]) \end{pmatrix} \tag{43}$$

Using the 2D relation from (21) provides

$$\begin{aligned} (\mathbf{h}_\theta \otimes \mathbf{I}_\chi)\mathrm{vec}(\mathbf{\Psi}_t[:,j,:]) &= \mathrm{vec}(\mathbf{I}_\chi \mathbf{\Psi}_t[:,j,:]\mathbf{h}_\theta^T) \\ &= \mathrm{vec}(\mathbf{\Psi}_t[:,j,:]\mathbf{h}_\theta^T) \end{aligned} \tag{44}$$

giving

$$(\mathbf{f}_\rho \otimes \mathbf{h}_\theta \otimes \mathbf{I}_\chi)\mathrm{vec}(\mathbf{\Psi}_t) =$$

$$= \begin{pmatrix} \mathrm{vec}\left(f_\rho[1,1]\mathbf{\Psi}_t[:,1,:]\mathbf{h}_\theta^T\right) \\ \mathrm{vec}\left(f_\rho[2,2]\mathbf{\Psi}_t[:,2,:]\mathbf{h}_\theta^T\right) \\ \vdots \\ \mathrm{vec}\left(f_\rho[n_\rho,n_\rho]\mathbf{\Psi}_t[:,n_\rho,:]\mathbf{h}_\theta^T\right) \end{pmatrix}$$

$$= \begin{pmatrix} \mathrm{vec}\left(\mathbf{Y}_\theta[:,1,:]\right) \\ \mathrm{vec}\left(\mathbf{Y}_\theta[:,2,:]\right) \\ \vdots \\ \mathrm{vec}\left(\mathbf{Y}_\theta[:,n_\rho,:]\right) \end{pmatrix} \tag{45}$$

$$= \mathrm{vec}(\mathbf{Y}_\theta).$$

Then, $\mathbf{Y}_\theta$ is given by

$$\mathbf{Y}_\theta[:,j,:] = f_\rho[j,j]\mathbf{\Psi}_t[:,j,:]\mathbf{h}_\theta^T. \tag{46}$$

Obtaining $\mathbf{Y}_\theta$ using (46) requires $n_\rho(n_\theta^2 n_\chi + 1)$ multiplications, and this is the most efficient form.

$\mathbf{Y}_\rho$: Finally, we will find $\mathbf{Y}_\rho$ using (36):

$$(\mathbf{h}_\rho \otimes \mathbf{I}_\theta \otimes \mathbf{I}_\chi)\mathrm{vec}(\mathbf{\Psi}_t) = (\mathbf{h}_\rho \otimes \mathbf{I}_{\chi\theta})\mathrm{vec}\begin{pmatrix} \mathbf{\Psi}_t[:,:,1] \\ \mathbf{\Psi}_t[:,:,2] \\ \vdots \\ \mathbf{\Psi}_t[:,:,n_\theta] \end{pmatrix} \tag{47}$$

where $\mathbf{I}_{\chi\theta}$ is an $(n_\theta n_\chi \times n_\theta n_\chi)$ identity matrix obtained from

$$\mathbf{I}_\theta \otimes \mathbf{I}_\chi = \mathbf{I}_{\chi\theta} \tag{48}$$

and

$$\text{vec}(\boldsymbol{\Psi}_t) = \text{vec}\begin{pmatrix} \boldsymbol{\Psi}_t[:,:,1] \\ \boldsymbol{\Psi}_t[:,:,2] \\ \vdots \\ \boldsymbol{\Psi}_t[:,:,n_\theta] \end{pmatrix} \tag{49}$$

as shown in (19). Using the 2D relation from (21) provides

$$(\mathbf{h}_\rho \otimes \mathbf{I}_{\chi\theta})\text{vec}\begin{pmatrix} \boldsymbol{\Psi}_t[:,:,1] \\ \boldsymbol{\Psi}_t[:,:,2] \\ \vdots \\ \boldsymbol{\Psi}_t[:,:,n_\theta] \end{pmatrix} = \text{vec}\left\{ \mathbf{I}_\chi \begin{pmatrix} \boldsymbol{\Psi}_t[:,:,1] \\ \boldsymbol{\Psi}_t[:,:,2] \\ \vdots \\ \boldsymbol{\Psi}_t[:,:,n_\theta] \end{pmatrix} \mathbf{h}_\rho^T \right\}$$

$$= \text{vec}\begin{pmatrix} \boldsymbol{\Psi}_t[:,:,1]\mathbf{h}_\rho^T \\ \boldsymbol{\Psi}_t[:,:,2]\mathbf{h}_\rho^T \\ \vdots \\ \boldsymbol{\Psi}_t[:,:,n_\theta]\mathbf{h}_\rho^T \end{pmatrix} \tag{50}$$

Then,

$$(\mathbf{h}_\rho \otimes \mathbf{I}_\theta \otimes \mathbf{I}_\chi)\text{vec}(\boldsymbol{\Psi}_t) = \text{vec}\begin{pmatrix} \boldsymbol{\Psi}_t[:,:,1]\mathbf{h}_\rho^T \\ \boldsymbol{\Psi}_t[:,:,2]\mathbf{h}_\rho^T \\ \vdots \\ \boldsymbol{\Psi}_t[:,:,n_\theta]\mathbf{h}_\rho^T \end{pmatrix} = \begin{pmatrix} \text{vec}\,(\mathbf{Y}_\rho[:,:,1]) \\ \text{vec}\,(\mathbf{Y}_\rho[:,:,2]) \\ \vdots \\ \text{vec}\,(\mathbf{Y}_\rho[:,:,n_\theta]) \end{pmatrix} = \text{vec}(\mathbf{Y}_\rho). \tag{51}$$

Then, $\mathbf{Y}_\rho$ is given by

$$\mathbf{Y}_\rho[:,:,k] = \boldsymbol{\Psi}_t[:,:,k]\mathbf{h}_\rho^T. \tag{52}$$

Obtaining $\mathbf{Y}_\rho$ using (52) requires $n_\rho^2 n_\theta n_\chi$ multiplications, and this is the most efficient form.

In summary, the 3D kronecker form can be expressed as

$$(\mathbf{f}_\rho \otimes \mathbf{f}_\theta \otimes \mathbf{h}_\chi + \mathbf{f}_\rho \otimes \mathbf{h}_\theta \otimes \mathbf{I}_\chi + \mathbf{h}_\rho \otimes \mathbf{I}_\theta \otimes \mathbf{I}_\chi + \tilde{\mathbf{V}})\text{vec}(\boldsymbol{\Psi}_t) =$$
$$\text{vec}(\mathbf{Y}_\chi) + \text{vec}(\mathbf{Y}_\theta) + \text{vec}(\mathbf{Y}_\rho) + \tilde{\mathbf{V}}\text{vec}(\boldsymbol{\Psi}_t) \tag{53}$$

with

$$\mathbf{Y}_\chi[:,j,k] = f_\rho[j,j]f_\theta[k,k]\mathbf{h}_\chi\boldsymbol{\Psi}_t[:,j,k] \tag{54}$$

$$\mathbf{Y}_\theta[:,j,:] = f_\rho[j,j]\boldsymbol{\Psi}_t[:,j,:]\mathbf{h}_\theta^T \tag{55}$$

$$\mathbf{Y}_\rho[:,:,k] = \boldsymbol{\Psi}_t[:,:,k]\mathbf{h}_\rho^T. \tag{56}$$

The kronecker form requires $2n_\rho n_\theta n_\chi^2 + 2n_\rho n_\theta^2 n_\chi + 2n_\rho^2 n_\theta n_\chi + n_\rho n_\theta n_\chi$ multiplications, while the matrix multiply form requires $n_\rho n_\theta(n_\chi^2 + 2) + n_\rho(n_\theta^2 n_\chi + 1) + n_\rho^2 n_\theta n_\chi + n_\rho n_\theta n_\chi$ multiplications. The matrix multiply form requires $n_\rho^2 n_\theta n_\chi + n_\rho n_\theta^2 n_\chi + n_\rho n_\theta n_\chi^2 - n_\rho n_\theta - n_\rho$ less multiplications than the kronecker form.

4   Parallelization

4.1   Parallelizing the Hamiltonian

Using (53 - 56) we can begin to parallelize the application of the APH Hamiltonian. First we separate each component part:

$$\mathcal{H}\boldsymbol{\Psi}_t = \mathrm{vec}(f_\rho[j,j]f_\theta[k,k]\mathbf{h}_\chi\boldsymbol{\Psi}_t[:,j,k]) + \mathrm{vec}(f_\rho[j,j]\boldsymbol{\Psi}_t[:,j,:]\mathbf{h}_\theta^T) +$$
$$\mathrm{vec}(\boldsymbol{\Psi}_t[:,:,k]\mathbf{h}_\rho^T) + \tilde{\mathbf{V}}\mathrm{vec}(\boldsymbol{\Psi}_t) \tag{57}$$

Now we see that none of these parts depends on the value of the others, so we can compute them individually and then simply add the resultant vectors after all calculations are completed. This was done with OpenMP in Fortran 90, as in the following code segment. (This is not an actual code segment, but rather an illustrative representation not dissimilar from the actual code.)

```
!=========
!APH Hamiltonian applicaton
!BEGIN T-Parallelization
!========
!$OMP PARALLEL DEFAULT(SHARED)&
!$OMP SECTIONS
!$OMP SECTION
! Apply T_rho:
 CALL trho_apply(psi3,rho3)
!$OMP SECTION
! Apply T_theta:
 CALL ttheta_apply(psi3,theta3)
!$OMP SECTION
! Apply T_chi
 CALL tchi_apply(psi3,chi3)
!$OMP SECTION
!Apply V
 CALL v_apply(nchir,nlength,psi3,vpsi)
!$OMP END SECTIONS
!$OMP END PARALLEL
 DO irho=1,nrho
   DO itheta=1,ntheta
     DO ichi=1,nchir
       i = aphindex(ntheta,nchir,irho,itheta,ichi)
       wavefunction(i)= &
       rho3(ichi, irho, itheta) + theta3(ichi, irho, itheta) &
   + chi3(ichi, irho, itheta) + vpsi(i)
     ENDDO
   ENDDO
ENDDO
```

Here, we open the parallelization and specify that all variables are shared–this ensures that all processes have access to the same initial/input wavefunction "psi3." Each section directive then initiates a new process which simply calls the relevant subroutine for each component ρ, θ, χ. Note that the input wavefunction and (for the kinetic energies) the output matrices are three-dimensional and specified by component. For instance, "rho3" is the matrix for the application of the ρ component of the kinetic energy operator onto Ψ_t. The potential energy subroutine produces a vector, however. After the parallel region is ended, the final series of DO loops uses a subroutine to determine the proper vector index corresponding to each three-dimensional element (calculated by function aphindex) and the three-dimensional matrices and $\tilde{V}\mathrm{vec}(\Psi_t)$ are vectorized as they are added together into "wavefunction", which corresponds to $\mathcal{H}\Psi_t$.

As noted at the end of the vectorization section, the matrix multiply form required $n_\rho n_\theta (n_\chi^2 + 2) + n_\rho (n_\theta^2 n_\chi + 1) + n_\rho^2 n_\theta n_\chi + n_\rho n_\theta n_\chi$ multiplications. The parallelized form, because it exectues these terms simultaneously, takes only as much time as the longest term (not counting the computational overhead involved in forking and merging processes). Since n_χ is nearly always the largest factor in these terms, the term for which n_χ is quadratic will usually be the largest and therefore determine the total computation time required, meaning that the calculation will be able to perform $n_\rho n_\theta (n_\chi^2 + 2) + n_\rho (n_\theta^2 n_\chi + 1) + n_\rho^2 n_\theta n_\chi + n_\rho n_\theta n_\chi$ multiplications in as much time as it takes to execute $n_\rho n_\theta (n_\chi^2 + 2)$ multiplications. Other considerations can reduce the number of multiplications required. For example, our matrices are banded, and therefore require less multiplications than a full matrix would.

4.2   Other Parallelizations

Besides the application of $\mathcal{H}$ on Ψ_t, there are several other places the method could be (further) parallelized. As we propagate the wavefunction's components in irreducible representation Γ, parallelization could be used to propagate these separately before combining them for final analysis. The benefit gained by this would depend on initial quantum numbers and parity, as well as the precise system–for AAA systems (belonging to the C_{6v} point group) with no initial angular momentum and even parity, the irreducible representations are A1 and E2. Here, E2 requires twice the coordinate space in n_χ as A1, and so would representing the limiting factor in time for this parallelization. This would suggest the time required to complete the full propagation of all irreducible representations would be reduced by a third if this were implemented. For other systems and other initial quantum states, the time gained by this parallelization would vary. In addition, the wavefunction could be split into its real and imaginary parts:

$$\mathcal{H}\Psi_t = \mathcal{H}\Psi_{real} + i\mathcal{H}\Psi_{imag} \tag{58}$$

Since both of these components should take the same amount of time, this could in theory reduce the time required for each application of the Hamiltonian by a factor of two if the Hamiltonian were applied simultaneously to both.

The resource expended for these improvements is additional processor usage, and their nested nature makes stacking them potentially impractical on many machines. Parallelizing the Hamiltonian application as shown in the code segment requires four processors. If the wavefunction is split into real and imaginary parts, eight processors are used. If–for an AAA system as discussed in the previous paragraph–both irreducible representations are propagated simultaneously, then sixteen processors would be required to achieve the proper speed-up. Obviously, many machines do not have that many cores available, making that level of parallelization infeasible.

5 Results

Parallelized code was compared with the same code compiled in single-threaded mode and the times required to complete the full propagation were noted. These results only account for the parallelization of the application of the Hamiltonian; the additional possible parallelizations proposed in section 4.2 are not implemented. First, in Table 1, we see the effects on a variety of different systems–$H + H_2$, $F + H_2$, and Li_3. In all systems, the parallel times were significantly lower than the serial code, giving a reduction of about 50% in total time required for these systems on appropriately dense grids (those dense enough to yield results).

Table 1. Parallelization Time Results

System	$H + H_2$	$F + H_2$	Li_3
$(n_\rho, n_\theta, n_\chi)$	(160,45,360)	(200,75,600)	(135,65,360)
Serial Time (minutes)	5.89	71.66	53.45
Parallel Times (minutes)	3.68	35.27	29.26
Percent Reduction (minutes)	39.0%	50.8%	45.3%

It is also interesting to look at the same system as the grid density becomes larger. In Fig. 1 we see the behavior of both single- and multi-threaded code as grid density increases in each coordinate. Clearly, the multi-threaded code outperforms the single-threaded variety. In addition, the multi-threaded code scales better as density increases. A quadratic fit applied to the single-threaded code for the n_χ plot gives a scaling of of $0.22n_\chi + 0.00064n_\chi^2$ while the multi-threaded code scales as $0.15n_\chi + 0.00040n_\chi^2$. For n_θ we also have an improvement of a different nature, with the single-threaded fit being $2.98n_\theta + 0.024n_\theta^2$ and a parallel fit of $0.19n_\theta + 0.030n_\theta^2$. Interestingly, here the speedup occurs in the linear scaling. This may be a result of the quadratic n_θ term dominating, becoming the determining factor in the time required for the parallel application of the Hamiltonian. The other terms–which are linear in n_θ do not contribute to the time required. Finally, we can look at the scaling in n_ρ. A linear fit applied to this data shows a slope of 1.19 for the serial code and .66 for the parallel code. On higher-density grids, then, the parallel code has an increasing advantage over the single-threaded version, with the possible exception of n_θ. However, if the fitted curves are valid for larger values of n_θ, the parallel version would remain faster until n_θ became as large as 467, a number far larger than usually used. It is also important to note that no difference in results is observed between the parallel and single-threaded codes.

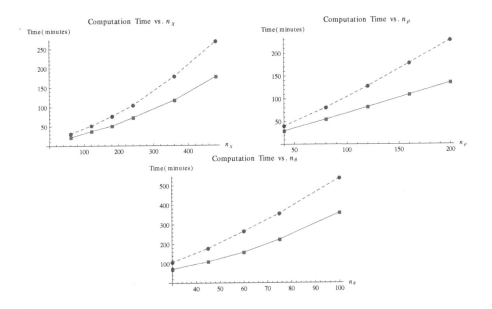

Fig. 1. Single-threaded times (blue, dashed) in comparison with the parallelized (purple, solid) times. Programs were run for $H + H_2$ with grid parameters at $n_\rho = 160$, $n_\theta = 45$, $n_\chi = 360$, with one of these parameters varying.

References

[1] Kuppermann, A.J., Schatz, G.C.: J. Chem. Phys. 62, 2502 (1975)
[2] Elkowitz, A.B., Wyatt, R.E.: J. Chem. Phys. 62, 2504 (1975a)
[3] Elkowitz, A.B., Wyatt, R.E.: J. Chem. Phys. 63, 702 (1975b)
[4] Pack, R.T., Parker, G.A.: J. Chem. Phys. 87, 3888 (1987), and references therein
[5] Crawford, J., Parker, G.A.: J. Chem. Phys. "State-to-State Three-Atom Time-Dependent Reactive Scattering in Hyperspherical Coordinates" (unpublished)
[6] Balakrishnan, N., Kalyanaraman, C., Sathyamurthy, N.: Phys. Rep. 280, 79 (1997)
[7] Colavecchia, F.D., Mrugala, F., Parker, G.A., Pack, R.T.: J. Chem. Phys. 118, 10387 (2003), and references therein

[8] Tal-Ezer, H., Kosloff, R.: J. Chem. Phys. 81, 3967 (1984)

[9] Cotton, F.A.: Chemical Applications of Group Theory, 3rd edn. Wiley, New York (1990)

[10] Iyengar, S.S., Parker, G.A., Kouri, D.J., Hoffman, D.K.: J. Chem. Phys. 110, 10283 (1999)

[11] Abadir, K.M., Magnus, J.R.: Matrix Algebra. Cambridge University Press, New York (2005)

Multi Reference versus Coupled Cluster ab Initio Calculations for the N2 + N2 Reaction Channels

Leonardo Pacifici, Marco Verdicchio, and Antonio Laganà

Department of Chemistry, University of Perugia, via Elce di Sotto, 8,
06123 Perugia (Italy)

Abstract. By making use of a Grid enabled *ab initio* molecular simulator we have tackled the *a priori* study of the $N_2(^1\Sigma_g^+) + N_2(^1\Sigma_g^+)$
process. A detailed analysis of the results obtained from high level *ab
initio* (Coupled Cluster) calculation of the electronic structure of N_4 for
a large number of nuclear geometries has singled out the fact that Coupled Cluster calculations are insufficiently accurate when the internuclear
distances of the approaching N_2 diatoms are stretched, because in such
cases the wavefunction of the N_4 system cannot be properly described by
a single determinant. For this reason we have carried out Multi Reference
calculations (using the same basis set) for a large number of the nuclear
geometries considered for the Coupled Cluster study. Then, a 4-atoms
global potential energy surface has been worked out for use in dynamics
calculations.

1 Introduction

In recent years, important advances in theoretical and experimental investigations on weakly bound molecular clusters and complexes have been reported [1,
2]. A large fraction of these studies has been carried out to better characterize
the dynamics of the formation of simple molecular clusters (such as the $(H_2)_2$
dimer [3]) as well as other important properties. This has allowed the assemblage of a potential energy surface (PES) suitable to describe, for example, H_2
+ H_2 collision processes [4]. Recently, growing efforts have been devoted also to
the determination of the interaction between two nitrogen molecules, the $(N_2)_2$
dimer. This system is important in high temperature atmospheric chemistry of
Earth (like that occurring in spacecraft reentry [5, 6]) and in low temperature
astrochemistry as well, in which the N_2 dimer plays an important role (like in
the atmospheric processes of Titan). For this reason related investigations have
been extended also to low temperature [7].

A first attempt to model an empirical potential for the nitrogen dimer was
carried out some time ago by fitting a Lennard-Jones potential to the second
virial coefficient and solid state data [8–12]. In more recent years, *ab initio* and
experimental data were combined to the end of improving the empirical formulation of the $(N_2)_2$ PES [13–15] and a partially *ab initio* PES was assembled by Van

B. Murgante et al. (Eds.): ICCSA 2013, Part I, LNCS 7971, pp. 31–46, 2013.
© Springer-Verlag Berlin Heidelberg 2013

der Avoird, Wormer and Jansen (AWJ) [16] by combining *ab initio* estimates of the short range $(N_2)_2$ interaction and long range electrostatic and dispersion terms. Still it was impossible to reproduce most of the available experimental and theoretical data [17, 18].

The first full *ab initio* PES of $(N_2)_2$ was developed by Stallcop and Partridge [19] by combining short and intermediate range potential energy values calculated at the CCSD(T) (Coupled-Cluster with Single and Double and perturbative Triple excitations) level of theory, with those obtained for the Van der Waals region from the evaluation of the second virial coefficients. Subsequently, further theoretical investigations were performed in order to better calibrate the PES [20–23].

However, despite the above mentioned efforts, the intermolecular interaction of $(N_2)_2$ is far from being well characterized. The fact that the system is unsuitable for experimental investigations (due to its optical inactivity) and that the potential strongly depends on both the distance and the mutual orientation of the two molecules, makes its accurate investigation quite difficult. As a matter of fact, *ab initio* calculations of the electronic structure of $(N_2)_2$ require both HTC (High Throughput Computing) and HPC (High Performance Computing). For this reason in our work within COMPCHEM [24] (the EGI (European Grid Infrastructure) [25] Virtual Organization (VO) of the Chemistry and Molecular & Material Science and Technology Community (CMMST)), aimed at designing the so called Grid Empowered Molecular Simulator (GEMS) [26–28], we chose as a case study the *ab initio* calculation of the geometries of the $N_2(^1\Sigma_g^+)$ + $N_2(^1\Sigma_g^+)$ system relevant to the related nitrogen atom exchange reaction [29].

Accordingly, in Section 2 we illustrate the coordinate system used and the investigated arrangements, in Section 3 we present the results of Coupled Cluster *ab initio* calculations and in Section 4 we discuss the related fit to assemble a proper PES. Then, in Section 5 we describe the corresponding higher level Multi Reference *ab initio* calculations. In Section 6 the fit of the mixed Multi Reference CCSD(T) set of energy points is reported and in Section 7 our conclusions are given.

2   The Coordinates Used

As discussed in ref [30] in order to build an initial *ab initio* picture of the $(N_2)_2$ intermolecular interaction we have already carried out *ab initio* calculations at a MP2 (Second order Møller-Plesset perturbation theory) [31] *ab initio* level by correcting the BSSE (Basis Set Superposition Error) via the full counterpoise procedure (FCP) [32], using a relatively small basis set (called cc-pVTZ (correlation consistent polarized triple valence)) involving 140 gaussian functions retrieved from the EMSL public database [33, 34]. We have also already performed CCSD(T) [35–37] *ab initio* calculations. Both calculations were performed using the serial version of GAMESS-US [38, 39] offered as a Grid service by the COMPCHEM Virtual Organization [24], that was running on typical Grid single-core machines.

For the definition of the geometry of the system a space fixed axis frame with the z axis passing through the centers of mass of the two nitrogen molecules was adopted. Using this kind of coordinate frame the relative positions of the nuclei

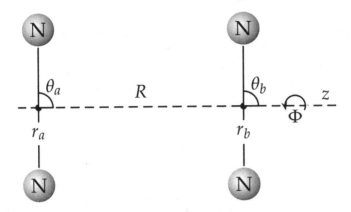

Fig. 1. Scheme of the radial and angular variables adopted to define the geometry of the $(N_2)_2$ system

of the N_4 4 atom system considered in our work in order to create the grid of points (geometries) of the discrete representation of the PES was uniquely defined in terms of the 6 variables (three radial and three angular, as shown in Figure 1). Accordingly, the orientation of the two diatomic internuclear vectors $\mathbf{r}_a$ and $\mathbf{r}_b$ with respect to the z axis of the space fixed frame is given by the angles θ_a and θ_b; the relative orientation of the planes formed by diatom a and the z axis and diatom b and the same axis is given by the dihedral angle Φ; the distance between the centers of mass of the two nitrogen molecules is given by R; the two intramolecular distances are given by the moduli r_a and r_b of the $\mathbf{r}_a$ and $\mathbf{r}_b$ vectors. The different reaction channels of the PES were therefore identified by five different θ_a, θ_b, Φ triples for R varying from 1 to 8 Å at fixed value of the bond length of the two monomers. Each of these (θ_a, θ_b, Φ; R) combinations (called arrangements, hereinafter) belongs to a different symmetry group (namely: D_{2h} (90, 90, 0; R), D_{2d} (90, 90, 90; R), $D_{\infty h}$ (0, 0, 0; R), C_{2v} (0, 90, 0; R), C_{2h} (45, 45, 0; R)). The different arrangements are labeled as H (parallel), X (X-shaped), L (linear), T (T-shaped) and Z (Z-shaped), respectively, as illustrated in Figure 2. A first set of calculations was performed by setting $r_a = r_b = 1.094$ Å (the N_2 equilibrium distance [40]). Other sets of calculations were performed by varying the Φ angle from $0°$ (H shape) to $90°$ (X shape), by a step of $15°$ in order to fully characterize the effect on the potential energy of distorting gradually the parallel arrangement into a crossed one (P15, P30, P45, P60, P75). At the same time, to sample the effect of stretching the N_2 bonds, further calculations were performed by varying the internuclear distances r_a and r_b as follows: $r_a = 1.094$ Å, $r_b = 1.694$ Å; $r_a = 1.494$ Å, $r_b = 1.694$ Å; $r_a = 1.694$ Å, $r_b = 1.694$ Å.

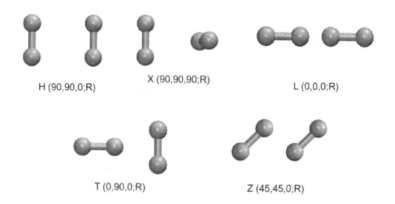

H (90,90,0;R)   X (90,90,90;R)   L (0,0,0;R)

T (0,90,0;R)   Z (45,45,0;R)

Fig. 2. Summary of the five investigated arrangements

3   The High-Level Coupled Cluster Calculations

In order to improve on the results obtained from the low level of theory *ab initio* calculations performed using the computational machinery illustrated above (the MP2 [31] method and the cc-pVTZ small basis using 140 gaussian functions) we resorted into using a CCSD(T) method and a cc-pV5Z [33, 34] basis set made of a total of 420 gaussian contractions. This eliminated the large BSSE that unphysically lowers the energy in the short range region. Using such basis set we carried out the calculations for the H, X, L, T, Z, P15, P30, P45, P60 and P75 arrangements. Obviously, the use of the larger basis set while improving the accuracy increased also the size of the matrices used in the calculations and made the computational cost three orders of magnitude larger (when compared with the smaller cc-pVTZ basis set). This prompted the use of HPC platforms through a cross submission from the EGI Grid (using gLite middleware) to the supercomputer platform of CINECA [41] on which parallel versions of the electronic structure *ab initio* package GAMESS-US 2009 [38] are available.

Due to their importance for reactive exchange our efforts focused on the calculation of the electronic energy values for the parallel (H-shape) and crossed (X-shape) arrangements. In Fig. 3 the long range CCSD(T) potential energy values are plotted as function of R, for the H, X, P30 and P60 arrangements. H and X arrangements show in the long range region a well deep 9.38 meV (with the minimum located at R=3.74 Å) and 11.55 meV (with the minimum located at R=3.64 Å), respectively. Moreover, in going from the H to the X arrangement the location on R of the minimum lowers gradually while the depth deepens. The pseudo 3D sketch given in the lower right hand side corner of FIg. 3 illustrates such variation in going from $\Phi = 0°$ to $\Phi = 90°$. No evidence has been found for the formation of stable structures at short range, whereas the opening of a reactive exchange channel when moving from a X to a H arrangements looks likely.

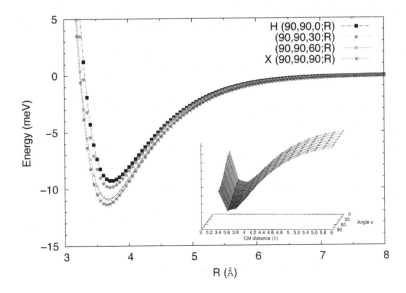

Fig. 3. Long range CCSD(T) potential energy curves plotted as a function of R at different values of Φ (a pseudo 3D representation is given in the right lower corner)

In order to better investigate this aspect we also calculated the evolution of the potential energy of the H and X arrangements when one or both intramolecular distances are stretched with respect to the N_2 equilibrium distance ($R_{eq} = 1.094$ Å). In particular, the internuclear distances were varied from 0.4 to 0.6 Å (see Figure 4 for the plot of the potential energy curves of the H and X arrangements). The analysis showed, however, that when the intramolecular distance is increased from equilibrium the role of non-dynamical correlation energy becomes important and the wavefunction of the system can not be properly described by a single determinant. Accordingly, for N_2 internuclear distances differing from the equilibrium value CCSD(T) values could not be trusted anymore, especially at short intermolecular distance. For this reason they were discarded with the exception of those that could be adjusted to fit some neighbor valid points.

4   The Fit of the *ab Initio* Potential Energy Values

In order to carry out the fitting of the calculated CCSD(T)/cc-pV5Z results we used the computational procedure of Paniagua et al. based on the MBE method [42] by embodying the gfit4c package [43] into GEMS. Three different sets of *ab initio* points (1440 two-body values, 4320 three-body values and 1917 four-body values) were used to fit the $N_2(^1\Sigma_g^+) + N_2(^1\Sigma_g^+)$ PES. *Ab initio* data and experimental information (when available) were used to tune both the two and three body terms thanks to the results of the work previously carried out for the assemblage of the N + N_2 PES [44, 45]. Moreover, the zero of all the *ab*

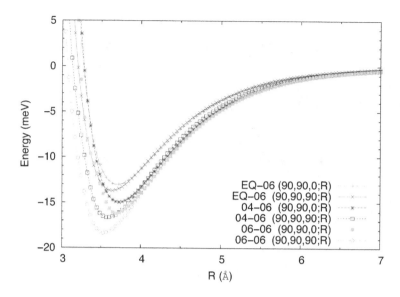

Fig. 4. CCSD(T) Van der Waals potential energy curves plotted as a function of R of the three stretched geometries for the H and X arrangements. The plus and cross symbols refer to the arrangements with one internuclear distance at the equilibrium and the other one stretched by 0.6 Å (EQ-06); the star and white-square symbols refer to the arrangements with one distance stretched by 0.4 Å and the other one by 0.6 Å (04-06) whereas the circle and the black-square symbols refer to the arrangements with both internuclear distances stretched by 0.6 Å (06-06).

initio values was set at the dissociation energy of the dimers. The scaling value has been calculated at the CCSD(T) level of theory using the same basis set. In particular, we calibrated the calculated potential energy values with respect to the energy point obtained by increasing R to a very large distance (20 Å) and by matching the obtained value with the one obtained for the isolated nitrogen molecules (the value used is -218.82612523 Hartree). The fitted two and three body components of the potential (both fitted to a polynomial of order 5) show a root mean square deviation of 0.005 and 0.05 eV, respectively, while that of the four body term is 0.2 eV with the largest deviation (up to 2 eV) occurring in highly repulsive regions.

The ability of the fitted PES to describe the N_2 dissociation was tested. The fitted PES gives a dissociation energy of 0.361 hartree at an equilibrium distance of 1.099 Å in good agreement with the experimental values (1.098 Å and 0.358 hartree [46, 47], respectively). In order to give a more global view of the fitted PES we show its isoenergetic contours in Figure 5. The contours of Figure 5 refer to the fixed R H arrangement and variable intramolecular r_a and r_b distances. The plots show that when the intermolecular distance is short (the $R = 3$ Å case of the left hand side panel), the interaction between the two stretching molecules leads clearly to the formation of two dissociating away channels separated by a

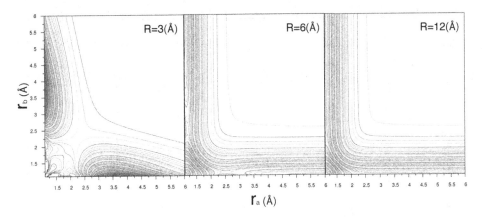

Fig. 5. Isoenergetic contour plots for the H arrangement at R=3 (left hand side panel), 6 (central panel) and 12 (right hand side panel) Å. Energy contours are drawn every 6 Kcal/mol.

quite large barrier. When R increases (the $R = 6$ Å case of the central panel) the interaction between the two molecules decreases. Yet at intermediate R values the two dissociation channels are separated from a quite stable insertion complex by two symmetric small barriers. On the other hand, at large R values (the $R = 12$ Å case of the right hand side panel) the two dissociation channels smoothly (with no barriers) conflue into the insertion minimum.

In order to better characterize the role played by the features of the short range region in driving the system to react via an exchange of the N atoms, we plot in Fig. 6 the isoenergetic contours for the H arrangement by varying $r_a = r_b$ while increasing R. The figure shows the occurrence of sideway barriers at short separation of the nuclei. In particular, for R, r_a and r_b all equal to about 2.5 Å the PES shows a deep well of -1.77 eV opening reactive path involving a clear change of arrangement.

5   The Multi Reference Calculations of the Potential Energy Surface

As already mentioned, when the interatomic distances are stretched from their equilibrium value, the N_4 system wavefunction is not properly described by a single determinant. When a triple bond, like that of the N_2 dimer, is stretched non-dynamical correlation becomes important and the electronic structure needs to be calculated using a Multi Reference (MR) method. This is particularly important when calculating the short range region of the PES where the two N_2 molecules are very close and the stretched configurations may become important for the description of bond breaking and formation.

In order to have a higher level of theory description of the interactions governing the dynamics of possible dissociation and recombination processes, we

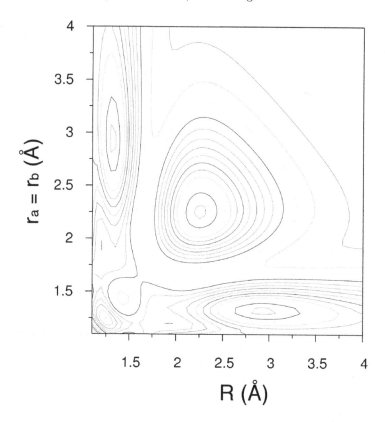

Fig. 6. Isoenergetic contour plots for the H arrangement with $r_a = r_b$. Energy contours are drawn every 50 Kcal/mol.

have repeated the calculations using the MRPT2 computational technique implemented in GAMESS-US [48, 49]. This method makes use, in fact, of a CASSCF reference wavefunction that allows the recovery of the static correlation energy and of the second order perturbation theory dynamical one. In particular, for the $N_2 + N_2$ case, it was chosen to correlate the two degenerate π_{2p} orbitals and the σ_{2p} orbital of each molecule while keeping the inner σ orbitals frozen at the SCF level. At this level of theory we repeated the calculations for the following arrangements:

- H, X, P15, P30, P45, P60 and P75 for $r_a = r_b = 1.094$ Å
- H, X, P15, P30, P45, P60 and P75 for $r_a = 1.094$ Å $r_b = 1.694$ Å
- H, X, P15, P30, P45, P60 and P75 for $r_a = 1.494$ Å $r_b = 1.694$ Å
- H, X, P15, P30, P45, P60 and P75 for $r_a = 1.694$ Å $r_b = 1.694$ Å

for values of R varying from 1 Å to 5 Å (though, in some cases, up to 10 Å). The outcomes of these calculations are shown in Figures 7, 8, 9 and 10 in which the potential energy zero is set at the complete separation of the two N_2 monomers.

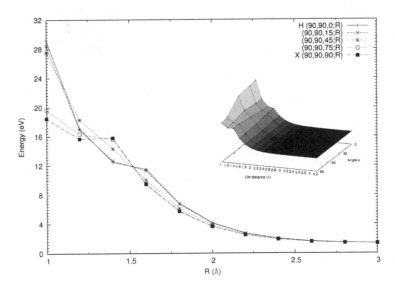

Fig. 7. Short range MRPT2 potential energy values plotted as a function of R for different values of Φ. The interatomic distances of each N_2 molecule are kept fixed at the equilibrium value ($r_a = r_b = 1.094$ Å).

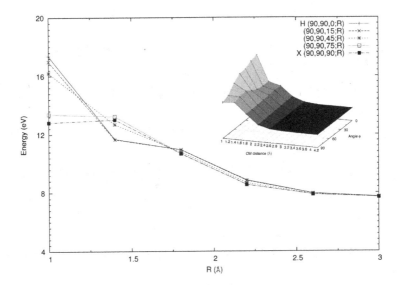

Fig. 8. Short range MRPT2 potential energy values plotted as a function of R for different values of Φ. One interatomic distance is stretched by 0.6 Å and the other is kept fixed at its equilibrium value ($r_a = 1.094$ Å , $r_b = 1.694$ Å).

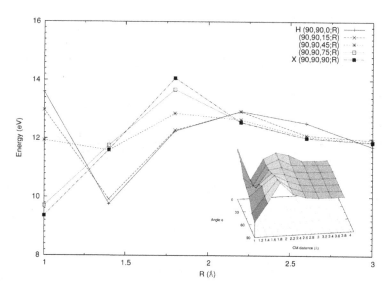

Fig. 9. Short range MRPT2 potential energy values plotted as a function of R for different values of Φ. One interatomic distance is asymmetrically stretched by 0.6 Å and the other one by 0.4 Å ($r_a = 1.494$ Å , $r_b = 1.694$ Å).

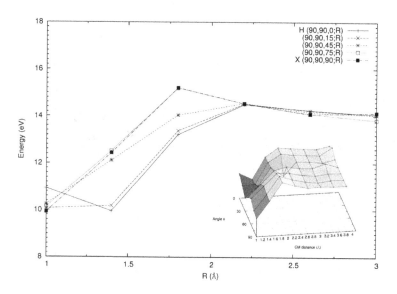

Fig. 10. Short range MRPT2 potential energy values plotted as a function of R for different values of Φ. Both interatomic distances are symmetrically stretched by 0.6 Å ($r_a = 1.694$ Å , $r_b = 1.694$ Å).

A pseudo three-dimensional sketch of the PES is also reported in rhs of the figures for a better understanding of its topology. As shown by Figure 7 the profile of the curves when both N_2 bonds are at the equilibrium value is close to the one obtained when using the CCSD(T) method. The positions and heights of the energy barriers for the two methods are also comparable, confirming the good quality of the CCSD(T) calculations for these arrangements. Figure 7, however, indicates that potential energy does not tend to the CCSD(T) value as R increases. This is due to the different method adopted for the calculation of the energy zero in CCSD(T) and MRPT2. Under optimal conditions, in fact, CCSD(T) methods are able to recover a fraction of the dynamical correlation larger than that of MRPT2. Such difference (1.3 eV) was estimated by a comparison of the values calculated using the two methods in the long range region of non stretched N_2 configurations.

The potential energy values calculated using the two levels of theory differ significantly when stretching the N_2 internuclear distances. As mentioned above, CCSD(T) calculations are inadequate to describe these configurations and the adoption of a Multi Reference picture is essential to have a correct description of the electronic structure of the system. Short range MRPT2 potential energy curves for the arrangement with one distance stretched by 0.6 Å (while the other is kept at equilibrium) are plotted in Figure 8. In the figure the potential energy calculated at $R = 3$ Å (the rhs side of the plot) is clearly higher than that of the non stretched arrangement, in agreement with the fact that the energy of the stretched molecule is larger than that at the equilibrium. On the other hand, the potential energy calculated at very short distances ($1 < R < 2$) is lower than that of the non stretched arrangement resulting in a decrease of the height of the barrier to exchange. Such difference is particularly large for the crossed (X) arrangement (at $R = 1$ it is about 7 eV) and supports the need for a change of arrangements along the reaction path.

This effect becomes even more pronounced when both the N_2 intermolecular distances are stretched. Figures 9 and 10 show the corresponding potential energy values for the asymmetric ($r_a = 1.494$ Å, $r_b = 1.694$ Å) and the symmetric ($r_a = 1.694$ Å, $r_b = 1.694$ Å) stretching of the two intramolecular bonds, respectively. As commented above, at $R = 3$ Å the energy is larger than that for the equilibrium distance because the related asymptote is more energetic. On the other hand, for $1 < R < 2$ the potential energy values decrease almost linearly with the shortening of R (the energy goes up again only for the H and P15 arrangements) further supporting the indication of the opening of a possible reactive channel in this region via an arrangement change.

6   The Fitting of the MRPT2/CCSD Potential Energy Surface

Also for the MRPT2 potential energy values, we carried out a first attempt to fit a global PES using the Paniagua suite of programs, in order to obtain a Fortran routine to be used in dynamics calculations. In practice, for the new fitting we

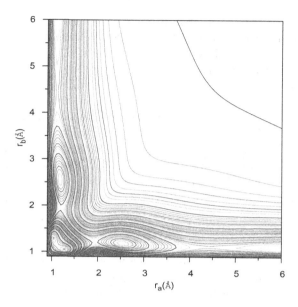

Fig. 11. Isoenergetic contour plots for the H arrangement at $R=3$ Å. Energy contours are drawn every 6 Kcal/mol.

used the CCSD(T) results for arrangements characterized by both N_2 in their equilibrium distance. On the contrary, for arrangements characterized by one or both N_2 internuclear distances stretched from their equilibrium value we replaced the CCSD(T) *ab initio* potential energy values, with the corresponding MRPT2 ones. These points were also scaled by taking into account the energy difference between the CCSD(T) energy values for the equilibrium arrangement and the corresponding MRPT2 ones (for the considered arrangement). A fit of these potential energy values, which were necessarily calculated only for a subset of the arrangements considered for the CCSD(T) calculations because of the associated heavy demand in terms of computer resources, was made of 1440 two-body points (fitted to a polynomial of order 5), 4320 three-body values (fitted to a polynomial of order 6) and 1361 four-body values (fitted to a polynomial of order 8). This led to a PES characterized by an average standard deviation of 0.27 eV and a maximum error of 1.52 eV (mainly located in the strongly repulsive regions). The resulting improved (with respect to that of Ref. [17] and Ref. [30]) fitted PES is illustrated in Fig. 11, where the isoenergetic contour plot for the H arrangements at $R = 3$ Å is shown. The figure clearly shows that at fixed intermolecular distance a reactive channel leading to the exchange of N atoms from reactants to products is open and bears a quite structured nature so that the Minimum Energy Path goes through various saddle points and wells, making the interchange of internal energy quite effective. This feature helps the rationalization of the path to reaction shown in Fig. 12, where the isoenergetic contours for the H arrangement obtained when varying R but constraining the

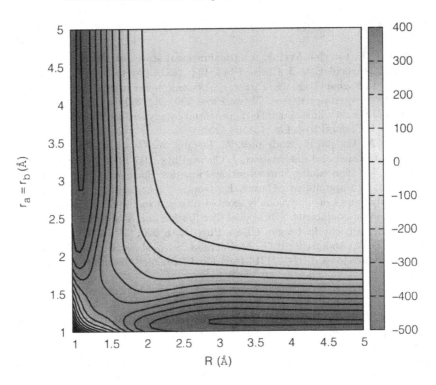

Fig. 12. Isoenergetic contour plots for the H arrangement with $r_a = r_b$

two N_2 intramolecular distances to be the same ($r_a = r_b$) are drawn. The contours given in Fig. 11 clearly show, in fact, that the fixed R internal energy exchange may be so structured that despite the fact that the $r_a = r_b$ path (see Fig. 12) has a high single transition state separating reactants from products, an efficient internal energy exchange may open alternative paths to reaction.

7   Conclusions

The calculations reported in this paper for the $N_2 + N_2$ system show, for the first time, that high level *ab initio* electronic structure calculations indicate the opening at fairly high values of a reactive channel enabling the exchange of N atoms. To this end, however, one has to resort to MRPT2 techniques and carry out appropriate fitting of the calculated potential energy values. The key feature of the reactive channel is a clear interplay of different internal degrees of freedom. Obviously, to find out the actual contribution of such interplay to the mechanisms governing the reactive process dynamical calculations need to be performed on the fitted PES as we have planned to do in the near future.

References

1. Nesbitt, D.J., van der Avoird, A.: Rovibrational states of the H_2O-H_2 complex: An ab initio calculation. J. Chem. Phys. 134, 44314 (2011)
2. Albertí, M., Faginas Lago, N., Pirani, F.: Benzene water interaction: From gaseous dimers to solvated aggregates. Chem. Phys. 339, 232–239 (2012)
3. Hinde, R.J.: A six-dimensional H_2H_2 potential energy surface for bound state spectroscopy. J. Chem. Phys. 128, 154308 (2008)
4. Ceballos, A., Garcia, E., Rodriguez, A., Laganà, A.: Quasiclassical kinetics of the $H_2 + H_2$ reaction and dissociation. J. Chem. Phys. 105(10), 1797–1804 (2001)
5. Capitelli, M.: Non-equilibrium vibrational kinetics. Springer, Berlin (1986)
6. Armenise, I., Capitelli, M., Garcia, E., Gorse, C., Laganà, A., Longo, S.: Deactivation dynamics of vibrationally excited nitrogen molecules by nitrogen atoms. Effects on non-equilibrium vibrational distribution and dissociation rates of nitrogen under electrical discharges. Chem. Phys. Lett. 200, 597 (1992)
7. Knauth, D.C., Andersson, B.G., McCandliss, S.R., Moos, H.W.: The Interstellar N_2 Abundance toward HD 124314 from Far-Ultraviolet Observations. Nature 429, 636 (2004)
8. Raich, J.C., Gillis, N.S.: The anisotropic interaction between nitrogen molecules from solid state data. J. Chem. Phys. 66, 846 (1977)
9. MacRury, T.B., Steele, W.A., Berne, B.J.: Intermolecular potential models for anisotropic molecules, with applications to N_2, CO_2, and benzene. J. Chem. Phys. 64, 1288 (1976)
10. Cheung, P.S.Y., Powles, J.G.: The properties of liquid nitrogen V. Computer simulation with quadrupole interaction. Mol. Phys. 32, 1383 (1976)
11. Cheung, P.S.Y., Powles, J.G.: The properties of liquid nitrogen. Mol. Phys. 30, 921 (1975)
12. Evans, D.J.: Transport properties of homonuclear diatomics I. Dilute gases. Mol. Phys. 34, 103 (1977)
13. Cappelletti, D., Vecchiocattivi, F., Pirani, F., Heck, E.L., Dickinson, A.S.: An Intermolecular potential for Nitrogen from a multi-property analysis. Mol. Phys. 93, 485 (1998)
14. Aquilanti, V., Bartolomei, M., Cappelletti, D., Caramona-Novillo, E., Pirani, F.: The N_2-N_2 system: An experimental potential energy surface and calculated rotovibrational levels of the molecular nitrogen dimer. J. Chem. Phys. 93, 485 (1998)
15. Gomez, L., Bussery-Honvault, B., Cauchy, T., Bartolomei, M., Cappelletti, D., Pirani, F.: Global fits of new intermolecular ground state potential energy surfaces for N_2-H_2 and N_2-N_2 van der waals dimers. Chem. Phys. Lett. 445, 99–107 (2007)
16. van der Avoid, A., Wormer, P.E.S., Jansen, A.P.J.: An improved intermolecular potential for nitrogen. J. Chem. Phys. 84, 1629–1635 (1986)
17. Cappelletti, D., Vecchiocattivi, F., Pirani, F., McCourt, F.R.W.: Glory structure in the N_2-N_2 total integral scattering cross section. A test for the intermolecular potential energy surface. Chem. Phys. Lett. 248, 237–243 (1996)
18. Huo, S.W.M., Green, S.: Quantum calculations for rotational energy transfer in nitrogen molecule collisions. J. Chem. Phys. 104, 7572–7589 (1996)
19. Stallcop, J.R., Partridge, H.: The N_2-N_2 potential energy surface. Chem. Phys. Lett. 281, 212–220 (1997)
20. Wada, A., Kanamori, H., Iwata, S.: Ab Initio MO studies of van der waals molecule $(N_2)_2$: Potential energy surface and internal motion. J. Chem. Phys. 109, 9434–9438 (1998)

21. Couronne, O., Ellinger, Y.A.: An ab initio and DFT study of $(N_2)_2$ dimers. Chem. Phys. Lett. 306, 71–77 (1999)
22. Leonhard, K., Deiters, U.K.: Monte Carlo Simulations of Nitrogen Using an Ab Initio Potential. Mol. Phys. 100, 2571–2585 (2002)
23. Karimi Jafari, M.H., Maghari, A., Shahbazian, S.: An improved ab initio potential energy surface for N_2-N_2. Chem. Phys. 314, 249–262 (2005)
24. Laganà, A., Riganelli, A., Gervasi, O.: On the structuring of the computational chemistry Virtual Organization COMPCHEM. In: Gavrilova, M.L., Gervasi, O., Kumar, V., Tan, C.J.K., Taniar, D., Laganá, A., Mun, Y., Choo, H. (eds.) ICCSA 2006. LNCS, vol. 3980, pp. 665–674. Springer, Heidelberg (2006)
25. The European Grid Initiative, http://www.egi.com (last access January 13, 2013)
26. Laganà, A.: Towards a grid based universal molecular simulator. In: Laganà, A., Lendvay, G. (eds.). Kluwer (2004)
27. Costantini, A., Gervasi, O., Manuali, C., Faginas Lago, N., Rampino, S., Laganà, A.: COMPCHEM: progress towards gems a grid empowered molecular simulator and beyond. Journal of Grid Computing 8, 571–586 (2010)
28. Rampino, S., Monari, A., Rossi, E., Evangelisti, S., Laganà, A.: Chem. Phys. 398, 192 (2012)
29. Laganá, A., Balucani, N., Crocchianti, S., Casavecchia, P., Garcia, E., Saracibar, A.: An extension of the molecular simulator GEMS to calculate the signal of crossed beam experiments. In: Murgante, B., Gervasi, O., Iglesias, A., Taniar, D., Apduhan, B.O. (eds.) ICCSA 2011, Part III. LNCS, vol. 6784, pp. 453–465. Springer, Heidelberg (2011)
30. Verdicchio, M., Pacifici, L., Laganà, A.: Grid enabled high level ab initio electronic structure calculations for the $N_2 + N_2$ exchange reaction. In: Murgante, B., Gervasi, O., Misra, S., Nedjah, N., Rocha, A.M.A.C., Taniar, D., Apduhan, B.O. (eds.) ICCSA 2012, Part I. LNCS, vol. 7333, pp. 371–386. Springer, Heidelberg (2012)
31. Møller, C., Plesset, M.S.: Note on an Approximation Treatment for Many-Electron Systems. Phys. Rev. 46, 618 (1934)
32. Boys, S.F., Bernardi, F.: The calculation of small molecular interactions by the differences of separate total energies. Some procedures with reduced errors. Mol. Phys. 19, 553–566 (1970)
33. Feller, D.: The Role of Databases in Support of Computational Chemistry Calculations. J. Chem. Phys. 17, 1571–1586 (1996)
34. Schuchardt, K., Didier, B., Elsethagen, T., Sun, L., Gurumoorthi, V., Chase, J., Li, J., Windus, T.: Basis Set Exchange: A Community Database for Computational Sciences. J. Chem Inf. Model. 47, 1045–1052 (2007)
35. Piecuch, P., Kucharski, S.A., Kowalski, K., Musial, M.: Efficient computer implementation of the renormalized coupled-cluster methods: The R-CCSD[T], R-CCSD(T), CR-CCSD[T], and CR-CCSD(T) approaches. Comput. Phys. Comm. 149, 71–96 (2002)
36. Bentz, J.L., Olson, R.M., Gordon, M.S., Schmidt, M.W., Kendall, R.A.: Coupled cluster algorithms for networks of shared memory parallel processors. Comput. Phys. Comm. 176, 589–600 (2007)
37. Olson, R.M., Bentz, J.L., Kendall, R.A., Schmidt, M.W., Gordon, M.S.: A novel approach to parallel coupled cluster calculations: Combining distributed and shared memory techniques for modern cluster based systems. J. Comput. Theo. Chem. 3, 1312–1328 (2007)

38. Schmidt, M.W., Baldridge, K.K., Boatz, J.A., Elbert, S.T., Gordon, M.S., Jensen, J.J., Koseki, S., Matsunaga, N., Nguyen, K.A., Su, S., Windus, T.L., Dupuis, M., Montgomery, J.A.: General atomic and molecular electronic structure system. J. Comp. Chem. 14, 1347–1363 (1993)
39. Gordon, M.S., Schmidt, M.W.: Theory and Applications of Computational Chemistry, the first forty years (2005)
40. Hay, P.J., Pack, R.T., Martin, R.L.: Electron correlation effects on the N_2-N_2 interaction. J. Chem. Phys. 81, 1360–1372 (1984)
41. http://www.cineca.it (last access January 21, 2013)
42. Sorbie, K.S., Murrell, J.N.: Analytical potentials for triatomic molecules from spectroscopic data. Mol. Phys. 52, 1387 (1975)
43. Aguado, A., Tablero, C., Paniagua, M.: Global fit of ab initio potential energy surfaces: II.1. tetraatomic systems ABCD. Comput. Phys. Comm. 134, 97 (2001)
44. Garcia, E., Saracibar, A., Gomez-Carrasco, S., Laganà, A.: Modelling the global potential energy surface of the $N + N_2$ reaction from ab initio data. Phys. Chem. Chem. Phys. 10, 2552–2558 (2008)
45. Caridade, P.J.S.B., Galvao, B.R.L., Varandas, A.J.C.: Quasiclassical Trajectory Study of Atom-Exchange and Vibrational Relaxation Processes in Collisions of Atomic and Molecular Nitrogen. J. Phys. Chem. A 114, 6063–6070 (2010)
46. Baerends, E.J., Ellis, D.E., Ros, P.: Chem. Phys. 2, 41 (1973)
47. Ziegler, T., Snijders, J.G., Baerends, E.J.: J. Chem. Phys. 74, 1271 (1981)
48. Hirao, K.: Chem. Phys. Lett. 190, 374 (1992)
49. Hirao, K.: Int. J. Quant. Chem. S26, 517 (1992)

A Theoretical Study of Formation Routes and Dimerization of Methanimine and Implications for the Aerosols Formation in the Upper Atmosphere of Titan

Marzio Rosi[1,2,*], Stefano Falcinelli[1], Nadia Balucani[3],
Piergiorgio Casavecchia[3], and Dimitrios Skouteris[3]

[1] Department of Civil and Environmental Engineering, University of Perugia, Via Duranti 93,
06125 Perugia, Italy
{marzio,nadia.balucani}@unipg.it, {stefano,piero}@dyn.unipg.it
[2] ISTM-CNR
[3] Department of Chemistry, University of Perugia, Italy
dimitris@impact.dyn.unipg.it

Abstract. Methanimine is a molecule of interest in astrobiology, as it is considered an abiotic precursor of the simplest amino acid glycine. Methanimine has been observed in the interstellar medium and in the upper atmosphere of Titan. In particular, it has been speculated that its polymerization can contribute to the formation of the haze aerosols that surround the massive moon of Saturn. Unfortunately, such a suggestion has not been proved by laboratory experiments. Methanimine is difficult to investigate in laboratory experiments because it is a transient species that must be produced *in situ*. To assess its potential role in the formation of Titan's aerosol, we have performed a theoretical investigation of possible formation routes and dimerization. The aim of this study is to understand whether formation and dimerization of methanimine and, eventually, its polymerization, are possible under the conditions of the atmosphere of Titan. Results of high-level electronic structure calculations are reported.

Keywords: ab initio calculations, potential energy surfaces, atmospheric models.

1 Introduction

Methanimine is an important molecule in prebiotic chemistry since it is considered a possible precursor of the simplest amino acid, glycine, via its reactions with HCN (and then H_2O) or with formic acid (HCOOH) [1]. According to this suggestion, the simplest amino acid can be formed 'abiotically' starting from simple molecules relatively abundant in extraterrestrial environments, such as the interstellar medium, and, possibly, primitive Earth. Interestingly, methanimine has been observed in the upper atmosphere of Titan, the massive moon of Saturn [2]. The atmosphere of Titan is believed to be somewhat reminiscent of the primeval atmosphere of Earth [3,4] and it is mainly composed of dinitrogen (97%), methane, simple nitriles such as HCN, and also other

---

* Corresponding author.

B. Murgante et al. (Eds.): ICCSA 2013, Part I, LNCS 7971, pp. 47–56, 2013.
© Springer-Verlag Berlin Heidelberg 2013

small hydrocarbons, like C_2H_6 [4]. Methanimine can be produced in the atmosphere of Titan by the reactions of N (^{2}D) with both methane and ethane [5], as well as by other simple processes, including the reaction between NH and CH_3 or reactions involving ionic species [2,6]. The recent model by Lavvas *et al.* [6] derived a larger quantity of methanimine than that inferred by the analysis of the ion spectra recorded by Cassini Ion Neutral Mass Spectrometer (INMS) [2]. However, there is a lot of uncertainty on the possible fate of methanimine in the upper atmosphere of Titan because of a severe lack of knowledge on the possible chemical loss pathways of this species [6]. CH_2NH is known to absorb in the UV region [7] and the possible photodissociation products are HCN/HNC + H_2 or H_2CN + H [8]. Also, having a C=N double bond, methanimine is a highly reactive molecule that can react with atomic/radical species and also ions, present in the upper atmosphere of Titan. Growing evidence suggests that nitrogen chemistry contributes to the formation of the haze aerosols in the Titan upper atmosphere [9-11]. In this respect, since imines are well-known for their capability of polymerizing, CH_2NH is an excellent candidate to account for the nitrogen-rich aerosols of Titan through polymerization and copolymerization with other unsaturated nitriles or unsaturated hydrocarbons. Lavvas et al. [6] suggested that polymerization of methanimine provides an important contribution to the formation of the nitrogen-rich aerosols, but a quantitative inclusion of this process in the model could not be obtained as there is no information (either experimental or theoretical) on methanimine polymerization. Since the first step of polymerization is dimerization, in this contribution we report on a theoretical characterization of methanimine dimerization. Electronic structure calculations of the potential energy surfaces representing the reactions of electronically excited atomic nitrogen, N(^{2}D), with methane and ethane are also presented, as they are possible formation routes of methanimine under the conditions of the upper atmosphere of Titan.

2   Computational Details

The potential energy surface of the species of interest was investigated by locating the lowest stationary points at the B3LYP [12] level of theory in conjunction with the correlation consistent valence polarized set aug-cc-pVTZ [13]. At the same level of theory we have computed the harmonic vibrational frequencies in order to check the nature of the stationary points, *i.e.* minimum if all the frequencies are real, saddle point if there is one, and only one, imaginary frequency. The assignment of the saddle points was performed using intrinsic reaction coordinate (IRC) calculations [14]. In selected cases, the energy of the main stationary points was computed also at the higher level of calculation CCSD(T) [15] using the same basis set aug-cc-pVTZ. Both the B3LYP and the CCSD(T) energies were corrected to 0 K by adding the zero point energy correction computed using the scaled harmonic vibrational frequencies evaluated at B3LYP/aug-cc-pVTZ level. In the evaluation of the binding energies, the energy of nitrogen atoms in their excited ^{2}D state was estimated by adding the experimental [16] separation N(^{4}S) – N(^{2}D) of 230.0 kJ/mol to the energy of N(^{4}S) at all levels of calculation. Thermochemical calculations were performed at the W1 [17] level of theory for selected processes. All calculations were done using Gaussian 09 [18] while the analysis of the vibrational frequencies was performed using Molekel [19].

3   Results and Discussion

3.1   Formation of Methanimine in the Atmosphere of Titan

Dinitrogen is characterized by a strong bond and is difficult to chemically fix in compounds [3,5]. However, the observation of nitriles in trace amounts in the atmosphere of Titan indicates that some forms of active nitrogen are produced by several processes, as for instance EUV absorption ($60 < \lambda < 82$ nm), electron impact induced dissociation, dissociative ionization and N_2^+ dissociative recombination [4,5]. Those processes produce N (^{4}S) and N (^{2}D) states in similar amounts [5]. The excited ^{2}D state is metastable with a very long radiative lifetime (~ 48 hours) [4,5]. The generation of atomic nitrogen in the first electronically excited state, ^{2}D, is extremely relevant in assessing the role of neutral nitrogen chemistry in the atmosphere of Titan because N(^{4}S) atoms exhibit very low reactivity with closed-shell molecules and the probability of collision with an open-shell radical is small [5]. The reaction of N (^{2}D) with methane and ethane could produce methanimine [20-22], while other imines and N-containing species are produced in the reactions of N(^{2}D) with acetylene and ethylene [23,24].

Let us analyze the minimum energy path, as obtained by high-level *ab initio* calculations, of the reactions N(^{2}D)+CH$_4$ and N(^{2}D)+C$_2$H$_6$.

3.1.1   N(^{2}D) + CH$_4$

The lowest stationary points localized on the potential energy surface of N(^{2}D)+CH$_4$ have been reported in Figure 1, where the main geometrical parameters (distances/Å and angles/°) are shown together with the energies computed at B3LYP/aug-cc-pVTZ, CCSD(T)/aug-cc-pVTZ and W1 level, relative to that of N(^{2}D) + CH$_4$. Many of the stationary points which are of interest in this work have been previously reported by Kurosaki *et al.* [25] and Jursic [26]. Kurosaki *et al.* optimized the geometries at MP2(full)/cc-pVTZ level of theory and computed the energies at PMP4(full,SDTQ)/cc-pVTZ level, while Jursic performed mainly CBS-Q calculations. The agreement of our work with their results is generally good, the small differences being due to the different methods employed. In the following discussion for simplicity we will consider only the CCSD(T) energies. The interaction of N(^{2}D) with CH$_4$ gives rise to the species CH$_3$NH which is more stable than the reactants by 427.6 kJ/mol. For this insertive approach we have found a saddle point which is, however, below the reactants both at B3LYP and CCSD(T) level of calculation. The presence of saddle points below the reactants has already been observed for very exothermic reactions, in particular involving methane [27]. However, since we were not successful in localizing a stable initial complex, the energy of this saddle point does not seem to be very meaningful. We tried also to localize this saddle point at MP2 level of calculations, but we did not succeed. Notably, this saddle point was computed at   several levels of calculations by Takayanagi *et al.* [25, 28] and their best estimate for the barrier height was calculated to be only 5.3 kJ/mol above the reactants. More recently, Ouk *et al.* [29] computed a barrier of 3.86 ± 0.84 kJ/mol with a multi-state multi-reference configuration interaction method. Once formed after N(^{2}D) insertion into a C-H bond, CH$_3$NH can either isomerize to species CH$_2$NH$_2$, which is the most stable isomer of the CH$_4$N PES, with a barrier of 154.2 kJ/mol or can directly dissociate to CH$_2$NH and H products through a slightly lower barrier of 141.4 kJ/mol. The other CH$_3$NH dissociation channels, *i.e.* those leading to CH$_3$+NH or CH$_3$N+H, are both barrierless, but they are very endothermic with respect to CH$_3$NH, being the dissociation energies 306.9 and 343.0 kJ/mol, respectively. The products CH$_3$ + NH can be reached also directly from N(^{2}D) + CH$_4$ through an hydrogen abstraction process with a saddle point which lies 60.4 kJ/mol below the reactants. Also in this case we were not able to localize any stable initial complex. The geometry of this saddle point is comparable to that obtained by Jursic [26]. In principle, CH$_3$N can be formed either in a triplet or in a singlet state. According to the present calculations, however, only the triplet state can be formed, because the singlet methylnitrene ($\tilde{a}$ ^{1}E) is too high in energy (ΔE=130.4±1.1 kJ/mol) [30]. Once formed by CH$_3$NH isomerization, CH$_2$NH$_2$ can also lose one H atom giving rise to the isomers CH$_2$NH (with an exit barrier of 163.8 kJ/mol) or CHNH$_2$ (in a very endothermic, 299.0 kJ/mol, barrierless process). CH$_2$NH$_2$ can also dissociate into CH$_2$(^{3}B$_1$,^{1}A$_1$)+NH$_2$: these barrierless dissociation channels are endothermic by 404.5 and 443.6 kJ/mol for the triplet and singlet CH$_2$ formation, respectively. Finally we have located a small barrier (34.2 kJ/mol) for the reaction of the CH$_2$NH product with its co-fragment H to produce CHNH + H$_2$. This reaction is exothermic by 34.7 kJ/mol. In Figure 2 we have reported a schematic representation of the two main reaction paths leading to methanimine which are the most relevant processes in this context.

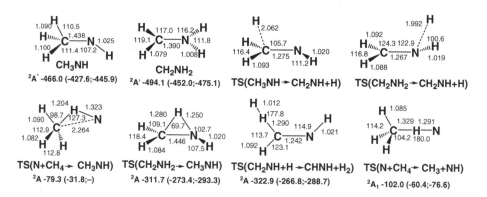

Fig. 1. B3LYP optimized geometries (Å and °) and relative energies with respect to N(^{2}D) + CH$_4$ (kJ/ mol) at 0 K of minima and saddle points localized on the PES of N(^{2}D) + CH$_4$; CCSD(T) and W1 relative energies are reported in parentheses

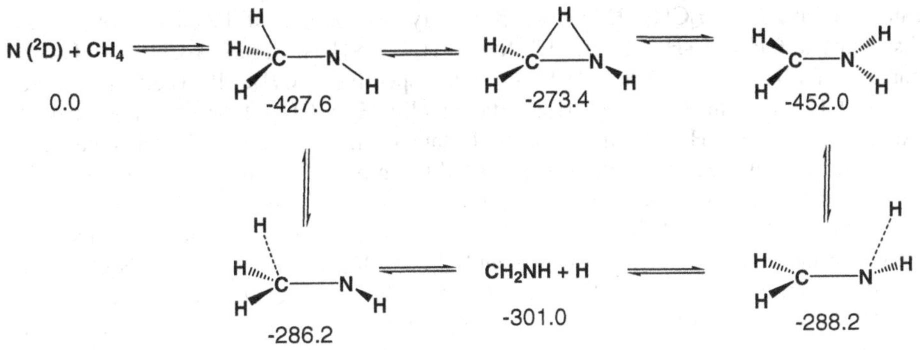

Fig. 2. Schematic representation of the two main reaction paths leading to methanimine. Relative energies (kJ/mol) computed at CCSD(T)/aug-cc-pVTZ level with respect to the energy of $N(^2D) + CH_4$.

3.1.2 $N(^2D) + C_2H_6$

The lowest stationary points localized on the potential energy surface of $N(^2D) + C_2H_6$ have been reported in Figure 3, where the main geometrical parameters (distances/Å and angles/°) are shown together with the energies computed at B3LYP/aug-cc-pVTZ and CCSD(T)/aug-cc-pVTZ (in parentheses), relative to that of $N(^2D) + C_2H_6$. For simplicity we will consider only CCSD(T) values in the following discussion. The interaction of $N(^2D)$ with C_2H_6 gives rise to the species CH_3CH_2NH which is more stable than the reactants by 447.0 kJ/mol. For this reaction we have found a saddle point which is, however, below the reactants both at B3LYP and CCSD(T) level of calculation. The presence of saddle points below the reactants has already been observed for very exothermic reactions [27]. However, since we were not successful in localizing a stable initial complex, the energy of this saddle point does not seem to be very meaningful. CH_3CH_2NH can isomerize to species CH_3NHCH_2, with a relatively high barrier of 253.6 kJ/mol or to species CH_3CHNH_2, which is the most stable isomer on the $N(^2D)+C_2H_6$ PES, with a barrier of 146.3 kJ/mol. CH_3CH_2NH can also dissociate to CH_3CHNH and H with a slightly lower barrier of 131.0 kJ/mol or to the more stable products CH_3 and CH_2NH with an even lower barrier of 113.6 kJ/mol. The other channels, *i.e.* the dissociation of CH_3CH_2NH to $CH_3CH_2 + NH$, $CH_3CH_2N + H$, $CH_3 + CH_2NH$ in its excited triplet state, and CH_2CH_2NH in its excited triplet state + H are barrierless, but they are very endothermic, being the dissociation energies 311.4, 348.2, 352.6, and 415.4 kJ/mol, respectively. CH_3NHCH_2 can isomerize to CH_3NCH_3 with a barrier of 172.2 kJ/mol or can dissociate to $CH_3 + CH_2NH$, $CH_3NCH_2 + H$, c-$CH_2(NH)CH_2 + H$, or $CH_2NHCH_2 + H$ with barriers of 125.2, 140.2, 268.7, 276.7 kJ/mol, respectively. CH_3NHCH_2 can dissociate also to $CH_3 + CH_2NH$ in its excited triplet state, $CH_3NH + CH_2$, CH_2NHCH_2 in its excited triplet state + H, and CH_3NCH_2 in its excited triplet state + H. All these processes are barrierless but they are very endothermic, being the dissociation energies 345.4, 396.4, 401.6, and 402.1 kJ/mol, respectively. The most stable isomer CH_3CHNH_2

can isomerize to $CH_2CH_2NH_2$ with a relatively low barrier of 193.6 kJ/mol or can dissociate to $CH_3CHNH + H$, $CH_2CHNH_2 + H$, $CHNH_2 + CH_3$, $CH_3CNH_2 + H$ with barriers of 148.3, 154.1, 258.9, 280.9 kJ/mol, respectively. CH_3CHNH_2 can dissociate also to CH_3CNH_2 in its excited triplet state + H, $CH_3 + CHNH_2$ in its excited triplet state, $NH_2 + CH_3CH$ both in its ground state or in its excited singlet state, and CH_2CHNH_2 in its excited triplet state + H; all these processes are barrierless but very endothermic being the dissociation energies 376.6, 395.8, 402.4, 416.7, and 419.2 kJ/mol, respectively. $CH_2CH_2NH_2$ can dissociate to $C_2H_4 + NH_2$ or to $CH_2CHNH_2 + H$ with relatively low barriers of 93.7 and 140.1 kJ/mol, respectively. $CH_2CH_2NH_2$ can dissociate also to $CH_2 + CH_2NH_2$, CH_2CHNH_2 in its excited triplet state + H, $CH_2CH_2NH + H$, and $CHCH_2NH_2 + H$; all these processes are barrierless but very endothermic being the dissociation energies 368.8, 380.3, 404.9, and 445.7 kJ/mol, respectively. CH_3NCH_3 can dissociate to $CH_3NCH_2 + H$ with a barrier of 148.6 kJ/mol or to $CH_3N + CH_3$ in a barrierless process endothermic by 287.0 kJ/mol. The dissociation of CH_3NCH_3 to CH_3NCH_2 in its excited triplet state + H is much more endothermic (399.3 kJ/mol). Therefore, the reaction is very complex with many possible products. However, RRKM calculations [19, 20] suggest that only some of the exit channels are really important; in particular the main channel, which accounts for the 78.8%, is the one leading to methanimine and methyl, with the second one, which accounts for the 12.4%, being the one leading to $CH_2CHNH_2 + H$, and the third one, which accounts for the 5.5%, being the one leading to $CH_3CH_2 + {}^3NH$. In conclusion, the main product of the reaction between $N\,({}^2D) + C_2H_6$ is methanimine. In Figure 4 we have reported a schematic representation of the two main reaction paths leading to methanimine.

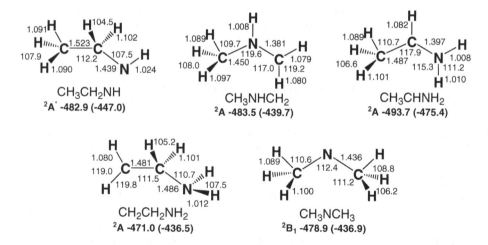

Fig. 3. B3LYP optimized geometries (Å and °) and relative energies with respect to $N({}^2D) + C_2H_6$ (kJ /mol) at 0 K of minima localized on the PES of $N({}^2D) + C_2H_6$; CCSD(T) relative energies are reported in parentheses

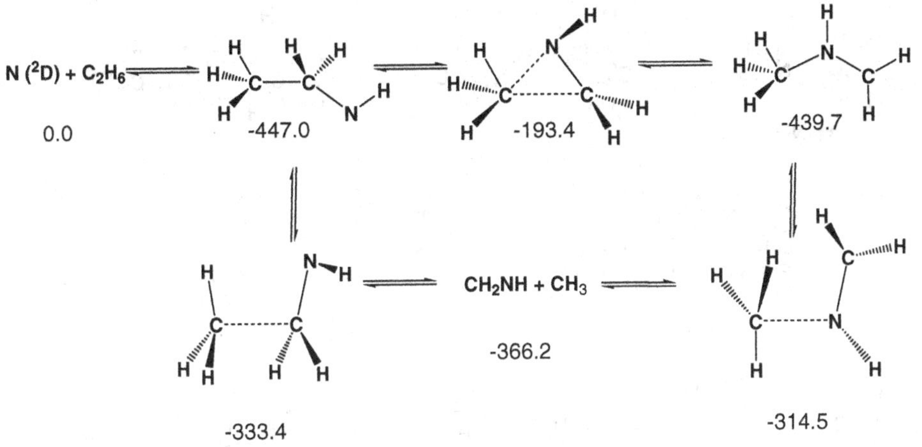

Fig. 4. Schematic representation of the two main reaction paths leading to methanimine. Relative energies (kJ/mol) computed at CCSD(T)/aug-cc-pVTZ level with respect to the energy of $N(^2D) + C_2H_6$.

3.2   Dimerization of Methanimine

The dimerization of methanimine leads to several minima whose structures and relative energies are reported in Figure 5. The approach of two methanimine molecules leads to the formation of an adduct (1) slightly bound (8.9 kJ/mol) in a barrierless process. Species (1) can isomerize to the compound (2), which shows an N—N bond, only slightly less stable than (1). The transfer of a hydrogen atom gives rise to species (3) less stable than (2) by 71.9 kJ/mol. The less stable species (3), however, can give rise to species (4) which is bound with respect to two molecules of methanimine by 8.7 kJ/mol at B3LYP level. The problem is that these isomerization processes show very high barriers. Also the transition states for the 1 → 2, 2 → 3, and 3 → 4 isomerization processes are reported in Figure 5. We can notice that the first reaction has a barrier of 343.5 kJ/mol, the second one shows a barrier of 257.7 kJ/mol, while the third one shows a slightly lower barrier of 181.0 kJ/mol. In Figure 6 we have summarized these results reporting a schematic representation of the reaction path leading to the dimerization of methanimine.

The presence of these high barriers suggests that these reactions cannot be important under the conditions of Titan, *i.e.* a surface temperature of 94 K and a stratospheric temperature of 175 K. Different processes, as co-polymerization or reactions with radicals or ions, should therefore be investigated in order to explain the discrepancy between the abundance of methanimine in the atmosphere of Titan derived   by the recent model of Lavvas *et al.* [6] and that inferred by the ion spectra recorded by Cassini Ion Neutral Mass Spectrometer (INMS) [2].

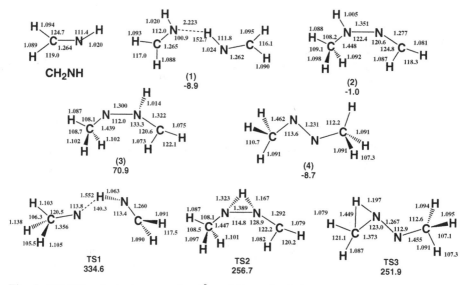

Fig. 5. B3LYP optimized geometries (Å and °) and relative energies with respect to 2 x CH₂NH (kJ /mol) at 0 K of minima and saddle points localized on the PES of the dimer of CH₂NH

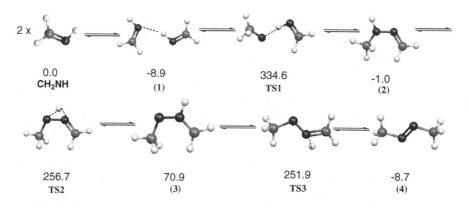

Fig. 6. Schematic representation of the reaction path leading to a stable methanimine dimer. Relative energies (kJ/mol) computed at B3LYP/aug-cc-pVTZ level with respect to the energy of two methanimine molecules are reported.

4   Conclusions

In summary, numerous elementary reactions in the upper atmosphere of Titan lead to the formation of methanimine, an important prebiotic molecule. Among them, the reaction of electronically excited nitrogen atoms, N(^{2}D), and abundant hydrocarbons have been considered and found to be efficient. Polymerization of methanimine in the gas phase has been invoked to explain the difference between model predictions and Cassini measurements [6]. According to the present results, dimerization of methanimine (the first step toward polymerization) is a process characterized by very high energy barriers and is difficult to believe it can be important under the conditions of Titan. The evaluation of the energies of the transition states, however, will be performed in future work at a higher level of accuracy and rate constant calculations will be derived on the basis of the results obtained.

Acknowledgment. Prof. Giorgio Liuti died during the preparation of the manuscript. This work and our future endeavours are dedicated to him, whose intellectual honesty and genuine love of chemistry continue to inform our scientific life. We acknowledge financial support from the Italian MIUR (Ministero Istruzione Università Ricerca): M.R. for project PRIN 2009 2009SLKFEX_004; P.C. for project PRIN 2010-2011 2010ERFKXL.

References

1. Balucani, N.: Int. J. Mol. Sci. 10, 2304 (2009)
2. Vuitton, V., Yelle, R.V., Anicich, V.G.: Astrophys. J. 647, L175–L178 (2006)
3. Balucani, N.: Chem. Soc. Rev. 41, 5473 (2012)
4. Vuitton, V., Dutuit, O., Smith, M.A., Balucani, N.: Chemistry of Titan's atmosphere. In: Mueller-Wodarg, I., Griffith, C., Lellouch, E., Cravens, T. (eds.) Titan: Surface, Atmosphere and Magnetosphere. Cambridge University Press (2013)
5. Dutuit, O., Carrasco, N., Thissen, R., Vuitton, V., Alcaraz, C., Pernot, P., Balucani, N., Casavecchia, P., Canosa, A., Le Picard, S., Loison, J.-C., Herman, Z., Zabka, J., Ascenzi, D., Tosi, P., Franceschi, P., Price, S.D., Lavvas, P.: Astrophys. J. Suppl. Ser. 204, 20 (2013)
6. Lavvas, P.P., Coustenis, A., Vardavas, I.M.: Planet. Space Sci. 56, 27 (2008), Lavvas, P.P., Coustenis, A., Vardavas, I.M.: Planet. Space Sci. 56, 67 (2008)
7. Teslja, A., Nizamov, B., Dagdigian, P.J.: J. Phys. Chem. A 108, 4433 (2004)
8. Larson, C., Ji, Y.Y., Samartzis, P., Wodtke, A.M., Lee, S.H., Lin, J.J.M., Chaudhuri, C., Ching, T.T., Chem, J.: Phys. 125, 133302 (2006)
9. Israel, G., Szopa, C., Raulin, F., Cabane, M., Niemann, H.B., Atreya, S.K., Bauer, S.J., Brun, J.F., Chassefiere, E., Coll, P., Conde, E., Coscia, D., Hauchecorne, A., Millian, P., Nguyen, M.J., Owen, T., Riedler, W., Samuelson, R.E., Siguier, J.M., Steller, M., Sternberg, R., Vidal-Madjar, C.: Nature 438, 796 (2005)
10. Imanaka, H., Smith, M.A.: Proc. Natl. Acad. Sci. 107, 12423 (2010)
11. Gu, X., Kaiser, R.I., Mebel, A.M., Kislov, V.V., Klippenstein, S.J., Harding, L.B., Liang, M.C., Yung, Y.L.: Astrophys. J. 701, 1797 (2009)
12. Becke, A.D.: J. Chem. Phys. 98, 5648 (1993), Stephens, P.J., Devlin, F.J., Chabalowski, C.F., Frisch, M.J.: J. Phys. Chem. 98, 11623 (1994)
13. Dunning Jr., T.H.: J. Chem. Phys. 90, 1007 (1989), Woon, D.E., Dunning Jr., T.H.: J. Chem. Phys. 98, 1358 (1993), Kendall, R.A., Dunning Jr., T.H., Harrison, R.J.: J. Chem. Phys. 96, 6796 (1992)
14. Gonzales, C., Schlegel, H.B.: J. Chem. Phys. 90, 2154 (1989), J. Phys. Chem. 94, 5523 (1990)

15. Bartlett, R.J.: Annu. Rev. Phys. Chem. 32, 359 (1981), Raghavachari, K., Trucks, G.W., Pople, J.A., Head-Gordon, M.: Chem. Phys. Lett. 157, 479 (1989), Olsen, J., Jorgensen, P., Koch, H., Balkova, A., Bartlett, R.J.: J. Chem. Phys. 104, 8007 (1996)
16. Moore, C.E.: Atomic Energy Levels. Natl. Bur. Stand (U.S.) Circ. N. 467, U.S., GPO, Washington, DC (1949)
17. Martin, J.M.L., de Oliveira, G.: J. Chem. Phys. 111, 1843 (1999), Parthiban, S., Martin, J.M.L.: J. Chem. Phys. 114, 6014 (2001)
18. Frisch, M.J., Trucks, G.W., Schlegel, H.B., Scuseria, G.E., Robb, M.A., Cheeseman, J.R., Scalmani, G., Barone, V., Mennucci, B., Petersson, G.A., Nakatsuji, H., Caricato, M., Li, X., Hratchian, H.P., Izmaylov, A.F., Bloino, J., Zheng, G., Sonnenberg, J.L., Hada, M., Ehara, M., Toyota, K., Fukuda, R., Hasegawa, J., Ishida, M., Nakajima, T., Honda, Y., Kitao, O., Nakai, H., Vreven, T., Montgomery Jr., J.A., Peralta, J.E., Ogliaro, F., Bearpark, M., Heyd, J.J., Brothers, E., Kudin, K.N., Staroverov, V.N., Kobayashi, R., Normand, J., Raghavachari, K., Rendell, A., Burant, J.C., Iyengar, S.S., Tomasi, J., Cosi, M., Rega, N., Milla, J.M., Klene, M., Knox, J.E., Cross, J.B., Bakken, V., Adamo, C., Jaramillo, J., Gomperts, R., Stratmann, R.E., Yazyev, O., Austin, A.J., Cammi, R., Pomelli, C., Ochterski, J.W., Martin, R.L., Morokuma, K., Zakrzewski, V.G., Voth, G.A., Salvador, P., Dannenberg, J.J., Dapprich, S., Daniels, A.D., Farkas, O., Foresman, J.B., Ortiz, J.V., Cioslowski, J., Fox, D.J.: Gaussian, Inc., Wallingford CT (2009)
19. Flükiger, P., Lüthi, H.P., Portmann, S., Weber, J.: MOLEKEL 4.3: Swiss Center for Scientific Computing, Manno (Switzerland) (2000-2002), Portmann, S., Lüthi, H.P.: Chimia 54, 766 (2000).
20. Balucani, N., Bergeat, A., Cartechini, L., Volpi, G.G., Casavecchia, P., Skouteris, D., Rosi, M.: J. Phys. Chem. 113, 11138 (2009)
21. Balucani, N., Leonori, F., Petrucci, R., Stazi, M., Skouteris, D., Rosi, M., Casavecchia, P.: Faraday Discussions 147, 189 (2010)
22. Rosi, M., Falcinelli, S., Balucani, N., Casavecchia, P., Leonori, F., Skouteris, D.: Theoretical study of reactions relevant for atmospheric models of titan: Interaction of excited nitrogen atoms with small hydrocarbons. In: Murgante, B., Gervasi, O., Misra, S., Nedjah, N., Rocha, A.M.A.C., Taniar, D., Apduhan, B.O. (eds.) ICCSA 2012, Part I. LNCS, vol. 7333, pp. 331–344. Springer, Heidelberg (2012)
23. Balucani, N., Alagia, M., Cartechini, L., Casavecchia, P., Volpi, G.G., Sato, K., Takayanagi, T., Kurosaki, Y.: J. Am. Chem. Soc. 122, 4443 (2000)
24. Balucani, N., Skouteris, D., Leonori, F., Petrucci, R., Hamberg, M., Geppert, W.D., Casavecchia, P., Rosi, M.: J. Phys. Chem. A 116, 10467 (2012)
25. Kurosaki, Y., Takayanagi, T., Sato, K., Tsunashima, S.: J. Phys. Chem. A 102, 254 (1998)
26. Jursic, B.S.: Int. J. Quantum Chem. 71, 481 (1999)
27. Fokin, A.A., Shubina, T.E., Gunchenko, P.A., Isaev, S.D., Yurchencko, A.G., Schreiner, P.R.: J. Am. Chem. Soc. 124, 10718 (2002), Schreiner, P., Fokin, A.A., Schleyer, P.V.R., Schaefer III, H.F.: Fundamental World of Quantum Chemistry. In: Brändas, E.J., Kryachko, E.S. (eds.), vol. II, pp. 349–375. Kluwer, The Netherlands (2003)
28. Takayanagi, T., Kurosaki, Y., Yokoyama, K.: Int. J. Quantum Chem. 79, 190 (2000)
29. Ouk, C.-M., Zvereva-Loëte, N., Bussery-Honvault, B.: Chem. Phys. Letters 515, 13 (2011)
30. Travers, M.J., Cowles, D.C., Clifford, E.P., Ellison, G.B.: J. Chem. Phys. 111, 5349 (1999)

Bonding Configurations and Observed XPS Features at the Hydrogen Terminated (100) Si Surface: What Can We Gain from Computational Chemistry

Paola Belanzoni[1], Giacomo Giorgi[2], and Gianfranco Cerofolini*

[1] Dipartimento di Chimica, Università di Perugia,
via Elce di Sotto 8, 06123 Perugia, Italy,
paola@thch.unipg.it
[2] Department of Chemical System Engineering, School of Engineering,
The University of Tokyo, 7-3-1 Hongo, Bunkyo-ku, Tokyo 113-8656, Japan
Research Center for Advanced Science and Technology (RCAST),
The University of Tokyo, 4-6-1 Komaba, Meguro-ku, Tokyo 153-8904, Japan
giacomo@tcl.t.u-tokyo.ac.jp

Abstract. Density functional (DFT) calculations for different size cluster models of the hydrogen-terminated HF_{aq}-etched (100) Si surface have been performed to verify that the quantities of interest (namely, atomic net charges and interatomic distances) in assigning the lines observed by X-ray photoelectron spectroscopy (XPS) vary weakly with cluster size. Net charge analysis based on Voronoi Deformation Density (VDD) method and accurate DFT geometry optimization calculations involving the smallest clusters as local models of various surface silicon atoms are used to assign chemical species to the features observed in the XPS spectra through evaluation of the chemical shifts, which are controlled by both the net charge and the Madelung potential truncated to nearest neighbours of the considered atoms.

Keywords: DFT calculations, hydrogen-terminated HF_{aq}-etched (100) Si surface, cluster modelling, X-ray photoelectron spectroscopy (XPS).

1 Introduction

The (100) surface of silicon has been subjected to intense experimental and theoretical study not only for its applications (it is the almost unique substrate used for the construction of integrated circuits), but also for its conceptual interest [1]-[4]. Of the procedures involved for the preparation of chemically homogeneous, atomically flat, nearly ideal, hydrogen terminated (100) Si surfaces, the oldest one (the HF_{aq} etching of the native oxide) has remained unrivalled in practice. As shown by the extended analysis based on infrared (IR) spectroscopy, the HF_{aq} etching results in prevailing SiH_2 terminations, but contains also SiH and SiH_3

---

* Author deceased.

B. Murgante et al. (Eds.): ICCSA 2013, Part I, LNCS 7971, pp. 57–68, 2013.

terminations, and siloxo (SiOH and SiOSi) defects [5]. The wide heterogeneity of the HF_{aq}-etched (100) Si surface is also confirmed by X-ray photoelectron spectroscopy (XPS). An angle-resolved XPS analysis of device-quality (100) Si substrates gave indeed evidence for eight superficial lines (in addition to that of elemental silicon Si^0), hereinafter referred to as follows: $Si^{-'}$, $Si^{k'}$ (with $k = 1,2,3$), and Si^n (with $n = 1, 2, 3, 4$). Table 1 lists the corresponding chemical shifts $\Delta\xi$ of 2p core electrons (with respect to the binding energy of elemental silicon) and line widths w [6].

Table 1. Chemical shift $\Delta\xi$ and line width w of all considered features for the HF_{aq}-etched surface

energy	$Si^{-'}$	Si^0	Si'	$Si^{2'}$	$Si^{3'}$	Si^1	Si^2	Si^3	Si^4
$\Delta\xi$ (eV)	-0.27	0.00	0.13	0.28	0.47	1.01	1.84	2.86	3.63
w (eV)	0.42	0.35	0.26	0.28	0.29	0.72	0.65	0.72	1.03

XPS is a powerful tool of surface analysis since, with the remarkable exception of hydrogen, it is sensitive to all elements and the survey spectra provide sufficiently accurate knowledge of the in-depth elemental composition and chemical configurations of the various atoms [7]-[9]. Atoms of the same element may however have different bonding configurations. The different chemical configurations are expected to result in different chemical shifts $\Delta\xi$ with respect to the atom in elemental state. In the absence of more detailed information, the configurations are tentatively given assuming the *naive assignation*: the chemical shift of an atom increases with its net charge Q as determined by its nearest neighbours in the hypothesis that the charge is the algebraic sum of the charges transferred along each bond due to the local electronegativity difference. With this attribution $\Delta\xi$ scales fairly linearly with Q ($\Delta\xi$ (eV) $= 1.94Q + 0.18$) [10]. Elementary electrostatic considerations, however, show that $\Delta\xi$ depends not only on Q but also on the Madelung potential U (in turn controlled by charge distribution on all nearby atoms), so that the attribution of XPS lines to bonding configurations guided by charge (or electronegativity distribution) alone may be invalidated. The attribution of XPS lines to surface configurations requires thus a knowledge of both Q and U. However, U can be determined only if the surface structure is known, that would render impossible any guess on surface structure unless U is controlled (although not completely determined) by the nearest neighbours of the considered atoms. The validation of this hypothesis is based on the ability to assign net charges to atoms in assigned chemical configurations. In turn, this requires: (i) the choice of a small cluster mimicking the local neighbourhood of the atom, (ii) an accurate ab initio modelling of the cluster, and (iii) a reliable criterion for assigning net charges to atoms in a molecule. This work is addressed to assess the adequacy of the cluster models to provide the partial charge and Madelung potential used in the description of the XPS chemical shifts and thus to assign chemical configurations to the XPS features observed at (100) Si surface resulting after etching in HF aqueous solution. Calculations on cluster models have been performed in the framework of density functional theory (DFT) and the Voronoi Deformation Density (VDD) method as implemented in the ADF package has been used to calculate the atomic charge, a fundamental quantity for rationalization of the XPS chemical shift data. We will show that the unique feature with negative chemical shift observed by XPS (Si$^{-\prime}$ in Table 1) can be attributed to the occurrence of silylene defect at the HF$_{aq}$-etched (100) Si surface.

1.1 Assigning XPS Features to Chemical Species-The Basic Equation

XPS lines could unambiguosly be assigned to bonding configurations if both the net charge Q and the Madelung potential U were known. Assume that one is able to assign to each atom its net charge Q; for the ith atom the dependence of the chemical shift $\Delta\xi_i$ on Q_i (hereinafter measured in units of proton charge e) is given by the following equation [11,12]

$$\Delta\xi_i = \epsilon_i Q_i + U_i + \mathcal{E} \tag{1}$$

where ϵ_i is the core-frontier coupling constant (an atomic property), $\mathcal{E}$ is the polarization energy (due to the polarization of the substrate when a core hole is formed after the photoemission, a substrate property), and U_i is the Madelung potential

$$U_i = 2E_0 a_0 \sum_{j\neq i} \frac{Q_j}{\mid \mathbf{r}_j - \mathbf{r}_i \mid} \tag{2}$$

where $2E_0$ and a_0 are the atomic units of energy and length (E$_0$ = 13.6 eV and a_0 = 0.53 Å), $\mathbf{r}_j$ is the position of the jth atom, and the sum is extended to all atoms but the ith. The value of ϵ_i can be estimated from elementary electrostatic: assuming for simplicity that the charge basin is spherical with radius R$_i$ and core electrons are distributed very close to the nucleus; if the net charge is uniformly distributed inside the sphere, the core-frontier coupling constant is given by

$$\epsilon_i = \frac{3E_0 a_0}{R_i} ; \tag{3}$$

otherwise, if the net charge is uniformly distributed on the spherical surface, one has

$$\epsilon_i = \frac{2E_0 a_0}{R_i} . \tag{4}$$

Since electrostatic repulsion spreads as much as possible the net charge, equation (3) is expected to be an overestimate of ϵ_i; on another side, ϵ_i cannot be lower than predicted by equation (4), so that one has

$$\frac{2E_0 a_0}{R_i} < \epsilon_i < \frac{3E_0 a_0}{R_i} \tag{5}$$

However, to be an atomic property R_i cannot be larger than the atomic radius a_i, so that

$$\frac{2E_0 a_0}{a_i} < \epsilon_i < \frac{3E_0 a_0}{a_i} \tag{6}$$

For silicon $a_{Si} = 1.17$ Å so that equation (6) gives that ϵ_{Si} may range in interval 12.3-18.5 eV. The major difficulty with equation (1) is due to the fact that it requires a knowledge of all the net charges Q_j and of the structure of the solid generating the Madelung potential U_i. The solution to the problem would be facilitated if the chemical shift of the ith atom were controlled by its first N nearest neighbours (referred to as $I^{(N)}(i)$) and the Madelung potential generated by the other atoms did not depend on the state of the considered atom. Rewrite equation (1) in the following form:

$$\Delta \xi_i = \epsilon_i Q_i + U_i^{(N)} + V_i^{(\tilde{N})} \tag{7}$$

where $U_i^{(N)}$ is the Madelung potential generated by the first, second, $\cdots$, Nth nearest neighbours to atom i,

$$U_i^{(N)} = 2E_0 a_0 \sum_{j \neq i, j \in I^{(N)}(i)} \frac{Q_j}{|\mathbf{r}_j - \mathbf{r}_i|} \tag{8}$$

and $V_i^{(\tilde{N})}$ is the sum of the polarization energy with the Madelung potential generated by the remaining atoms

$$V_i^{(\tilde{N})} = 2E_0 a_0 \sum_{j \neq i, j \notin I^{(N)}(i)} \frac{Q_j}{|\mathbf{r}_j - \mathbf{r}_i|} + \mathcal{E} \tag{9}$$

where the upper index '$(\tilde{N})$' denotes that it includes all atoms but the first, second, $\cdots$, Nth nearest neighbours to i, $\{j \mid j \notin I^{(N)}(i)\}$. The interest in form (7) of equation (1) resides in the hope that $V_i^{(\tilde{N})}$ is nearly independent of i

$$\forall i (V_i^{(\tilde{N})} \simeq V^{(\tilde{N})}) \tag{10}$$

so that equation (7) becomes

$$\Delta \xi_i \simeq \epsilon_i Q_i + U_i^{(N)} + V^{(\tilde{N})}. \tag{11}$$

Equation (11) is trivially rigorous for N $\rightarrow \infty$ and is expected to provide an adequate description of the chemical shift taking for N the order of neighbours over which the surface may be viewed as homogeneous. The simplest case of equation (11) is obtained taking N=0:

$$\Delta \xi_i = \epsilon Q_i + V^{(\tilde{0})} \tag{12}$$

This approximation gives the *naive attribution* based on the electronegativity distribution alone in determining the chemical shift, i.e. XPS lines are ordered for increasing chemical shift assigning each of them to the candidate species ordered for net charge. The naive attribution is however unsatisfactory as discussed in ref. [13] , and it cannot be accepted on the grounds of the insufficient quality and lack of physico-chemical consistency of description (12). These considerations suggest thus to go to the case N=1. Since in the case N=1 the chemical shift is controlled not only by the net charge but also by the Madelung potential truncated to nearest neighbours, the analysis of this case requires a sufficiently accurate, small cluster, model of the surface to produce the local distribution of atoms and bonds, allowing the extraction of realistic charges and internuclear distances.

2   Computational Details

Three different size cluster models of hydrogen-terminated silicon atoms forming fused cyclohexasilanes [14] have been considered to monitor the convergence of the quantities of interest in assigning the XPS chemical features of the silicon bonding configurations, including the silylene defect, with increasing cluster size. The smallest cluster Si_9H_{14} has a single surface dimer, the intermediate sized $Si_{16}H_{22}$ cluster model has three surface Si atoms along a single row, and the largest $Si_{58}H_{62}$ cluster model has three rows of three Si atoms in the surface. For each cluster model, four different geometrical structures were considered, corresponding to different configurations: the dihydride SiH_2 1×1 (100); the monohydride $(SiH)_2$ 2×1 (100); the silylene Si 1×1 (100) defect; and the clean Si_2 2×1 (100) configuration. Two additional small clusters, $Si_9H_{15}(OH)$ and $Si_9H_{15}(SiH_3)$, have been used to model the SiOH and the SiH_3 centre, respectively, that occur at the HF_{aq}-etched (100) Si surface. DFT calculations with the Becke-Perdew (BP86) exchange-correlation functional [15]-[17] were performed using the Amsterdam Density Functional (ADF) program package 2007.1 [18]-[22]. The molecular orbitals were expanded in a Slater-type STO basis set of triple-ζ doubly polarized TZ2P quality for all atoms. The core orbitals were kept frozen up to 2p for Si, or 1s for O. Convergence criteria for full geometry optimizations were 1×10^{-3} hartrees in the total energy, 5×10^{-4} hartrees $\mathring{A}^{-1}$ in the gradients, 1×10^{-2} $\mathring{A}$ in bond lengths, and $0.20°$ in bond angles. In all cases no symmetry constraint was imposed and the stability of the structures was checked by performing a normal-mode analysis and checking that all vibration frequencies are positive for the Si_9H_{14} and $Si_{16}H_{22}$ cluster models; the $Si_{58}H_{62}$ clusters were too large for feasible frequency calculations. The calculated equilibrium structures for all the Si_9H_{14} and $Si_{16}H_{22}$ cluster models are true minima. The ground state spin configuration of all clusters, including the silylene defect, is a singlet. However, for the silylene defect, a triplet spin state has been calculated that is above the singlet ground state for the three models by 0.8-0.9 eV and corresponds to a geometry structure where the two hydrogen atoms are strongly symmetric; i.e. both hydrogen atoms are in the usual silanic configuration, with the two unpaired electrons localized on the silylene Si. The Voronoi Deformation Density (VDD) method, as implemented in the ADF package, was chosen for computing atomic charges in the clusters, thus avoiding the

unwanted dependence on the basis set suffered by Mulliken population analysis
[23].

3   Results and Discussion

3.1   Cluster Modelling of Surface Sites

The use of small clusters is crucial for consistency with a description, like that
of equation (11), where the Madelung potential is truncated at the first neighbours. Of course, this description has a meaning only if the resulting charge
and the interatomic distance vary weakly with cluster size. Figure 1 provides
stick-and-ball views of the structures of the considered clusters as result from
optimization calculations. The three different size cluster models, in the four
different geometrical structures mimicking the considered superficial configurations, are depicted; main geometrical parameters are also shown. We find that
the bond lengths do not substantially change by increasing the cluster size.

VDD atomic charges on superficial silicon and hydrogen atoms are reported
in Table 2 for all different size clusters.

Table 2. Partial charges (VDD) on superficial silicon and hydrogen atoms for different size cluster models in the dihydride, monohydride, silylene defect, and clean
configurations.

cluster size	Si_{sil} H_{bridge}	Si	H
Dihydride			
Si_9H_{16}		0.11	-0.05
$Si_{16}H_{24}$		0.12	-0.03
$Si_{58}H_{64}$		0.12	-0.03
Monohydride			
Si_9H_{14}		0.03	-0.04
$Si_{16}H_{22}$		0.06/0.03	-0.05
$Si_{58}H_{62}$		0.05/0.04	-0.05
Silylene defect			
Si_9H_{14}	-0.05 0.04	0.11	-0.05
$Si_{16}H_{22}$	-0.06 0.03/0.04	0.09	-0.05
$Si_{58}H_{62}$	-0.08 0.03	0.11	-0.04
Clean			
Si_9H_{12}		-0.11(up)-0.01(down)	
$Si_{16}H_{20}$		-0.13(up) 0.05(down)	
$Si_{58}H_{60}$		-0.14(up) 0.07(down)	

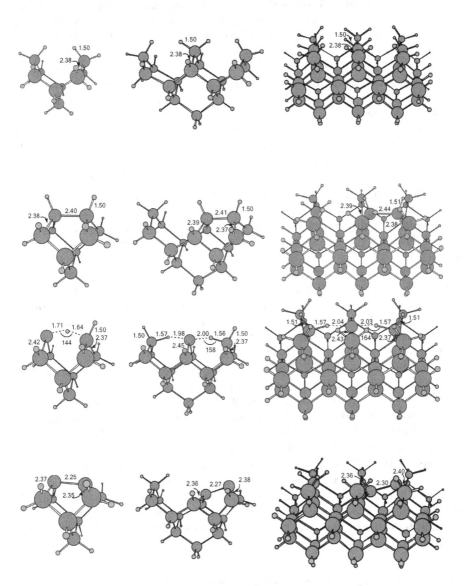

Fig. 1. Optimized structures and relevant geometrical parameters ($\mathring{A}$) in the considered cluster models of different sizes in different superficial bonding configurations: dihydride (top), monohydride and silylene defect (middle), clean (bottom)

The calculated charges are found to slightly change while increasing the cluster size. Our results show that the quantities of interest in assigning the XPS chemical shift (namely, Voronoi charges and interatomic distance) vary weakly with cluster size. In a previous work by us [24], we proposed that the occurrence of silylene centers with a partial negative charge could be considered as being responsible for the XPS feature with negative chemical shift observed at the hydrogen-terminated (100) Si surface. The cluster picture of the silylene defect shows that the hydrogen atoms in both nearby silicon dihydrides are notably asymmetric: whereas in each silicon dihydride one hydrogen is in the usual silanic configuration, the other is at a bond distance of silylene silicon too (see Figure 1). The configuration of each hydrogen coordinated to silylene silicon is reminiscent of that in electron-deficient compounds like diborane; alternatively, the hydrogen may be viewed as an off-axis bond-centered proton in a strongly relaxed Si-Si bond. Silylene is a configuration of silicon (with 6 electrons in the outer shell) where the silicon atom is covalently bonded with single bonds to two other atoms. Clearly enough, the 6-electron configuration of silylene makes this structure highly reactive. At the (100) Si surface, where silylene could be formed during HF_{aq} etching, the 6-electron configuration can be stabilized via interaction with species unavoidably present in the HF_{aq}-etching solution, due to the silylene zwitterionic nature (see Ref. [24]). However, what is interesting in the present work, is that also the negative charge on silylene Si (denoted as Si_{sil}) hardly varies with cluster size. We can conclude that since the internuclear distances (Figure 1) and net charges (Table 2) are found to slightly change while increasing the cluster size for all the different considered superficial silicon configurations, it is reasonable to take data resulting from the smallest cluster modelling for assigning XPS features. To this aim, we used the two small clusters whose optimized geometries are shown in Figure 2.

Both clusters may be used for modelling the SiH_2 centre; the top cluster is used to model the SiH_3 centre; the bottom cluster is used to model the $Si(H)OH$ centre. Modelling the SiH centre is less trivial: the top cluster shows indeed not only a silicon monohydride resulting from the substitution of a silyl group for hydrogen at the otherwise perfect 1×1 (100) surface site (referred to as (100) SiH), but also two silicon monohydrides with symmetry and orientation characteristic of silicon monohydrides at the (111) surface (referred to as (111)SiH). In addition, the smallest cluster Si_9H_{14} modelling the silylene defect (see Figure 1 for interatomic distances, and Table 2 for charges) has been also used to assign XPS features.

3.2 Attribution of XPS Features Based on Theoretical Data

The analysis of the case $N = 1$ of equation (11) requires a sufficiently accurate, small cluster, model of the bonding configurations to produce the local distribution of atoms and bonds, allowing the extraction of realistic charges and internuclear distances. The smallest $Si_9H_{15}(OH)$ and $Si_9H_{15}(SiH_3)$ clusters shown in Figure 2 have been used to calculate the expected chemical shifts of the considered surface centers. They are summarized in Table 3 where the Voronoi

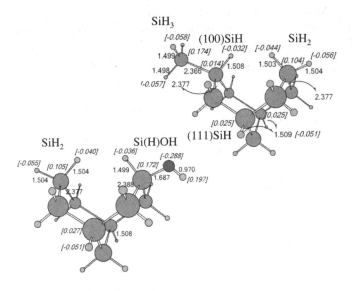

Fig. 2. Clusters used for assigning XPS features. Calculated Voronoi atomic charges on the considered hydrogen-terminated silicon atoms and relevant geometrical parameters (Å) for truncated Madelung potential evaluation are reported.

net charges and the calculated Madelung potential $U^{(1)}$ (eV) are also reported for $\epsilon = 12.3$ eV and $\epsilon = 18.5$ eV.

We observe that, irrespective of the assumed value of ϵ in the considered interval 12.3-18.5 eV, the putative chemical shift $\Delta\xi$ ($= \epsilon Q + U^{(1)}$, except for the additive term $V^{(\bar{1})}$) increases in the order

$$\Delta\xi \text{ [SiH]} < \Delta\xi \text{ [SiH}_2] < \Delta\xi \text{ [SiH}_3]$$

Rather, the value of ϵ affects only the position of Si(H)OH in the sequence: the order shifts from

$$\Delta\xi \text{ [Si(H)OH]} < \Delta\xi \text{ [SiH]} < \Delta\xi \text{ [SiH}_2] < \Delta\xi \text{ [SiH}_3]$$

for $\epsilon = 12.3$ eV, to

$$\Delta\xi \text{ [SiH]} < \Delta\xi \text{ [Si(H)OH]} < \Delta\xi \text{ [SiH}_2] < \Delta\xi \text{ [SiH}_3]$$

for $\epsilon = 18.5$ eV. We may thus limit to consider the tentative assignments in Table 4, where we consider four other counterintuitive assignments (CIA_1, CIA_2, CIA_3, and CIA_4) characterized by four different positions of Si(H)OH in the silicon-hydride alignment.

For them the core-frontier coupling constant ϵ, surface bias $V^{(\bar{1})}$, and correlation coefficient r are given by

$\epsilon = 16.19$ eV and $V^{(\bar{1})} = +0.01$ eV, with r = 0.938 for CIA_4

$\epsilon = 15.32$ eV and $V^{(\bar{1})} = +0.11$ eV, with r = 0.967 for CIA_3

$\epsilon = 13.84$ eV and $V^{(\bar{1})} = +0.29$ eV, with r = 0.968 for CIA_2

$\epsilon = 11.54$ eV and $V^{(\bar{1})} = +0.24$ eV, with r = 0.956 for CIA_1

The statistical descriptions of CIA_2 and CIA_3 are nearly the same (with correlation coefficients closer to 1) and they provide the optimal description. Deciding

Table 3. Charge, Madelung potential and expected chemical shifts (apart from the additive term $V^{(1)}$ and for ϵ at the extremes of its expected validity interval) of the considered surface centres

Centre	Q	$U^{(1)}$ (eV)	$\epsilon Q + U^{(1)}$ (eV)	
			$\epsilon = 12.3$ eV	$\epsilon = 18.5$ eV
(100)SiH	0.014	+0.753	+0.952	+1.012
(111)SiH	0.025	-0.487	-0.180	-0.025
SiH$_2$	0.104	-0.958	+0.321	+0.966
Si(H)OH	0.172	-2.804	-0.688	+0.378
SiH$_3$	0.174	-1.577	+0.563	+1.642

Table 4. The considered assignation schemes

Line	Si$^{-\prime}$	Si$^{\prime}$	Si$^{2\prime}$	Si$^{3\prime}$	Si^{1}
Assignation					
CIA$_4$		SiH	SiH$_2$	Si(H)OH	SiH$_3$
CIA$_3$		SiH	Si(H)OH	SiH$_2$	SiH$_3$
CIA$_2$		Si(H)OH	SiH	SiH$_2$	SiH$_3$
CIA$_1$	Si(H)OH	SiH	SiH$_2$	SiH$_3$	

which between them is a better model of reality cannot be done on statistical grounds but on physical grounds combining the synchroton-radiation data of the (111) SiH surface and the relative abundances of the various species resulting from IR analysis (see Ref. [13]). Both statistical and physical considerations support CIA$_3$ as the optimal assignment that, however, leaves Si$^{-\prime}$ unassigned. Accounting for feature Si$^{-\prime}$ is not trivial. We consider the hypothesis that Si$^{-\prime}$ is due to some form of silylene. Then the calculated chemical shift of silylene modelled by Si$_9$H$_{14}$ cluster (Q=-0.05, $U^{(1)}$=+0.56 eV, $\Delta\xi$=-0.10 eV) is consistent with the one of Si$^{-\prime}$. Assuming that the attributions of CIA$_3$ are correct and that Si$^{-\prime}$ can actually be assigned to silylene defect, the whole data can be described by the following equation:

$$(\Delta\xi - U^{(1)})(eV) = 16.00Q + 0.01 \qquad (13)$$

with a correlation coefficient of 0.987, better than the one obtained ignoring Si$^{-\prime}$. According to equation (13) the chemical shift of silylene centre would be -0.23 eV, in good agreement with experiment.

4   Conclusion

Density functional calculations for different size cluster models of the hydrogen terminations at the HF$_{aq}$-etched (100) Si surface have demonstrated that their mean properties are determined by really few neighbours and can thus be modelled with small clusters. On the basis of the optimized geometries and of the Voronoi charges deduced from DFT calculations for such clusters, the four lines (in addition to that of elemental silicon S^0) detected in the spectral region with chemical shift in the range between -0.3 and +1 eV of the XPS spectra of HF_{aq}-etched (100) Si have had the following attributions:

$Si^{'}$, silicon monohydrides at (111) facets

$Si^{2'}$ silicon terminated with one hydrogen and one silanol group

$Si^{3'}$ silicon dihydrides

Si^1 silicon trihydrides or its associated silicon monohydride defects on facets with (100) orientation

In addition, the chemical configuration associated with $Si^{-'}$ has been attributed to fully dehydrated silylene defects at 1×1 (100) SiH_2 surface. In conclusion, the considered example has shown that the assignment of spectral lines basing on the differences of electronegativities alone, though of common use when dealing with the inverse problem in XPS, may lead to erroneous attributions and must thus taken with caution. Theoretical calculations performed on suitable, small cluster models of accurate atomic net charge (VDD) and bond lenghts, that are needed for the description of the chemical shift in terms of partial charge and truncated Madelung potential, can provide information even in complicate situations, like that characterizing the hydrogen-terminated (100) Si prepared by HF_{aq} etching of the native oxide.

References

1. Tsumuraya, T., Batcheller, S.A., Masamune, S.: Angew. Chem. Int. Ed. Engl. 30, 902 (1991)
2. Nergaard Waltenburg, H., Yates, J.T.: Chem. Rev. 95, 1589 (1995)
3. Aoyama, T., Goto, K., Yamazaki, T., Ito, T.: J. Vac. Sci. Technol. A 14, 2909 (1996)
4. Cerofolini, G.F., Galati, C., Reina, S., Renna, L., Spinella, N., Jones, D., Palermo, V.: Phys. Rev. B 72, 125431 (2005)
5. Niwano, M., Kageyama, J., Kinashi, K., Takashi, I., Miyamoto, N.: J. Appl. Phys. 76, 2157 (1994)
6. Cerofolini, G.F., Giussani, A., Carone Fabiani, F., Modelli, A., Mascolo, D., Ruggiero, D., Narducci, D., Romano, E.: Surf. Interface Anal. 39, 836 (2007)
7. Seah, M.P., Briggs, D.: Practical Surface Analysis: Auger and X-ray Photoelectron Spectroscopy. IM Publications, Chichester (2003)
8. Briggs, D., Grant, J.T.: Surface Analysis by Auger and X-ray Photoelectron Spectroscopy, 2nd edn. Wiley, Chichester (1992)
9. Cumpson, P.J.: Appl. Surf. Sci. 16, 144 (1999)
10. Cerofolini, G.F., Giussani, A., Modelli, A., Mascolo, D., Ruggiero, D., Narducci, D., Romano, E.: Appl. Surf. Sci. 254, 5781 (2008)
11. Nefedov, V.I., Yarzhemsky, V.G., Chuvaev, A.V., Trishkina, E.M.: J. Electron Spectrosc. Relat. Phenom. 46, 381 (1988)
12. Cerofolini, G.F., Meda, L., Re, N.: Appl. Phys. A 72, 603 (2001)
13. Cerofolini, G.F., Romano, E., Narducci, D., Belanzoni, P., Giorgi, G.: Appl. Surf. Sci. 256, 6330 (2010)

14. Konecny, R., Hoffman, R.: J. Am. Chem. Soc. 121, 7918 (1999)
15. Becke, A.D.: Phys. Rev. A 38, 3098 (1988)
16. Perdew, J.P.: Phys. Rev. B 33, 8822 (1986)
17. Perdew, J.P.: Phys. Rev. B 34, 7406 (1986)
18. ADF2007.01, SCM, Theoretical Chemistry, Vrije Universiteit Amsterdam, The Netherlands (2007), http://www.scm.com
19. Baerends, E.J., Ellis, D.E., Ros, P.: Chem. Phys. 2, 42 (1973)
20. te Velde, G., Bickelhaupt, F.M., van Gisbergen, S.J.A., Fonseca Guerra, C., Baerends, E.J., Snijders, J.G., Ziegler, T.: J. Comput. Chem. 22, 931 (2001)
21. Fonseca Guerra, C., Snijders, J.G., te Velde, G., Baerends, E.J.: Theor. Chem. Acc. 99, 391 (1998)
22. te Velde, G., Baerends, E.J.: J. Comput. Phys. 99, 84 (1992)
23. Fonseca Guerra, C., Handgraaf, J.W., Baerends, E.J., Bickelhaupt, F.M.: J. Comput. Chem. 25, 189 (2004)
24. Belanzoni, P., Giorgi, G., Sgamellotti, A., Cerofolini, G.F.: J. Phys. Chem. A 113, 14375 (2009)

Modeling the Intermolecular Interactions and Characterization of the Dynamics of Collisional Autoionization Processes

Stefano Falcinelli[1,*], Marzio Rosi[1,2], Pietro Candori[1], Franco Vecchiocattivi[1],
Alessio Bartocci[3], Andrea Lombardi[3], Noelia Faginas Lago[3], and Fernando Pirani[3]

[1] Department of Civil and Environmental Engineering,
University of Perugia,
Via Duranti 93, 06125 Perugia, Italy
{stefano,atomo,vecchio,pirani}@dyn.unipg.it, marzio@unipg.it
[2] ISTM-CNR, University of Perugia,
Via Elce di Sotto 8, 06123 Perugia, Italy
[3] Department of Chemistry, University of Perugia,
Via Elce di Sotto, 8, 06123 Perugia, Italy
{ebiu2005,piovro}@gmail.com

Abstract. The autoionization dynamics of triatomic molecules induced by $He^*(2^{3,1}S_{1,0})$ and $Ne^*(^3P_{2,0})$ collisions has been discussed. The systems are analyzed by using an optical potential model within a semiclassical approach. The real part of the potential is formulated applying a semiempirical method, while the imaginary part has been used in the fitting procedure of the data adjusting its pre-exponential factor. The good agreement between calculations and experiment confirms the attractive nature of the potential energy surface driving the He^* and Ne^*-H_2O dynamics.

Keywords: intermolecular potentials, collisional autoionization, semiempirical method, Penning ionization.

1   Introduction

Autoionization processes in slow molecular collisions can occur because the autoionization time is usually shorter than the characteristic molecular collision time at thermal energies (~10^{-12} s). The basic requirement is that the two partners should have enough internal energy to produce a collision complex degenerate with the ionization continuum. Collisional autoionization must present several analogies with photoionization. For example, in both cases when the electron is ejected, the measurement of its energy and momentum provides a detailed spectroscopy of the ionic product, which can then be correlated with its subsequent dynamical evolution as showed by a number of studies performed by our group by using synchrotron radiation [1-9].

---

[*] Corresponding author.

B. Murgante et al. (Eds.): ICCSA 2013, Part I, LNCS 7971, pp. 69–83, 2013.
© Springer-Verlag Berlin Heidelberg 2013

However, while in photoionization the conditions of the autoionized molecules are determined by the energy and polarization of the photon [10-16], in the collisional autoionization process these conditions are determined by the collision characteristics of the system under consideration, such as relative velocity, internal states of the reactants, relative orientation, etc. [17-21].

A collisional autoionization process can be schematically written as follows:

$$X + Y \rightarrow [X \cdots Y]^*$$

where X and Y are atoms or molecules and $[X \cdots Y]^*$ the collision complex in an autoionizing state,

$$[X \cdots Y]^* \rightarrow [X \cdots Y]^+ + e^-$$

After an $[X \cdots Y]^+$ ionic complex is formed, the collision continues towards the final ionic products

$$[X \cdots Y]^+ \rightarrow \text{ion products.}$$

In the literature this collisional autoionization process is often called "Penning" ionization after the early observation in 1927 by F.M.Penning [22]. This process has long attracted the attention of the scientific community, as shown by the large number of papers and review articles on this topic [23-26]. Specific features make these processes very interesting from a fundamental point of view. However many applications of collisional autoionization to important fields like radiation chemistry, plasma physics and chemistry, combustion processes, and the development of laser sources, are also possible [24-25].

Several experimental techniques are used to study the microscopic dynamics of these collisional autoionization processes. It is well established that the most valuable information about the dynamics of a collisional process is provided by molecular beam scattering experiments. In such cases one can study the process by detecting the metastable atoms, the electrons or the product ions [17-21]. In general elastic scattering experiments provide information mainly about the pre-ionization dynamics, electron energy spectra provide information about the dynamics of the autoionization event, and ion mass spectrometric experiments mainly about the post-ionization dynamics [24-25]. The rare gas atoms, when excited to their first metastable levels, are very suitable for these experimental studies because of their high energy content and relatively long life-time, which allows them to survive along beam path lengths typical of the laboratory molecular beam experiments. The electronic energy and life-time values of metastable rare gas atoms are listed in Table 1. The energy values are large enough for an autoionizing complex to be formed in most cases. Metastable helium atoms have enough energy to ionize all known atomic and molecular species, except ground state helium and neon atoms, while in the case of metastable neon atoms the exceptions also include fluorine atoms.

In general, the role of metastable rare gas atoms in Earth's atmospheric reactions is of interest for a number of reasons: (i) the quantity of rare gas atoms in the atmosphere is not negligible (argon is the third component of the air, after nitrogen and oxygen molecules: 0.934 % ± 0.001% by volume compared to 0.033 % ± 0.001% by volume of CO_2 molecules); (ii) recent studies have shown that the exosphere of our planet is rich in $He^*(2^1S_0)$ metastable atoms [27]; and (iii) rate constants for ionization processes induced by metastable rare gas atoms are generally larger than those of common bimolecular chemical reactions of atmospheric interest.

Table 1. Some characteristics of metastable rare gas atoms [24,25]

Atoms	Excitation energy (eV)	Lifetime (s)
$He^*(2^1S_0)$	20.6158	0.0196
$He^*(2^3S_1)$	19.8196	9000
$Ne^*(^3P_0)$	16.7154	430
$Ne^*(^3P_2)$	16.6191	24.4
$Ar^*(^3P_0)$	11.7232	44.9
$Ar^*(^3P_2)$	11.5484	55.9
$Kr^*(^3P_0)$	10.5624	0.49
$Kr^*(^3P_2)$	9.9152	85.1
$Xe^*(^3P_0)$	9.4472	0.078
$Xe^*(^3P_2)$	8.3153	150

For example, the rate constants for Penning ionization processes are of the same order of magnitude of gas phase bimolecular reactions of $O(^1D)$, Cl, or Br, which are known to be relevant in atmospheric chemistry, whereas other reactions exhibit rate constants that are at least one order of magnitude smaller [29].

Considering the planetary atmospheric compositions, we can suppose that rare gas atoms are involved in several atmospheric phenomena not only on Earth, but also on other planets of the Solar System, like Mars and Mercury. In fact, argon is the third component of the martian atmosphere with its 1.6% and helium is a relatively abundant species (about 6%) in the low density atmosphere of Mercury. As mentioned in the previous section, a basic step in chemical evolution of planetary atmospheres and interstellar clouds (where the 89% of atoms are hydrogen and 9% are helium) is the interaction of atoms and molecules with electromagnetic waves (γ and X rays, UV light) and cosmic rays. Therefore, He^* and Ar^* formation by collisional excitation with energetic target particles (like electrons, protons or alpha particles) and the subsequent Penning ionization reactions could be of importance in these environments. In particular, considering the tenuous exosphere of Mercury, it is interesting to note that in 2008 the NASA spacecraft MESSENGER discovered surprisingly large amounts of water and several atomic and molecular ionic species including O^+, OH^-, H_2O^+ and H_2S^+ (NASA and the MESSENGER team will issue periodic news releases and status reports on mission activities and make them available online at http://messenger.jhuapl.edu and http://www.nasa.gov/messenger). Taking into account the suggested polar deposits of water ice in this planet, McClintock and Lankton (2007) have suggested impact vaporization mechanisms [30]. We argue that in these parts of the planetary surface, the presence of He* and its collisions can cause subsequent ions formation. In this respect, the study of the Penning ionization of water by He* and Ne* metastable rare gas atoms producing H_2O^+, OH^+ and O^+ (see the next section), could help in explaining the possible routes of formation for these ionic species in Mercury's exosphere. Main purpose of this work is to demonstrate that the measure under high resolution conditions of microscopic quantities as collisional ionization cross sections and the proper description of the intermolecular interaction, that is its accurate analytical formulation, represent crucial steps for the detailed characterization of the dynamic of the elementary involved processes.

2   Mass Spectrometric Determinations and Cross Section Measurements

The experimental apparatus, used for the mass spectrometric determinations and for cross section measurements here reported, has been previously employed and is described in detail in a recent paper [27]. Basically, it consists of an effusive or, alternatively, a supersonic metastable neon atom beam, which crosses at right angles an effusive secondary beam of water molecules, produced by a microcapillary array. The ions produced in the collision zone are extracted, focused, mass analyzed by a quadrupole filter and then detected by a channel electron multiplier.

In order to cover a wide collision velocity range, the neon beam is produced by two sources, which can be used alternately. The first one is a standard effusive source mantained at room temperature, coupled with an electron bombardment device, while the second one is a microwave discharge beam source operating with pure neon at a pressure of ~10^{-3} atm. Together with metastable atoms, the discharge source produces a large number of Ne(I) photons. These allow comparative studies of Penning ionization and photoionization. The supersonic beam is used when we need to maximize the intensity of metastable atom production.

The Ne* atom velocity is analyzed by a time-of-flight (TOF) technique: the beam is pulsed by a rotating slotted disk and the metastable atoms are counted, using a multiscaler, as a function of the delay time from the beam opening. By using this technique, the velocity dependence of the cross section is obtained. Time delay spectra of the metastable atom arrival at the collision zone are recorded, as well as the time spectra of the product ion intensity. Then the relative cross sections, σ, as a function of the collision energy, E, are obtained, for a given delay time τ, according to the equation

$$\sigma(E) = \frac{I^+(\tau)v_1}{I^*(\tau)g}$$

(1)

where I^+ and I^* are the intensities of the product ions and metastable atoms, respectively, v_1 is the laboratory velocity of the Ne*, and g is the relative collision velocity. By the TOF technique, the separation of photo-ions and Penning-ions is very easy, the former being detected at practically zero delay time.

Absolute values of the total ionization cross section for different species can be obtained by the measurement of relative ion intensities in the same conditions of metastable atom and target gas density in the crossing region. This allows the various systems to be put on a relative scale which can be normalized by reference to a known cross section such as, in the present case, the Ne^*–Ar absolute total ionization cross section by West et al. [31].

Following the considerations made on the importance of Penning ionization of water by $He^*(2^3S_1, 2^1S_0)$ and $Ne^*(^3P_{2,0})$ in the understanding of the chemical composition of Mercury's atmosphere, we report here the mass spectrometric determination of the channel branching ratios for both systems as obtained at an averaged collision energy of 70 meV:

$$He^*(2^3S_1, 2^1S_0) + H_2O \rightarrow He + H_2O^+ + e^- \qquad 77.8\%$$
$$\rightarrow He + H + OH^+ + e^- \qquad 18.4\%$$
$$\rightarrow He + H_2 + O^+ + e^- \qquad 3.8\%$$

$$Ne^*(^3P_{2,0}) + H_2O \rightarrow Ne + H_2O^+ + e^- \qquad 96.7\%$$
$$\rightarrow Ne + H + OH^+ + e^- \qquad 2.9\%$$
$$\rightarrow Ne + H_2 + O^+ + e^- \qquad 0.4\%$$

No experimental evidence has been recorded for the HeH^+, NeH^+ and HeH_2O^+, NeH_2O^+ ion production, at least under our experimental conditions.

Total ionization cross sections for $Ne^*(^3P_{2,0})$-H_2O collisions are reported in Figure 1 as a function of the relative collision energy. The recorded data show a decreasing trend as a function of the collision energy, with an absolute value of ≈ 47.5 $Å^2$ at 0.070 eV.

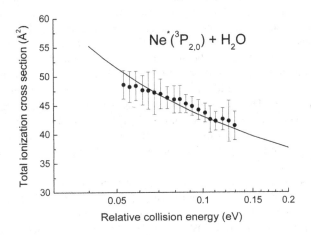

Fig. 1. The total ionization cross sections of the $Ne^*(^3P_{2,0})$-H_2O system as a function of the relative collision energy. The curve represent the calculation of the cross section by the use of the present semiempirical potential and adjusting only the pre-exponential parameter of the imaginary part of the optical potential.

An analysis of present cross section data, in terms of the intermolecular interaction between the two collision partners, requires the use of an optical potential model [24,25], defined as a combination of a real part with an imaginary part. While the real part affects the approach dynamics of the two partners, the imaginary one controls the ionization probability, being related to the reciprocal of the lifetime of the system. The interaction between Ne[*] and H_2O, as discussed below, results to be strongly anisotropic. However, for the present calculation we have used the spherical average of such a potential, because, in the collision events of the present experiment, the rotational period of water molecules is rather fast when compared with typical collision time.

In the following section a description of the optical potential that we have used is reported and the semiclassical calculation of the total ionization cross section is also described.

3   Theoretical Approaches: The Optical Potential Model

From a theoretical point of view, collisional autoionization phenomena in atom-atom systems are generally treated on the basis of a local complex potential,

$$W(R) = V(R) - \frac{i}{2}\Gamma(R) \tag{2}$$

called also "optical potential", whose real part $V(R)$ represents the interaction controlling the approach of the colliding particles and the imaginary part $\Gamma(R)$, the "potential width", is related to the decay (ionization) probability of the system at the intermolecular distance R [24-28]. From a general point of view, the real part of the interaction potential defines, at any intermolecular distance R, the energy of the bound wave function of the system. Instead, the imaginary part, which controls the decay, depends more specifically on the characteristics of the electron wave function directly involved in the charge transfer and ionization. In the simple atom-molecule case, both the real and imaginary parts are expected to be anisotropic, because the real part of the potential depends on the angle of approach of the metastable X[*] atom towards the target Y molecule, while the potential width varies with the geometry of the autoionizing collisional complex [X⋯Y][*]. This latter point can be easily understood when the commonly accepted electron exchange mechanism, originally proposed by Hotop and Niehaus [23], is taken into account. According to this model, autoionization occurs through a transfer of an outer-shell electron of the target Y into the inner-shell vacancy of the excited atom X[*], which subsequently ejects an external electron. In an orbital model, such process depends strongly on the overlap between the orbitals involved in the electron transfer. The ionization probability is expected to be maximum at those distances where the relative motion is slowest (i.e., at the neighboring of the turning point) and for approaches of the metastable atom along the directions where the electron density of the orbital to be ionized is highest.

Studies on anisotropy effects in Penning ionization are still rather scarce [26,32-34]; due to the weakness of the interaction there is a difficulty in the proper characterization of the optical potential for many configurations of the system. Only few rigorous atom-molecule treatments are present in the literature, in particular, for the systems $He^*(2\,^3S_1)+H_2$, N_2, H_2O [24,25,35,36]. In these cases, *ab initio* complex potentials have been calculated and the ionization dynamics investigated by using quantum approaches or quasiclassical trajectory methods. In many other cases, semiempirical potentials and estimated energy widths have been used to analyze experimental data [27,28,37,38]. In this work, we study the collisional autoionization dynamics of the triatomic target molecules with excited $He^*(2\,^{3,1}S_{1,0})$ and $Ne^*(^3P_{2,0})$ atoms. Our aim is a more quantitative interpretation of experimental results concerning H_2O target molecules, also in the attempt to give a better account of how interaction anisotropies play a selective role in the autoionization process. In order to accomplish this task, we use a method based on the identification, modeling, and combination of the leading components of the interaction potential. As we will see in the following, our model potential can be expressed in a simple analytical form and intended to be of completely general validity. Furthermore, present results can also stimulate accurate *ab initio* calculations which can provide further crucial information on other basic features of the full intermolecular potential energy surfaces.

3.1   The Optical Potential for Triatomic Molecules

From a general and quantitative point of view, a system formed by a metastable atom interacting with a triatomic molecule, as H_2O, dynamically evolves on a multidimensional potential energy surface (PES). In the thermal energy range, as the case here discussed, effects of internal degree of freedom, arising from molecular deformations induced by collision events, can be neglected and therefore the optical potential W, controlling ionization processes, can be written as the combination of a real V and an imaginary Γ part [27,28]:

$$W(R,\vartheta,\varphi) = V(R,\vartheta,\varphi) - \frac{i}{2}\Gamma(R,\vartheta,\varphi)$$

$$(3)$$

where R is the distance of the metastable atom from the center of mass of the molecule, ϑ is the polar angle between the C_{2v} axis of H_2O and R ($\vartheta=0$ refers to the oxygen side) and φ is the azimuthal angle ($\varphi=\pi/2$ describes the atom located on the water plane). The real part, $V(R, \vartheta, \varphi)$, has been assumed to be mainly characterized by taking into account the interactions in six limiting basic configurations. Such interactions are indicated as V_{C2v-O} and V_{C2v-H} (with $\vartheta=0$ and $\vartheta=\pi$ respectively), $V_\perp$ ($\vartheta=\pi/2$, $\varphi=0$) and V_{planar} ($\vartheta=\pi/2$, $\varphi=\pi/2$), where $V_\perp$ and V_{planar} are doubly degenerated. As we have done previously [27,28], their strength and their radial dependence have been described by a semiempirical method developed in our laboratory [39,40], which is founded on the identification and description of some leading interaction components, to be considered "effective", since indirectly including the role of less important contributions. The method exploits correlation and perturbative formulas, providing range and strength of such components in terms of fundamental physical properties of the interacting partners, as the value of Ne^* and H_2O polarizability (27.8 $Å^3$ and 1.47 $Å^3$ respectively) and that of H_2O dipole moment (1.85 D) [28,41]. In particular, five leading-effective interaction components are taken into account:

1. The dispersion attraction, dominant at large R, combines with the size repulsion, prevalent at short R, providing a global component called, as previously suggested [39], van der Waals interaction (V_{vdw}).

2. A weak charge transfer contribution, V_{CT}, arises from the perturbative electron exchange between the outermost occupied orbital of the metastable and the LUMO of the H_2O molecule. V_{CT} affects only V_{C2v-H}, since the involved configuration (see Figure 2) is the only one promoting the formation of a weak hydrogen bond [42,43].

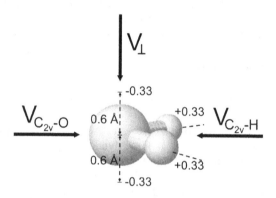

Fig. 2. A schematic structure of the water molecule, where the direction for the V_{C2v-H}, V_{C2v-O} and $V_\perp$ interactions are indicated. In the figure also the partial electrostatic charges located on the two hydrogen atoms and below and above the oxygen atom (at 0.6 Å from the oxygen nucleus) are indicated (for details and references see the text).

3. The asymptotic long range induction, due to permanent dipole of water interacting with the induced multipole on the strongly polarizable Ne^*, increases the long range attraction. In this paper its averaged contribution has been enclosed in V_{vdw} through a modification of the parameters involved in its formulation [39,42,43] (see also below).

4. As it has been done in Ref. 28, V_{qm} describes the balancing of induction attraction with size repulsion associated to the interaction, emerging at short R, when the metastable atom assumes a partial and "effective" charge q. This charge arises from the strong polarization effects of the metastable atom, which is a big "floppy" species with a radius of some angstroms, and is promoted by the electron lone pairs of water [28,30]. These effects, emerging as the intermolecular field increases, originate positive and negative charge centers, respectively located on the nucleus and in the outside region and separated by a distance comparable or larger than the intermolecular separation R. This component plays a crucial role at intermediate and short R in determining the behavior of V_{C2v-O} and $V_\perp$. Following the same guidelines of

previous studies [28], we have found appropriate to represent V_{qm}, in the neighborhood of the equilibrium distance of the collision complex, by the use of an effective charge q=0.75 a.u., since the positive charge center on polarized Ne* is significantly closer to the interacting partner [28], while the negative charge is repelled outside.

5. The electrostatic component, V_{electr}, emerging again at intermediate and short R, depends on the interaction between q and the charge distribution on water. The latter, confined in a restricted space (see Figure 2), is assumed to be consistent with the water dipole and also provides reasonable values of the quadrupole moment components. Under these conditions V_{electr} can be represented as a sum of coulomb terms.

3.2   Formulation of the Intermolecular Interaction

According to these considerations, in the chosen basic configurations the semiempirical formulation of V(R) takes the form:

$$V_{C2v-H}(R) = V_{vdW}(R) + V_{CT}(R)$$
$$V_{C2v-O}(R) = V_{vdW}(R)f(R) + \left(V_{qm}(R) + V_{electr}(R)\right)\left(1 - f(R)\right)$$
$$V_{\perp}(R) = V_{vdW}(R)f(R) + \left(V_{qm}(R) + V_{electr}(R)\right)\left(1 - f(R)\right)$$
$$V_{planar}(R) = V_{vdW}(R)f(R) + \left(V_{qm}(R) + V_{electr}(R)\right)\left(1 - f(R)\right)$$

$$(4)$$

V_{vdW} and V_{qm} have been represented by the improved Lennard Jones function, recently introduced [38,44] and found to be able to describe weak interactions of different nature and in a wide range of configurations. Its general form is

$$V_{ILJ}(R) = \varepsilon \left[\frac{m}{n(R) - m} \left(\frac{R_m}{R}\right)^{n(R)} - \frac{n(R)}{n(R) - m} \left(\frac{R_m}{R}\right)^{m} \right]$$

$$(5)$$

where

$$n(R) = \beta + 4\left(\frac{R}{R_m}\right)^2$$

and ε is the well depth while R_m is its location and β is a shape parameter. For neutral-neutral systems m=6, while for ion-neutral cases m=4.

V_{CT} has an exponential form, being related to the overlap integral between orbitals exchanging the electron, whose radial dependence is modulated by the ionization potential and electron affinity of the partners [27,45]:

$$V_{CT}(R) = A_H e^{-\gamma_H R}$$

$$(6)$$

where the pre-exponential A_H and the exponent γ_H, are listed in Table 2 and have been determined extending the phenomenology recently obtained for water [42].

Finally the "switching function" f(R), which defines the relative weight in eq.(4), describes the transition of the metastable atom from the weakly perturbed state at large R, where it is assumed to maintain a nearly spherical symmetry, to the strongly polarized one at short R. The f(R) function modulates the contribution of $V_{vdW}(R)$, $V_{qm}(R)$ and $V_{electr}(R)$ when R varies and guaranties the smoothness of the analytical function representing the global V(R).

As in previous case [28] we define f(R) as follows

$$f(R) = \frac{1}{1 + e^{\left(\frac{R_0 - R}{d}\right)}} \tag{7}$$

where R_0 represents the distance where the combined potential forms have equal weight, while d describes how fast the passage occurs. All the used potential parameters are reported in Table 2.

Table 2. Potential parameters used in the formulation of the various interaction components determining the real and imaginary part of the optical potential

Component	Parameters		
V_{vdW}	$\varepsilon = 9.29$ meV	$R_m = 5.000$ Å	$\beta = 7.0$
V_{qm}	$\varepsilon = 69.13$ meV	$R_m = 3.989$ Å	$\beta = 9.0$
V_{CT}	$A_H = 4277$ meV	$\gamma_H = 1.1$ Å^{-1}	
f(R)	$R_0 = 4.5$ Å	$d = 0.5$ Å	
Γ	$A_\Gamma = 3371$ meV	$\gamma_\Gamma = 1.95$ Å^{-1}	

Following the suggestions of a "minimal" model [46], concerning the use of a defined number of spherical harmonics to represent the PES in water-atom systems, the spherical component $V_{sph}(R)$, relevant for the present analysis, has been obtained at each R through the following weighted average expression

$$V_{sph}(R) = \frac{1}{6}\left[V_{C2v-H}(R) + V_{C2v-O}(R) + 2V_{\perp}(R) + 2V_{planar}(R)\right] \tag{8}$$

The dependence on R of the V_{C2v-O}, V_{C2v-H} and $V_{\perp}$ components are shown in Figure 3, while V_{sph} is plotted in the lower part of Figure 4. These potential energy results, which refer to the limit configurations of the $[Ne \cdots H_2O]^*$ autoionizing collision complex, appear to be in agreement with an early *ab initio* calculation by Bentley [47]. Our results could stimulate new accurate *ab initio* calculations and allow to fill the lack of computational approaches and of dynamic trajectories calculations for such kind of systems.

The imaginary component $\Gamma(R)$ accounts for the disappearance of the system because of ionization. In particular, the quantity $h/2\pi\Gamma(R)$ can be seen as the life time of the system at the intermolecular distance R. For the case here discussed, $\Gamma(R)$ promotes the formation of H_2O^++Ne and can be related, following the exchange mechanism proposed for the Penning process, to the coupling element by charge transfer between water molecule and Ne^+ ionic core [45]. Such coupling, controlled again by the overlap integral between orbitals exchanging the electron, has been represented by an exponential function as follows:

$$\Gamma(R) = A_\Gamma e^{-\gamma_\Gamma R} .$$
(9)

The exponent γ_Γ has been fixed as suggested in [48-49], and found in the right scale of previous evaluations [24,36]. The pre-exponential factor A_Γ has been the only parameter adjusted during the analysis of the experimental data, in order to reproduce the velocity dependence and the absolute value of the total ionization cross section. Also the parameters of Γ are given in Table 2 and the function is also plotted in the upper part of Figure 4.

3.3   The Semiclassical Cross Section Calculation

By the use of the spherically averaged potential described above, we have calculated, for each collision energy, the center of mass total ionization cross sections within the semiclassical method, that appears to be adequate for the present system and for this range of collision energies [17-21,24,25]. In such an approximation, the cross section at the relative collision energy E is given by

$$\sigma_{CM}(E) = 2\pi \int_0^\infty P(b)b\,db$$
(10)

where b is the classical impact parameter and P(b) is the ionization probability as a function of b, which is determined by

$$P(b)=1-\exp\left\{-\frac{2}{\hbar g}\int_{R_c}^\infty\left[1-\frac{V_{sph}(R)}{E}-\frac{b^2}{R^2}\right]^{-\frac{1}{2}}\Gamma(R)dR\right\}$$
(11)

being R_c the classical turning point.

In the Figure 1, the cross sections so calculated and convoluted over the distributions of the relative velocities are compared with experimental results. As evident, the agreement is very good, especially considering that we have empirically adjusted only one parameter, the pre-exponential term, A_Γ, being the entire real potential the one evaluated independently, as described above, from the properties of the two separated collision partners.

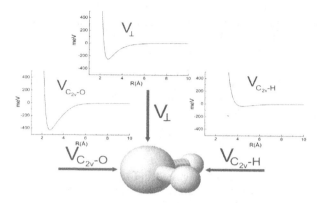

Fig. 3. Plot of the V_{C2v-H}, V_{C2v-O} and $V_\perp$ potential energy curves (see the text)

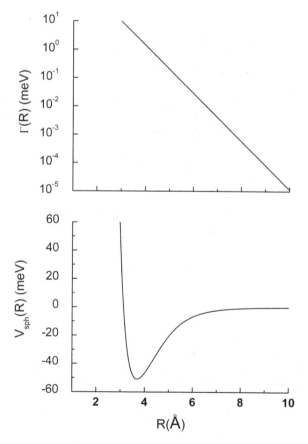

Fig. 4. The spherical optical potential used for the cross section calculation. Lower part represents the real part, while the upper part the imaginary part (see the text).

4   Conclusions

The measured cross sections that are reported in Figure 1, exhibit a decrease when the relative collision energy increases. As emphasized in the previous section, this behavior is well known to indicate an effective attractive interaction between the two collision partners, for those distances and configurations mainly responsible for the ionization [17-21].

The quality of the cross section fit shown in Figure 1, obtained by adjusting only one parameter of Γ, clearly supports the reliability of the strength and range of the spherical V_{sph} component, here generated by an average of selected limiting configurations. It must be noted that while the ionization mainly occurs when the metastable neon atom is approaching the H_2O molecule along the directions of the two lone pairs, the overall collision dynamics, which involves randomly oriented and fast rotating water molecules, is mainly controlled by the spherical average of the real part of the intermolecular potential.

A more direct probe of the anisotropy of the PES can be provided by an analysis of the features of electron energy spectra recently measured [50]. The electron spectrum appeared to be composed of two bands: one for the formation of H_2O^+ ion in the ground $X(^2B_1)$ electronic state and another one for the formation of H_2O^+ in the first excited $A(^2A_1)$ electronic state. These two states are obtained by removing one electron from the perpendicular and C_{2v} lone pair, respectively. Therefore the spectrum features can be in principle fitted by using defined cuts of the full PES which also includes an anisotropic imaginary part, and such an effort is in progress in our laboratory. Note that the potentials obtained in this work, for the selected limiting configurations, correspond to cuts of the real part of the full PES and, as mentioned in the previous section, they are in good agreement with the *ab initio* calculations performed by Bentley [47]. These cuts of the PES are here given in an analytical form and this is crucial, together to a formulation of the ionic interaction over the $[Ne \cdots H_2O]^+$ exit channel that we are able to formulate by using the same approach, to try a quantitative analysis of the electron energy spectra already recorded for such system [50].

Since the microscopic exchange mechanism of the Penning ionization involves an electron jump of the outer shell electron of the molecule, essentially from one of the two lone pairs, towards the inner shell vacancy of the Ne^* atom [23-26], which shows a structure isoelectronic with that one of a fluorine atom, this phenomenon can be seen as the formation of an halogen-like bond, and this is presently of great interest both for applications and fundamental points of view [51,52].

Acknowledgments. Prof. Giorgio Liuti died during the preparation of the manuscript. This work and our future endeavours are dedicated to him, whose intellectual honesty and genuine love of chemistry continue to inform our scientific life. Financial contributions from the MIUR (Ministero dell'Istruzione, dell'Università e della Ricerca) are gratefully acknowledged.

References

1. Moix-Teixidor, M., Pirani, F., Candori, P., Falcinelli, S., Vecchiocattivi, F.: Chem. Phys. Lett. 379, 139 (2003)
2. Candori, P., Falcinelli, S., Pirani, F., Tarantelli, F., Vecchiocattivi, F.: Chem. Phys. Lett. 436, 322 (2007)
3. Alagia, M., Brunetti, B.G., Candori, P., Falcinelli, S., Moix-Teixidor, M., Pirani, F., Richter, R., Stranges, S., Vecchiocattivi, F.: J. Chem. Phys. 120, 6985 (2004)
4. Alagia, M., Biondini, F., Brunetti, B.G., Candori, P., Falcinelli, S., Moix-Teixidor, M., Pirani, F., Richter, R., Stranges, S., Vecchiocattivi, F.: J. Chem. Phys. 121, 10508 (2004)
5. Alagia, M., Boustimi, M., Brunetti, B.G., Candori, P., Falcinelli, S., Richter, R., Stranges, S., Vecchiocattivi, F.: J. Chem. Phys. 117, 1098 (2002)
6. Alagia, M., Brunetti, B.G., Candori, P., Falcinelli, S., Moix Teixidor, M., Pirani, F., Richter, R., Stranges, S., Vecchiocattivi, F.: J. Chem. Phys. 124, 204318 (2006)
7. Alagia, M., Candori, P., Falcinelli, S., Lavollée, M., Pirani, F., Richter, R., Stranges, S., Vecchiocattivi, F.: Chem. Phys. Lett. 432, 398 (2006)
8. Alagia, M., Brunetti, B.G., Candori, P., Falcinelli, S., Moix-Teixidor, M., Pirani, F., Richter, R., Stranges, S., Vecchiocattivi, F.: J. Chem. Phys. 120, 6980 (2004)
9. Alagia, M., Candori, P., Falcinelli, S., Mundim, K.C., Mundim, M.S.P., Pirani, F., Richter, R., Stranges, S., Vecchiocattivi, F.: Chem. Phys. 398, 134 (2012)
10. Alagia, M., Candori, P., Falcinelli, S., Lavollée, M., Pirani, F., Richter, R., Stranges, S., Vecchiocattivi, F.: J. Chem. Phys. 126, 201101 (2007)
11. Alagia, M., Candori, P., Falcinelli, S., Lavollée, M., Pirani, F., Richter, R., Stranges, S., Vecchiocattivi, F.: J. Phys. Chem. A 113, 14755 (2009)
12. Alagia, M., Candori, P., Falcinelli, S., Lavollée, M., Pirani, F., Richter, R., Stranges, S., Vecchiocattivi, F.: Phys. Chem. Chem. Phys. 12, 5389 (2010)
13. Alagia, M., Candori, P., Falcinelli, S., Mundim, M.S.P., Pirani, F., Richter, R., Rosi, M., Stranges, S., Vecchiocattivi, F.: J. Chem. Phys. 135, 144304 (2011)
14. Alagia, M., Candori, P., Falcinelli, S., Pirani, F., Pedrosa Mundim, M.S., Richter, R., Rosi, M., Stranges, S., Vecchiocattivi, F.: Phys. Chem. Chem. Phys. 13, 8245 (2011)
15. Alagia, M., Callegari, C., Candori, P., Falcinelli, S., Pirani, F., Richter, R., Stranges, S., Vecchiocattivi, F.: J. Chem. Phys. 136, 204302 (2012)
16. Alagia, M., Bodo, E., Decleva, P., Falcinelli, S., Ponzi, A., Richter, R., Stranges, S.: Phys. Chem. Chem. Phys. 15, 1310 (2013)
17. Brunetti, B., Cambi, R., Falcinelli, S., Farrar, J.M., Vecchiocattivi, F.: J. Phys. Chem. 97, 11877 (1993)
18. Brunetti, B.G., Falcinelli, S., Giaquinto, E., Sassara, A., Prieto-Manzanares, M., Vecchiocattivi, F.: Phys. Rev. A 52, 855 (1995)
19. Brunetti, B., Falcinelli, S., Sassara, A., De Andres, J., Vecchiocattivi, F.: Chem. Phys. 209, 205 (1996)
20. Brunetti, B., Candori, P., Falcinelli, S., Vecchiocattivi, F., Sassara, A., Chergui, M.: J. Phys. Chem. A 104, 5942 (2000)
21. Brunetti, B.G., Candori, P., Falcinelli, S., Kasai, T., Ohoyama, H., Vecchiocattivi, F.: Phys. Chem. Chem. Phys. 3, 807 (2001)
22. Penning, F.M.: Naturwissenschaflen 15, 818 (1927)
23. Hotop, H., Niehaus, A.A.: Z. Phys 228, 68 (1969)
24. Siska, P.E.: Rev. Mod. Phys. 65, 337 (1993)
25. Brunetti, B., Vecchiocattivi, F.: Cluster Ions, pp. 359–445. Wiley&Sons (1993), Ng, C.Y. (ed.)

26. Ogawa, T., Ohno, K.: J. Chem. Phys. 110, 3733 (1999)
27. Biondini, F., Brunetti, B.G., Candori, P., De Angelis, F., Falcinelli, S., Tarantelli, F., Teixidor, M.M., Pirani, F., Vecchiocattivi, F.: J. Chem. Phys. 122, 164307 (2005)
28. Biondini, F., Brunetti, B.G., Candori, P., De Angelis, F., Falcinelli, S., Tarantelli, F., Pirani, F., Vecchiocattivi, F.: J. Chem. Phys. 122, 164308 (2005)
29. Atkinson, R., Baulch, D.L., Cox, R.A., Hampson, R.F., Kerr Jr., J.A., Troe, J.: J. Phys. Chem. Ref. Data 21, 1125 (1989)
30. McClintock, W.E., Lankton, M.R.: Space Sci. Rev. 131, 481 (2007)
31. West, W.P., Cook, T.B., Dunning, F.B., Rundel, R.D., Stebbings, R.F.: J. Chem. Phys. 63, 1237 (1975)
32. Brunetti, B., Candori, P., De Andres, J., Pirani, F., Rosi, M., Falcinelli, S., Vecchiocattivi, F.: J. Phys. Chem. A 101, 7505 (1997)
33. Ben Arfa, M., Lescop, B., Cherid, M., Brunetti, B., Candori, P., Malfatti, D., Falcinelli, S., Vecchiocattivi, F.: Chem. Phys. Lett. 308, 71 (1999)
34. Brunetti, B., Candori, P., Ferramosche, R., Falcinelli, S., Vecchiocattivi, F., Sassara, A., Chergui, M.: Chem. Phys. Lett. 294, 584 (1998)
35. Ishida, T.: J. Chem. Phys. 105, 1392 (1996)
36. Dunlavy, D.C., Siska, P.E.: J. Phys. Chem. 100, 21 (1996)
37. Falcinelli, S., Fernandez-Alonso, F., Kalogerakis, K.S., Zare, R.N.: Mol. Phys. 88, 663 (1996)
38. Cappelletti, D., Candori, P., Falcinelli, S., Albertì, M., Pirani, F.: Chem. Phys. Lett. 545, 14 (2012)
39. Cambi, R., Cappelletti, D., Liuti, G., Pirani, F.: J. Chem. Phys. 95, 1852 (1991)
40. Cappelletti, D., Liuti, G., Pirani, F.: Chem. Phys. Lett. 183, 297 (1991)
41. Radzig, A.A., Smirnov, B.M.: Reference data on atoms, molecules and ions. Springer, Berlin (1985)
42. Cappelletti, D., Ronca, E., Belpassi, L., Tarantelli, F., Pirani, F.: Acc. Chem. Res. 45(9), 1571 (2012)
43. Roncaratti, L.F., Belpassi, L., Cappelletti, D., Pirani, F., Tarantelli, F.: J. Phys. Chem. 113, 15223 (2009)
44. Pirani, F., Brizi, S., Roncaratti, L.F., Casavecchia, P., Cappelletti, D., Vecchiocattivi, F.: Phys. Chem. Chem. Phys. 10, 5489 (2008)
45. Brunetti, B., Candori, P., Falcinelli, S., Lescop, B., Liuti, G., Pirani, F., Vecchiocattivi, F.: Eur. Phys. J. D 38, 21 (2006)
46. Barreto, P.R.P., Albernaz, A.F., Capobianco, A., Palazzetti, F., Lombardi, A., Grossi, G., Aquilanti, V.: Comp. Theor. Chem. 990, 53 (2012)
47. Bentley, J.: J. Chem. Phys. 73, 1805 (1980)
48. Grice, R., Herschbach, D.: Mol. Phys. 98, 159 (1974)
49. Pirani, F., Giulivi, A., Aquilanti, V.: Mol. Phys. 98, 1749 (2000)
50. Brunetti, B., Candori, P., Cappelletti, D., Falcinelli, S., Pirani, F., Stranges, D., Vecchiocattvivi, F.: Chem. Phys. Lett. 539-540, 19 (2012)
51. Legon, A.C.: Phys. Chem. Chem. Phys. 12, 7736 (2010)
52. Murray, J.S., Lane, P., Clark, T., Politzer, P.: J. Mol. Model. 13, 1033 (2007)

Implementation of the ANSYS® Commercial Suite on the EGI Grid Platform

Alessandro Costantini[1], Diego Michelotto[1], Marco Bencivenni[1],
Daniele Cesini[1], Paolo Veronesi[1], Emidio Giorgio[1], Luciano Gaido[1],
Antonio Laganà[2], Alberto Monetti[3],
Mattia Manzolaro[3], and Alberto Andrighetto[3]

[1] INFN-IGI, Italy
`alessandro.costantini@pg.infn.it`, {`diego.michelotto,marco.bencivenni,`
`daniele.cesini,paolo.veronaesi`}`@cnaf.infn.it`
`emidio.giorgio@ct.infn.it`, `luciano.gaido@to.infn.it`
[2] Dep. of Chemistry, University of Perugia, Italy
`lag@dyn.unipg.it`
[3] LNL - INFN, Italy
{`alberto.monetti,mattia.manzolaro,`
`alberto.andrighetto`}`@lnl.infn.it`

Abstract. This paper describes and discusses the implementation, in a high-throughput computing environment, of the ANSYS® commercial suite. ANSYS® implements the calculations in a way which can be ported onto parallel architectures efficiently and for this reason the User Support Unit of the Italian Grid Initiative (IGI) and the INFN-Legnaro National Laboratories (INFN-LNL) worked together to implement a Grid enabled version of the ANSYS® code using the IGI Portal, a powerful and easy to use gateway to distributed computing and storage resources. The collaboration focused on the porting of the code onto the EGI Grid environment for the benefit of the involved community and for those communities interested in exploiting production Grid infrastructures in the same way.

1 Introduction

The increasing availability of computer power on Grid platforms has prompted the implementation of complex computational codes on distributed systems and, at the same time, the development of appropriate visual interfaces and tools able to minimize the skills requested to the final user to carry out massive Grid calculations. The work has been carried out within a collaboration with the User Support Unit of the Italian Grid Infrastructure (IGI) [1] and the INFN-Legnaro National Laboratories (INFN-LNL) [2] aimed at implementing a complex engineering simulation software use-case on distributed systems making use of the IGI web portal is here presented and discussed.

IGI is a Joint research Unit (JRU) made of 18 Italian academic and research institutions that is based on a Memorandum of Understanding (MoU) [3] signed

B. Murgante et al. (Eds.): ICCSA 2013, Part I, LNCS 7971, pp. 84–95, 2013.
© Springer-Verlag Berlin Heidelberg 2013

on December 2007 and has been active in various national and European Grid projects [4–6]. IGI is actively participating to the EGI-InSPIRE project [7] and is one the largest National Grid Initiatives (NGI) of the European Grid Infrastructure (EGI) [8] with a recognized leadership both in the Grid technology development and in the management of the distributed computing infrastructure operations supporting research communities. The Italian Grid Infrastructure currently consists of more than 50 geographically distributed sites, providing about 33000 computing cores and 30PB of storage capacity and supports more then 50 Virtual Organizations with thousands of active users. The infrastructure implements a customized version of the gLite [9] middleware distributed by the European Middleware Initiative (EMI) project [10]. One of the role of IGI is to satisfy the compute and storage demand of various user communities such as high energy physics, computational chemistry, bioinformatics, astronomy and astrophysics, earth science.

In the present paper we describe the porting to the Grid of a software package related to the SPES experiment [11] carried out at the INFN Legnaro laboratories and concerning the electro-thermal design of high temperature devices for the production of Radioactive Ion Beams. The related numerical computing is strongly non-linear mainly because of the radiative heat transfer computation, and require a relevant computational power to obtain a solution.

The application chosen to be ported to the Grid environment, making use of the IGI resources is the ANSYS® commercial suite [12]. ANSYS® is an engineering simulation software (computer-aided engineering, or CAE) that offers a comprehensive range of engineering simulation solution sets providing access to virtually any field of engineering simulation that a design process requires. In the present work the ANSYS® suite has been installed and configured in some IGI sites. A web interface to run the simulations exploiting the production Grid services was provided through a dedicated graphical interface of the IGI Web Portal [13] which is a powerful and easy to use gateway to distributed computing and storage resources.

At present, for its features, the ANSYS® package has been used as a Grid computing test bed with the aim to extend the work done to other applications belonging to different domains such, for example, computational chemistry which is represented in EGI by an active community.

This paper introduces the original ANSYS® code and the way it was ported onto the IGI/EGI platform. The purpose of our work is twofold. We are aiming in fact both at creating a user friendly Grid application which is beneficial for the SPES community and to introduce a method that can be applied by other groups to port parameter study style applications based on commercial suites onto production Grids. Because of the method adopted and of the tools used we believe that several e-Science communities would be able to follow the same approach and exploit production Grid infrastructures in the same way. The method we chose is quite generic and resulted in a parameter study application that implements parallel execution of the code, using several Grid computing resources simultaneously. The IGI web Portal tool used to create the customized web interface can be used on all the major production Grids based on the European Middleware Initiative (EMI) [10] middleware stack.

This paper is organized as follows: in section 2 the articulation of the ANSYS® program is described; in section 3 a benchmark calculation is discussed; in section 4 the steps needed for the Grid porting process of the application are illustrated; in section 5 the results obtained using the Grid enabled version of the code and related performances are analyzed. Our conclusions are summarized in section 6 where we argue that thanks to the exploitation of Grid resources the user community work with ANSYS® ported into the Grid environment is more effective if compared with their usual workflows run on local resources.

Other application domains and research communities could benefit from the collaboration with the IGI User Support Unit to port applications to production Grids.

2   ANSYS Program Overview

The ANSYS® suite is a FEM (Finite Element Method) commercial program for simulations of models belonging to various physical environments to simulate problems concerning mechanical, thermal, electrical, magnetic, fluid dynamics matters. Depending on the model to simulate various packages of the ANSYS® suite are available, as the Mechanical, Fluid Dynamics, Electromagnetic and Multiphysics.

Accordingly, ANSYS® covers with its packages the following main tasks and disciplines:

1 *Mechanical*: this is the most used package of the ANSYS® suite. It allows to simulate static, transient, modal or harmonic structural analysis, also in large displacements or with non linear material behaviour, contact effects, static or transient thermal simulations with conduction, convection or radiation heat exchanges, static, harmonic or transient magnetic and electrostatic analysis. The coupling between the various fields is allowed only if the element used for the FEM analysis has the degrees of freedom of all the concerned fields. This package uses ANSYS® Parametric Design Language (APDL), that allows to launch the simulations without a graphic interface inputting the instructions by text files.

2 *Multiphysics*: this package is needed by those computational analysis that do not have in the "Mechanical" package corresponding elements with all the degrees of freedom of the concerned fields of the simulation. Moreover its use is recommended for weakly coupled physics fields where the results carried out in a specific field influence the others (but not viceversa) by limiting the use of two-way coupling simulations which take longer calculation time.

3 *Fluid Dynamics*: this environment allows the resolution of model flow, turbulence and heat transfer on fluids, with the possibility to take in account the combustion reactions, bubbles formation and multiphase systems. The simulations can be executed by using two different packages: FLUENT, the newest model flow package and CFX (Computational Fluid dynamiX), the oldest

model flow package. These packages have been integrated in the ANSYS® suite in the recent years and for this reason they are not using APDL commands;

4 *Electromagnetic*: this environment can be simulated by using Maxwell or HFSS (High Frequency Structural Simulator) packages according to the type of the analysis carried out by the user. The Maxwell package allows to simulate electromagnetic and electromechanical devices including motors, actuators, transformers, sensors and coils using the accurate finite element method to solve static, frequency-domain, and time-varying electromagnetic and electric fields. On the other hand the HFSS package is used for 3D full-wave electromagnetic field simulation for high-frequency and high-speed components.

Most often observable properties are the results of averaging (or integrating) over energies, time, etc. which means that ANSYS® runs have to be repeated a large number of times making the exploitation of the distributed resources available in Grids highly effective for this kind of analyses.

3   Heat Dissipation Simulation of the Radioactive Isotopes Production Target

The SPES project (Selective Production of Exotic Species) is a multi user project aimed to develop a Radioactive Ion Beam (RIB) facility to cover interdisciplinary applied physics in the fields of medical applications, material science and nuclear physics.

The core of the facility is the apparatus shown in Figure 1 where 7 coaxial discs made by uranium carbide (UC_2 + 2C, namely UC_x) are impinged by a 40 MeV proton beam yielding radioactive isotopes by nuclear fission. The power deposited by protons in the disks is dissipated mainly by thermal radiation. The spaces between the disks in the axial direction strongly influence the mechanical stresses and the temperatures of the disks which can reach $2300°C$. The disks are placed in a graphite box that is in turn hosted in a Tantalum tube. An electrical current is flowing through the tube and contributes to control the target temperature field and the thermal stress in the disks. Finally the isotopes produced in the target are collected by the transfer line and directed to the ion source, where they are ionized and accelerated by a 40 kV potential field.

To optimize the aforementioned components, simulations with the ANSYS® suite are performed studying in detail their thermal, electric and structural behavior thanks to FEM models.

For the electrical field, the gradient of the electric potential $V(x, y, z)$ defines the current density $j(x, y, z)$ according to Ohm's equation:

$$ j = -\frac{1}{\rho(T)}\nabla V \tag{1} $$

where $\rho(T)$ is the electrical resistivity dependent on the temperature. Since the electrical resistivity of materials is influenced by temperature and, on the other

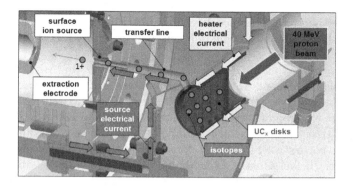

Fig. 1. Sketch of the core of the SPES facility and its components

hand, Joule heating (see eq. 2) affects the temperature field of the system, the thermal and the electrical problems are coupled.

$$\dot{q} = -\nabla V \cdot j \tag{2}$$

where $\dot{q}(x, y, z)$ is the heat power dissipation per unit volume and the dot symbol is the scalar product between the two vectors.

For this reasons with the ANSYS® mechanical package a two-way coupling simulations are performed, making use of elements characterized by two degree of freedom: temperature and voltage. In particularly for the thermal field, convection is not taken into account, considering that the components are closed in a high vacuum environment. The conduction and radiation, the heat transfer modes governing the thermal behavior of the system, are solved in conjunction: the radiative heat fluxes, coming from the solution of the radiative problem (see eq. 3), are assigned as boundary conditions to the conductive problem, whereas in the superficial temperature distribution, the solution of the conductive equations (see eq. 4) provides boundary data to compute the radiative heat fluxes.

$$\sum_{i=1}^{N} \left[\frac{\delta_{ij}}{\varepsilon_i} - F_{j-i} \left(\frac{1 - \varepsilon_i}{\varepsilon_i} \right) \right] \cdot q_i = \sum_{i=1}^{N} (\delta_{ij} - F_{j-i}) \cdot \sigma \cdot T_i^4 \tag{3}$$

In equation 3 [14] ε_i is the hemispherical total emissivity of surface i, δ_{ij} is the Kronecker delta ($\delta_{ij} = 1$ if $i = j; \delta_{ij} = 0$ if $i \neq j$), F_{j-i} is the radiation view factor, q_i is the net rate of energy loss per unit area by radiation from the surface i, σ is the Stefan-Bolzmann constant and T_i is the absolute temperature of surface i.

$$\frac{\partial}{\partial x} \left(k \frac{\partial T}{\partial x} \right) + \frac{\partial}{\partial y} \left(k \frac{\partial T}{\partial y} \right) + \frac{\partial}{\partial z} \left(k \frac{\partial T}{\partial z} \right) + \dot{q} = \rho\, c\, \frac{T}{t} \tag{4}$$

In equation 4 [15] $T(x, y, z)$ is the temperature field in the volume V, t is the time, ρ, c, k are respectively the density, the specific heat and the thermal conductivity of the material and $\dot{q}$ is the heat source per volume unit.

To calculate the stresses induced by thermal gradients, the *Multiphysics* package is used. The temperature's field obtained thanks to the thermal-electric simulation is assigned as a body load at the nodes in the structural analysis, as showed in eq. 5:

$$\varepsilon = [D]^{-1}\sigma + \alpha\Delta T \tag{5}$$

where ε is the total strain vector, D is the elastic stiffness matrix, σ is the stress vector, α the vector $[\alpha_x \ \alpha_y \ \alpha_z \ 0 \ 0 \ 0]^T$ containing the coefficients of thermal expansion, $\Delta T(x, y, z)$ is the temperature difference between the current temperature and the initial one.

The present simulation is a typical one-way coupling simulation because only one field strongly influence the other (but not vice versa).

In Fig. 2 an example of the temperature's results carried out from the adoption of the FEM method are showed. In the model a current (from 600 A to 1300 A with steps of 100 A) coming from the left lateral wing and going to the other, is simulated passing through the Tantalum tube, which has an external diameter of 50 mm and thickness of 0.2 mm. The temperature results are reported in Table 1.

Table 1. Results carried out from the adoption of the FEM model

I [A]	Volt [V]	T_1 [$°C$]	T_2[$°C$]
600	2.72	1144.46	1166.97
700	3.44	1271.93	1296.06
800	4.20	1391.09	1416.86
900	5.02	1503.64	1530.95
1000	5.87	1610.40	1639.18
1100	6.76	1712.18	1742.42
1200	7.69	1809.64	1841.39
1300	8.66	1903.41	1936.69

4   The Adopted Tool: IGI Web Portal

The already described suite ANSYS® has been installed and configured in some sites of the IGI domain and a web interface to run the simulation exploiting the production Grid services was provided through a dedicated portlet of the IGI Portal (see Fig. 3). The IGI web Portal is a powerful and easy to use gateway that enable the final user to access the Grid infrastructure, supporting the user in many tasks by hiding the inner complexity of Grid infrastructures usage (proxy credential handling, job submission, data management, error recovery, etc.). As the IGI portal is based on the Liferay technology [16], web Graphical User Interfaces (GUI) can be added as dedicated portlets [17] enhancing the flexibility and the possibilities of customizations.

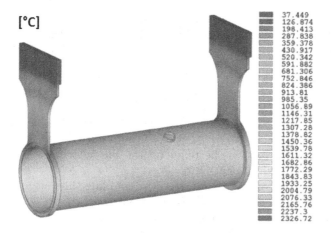

Fig. 2. Temperature map in the apparatus carried out from the adoption of the FEM model at $1300A$

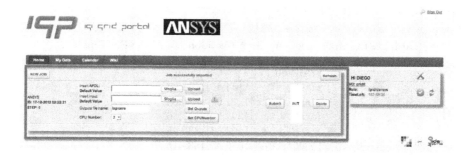

Fig. 3. Sketch of the main porlet developed for ANSYS® suite and integrated in the IGI Portal

The user can access the Portal functionalities by using (i) a federation membership (actually EduGAIN [18] and Idem [19] are supported); (ii) a personal certificate released by a valid Certification Authority and a membership to a proper Virtual Organization (VO) supported by the Infrastructure. Since the ANSYS® suite is a commercial package, to be compliant with the terms of license imposed by the seller, a license handling mechanism has been implemented based on standard Flex servers [20] and on the EMI VO Management Service (VOMS) [21]. This mechanism enable the ANSYS® runs only for those users that have been registered in a proper VO group managed by the local license owner. In the present case the group is called "ansys" and belongs to the GRIDIT VO.

In a typical usecase the user provides the initial input files and configuration parameters and waits for the results until the calculation is terminated. For this reason the developed graphical interface (see Fig. 4) enables the user to upload

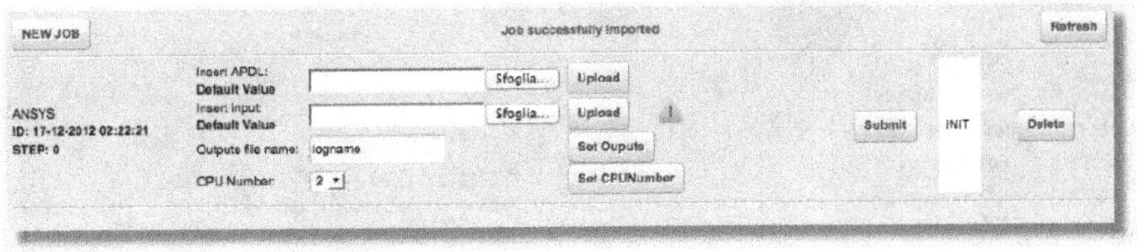

Fig. 4. Sketch of the dedicated porlet developed for ANSYS® suite. A proper graphical interface has been developed to set the needed parameters.

the needed input files and to set the relevant simulation parameters such the number of CPUs that have to be used for a single submission.

The above described execution process may take several hours, even days using the resources available to run this application. As the granted amount of CPU time is limited on Grid sites (from 12 hours to few days), special care was taken in handling the checkpointing of the calculation where a set of specialized bash components have been developed with the twofold purpose of setting the computational environment needed to run the application and monitoring the computation time allowing a safely interruption of the application. In the first version of the developed GUI interface, at the end of each calculation the user had to retrieve the related output files directly from the IGI Portal, analyze them and submit a new job to continue the current work.

To meet the requirements of the SPES community we developed a first pro-totypical workflow using the WSPgrade workflow engine [22] able to monitor a set of continuous runs in an automated way. In the automatic workflow (see Fig. 5) each step is conditioned by event-related dependences occurring at runtime making possible the execution of complex analysis involving both structural and electro-thermic models.

For security reasons and to avoid possible execution inner loops, the workflow has been equipped with a total amount of 10 consecutive job submissions (that is equivalent to an average of 10 days of continuative calculation for a single experiment). If the simulation time granted to the user by the adoption of the workflow is not enough, a new workflow can be submitted starting from the outcomes of the previous run. This approach increases the feeling of the users with the submission procedures with a consequent reduction of support requests.

As an added value, the combined use of the GUI interface and the implemented bash components make the whole Grid execution process completely transparent to the final user, requiring his/her evaluation only for those application-related failures which may occur during the calculations.

Another crucial aspect of such long time simulations is the evolution's audit of the calculations at runtime. In this case we made use of the SRM [23] client-server functionalities which enable to copy selected files from the Worker Node (WN) where the job is physically running to a Grid Storage Elements (SEs), where the file are stored in a temporary or permanent way. Using the same

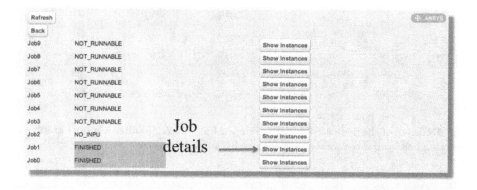

Fig. 5. Sketch of the dedicated workflow (and related components) developed for ANSYS® suite. The workfow enable the final user to perform an equivalent to 10 days run of continuous simulation limiting any intervention.

SRM functionalities, the files in the SE are made available for inspection at runtime and can be accessed by the user directly via the web GUI.

The main bash functions which interact with the GUI are here described:

- *SETENV* Checks the environment parameters (both provided by the GUI and set in the WN) needed to start the calculation
- *USERFOLDER* Checks if user-folder exists using the SRM-client functionalities and create it, if needed;
- *PREPARETOPUT* Uses the PreparetoPut function implemented in the SRM-client to copy a rewritable file(s) in the SE and store it for a limited amount of time (file lifetime has been set by default to 24 hours);
- *PREPAREINPUT* Retrieves the input file needed for the calculation depending on the value provided by the GUI (first run or resubmission after hang-up);
- *RUNNINGAPP* Runs the executable monitoring its activity for a fixed amount of time, that depends from the CPU time assigned by the batch system to the queue where the job is running, and gracefully kill it allowing a safe interruption of the application. This component implements a check procedure that, at fixed intervals, control the status of the running application and uploads the needed file(s) to the SE.
- *CHECKLOGS* Uploads the file(s) created by PREPARETOPUT function to the SE making use of specific commands available on the WNs.
- *PREPAREOUTPUT* The function manages the output files carried out from the calculations acting on both sides: from the WN selects the output files making a univocally named archive; from the SE (i) removes the oldest output file, if it exists, (ii) rename the output file carried out from the previous run assigning a proper name, (iii) copies the actual output file from the WN to the SE (in this case the file lifetime has been set by default to 7 days).

Moreover, an automated notification mechanism has been implemented in the IGI Portal. This mechanism enables the user to set the Portal to send customized e-mails when the simulation is completed or when the proxy lifetime is close to expire.

The features implemented in the IGI Portal enables the user to perform long time simulations on the Grid infrastructure in a completely transparent way and to retrieve the outcomes of the calculation directly via web. In this way the user can check the consistency of the output at runtime, evaluating possible strategies aimed at saving time, computing resources and at avoiding waste of license usage.

5   Performances

The workflow developed for the ANSYS® suite was implemented in the IGI web portal [13] and all the scripts and processes for the Grid execution were generated in order to meet their requirements. The executables derived from the ANSYS® Release 13.0 and compiled on Linux SL5 platform look for OpenMotif, OpenMP and Mesa libraries that should be already installed in a typical Linux installation.

Local compilation assures that the program is binary compatible with the Computing Elements of the IGI Grid and that the program will not run into incompatibility errors due to the lack of fundamental libraries.

The Grid enabled version of the ANSYS® is executed simultaneously on multiple IGI resources. Because the time spent in the submission stage is neglegible compared with that of the ANSYS® run, the latter is the dominant contribution to the overall execution duration. One execution of a single instance of ANSYS® on a Intel machine with 2.3 GHz CPU and 4 GB memory takes from 40 to 50 hours, depending on the chosen parameters values in the case-study. The main benefit of the grid implementation is the possibility of running ANSYS for different sets of parameters during the same time window on the resources of the VO. As the average execution time for a single instance of ANSYS® on the Grid is almost equivalent to the one on a dedicated local machine, the added value of Grid runs is the possibility to submit in parallel different sets of concurrent simulations. Accordingly, as soon as there are at least 6 ANSYS® jobs running simultaneously, the Grid based execution is advantageous because 6 jobs running simultaneously on 6 IGI Grid resources will take on average 2 days. Meanwhile, the same simulation would take about 10 days on a local machine. The more parameter study jobs are executed, the higher speedup can be achieved on the Grid.

As the present work is in its prototypical implementation, there are still some disadvantages that have to be addressed and related on the use of the ANSYS® suite. In fact, it is not yet possible to use inner approaches as the multi-load step with the automatic resubmission and some methods for the parametric optimization.

6   Conclusions

The paper describes the work done aimed at porting the ANSYS® suite onto the EGI Grid environment as a result of a collaboration between the User Support Unit of the Italian Grid Initiative and the INFN-Legnaro National Laboratories (INFN-LNL). The porting of legacy applications onto the Grid infrastructure, together with the development of the related workflows and gateways, is being carried out as part of a more general effort to build a solid platform, offered to users as a service, for assembling accurate multi scale realistic simulations. Although repeated submissions in the workflow can make the execution of a ANSYS® run slow if compared with a local machine, the overall execution time of a parameter study simulation is far shorter on the Grid environment than on a local CPU. For the SPES community case study the overall Grid execution time is smaller than that of the sequential execution for a parameter space larger than 2.

The implemented case study demonstrates not only the power of interdisciplinary group work, but also details the application porting process, providing a reusable example for other groups interested in porting their applications to production Grid systems. It is the case of the Computational Chemistry community operating in EGI which pursue the goal of designing user friendly Grid empowered versions of the molecular system simulation workflows SIMBEX [24] and GEMS [25].

Acknowledgments. Authors acknowledge A. Prevedello and M. Marin for their collaboration in testing the system and for the information given related to the observed problems. The research leading to the results presented in this paper has been possible thanks to the grid resources and services provided by the European Grid Infrastructure (EGI) and the Italian Grid Infrastructure (IGI). For more information, please reference the EGI-InSPIRE paper (http://go.egi.eu/pdnon) and the IGI web server (http://www.italiangrid.it)

References

1. IGI (Italian Grid Infrastructure), http://www.italiangrid.it/ (last seen February 2013)
2. INFN - Legnaro National Laboratories, http://www.lnl.infn.it/ (last seen February 2013)
3. IGI MoU, https://www.italiangrid.it/sites/default/files/ (last seen February 2013)
4. DataGRID project, http://eu-datagrid.web.cern.ch/eu-datagrid/ (last seen February 2013)
5. EGEE website, http://public.eu-egee.org (last seen February 2013)
6. INFN-GRID project, www.infn.it/globus (last seen February 2013)
7. EGI-Inspired project, http://www.egi.eu/about/egi-inspire/ (last seen February 2013)
8. European Grid Infrastructure, http://www.egi.eu (last seen February 2013)

9. gLite website, http://glite.web.cern.ch/glite (last seen February 2013)
10. European Middleware Initiative, http://www.eu-emi.eu (last seen February 2013)
11. Manzolaro, M., Meneghetti, G., Andrighetto, A.: Nucl. Instrum. Methods Phys. Res., Sect. A 623, 1061–1069 (2010)
12. ANSYS website, www.ansys.com/ (last seen February 2013)
13. IGI Portal, https://portal.italiangrid.it/ (last seen February 2013)
14. Siegel, J.H.: Thermal Radiation Heat Transfer, 4th edn. Taylor & Francis, New York (2002)
15. Incropera, F., Dewitt, D., Bergman, T., Lavine, A.: Fundamentals of Heat and Mass Transfer, 6th edn. Wiley, New York (2007)
16. Liferay technology, www.liferay.com (last seen February 2013)
17. JSR 286 reference, http://www.jcp.org/en/jsr/detail?id=286 (last seen February 2013)
18. eduGAIN website, www.edugain.org (last seen February 2013)
19. Idem website, https://www.idem.garr.it/ (last seen February 2013)
20. Macrovision, FLEXlm Reference manual. Printed in the USA (2004)
21. VOMS reference, http://vdt.cs.wisc.edu/components/voms.html (last seen February 2013)
22. Sipos, G., Kacsuk, P.: Multi-Grid, Multi-User Workflows in the P-GRADE Portal. Journal of Grid Computing 3, 221–238 (2005)
23. SRM website, http://storm.forge.cnaf.infn.it/home (last seen February 2013)
24. Gervasi, O., Laganà, A.: SIMBEX: a portal for the a priori simulation of crossed beam experiments. Future Gener. Comput. Syst. 20(5), 703–716 (2004)
25. Costantini, A., Manuali, C., Faginas Lago, N., Rampino, S., Laganà, A.: COMPCHEM: progress towards GEMS a Grid Empowered Molecular Simulator and beyond. Journal of Grid Computing 8(4), 571–586 (2010)

An Efficient Taxonomy Assistant for a Federation of Science Distributed Repositories: A Chemistry Use Case

Sergio Tasso[1], Simonetta Pallottelli[1], Giovanni Ciavi[1],
Riccardo Bastianini[1], and Antonio Laganá[2]

[1] Department of Mathematics and Computer Science, University of Perugia
via Vanvitelli, 1, 06123 Perugia, Italy
{sergio,pallottelli}@unipg.it, gciavi@yahoo.it,
riccardo@bastianini.org
[2] Department of Chemistry, University of Perugia
via Elce di Sotto, 8, 06123 Perugia, Italy
lagana05@gmail.com

Abstract. The paper describes the design and the implementation of a built-in assistant software module aimed at managing the metadata of a federation of collaborative repositories of learning objects. The paper focuses mainly on the standardization of the classification criteria and their application to Molecular sciences.

Keywords: repository, taxonomy, learning objects, G-Lorep.

1 Introduction

The progress made during the EGEE [1] and the EGI-inspire [2] European projects, has allowed the Chemistry, Molecular and Materials Science and Technologies (CMMST) community to set up a Virtual Team (VT) [3] of the European Grid Infrastructure (EGI) [4] devoted to grounding the assemblage of a homonymous Virtual Research Community (VRC). VRCs are groups of like-minded researchers, organised by discipline or computational model, which can draw benefits from having a partnership with EGI. For example, they can benefit from resources and support available within the National Grid Infrastructures (the main stakeholders of EGI.eu) as well as from the workshops and forums organised by EGI. VRCs can also receive support on resolving specific technical issues with EGI services and they will constitute a pillar of the user-focused evolution of EGI's production infrastructure. The mentioned Virtual team is committed to document the evolution of the community from the existing CMMST Virtual Organizations (like COMPCHEM [5] and GAUSSIAN [6]) into a VRC. Such VRC will represent the CMMST community in EGI, will identify the technologies, resources and services already existing within EGI and usable to satisfy the requirements of the users, will single out the

B. Murgante et al. (Eds.): ICCSA 2013, Part I, LNCS 7971, pp. 96–109, 2013.
© Springer-Verlag Berlin Heidelberg 2013

technologies to be developed or imported into EGI and integrated with the production infrastructure in order to allow the VRC members to efficiently manage the relevant tasks.

As a result, the main task of the CMMST VRC will be the collaborative exploitation of research and its application for innovation to fields like Chemical Engineering, Biochemistry, Chemometrics, Omic-sciences, Medicinal chemistry, Forensic chemistry, Food chemistry, Materials, Energy, etc. At the same time, however, increasing emphasis is being put on the collaborative development of research based education. Such challenge has been undertaken in Europe by the European Chemistry Thematic Network Association (ECTNA) [7] that has established for that purpose a Virtual Education Community (VEC) [8, 9]. By relying on the same technological ground as the CMMST VRC, the VEC takes care of supporting harmonization of Chemistry curricula (including related labels), exploiting the use of modern computing technologies in education (including electronic self evaluations tests) supporting institutions bearing Eurolabels (like the Erasmus Mundus consortium for the master in Theoretical Chemistry and Computational Modelling (TCCM) [10]).

Along this line, our laboratories have started working on tools storing, identifying, localizing and reusing CMMST educational materials. Such materials are often the result of a complex process requiring time consuming calculations and the use of sophisticated multimedia rendering products whose objective is the offer of support to students attempting to understand physical phenomena and chemistry processes at microscopic (nanometer) level. In our approach such a block of knowledge is packed into units (called Learning Objects or shortly LOs). LOs are self-consistent, modular, traceable, reusable and interoperable blocks of knowledge which do not only represent consistently a well defined topic but do also bear a specific pedagogical background and embody a significant amount of multimediality and interactivity. Their size usually corresponds to a content of 7(+/-2) concepts (as suggested by CISCO [11]) delivering the front teaching content ranging from ¼ to ½ ECTS credits. The development of Grid technologies has made available a robust platform for empowering LOs with distributed repositories assembled by federating local repositories on the Internet as we did when developing G-LOREP [12] a distributed and collaborative repository of LOs. After all, the building of a system of distributed LO repositories exploiting the collaborative use of metadata has in fact already shown to play a key role in the success of physical sciences teaching and learning.

Key features of G-LOREP are its native suitability for heterogeneous environments and materials as well as its decoupled and evolutionary structure. This has stimulated the development of an efficient tool for filing and retrieving distributed information and has suggested the following structuring of the paper:

In section 2 metadata and taxonomies of our taxonomy assistant are discussed, in section 3 related inherence criteria are analysed, in section 4 a CMMST use case is presented and in section 5 some conclusions are drawn together with indications for future work.

2   The Taxonomy Assistant

2.1   Metadata and Taxonomies

The non automatic tagging of a LO is a long, costly, and error prone subjective process. For this reason the adoption of metadata (data about data: a metadata record consists of a set of attributes or elements describing the considered set of data) standards and of related automatic tagging procedures is extremely important. During the last few years, various open metadata standards have become popular (e.g. IMS [13], SCORM Dublin Core [14] and IEEE LOM [15]). An analysis performed on the following criteria:

- Easing the acquisition and usage of learning instruments.
- Allowing both automated and user-driven content retrieval.
- Supporting LO re-use in multiple learning contexts.
- Fostering content interoperability (information share and exchange using whichever technology compatible with the learning system).

Made us adopt the Dublin Core standard that is developed by the workgroup (DCMI Education Community) [16]. The Dublin Core Metadata is made of 15 descriptive elements. It has been engineered to enable the LO authors to describe their content in a standardized way. Essential characteristics of the Dublin Core are simplicity, interoperability and flexibility. Its success is due to the easy intelligibility of the descriptive elements, to the universally accepted semantics and to its straightforward application to different languages.

However, the Dublin Core algorithm is often too general to appropriately describe specific LOs. For this reason the user has often to extend and personalize the metadata schema to the end of making them accommodate his/her specific educational needs. The (undesirable) consequence of this is the fact that a change of the metadata disrupts interoperability (unless a mapping of the application profiles is provided). Despite this, several projects adopt the Dublin Core and join related initiatives. Moreover, the Dublin Core metadata does not fully meet the attributes required to describe the pedagogic perspective of the LOs such as: beneficiaries, autoconsistency, didactic level, quality indicators, etc. aspects which are better addressed (and solved) by other metadata standards and in particular by the IEEE Learning Object Metadata (LOM) one. LOM is the standard stating the minimum set of properties necessary to LO management, allocation and evaluation. It has been approved by the IEEE (Institute of Electrical and Electronics Engineers) in July 2002 (code 1484.12.1-2002) and specifies a conceptual schema defining the structure of a metadata instance for LOs.

In particular, the IEEE section Learning Technology Standards Committee (LTSC) created a Learning Object Metadata standard (LOM). As shown in Figure 1 LOM is articulated into nine descriptive areas (**categories**) containing groups of attributes arranged in a tree structure and resulting in a total of 70 descriptive elements.

LOM Categories	Educational Category Element	Description
General Lifecycle Meta-metadata Technical Educational Rights Relation Annotation Classification	Interactivity Type	active: Active learning (e.g., learning by doing) is supported by content that directly induces productive action by the learner. expositive: Expositive learning (e.g., passive learning) occurs when the learner's job mainly consists of absorbing the content exposed to them. mixed: A blend of active and expositive interactivity types.
	Learning Resource Type	exercise, simulation, questionnaire, diagram, figure, graph, index, slide, table, narrative text, exam, experiment, problem statement, self assessment, lecture
	Interactivity Level	very low, low, medium, high, very high
	Semantic Density	very low, low, medium, high, very high
	Intended End User Role	teacher, author, learner, manager
	Context	school, higher education, training, other
	Typical Age Range	(range)
	Difficulty	very easy, easy, medium difficult, very difficult
	typical Learning Time	open text element
	description	open text element
	language	standardized def.

Fig. 1. The IEEE Learning Object Metadata

2.2 Taxonomy Assistant 2.0

The classification of educational objects (like the LO ones) is instrumental to the purpose of organizing entities of the knowledge domain in a way that enables their efficient re-use and their automatic handling. This is grounded on appropriate tree classifications (taxonomies) that are based on the subdivision of the set of related entities into more homogeneous subsets easier to handle for search and cataloguing. In our work we have adopted Taxonomy Assistant 2.0 (TA2.0). TA2.0 is a Drupal module that once installed interacts with the LOs via the existing linkableobject and dis_cat modules. TA2.0 analyses the related textual content entered by the user as LO description in order to help with the selection of the category better related to the object. For that purpose TA2.0 generates a message whose formulation varies depending on the proposed text and other contextual variables (like, for example, whether or not the user has selected a category). However, the user is not really bound to accept the suggested text. He/she can edit, for example, the suggested text by adding some words or changing the LO category into the one suggested to the end of increasing its inherence. The TA2.0 algorithm works, in fact, on pattern matching between keywords and the terms of the thesaurus. Therefore, it might happen that to catch the actual meaning of a word appearing in various places with slightly different connotations some integrations could become necessary.

In TA2.0 the different categories are arranged as a forest graph in which each tree represents a science area (e.g. Computer Science, Mathematics, Physics, Chemistry, Biology, etc.) that is composed of various sub-categories with the most specific ones being the tree leaves and the most generic ones the roots.

Such structure is compliant with Various classification schemes such as the Dewey Decimal Classification (DDC) [17] that has been used for describing our test federation. The flowchart of TA2.0 is sketched in Fig. 2.

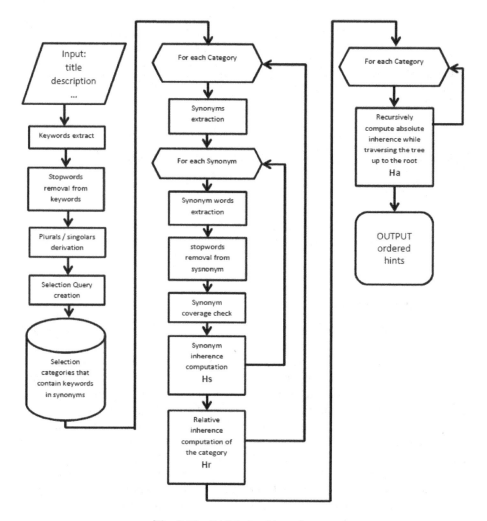

Fig. 2. The TA2.0 algorithm schema

To start the procedure the title, the description and (optionally) the category of the LO is prompted. Out of these input information TA2.0 works out the description text by applying the following procedure:

1- the text is converted to lower case
2- all non essential characters are pulled out (punctuation marks, parentheses, apostrophes, etc.)
3- all text words are checked against the list of stopwords[1] [18] in order to pull out them.

---

[1] Stopwords are words of negligible interest, usually considered as not particularly meaningful from searching engines. Among them one can name articles, pronouns, adverbs, and other words whose interest is lowered by frequent usage.

The goal of that is to end with a series of words (keywords) deprived of inessential terms and use them to query the database. In order to increase the probability of catching meaningful correspondences between the text keywords and the database, TA2.0 was enriched also with some elimination mechanisms based on standard grammar rules (like singular/plural extension).

Each category is associated with a thesaurus made of a set of terms (synonyms) having the same meaning in that category (synonyms may be composed of more than one word). The fraction given by the number X of keywords found in the description of the LO and the total number N of keywords defining the synonym is called **coverage.** The coverage is used to the end of comparing the user submitted keywords with those contained in the database. As to the already mentioned concept of inherence, it can be quantified as follows: for all the records obtained from the query each synonym is split into single words whose presence (as such) in the database is counted. The result is taken as a measure of the inherence considered as the similarity between the synonyms and the user text (and evaluate in this way the propriety of a category assignment).

3   Inherence Criteria

3.1   Synonym Inherence as an Efficiency Index

The **inherence of a synonym** H_s with a LO is the extent of relevance (or pertinence P) of the given synonym in describing a LO. Therefore H_s can be taken as an efficiency index if the following assumptions:

a) let the value of H_s be as high as the number N of words composing the synonym;
b) let the value of H_s for partial coverage depend on the recurrence of a word among those composing the synonym (a coverage of "1 out of 3" must result in a H_s value higher than the one associated with a coverage of "1 out of 4");

are made. To work out an algebraic formulation of H_s we make also the following positions:

- R_i is the number of occurrences of the words in synonym i,
- S_i is the number of words composing synonym i,
- K is the total number of valid keywords,
- $U_i = R_i/S_i$ is the ratio between the number of occurrences of the words R_i found in synonym i and the number of words S_i found in it,
- $P_i = S_i/K$ is the ratio between the words observed in synonym i and the total number of considered keywords K (that is called either pertinence or relevance).

At first the simple formulation of H_s was taken to be the following power-like one

$$H_s = R_i^{U_i} = R_i^{\left(\frac{R_i}{S_i}\right)}$$

that gives 1 in the case of a "1 out of N" coverage and gives N for a complete coverage "N out of N". Figure 3 shows the values of such H_s function when plotted against the number of occurrences for different values of the number of words composing the synonym.

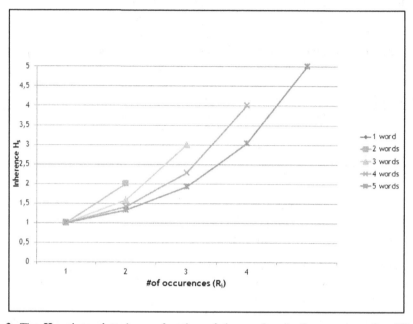

Fig. 3. The H_s values plotted as a function of the number # of occurrences for different numbers of words composing the synonym

As clearly shown by the figure, the value of H_s is the same in both the "1 out of N" and "1 out of 1" cases. This is in conflict with the requirement b) listed above. To correct such behaviour, the H_s function was then multiplied by U_i obtaining the revised form of the H_s function,

$$H_s = R_i^{\left(\frac{R_i}{S_i}\right)} \frac{R_i}{S_i}$$

whose behaviour is shown in Figure 4.

Moreover, we can exalt the inherence value of the most interesting synonyms by multiplying the revised H_s function by the relevance P_i. Accordingly, the H_s inherence value of synonym i (H_{s_i}) is:

$$H_{s_i} = R_i^{\left(\frac{R_i}{S_i}\right)} \frac{R_i}{S_i} P_i$$

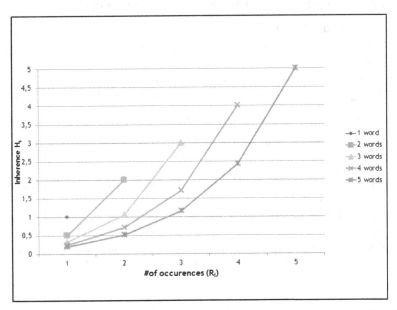

Fig. 4. The revised H_s values plotted as a function of the number # of occurrences for different P_i.numbers of words composing the synonym

3.2 Inherence of Categories

In order to work out more appropriate efficiency indices, the significance of repeated partial occurrences in synonyms within a category has been calibrated. To this end the quantities

- Relative inherence of a category, (H_r)
- Absolute inherence of a category (H_a)

have been defined as the sum of all the synonym inherences H_s of the considered category and as the sum of the relative inherences H_r associated to all the categories found in the path root/node, weighted by the related level d, respectively.

Accordingly, the *relative inherence H_r* is formulated as:

$$H_r = \sum_{i=1}^{\#of\ synonyms} H_{s_i}$$

By expanding the formula through the substitution of H_s we obtain:

$$H_r = \sum_{i=1}^{\#of.Synonyms} R_i^{U_i} U_i P_i \quad \Rightarrow \quad H_r = \frac{1}{K} \sum_{i=1}^{\#of.Synonyms} \left(R_i^{\left(\frac{R_i}{S_i}\right)} \frac{R_i^2}{S_i} \right)$$

To work out an algebraic formulation of H_a we make the following positions:

- d is the depth of the node,
- $R_{i,d}$ is the number of matching words found in the synonymous i at level d,
- $S_{i,d}$ is the number of words composing synonym i at level d,
- $U_{i,d} = \frac{R_{i,d}}{S_{i,d}}$ is the ratio between the number of occurrences and the number of words composing the synonymous i at level d,

We have further set specific positions in order to

a) avoid giving too much weight to a category with a H_r value obtained from many occurrences of partial coverage for different synonyms. In this case the inherence is still given a power formulation (by doing so we exalt the contribution of the complete coverage of one synonym);

b) account for the same partial coverage "X out of N" occurring more than one time in a single category (if a word belongs to more than one synonym). In this case the final H_r value of the node is obtained by adding the H_s inherences of the various synonyms. For example, in the case of one occurrence of "1 out of 2" coverage and two occurrences of "1 out of 3" coverage, the final inherence value is $H_r = H_s("1 \text{ out of } 2") + 2*H_s("1 \text{ out of } 3")$;

c) exalt more specific categories. In this case the inherence value of each node is weighted (by multiplying it by the level d in which the node is located in the tree) and we add the inherence values of the nodes preceding the one considered in the tree hierarchy.

In this way we work out the absolute inherence of a category obtained by adding the relative inherence value of each ancestor category i lying on the path from the root to the considered category multiplied by the level d to which the category belongs.

$$H_a = \sum_{d=1}^{depth} H_{r_d} d$$

or

$$H_a = \frac{1}{K} \left[\sum_{d=1}^{depth} \left(\sum_{i=1}^{\#of.Synonyms} \frac{R_{i,d}^{\left(\frac{R_{i,d}}{S_{i,d}}\right)+2}}{S_{i,d}} \right) d \right]$$

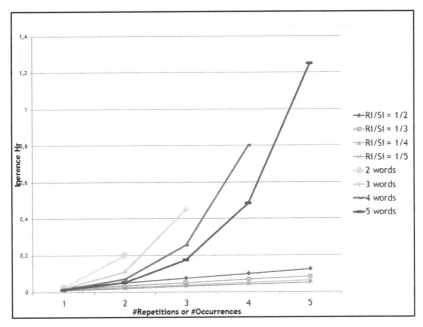

Fig. 5. The H_r values plotted as a function of the number # of repetitions (or occurrences) for different numbers of words composing the synonym

4   Use Case Procedure and Results

4.1   Tuning of the Procedure

Here a first comparison of the validity of the adopted procedure is performed by comparing the classification obtained by evaluating H_a for a set of publications taken from the literature using the taxonomy tree adopted in ref. 19 with the one of the official publisher. Then a second comparison is made after improving the description of some categories of the tree following the indications obtained from the first test. To perform the validity tests a database of title, abstract, and related classification of a randomly selected sample of 40 articles published on JAMS [20] was created. In order to carry out the comparison of the TA2.0 classification with that of JAMS [21] the latter was converted into the DDC one.

Results obtained are shown in Figure 6 in which the frequency of occurrence of the ranking deviation between our classification and that of JAMS is shown (deviation 0 means exact reproduction and deviation n means displacement of n places). As shown by figure 6 the prediction of our method is exact in 27.5 % of the cases considered whereas the remaining 65 % of cases our prediction deviates for less than 5 places. Yet, there has been a residual 15 % of case for which our method was unable to perform the classification.

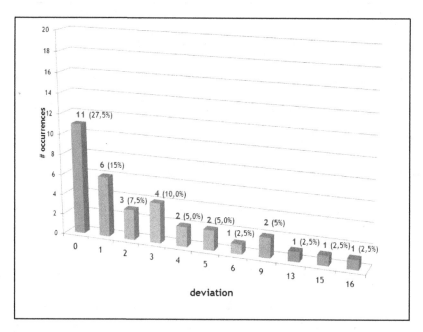

Fig. 6. The number of classification deviations plotted as a function of ranking difference (in brackets the percentage)

The results obtained after improving the description of the tree following the indications of the journal are illustrated in Figure 7. As apparent from the figure there is a tremendous improvement in the occurrence of exact prediction (50%) while in 65 % of cases our prediction still deviates for less than 5 places. At the same time the percentage of unclassified cases lowers down to 12%.

This shows the validity of our TA2.0 algorithm and points out the importance of an appropriate definition of the taxonomy tree.

4.2   Use Case Results

Before dealing with the application of TA2.0 on CMMST, we tackled a Mathematics use case by considering a JAMS paper as a LO that was filed under the Dewey category 515.73 (*Topological vector spaces*) and the paper's title and abstract were input into the TA2.0 module. TA2.0 reacts to the user's choice and displays a list of synonyms related to the selected category. The list can either be edited (if the user has enough knowledge of the subject and of DDC) in order to choose the appropriate category or the TA2.0 suggestion is followed and the user invited to integrate the solution offered.

A similar roadmap is followed for the Chemistry use case. The user input is listed in Figure 8.

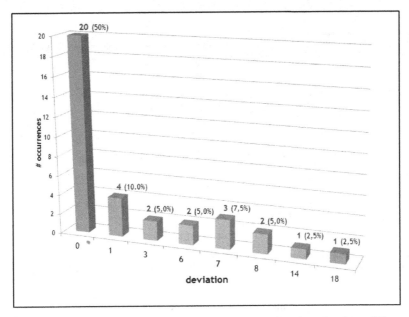

Fig. 7. The frequency of classification deviations plotted as a function of ranking difference

Title:

Coupling Quantum Interpretative Techniques: Another Look at Chemical Mechanisms in Organic Reactions

Description (20 words):

A cross ELF/NCI analysis is tested over prototypical organic reactions. The synergetic use of ELF and NCI enables the understanding of reaction mechanisms since each method can respectively identify regions of strong and weak electron pairing. Chemically intuitive results are recovered and enriched by the identification of new features. Noncovalent interactions are found to foresee the evolution of the reaction from the initial steps. Within NCI, no topological catastrophe is observed as changes are continuous to such an extent that future reaction steps can be predicted from the evolution of the initial NCI critical points. Indeed, strong convergences through the reaction paths between ELF and NCI critical points enable identification of key interactions at the origin of the bond formation. VMD scripts enabling the automatic generation of movies depicting the cross NCI/ELF analysis along a reaction path (or following a Born–Oppenheimer molecular dynamics trajectory) are provided as Supporting Information.

Fig. 8. User input for TA2.0 in an Organic Chemistry use case

The user can now continue by editing the displayed fields in order to describe the LO, and is encouraged to use the suggested synonyms. As soon as both the title and description of the LO have been filled in, or upon explicit user request, TA2.0 is activated.

If the user does not already know which category is right for the LO, he/she can enter the title and description of the LO in the appropriate fields, without choosing any category. The assistant will then employ the algorithm described above to suggest the most relevant categories to the user. The user can then choose the category for the LO. If he/she does not choose the best category selected by the assistant, it will invite the user to include more of the associated synonyms to improve the LO description (see Figure 9).

Categories suggested by Taxononomy assistant:

This is the list of categories that are compatible with the text and their value inherence (Hin value) and relevance (max for single term% | total %):

(Remember that you haven't yet selected a category from the vocabularies)

541.2 - Theoretical Chemistry *(keywords:*'**reaction**' *'molecular bond'* '**quantum**' *)(Hin Value: 100) Relevance: (max:2.9%) | (Tot:15.9%)*

541.36 - Thermochemistry & Thermodynamics *(keywords:*'**reaction**' *'formation'* '**point**' *)(Hin Value: 35.3) Relevance: (max:1.4%) | (Tot:7.2%)*

541.39 - Chemical reactions *(keywords:*'**reaction**' *)(Hin Value: 25.5) Relevance: (max:1.4%) | (Tot:5.8%)*

515.78 - Special topics of functional analysis *(keywords:*'**analysis**' *)(Hin Value: 17.6) Relevance: (max:1.4%) | (Tot:5.8%)*

515.73 - Topological vector spaces *(keywords:*'**topological**' '**continuous**' *)(Hin Value: 11.8) Relevance: (max:1.4%) | (Tot:2.9%)*

541.34 - Solutions Chemistry *(keywords:*'**point**' *)(Hin Value: 9.8) Relevance: (max:1.4%) | (Tot:2.9%)*

543.6 - Non-Optical Spectroscopy *(keywords:*'**electron**' '**analysis**' *)(Hin Value: 9.8) Relevance: (max:1.4%) | (Tot:4.3%)*

514.7 - Analytic Topology *(keywords:*'**analysis**' *)(Hin Value: 7.8) Relevance: (max:1.4%) | (Tot:1.4%)*

547.2 - Organic Chemical Reactions *(keywords:*'**reaction**' *)(Hin Value: 7.8) Relevance: (max:1.4%) | (Tot:1.4%)*

543.2 - Classical Methods *(keywords:*'**analysis**' *)(Hin Value: 7.8) Relevance: (max:1.4%) | (Tot:2.9%)*

512.5 - Linear Algebra *(keywords:*'**topological**' *)(Hin Value: 6.5) Relevance: (max:1.4%) | (Tot:2.9%)*

547 - Organic Chemistry *(keywords:*'**organic**' *)(Hin Value: 3.9) Relevance: (max:1.4%) | (Tot:1.4%)*

514.2 - Algebraic Topology *(keywords:*'**topological**' *)(Hin Value: 3.9) Relevance: (max:1.4%) | (Tot:1.4%)*

543.5 - Optical Spectroscopy (Spectrum Analysis) *(keywords:*'**molecular**' *)(Hin Value: 3.9) Relevance: (max:1.4%) | (Tot:1.4%)*

519.5 - Statistical Mathematics *(keywords:*'**analysis**' *)(Hin Value: 3.9) Relevance: (max:1.4%) | (Tot:1.4%)*

548.8 - Physical and Structural Crystallography *(keywords:*'**method**' *)(Hin Value: 3.9) Relevance: (max:1.4%) | (Tot:1.4%)*

543.8 - Chromatography *(keywords:*'**analysis**' *)(Hin Value: 3.9) Relevance: (max:1.4%) | (Tot:1.4%)*

515 - Analysis *(keywords:*'**analysis**' *)(Hin Value: 3.9) Relevance: (max:1.4%) | (Tot:1.4%)*

514.3 - Topology of Spaces *(keywords:*'**point**' *)(Hin Value: 2.6) Relevance: (max:1.4%) | (Tot:1.4%)*

541 - Physical Chemistry *(keywords:*'**molecular**' *)(Hin Value: 2) Relevance: (max:1.4%) | (Tot:1.4%)*

518 - Numerical Analysis *(keywords:*'**method**' *)(Hin Value: 1.3) Relevance: (max:1.4%) | (Tot:1.4%)*

Depending on your choices and the text entered in the fields 'title' and 'description' Taxonomy Assistant suggests the categories that are most relevant to them.

Fig. 9. Results obtained from TA2.0 for the Organic Chemistry use case

5   Conclusion and Future Work

We hereby described TA2.0, an assistant which helps a user during the delicate process of cataloguing a LO in a federation of distributed and collaborative repositories. A correct LO classification is fundamental for other aspects and features of G-LOREP a project based on distributed computing educational environments like the one promoted by the VEC committee of ECTNA. In this respect, the search process can greatly benefit from the usage of numerical descriptors for categories, allowing both to retrieve and to let users navigate amongst related content. TA2.0 suggests to the user the correct place for a LO inside the taxonomy tree. The reported test results have shown to satisfy the expectations of both mathematical and Chemistry communities even if the algorithm will need further testing. The work has shown also where future activities will have to be addressed: improving our taxonomy forest, including more categories and more synonyms, hopefully with help from Dewey and library science experts. The reason for this is that, during our tests, we noticed a great improvement in the algorithm outputs while expanding our taxonomy trees with more synonyms. In the meantime, the TA2.0 module will be adopted by our G-LOREP test federates and will offer its helpful advice to the members of ECTNA to allow them to upload new material to the federation. Further improvements will be planned in the future by including machine learning techniques to enrich the synonyms associated to each category in the taxonomy tree, by exploiting semantic-based and computer-assisted tools and users feedback.

Acknowledgements. The authors acknowledge ECTNA (VEC standing committee) and the EC2E2N 2 LLP project for stimulating debates and financial support. Thanks are due also to EGI and IGI and the related COMPCHEM VO.

References

1. http://www.eu-egee.org/ (last access February 2013)
2. http://www.egi.eu/about/egi-inspire/ (last access February 2013)
3. https://wiki.egi.eu/wiki/Virtual_team (last access February 2013)
4. http://www.egi.eu/ (last access February 2013)
5. http://compchem.unipg.it (last access February 2013)
6. http://egee.grid.cyfronet.pl/Applications/gaussian-vo/ (last access February 2013)
7. http://ectn-assoc.cpe.fr/ (last access February 2013)
8. Laganá, A., Riganelli, A., Gervasi, O., Yates, P., Wahala, K., Salzer, R., Varella, E., Froehlich, J.: ELCHEM: A metalaboratory to develop grid e-learning technologies and services for chemistry. In: Gervasi, O., Gavrilova, M.L., Kumar, V., Laganá, A., Lee, H.P., Mun, Y., Taniar, D., Tan, C.J.K. (eds.) ICCSA 2005. LNCS, vol. 3480, pp. 938–946. Springer, Heidelberg (2005)
9. Laganà, A., Manuali, C., Faginas Lago, N., Gervasi, O., Crocchianti, S., Riganelli, A., Schanze, S.: From Computer Assisted to Grid Empowered Teaching and Learning Activities in Higher Level Chemistry Education. In: Eilks, I., Byers, B. (eds.) Innovative Methods of Teaching and Learning Chemistry in Higher Education. RCS Publishing (2009)
10. http://www.emtccm.org/ec2e2n (last access February 2013)
11. http://www.cisco.com/ (last access February 2013)
12. Pallottelli, S., Tasso, S., Pannacci, N., Costantini, A., Lago, N.F.: Distributed and Collaborative Learning Objects Repositories on Grid Networks. In: Taniar, D., Gervasi, O., Murgante, B., Pardede, E., Apduhan, B.O. (eds.) ICCSA 2010, Part IV. LNCS, vol. 6019, pp. 29–40. Springer, Heidelberg (2010); Tasso, S., Pallottelli, S., Bastianini, R., Lagana, A.: Federation of Distributed and Collaborative Repositories and Its Application on Science Learning Objects. In: Murgante, B., Gervasi, O., Iglesias, A., Taniar, D., Apduhan, B.O. (eds.) ICCSA 2011, Part III. LNCS, vol. 6784, pp. 466–478. Springer, Heidelberg (2011)
13. http://www.imsglobal.org/ (last access February 2013)
14. http://www.adlnet.org/scorm/scorm-2004-4th/ (last access February 2013)
15. http://ltsc.ieee.org/wg12/files/ LOM_1484_12_1_v1_Final_Draft.pdf (last access February 2013)
16. http://dublincore.org/groups/education/ (last access February 2013)
17. Mitchell, J.S., et al.: Dewey Decimal Classification and relative index / devised by Melvil Dewey, 23th edn., vol. 4. OCLC, Dublin (2011)
18. http://www.ranks.nl/resources/stopwords.html (last access January 2013)
19. Tasso, S., Pallottelli, S., Ferroni, M., Bastianini, R., Laganà, A.: Taxonomy Management in a Federation of Distributed Repository: A Chemistry Use Case. In: Murgante, B., Gervasi, O., Misra, S., Nedjah, N., Rocha, A.M.A.C., Taniar, D., Apduhan, B.O. (eds.) ICCSA 2012, Part I. LNCS, vol. 7333, pp. 358–370. Springer, Heidelberg (2012)
20. http://www.ams.org/publications/journals/journalsframework/ jams (last access February 2013)
21. http://www.math.unipd.it/~biblio/msc-cdd/ (last access January 2013)

Evaluation of a Tour-and-Charging Scheduler for Electric Vehicle Network Services*

Junghoon Lee and Gyung-Leen Park**

Dept. of Computer Science and Statistics
Jeju National University, Republic of Korea
{jhlee,glpark}@jejunu.ac.kr

Abstract. This paper evaluates the waiting time for a visiting schedule created by a tour-and-charging scheduler for electric vehicles in Jeju city area. As a promising vehicle network application, this service alleviates the range anxiety of electric vehicles by finding an energy-efficient tour schedule, alternately considering tour and charging. For the field test on the real-life tour spot distribution and charger availability, 3 model tour courses are selected first, each of which has the tour length of 155, 166, and 148 *km*, respectively, and different number of chargers on the path. The evaluation process measures and compares the tour time and thus the waiting time on 3 courses for tour schedules generated by our scheduler and legacy traveling salesman problem solver. The experiment discovers that our scheme reduces the waiting time by up to 21.1 % and eliminates the waiting time if the stay time intervals are 90, 80, and 40 minutes for each course on the current distribution of spots and chargers.

Keywords: electric vehicle, rent-a-car business, tour-and-charging schedule waiting time, real-life distribution.

1 Introduction

Current wireless communication technologies are able to provide stable connections having reasonable bandwidth even for fast-moving vehicles. 5networks, VANETs (Vehicular Ad-hoc NETworks), and the like [1]. Particularly, cellular networks make it possible for vehicles to ubiquitously access the global network such as the Internet not restricted by space or time [1]. Moreover, as the telcos are continuously lowering the communication fee, this network will host more diverse vehicle applications. Here, connected vehicles can fully take advantage of various and intelligent information services interacting with high-performance servers. People can access the global network through in-vehicle computers such as telematics devices or their own smart phones. Especially for the users in vehicle, location-based services will be very useful and such services necessarily take the current location obtained by the embedded GPS module.

* This research was financially supported by the Ministry of Knowledge Economy (MKE), Korea Institute for Advancement of Technology (KIAT) through the Inter-ER Cooperation Projects.
** Corresponding author.

B. Murgante et al. (Eds.): ICCSA 2013, Part I, LNCS 7971, pp. 110–119, 2013.
© Springer-Verlag Berlin Heidelberg 2013

In the mean time, electric vehicles, or EVs, are expected to gradually replace gasoline-powered vehicles in the near future, as their energy efficiency is much better [2]. However, the critical drawback of EVs lies in short driving range and long charging time. The driving range denotes the distance a fully charged vehicle can drive without additional battery charging. As for slow charging, it takes about 6 ∼ 7 hours to fully charge an EV, but it can drive just about 100 km [3]. Moreover, this driving distance is further shortened by air-conditioner and brake operations. Short driving range is not a problem for EVs mainly used in our everyday lives as the daily driving distance hardly exceeds the driving range as long as the drivers didn't forget to charge their vehicles overnight. However, the daily driving distance of tour rent-a-cars and delivery vehicles is usually longer than the driving range.

Here, EVs can benefit from the vehicle network as many sophisticated services can be provided to them via the network for the sake of overcoming their problems in long charging time and short driving range [4]. For example, the information server can not only search the energy-efficient route but also allocate a charging station having the smallest waiting time [5]. Particularly, as battery charging can be done while the drivers are taking a tour, the waiting time is dependent on the visiting sequence. Here, waiting time is the time length a tourist must wait his or her EV battery to be charged enough to reach the next destination. Our previous work first has designed an estimation model of the waiting time for a given tour schedule accounting for this stay-and-charging [6]. Then, the backtracking-based search scheme finds a visiting schedule having the minimal waiting time. For this scheme, it is necessary to evaluate the performance in a real-life tour environment. It is obvious that spatial distribution and stay time of tour spots will affect waiting time.

In this regard, this paper tailors tour-and-charging scheduler and analyzes its performance in Jeju city. Jeju area, as a smart grid test bed and a well-known tourist place embracing plenty of natural attractions, possesses hundreds of EVs and also hundreds of chargers over the whole island. Here, EV rent-a-car business is about to start its service according to the ambitious vision of replacing whole vehicles with EVs by 2030. The deployment of EVs to rent-a-cars will prompt the society-wide penetration of EVs in delivery systems, public transportation systems, and the like. This paper is organized as follows: After issuing the problem in Section 1, Section 2 reviews related work on intelligent EV services. Section 3 describes the tour-and-charging scheduler and geographically represents model tour courses. Section 4 demonstrates the waiting time estimation results for 3 courses. Finally in Section 5, the study is summarized with a brief introduction of future work.

2   Related Work

As an example of EV operation planning, [7] presents a multiple ant colony algorithm to solve vehicle scheduling problems with route and fueling (or charging) time constraints. Considering the limited travel miles of EVs, this scheme tries to minimize the charging time by means of efficient transit scheduling. Such vehicle scheduling problems optimize fleet operations of public transportation systems but belong to NP-hard category, inherently rooted from the well-known traveling salesman problem, or TSP from now on [8]. Their algorithm is built upon a multiple objective function to minimize the number of tours as well as to minimize the total deadhead time, while the precedence is put on the first. Route construction and trail update procedures are defined to regulate two conflicting goals of fast convergence and prematurity avoidance. Particularly, a bipartite graph model combined with its optimization algorithm minimizes the number of required EVs to meet the charging time constraint.

[9] increases the driving range by finding an efficient route integrating the information from diverse cooperative transport infrastructure such as charging facilities and public transport. The main idea is to export all public transportation data to a graph, where the arc length is defined by time to move between its two end points. Here, the arc weight is a combination of several parameters such as time, cost, CO_2 emissions, and city traffic conditions. This graph model extensively includes the data from heterogeneous transports including car and bike sharing systems for multimodal planning in Lisbon area. The parameter orchestration will be completed in their final project year. In addition, some heuristics are designed to reduce the memory data size in integrating diverse information. Finally, a mobile application, running on on-board systems or smart phones, helps the driver to find an efficient plan for EV-based multimodal journey.

In the mean time, electric trucks are also introduced to the market place and large logistics companies, such as Fed Ex and Frito Lay, are testing and putting them into service [10]. To examine the competitiveness of electric delivery trucks, [10] builds a new analysis model integrating routing constraints, speed profiles, energy consumption, and vehicle ownership cost. Particularly, the authors exploit a continuous approximation of the vehicle routing problem to estimate the average cost of serving routes. This approximation is based on the spatial demand density and derives analytical insights about the relation between involved parameters to recognize key variables. Here, for the tour distance approximation of a TSP, both the number of customers and the number of trucks are integrated into the analysis model. Their research addresses that cost savings by the reduction of operation cost overcomes the high initial cost for purchasing electric trucks in the logistics business.

3   Tour and Charging Scheduler

3.1   EV Rent-a-Car Tour

The Republic of Korea, after being designated as a smart grid initiative country in 2009, launched a smart grid test-bed in Jeju area, an island located in the southernmost part of Korean territory. This enterprise is aiming at testing leading-edge technologies and developing business models in 5 major areas of smart power grid, smart place, smart transportation, smart renewables, and smart electricity services. In the mean time, Jeju province is one of the most famous tour places in East Asia, having many natural tourist attractions including beaches, volcanoes, cliffs, and so on. Hence, for environment preservation, its local government is ambitiously trying to accelerate the penetration of EVs to the entire island. As a step of this effort, an EV rent-a-car business is now about to begin its service. According to the tour pattern analysis, most daily driving distance is estimated to be a little bit longer than the driving range of EVs, so an efficient tour and charging schedule can possibly make the tour more convenient.

Figure 1 illustrates our system model. Charging stations and facilities are scattered over the island area. Some tour spots have a charger, while others not. Tourists renting out an EV select the tour spots they want to visit and submit to the remote server along with its current location via the vehicle network. They are not aware of whether a spot has chargers or not. The remote server, essentially capable of performing high speed computation and manipulating large data volume, searches the route having the minimal waiting time induced by EV charging. The tour schedule takes into account the current EV location, road network characteristics, battery capacity, and charging facility availability. Here, as contrast to the fuel consumption of gasoline-powered vehicles, the battery discharge model of EVs is very complex and affected by road shapes, slopes, and driving conditions. So, it is parameterized and processed in spatial database [11]. The result is sent back to the tourists. If they are not satisfied with the tour schedule, they will modify the selection and resubmit to the server.

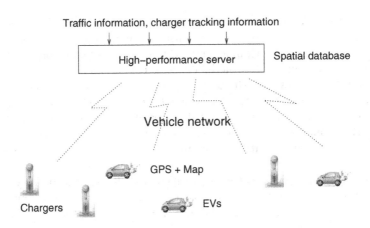

Fig. 1. Vehicle network service architecture

Tour scheduling is basically a process of deciding a visiting order for a given set of destinations. As a tour generally returns to the start position (rent-a-car stations or hotels), it is equivalent to TSP. However, the schedule must try to reduce waiting time not just on the driving distance. Waiting time takes place when battery remaining is not enough to reach the next destination when the EV is to depart the current spot. Our previous work has designed a tour scheduling scheme for EV rent-a-car tours based on the assumption that battery charging can be conducted at each tour spot while the tourists are taking a tour [6]. It models the battery amount gained when this charging-while-stay is feasible. Then, search space is traversed through backtracking-based methods or genetic algorithms. The schedule having the smallest waiting time is taken as final route recommendation. For more details, refer to [6]

3.2   Model Courses

The performance of this tour and charging scheduler is deeply dependent on the distribution of tour spots and the availability of chargers. If the tour length is less than the driving range of an EV, waiting time reduction is meaningless. On the contrary, when the tour length is too long, the waiting time cannot be tolerated by tourists even if it is much reduced by an efficient schedule. For its validation, we select 40 most commonly visited spots in Jeju area and check whether each of them installs EV chargers. Then, according to the geographical affinity, 3 model courses are selected. Each course fits for a daily tour, consists of 9 destinations, and different number of chargers. Model course 1 starts from the hotel area in Jeju city and covers tour spots in east area. It has 5 chargers along 155 km long route. Model course 2 covers the west area, has 2 chargers, and is 166 km long. Model course 3 embraces the south area, has 5 chargers, and is 148 km long. These courses are depicted in Figure 2.

The distance between each pair of 40 spots is calculated by means of the A* algorithm for the road network of Jeju area. As the power consumption model for each road is not yet completed, we just consider the distance of segments at this stage. Our scheduler can work independently of link weights as long as they are given as numerical values. Each model course creates its own cost matrix to run a TSP solver, which traverses the search space and evaluates feasible schedules. Actually, for 9 destinations, the execution time is not significant on average performance personal computers, even if all feasible schedules are investigated. Following the sequence alternately consisting of move and stay, the evaluation process basically decreases battery remaining for a move between two consecutive spots in the visiting sequence by the distance between them. At a stay, the process checks if the spot has a charging facility. If it has, the battery amount increases in proportion to the stay time. Here, every parameter is aligned to the distance credit, which denotes the distance with current battery remaining.

The background image of Figure 2 is the road network of Jeju area. In its center, there is a big mountain, so the road density is very low. The road network density coincides with the population density. Large rectangles mark the locations of tour spots. Spots having chargers are additionally marked with charger images. Each course is represented by a route calculated by a legacy TSP solver which minimizes the tour length in network distance. As the shape detail of each road segment is abbreviated, the actual distance can be larger than as it looks. Course 1 has relatively large number of chargers, but they are located along the coast area. The path across the mountain area looks charming to

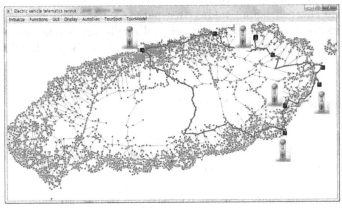

(a) Course 1

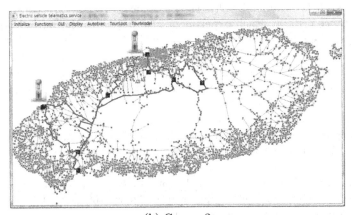

(b) Course 2

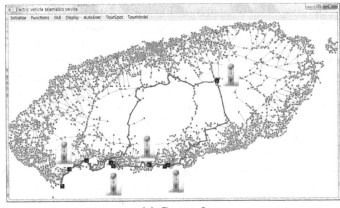

(c) Course 3

Fig. 2. Model course description

reduce the driving length, but it may increase the waiting time, as an EV must be sufficiently charged before taking this path. Course 2 includes just 2 spots having chargers, so the effect of our scheduler will not be significant. Course 3 has sufficient number of chargers and they are properly distributed over the route.

4   Estimation Result Analysis

This section conducts the waiting time estimation for 3 model tour courses selected in Section 3.2. A prototype version of the tour-and-charging scheduler has been implemented using Visual C++ 6.0. There are some assumptions on it. First of all, as the EV is charged overnight at a hotel, it is fully charged when passengers start their daily trip. We compare the tour time, which includes the waiting time for EV charging in addition to the pure driving time, with the classic TSP solver. It is obtained by $O(n!)$ search space traversal, just considering the driving distance criteria. Each experiment also measures the pure driving time to investigate the added waiting time brought by EV charging. The experiment assumes that the average EV speed is 60.0 kmh and an EV can drive 90 km with 6 hour charging. Hence, the distance of 30.0 km corresponds to 30.0 min drive while 1 minute charging earns 0.25 km credit. The initial battery amount, or the distance credit, is 90.0 km.

The first experiment measures the effect of the stay time in each tour spot having chargers. Even if the tour spot distribution is fixed, the stay time in each spot will be different according to personal preference, weather, and many other factors. The scheduling service can be more sophisticated if it combines such information. However, it's out of scope of this paper and we just focus on the effect of stay time, changing it from 30 to 100 min. Figure 3 shows the estimation result. Here, each course has its own pure driving time and it is plotted by a straight line labeled with *PureDrive*. The pure driving time is not affect by EV charging. For comparison, Figure 3 plots the tour time according to the TSP solver and our scheme, marked by TSP and $EvSched$, respectively. Actually, reducing the driving distance contributes to reducing the waiting time. The difference from the pure driving time will be the waiting time.

For Course 1, the difference between the TSP solver and our scheduling scheme can be observed during the interval where stay time is between 80 and 100 min. The waiting time linearly decreases according to the increase in the stay time for both schemes. With the given charger distribution, the path across the mountain can hardly be excluded in the route. Here, two schemes show quite similar waiting time. When the stay time is 90 min, our scheduler finds a schedule having no waiting time. Anyway, the performance gap between two schemes reaches 4.6 % when the stay time is 80 min. For Course 2, having the limited number of chargers, we can hardly expect the advantage of the tour-and-charging scheduler. Hence, both schemes show the same waiting time for the whole experiment range. Course 3 can most benefit from the overlapped charging and tour. Our scheme finds the schedule having no waiting time already when the stay time is 50 *min*, while the TSP solver cannot find such a schedule during the whole range. The performance gap reaches 21.1 % when the stay time is 40 *min*.

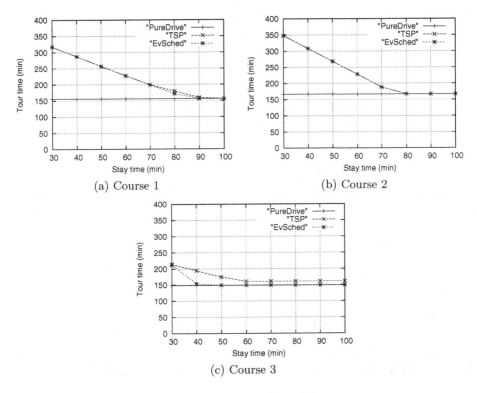

Fig. 3. Tour time for each model course

Next, charging facilities are sure to be installed in more tour spots according to the penetration of EVs [12]. Hence, the second experiment measures the effect of the number of chargers to the waiting time. In this experiment, stay time is fixed to 50 *min*. If n out of 9 spots have chargers and n is less than 9, they are selected randomly. Figure 4 plots the results. In Course 1, the tour-and-charging scheduler reduces waiting time by up to 34.1 % when the number of chargers is 4, compared with the TSP solver. Moreover, waiting time completely disappears when the number of chargers is 5. For the TSP case, 6 chargers are necessary. In Course 2, two schemes have the same waiting time for the whole range. Their waiting time reaches 0 with 5 chargers. Here, the shortest driving length minimizes the waiting time. Course 3 extends the performance gap to 35.2 % when the number of chargers is 4. It is because the tour spots are distributed evenly, so many tour schedules closer to the shortest path are available and they may have different waiting time. Then, we can just select the best of them.

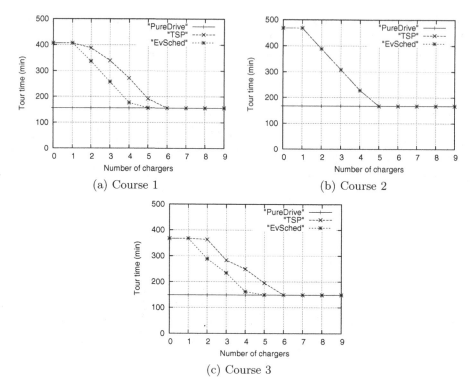

Fig. 4. Charging facility effects

5   Conclusion

As a major part of smart grids, smart transportation is aiming at achieving energy efficiency in transportation systems. It mainly focuses on the fast and large deployment of EVs into our daily lives. To overcome their problems in long charging time and short driving range, an intelligent information service is indispensable. This paper has presented first a tour scheduler for EVs, targeting at EV rent-a-car business which is preferred in tour areas possessing natural attractions. For the evaluation purpose, we have selected 3 model courses which are most commonly taken in Jeju city area and run the tour-and-charging scheduler to compare the waiting time with the legacy TSP solver. The experiment results reveal that our scheme reduces the waiting time by up to 21.1 % and eliminates the waiting time if the stay time intervals are 90, 80, and 40 min on each course. In addition, if more than 55.5 % of spots install charging facilities, the waiting time in the tour will be eliminated, indicating that EVs can be promisingly exploited by a tour rent-a-car business.

Actually, this paper has assumed that every charging facility is always available when an EV arrives, as the current power provision is sufficient considering the current number of EV rent-a-cars. However, this assumption will be invalid when more EVs are deployed in the near future. In this case, the scheduler must take into account the temporal availability of chargers and it will be built on a reservation-based scheduler. This will be the future work in our research.

References

1. Fallah, P., Huang, C., SenGupta, R., Krishnan, H.: Analysis of Information Dissemination in Vehicular Ad-Hoc Networks With Application to Cooperative Vehicle Safety Systems. IEEE Transactions on Vehicular Technology 60, 233–247 (2011)
2. Freire, R., Delgado, J., Santos, J., Almeida, A.: Integration of Renewable Energy Generation with EV Charging Strategies to Optimize Grid Load Balancing. In: IEEE Annual Conference on Intelligent Transportation Systems, pp. 392–396 (2010)
3. Botsford, C., Szczepanek, A.: Fast Charging vs. Slow Charging: Pros and Cons for the New Age of Electric Vehicles. In: International Battery Hybrid Fuel Cell Electric Vehicle Symposium (2009)
4. Frost & Sullivan: Strategic Market and Technology Assessment of Telematics Applications for Electric Vehicles. In: 10th Annual Conference of Detroit Telematics (2010)
5. Kobayashi, Y., Kiyama, N., Aoshima, H., Kashiyama, M.: A Route Search Method for Electric Vehicles in Consideration of Range and Locations of Charging Stations. In: IEEE Intelligent Vehicles Symposium, pp. 920–925 (2011)
6. Lee, J., Kim, H., Park, G.: Integration of Battery Charging to Tour Schedule Generation for an EV-Based Rent-a-Car Business. In: Tan, Y., Shi, Y., Ji, Z. (eds.) ICSI 2012, Part II. LNCS, vol. 7332, pp. 399–406. Springer, Heidelberg (2012)
7. Wang, H., Shen, J.: Heuristic Approaches for Solving Transit Vehicle Scheduling Problem with Route and Fueling Time Constraints. Applied Mathematics and Computation 190, 1237–1249 (2007)
8. http://www.tsp.gatech.edu/concorde.html
9. Ferreira, J., Moneiro, V., Afonso, J.: Cooperative Transportation System for Electric Vehicles. In: Annual Seminar on Automation, Industrial Electronics and Instrumentation, pp. 452–457 (2012)
10. Davis, B., Figliozzi, M.: A Methodology to Evaluate the Competitiveness of Electric Delivery Trucks. Transportation Research Part E 49, 8–23 (2013)
11. Ferreira, J., Pereira, P., Filipe, P., Afonso, J.: Recommender System for Drivers of Electric Vehicles. In: Proc. International Conference on Electronic Computer Technology, pp. 244–248 (2011)
12. Morrow, K., Karner, D., Francfort, J.: Plug-in Hybrid Electric Vehicle Charging Infrastructure Review. Battelle Energy Alliance (2008)

An Adaptive Connection Scheduling Method Based on Yielding Relationship in FlashLinQ*

Dong-Hyun Kim, Bum-Gon Choi,
Sueng Jae Bae, and Min Young Chung**

College of Information and Communication Engineering
Sungkyunkwan University
300, Chunchun-dong, Jangan-gu, Suwon, Gyeonggi-do, 440-746, Republic of Korea
{dhk1231,gonace,noooi,mychung}@skku.edu

Abstract. As device-to-device (D2D) communications enables direct communication between user equipments, it can alleviate traffic overload on base stations. Qualcomm has introduced a new OFDM-based synchronous frame architecture for MAC/PHY, which is called FlashLinQ. In FlashLinQ, D2D user equipments perform signal-to-interference ratio based connection scheduling in order to distributively access wireless medium. The connection scheduling scheme enables D2D user equipments to simultaneously transmit data through same wireless medium. However, in this scheme since D2D user equipments simultaneously perform the medium access, they unnecessarily may yield data communication. In this paper, we propose a scheme that D2D user equipments adaptively perform connection scheduling by ignoring the interference from D2D links which are expected to yield data transmission. We verify that the proposed scheme can improve the system performance through simulations.

Keywords: D2D communications, FlashLinQ, medium access, OFDM system.

1 Introduction

Recently, as smart mobile devices have been supplied and various multimedia applications have been widely used, data traffic which has to be disposed in mobile network has been sharply increased [1,2]. The phenomenon makes data traffic concentrated on base stations, and causes radio channels of base stations to be overloaded. In order to solve this problems, device-to-device (D2D) communications has been considered as one of techniques which can be adopted in the next generation cellular communication systems such as 3GPP LTE-Advanced.

---

* This research was supported by the MSIP(Ministry of Science, ICT&Future Planning), Korea, under the ITRC (Information Technology Research Center) support program supervised by the NIPA (National IT Industry Promotion Agency) (NIPA-2013-(H0301-13-1005)).
** Min Young Chung is corresponding author.

B. Murgante et al. (Eds.): ICCSA 2013, Part I, LNCS 7971, pp. 120–130, 2013.

Since D2D communications enables direct data transmission between D2D user equipments (DUEs) without relay of base stations, it can reduce data traffic concentrated on base stations [3,4,5].

Qualcomm has introduced FlashLinQ (FLQ) for distributed D2D communications [6,7]. In FLQ, DUEs distributively access wireless medium based on single-tone signaling. Because single-tone signals are transmitted through dedicated resources for each D2D link, DUEs can exchange the signals without interference. In order to access wireless medium, paired DUEs estimate the interference from/to neighboring links. As listening the single-tone signals from neighboring DUEs, DUEs can calculate signal-to-interference ratio (SIR) of them and neighboring links, and can determine whether to yield the medium access depending on the calculated SIR. This method is called connection scheduling in FLQ. The connection scheduling enables one or more links to transmit data through whole frequency band at the same time. However, in this method, because DUEs do not know the scheduling results of neighboring links, they may excessively yield medium access even though some links are actually impossible to transmit data at a traffic slot. The excessive give-up makes the system performance degraded.

This paper proposes an adaptive connection scheduling scheme by ignoring the interference from the links which are expected to yield the medium access. In order to know which links will yield the medium access, we introduce an additional time slot with the purpose of information collection about the scheduling results of neighboring links at each traffic slot. Based on the information, DUEs can calculate SIR with only the links which are possible to transmit data, and then determine whether to perform data transmission.

The rest of this paper is organized as follows. At Section 2, in order to explain our proposed scheme, we describe the background knowledge about FLQ. Section 3 and 4 explain the proposed scheme, and verify the improved performance by the proposed scheme through simulation results, respectively. Finally, Section 5 gives a conclusion.

2 Backgrounds

In order to distributively perform direct communication, FLQ proposed a frame structure composed of synchronization, discovery, paging, and traffic periods as shown in Fig. 1. In the synchronization period, DUEs synchronize with other DUEs. In discovery period, DUEs broadcast single-tone signals including their own information and identify neighboring DUEs by exchanging the signals with each other. At the paging period, DUE establishes a link with its corresponding DUE. A communication link established between DUEs is allocated locally unique connection identifier (CID). Then, DUEs perform connection scheduling in order to access wireless medium, and DUEs succeeding in the medium access perform data transmission in the traffic period. The traffic period consists of connection scheduling, rate scheduling, data transmission, and acknowledgement periods as presented in Fig. 1. The connection scheduling period is divided into Tx- and Rx-orthogonal frequency division multiplexing (OFDM) blocks where

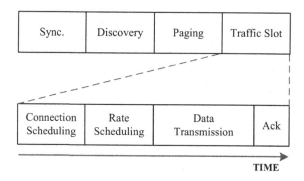

Fig. 1. The frame structures of FLQ

Tx- and Rx-blocks are used for single-tone signaling of transmitting DUE (Tx-DUE) and receiving DUE (Rx-DUE), respectively. In order to access wireless medium, DUEs exchange signals by using Tx- and Rx-blocks. With consideration of link priority and link quality, it is determined whether DUE accesses medium or not. DUEs try accessing wireless medium by performing the signal exchange. After the connection scheduling, Tx-DUEs of link which is successful in the medium access transmit pilot signals to their paired Rx-DUEs, and then the Rx-DUE responds to the pilot signals by sending signals including channel quality indicator (CQI) in the rate scheduling. Finally, at the data transmission and acknowledgement periods, the Tx-DUEs transmit data traffic to their Rx-DUEs and the Rx-DUEs acknowledge the reception results of the data.

2.1   Connection Scheduling in FLQ

The goal of SIR based connection scheduling is to find links which are able to simultaneously transmit data through whole frequency band while maintaining sufficiently large SIR. For the purpose, paired DUEs exchange single-tone signals through Tx- and Rx-blocks and they estimate SIR with considering other links which have higher priority than their own. If the calculated SIR is lower than a predetermined threshold, the DUEs yield their data transmission to the higher-priority links. However, if the SIR is higher than the threshold, the DUEs perform data transmission. In FLQ, the procedure is categorized as Rx- and Tx-yielding determination. While Rx-DUEs estimate SIR by considering the interference from higher-priority links at Rx-yielding determination, at Tx-yielding determination, Tx-DUEs estimate SIR by considering the interference from its link to higher-priority links.

In order to explain the procedure of the connection scheduling in FLQ, we assume that only two links participate in the connection scheduling as shown in Fig. 2, DUE A and C have data for DUE B and D, respectively and the upper link (link 1) has higher priority than the lower link (link 2) does. Also, h_{xy} means the channel gain between DUE X and Y.

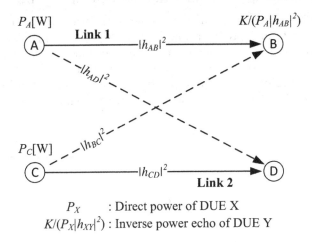

$$P_A[\text{W}] \qquad\qquad\qquad\qquad K/(P_A|h_{AB}|^2)$$

$$P_X \qquad : \text{Direct power of DUE X}$$
$$K/(P_X|h_{XY}|^2) : \text{Inverse power echo of DUE Y}$$

Fig. 2. The example of two geographically adjacent D2D links

Rx-Yielding Determination. At Tx-block, Tx-DUE A and C transmit the direct power (DP) signals with transmission power of P_A [W] and P_C [W], respectively. Then, Rx-DUE D receives the DP signals from Tx-DUE A and C, and based on the received signal power, calculates the SIR_D of link 2. If this SIR_D dissatisfies following Eq. 1,

$$\frac{P_C \times |h_{CD}|^2}{P_A \times |h_{AD}|^2} > \gamma_{Rx} \tag{1}$$

Rx-DUE D yields medium access due to the strong interference from link 1 at corresponding traffic slot (Rx-yielding). On the other hand, if the SIR_D is higher than the threshold (γ_{Rx}), Rx-DUE D transmits a inverse power echo (IPE) signal to Tx-DUE C.

Tx-Yielding Determination. Here, since link 1 has the first priority, Rx-DUE B does not perform SIR calculation, and transmits a IPE signal through Rx-block with transmission power of $K/(P_A \cdot |h_{AB}|^2)$ [W], where K means a system constant number. Then, Tx-DUE C receives the signal with transmission power of $(K \cdot |h_{BC}|^2)/(P_A \cdot |h_{AB}|^2)$, and estimates the SIR_B in order to consider interference effect from link 2 to link 1.

$$\left(\frac{K \times |h_{BC}|^2}{P_A \times |h_{AB}|^2} \times \frac{P_C}{K}\right)^{-1} = \frac{P_A \times |h_{AB}|^2}{P_C \times |h_{BC}|^2} > \gamma_{Tx} \tag{2}$$

If the calculated SIR_B dissatisfies Eq. 2, the DUE C yields data transmission to link 1 (Tx-yielding). If not, DUE C performs data transmission after following rate scheduling.

2.2   The Disadvantage of Connection Scheduling in FLQ

In the connection scheduling, some links may perform Rx- or Tx-yielding if their SIR is lower than a predetermined threshold γ_{Rx} or γ_{Tx}. This SIR based scheduling causes unnecessary yielding problem which means that any link yields data transmission because it considers interference from/to other higher-priority links which are actually impossible to access wireless medium. In order to explain the problem, we consider a environment as shown in Fig. 3.

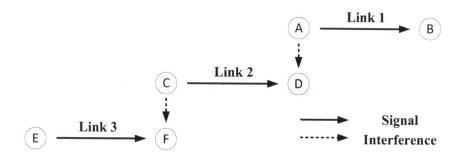

Fig. 3. The example of cascade yielding problem among three D2D links

We assume that Tx-DUE A, C, and E have data for Rx-DUE B, D, and F, link 1, 2, and 3 have priority numbers the same as their link numbers (e.g., link 1 has the highest priority.), and link 2 and 3 experience strong interference from link 1 and 2, respectively. In this case, if three links simultaneously attempt to access wireless medium, both link 2 and 3 perform Rx-yielding due to interference from link 1 and 2, respectively even though link 3 is possible to simultaneously transmit data with link 1. In other words, by considering interference from link 2 which is expected to yield the medium access to link 1, link 3 unnecessarily determines Rx-yielding. In this paper, we define this problem as cascade yielding problem.

M. Leconte et al. proposed a repeated connection scheduling scheme, where DUEs perform Rx- and Tx-yielding determination one and more times at a traffic slot [8]. In the connection scheduling in the basic FLQ, since DUEs attempt to perform connection scheduling only once per a traffic slot, there is high possibility that a D2D link yields to data transmission due to the cascade yielding problem. In this scheme, as DUEs repeatedly perform connection scheduling based on repeated Tx- and Rx-block structure at the same traffic slot. For example, if link X yields to data transmission as a result of Rx- or Tx-yielding, the link can re-try the connection scheduling through the second Tx- and Rx-blocks. Therefore, this scheme enables to mitigate the cascade yielding problem and thus each link has more opportunity for data transmission. However, in the scheme, since Tx- and Rx-blocks should be repeatedly deployed several times in order to completely solve cascade yielding problem, transmission time is decreased proportionally to the growing number of repeated Tx- and Rx-blocks.

3   Proposed Connection Scheduling Method

In this section, we propose a connection scheduling method that DUEs adaptively perform SIR based connection scheduling by considering yielding relationship where the relationship means that any link causes other links to yield data transmission. In order to know the yielding relationship, DUEs collect the information about scheduling results at each traffic slot. Then, based on the information, DUEs calculate SIR by considering the interference from/to the links which have higher-priority and are expected to access wireless medium. The proposed scheme consists of information collection and adaptive connection scheduling phases.

3.1   The Method of Gathering Yielding Relationship

In our scheme, DUEs collect information about scheduling results of neighboring D2D links. However, in the existing structure of traffic slot, DUEs are unable to know the results, since scheduled DUEs immediately perform rate scheduling with full band signaling after connection scheduling. For information collection of DUEs, we introduce a scheduling results broadcasting (SRB) OFDM block which has the same structure as the existing Tx- and Rx-block. The SRB block is located between connection scheduling and rate scheduling periods as shown in Fig. 4. Each Tx-DUE which finally succeeds in medium access broadcasts the single-tone signal through the SRB block. Then, as the remainder which does not broadcast signals senses the SRB block, it can collects information about which links are successful in medium access at every traffic slot.

Based on the scheduling results collected from SRB block, each DUE manages the information as a table as shown in Fig. 5(a), where '+1' and '-1' respectively mean success and failure in medium access of link j affected by link i, and '0'

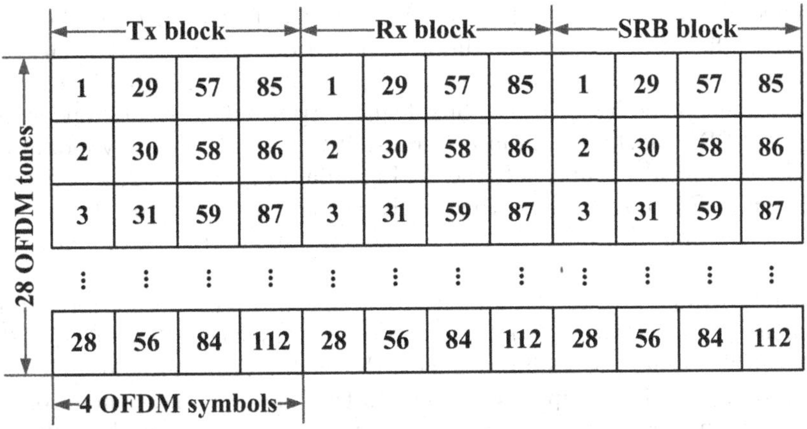

	Tx block				Rx block				SRB block		
1	29	57	85	1	29	57	85	1	29	57	85
2	30	58	86	2	30	58	86	2	30	58	86
3	31	59	87	3	31	59	87	3	31	59	87
⋮	⋮	⋮	⋮	⋮	⋮	⋮	⋮	⋮	⋮	⋮	⋮
28	56	84	112	28	56	84	112	28	56	84	112

28 OFDM tones (vertical, left axis) — 4 OFDM symbols

Fig. 4. The connection scheduling structure in the proposed scheme

means that the yielding relationship between link i and j is not identified. The initial value of each cell is set at '0' and the values are updated every traffic slot. In order to explain the way of updating the table, we consider the scheduling results as shown in Fig. 5(b) where only three neighboring links perform connection scheduling and Tx-DUEs of the links which are successful in the medium access broadcast single-tone signals through SRB block. Whether any Tx-DUE transmits a single-tone signal is described as ○ and × symbols, respectively. In

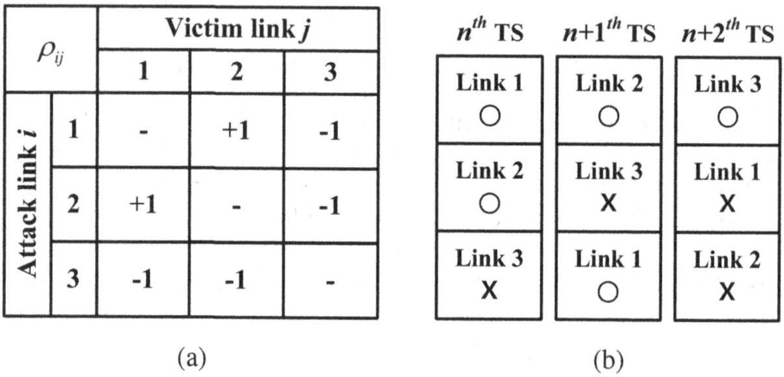

ρ_{ij}		Victim link j		
		1	2	3
Attack link i	1	-	+1	-1
	2	+1	-	-1
	3	-1	-1	-

n^{th} TS	$n+1^{th}$ TS	$n+2^{th}$ TS
Link 1	Link 2	Link 3
○	○	○
Link 2	Link 3	Link 1
○	×	×
Link 3	Link 1	Link 2
×	○	×

(a)   (b)

Fig. 5. The example for the method of collecting the yielding relationship information based on SRB block

Fig. 5(b), each column presents SRB blocks at traffic slot from n^{th} to $n + 2^{th}$ and a link which occupies the highest block has the first priority at the corresponding traffic slot. Also, for the simple explanation, we assume that all DUEs acquire the common information in this example. Through the SRB block of the nth traffic slot, while Tx-DUEs of link 1 and 2 transmitted single-tone signals through the SRB block, link 3 did not transmit the signal. Then, by monitoring the SRB block of nth traffic slot, the DUEs update (1, 2) and (2, 1) cells with '+1' shown in Fig 5(a), because this result means that link 1 and 2 can simultaneously transmit data. On the other hand, the values of the rest cell remain as '0'.

At $n + 1^{th}$ traffic slot, the DUEs determine that link 3 cannot perform simultaneous medium access with link 2, and then update the cell (2, 3) and (3, 2) as '-1'. At $n + 2^{th}$ traffic slot, as link 1 fails in the medium access due to link 3, the cells (1, 3) and (3, 1) are updated as '-1'. In this way, each DUE monitors the SRB blocks, and manages the scheduling results as a table at each traffic slot. This information is used for adaptive connection scheduling while continuously updated each traffic slot.

3.2   The Adaptive Connection Scheduling

In the adaptive connection scheduling, DUEs perform SIR based connection scheduling, considering both the collected information and priority of D2D links. Each link calculate SIR by excluding the interference from/to the links which have higher-priority and are expected to yield medium access. Fig. 6 shows an example in order to explain the procedure of the adaptive connection scheduling. In this example, we assume that there are four D2D links and their priority is the same as their link numbers. They simultaneously participate in connection scheduling and each DUE has the same yielding relationship table as shown in Fig. 6(a).

In Fig. 6(b), each column presents Rx- or Tx-yielding determination process of links where ○ and × symbols mean success and failure in medium access, respectively, and △ symbol presents that DUEs do not know whether the corresponding link accesses medium or not. At Rx-yielding, since link 1 has the highest priority, it succeeds in the medium access. Link 2 decides to Rx-yielding due to the interference from link 1. Link 3 expects that link 2 fails in the medium access based on the table, and calculates SIR by excluding the interference from link 2. Based on the table, link 4 performs connection scheduling by considering the interference only from link 1 and 3, because it can knows that link 2 should yield to data transmission. At Tx-yielding determination, likewise as above, link 1 accesses medium, and link 2 yields to medium access due to the interference from itself to link 1. Link 3 performs connection scheduling only considering the interference to link 1. Finally, link 4 considers the interference to link 1 and link 3 at the SIR calculation. If the scheduling scheme of the basic FLQ is applied to this environment, since link 3 and 4 should consider the interference from/to link 2 which is expected to yield the medium access, the probability that they

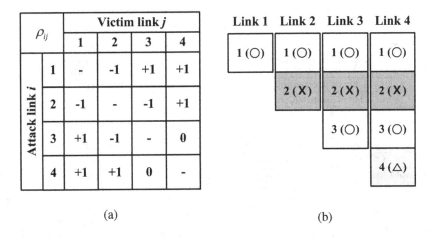

(a)   (b)

Fig. 6. The example for the adaptive connection scheduling of D2D links

perform Rx- or Tx-yielding may become higher. However, in the proposed scheme, since DUEs do not calculate SIR with the links which are expected to yield medium access, their probability of medium access can be increased.

4 Performance Evaluation

In order to evaluate the performance of the proposed scheme, we performed simulations based on C programming. we consider nine square sectors are considered, where center is target and the others are adjacent. D2D pairs are uniformly distributed in each square plane and distance between Tx- and Rx-DUE is randomly determined within 500 meters. We simulate three types of schemes which are a basic FLQ, the existing scheme based on repeated Tx- and Rx-block, and the proposed scheme in order to compare the system performance. In the simulations for the basic FLQ and the existing scheme, the connection scheduling is performed respectively once (N=1) and three times (N=3) per a traffic slot. The numeric values of the simulation parameters are shown in Table. 1.

Fig. 7 shows total throughput according to the number of D2D links. The total throughput of the scheme that Tx- and Rx-block is repeated three times in a traffic slot lower than that of the basic FLQ because the additional Tx- and Rx-blocks decreases the proportion of data transmission period. On the other hand, the proposed scheme has the highest throughput compared with those of other schemes. Since at SIR calculation the proposed scheme exclude the interference from the links expected to yield data transmission, more links can succeed in simultaneous medium access and the total throughput is improved.

Fig. 8 presents the average number of concurrent transmission links in a traffic slot. The Tx- and Rx-block repeated scheme almost does not improve the average number of concurrent transmission links of the basic FLQ. Since the scheme simply repeated the same signaling structure, the average number of concurrent

Table 1. Simulation Parameters

Parameters	Value
A network size	1 km × 1 km
Carrier frequency	2.4 GHz
Total bandwidth	5 MHz
Length of a traffic slot	2.08 msec
Size of Tx- and Rx-block	28 OFDM tone × 4 OFDM symbol
Bandwidth of a OFDM tone	78 kHz
OFDM symbol duration	12.8 μsec
Threshold for Tx- and Rx-yielding (γ_{Tx}, γ_{Rx})	9 dB
Maximum number of CID	112
Transmission power of DUEs	20 dBm
Path loss model	ITU-1411 LOS model [9]
Traffic model	Full buffer best effort traffic [10]
Spectral efficiency	Spectral efficiency lookup table [11]

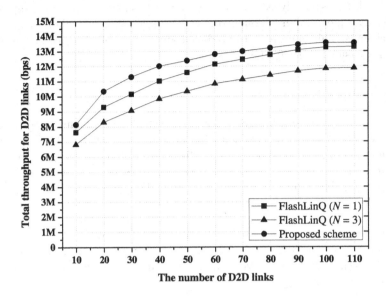

Fig. 7. Total throughput with variation of the number of D2D link

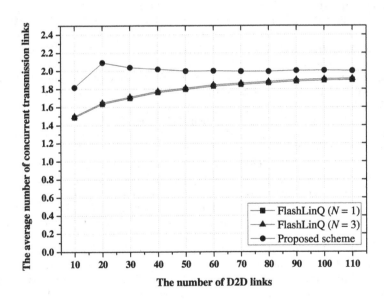

Fig. 8. The average number of concurrent transmission link with variation of the number of D2D link

transmission links is nearly equal to that of the basic FLQ. At the proposed scheme, the average number of concurrent transmission links is higher than those of other schemes. This is because the reduced amount of interference makes SIR of D2D links easier to satisfy the threshold.

5   Conclusion

This paper proposed the adaptive connection scheduling scheme in order to mitigate the cascade yielding problem. In the proposed scheme, DUEs collect the scheduling results of neighboring D2D links at each traffic slot, and at SIR calculation, exclude the interference from the links expected to yield data transmission. Thus, as the amoung of interference which D2D links consider decreases, SINR of D2D links becomes easy to satisfy the threshold. Through simulation results we confirm that the proposed scheme increases the number of concurrent transmission links, and so the system performance is improved. However, in the proposed scheme, the introduction of the SRB block generates additional redundancy. As a further work, optimized frame structure will be studied in order to minimize the redundancy.

References

1. UMTS Forum Report 44: Mobile Traffic Forecasts 2010-2020 Report (2011)
2. OVUM: Mobile Broadband Users and Revenues Forecast Pack to 2014 (2009)
3. Fodor, G., Dahlman, E., Mildh, G., Parkvall, S., Reider, N., Miklos, G., Zoltan, Z.: Design Aspects of Network Assisted Device-to-Device Communications. IEEE Communications Magazine 50, 170–177 (2012)
4. Lei, L., Zhong, Z., Lin, C., Shen, X.: Operator Controlled Device-to-Device Communications in LTE-Advanced Networks. IEEE Wireless Communications Magazine 19, 96–104 (2012)
5. Doppler, K., Rinne, M., Wijting, C., Ribeiro, C., Hugl, K.: Device-to-Device Communication as an Underlay to LTE-Advanced Networks. IEEE Communications Magazine 47, 42–49 (2009)
6. Xinzhou, W., Tavildar, S., Shakkottai, S., Richardson, T., Junyi, L., Laroia, R., Jovicic, A.: FlashLinQ: A Synchronous Distributed Scheduler for Peer-to-Peer Ad Hoc Networks. In: IEEE Allerton Conference, pp. 514–521 (2010)
7. Corson, M.S., Laroia, R., Junyi, L., Park, V., Richardson, T., Tsirtsis, G.: Toward Proximity-Aware Internetworking. IEEE Wireless Communications 17, 26–33 (2010)
8. Junyi, L., Rajiv, L., Thomas, R., Xinzhou, W., Saurabh, T.: Methods and Apparatus Related to Scheduling Traffic in a Wireless Communications System Using Shared Air Link Traffic Resources. US Patent App. Pub., US 7,983,165 B2 (2011)
9. Cichon, D.J. and Kerner, T.: Propagation Prediction Models. COST 231. Final Rep. (1995)
10. Report ITU-R M.2135-1, Guidelines for Evaluation of Radio Interface Technologies for IMT-Advanced (2009)
11. 3GPP, Evolved Universal Terrestrial Radio Access (E-UTRA); Radio Frequency (RF) System Scenarios (Release 10). TR 36.942 V10.2.0 (2011)

A Probabilistic Medium Access Scheme for D2D Terminals to Improve Data Transmission Performance of FlashLinQ*

Hee-Woong Yoon, Jungha Lee, Sueng Jae Bae, and Min Young Chung**

College of Information and Communication Engineering
Sungkyunkwan University
300, Chunchun-dong, Jangan-gu, Suwon, Gyeonggi-do, 440-746, Republic of Korea
{hiwoong,jhlee88,noooi,mychung}@skku.edu

Abstract. In existing wireless communication systems such as cellular network, base station can become bottleneck since terminals have to exchange data traffic each other only through a base station. In order to solve this problem, device-to-device (D2D) communication has been considered. Recently, Qualcomm Inc. has introduced FlashLinQ (FLQ) for D2D communication, which has a radio frame based on orthogonal frequency division multiplex (OFDM). In FLQ, terminals distributively access to medium based on their D2D link qualities and interference from other terminals. However, terminal with low link quality can cause that other terminals excessively give up their medium access. In order to solve this problem, we propose a probabilistic medium access scheme for D2D terminals in FLQ. In the proposed scheme, terminal stochastically tries accessing to medium based on its link quality, and the excessive yieldings of other terminals are reduced. Through simulation, we evaluate performance of the proposed scheme by comparing that of FLQ. We show that the proposed scheme can improve performance of FLQ by simulation results.

Keywords: FlashLinQ, Medium Access Scheme, Device to Device, Connection Scheduling, Yielding.

1 Introduction

As use of smart devices has been expanded, wireless traffic which has to be processed in wireless networks has rapidly increased [1]. In infra-based wireless network such as cellular network, terminals can exchange data traffic with each other only through a base station (BS) [2]. Increase of data traffic disposed by BS causes heavy loads on BS. As one of technologies which can offloads burden of

---

* This research was supported by the MSIP(Ministry of Science, ICT&Future Planning), Korea, under the ITRC (Information Technology Research Center) support program supervised by the NIPA (National IT Industry Promotion Agency) (NIPA-2013-(H0301-13-1005)).
** Corresponding author.

B. Murgante et al. (Eds.): ICCSA 2013, Part I, LNCS 7971, pp. 131–141, 2013.

BS, device-to-device (D2D) communication which enables terminals to directly communicate with each other has been considered [3]. Since terminals exchange data with their corresponding terminals through direct communication link in D2D communication, BS does not have to relay data for them as shown in Fig. 1. Additionally, D2D terminals can communicate with each other when their infrastructure is not available [4].

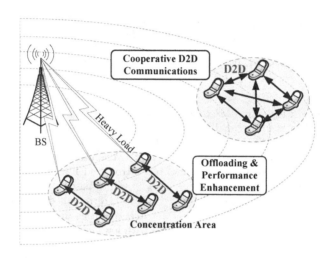

Fig. 1. Scenario of D2D communication

For supporting D2D communication, Qualcomm has introduced a medium access technology, called FlashLinQ (FLQ) [5][6]. In FLQ, D2D links which are direct links between D2D terminals and their corresponding D2D terminals have their own connection identification (CID). CID is locally unique, and priority of link is determined based on CID of the link. Each D2D terminal distributively tries accessing to medium by exchanging single-tone signals based on orthogonal frequency division multiplex (OFDM) structure.

In connection scheduling period for D2D communication, D2D terminal considers priority of link and link quality between its corresponding D2D terminal and itself. D2D terminals determine whether they conduct medium access or not by considering link qualities of themselves and interference from/to D2D links which have higher priorities. For example, D2D link with low quality does not access to medium with high probability, because it expects that its D2D communication is not successful through D2D link due to its low link quality. If a D2D link which has low priority is expected to severely interfere with D2D link which has higher priority, it gives up medium access and yields the medium to D2D link with higher priority.

D2D terminals of D2D link with low link quality cannot successfully communicate with their pair even though they obtain opportunities of medium access.

In addition, if D2D terminal with low D2D link quality participates in connection scheduling, it may cause that D2D terminals with lower link priority but relatively higher link quality yields medium access. In spite of having good link quality, D2D terminal gives up transmission due to interference from D2D link with higher priority but low link quality. This may cause degradation of system performance.

In order to reduce the excessive yielding of D2D terminals with good link quality, we propose a probabilistic medium access scheme for D2D terminals in this paper. The rest of this paper is organized as follows. In Section 2, we briefly introduce frame structure of FLQ and connection scheduling procedure of terminals for their medium access. The proposed scheme is introduced in Section 3. In Section 4, we analyze performance of the proposed scheme. Finally, Section 5 gives a conclusion of this paper.

2 Preliminary

2.1 Frame Structure of FlashLinQ

For D2D communication, FLQ defines a periodic super-frame composed of synchronization period, discovery period, paging period, and traffic period. In synchronization period, D2D terminals basically synchronize time and frequency with neighbor terminals through geographic information system such as global positioning system (GPS). In discovery period, D2D terminals broadcast beacon signals including their information and detect the existence of neighbor D2D terminals through beacon signals transmitted by the neighbor ones. In paging period, D2D terminals inquiry their neighborhood to find available CIDs. If D2D terminals find available CIDs in their neighborhood, they respectively form their own list of available CIDs. After forming the list of available CIDs, D2D terminals exchange their own list of the CIDs with their pair D2D terminals by signaling based on OFDM. By exchanging the list of available CIDs and selecting a CID in the list, they can obtain a locally unique CID in their neighborhood. D2D terminal and its pair D2D terminal can establish D2D link with the selected CID. In traffic period, the established D2D links perform connection scheduling and transmit or receive their data traffic with consideration of their priorities and their D2D link qualities.

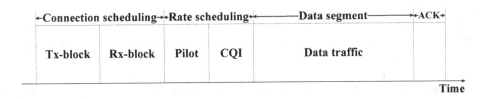

Fig. 2. Structure of a traffic slot

In traffic period, there are traffic slots for data transmission. A traffic slot is composed of connection scheduling, rate scheduling, data segment, and ACK as in illustrated Fig. 2. In connection scheduling, D2D terminal decides whether it accesses to medium or not based on its priority and signal to interference ratio (SIR). At rate scheduling, the scheduled D2D terminal and its pair D2D terminal in the connection scheduling exchange a wide-band pilot signal and channel quality indicator (CQI) to determine their own code rate and modulation scheme for data segment based on the quality of their link. At data segment, D2D terminal transmits data traffic to its pair D2D terminal with the determined rate. As a response to successful reception of data traffic the receiving D2D terminal transmits ACK to its pair D2D terminal.

2.2   Connection Scheduling Procedure

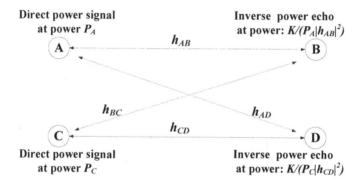

Fig. 3. Scenario of D2D signals and interference

FLQ defines a procedure for medium access of D2D terminals, called connection scheduling. In connection scheduling, D2D terminals exchange single-tone signals with their pair for medium access. By exchanging the single-tone signals based on OFDM structure, D2D terminals can obtain information which contains quality of their D2D links and interference from neighboring D2D links. By using the information, each D2D terminal decides whether it communicate with its pair or not. Fig. 3 shows an example of the connection scheduling between two D2D links. Transmitter A and receiver B have been established a D2D link with first priority, and transmitter C and receiver D have been established a D2D link with second priority. The connection scheduling is composed of two steps; Rx-yielding decision and Tx-yielding decision. For Rx-yielding decision, A and C transmit direct power signal (DPS) to their pair terminals with power P_A and P_C respectively. As the channel gain between A and D and the channel gain between C and D are $|h_{AD}|^2$ and $|h_{AC}|^2$, respectively, D receives two DPSs from A and C with power $P_A|h_{AD}|^2$ and $P_C|h_{CD}|^2$. D regards the DPS from

A as interference signal, and the DPS from C as oriented signal. D measures SIR of the received signals based on the strength of two DPSs. D compares the measured SIR with a predefined Rx-yielding threshold γ_{Rx} as follows,

$$\frac{P_C \times |h_{CD}|^2}{P_A \times |h_{AD}|^2} > \gamma_{Rx} \tag{1}$$

γ_{Rx} is a minimum SIR required to successful data traffic reception of a receiver from its pair transmitter, when a transmitter with higher priority than that of the receiver transmits data traffic to its pair receiver. If the measured SIR is higher than γ_{Rx}, D make decision that it can successfully receive data traffic from C even though the interference from A is generated. On the other hand, if the measured SIR is lower than γ_{Rx}, D decide that it cannot receive data traffic from C due to the interference from data traffic transmission from A to B. Then D immediately cease the connection scheduling procedure and does not participate in following Tx-yielding decision procedure (Rx-yielding decision).

If B and D decide to be able respectively to receive data traffic from their pairs, B and D transmit inverse power echo (IPE) signals to their pair as response to the received DPS at power $K/(P_A|h_{AB}|^2)$ and $K/(P_C|h_{CD}|^2)$ respectively. K is a constant and can be changeable to adapt to systems. C receives the IPE signal from B with power $K|h_{AC}|^2/(P_A|h_{AB}|^2)$. As multiplying the strength of received IPE by P_C/K and inverting it, C can calculates SIR of B, when it transmits data traffic to D as follows,

$$\frac{P_A \times |h_{AB}|^2}{P_C \times |h_{BC}|^2} > \gamma_{Tx} \tag{2}$$

γ_{Tx} is a minimum SIR required to successful data traffic transmission of a transmitter to its pair receiver. If the calculated SIR is more than a predefined Tx-yielding threshold γ_{Tx}, C decides that it can transmit data traffic to its pair. On the other hand, if the calculated SIR is less than γ_{Tx}, C decides that its transmission to D causes much interference to reception of B from A. Thus it decides not to participate in following rate-scheduling (Tx-yielding decision).

In connection scheduling of FLQ, D2D terminals distributively decide whether to access to medium based on their SIRs and priorities. However, FLQ has a problem that terminals excessively yield its medium access as shown in Fig. 4. We consider a case of that transmitter A, C and E have respectively established a D2D link 1, 2 and 3 with receiver B, D and F. Link index represents a priority of the link. Link 1 and 3 have high link quality while link 2 has low link quality. D and F have received stronger interference from A and C than oriented signal strength of its corresponding terminal C and E.

Link 1 can access to medium regardless of existence of link 2 and 3, since it has the highest priority. On the other hand, link 2 and 3 distributively decide whether to access to medium with consideration of interference from/to link(s) with higher priority. At the same time, t_0, A, C, and E respectively transmit DPSs to their corresponding terminals, B, D, and F. At the time, $t_0+\delta_1$, D with low link quality may give up medium access to link 1 due to interference

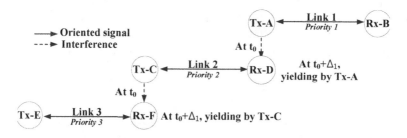

Fig. 4. Excessive yielding problem

from A. As a result, link 2 may yield the medium access to link 1. Simultaneously, F receives interference from A and C. F may also yield medium access for link 1 and 2, even though link 3 has a high link quality. Despite link 1 and 3 can simultaneously conduct medium access, link 3 yields the medium access because of interference from link 2. In addition, if link 2 does not yield the medium access, it may cause Tx-yielding of link 3. Link 2 cannot efficiently conduct D2D communication than that of link 3 in terms of data transmission rate, since link 2 has lower link quality than link 3.

3   Probabilistic Medium Access Scheme (PMAS) for FLQ

In order to solve a problem that D2D links with low link qualities cause excessive yieldings of neighbor D2D links with lower priorities in connection scheduling of FLQ, we propose a probabilistic medium access scheme for D2D terminals. In the proposed scheme, a D2D terminal stochastically tries accessing to medium based on its D2D link quality. The proposed scheme can reduce excessive Rx- or Tx-yielding of D2D terminals by coordinating the probability of medium access tried by D2D terminals. The method of determining the probability of medium access is as follows.

We consider a scenario as shown in Fig. 5. Tx and Rx refer to D2D transmitter and D2D receiver, respectively. Tx and Rx 1 compose a D2D pair, and 5 neighbor D2D Rxs are deployed around Tx 1. Tx 1 can measure link gain between Rxs and itself by using strength of the received signal since Rxs transmit single-tone signals to Txs in paging period. Based on the link gains, Tx 1 can estimate the amount of interference from itself to each Rx. With considering the amount of interference from Tx 1 to each neighbor Rxs, Tx 1 divides its neighbor Rxs into two groups: **S** and **W**. **S** is the group of Rxs expected to be severely interfered by the transmission of Tx 1. Rxs which are not expected to be relatively less interfered by Tx 1 are grouped in **W**. In the proposed scheme, we set the criterion of determining whether Rx is severely interfered or not as strength of the signal received at Tx 1 from its corresponding D2D pair, Rx 1. For example, Rxs 2 and 3 which receive stronger interference from Tx 1 than the strength of signal received by Rxs 4, 5, and 6.

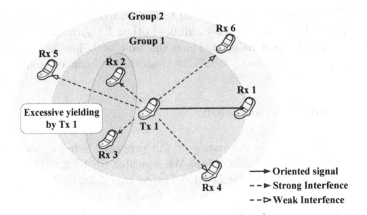

Fig. 5. Scenario of D2D signals and interference in the proposed scheme

By grouping the Rxs, Tx 1 can know how many Rxs receive strong interference from itself. For determining the probability of trying medium access, Tx 1 calculates α which is the ratio of severely interfered Rx to total Rxs as follows,

$$\alpha = \frac{N_{\mathbf{S}}}{N} \tag{3}$$

where $N_{\mathbf{S}}$ and N denote the number of Rx in $\mathbf{S}$, the total number of Rxs around Tx 1, respectively. If Tx and Rx 1 conduct D2D communication even though the link between Tx and Rx 1 has low link quality, they cannot sufficiently achieve high performance in terms of data transmission rate. In addition, a lot of neighbor link may excessively yield its medium access due to interference from/to the link.

In order to reduce the excessive yieldings, Tx 1 stochastically tries accessing to medium by considering its link quality. Based on α, Tx 1 determines probability of trying accessing to medium. We consider two scenarios according to priority of Tx 1. If Tx 1 has a highest priority, since it does not yield its medium access to other D2D link, it tries accessing to medium by transmitting DPS to its pair regardless of the α. This guarantees an opportunity for a link's medium access, even though it has low link quality. If priority of a Tx 1 is not highest, since it may yield its medium access to a D2D link with higher priority than its priority, it stochastically tries accessing to medium with a probability based on the α as follows,

$$P_{DPS} = 1 - \alpha \tag{4}$$

where P_{DPS} denotes the probability of trying medium access for Tx 1. This means that link quality of Tx 1 determines a probability that Tx 1 transmits DPS to its pair. If Tx 1 has low link quality, it transmits DPS to Rx 1 with low probability. Otherwise, it tries accessing to medium with high probability. Interference generated by Tx 1 with low link quality can be reduced, and more neighbor D2D receivers can try accessing to medium than those of conventional FLQ. In addition, the excessive Tx-yielding of other Tx with lower priority than that of Tx 1 can be also reduced, since the link with low link quality tries accessing to medium with less probability than those of conventional FLQ.

4 Performance Evaluation

In order to evaluate the performance of the proposed scheme, we performed simulations based on C programming. We consider up to 220 D2D terminals (110 D2D links), they are uniformly distributed in a 1 km × 1 km rectangular area. A role of each D2D terminal is predefined as a transmitter or a receiver before simulation. Detailed simulation parameters are given in 1.

Table 1. Simulation parameters

Parameters	values
Dimension	1 km × 1 km
Maximum number of terminals	220
Carrier frequency	2.4 GHz
Total bandwidth	5 MHz
Size of Tx/Rx block	28 OFDM tone × 4 OFDM symbol
Threshold for Tx-/Rx-yielding (γ_{Tx}, γ_{Rx})	9 dB
Transmission power	20 dBm
Path loss model	ITU-R P1411 Outdoor [7]
Traffic model	Full buffer best effort traffic [8]
Spectral efficiency	Spectral efficiency lookup table [9]
Noise density	-174 dBm/Hz

The average number of concurrent transmission as increasing the number of D2D links is shown in Fig. 6. The average number of concurrent transmission is defined as the average number of D2D links which simultaneously access to medium at the same time. Since more D2D links try accessing to medium as the number of D2D links increases, more D2D links simultaneously conduct D2D communication at a data traffic slot. Thus, the number of concurrent transmission also increases. The proposed scheme achieves a higher performance than FLQ in terms of the average number of concurrent transmission. A link with low link quality tries accessing to medium with less probabilities than link with high link quality since a D2D link of the proposed scheme stochastically tries accessing to medium based on its link quality. Overall interference of network can be reduced and D2D links can efficiently perform connection scheduling with low interference. As a result, the number of D2D links which excessively yield its medium access decreases and the average number of D2D links that can simultaneously access to medium at a data traffic slot increases.

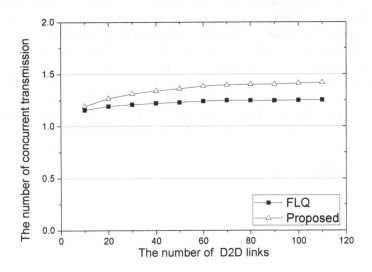

Fig. 6. The average number of concurrent transmission as increasing the number of D2D links

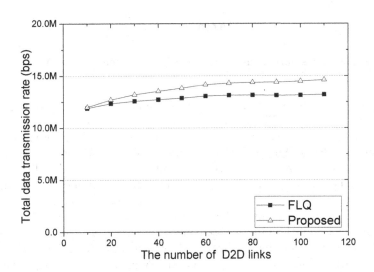

Fig. 7. Data transmission rate as increasing the number of D2D links

Total data transmission rate as increasing the number of D2D links is shown in Fig. 7. Total data transmission rate is defined as that sum rate of data transmission of each D2D link accessing to medium. As the number of D2D links increases, total data transmission rate also increases. In addition, total data transmission rate of the proposed scheme is higher than that of FLQ. The proposed scheme has higher performance in terms of the average number of concurrent transmission than conventional FLQ shown as Fig. 6, it can enable more terminals to access to medium. Terminals of the proposed scheme can exchange more data traffic than those of conventional FLQ. Thus, proposed scheme can improve performance of FLQ in terms of data traffic transmission rate.

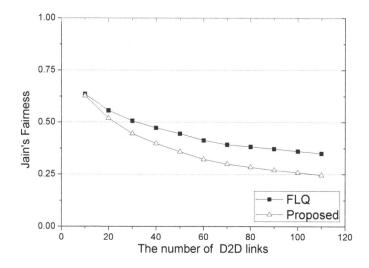

Fig. 8. Jain's fairness index as increasing the number of D2D links

In order to evaluate fairness among user throughput, we analyze Jain's fairness index of two schemes in Fig. 8. Jain's fairness is well-known index to evaluate fairness among users [10]. Fairness of the proposed scheme is lower than that of conventional FLQ. The proposed scheme does not fairly give opportunities of medium access for all D2D links. D2D link with high link quality may obtain many opportunities of medium access for D2D communication, while D2D links with low quality may not. As a result, the proposed scheme achieves lower fairness among user throughput than that of conventional FLQ.

A main purpose of the proposed scheme is to reduce the excessive yielding of terminals and to increase the number of links which perform D2D communication at the same time. The link with low link quality cannot achieve high data rate even though it obtains a opportunity of medium access. In addition, the link can cause the excessive Tx-yielding of other terminals with lower priority than that of the link. As a link with low link quality gives up its medium access, the more links may simultaneously access to medium for their D2D communication as shown in Fig. 6. The proposed scheme may guarantee medium access of more links, even though the fairness among user throughput of the proposed scheme is lower than that of conventional FLQ.

5   Conclusion

In this paper, we propose a probabilistic medium access scheme of D2D terminal for performance improvement in FLQ. In the proposed scheme, each D2D terminal stochastically tries accessing to medium for its D2D communication based on its link quality. The proposed scheme can reduce the overall interference and the excessive yieldings of terminals. Terminals of the proposed may efficiently perform connection scheduling with low interference. Simulation results show that the proposed scheme improves performance of FLQ in terms of the number of concurrent data transmission and total data transmission rate, while fairness of the proposed scheme deteriorates in comparison with FLQ. It means that terminals of the proposed scheme cannot fairly obtain opportunities for their D2D communication. A method to improve the fairness of the proposed scheme is required in further study.

References

1. Yeh, S., Talwar, S., Wu, G., Himayat, N., Johnsson, K.: Capacity and Coverage Enhancement in Heterogeneous Networks. IEEE Wireless Communications 18(3), 32–38 (2011)
2. Hakola, S., Chen, T., Lehtomaki, J., Koskela, T.: Device-to-Device (D2D) Communication in Cellular Network - Performance Analysis of Optimum and Practical Communication Mode Selection. In: IEEE Wireless Communication and Networking Conference (WCNC), pp. 1–6 (2010)
3. Doppler, K., Rinne, M., Wijting, C., Ribeiro, C., Hugl, K.: Device-to-Device Communication as an Underlay to LTE-Advanced Networks. IEEE Communications Magazine 47(12), 42–49 (2009)
4. Fodor, G., Dahlman, E., Mildh, G., Parkvall, S., Reider, N., Miklos, G.: Design Aspects of Network Assisted Device-to-Device Communications. IEEE Communications Magazine 50(30), 170–177 (2012)
5. Wu, X., Tavildar, S., Shakkottai, S., Richardson, T., Li, J., Laroia, R., Jovicic, A.: FlashLinQ: A Synchronous Distributed Scheduler for Peer-to-Peer Ad Hoc Networks. In: IEEE Allerton Conference, pp. 514–521 (2010)
6. Corson, M.S., Laroia, R., Junyi, L., Park, V., Richardson, T., Tsirtsis, G.: Toward Proximity-Aware Internetworking. IEEE Wireless Communications 17, 26–33 (2010)
7. Cichon, D.J., Kerner, T.: Propagation Prediction Models. COST 231. Final Rep. (1995)
8. Report ITU-R M.2135-1, Guidelines for Evaluation of Radio Interface Technologies for IMT-Advanced (2009)
9. 3GPP, Evolved Universal Terrestrial Radio Access (E-UTRA); Radio Frequency (RF) System Scenarios (Release 10). TR 36.942 V10.2.0 (2011)
10. Jain, R., Chiu, D.M., Hawe, W.: A Quantitative Measure of Fairness and Discrimination for Resource Allocation in Shared Computer Systems

Toward Smart Microgrid with Renewable Energy: An Overview of Network Design, Security, and Standards

Mihui Kim[*]

Department of Computer & Web Information Engineering, Computer System Institute,
Hankyong National University, Anseung, Korea
mhkim@hknu.ac.kr

Abstract. An increasing demand for efficient and safe electricity management has motivated the development of smart grid and accelerated the technical research for smart grid. On one side, a smart microgrid has been recently given an interest as a relatively small-scale, self-contained, medium/low voltage electric power system (EPS); it houses various distributed energy resources (DERs) with renewable energy and controllable loads in a physically close location. In this paper, we overview the research trend for the network design and security issues of smart microgrid, different from smart grid, and survey the effort of standardization for them. Through the survey, we derive the novel research issues to address for the successful realization of smart microgrid.

Keywords: Smart Microgrid, Renewable Energy, Distributed Energy Resources (DER), State of the Art of Research and Standard Trend, Network Design, Security.

1 Introduction

Smart grid is to expand the current capabilities and efficiency of the grid's generation, transmission, and distribution systems for autonomous power distribution, efficient electricity management and safety. Moreover, the next-generation electric power systems will not only address the existing problems in the current power systems, but also add in advanced new features as follows: support for diverse devices, superior power quality, operation efficiency and estimation, grid security, consumer participation, grid self-correction, and market boost [1].

Comparing with the smart grid, a microgrid is a relatively small-scale, self-contained, medium/low voltage electric power system (EPS) that houses various distributed energy resources (DERs) (i.e., solar panels or wind turbines) with renewable energy and controllable loads in a physically close location. The microgrid can benefit from less transmission losses and less cable costs because of in vicinity of generator and consumption. Moreover, it can decrease carbon emissions, and increase the resilience of the utility grid [2,3].

---

[*] Corresponding author.

B. Murgante et al. (Eds.): ICCSA 2013, Part I, LNCS 7971, pp. 142–156, 2013.

Smart microgrid is a relatively new concept and paid attention in research and industrial fields because of the benefits. In this paper, we present the state-of-the-art research and standardization trends of smart microgrid. The rising number of DERs in smart microgrid brings up new research issues. They should optimally connect with each other to share or route the energy. Network design issue thus becomes important. Sharing the energy provides a new economic model, and it emphasizes security in smart in addition to basic safety in grid. Moreover, the exposure of sensitive information, i.e., all kinds of personal attributes and activities, makes the privacy problem more significant. To optimally share the energy, real-time management is required, and on the other hand it could become a palatable target of adversary from a viewpoint of availability threat.

The new issues in smart microgrid affect standardization. At first, to deal with the peculiarities of different DESs with sophisticated sensing and actuating units, the IEEE Std. 1451 is suggested as a system design model of energy gateways or nodes with uniform interfaces [4]. These nodes have an energy interface not only to the power distribution grid but also to active sources or loads. Moreover, communication standards focus on the smart microgrid. Cognitive radio is a candidate technique for smart microgrid networks (SMGNs) [5], and also WiMAX and WiFi are considered [6]. In this paper, we introduce the standards for smart microgrid.

The remainder of the paper is organized as follows. Section 2 clarifies network architecture of smart grid and microgrid, and presents the research issues appearing because of the architectural features. Section 3 provides the research trend and Section 4 introduces the standardization contents. Section 5 concludes this paper with still-open research issues.

2   Network Architecture of Smart Grid and Smart Microgrid, and Research Issues

In this chapter, we compare the network architecture of smart grid and smart microgrid with figures and explain their relationship. Different network architectures give novel research issues. We also introduce research issues with recent attention, related with smart microgrid.

2.1   Network Architecture

Smart grid has a huge infrastructure and communication networks, and thus the network has hierarchical architecture: home area network (HAN), neighborhood area network (NAN), and wide area network (WAN), as shown in Figure 1. The HAN provides access to in-home appliances while the NAN connects smart meters to local access points, and WAN provides communication links between the grid and core utility systems. Figure 1 shows a basic illustration of the electrical power grid and the smart grid multitier communications network. As the characteristics of each tier network, different wireless communication techniques can be adapted, i.e., WiFi or Zig-Bee for HAN in indoor small area, WiMAX, cognitive radio (CR) WiFi or 4G for NAN with wireless mesh topology, and WiMAX, 4G or CR for WAN [7].

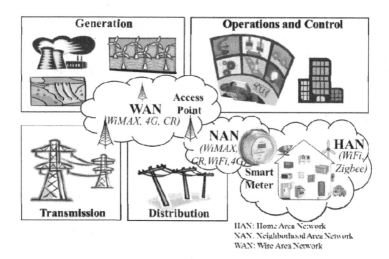

Fig. 1. Hierarchical Architecture of Smart Grid [7]

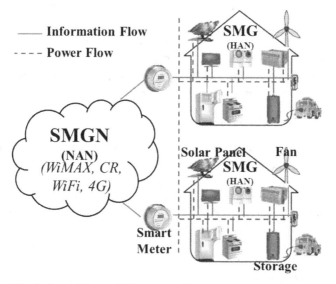

Fig. 2. Smart Microgrid Network with Renewable Energy in Smart Grid

Meanwhile, smart microgrid (SMG) is reminiscent of a small scale smart grid; it has abundant distributed generators and consumers for medium/low voltage electric power in close location, but requires the control such as advanced metering infrastructure (AMI) of smart grid in order to efficiently share the energy. Therefore, the generators and consumers is the same as HAN in smart grid, and smart microgrid network (SMGN) connecting them play a similar role with the NAN, as shown in Figure 2.

As one of main NAN/SMGN communication technologies, 3GPP and WiMAX (IEEE 802.16) have been doing research for the architecture of smart grid, i.e., an important application of machine to machine (M2M) communication [8]. User equipment (UE) with machine type communication (MTC) application in M2M may become the smart meters, and MTC users may be authorized someone requiring information in smart microgrid, e.g., other smart meters or utility company.

At first, 3GPP introduces architectural reference model to provide smart grid/microgrid services in [9]. Figure 2 shows the architectural reference model consisting of UE using MTC application, MTC server, and 3GPP network entities involved in MTC. The model covers three architecture models, direct model, indirect models, and hybrid model. In the direct model, the 3GPP operator provides direct communication where MTC application directly connects to the operator network without the use of any MTC server. In the indirect model, communication is controlled by MTC server provider or 3GPP operator. In the hybrid model, the data transmission in the user plan is used for the direct model while the signaling in the control plan is used for the indirect model.

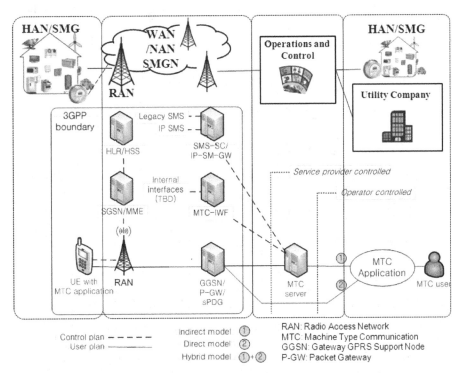

Fig. 3. Architectural reference model for smart microgrid in the 3GPP [9]

IEEE 802.16 introduces high level system architecture for M2M communications as shown in Figure 4 [10]. It can be also applied to SMGN. The system architecture consists of IEEE 802.16 M2M devices with M2M functionality, M2M server commuting to one or more IEEE 802.16 M2M devices (e.g., smart meters), and M2M service consumer (e.g., other meters or utility company). The M2M server may be located inside or outside the connectivity service network (CSN). The M2M application operates on IEEE 802.16 M2M devices and the M2M server. The system architecture provides the communication between one or more IEEE 802.16 M2M devices and an M2M server, as well as point-to-multipoint communication between IEEE 802.16 M2M devices and IEEE 802.16 base station. In the basic system architecture, IEEE 802.16 M2M devices can perform as aggregation points for non IEEE 802.16 M2M devices with different radio interfaces. In the advanced system architecture, IEEE 802.16 M2M devices can act like an aggregation points for other IEEE 802.16 M2M devices.

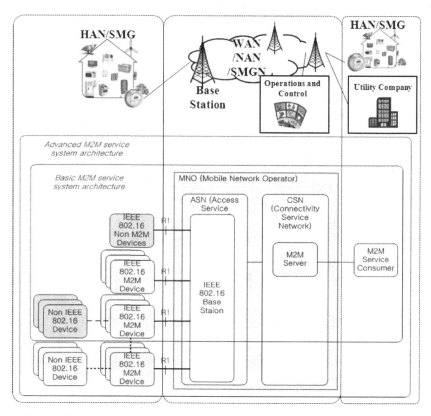

Fig. 4. M2M service system architecture for smart microgrid in IEEE 802.16 [10]

Meanwhile, smart microgrid has the important features in the high penetration level of distributed renewal power generators and it has to have the close relationship with consumers. To keep systematical connection, networking or topology issue are recently introduced. Power sharing and routing emerges as a novel feature and survivability, reliability and availability are highlighted. Table 1 summarizes the characteristic of smart microgrid in comparison with one of smart grid. We will introduce the research trend for these features in detail in next section.

Table 1. Comparison of smart microgrid and smart grid

Element	Smart Microgrid	Smart Grid
Generator	• small scale • gathered in vicinity (neighborhood) • limited energy sources, e.g., solar panels or wind turbines • medium/low voltage electric power system	• large scale • distributed in wide area • various energy sources, e.g., water (hydroelectric), wood, wind, organic waste, geothermal. • very high voltage electric power system
Consumer	• small scale • located together in generators • should compare with costs of between smart microgrid and smart grid	• large scale • separated from generators • should compare with costs in time slots
Network	• small scale • two tier networks (HAN/NAN)	• large scale • three tier networks (HAN/NAN/WAN)

2.2   Research Issues

To support such new features of SMG, recently researches focus on the follows: *network design*, *security*, *control algorithms*, and *system architecture*. At first, optimally sharing the energy or routing [11,12], and overlay topology [2] between DERs. The security, as a noticeable issue on SMG, emphasizes the privacy [13-16] and attack defense [17,18] because the data of SMG is mostly private data and inviting data to adversaries. The control (e.g., handling and fault isolation) of a high number of distributed generators (i.e., rooftop photovoltaic (PV) panels) is challenging. SMGN has scalability issue due to such many generators, should do real-time monitoring and quick response. Moreover, novel algorithms or new information-based control mechanisms are required in SMGN, such as scheduling based on charged energy [19], or energy routing [2,11]. System architecture for SMG is newly designing, e.g., connection of smart meters through cognitive radio [5] and dynamic energy-oriented scheduling [19]. Table 2 summarizes the solving problems in recent research and their solutions, and the detail contents describes in Section 3.

Table 2. Recent research issues for smart microgrid

	Solving problems in recent research	Solutions
Network design [2,12]	• network to provide reliable service, faster restoration, sus-	• overlay topology design maximizing the usage of re-

	tainable grid operation and platform for exporting or importing power energy routing • topology between DERs	newable energy [2] • cost-aware network design [12]
Security [5,11,13-18]	• attack defense • secure routing • privacy	• detection method using Kernel GLRT for malicious data attack in state estimation [5] • secure key management using a PKI, and a secure routing procedure [11] • privacy-preserving protocols in computation for power consumption or billing [13-14]
Control algorithms [11,19]	• scalability • real-time monitoring and quick response • energy routing • scheduling	• energy routing to increase the utilization of renewable resources and reduce the dependency of the SMG to the utility grid [11] • scheduling based on charged energy/cost [19-21]
System architecture [5,19]	• connection of smart meters • scheduling	• utilize cognitive radio in white spaces [5] • dynamic energy-oriented scheduling [19]

3   Research Trend of Smart Microgrid

3.1   Network Design Issues

To share renewable energy efficiently between DERs, energy routing (i.e., setting up energy efficient path) is brought up as a novel feature in SMGNs [8,20-22]. The authors in [20] propose a novel stochastic framework, leveraging distributed storage that alleviates many of the problems of the current grid, e.g., difficulty of the grid in routing the renewable sources due to their stochastic and often volatile nature. In [21], to maximally utilize the distributed energy resources and minimize the energy transmission overhead, the authors develop the distributed energy routing protocols for smart grid; it can be also applied to smart microgrid. The authors in [11] propose a secure energy routing mechanism, and the authors in [22] show that false data injection attacks against distributed energy routing can effectively disrupt the effectiveness of energy distribution process, posing significant supplied energy loss, energy transmission cost and the number of outage users, through simulation.

In [2,12], SMG is introduced as a technology to provide reliable service, faster restoration, sustainable grid operation, and a platform for exporting or importing power,
in order to increase the utilization of renewable resources, reduces the dependency of
SMG to the utility grid, and consequently reduces the load on the grid. To increase
reliability and survivability in SMGNs, the overlay topology design maximizing the
usage of renewable energy is presented in [2]. The topology is made with the minimized number of SMGs in each cluster through integer linear programming (ILP)
formulation. SMGs in a cluster share the energy with each other. As a next version,
the authors in [12] upgrade to cost-aware network design, enabling economic power
transaction with ILP to match the excess energy of each SMG to demands of other
SMGs. They give an example of different SMGN topologies formed by their proposed scheme (i.e., cost-aware smart microgrid) for two different time periods which
are 01:00-03:00 and 19:00-21:00. Each ring depicts a cluster. In Figure 5(a), since
demand is not high during overnight, several SMG clusters can form survivable rings.
Meanwhile, as shown in Figure 5(b), during mid-peak hours, the overlay topology
design scheme tends to form larger ring in order to fulfill the increasing power demand and during peak hours only one ring is formed since the only feasible solution is
clustering all SMGs in a ring in order to fulfill the load

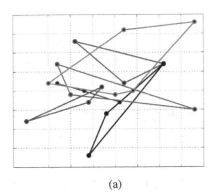
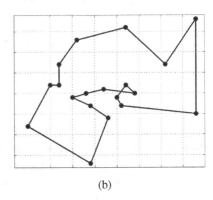

(a)   (b)

Fig. 5. An example of different SMGN topologies between a) 01:00-03:00, b)19:00-21:00 [9]

Cognitive radio in white spaces (e.g., unused local TV broadcast spectra) can support physical-layer security with detected low-latency communication links for smart
microgrid. Thus, the authors in [5] systematically design the system architecture,
control algorithms (e.g., price-based utility function with sophisticated control strategy), and security for microgrid. Meanwhile, in [23] the authors model the micro grids
using the graph theory and developed the optimization solution to determine location
where the new transmission, generation, and storage facility will be installed.

3.2   Security Issues

Security is as important as or more important than any other performance of interest
for the microgrid. Thus, several recent researches have considered the security aspect,

especially *secure routing* [8,22], *attack defense* [5,24], and *privacy preservation* [13-15].

As we introduced, energy routing between DERs is a novel research issue in SMGN. The energy routing can be vulnerable against false energy sharing information, failing to report security violation and so on, and thus the authors in [11] proposes a secure key management using a public key infrastructure (PKI), and a routing procedure based on transferring securely routing messages. The authors in [22] show impact of false data injection attacks against distributed energy routing .

The authors in [5] emphasis the importance of security in the both information and power flows in SMGN. Especially autonomous recovery against unpredicted faults or contrived attacks should be considered for secure power flows, they propose a detection method using Kernel GLRT for malicious data attack in state estimation of SMG. The authors in [24] introduce several intelligent attacks in smart grid communication, e.g., *vulnerability attack*, *data injection attack*, and *intentional attack*, and present Even though all of those attacks are based on smart grid communication, they also can be launched in SMGN. Table 3 explains the attacks and countermeasures.

Table 3. Practical attacks in smart microgrid and countermeasures

Attack	Description	Countermeasure
vulnerability attack	• attack method: by the malfunction of a device or communication channel, or the de-synchronization of feedback information • attack effect: an incorrect control process at the control center	• can be prevented by introducing the fault diagnosis scheme [25] to infer the fault detection and localization
data injection attack	• attack method: to alter the measurements of some meters in order to manipulate the operations of the smart grid • attack effect: adversary with full understanding of the network topology disrupts the network operations by paralyzing some fraction of nodes with the highest degree	• it is possible to defend against malicious data injection if a small subset of measurements can be made immune to the data injection attacks [26]
intentional attack	• attack method: with full understanding of the network topology, the adversary can fully utilize the network structure to disrupt the network operations by paralyzing some fraction of nodes with the highest degree • attack effect: network disruption due to node disconnections in the communication network.	• a fusion-based defense mechanism is proposed in [27] to defend intentional attack by utilizing the feedback information from each node for attack inference and defense reaction.

Among several security issues, the privacy preservation is especially important in SMGN, because meter readings imply a sketch of daily activities of households, as shown in Figure 6. Through the electrical usage values, the adversary can

•Detect how many people live at your house by watching the number of cycles of your hot water heater (not accounting for bad hygiene);
•Know when you're home by the energy cycle of the TV;
•Know when you're awake by the energy signature of the coffee pot or the toaster;
•Know whether you've got a hydroponics 'project' in the house.

Some privacy-preserving protocols in computation for power consumption or billing are proposed [13-14]. The scheme in [14] allows an electricity service provider obtain sums of meter readings over a time period and a monitoring center obtain sums of meter readings from meters in an area at some recent time unit while keeping individual meter reading private.

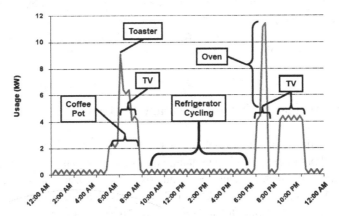

Fig. 6. An example of electrical usage in a home [28]

3.3   Control Issues

As we explain the characteristic of smart microgrid in Table 1, a high number of distributed generators or consumers are located in neighborhood. The control server should handle them and guarantee their fault isolation. Moreover, the control server should compare the real-time cost information between smart microgrid and smart grid for energy routing, and thus it is required doing real-time monitoring and quick response. Authors in [19] propose dynamic energy-oriented scheduling method for sustainable wireless sensor network, which treats energy as a first-class schedulable resource and dynamically schedules. This scheduling can be also applied for SMGN with the dynamic change of the SMG's available energy. The authors [29] observe that microgrid technology provides an opportunity and a desirable infrastructure for improving the efficiency of energy consumption in buildings. To improve building energy efficiency in operation is to coordinate and optimize the operation of various energy sources and loads, they tend to address the scheduling problem of building energy supplies. Meanwhile, a novel and important control is energy routing. Authors in [11] focus secure energy routing.

3.4   System Architecture Issues

Systems for SMG should be newly designed to support novel features as we explained in Section 2. Authors in [5] systematically investigate the novel idea of applying the next generation wireless technology, cognitive radio network, for the smart microgrid. Cognitive radio is originally proposed as a solution to rationalize the concept of recycling the spectrum in today's spectrum hungry scenario. Unlicensed users utilize the licensed frequency and when that particular band is not in use. They design the system architecture for SMG adapting cognitive radio.

Moreover, systems in SMGN, e.g., smart meter and EPS, should perform their operation according to new factors, e.g., available energy in their HAN, or costs for buying or selling energy in smart microgrid or smart grid. Energy is an important factor for operation. Thus system operations should be designed in considering the new factors. The dynamic energy-oriented scheduling [19] can be applied for system design in SMGN.

4   Standardization Trend of Smart Microgrid

The standardization is an important factor to realize the promising techniques, i.e., SMG. The authors in [9] provide a contemporary look at the current state of the art in smart grid, including technologies and standards. As smart grid communication technologies, GSM, GPRS, 3G, WiMAX, PLC, WiFi and ZigBee are compared in [6]. Cognitive radio is a candidate technique for SMGNs [5]. Table 4 overviews the characteristics with these standard technologies for HAN/NAN/WAN. All are related with SMGN because it has the same as architecture of NAN

Table 4. Standard technologies for HAN/NAN/WAN [6,30]

Standard	Spectrum	Data Rate	Coverage	Applications	Limitation
GSM	900-1800 MHz	Up to 14.4 Kbps	1-10km	HAN,NAN, WAN	Low data rate
GPRS	900-1800 MHz	Up to 170 Kbps	1-10km	HAN,NAN, WAN	Low data rate
3G	1.92-1.98 GHz 2.11-2.17 GHz (licensed)	384 Kbps – 2Mbps	1-10km	HAN,NAN, WAN	Costly spectrum fees

Table 4. *(Continued)*

Wi-MAX	2.5 GHz, 3.5 GHz, 5.8 GHz	Up to 75 Mbps	10-50 km (LOS) 1-5 km (NLOS)	NAN, WAN	Not wide-spread
PLC	1-30 MHz	2-3 Mbps	1-3 km	HAN, NAN	Harsh, nosy channel environment
ZigBee	2.4 GHz, 868-915 MHz	250 Kbps	30-50m	HAN	Low data rate, short range
WiFi	2.4 GHz, 5 GHz	Up to 600 Mbps	Up to 50m	HAN, NAN	Vulnerable in security
Cognitive Radio	700 Mhz, 2.4 GHz and 5.1 Ghz	Up to 100 Mbps	Up to 30 Km	NAN, WAN	Practical difficulty

The following standards for HAN in SMGNs are introduced in [6,31]: HomgPlug (powerline technology to connect the smart appliance), HomgPlug Green PHY (specification developed as a low power, cost-optimized power line networking specification standard), U-SNAP (providing many communication protocol to connect HAN devices to smart meters), Z-Wave (alternative solution to ZigBee that handles the interference with 802.11b/g), and openHAN (home area network device communication, measurement, and control). As standards related with DER, smart home, E-storage, and E-mobility, IEC 62056 and IEC 62051-54/58-59 are introduced in [31], and IEC 61850-7-410/420 is relevant to hydro/distributed energy communication, DER, and EMS.

As a novel aspect in SMGNs, to support deal with the peculiarities of different DESs with sophisticated sensing and actuating units, the IEEE Std. 1451 is suggested as a system design model of energy gateways or nodes with uniform interfaces [4]. These nodes have an energy interface not only to the power distribution grid but also to active sources or loads. As shown in Figure 7, in the IEEE 1451 architecture two different network entities are defined: the network capable application processor (NCAP) and the transducer interface module (TIM). NCAP and TIM nodes communicate through the transducer independent interface (TII). Different transducer electronic data sheets (TEDSs) are distributed throughout a TIM, supporting the documentation and configurability of several different features. Figure 8 shows the microgrid networking protocol performing in energy gateways or nodes based on IEEE Std. 1451.

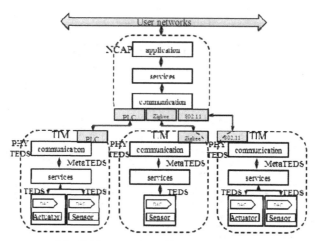

Fig. 7. Elements in an IEEE1451 network

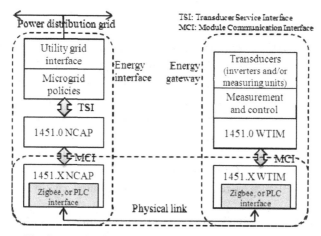

Fig. 8. Micrgrid networking protocols based on IEEE1451

5   Conclusions

In this paper, we have provided the state-of-the-art research and standardization in SMGN with potential advantages and research challenges, which is a little different architecture and issues from smart grid. As noticeable points, sharing renewal energy among DERs brings out a novel research issue in SMGN, i.e., *energy sharing*, and thus *novel network design* and *control mechanisms* in such architecture should be planned. Security issues, i.e., *secure routing*, *attack defenses*, and *privacy preservation*, have been researched. Various *standardizations* for communications or devices have been processing for successful realization of

SMGN. However, the results are still in early stages, and thus the still-open research issues should be discovered and addressed.

Acknowledges. This research was supported by Basic Science Research Program through the National Research Foundation of Korea (NRF) funded by the Ministry of Education, Science and Technology (2012-0004279). The author would also like to thank Dr. Douglas S. Reeves in North Carolina State University for his helpful comments and support.

References

1. Wang, W., Xu, Y., Khanna, M.: Survey Paper: A survey on the communication architectures in smart grid. Comput. Netw. 55(15) (2011)
2. Erol-Kantarci, M., Kantarci, B., Mouftah, H.T.: Reliable overlay topology design for the smart microgrid network. IEEE Network 25(5) (2011)
3. Alwala, S., Feliachi, A., Choudhry, M.A.: Multi Agent System based fault location and isolation in a smart microgrid system. In: Proceedings of Innovative Smart Grid Technologies (January 2012)
4. Bertocco, M., Giorgi, G., Narduzzi, C., Tramarin, F.: A case for IEEE Std. 1451 in smart microgrid environments. In: Proceedings of IEEE International Conference on Smart Measurements for Future Grids (SMFG) (November 2011)
5. Qiu, R.C., Hu, Z., Chen, Z., Guo, N., Ranganathan, R., Hou, S., Zheng, G.: Cognitive Radio Network for the Smart Grid: Experimental System Architecture, Control Algorithms, Security, and Microgrid Testbed. IEEE Transactions on Smart Grid 2(4) (2011)
6. Gungor, V.C., Sahin, D., Kocak, T., Ergut, S., Buccella, C., Cecati, C., Hancke, G.P.: Smart Grid Technologies: Communication Technologies and Standards. IEEE Transactions on Industrial Informatics 7(4) (2011)
7. Kim, M.: A survey on guaranteeing availability in smart grid communications. In: Proceedings of International Conference on Advanced Communication Technology (ICACT) (February 2012)
8. Kwon, Y.M., Kim, J.S., Chung, M.Y., Choo, H., Lee, T.J., Kim, M.: State of the Art 3GPP M2M Communications toward Smart Grid. KSII Transactions on Internet and Information Systems 6(2), 468–479 (2012)
9. 3GPP TR 23.888 v1.3.0: System Improvements for Machine-Type Communications (June 2011)
10. IEEE 80216p-10_0005: Machine to Machine (M2M) Communications Technical Report (November 2010)
11. Zhu, T., Xiao, S., Ping, Y., Towsley, D., Gong, W.: A secure energy routing mechanism for sharing renewable energy in smart microgrid. In: Proceedings of IEEE International Conference on Smart Grid Communications (SmartGridComm) (October 2011)
12. Erol-Kantarci, M., Kantarci, B., Mouftah, H.T.: Cost-Aware Smart Microgrid Network design for a sustainable smart grid. In: Proceedings of GLOBECOM Workshops (GC Wkshps) (December 2011)
13. Erkin, Z., Tsudik, G.: Private computation of spatial and temporal power consumption with smart meters. In: Bao, F., Samarati, P., Zhou, J. (eds.) ACNS 2012. LNCS, vol. 7341, pp. 561–577. Springer, Heidelberg (2012)

14. Lin, H.-Y., Tzeng, W.-G., Shen, S.-T., Lin, B.-S.P.: A Practical Smart Metering System Supporting Privacy Preserving Billing and Load Monitoring. In: Bao, F., Samarati, P., Zhou, J. (eds.) ACNS 2012. LNCS, vol. 7341, pp. 544–560. Springer, Heidelberg (2012)

15. Rottondi, C., Verticale, G., Capone, A.: A security framework for smart metering with multiple data consumers. In: Proceedings of IEEE Conference on Computer Communications Workshops (INFOCOM WKSHPS) (March 2012)

16. He, D., Chen, C., Bu, J., Chan, S., Zhang, Y., Guizani, M.: Secure service provision in smart grid communications. IEEE Communications Magazine 50(8) (2012)

17. Li, X., Liang, X., Lu, R., Shen, X., Lin, X., Zhu, H.: Securing smart grid: cyber attacks, countermeasures, and challenges. IEEE Communications Magazine 50(8) (2012)

18. Kolesnikov, V., Lee, W., Hong, J.: MAC aggregation resilient to DoS attacks. In: Proceedings of IEEE International Conference Smart Grid Communications (Smart-GridComm) (October 2011)

19. Zhu, T., Mohaisen, A., Ping, Y., Towsley, D.: DEOS: Dynamic energy-oriented scheduling for sustainable wireless sensor networks. In: Proceedings of IEEE INFOCOM (March 2012)

20. Baghaie, M., Moeller, S., Krishnamachari, B.: Energy routing on the future grid: A stochastic network optimization approach. In: Proceedings of International Conference on Power System Technology (POWERCON), pp. 1–8 (October 2010)

21. Lin, J., Xu, G., Yu, W., Yang, X., Bhattarai, S.: On distributed energy routing protocols in smart grid (2011),
http://pages.towson.edu/wyu/lxyybReportJuly2011.pdf

22. Lin, J., Yu, W., Yang, X., Xu, G., Zhao, W.: On False Data Injection Attacks against Distributed Energy Routing in Smart Grid. In: Proceedings of the 2012 IEEE/ACM Third International Conference on Cyber-Physical Systems (ICCPS 2012), pp. 183–192 (2012)

23. Marinescu, C., Deaconu, A., Ciurea, E., Marinescu, D.: From microgrids to smart grids: modeling and simulating using graphs: part I active power flow. In: Proceedings of International Conference on Optimization of Electrical and Electronic Equipment, OPTIM (2010)

24. Chen, P.-Y., Cheng, S.-M., Chen, K.-C.: Smart attacks in smart grid communication networks. IEEE Communications Magazine 50(8) (August 2012)

25. He, M., Zhang, J.: A Dependency Graph Approach for Fault Detection and Localization Towards Secure Smart grid. IEEE Trans. Smart Grid 2(2), 342–351 (2011)

26. Kim, T.T., Poor, H.V.: Strategic Protection Against Data Injection Attacks on Power Grids. IEEE Trans. Smart Grid 2(2), 326–333 (2011)

27. Chen, P.-Y., Chen, K.-C.: Intentional Attack and Fusion-based Defense Strategy in Complex Networks. In: Proceedings of IEEE GLOBECOM (December 2011)

28. Seward, M.: Security: Smart Grid Data – the 'wild west' of privacy rights. (2011),
http://blogs.splunk.com/2011/05/27/smart-grid-data-the-wild-west-of-privacy-rights/

29. Guan, X., Xu, Z., Jia., Q.-S.: Energy-efficient buildings facilitated by microgrid. IEEE Transactions on Smart Grid 1(3) (December 2010)

30. Ackland, B., et al.: High Performance Cognitive Radio Platform with Integrated Physical and Network Layer Capabilities. In: Berkeley Wireless Res. Center Cognitive Radio Workshop (2004)

31. Rohjans, S., Uslar, M., Bleiker, R., González, J., Specht, M., Suding, T., Weidelt, T.: Survey of Smart Grid Standardization Studies and Recommendations. In: Proceedings of IEEE International Conference on Smart Grid Communications (SmartGridComm) (October 2010)

Cross-Layered OFDMA-Based MAC and Routing Protocol for Multihop Adhoc Networks

Khanh Quang, Van Duc Nguyen,
Trung Dung Nguyen, and Hyunseung Choo

School of Electronics and Telecommunications, Hanoi University of Science
and Technology, Viet Nam
Sungkyunkwan University School of Information & Communication Engineering,
Korea
{khanhnq1@vms.com.vn,ducnv-fet@mail.hut.edu.vn,dungnt@mail.hut.edu.vn,
choo@ece.skku.ac.kr}

Abstract. In this paper, a dynamic Sub-channel Assignment Algorithm (DSA) based on orthogonal frequency division multiple access (OFDMA) technology operating in the time division duplexing (TDD) and a new routing protocol are proposed. The proposed of dynamic Sub-channel Assignment Algorithm solves several drawbacks of existing radio resource allocation techniques in OFDM system used in ad-hoc and multihop networks, such as the hidden and exposed node problem, mobility, co-channels interference in frequency (CCI). An interference avoidance mechanism allows the system to reduce CCI and to operate with full frequency re-use. The proposed routing protocol is jointed with the MAC protocol based the algorithm to ensure the mobility and multi-hop, thus the quality of service in ad-hoc and multi-hop networks is significantly improved.

Keywords: OFDMA, MAC protocol, routing protocol, ad-hoc networks, multi-hop networks.

1 Introduction

A major challenge in wireless networks for multi hop communications is the co-channel interference (CCI). This interference is introduced when two different radio stations simultaneously use the same frequency. It is mainly caused by the spectrum allocated for the system being reused multiple times in TDMA network. CCI is one of the major limitations in cellular and Personal Communication Services wireless telephone networks since it significantly decreases the carrier-to-interference ratio. In addition, it makes the diminished system capacity, more frequent handoffs, and dropped calls. IEEE 802.11 Distributed Coordination function operation is based on conventional carrier mechanism (CSMA/CA) in order to prevent channel collisions, CCI and provide the communication between multiple pairs of independent mobile nodes without access points or base stations such as mobile adhoc networks [1-2].

B. Murgante et al. (Eds.): ICCSA 2013, Part I, LNCS 7971, pp. 157–172, 2013.
© Springer-Verlag Berlin Heidelberg 2013

Recently, orthogonal frequency division multiplexing (OFDM) has been intensively investigated for wireless data transmission in broadband cellular and ad-hoc networks. The multiple access technique for these networks is OFDMA [3]. The concept of this technique is to assign different users to different sub-channels in order to avoid interferences.

The paper is organized as follows: In section 2, we briefly review sub-channels allocation methods. Session 3 describe the proposed DSA algorithm. In session 4, the proposed routing protocol is presented. Simulation schemes, numerical results, are discussed in section 5. Finally, conclusions are drawn in session 6.

2   Review of Sub-channels Allocation Methods

2.1   OFDM-FDMA Fix Allocation

The fixed allocation method of OFDM-FDMA for multiuser communications was proposed [8]. In such method, different users will be fixedly assigned to different sub-channels. Therefore, this method has not any anti-interference mechanism.

2.2   OFDM-FDMA Random Allocation

The OFDM-FDMA random allocation method is based on the idle and busy of sub-channels allowing users to accounts different sub-channels [9]. However, it does not have any attention to the network interference. Once a sub-channel is selected, a user starts transmitting using the selected sub-channel. During a transmission process, if a sub-channel does not meet the required QoS, it will be released and assigned to new user. Although the method is simple and it offers an adaptive mechanism, it does not provide CCI avoidance.

3   Proposed DSA Algorithm

Co-channel interference (CCI) is crosstalk from more than one different radio transmitter using the same frequency in wireless networks. Reducing CCI is very important since it makes the poor throughput performance. To avoid CCI and collisions, we propose a novel channel allocation algorithm called DSA which supports simultaneous transmissions in Vehicle Ad Hoc Network among nodes. In this section, the problem of CCI in OFDMA/TDD in wireless networks is discussed in detail. Then, we present the proposed Dynamic Subchannel Assignment (DSA) algorithm.

3.1   CCI in OFDMA/TDD System

To illustrate the problem of CCI, a simple scenario consisting of two base stations (BSs) and four mobile stations (MSs) is depicted in Fig. 1. Let us assume an example of exchanging data among BSs and MSs as following. The mobile station MS_1^{Rx}, MS_2^{Rx} and MS_3^{Rx} receive data from base station BS_1^{Tx}, while at the same time the mobile station BS_2^{Tx} transmits its data to the base station MS_4^{Rx}. In such scenario, BS_2^{Tx} causes CCI to receiving data of MS_1^{Rx}, MS_2^{Rx} and MS_3^{Rx}; the base station BS_1^{Tx} causes CCI to receiving data of MS_4^{Rx}. Note that CCI only exists in TDD mode.

If a node can select appropriate sub-channels in available sub-channel set before the data transmission, CCI is minimized, thus increase the throughput of the network.

3.2 Sub-channel Selection Based on Busy-Signals

In the previous example, CCI was introduced and it badly affects to the receiving data of MS_1^{Rx}, MS_2^{Rx} and MS_3^{Rx}. To avoid this interference, MS_1^{Rx}, MS_2^{Rx} and MS_3^{Rx} have to broadcast a busy signal when they receive data. Before data transmission, BS_2^{Tx} will first hear busy signals in all channels and then compares the received signal power with a predetermined threshold. If the power of busy signal in a channel is lower than the threshold, BS_2^{Tx} can select the channel for data transmission.

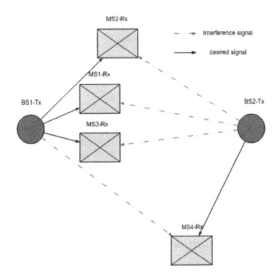

Fig. 1. Co-channel interference in OFDMA/TDD system

3.3 Dynamic Sub-channel Assignment Algorithm

In this section, we propose MAC frame structure for downlink and uplink transmission based on incorporation the busy tone in OFDMA/TDD network. The proposed MAC frame is illustrated in Fig.2. The busy signal is sent an in-band signaling before data transmission. Data sub-channels are mutually orthogonal and used for transmitting OFDM symbols.

Fig. 2 shows data transmitting and receiving in view point from BS and MS, respectively. The structure of the uplink frame is similar to those of the downlink frame. Each sub-frame includes a busy signal channel (an OFDM symbol) used for signaling busy tone. MAC frame length is chosen having corresponding to the time correlation of the channel. Initially, the sub-channel is selected by only the transmitter. On the other hand, the adaptive period is adjusted by both transmitter and receiver. We define two kinds of signal:

– Busy signal: transmitted only in a channel by MS receiver.
– Data signal: transmitted only by BS transmitter.

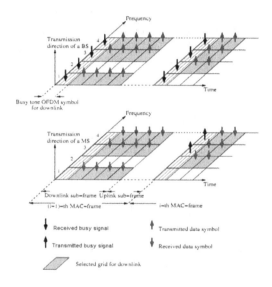

Fig. 2. Structure of the MAC frame for proposed Algorithm

In Fig. 3, the proposed algorithm includes two following steps:

Link Initialization. Link initialization occurs when m^{th} BS need to transmit data to k^{th} MS in the network. Firstly, m^{th} BS listens to all busy signals and compares each of these busy signals with a given threshold to obtain a set of available sub-channel for data transmission called set A. A sub-channel is available if the power of received busy signal is below the threshold. The mth BS will start transmit its data to receiver using a set of available channel at the $(i-1)^{\text{th}}$ MAC frame. We propose a mathematical model 1 where $a_{l,i-1}^{k,m}$ denotes the sub-channel assignment for l^{th} sub-channel at the $(i-1)^{\text{th}}$ MAC frame of the link between m^{th} BS and k^{th} MS. If the sub-channel is assigned to the link between m^{th} BS and k^{th} MS, then $a_{l,i-1}^{k,m} = 1$, otherwise $a_{l,i-1}^{k,m} = 0$. The outcome of this channel assignment is obtained by comparing the busy signal with the threshold as following:

$$a^{k,m}_{l,i-1} = \begin{cases} 1 \text{ if } (|\hat{B}^m_{l,i-1}| \leq I_{\text{thr}}) \\ 0 \text{ otherwise} \end{cases} \tag{1}$$

I_{thr} is a threshold being a measure for the interference that this transmission would effect to other co-existing transmission. Set A is all sub-channels having channel assignment by model 1.

Dynamic Subchannel Assignment. At the $(i-1)^{\text{th}}$ MAC frame, k^{th} MS estimates the SINR on each sub-channel of set A. Based the given QoS requirement for the transmission, k^{th} MS will decides to maintain or release the respective sub-channel of set A. k^{th} MS will only broadcasts the busy signal on the remained sub-channels. If the sub-channel is specified, then $b^{k,m}_{l,i-1}$ is assign 1, otherwise $b^{k,m}_{l,i-1} = 0$.

We propose a mathematical model 2 where $b^{k,m}_{l,i-1}$ denotes the reservation of lth sub-channel for the ith MAC frame. $\breve{\gamma}^k_{l,i-1}$, γ_{req} are SINR of estimated by k^{th} MS and required QoS.

$$b^{k,m}_{l,i-1} = \begin{cases} 1 \text{ if } (|\hat{B}^m_{l,i-1}| \leq I_{\text{thr}}) \ \& \ (\breve{\gamma}^k_{l,i-1} \geq \gamma_{\text{req}}) \\ 0 \text{ otherwise} \end{cases} \tag{2}$$

Since the kth MAC frame, the condition for the sub-channel assignment on the desired or new link between m^{th} BS and k^{th} MS for subsequent MAC frames is given as follows:

$$a^{k,m}_{l,i} = \begin{cases} 1 \text{ if } (\bar{a}^m_{l,i-1} | \hat{B}^m_{l,i}| \leq I_{\text{thr}}) \text{ or } (b^{k,m}_{l,i-1} = 1) \\ 0 \text{ otherwise} \end{cases} \tag{3}$$

Where $\hat{B}^m_{l,i}$ is the busy signal received at the mth BS on the lth sub-channel at the ith MAC frame and

$$\bar{a}^{k,m}_{l,i-1} = \begin{cases} 1 \text{ if } a^{k,m}_{l,i-1} = 0 \\ 0 \text{ otherwise} \end{cases} \tag{4}$$

Since the ith MAC frame, set A includes all maintained sub-channels (set B of $(i-1)^{\text{th}}$ MAC frame) and new respective sub-channels if the power of their received busy signal are below the threshold.

4   Proposed Routing Protocol Description

In ad-hoc and multi-hop networks, a node transmits its data to a destination node via an optimal route. In order to find the optimal route, the routing is implemented. In this section, we propose a new routing protocol which can be easily jointed to the MAC protocol based the DSA algorithm. The corporation

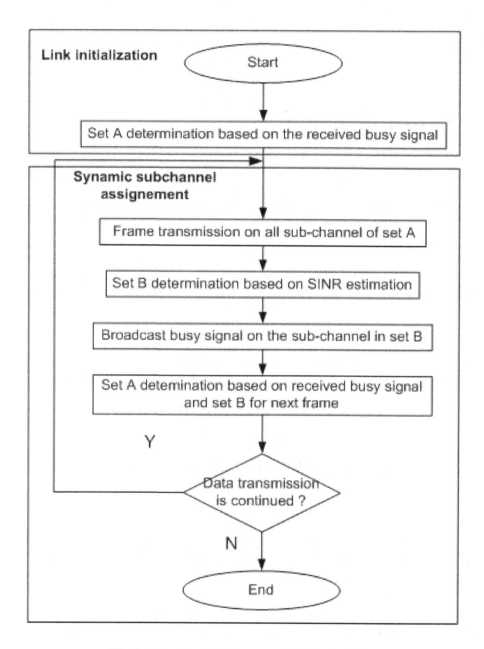

Fig. 3. Flow chart of the proposed DSA algorithm

between the routing and the MAC protocol not only ensures the mobility and multi-hop but also gains the performance in term of throughput in the ad-hoc multi-hop networks.

The routing protocols working in wireless networks depend on the radius of coverage of nodes. The connection among all nodes changes fast over time. The principal cost of the proposed routing protocol is the combination of the distance among nodes and the number of selected channel of a route. The finding the shortest route in the proposed protocol is based on the shortest route finding algorithms and the radius of coverage of nodes in the network. To ensure the mobility and the performance in term of throughput of the network, the MAC layer provides the number of selected channels of a route to the routing protocol. This value is compared with a proposed threshold called channel_threshold in order to update the optimal route for the transmission.

Before describing the flow chart of this protocol, we define:

- R_i is the radius of coverage of ith node in network.
- The network is supposed to be a graph with nodes and edges.
- The connection among nodes in the coverage is described in a connection matrix.

$$M_{ij} = \begin{cases} \text{distance}(i,j) & \text{if } distance(\text{i,j}) \leq \min(R_i, R_j) \\ 0 & \text{otherwise} \end{cases} \tag{5}$$

Where, M_{ij} and distance(i,j) are the connection state and the distance between the ith node and the jth node, respectively.

The proposed routing protocol finds the route using the three following steps in Fig. 4:

1. The input of this step is the radius of coverage of all nodes and the distances between all node-pairs in the network. The Connection matrix is build based on the proposed mathematical model 5. The value of init_loop is initially set 0.
2. In this step, the shortest route and the number of available routes (iNP) are obtained by using the dijkstra algorithm shortest route algorithm, see [11]. The inputs of the step are the positions of the destination and source node, and the Connection matrix.
3. This step aims to find the optimal route for the transmitting data. Firstly, iNP is checked. If iNP is non-zero, there is at least one available route between the source node and the destination node. The number of selected sub-channel (iSC) of the shortest route is derived by using the proposed DSA algorithm. If iSC is larger than the given channel_threshold, it means the route ensures the requirement of throughput of the system. Consequently, the shortest route is selected as the optimal route for the transmission data. Otherwise, the route does not satisfy the requirement of the system. We, therefore, have to find another route. The init_loop is set 1 which points that there is at least one available route being not satisfy requirement of

throughput of the network. The step is directed to the beginning of step 2 and the shortest route is excluded in the dijkstra algorithm. At the end of step 1, if iNP is zero, these are two cases can be happen. The first case is that there is no available route between the source node and the destination node (init_loop is 0), hence, the transmission is stopped and it has to delay for some time to transmit the data. In the other case, these are at least one route available. However, all of the nodes do not satisfy the requirement of throughput of the network. In this case the shortest node is selected for the transmission data.

5 Simulation Model

In the simulation scheme, OFDM parameters are selected as in table 1. 16-QAM modulation is selected for all sub-channels. We will present the performance in term of throughput of the network of our proposed method in three scenarios. In the first scenario, the performance of the proposed DSA is demonstrated in single-hop adhoc networks without the proposed routing algorithm. Then, the proposed DSA algorithm and the proposed routing protocol are applied in the scenario of multi-hop adhoc networks. However, in this case, the routing is implemented without selecting the optimal route in term of the number of selected channel of a route. Finally, the proposed routing protocol cooperated with the proposed MAC protocol based on DSA algorithm is fully carried out in multi-hop adhoc networks.

Table 1. OFDM SYSTEM PARAMETERS

Parameters	Values
Bandwidth (B)	20 MHz
Sampling Interval ($t_a = 1/B$)	50 ns
FFT length (N_{FFT})	256
OFDM symbol duration (T_s)	12.8 μs
Guard interval (T_{G})	2 μs
Frequency (f_c)	1.9 GHz
Modulation scheme	16-QAM
SINR$_{\min}$ (γ_{req})	16 dB
Channel_threshold	42

5.1 Throughput of Multihop Adhoc Networks Based on Only the Proposed DSA Algorithm and MAC Structure

Analysis Model. In this scenario, the performance of the proposed DSA algorithm and the MAC structure is evaluated in a multi-hop adhoc network considering the effect of the length of selected routes and I_{thr}. Therefore, the routing

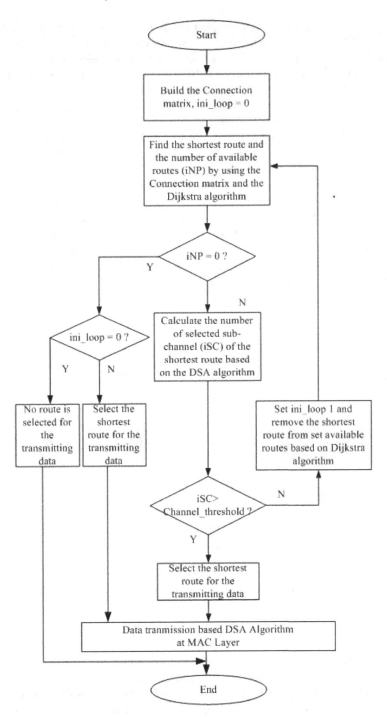

Fig. 4. Flow chart of the proposed routing protocol

protocol used in this model is not the proposed routing protocol presented in section IV. It is just the principle routing which considers only the distance of a route as the cost of this route. The shortest route being selected as the optimal route for transmitting data is obtained by using the dijkstra algorithm. The input of the dijkstra algorithm is the Connection matrix of the network. Before each transmission section, the Connection matrix is built as proposed in section IV. Then, the shortest route outputted from the dijkstra algorithm is selected. The data will be transmitted through the selected route. We consider a vehicles network VANET includes 6 mobile stations MS indexed from 1 to 6 with the initial state depicted in Fig. 5. MS_1, MS_5, and MS_6 are running step by step while the other mobile stations are fixed. The running step length is set 5 meters. MS_5 and MS_6 run in the same direction being opposite the running direction of MS_1. While the mobile stations are running, MS_1 is transmitting its data to MS_5. After each step, throughput of whole network is obtained. Fig. 6 gives an example of the connection state of the network. It can be seen from Fig. 6 that the shortest route (MS_1,MS_2,MS_4,MS_5) is selected for the transmitting data.

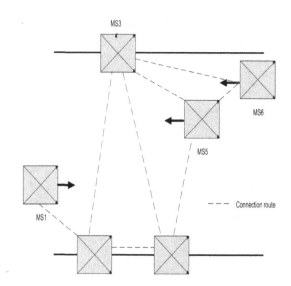

Fig. 5. Initial state of networks used in simulation in the case of ad-hoc multihop network

Numerical Results. Fig. 7 plots the length of route from MS_1 to MS_5 with respect to running steps. Since the routing protocol is not based on I_{thr}, the length of routes for three cases of I_{thr} are the same. We truncate the process into three groups A, B and C. In each group, we can see that lengths of routes seem to be the same. When the length of a route is high, it means that the number of intermediate nodes also high. Therefore, the data is transmitted through many

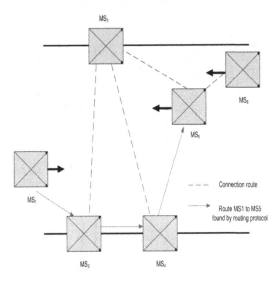

Fig. 6. An example of the route from MS_1 to MS_5

sections. In some sections, some sub-channels do not meet the QoS. Hence, the channel is not full-fill, thus making the decrease of throughput. This conclusion is backed-up from group A and B where lengths of routes are high and throughput is small in Fig. 8. On the other hand, when length of a route is small, the number of intermediate nodes is also small. Data can be directly transmitted from source to destination or through few intermediate nodes. In such the case, the effect of interferences, expose node, and hidden node is reduced, and channel is full-fill. Consequently, throughput of the network is significantly increased. It can be seen from group C, the length of route at each step is small and throughputs would, therefore, be high. Throughput is also depends on the I_{thr} which decides a sub-channel be selected for transmission data or not. When I_{thr} is small, the probability that a sub-channel is selected is high. Hence, throughput is also high. The effect of I_{thr} is clear when the length of a route is high (in A and B group), since it affects to the selection of sub-channel in many intermediate sections. On the other hand, when mobile stations are close (in group C), the role of I_{thr} is not so important. Therefore, throughputs seem to be similar for three cases of I_{thr} of -20, -40, -60 dBm. It is clear that the throughput of the network depends on the distance between the source and destination nodes as well as I_{thr}.

5.2 Throughput of Multihop Adhoc Networks Based on the Proposed Routing Protocol Jointed with the MAC Layer

Analysis Model. The performance of the proposed routing protocol is evaluated in this section. We implemented the routing protocol jointed with the proposed MAC layer on the network as one presented in section B. However, we

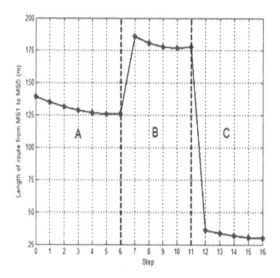

Fig. 7. Length of route from MS_1 to MS_5

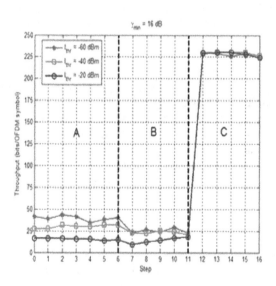

Fig. 8. Throughput of the network when MS_1 sends data to MS_5

consider the transmission the data from MS_1 to MS_6. The channel allocation for the transmission data between nodes in the network is based on the proposed DSA algorithm. The threshold Channel_threshold is experimentally set 42. A mathematical method of deriving the optimal threshold will be proposed in our future work. γ_{min} and I_{thr} are set -80 dBW and 16 dB, respectively. An example of a route found by using the proposed routing protocol is drawn in Fig. 9. In this state, the shortest route was found by the dijkstra algorithm. The route, however, does not meet the requirement of the number of selected sub-channels. Hence, it was not selected for the transmitting data. On the other hand, our proposed protocol derived another route which has a longer length than the shortest route. Since the number of selected sub-channel of this route is larger than Channel_threshold, it was selected as the optimal route.

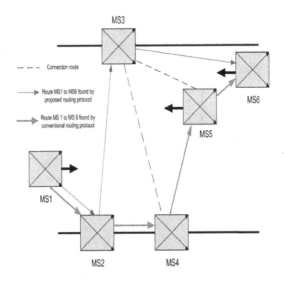

Fig. 9. An example of a route from MS_1 to MS_6 by using the proposed routing protocol and the conventional routing protocol

Numerical Results. The performance of the proposed routing protocol is depicted in Fig. 10 and Fig. 11 where Conventional routing protocol is the principle routing based on only the dijkstra algorithm. It can be seen from these results that the throughput of the network depends on the distance between the source node and the destination node as the conclusion in section B. In the group A where the destination and the source node are far, the number of intermediate nodes is high. Hence, the transmission is significantly suffered from the CCI as well as the hidden and expose node problem. Consequently, the number of selected sub-channels of all available routes each is smaller than Channel_threshold. Therefore, the proposed routing selected the shortest route as the optimal route to transmit the data and there is no difference of throughput of the proposed routing protocol and the conventional one. On the other hand in group C, since the source and the destination are very close, they transmit its data directly or via a few of intermediate node. The effect of the CCI and the hidden and expose node problem is considerably fell, thus making the increase of the number of selected sub-channels. The number of selected sub-channel for all routes each is larger than Channel_threshold. Therefore, as in group A, the proposed routing protocol selected the shortest route as the optimal route. In this group, the proposed routing and conventional routing protocol demonstrate the same performance. In the group B, some routes have the number of selected sub-channel being larger than threshold, while one of other routes each is smaller than threshold. In some case, the shortest route does not meet the requirement of the selected sub-channels; hence it is not selected as the optimal route. Our proposed find another route satisfy the requirement with longer length of route. Therefore, in this group, the proposed method outperforms the conventional routing protocol in term of throughput of the network.

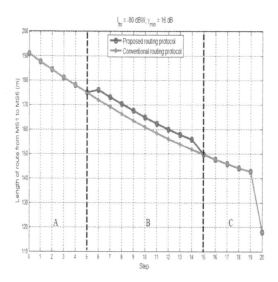

Fig. 10. Length of the route from MS_1 to MS_6 using the proposed routing protocol and the conventional routing protocol

6   Conclusion

In this paper, we present a dynamic Sub-channel Assignment Algorithm base OFDM and a new routing protocol. A concept of ad-hoc and multi-hop networks based on the DSA algorithm is simulated with the proposed routing protocol. The proposed algorithm has an interference avoidance mechanism, so the throughput of the network can be maximized. The problems of cellular networks

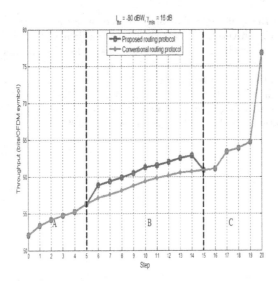

Fig. 11. Throughput of the network using the proposed routing protocol and the conventional routing protocol

such as hidden and expose node are solved. A routing protocol which is joint to the MAC protocol is proposed. The proposed routing protocol improves the throughput of ad-hoc and multi-hop networks as well as ensures the mobility, multi-hop of ad-hoc and multi-hop networks. In the future, we will study mathematic model for Channel_threshold to improve and optimize QoS of ad-hoc multi-hop network.

References

1. IEEE 802.11 Working Group, Wireless LAN Medium Access Control (MAC) and Physical Layer (PHY) Specification (1997)
2. Chen, K.C.: Medium access protocols of wireless LANs for mobile computing. IEEE Network 8(5), 50–63 (1994)
3. Rohling, H., Grünheid, R.: Performance comparison of different multiple access schemes for the downlink of an OFDM communication system. In: IEEE 47th Vehicular Technology Conference (May 1997)
4. Kulkarni, G., Srivastava, M.: Subcarrier and bit allocation strategies for ofdma based wireless ad hoc networks. In: IEEE GLOBECOM (2002)
5. Venkataraman, V., Shynk, J.J.: Adaptive algorithms for ofdma wireless ad hoc networks with multiple antennas. In: Conference Record of the Thirty-Eighth Asilomar Conference on Signals, Systems and Computers (2004)
6. Kim, S.W., Kim, B.: OFDMA-Based reliable multicast MAC protocol for Wireless Ad-hoc Network. ETRI Journal 31(1) ·(February 2009)
7. Nguyen, V.-D., Hass, H., Kyamakya, K., Chedjou, J.-C., Nguyen, T.-H., Yoon, S., Choo, H.: Decentralized dynamic sub-carrier assignment for OFDMA-based adhoc and cellular networks. IEICE Trans. Commun. E92-B(12), 3753–3764 (2009)

8. Rahman, M.I.: Basic about multi-carrier based multiple access techniques. TT R-04-1001, Aalborg University, Denmark, Elsevier Science (2005)
9. Stemick, M., Rohling, H.: AOFDM-FDMA scheme for the uplink of a mobile communication system. Wireless Personal Communications 40(2), 157–170 (2007)
10. Torrance, J.M., Hanzo, L.: Optimization of switching levels for adaptive modulation in slow Rayleigh fading. IEEE Elec. Lett. 32(13), 1167–1169 (1996)
11. Skiena, S.: Implementing Discrete Mathematics: Combinatorics and Graph Theory with Mathematica, pp. 225–227. Addison-Wesley (1990)
12. Van, T.P., Nguyen, V.D.: Location-aware and load-balanced data delivery at roadside units in vehicular Ad hoc networks. In: IEEE 14th International Symposium on Consumer Electronics (2010)

Content-Based Chunk Placement Scheme for Decentralized Deduplication on Distributed File Systems

Keonwoo Kim[1], Jeehong Kim[1], Changwoo Min[1,2], and Young Ik Eom[1]

[1] College of Information and Communication Engineering, Sungkyunkwan University
Suwon, Korea
{kkw0528,jjilong,multics69,yieom}@skku.edu
[2] Samsung Electronics Co., Ltd., Suwon, Korea
multics69@skku.edu

Abstract. The rapid growth of data size causes several problems such as storage limitation and increment of data management cost. In order to store and manage massive data, Distributed File System (DFS) is widely used. Furthermore, in order to reduce the volume of storage, data deduplication schemes are being extensively studied. The data deduplication increases the available storage capacity by eliminating duplicated data. However, deduplication process causes performance overhead such as disk I/O. In this paper, we propose a content-based chunk placement scheme to increase deduplication rate on the DFS. To avoid performance overhead caused by deduplication process, we use *lessfs* in each chunk server. With our design, our system performs decentralized deduplication process in each chunk server. Moreover, we use consistent hashing for chunk allocation and failure recovery. Our experimental results show that the proposed system reduces the storage space by 60% than the system without consistent hashing.

Keywords: Deduplication, Distributed file system, Chunk placement, Consistent hashing.

1 Introduction

The amount of digital information is rapidly increasing all over the world. According to the forecast by IDC, the amount of digital information will grow by a factor of 50 over the next decade [1, 2]. Also, most of data in the digital universe is unstructured one such as image, video, audio, and document files [1]. Moreover, an influx of data is rapidly growing in cloud storage [1]. This rapid growth of data size causes several problems such as storage limitation, increment of data management cost, and network traffic congestion [3, 4]. For storing and managing massive data in cloud storage, cloud storage service provider generally uses Distributed File System (DFS). However, despite wide adoption of DFS, rapidly growing data size causes additional storage and management cost. Therefore, data size optimization techniques are required for cloud storage systems. Among

B. Murgante et al. (Eds.): ICCSA 2013, Part I, LNCS 7971, pp. 173–183, 2013.
© Springer-Verlag Berlin Heidelberg 2013

data size optimization studies [3–5], data deduplication is a representative research issue. Data deduplication eliminates duplicated data, by which available storage space is increased and additional storage cost is reduced.

Duplicated data reduces available storage space. Most of cloud storage services use data deduplication to solve lack of available storage space. However, previous research [6–11] on DFS do not support data deduplcation because of performance overheads that are caused by additional computation cost and disk I/O. In this paper, we propose a content-based chunk placement for the data deduplication on the DFS. In order to distribute deduplication processes, we integrate *lessfs* that is an inline deduplication file system with each chunk server of MooseFS. Thus, we achieve decentralized data deduplication process, while maintaining high performance. In MooseFS [6], a file is divided into multiple chunks and each of them is stored in chunk server dispersedly. So, by using consistent hashing, the chunk placement module gather same chunks in one chunk server, by which we enhance the deduplication rate. We make following contributions in this paper:

- We introduce a new content-based chunk placement for deduplication system on MooseFS. Furthermore, we design a content-based chunk placement for deduplication system that replaces the file system of chunk server in MooseFS with *lessfs*.
- We coordinate consistent hashing with a content-based chunk placement for deduplication system for enhancing the deduplication rate. By performing deduplication process in each chunk server, our system can avoid bottleneck of the master server.
- We evaluate our chunk placement for deduplication system, and show its effectiveness.

The rest of the paper is organized as follows: We review background in Section 2. Section 3 describes the key ideas and implementation details. In Section 4, we evaluate our system and show the results. Finally, we conclude in Section 5.

2   Background and Related Work

2.1   DFS

A DFS is file system that allows access to files from multiple hosts via a communication network [12]. With DFS, a client can access remote files in the same way that it accesses local files. DFS generally consists of single master server and multiple storage servers. Master server manages file metadata and chunk server keeps chunk data of files. Typical open-source based DFSs are MooseFS [6], XtreemFS [7, 8], Ceph [10], Google file system[11], and GlusterFS [9]. We choose MooseFS as our DFS platform because it is general-purpose and good performance file system. MooseFS consists of single master server, multiple metadata backup servers, and multiple chunk servers. Master server manages whole file system and stores metadata for each file. Chunk servers store chunk data of each file. Metadata backup servers store metadata changelogs and periodically download metadata files from master server [6].

2.2   Data Deduplication in Local File System

A data deduplication technique is used to increase the storage capacity by detecting and eliminating duplicated data. With the scheme, each chunk can have only a single copy in the system. The data deduplication exists in the following types: post-process and inline data deduplication. Post-process data deduplication occurs after data has been written, whereas inline data deduplication occurs before the data has been written [5]. Therefore, inline deduplication causes more network traffic than post-process deduplication. On the other hand, post-process deduplication requires storage space to store the duplicated data before deduplication is performed. Typical open-source based deduplication file systems are *lessfs*, zfs, and sdfs [4, 5]. We use *lessfs* because it is a good performance deduplication file system. *Lessfs* is an inline block-level deduplication in Linux file system and uses data compression (e.g., LZO, QuickLZ, Snappy, bzip, and gzip). Also, block size can be defined as 4, 8, 16, 32, 64, and 128 KB. Through configuring block size, we can adjust tradeoff between throughput and deduplicated size. To store metadata, *lessfs* uses low-level FUSE API and database (e.g., BerkeleyDB, hamsterdb, and Tokyo Cabinet). Also, by using cache, *lessfs* reduces disk I/O.

2.3   Related Work

As the data size rapidly increases, the data deduplication is extensively studied to optimize storage capacity. The data deduplication technique is divided into two categories [3]: primary data deduplication and secondary data deduplication. Primary data storage directly interacts with application. In other words, application directly affects data. Thus, primary data deduplication systems are latency-aware and use RPC based protocols [13]. On the other hand, secondary data storage copies and archives data to recover from data loss and corruption. Secondary data deduplication systems are throughput-aware and use streaming protocols [13] because this storage processes large amounts of data. Mayer et al. [14] examined in primary data and secondary data. They found that block-level deduplication saves just about 10% more space than the whole-file deduplication. However, experimental result of El-Shimi et al. [15] showed that block-level deduplication saves from 2.3 times to 15.8 times more storage space than whole-file deduplcation. This difference of experimental results is due to the difference of Mayer's data set and El-Shimi's data set. HYDRAstor [16] is a secondary data storage and block-level deduplication system, and uses distributed hash table for scalability. iDedup [13] is a inline data deduplication for the primary storage, and uses in-memory indexing and metadata cache. Lillibridge et al. [17] propose a inline deduplication system and use parse indexing which is in-memory index. Wei et al. [18] use bloom filter and dual cache. Zhu et al. [19] use summary vector and locality preserved caching.

Previous research [13–20] regards deduplication process as performance degradation factor because of disk I/O. Therefore, in-memory indexing or cache is used to reduce deduplication overhead. In-memory indexing such as Bloom filter can reduce disk I/O and quickly searches chunk or block which is a unit of deduplication. Cache is also used to reduce disk I/O.

However, most of studies [13–17, 19] perform centralized data deduplication. This causes bottleneck of the central server and increases I/O latency. For this reason, we use *lessfs* in each chunk server of MooseFS to decentralize deduplication.

3   Design and Implementation

3.1   Chunk Placement of Existing MooseFS

When the client of MooseFS executes the write operation, write operation steps of existing MooseFS are as follows: (1) The client requests a chunk server list to the master server to store chunks in chunk servers. (2) The master server returns a chunk server list that consists of information of chunk servers which have more available space than the other chunk servers. (3) The client sends the chunk data to chunk servers that correspond with a chunk server list. (4) Chunk servers receive and store chunk data.

Therefore, chunks are stored on the chunk server that has more available space than the other servers. This approach is fitted for a system requiring storage load balancing.

3.2   Deduplication on MooseFS

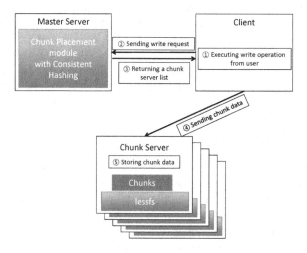

Fig. 1. Overall architecture

We propose the deduplication on MooseFS. MooseFS is chunk-based DFS. If a file size is larger than 64MB, a file is divided up into 64MB chunks. Otherwise, a chunk size becomes the same as a file size. MooseFS can consist of a single master server and the multiple chunk servers as Fig. 1 shows. Chunks are dispersedly stored in chunk servers that are distributed on the network. Chunk server can use many file systems but ext4 is generally used. To implement the deduplication on DFS, we use a *lessfs* that is a block-level and inline deduplication file system. In chunk server of Fig. 1, we mount *lessfs* in the chunk server for the data deduplication. Therefore, all chunks of chunk server are stored in a mount point of *lessfs* and the duplicated chunk data are eliminated by *lessfs*. The deduplication process is regarded as causing performance overheads such as the additional computation cost and the disk I/O. If deduplication process is performed in a master server, those overheads cause bottleneck of a master server. Therefore, the data deduplication process is performed in each chunk server so we can avoid the bottleneck to deduplicate in a master server. Also, the disk I/O can be reduced because *lessfs* uses cache and in-memory database.

3.3   Consistent Hashing for MooseFS

Consistent Hashing. Consistent hashing [21, 22] is used in a changing population of web server environment. If a sever node is added or removed, all objects of web server nodes have to be relocated. We use consistent hashing to solve this problem. Each node is mapped on a hash ring and has a hash value range. Moreover, each data object has hash value and belongs to each node. When data object is stored, the system finds a node by comparing hash value range of node with hash value of data object. After the system find a node, the data object is stored in the found node. By using consistent hashing, we can avoid all object data relocation. Consistent hashing is used in many distributed system such as Dynamo [23], Cassandra [24], and Memcached [25].

Consistent Hashing for MooseFS. When communicating with servers, Moose-FS uses 32-bit CRC that is a hash function to detect error. For identifying chunk, we utilize 32-bit CRC. In master server, we make a data structure for the chunk server node in consistent hashing. Each node also has start and end of 32-bit CRC value, size of range, and next node pointer and consistent hahsing is implemented as a circular linked list. When chunk server is registered, new node is created and included in a circular linked list. The range of consistent hashing is from 0 to 2^{32}-1 because CRC is 32-bit. Each chunk server has 32-bit CRC hash range and each chunk is stored in the corresponding chunk server. To find the chunk server, the master server compares 32-bit CRC value of chunk with a hash range of the chunk server. In Fig. 2-(a), consistent hashing include the chunk servers and chunks. Chunk 1 belongs to chunk server A, chunk 2 belongs to chunk server B, and chunk 3, chunk 4, and chunk 5 belong to chunk server C. Subsequently, in Fig. 2-(b), if chunk server D is added in consistent hashing, range of chunk server C is divided into two halves. So, range

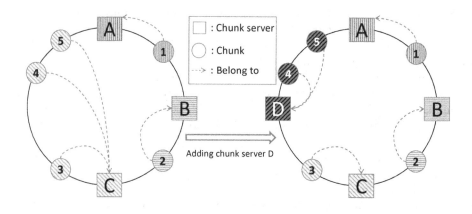

Chunk server	Hash range	Stored chunks
A	$[0, 2^{30})$	1
B	$[2^{30}, 2^{31})$	2
C	$[2^{31}, 2^{32})$	3, 4, 5

(a) Before adding chunk server D

Chunk server	Hash range	Stored chunks
A	$[0, 2^{30})$	1
B	$[2^{30}, 2^{31})$	2
C	$[2^{31}, 3 \times 2^{30})$	3
D	$[3 \times 2^{30}, 2^{32})$	4, 5

(b) After adding chunk server D

Fig. 2. Consistent Hashing for MooseFS

of chunk server C becomes from 2^{31} to 3×2^{30}-1 and range of chunk server D becomes from 3×2^{30} to 2^{32}-1. Therefore, chunk 4 and chunk 5 are relocated from chunk server C to chunk server D because of changing range of chunk server C. When the client executes a write operation, the write operation steps of the proposed chunk placement are as follows: (1) The client generates 32-bit CRC value that corresponds with foremost 4KB of the chunk data. (2) The client sends a write request message to the master server with 32-bit CRC value. (3) To find a appropriate chunk server, the master server travels a circular linked list with comparing 32-bit CRC value with start and end value of each chunk server node. (4) MooseFS not use replication, the master server stores node information to a chunk server list. (5) Otherwise, replicas of chunk data are stored other chunk servers. Therefore, the master server chooses found node of step (4) and its next nodes. (6) The master server returns a chunk server list to the client. (7) The client sends chunk data to chunk servers that correspond with a chunk servers list. (8) Chunk servers receive and store chunk data. For content-based chunk placement, master server compares 32-bit CRC range of chunk server and 32-bit CRC value of chunk. However, the larger size data is input to 32-bit CRC, the more time is spent. Therefore, 32-bit CRC computation of all 64 MB chunks causes the bottleneck in the master server. If the master server performs 32-bit CRC computation and comparing with 32-bit CRC values, the performance of whole system is degraded. Chen et al. [26] propose content-based sampling which produces 32-bit CRC value that corresponds with the first four bytes of each page. Moreover, they examined choosing other bytes and found that using the first four bytes performs well. For this reason, we hash a foremost 4 KB of a 64 MB chunk.

4 Evaluation

In this section, we evaluate how much our deduplication system reduces the data. We installed client, master server, and chunk server in one PC and installed chunk server in 17 PCs. Also, we mount *lessfs* on all PCs. Therefore, the total number of chunk servers is 18. We used multimedia data set (about 936 MB) which includes movie files, audio files, and document files. To identify what amount of duplicated data is eliminated, we make data *set-1*, *set-2*, and *set-3* that include same data and have different name.

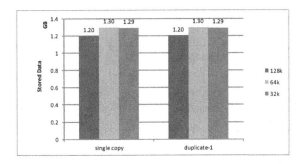

Fig. 3. Stored data size according to block size variation

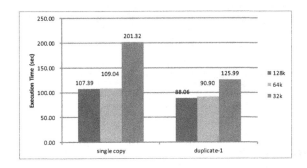

Fig. 4. Execution time according to block size variation

First of all, we experiment to find suitable block size of *lessfs*. In Fig. 3, single copy of x-axis means that unique *set-1* is stored in MooseFS, duplicate-1 of x-axis means that both *set-1* and *set-2* were stored in MooseFS. Y-axis means stored data size on all chunk servers. 128 KB, 64 KB, and 32 KB are block size of *lessfs*. Stored data size of 128 KB is smaller than 64 KB and 32 KB block size.

In Fig. 4, execution time of 128 KB is the fastest but execution time of 32 KB is almost twice as execution time of 128 KB and 64 KB in single copy. The reason of this result is that block size is too small. 32 KB block size takes long time to execute, because small block size occurs to create more metadata and many comparision to detect data duplication. Therefore, optimal deduplication block size is 128 KB, and we experiment our system using 128 KB block size.

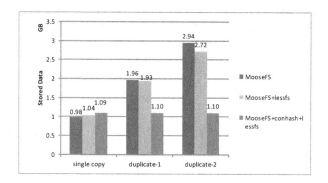

Fig. 5. Stored data size in MooseFS, MooseFS with *lessfs*, and MooseFS with consistent hashing and *lessfs*

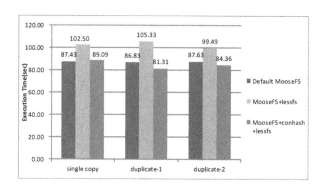

Fig. 6. Execution time in MooseFS, MooseFS with *lessfs*, and MooseFS with consistent hashing and *lessfs*

In Fig. 5, `Default MooseFS` means stored data size in default MooseFS which not use *lessfs*. `MooseFS+lessfs` means stored data size in deduplication on default MooseFS that provide default chunk placement. `MooseFS+conhash+lessfs` means stored data size in deduplication on MooseFS that provide chunk placement using consistent hashing. Fig. 5 shows that our chunk placement scheme is more effective than chunk placement scheme of default MooseFS. Data size of single copy is more than 936 MB because *lessfs* creates metadata to manage data. We define deduplication rate as 100 - ((single copy size × a number of duplicate) ÷ actual stored data size × 100). Deduplication rate of `MooseFS+lessfs` is about 12%. In this regard, chunk placement scheme is needed for improving deduplication rate. Therefore, we experiment applying chunk placement with consistent hashing. As a result, deduplication rate is about 99%. This result shows that proposed system reduces the storage space by 60% than the system without consistent hashing. All data is not eliminated because *lessfs* creates metadata about duplicated data.

In single copy of Fig. 6, execution time of `MooseFS+conhash+lessfs` is slightly slower than `Default MooseFS` because of chunk hashing and deduplication overhead. However, duplicated file copy causes fewer write operation due to eliminating duplicated data. Because `MooseFS+lessfs` does not use content-based chunk placement, `MooseFS+lessfs` rerely performs deduplication. Therefore, `MooseFS+lessfs` is slower than the others. Duplicated file copy of `MooseFS+con hash+lessfs` is faster than single copy and duplicated copy of `Default MooseFS`.

5   Conclusion

In this paper, we study content-based chunk placement for decentralized deduplication on the DFS. We describe how we design our system in detail. We utilize open-source based DFS (MooseFS) and deduplication file system (*lessfs*). We integrate the chunk server of MooseFS with *lessfs*. Therefore, contrary to general deduplication studies, our system can avoid the bottleneck of master server because each chunk server performs decentralized deduplication processes. Moreover, in order to reduce chunk hash overhead, we design content-based chunk placement with consistent hashing that hash foremost 4 KB of chunk data. As a result, experimental results show that proposed system reduces the storage space by 60% than the system without consistent hashing and execution time of proposed system is similar to execution time of default MooseFS.

Acknowledgements. This work was supported by the IT R&D program of MKE/KEIT. [10041244, SmartTV 2.0 Software Platform].

References

1. Gantz, J., Reinsel, D.: 2011 Digital Universe Study: Extracting Value from Chaos. Technical report, IDC (2011)
2. Gantz, J., Reinsel, D.: The Digital Univers. In: 2020: Big Data, Bigger Digital Shadows, and Biggest Growth in the Far East. Technical report, IDC (2011)
3. DuBois, L., Amaldas, M.: IDC key-considerations deduplication. Technical report, IDC (2010)
4. Webb, N.: Open Source Data Deduplication. In: Linuxfest Northwest, Bellingham, WA, USA (April 2011)

5. Koutoupis, P.: Data Deduplication with Linux. 7 (2011)
6. MooseFS, http://www.moosefs.org
7. Hupfeld, F., Cortes, T., Kolbeck, B., Stender, J., Focht, E., Hess, M., Malo, J., Marti, J., Cesario, E.: The XtreemFS architecture—a case for object-based file systems in Grids. Concurrency and Computation: Practice and Experience 20(17), 2049–2060 (2008)
8. XtreemFS, http://www.xtreemfs.org
9. GlusterFS, http://www.gluster.org
10. Weil, S., Brandt, S., Miller, E., Long, D., Maltzahn, C.: Ceph: A scalable, high-performance distributed file system. In: Proceedings of the 7th Symposium on Operating Systems Design and Implementation (OSDI), pp. 307–320 (2006)
11. Ghemawat, S., Gobioff, H., Leung, S.: The Google file system. ACM SIGOPS Operating Systems Review 37, 29–43 (2003)
12. Thanh, T., Mohan, S., Choi, E., Kim, S., Kim, P.: A taxonomy and survey on distributed file systems. In: Fourth International Conference on Networked Computing and Advanced Information Management, NCM 2008, vol. 1, pp. 144–149. IEEE (2008)
13. Srinivasan, K., Bisson, T., Goodson, G., Voruganti, K.: iDedup: Latency-aware, inline data deduplication for primary storage. In: Proceedings of the Tenth USENIX Conference on File and Storage Technologies (FAST 2012), San Jose, CA (2012)
14. Meyer, D., Bolosky, W.: A study of practical deduplication. ACM Transactions on Storage (TOS) 7(4), 14 (2012)
15. El-Shimi, A., Kalach, R., Kumar, A., Oltean, A., Li, J., Sengupta, S.: Primary Data Deduplication-Large Scale Study and System Design. In: Proccedings of the USENIX Annual Technical Conference 2012 (2012)
16. Dubnicki, C., Gryz, L., Heldt, L., Kaczmarczyk, M., Kilian, W., Strzelczak, P., Szczepkowski, J., Ungureanu, C., Welnicki, M.: Hydrastor: A scalable secondary storage. In: Proccedings of the 7th Conference on File and Storage Technologies, pp. 197–210. USENIX Association (2009)
17. Lillibridge, M., Eshghi, K., Bhagwat, D., Deolalikar, V., Trezise, G., Camble, P.: Sparse indexing: large scale, inline deduplication using sampling and locality. In: Proccedings of the 7th Conference on File and Storage Technologies, pp. 111–123 (2009)
18. Wei, J., Jiang, H., Zhou, K., Feng, D.: Mad2: A scalable high-throughput exact deduplication approach for network backup services. In: 2010 IEEE 26th Symposium on Mass Storage Systems and Technologies (MSST), pp. 1–14. IEEE (2010)
19. Zhu, B., Li, K., Patterson, H.: Avoiding the disk bottleneck in the data domain deduplication file system. In: Proceedings of the 6th USENIX Conference on File and Storage Technologies, vol. 18 (2008)
20. Clements, A., Ahmad, I., Vilayannur, M., Li, J., et al.: Decentralized deduplication in SAN cluster file systems. In: Proceedings of the 2009 Conference on USENIX Annual Technical Conference, p. 8. USENIX Association (2009)
21. Karger, D., Sherman, A., Berkheimer, A., Bogstad, B., Dhanidina, R., Iwamoto, K., Kim, B., Matkins, L., Yerushalmi, Y.: Web caching with consistent hashing. Computer Networks 31(11), 1203–1213 (1999)
22. Karger, D., Lehman, E., Leighton, T., Panigrahy, R., Levine, M., Lewin, D.: Consistent hashing and random trees: Distributed caching protocols for relieving hot spots on the World Wide Web. In: Proceedings of the Twenty-Ninth Annual ACM Symposium on Theory of Computing, pp. 654–663. ACM (1997)

23. DeCandia, G., Hastorun, D., Jampani, M., Kakulapati, G., Lakshman, A., Pilchin, A., Sivasubramanian, S., Vosshall, P., Vogels, W.: Dynamo: amazon's highly available key-value store. ACM SIGOPS Operating Systems Review 41, 205–220 (2007)
24. Cassandra, http://cassandra.apache.org
25. Memcached, http://memcached.org/
26. Chen, F., Luo, T., Zhang, X.: CAFTL: A content-aware flash translation layer enhancing the lifespan of flash memory based solid state drives. In: Proceedings of the 9th USENIX Conference on File and Stroage Technologies, p. 6. USENIX Association (2011)

Protecting Wireless Sensor Networks from Energy Exhausting Attacks

Vladimir V. Shakhov

Institute of Computational Mathematics and Mathematical Geophysics of SB RAS
Novosibirsk 630090, Russia
shakhov@rav.sscc.ru

Abstract. One of the critical issue of wireless sensor networks is a poor sensor battery. It leads to sensor vulnerability for battery exhausting attacks. Quick depletion of battery power is not only explained by intrusions but also by a malfunction of networks protocols. In this paper, a special type of denial-of-service attack is investigated. The intrusion effect is the depletion of sensor battery power. In contrast to general denial-of-service attack, quality of service under the considered attack is not necessarily degraded. Therefore, the application of traditional defense mechanism against this intrusion is not always possible. Taking into account proprieties of wireless sensor networks, a model for evaluation of energy consumption under the attack is described. Using this model, the attack detection method is offered.

Keywords: wireless sensor networks, intrusion detection, energy exhausting attack.

1   Introduction

Wireless sensor networks (WSNs) have evolved over the last decade from a stage where these networks were designed in a technology-dependent manner to a stage where some broad conceptual understanding issues now exist. Significant progress in the area micro-electro-mechanical technologies have enabled the development of inexpensive sensor nodes that communicate wirelessly. Wireless sensors can be used in a wide range of applications, such as, air pollution monitoring [1], landslide detection [2], military applications and tracking [3], health care [4]-[7], smart home [8]-[10] etc. As WSNs based applications are deployed, security emerges as an essential requirement. However, wireless sensors are vulnerable for malefactor due to the following reasons. Sensor resources are limited. Usually, WSNs nodes facilities do not allow the application of an efficient defense scheme against intrusions. WSNs technologies are relatively new and so, the corresponding defense tools are poor. Wireless sensor uses unreliable channels. In wireless communications, the signal transmission is disturbed by a noise. A sensor has to contact with other sensors. It is a facility for infection extension. Thus, a security problem is the critical issues of wireless sensor technology. Hence, it is very important to investigate on the potential attacks against wireless terminals.

B. Murgante et al. (Eds.): ICCSA 2013, Part I, LNCS 7971, pp. 184–193, 2013.

There are many sources for the material on security issues in wireless sensor networks (WSNs). Surveys of attacks against WSNs can be found in [11]-[14]. Particularly, taxonomy of denial of service (DoS) attacks in WSNs is described in [15]. Several recent contributions to the literature have addressed privacy issues in sensor systems [16]. Generally attacks against sensors can be classified into attacks on physical, medium access control, network, transportation, and application layers.

Typical attacks on sensor networks are as follows: jamming, lower duty cycle tampering, manipulating routing information, selective forwarding attack, Sybil attack etc. Possible defense techniques include spread-spectrum, tamper-proofing, effective key management schemes, error correcting code, rate limitation, authentication, encryption, redundancy, probing. However, this technique does not protect against the following intrusions. First, a malware at infective node may amplify the transmission range. High illegal electromagnetic emissions cause an exhaustion of a sensor battery. It can also be a reason for noise and degradation of the electromagnetic compatibility of the neighbor sensors. Next, some data aggregation protocols allow to change a packets size. Hence, an intruder can drastically increase a size of legal packets. The mentioned intrusions lead to quick battery exhausting. In this paper we consider a threat of rash depletion of sensor node battery. This type of threat is distinguishing characteristic of wireless sensors technology.

The rest of the paper is organized as follows. In Section 2 the author discusses the reasons, which can lead to the effect of fast battery depletion. Then, we propose a model for DoB attack, provide a method for sensor lifetime calculations, discuss the concept of protection against DoB. Some particular protection strategies are compared in Section 4. Finally, we conclude the paper in Section 5.

2   Energy Exhausting Attack

2.1   Attack Description

It is a well known fact that a cost of sensor components is a critical consideration in the design of practical sensor networks. A cost of sensor network increases with sensor battery power. It is often economically advantageous to discard a sensor rather than sensor recharging. By this reason a battery power is usually a scare component in wireless devices. On the other hand, sensor lifetime depends on battery lifetime. High illegal electromagnetic emissions cause an exhaustion of a sensor battery. It can also be a reason for noise and degradation of the electromagnetic compatibility of the neighbor sensors. The goal of intrusions is the sensors lifetime degradation. In contrast to common DoS attack, quality of service under the considered attack is not necessary degraded.

The considered situation arises in the following cases. The attack affect can be caused by external attack. The intrusion, exemplified by relaxed jamming attack, is based on noise generating. If a sensor cannot adaptively change the power of transmission then, the intrusion is equivalent to the well-known DoS (or DDoS) attack on WSNs which is named as jamming. Jamming can disrupt wireless connections and can occur either unintentionally in the form of interference or collisions. If a sensor can counteract against a noise by an increase in the power of transmission then, the object of attack is the battery of a sensor. A jamming attack is usually effective since no expensive hardware is needed in order to be launched, it can be implemented by simply listening to the open medium and broadcasting in the same frequency band as the network and if launched wisely, it can lead to significant effects with small incurred cost for the intruder.

With regard to the system and impact of jamming attacks, they usually aim at the physical layer and are realized by means of a high transmission power signal that corrupts a communication link or an area. In the case of relaxed jamming the generated noise does not disturb the signal transmission. But it stimulates sensors to increase the power gain. All applications can get the required level of service and do not feel the intrusion.

The second, exemplified by packets flooding, is based on spoofed traffic generating. Sensors have to spend energy for the spoofed packets transmission. To make difficulties for the attack detection, an attacker can keep all legal traffic and often retransmit legal packets only.

The third, exemplified by modifications of Hello flooding attack, is based on illegal instructions to increase the signal power. In the Hello flooding attack, sensors try to get an initiator of illegal Hello message, retransmit packets and waste energy. In general, an intruder can send illegal instructions to sensors as well as successful delivery notifications.

The energy exhausting attack is not limited by the relaxed jamming attack or flooding attacks. For example, sybil or wormhole can get the same effect. In same cases, the denial of sleeping attack can be organized by external way.

The next reason of the threat is an inner attack or malware. WSNs are prone to the spread of self-replicating malicious codes known as malware. The malware can be used to initiate different forms of attacks ranging from the less intrusive eavesdropping of the sensed data to the more virulent disruption of node functions such as relaying and establishing end-to-end routes, or even destroying the integrity of the in transit sensed data, as in unauthorized access and session hijacking attacks. Malware can deplete the energy reserves of the sensor nodes and render them dysfunctional either deliberately or as a result of aggressive media access actions in attempt to infect others.

The economic viability of the investments on the sensing infrastructure is therefore contingent on the design of effective security countermeasures. The energy exhausting effect can be generated by unintentional actions as well. It can be failure of inner sensor protocols or inefficient function of network protocol.

2.2   Related Works

In previous works there have been several notions of energy exhausting attacks. In [17] authors investigate reliability of transport protocols for WSNs. They demonstrate that the method of end-to-end communications reliability supporting based on control packet injections and packets replication opens opportunities for energy depletion attacks. It is concluded that it is impossible to fully protect a protocol against energy depleting attacks without authentication (using cryptographic solutions). However, authors do not consider any tools to counteract against the intrusion. In [18], the battery depletion effect through reduction of sleep cycles of is described. We mentioned it above.

In [19], [20] authors investigate malware attacks in wireless networks. They represent the propagation of malware in a battery-constrained wireless network by an epidemic model, provide technique to quantify the damage that the malware can inflict on the network, describe conditions for optimal attack policy. However, specificity of WSANs is not properly used. Details of intrusion and attack detection methods are not considered as well. It seems the provide results cannot be used for WSNs protection against energy depleting attacks.

Thus, most of the existing work on security control of wireless sensor networks provide a general notion of energy exhausting attacks possibility. In particular cases the attack impact had been evaluated by simulation technique. Comprehensive methods for performance analysis of energy exhausting attacks and estimations of defense approaches efficiency have not been offered. Related mathematical models have not been developed.

3   Attack Model

3.1   Assumptions and Designations

Assume, a firmware of sensors that supports sleeping mode. It is a widely used approach on energy consumption optimization technique. A sensor can get the following stages:

- Sleep stage - A sensor does not transmit any data and spends minute amount of energy.
- Idle stage - A sensor works but it does not transmit any data.
- Transmit stage - A sensor transmits some data.

Assume, the sensor stay durations in the states are independent and exponentially distributed. Because of these assumptions, the system can be modeled with Markov chains. We could analyze the system in terms of continuous-time Markov chain technique. It is sufficient, however, for our purposes in this paper to use the simpler theory of discrete-time Markov chains.

Assume that the intensities of sensors activating are λ_{si} and λ_{st}. This values are defined by the MAC protocols for lifetime prolongation. Let the intensities of the sensors deactivating and placing in the sleep mode be λ_{is} and λ_{ts}.

Following some protocols, the sensor transmits data at the end of an active period (directly before sleeping). In this cases, $\lambda_{is} = 0$. Other intensities are defined by the target behavior, an offered load and the current capacity of a sensor battery. Let us make designations for sensors energy consumption per a time unit in the mentioned stages: E_{sleep}; E_{idle}; $E_{transmit}$. The expected sensor energy consumption is as follows.

$$E = P_{sleep}E_{sleep} + P_{idle}E_{idle} + P_{transmit}E_{transmit} \qquad (1)$$

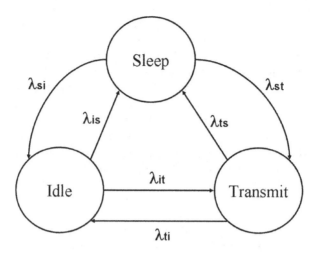

Fig. 1. The state diagram of sensor behavior

Here, P_{sleep}; P_{idle}; $P_{transmit}$ are the steady-state occupancy probabilities for Sleep states, Idle state, and Transmit state correspondingly. If the current capacity of a sensor battery equals C then the sensor lifetime LT can be estimated as follows

$$LT = C/E$$

Let us remark that a sensor is under the attack in the Transmit state. Therefore, we have to know probabilities of other states to evaluate counteracting strategies against the attack.

3.2 Sensor Lifetime Estimation

From the state diagram above, we can derive a set of equations that can be solved for steady-state occupancy probabilities. Application of equality (1) then yields the estimation of sensor lifetime. By writing down the equilibrium equations for the steady-state probabilities, we obtain

$$P_{sleep}(\lambda_{si} + \lambda_{st}) = P_{idle}\lambda_{is} + P_{transmit}\lambda_{ts};$$

$$P_{idle}(\lambda_{is} + \lambda_{it}) = P_{sleep}\lambda_{si} + P_{transmit}\lambda_{ti};$$

$$P_{transmit}(\lambda_{ts} + \lambda_{ti}) = P_{sleep}\lambda_{st} + P_{idle}\lambda_{it}.$$

The received equations are not independent. Using the normalization condition,

$$P_{sleep} + P_{idle} + P_{transmit} = 1,$$

the system can be rewritten as,

$$P_{sleep}(\lambda_{si} + \lambda_{st} + \lambda_{is}) + P_{transmit}(\lambda_{ts} - \lambda_{is}) = \lambda_{is},$$

$$P_{sleep}(\lambda_{it} - \lambda_{st}) + P_{transmit}(\lambda_{ts} + \lambda_{ti} + \lambda_{it}) = \lambda_{it}.$$

Now we get the system of independent equations. It is convenient to rewrite the equations as follows

$$a_1 P_{sleep} + b_1 P_{transmit} = \lambda_{is};$$

$$a_2 P_{sleep} + b_2 P_{transmit} = \lambda_{it};$$

where

$$a_1 = \lambda_{si} + \lambda_{st} + \lambda_{is};$$

$$b_1 = \lambda_{ts} - \lambda_{is};$$

$$a_2 = \lambda_{it} - \lambda_{st};$$

$$b_2 = \lambda_{ts} + \lambda_{ti} + \lambda_{it}.$$

From these equations, we obtain

$$P_{transmit} = \frac{\lambda_{is} a_2 - \lambda_{it} a_1}{a_2 b_1 - a_1 b_2},$$

$$P_{sleep} = \frac{\lambda_{is} - b_1 P_{transmit}}{a_1},$$

$$P_{idle} = 1 - P_{transmit} - P_{sleep}.$$

Using the formulas for steady-state occupancy probabilities, we can calculate a sensor lifetime LT. Note that the formula (1) represents the criterion for comparison of different particular defense scenarios.

4   Counteracting Method

Generally, the defense method against the considered attack in wireless sensor networks can be implemented by the following way,

- Step 1 - To detect the attack.
- Step 2 - To localize the attacked sensors.
- Step 3 - To place the attacked sensors in sleeping mode.

A sensor can be equipped with a firmware for sleeping mode activation depending on the observed Bit Error Rate and battery capacity. From the above consideration, it is clear that to effectively counteract intrusions, a protection mechanism has to be activated in the attack stage and deactivated at the normal stage. If the defense scheme is used in the normal stage then, the system functionality degrades. Therefore, it is important to detect an intrusion presence. For the detection of the considered intrusion an algorithm for discord detection can be used. Let us consider a random sequence of transmitted bits. Assume that the probability of the error bit is constant. The probability of error in the normal state differs from the probability of error under intrusions.

Thus, we have to detect a change point in a sequence of random variables distributed by Gauss CDF. Here, the change point is the star time of an attack. It needs to minimize the difference between the change point and the alarm signal generating moment. A change-point detection procedure based on cumulative sums (CUSUM) can be used. It can be remarked that an attacked system can work without significant losses during some time period. If a false alarm probability is given then, we can use the efficient algorithm for the detection of change point with an admissible lag [21].

The detection procedure in the common case is as follows. Let us consider a random sequence of consumed energy units. Assume that the probability of the error bit is constant. The probability of error in the normal state differs from the probability of error under intrusions. Thus, we have to detect a change point t in a sequence of random variables $\{e_i\}$ distributed by using normal CDF. The change point t is the star time of an attack (moment of discord).

Let A be a moment of time when a discord is detected (an alarm is generated). It needs to minimize the difference $A - t$. If $A < t$ then, a false alarm is generated. This implies that the defense mechanism is used during a normal stage. It needs to avoid this situation. An efficient point-of-change detection method seldom generates false alarms. A conventional change-point detection procedure based on cumulative sums (CUSUM) makes an alarm at that time

$$A = \arg \inf_i \{i \leq 1; S_i = max(0, S_{i-1} + \Delta_i) > h\},$$

$$S_0 = 0.$$

Here, h is a threshold value. Optimal threshold calculation is a nontrivial problem. It can be defined by simulation. Δ_i is a function of $\{e_i\}$. Generally, it is a logarithm of likelihood ratio.

Taking into account the results above, let us compare sensor survivability for two defense scenarios. Let $\lambda_{st} = \lambda_{is} = \lambda_{ti} = 0$. It means a sensor transmits data at the end of an active slot of duty cycle. After transmission session, sensor gets the sleep mode. A sensor does not transmit any data directly after awakening. Assume, $\lambda_{it} = 1$. Let λ_{si} and λ_{ts} be controlled parameters. In this case the formulas above are simplified.

Remark, $E_{sleep} << E_{idle} << E_{transmit}$ under DoB intrusions. If we reduce λ_{si} or λ_{ts} then sensors lifetime is increased. Let us consider two scenario. In the first, both x and y parameters are smoothly sequentially reduced, step by step. In the second, λ_{ts} is quickly reduced. It means a sensor increases sleeping time and packets gathering happen. So, sometimes the parameter λ_{si} gets stepwise changes and increases the value. The second strategy seems preferable. But as it was shown on the Figure 2 both strategies can be applied. A scenario advantage depends on the concrete situation.

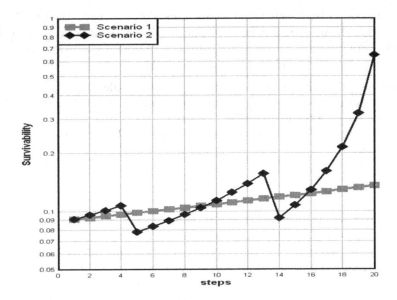

Fig. 2. The state diagram of sensor behavior

Let us consider the effect of attack. We use the assumptions above and let $\lambda_{it} = \lambda_{si} = \lambda_{ts} = \lambda$ in the normal case, where λ is a constant. Let $\lambda_{it} = A$ under the attack and for the constant A it is true the inequality $A \gg \lambda$. The degradation of network lifetime under attack can be estimated by the ration of transmission state probabilities, which are calculated in normal case and in the case of intrusion. It need to take into account that an attacked sensor spends additional energy in the transmission stage. Assume, the intensity of energy consumption is proportional to the attack intensity. The results of network lifetime degradation (fraction) is shown in the second column of Table 1.

Table 1.

ratio A/λ	LF degradation	sleep mode increasing
2	0.42	1.5
5	0.15	2.14
10	0.07	2.5
50	0.01	2.89
100	0.007	2.94
200	0.003	2.97
500	0.001	2.98

If the intrusion is detected and sensor gets sleeping mode then the attack effect is reduced. Assume, if defense mechanism is activated then $\lambda_{ts} = A$. It is clear that here

$$P_{idle} = P_{transmit} << P_{sleep}.$$

In this situation the attack is prevented, however sleeping time has to be increased and the sensor duty cycle has to be revised. The third column of Table 1 shows the sleeping mode increasing (times).

Thus, the energy exhausting attack can be efficiently prevented by the proposed approach.

5   Conclusion

In this paper we investigated a survivability of WSNs nodes under the attack named Denial of Battery. A corresponding mathematical models based on stochastic processes have been developed. The considered intrusion can be classified as a novel type of attack in wireless sensor networks. In contrast to the previously considered attacks, the object of a novel potential attack is sensors batteries. Preferable counteracting technologies can be implementation of multilevel independent power control and also, effective monitoring based on the methods of discord detection.

References

1. Ma, Y., Richards, M., Ghanem, M., Guo, Y., Hassard, J.: Air Pollution Monitoring and Mining Based on Sensor Grid in London. Sensors 8, 3601–3623 (2008)
2. Zhao, Y., Shouzhi, X., Shuibao, Z., Xiaomei, Y.: Distributed detection in landslide prediction based on Wireless Sensor Networks. In: Proceedings of World Automation Congress, Puerto Vallarta, Mexico, June 24-28, pp. 235–238 (2012)
3. Akyildiz, I.F., Su, W., Sankarasubramaniam, Y., Cayirci, E.: Wireless Sensor Networks: A Survey. Computer Networks 38, 393–422 (2002)
4. Istepanian, R., Jovanov, E., Zhang, Y.: Guest editorial introduction to the special section on M-Health: Beyond seamless mobility and global wireless Health-Care connectivity. IEEE Trans. Inf. Technol. Biomed. 8, 405–414 (2004)
5. Milenkovic, A., Otto, C., Jovanov, E.: Wireless sensor network for personal health monitoring: Issues and an implementation. Comput. Commun. 29, 2521–2533 (2006)
6. Junnila, S., Kailanto, H., Merilahti, J., Vainio, A.-M., Vehkaoja, A., Zakrzewski, M., Hyttinen, J.: Wireless, Multipurpose In-Home Health Monitoring Platform: Two Case Trials. IEEE Trans. Inf. Technol. Biomed. 14, 447–455 (2010)
7. Bachmann, C., Ashouei, M., Pop, V., Vidojkovic, M., Groot, H.D., Gyselinckx, B.: Low-power wireless sensor nodes for ubiquitous long-term biomedical signal monitoring. IEEE Commun. Mag. 50, 20–27 (2012)
8. Han, K., Shon, T., Kim, K.: Efficient mobile sensor authentication in smart home and WPAN. IEEE Trans. Consum. Electr. 56, 591–596 (2010)
9. Byun, J., Jeon, B., Noh, J., Kim, Y., Park, S.: An intelligent self-adjusting sensor for smart home services based on ZigBee communications. IEEE Trans. Consum. Electr. 58, 591–596 (2012)

10. Nakamura, M., Igaki, H., Yoshimura, Y., Ikegami, K.: Considering Online Feature Interaction Detection and Resolution for Integrated Services in Home Network System. In: Proceedings of the 10th International Conference on Feature Interactions in Telecommunications and Software Systems, Lisbon, Portugal, June 11-12, pp. 191–206 (2009)
11. Zhou, Y., Fang, Y., Zhang, Y.: Securing wireless sensor networks: a survey. IEEE Commun. Surv. Tutor. 10, 6–28 (2008)
12. Young, M., Boutaba, R.: Overcoming Adversaries in Sensor Networks: A Survey of Theoretical Models and Algorithmic Approaches for Tolerating Malicious Interference. IEEE Commun. Surv. Tutor. 13, 617–641 (2011)
13. Chen, X., Makki, K., Yen, K., Pissinou, N.: Sensor network security: a survey. IEEE Commun. Surv. Tutor. 11, 52–73 (2009)
14. Mpitziopoulos, A., Gavalas, D., Konstantopoulos, C., Pantziou, G.: A survey on jamming attacks and countermeasures in WSNs. IEEE Commun. Surv. Tutor. 11, 42–56 (2009)
15. Raymond, D., Midkiff, S.: Denial-of-Service in Wireless Sensor Networks: Attacks and Defenses. IEEE Pervasive Computing 7, 74–81 (2007)
16. Sun, J., Fang, Y., Zhu, X.: Privacy and emergency response in e-healthcare leveraging wireless body sensor networks. IEEE Wireless Communications 17, 66–73 (2010)
17. Buttyan, L., Csik, L.: Security analysis of reliable transport layer protocols for wireless sensor networks. In: Proceedings of the 8th IEEE International Conference on Pervasive Computing and Communications Workshops (PERCOM Workshops 2010), pp. 419–424 (April 2010)
18. Brownfield, M., Gupta, Y., Davis, N.: Wireless sensor network denial of sleep attack. In: Proc. 6th Annu. IEEE SMC Inform. Assurance Workshop, pp. 356–364 (2005)
19. Khouzani, M., Sarkar, S.: Maximum damage battery depletion attack in mobile sensor networks. IEEE Trans. Autom. Control 56(10), 2358–2368 (2011)
20. Khouzani, M., Sarkar, S., Altman, E.: Maximum damage malware attack in mobile wireless networks. IEEE/ACM Transactions on Networking 20(5), 1347–1360 (2012)
21. Shakhov, V.V., Choo, H., Bang, Y.: Discord model for detecting unexpected demands in mobile networks. Future Generation Comp. Syst. 20(2), 181–188 (2004)

Semantic Annotation of the CEREALAB Database by the AGROVOC Linked Dataset

Domenico Beneventano, Sonia Bergamaschi, and Serena Sorrentino

Department of Engineering "Enzo Ferrari" - via Vignolese 905, 41125 - Modena
University of Modena and Reggio Emilia - Italy
firstname.lastname@unimore.it

Abstract. The objective of the CEREALAB database is to help the breeders in choosing molecular markers associated to the most important traits. Phenotypic and genotypic data obtained from the integration of open source databases with the data obtained by the CEREALAB project are made available to the users. The CEREALAB database has been and is currently extensively used within the frame of the CEREALAB project.

This paper presents the main achievements and the ongoing research to annotate the CEREALAB database and to publish it in the Linking Open Data network, in order to facilitate breeders and geneticists in searching and exploiting linked agricultural resources. One of the main focus of this paper is to discuss the use of the AGROVOC Linked Dataset both to annotate the CEREALAB schema and to discover schema-level mappings among the CEREALAB Dataset and other resources of the Linking Open Data network, such as NALT, the National Agricultural Library Thesaurus, and DBpedia.

1 Introduction

The CEREALAB database [15] is a tool realized to help the breeders in choosing molecular markers associated to the most important economically phenotypic traits. The CEREALAB database can help breeders and geneticists to unravel the genetics of economically important phenotypic traits, to identify and to choose molecular markers associated to key traits, and finally to choose the desired parentals for breeding programs. The CEREALAB database development was one of the objectives of the CEREALAB projects and of the BIOGEST-SITEIA laboratory[1] funded by Emilia-Romagna regional government (Italy), aiming to increase the competitiveness of Regional seed companies through the use of modern selection technologies such as the Marker-Assisted Selection (MAS).

The key feature of the CEREALAB database is that phenotypic and genotypic data are obtained from the integration of open source databases with the data obtained by the CEREALAB project. Data integration is obtained by using the MOMIS[2] system (Mediator envirOnment for Multiple Information Sources), a framework to perform integration of structured and semi-structured data sources [2,4]. MOMIS is characterized

---

[1] www.biogest-siteia.unimore.it

[2] http://www.dbgroup.unimore.it/Momis/

B. Murgante et al. (Eds.): ICCSA 2013, Part I, LNCS 7971, pp. 194–203, 2013.
© Springer-Verlag Berlin Heidelberg 2013

by a classical wrapper/mediator architecture: the local data sources contain the real data, while a Global Schema (GS) provides a *reconciled, integrated, read-only view* of the underlying sources. The GS and the mappings between the GS and the local sources are semi-automatically defined at design time by the Integration Designer component of the system [2]. After GS creation end-users can pose queries over the GS in a transparent way w.r.t. the local sources. An open source version of the MOMIS system is delivered and maintained by the academic spin-off DataRiver[3].

The *Linked Open Data* (LOD) project is a community effort (founded by the W3C Semantic Web Education and Outreach (SWEO) group[4]) which aims to extend the Web with data by publishing various open data sets as RDF on the Web and by setting RDF links between data items from different data sources[7]. Nowadays, LOD includes more than 300 data sets from different domains of knowledge: among them, we recall general domain data sets like Wikipedia[5] and WordNet[6], and agriculture data sets like AGROVOC[7] and NALT[8]. The great majority of agricultural resources is typically accessed only by closed communities and even when they are make available on the Web, they look more as a sets of disconnected information units than as an integrated information space [16].

As a consequence, this paper presents the main achievements and the ongoing research to annotate the CEREALAB database and to discover semantic mappings between it and other LOD datasets in the network, in order to facilitate breeders and geneticists in searching and exploiting linked agricultural resources. To this aim, we propose the following process consists of three main steps:

RDF-ization for converting the data into RDF.
   To publish and annotate CEREALAB in the LOD network, first of all, we need to translate the relational CEREALAB database into an RDF database (this process is called *RDF-ization* [17]).
Semantic Enrichment for understanding the semantics of source schemata.
   Even if the LOD community is constantly growing, there is still a few applications making use of its data sets. This is mainly due to the fact that links between LOD data sets are almost exclusively on the instance level, while schema level information is almost ignored [11]. To efficiently use LOD data sets, consumers need to deeply understand the semantics of source schemata; moreover, the hidden meanings associated to schema elements can be exploited in order to discover semantic mappings and thus to perform integration of different LOD data sets [18].
Linking and mapping for discovering instance level-links and semantic mapping between the public data sets and other LOD resources;
   After having published the data set to the LOD, we need to link the data set to other LOD resorces; creating links and mappings between the published resources is a

---

[3] http://www.datariver.it
[4] http://www.w3.org/wiki/SweoIG/TaskForces/CommunityProjects/
   LinkingOpenData
[5] http://it.wikipedia.org/
[6] http://wordnet.princeton.edu/
[7] http://aims.fao.org/standards/agrovoc/about
[8] http://agclass.nal.usda.gov/

key part of the Linked Open Data paradigm. We can identify two kinds of connection: *Instance-Level links* which are established between LOD data set instances and *Schema-level mappings* which are established between schema concepts. In the LOD cloud, instance-Level links represent the great majority of links; on the contrary, schema-level mappings are almost absent in the LOD cloud even if, as previously described, they represents a fundamental means to integrate different LOD resources. Integrating LOD data sets from different organizations can yield high value information [9]; in particular, schema-level integration of LOD data sets is an issue which has been pointed out in [7] as a core challenge.

In this paper we focus on the *Semantic Enrichment* and mapping phases: to describe the semantics at the schema level, we use *semantic annotation*, i.e. the explicit association of one or more *meanings* to schema element labels (classes and attributes names) with respect to reference knowledge sources. Then, we will show how semantic annotation is a key tool in the context of LOD integrations since starting from semantic annotations it is possible to discover schema-level mappings, i.e. semantic correspondences at the schema-level.

One of the main focus of this paper is to discuss the use of AGROVOC Linked Dataset either to annotate the CEREALAB schema and to discover schema-level mappings among the CEREALAB Dataset and other resorces of the Linking Open Data network. The choice to use AGROVOC is motivated in section 2, after a brief synthesis on the state-of-the-art. In particular, AGROVOC will be first used in the *Semantic Enrichment* step: in Section 3 we will discuss how to annotate the CEREALAB schema with respect to AGROVOC. Then in Section 4, we will show how AGROVOC is used as background knowledge to discover schema-level mappings among the CEREALAB Dataset and other LOD resources. Finally in Section 5, we give our concluding remarks and describe future work.

2   State of the Art and Motivations

Biological ontologies are essential tools for accessing and analyzing the rapidly growing pool of plant genomic and phenomic data, since they offer a flexible framework for comparative plant biology, based on common botanical understanding [21]; in particular, the authors highlighted that as genomic and phenomic data become available for more species, that the annotation of data with ontology terms will become less centralized, while at the same time, the need for cross-species queries will become more common, causing more researchers in plant science to turn to ontologies.

As stated in [21], one of the most relevant ontologies for the plant sciences (excluding specialized ontologies used by crop breeders and agronomists) is the Plant Ontology[9], which covers gross plant anatomy and morphology at the level of the cell and higher, as well as plant development stages. In [10] a retrieval engine which exploits the content and structure of available domain ontologies to expand and enrich retrieval results in major plant genomic databases is proposed. A more sophisticated semantic application

---

[9] http://www.plantontology.org/

is Semantic J-SON [12] which can be used for biological applications including genome design, sequence processing, inference over phenotype databases, and full-text search indexing.

In the context of the Linked Open Data effort, the main agriculture data set is AGROVOC[10], a multilingual agricultural thesaurus developed by the United Nations Food and Agricultural Organization and mainly used for controlled-vocabulary indexing and retrieving data in agricultural information systems; it was developed with the aim of standardizing the indexing process of agricultural resources. In particular, AGROVOC includes the AGRIS network[11], an information system for indexing and retrieval, based on a global public domain Database with 4190822 structured bibliographical records on agricultural science and technology. AGROVOC is published as Linked Open Data, containing links and references to many other Linked Datasets in the LOD cloud, including EUROVOC[12], a multilingual, multidisciplinary thesaurus covering the activities of the EU, NALT[13], the National Agricultural Librarys Agricultural Thesaurus, and DBpedia[14].

As stated in [13], AGROVOC defines the stage for organizations around the world to start publishing their agriculture knowledge models by linking them to AGROVOC, as well as utilizing AGROVOC for resource management.

Starting from this consideration, one of the main focus of this paper is to discuss the use of AGROVOC in the process of publishing and linking the CEREALAB database to the LOD cloud. Another important reason behind the choice of using AGROVOC as semantic resource for the CEREALAB database is that it covers the great majority of the CERELAB names (indicatively a 80%) while Plant Ontology, another relevant resource, only contains the 60% of the names.

Among the other tools for agricultural data management where AGROVOC is being used, one of the most significant is VocBench[15]; developed by FAO and its partners, VocBench provides tools and functionalities that facilitate both collaborative editing and multilingual terminology. the system architecture and significant set of features available in the VocBench are discussed in [19]; in particular, it facilitates as a collaborative tool allowing experts to manage multilingual terminology and semantic concept; moreover, it provides a workflow for the maintenance, validation, and consistency checks. In our framework, we use AGROVOC mainly to to automatically discover mappings between CEREALAB and other Liked Open Datasets.

3   Semantic Annotation of the CEREALAB Database

A preliminary idea to annotate the CEREALAB schema with respect to a reference knowledge source was presented in [1]. As described above, we decided to exploit AGROVOC as semantic resource for annotating the CEREALAB database. AGROVOC

---

[10] http://aims.fao.org/website/AGROVOC-Thesaurus/sub
[11] http://agris.fao.org/
[12] http://eurovoc.europa.eu/
[13] http://agclass.nal.usda.gov/
[14] http://dbpedia.org
[15] http://aims.fao.org/tools/vocbench-2

is a thesaurus made up of agricultural terms consisting of one or more words. Each term is related to other terms via a set of relationships including *BT* (broader term), *NT* (narrower term) and *RT* (related term).

Figure 1 shows an excerpt of the AGROVOC semantic network, for instance, for the term "Lodging" we can easily derive that it is a kind of "plant damage" (as they are connected by a BT relationship) and that it is related to other terms such as "Lodging resistance", "Rain" etc. Moreover, AGROVOC provides term definitions in different languages (e.g., . This property is particularly relevant as, starting from the annotations, we can automatically enrich the CEREALAB schema with the natural language descriptions of its elements thus helping even not skilled users to understand the meaning of specific agricultural terms.

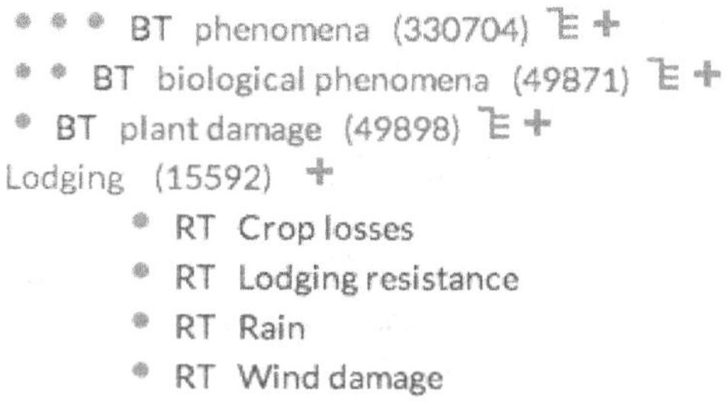

```
* * * BT  phenomena  (330704)  E +
  * * BT  biological phenomena  (49871)  E +
    * BT  plant damage  (49898)  E +
Lodging  (15592)  +
              * RT  Crop losses
              * RT  Lodging resistance
              * RT  Rain
              * RT  Wind damage
```

Fig. 1. An excerpt of the AGROVOC semantic network for the term "Lodging"

However, CEREALAB contains several names that are not specific of the agricultural domain and thus they are not present in AGROVOC (e.g. "index" and "ratio"). For these terms, we decided to employ in association with AGROVOC the WordNet general domain thesaurus. Since WordNet terms may have more than one possible meaning (e.g., the term "index" may mean a numerical scale or the finger next to the thumb), a Word Sense Disambiguation (WSD) algorithm is needed; we decided to use our CWSD (Combined Word Sense Disambiguation) algorithm [6]; however, other WSD algorithms might be applied.

The main problem we encountered in the application of WSD algorithms to our context was the presence in CEREALAB of several compound nouns (i.e., nouns composed by more than one term, e.g., "Septoria Tritici"). To annotate these terms we employed the normalization tool described in [20] which allows to annotate compound names by considering the meaning of its constituent terms. Moreover, it provides additional functionalities to semi-automatically expand abbreviations and acronyms. Starting from the semantic annotations, it is possible to automatically discover semantic mappings among schema elements.

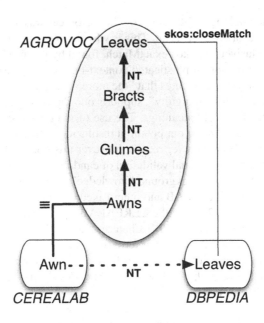

Fig. 2. Schema mapping discovery between CEREALAB and DBPedia by using AGROVOC as background knowledge

4   Schema Mapping Discovery between CEREALAB and Linked Open Datasets

In this section, we describe how it is possible to discover semantic mappings among CEREALAB and other LOD resources starting from the previous obtained semantic annotations.

As described above, AGROVOC is published as Linked Open Data, containing links and references to many other Linked Datasets in the LOD cloud. Among them there is the general scope resources DBpedia[16] (with 1099 matches to AGROVOC) and the specific agricultural domain resource NALT (National Agricultural Librarys Agricultural Thesaurus). As discussed in [8], the mappings between AGROVOC and the other LOD resources are expressed by using SKOS (Simple Knowledge Organization system)[17] a language built upon RDF designed to represent thesauri and taxonomies. In particular, two kind of mappings are considered: **skos:exactMatch** defined between an AGROVOC terms and the terms of another specific agricultural LOD resource (e.g., NALT): these mappings were found by applying string similarity matching algorithms to pairs of labels; then, some of these candidate mappings were validated by domain experts; **skos:closeMatch** defined between AGROVOC terms and the terms of a general domain LOD resource (e.g., DBPedia).

---

[16] http://dbpedia.org
[17] http://www.w3.org/2004/02/skos/

The objective when linking AGROVOC to other resources was to provide only main anchors, privileging accuracy over recall. This is why they (mostly) rejected skos:close Match and relied exclusively on skos:exactMatch, found by means of string-similarity techniques as opposed to more sophisticated context-based approaches. Also, the One Sense per Domain assumption specifies that "the more specialized a domain is, the less is the influence of word sense ambiguity", supports our claim that (in our case) similar strings correspond to equivalent meanings. The use of more sophisticated approaches might have contributed to filtering out potential results more than widening their number (thus incrementing precision over recall), however this potential loss of precision was well compensated by the manual validation of candidate links by a domain expert.

We can use AGROVOC as "background knowledge" in order to automatically discover mappings between CEREALAB and other Liked Open Datasets. Let us consider, the example shown in figure 2: given the CEREALAB term "Awn" annotated with the AGROVOC concept "Awns" and the DBPedia term "Leaves" linked to the AGROVOC concept "Leaves", as "Awns" is a narrower term of "LeavesAwns" in AGROVOC (i.e., they are connected by a chain of NT relationships), we can automatically derive and NT/BT relationship between the AGROVOC term "Awn" and the DBPedia term "Leaves".

In the same way, we can address the problem of mapping discovery in case of compound nouns. Let us consider the example shown in Figure 3 where we have annotated the compound noun "Frost Damage" with respect to AGROVOC. However, this compound noun is not present in DBPedia but there exist only its constituent terms "Frost" and "Damage". By exploiting the correspondences between DBPedia and AGROVOC we can discover that the CEREALAB "Frost damage" term is connected to DBPedia by an RT relationship to "Frost" and an NT/BT relationship to "Damage".

The same mapping disocvery process can be applied to all the other LOD resources aligned with AGROVOC. Figure 4 shows an example where we discover mappings between CEREALAB and DBPedia and CEREALAB and NALT, starting from the annotation of the CEREALAB term "Awn" and by exploiting the semantic network of AGROVOC together with the LOD mappings.

In the above examples the alignement are based on SKOS relationship (NT/BT); however, it is possible to discover other types of alignments like equivalence (the semantic driven mapping discovery process is described in [6,5]).

5   Conclusion and Future Work

This paper presented the main achievements and the ongoing research to publish and link CEREALAB to the LOD cloud in order to facilitate breeders and geneticists in searching and exploiting linked agricultural resources.

One of the main directions for our future research will be the extension of the CE-REALAB LOD with *provenance*, which is crucial in enriching the context surrounding open data and can be used in the process of trustworthiness and quality assessment. The Data Provenance in the Web of Data is a new technique which enabled the publishers and consumers of Linked Data to assess the quality and trustworthiness of the data [14]. In [3], we discussed the ongoing effort on the design and development of a Provenance

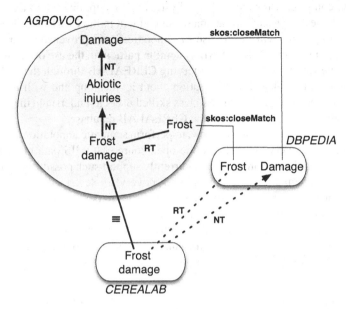

Fig. 3. Schema mapping discovery between CEREALAB and DBPedia starting from compound noun annotations

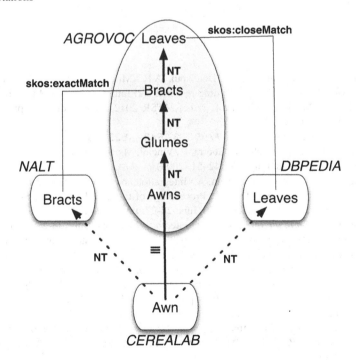

Fig. 4. Schema mapping discovery between CEREALAB and DBPedia and between CERE-ALAB and NALT

Management component for the MOMIS System and its application to the domain of genotypic and phenotypic cereal-data management within the CEREALAB project.

Another future work will be the use of annotation (and provenance) to enrich the GUI functionalities of the CEREALAB system; in particular, the use of semantic annotation techniques might help users in querying CEREALAB through the GUI: we can associate each GUI label with a description about its meaning and with is-a relationships with other terms, thus helping also less skilled users to understand the meaning of the specific agricultural terms used in the CEREALAB database.

From an implementation point of view, to perform semantic annotation and semantic driven mapping discovery, we will use the open-source MOMIS data integration system; the MOMIS tool functionalities that currently support such possibilities for semantic enrichment and semantic driven mapping discovery are described in [6,5]; however, MOMIS will need to be reviewed and adapted in order to deal with LOD resources. In this way the proposed approach could be properly evaluated; in particular we will evaluate whether is it possible to annotate all the entries of CEREALAB, what are the entries which put difficulties and how the performances of the proposed approach depend on the quality of the used sources.

Acknowledgements. This work is partially supported by the BIOGEST-SITEIA laboratory (www.biogest-siteia.unimore.it), funded by Emilia-Romagna (Italy) regional government.

References

1. Beneventano, D., Bergamaschi, S., Dannaoui, A.R., Milc, J., Pecchioni, N., Sorrentino, S.: The CEREALAB database: Ongoing research and future challenges. In: Dodero, J.M., Palomo-Duarte, M., Karampiperis, P. (eds.) MTSR 2012. CCIS, vol. 343, pp. 336–341. Springer, Heidelberg (2012),
   http://dx.doi.org/10.1007/978-3-642-35233-1_32
2. Beneventano, D., Bergamaschi, S., Guerra, F., Vincini, M.: Synthesizing an integrated ontology. IEEE Internet Computing 7(5), 42–51 (2003)
3. Beneventano, D., Dannoui, A.R., Sala, A.: Integration and provenance of cereals genotypic and phenotypic data. In: SEBD 2011, Proceedings of the 20th Italian Symposium on Advanced Database Systems, Venice, Italy, June 24-27, pp. 84–94 (2012); poster presented at the *Data Integration in the Life Sciences* Conference, Maryland, June 28-29, (2012)
4. Bergamaschi, S., Beneventano, D., Guerra, F., Orsini, M.: Data integration. In: Handbook of Conceptual Modeling: Theory, Practice and Research Challenges. Springer (2011)
5. Bergamaschi, S., Beneventano, D., Po, L., Sorrentino, S.: Automatic normalization and annotation for discovering semantic mappings. In: Ceri, S., Brambilla, M. (eds.) Search Computing II. LNCS, vol. 6585, pp. 85–100. Springer, Heidelberg (2011),
   http://dl.acm.org/citation.cfm?id=1983774.1983785
6. Bergamaschi, S., Po, L., Sorrentino, S.: Automatic annotation in data integration systems. In: Meersman, R., Tari, Z. (eds.) OTM 2007 Ws, Part I. LNCS, vol. 4805, pp. 27–28. Springer, Heidelberg (2007)
7. Bizer, C., Heath, T., Berners-Lee, T.: Linked data - the story so far. Int. J. Semantic Web Inf. Syst. 5(3), 1–22 (2009)

8. Caracciolo, C., Morshed, A., Stellato, A., Johannsen, G., Jaques, Y., Keizer, J.: Thesaurus maintenance, alignment and publication as linked data: The agrovoc use case. International Journal of Metadata, Semantics and Ontologies 7(1), 65–75 (2012),
   http://eprints.rclis.org/17734/
9. Coletta, R., Castanier, E., Valduriez, P., Frisch, C., Ngo, D., Bellahsene, Z.: Public data integration with websmatch. CoRR abs/1205.2555 (2012)
10. Green, J.M., Harnsomburana, J., Schaeffer, M.L., Lawrence, C.J., Shyu, C.R.: Multi-source and ontology-based retrieval engine for maize mutant phenotypes. Database (2011),
    http://database.oxfordjournals.org/content/2011/
    bar012.abstract
11. Jain, P., Hitzler, P., Sheth, A.P., Verma, K., Yeh, P.Z.: Ontology alignment for linked open data. In: Patel-Schneider, P.F., Pan, Y., Hitzler, P., Mika, P., Zhang, L., Pan, J.Z., Horrocks, I., Glimm, B. (eds.) ISWC 2010, Part I. LNCS, vol. 6496, pp. 402–417. Springer, Heidelberg (2010)
12. Kobayashi, N., Ishii, M., Takahashi, S., Mochizuki, Y., Matsushima, A., Toyoda, T.: Semantic-JSON: a lightweight web service interface for semantic web contents integrating multiple life science databases. Nucleic Acids Research 39(Web Server) (June 2011),
    http://www.nar.oxfordjournals.org/cgi/doi/10.1093/nar/gkr353
13. Lukose, D.: World-wide semantic web of agriculture knowledge. Journal of Integrative Agriculture 11(5), 769–774 (2012),
    http://www.sciencedirect.com/science/article/pii/
    S2095311912600665
14. Marjit, U., Sharma, K., Biswas, U.: Provenance representation and storage techniques in linked data: A state-of-the-art survey. International Journal of Computer Applications 38(9), 23–28 (2012); published by Foundation of Computer Science, New York, USA
15. Milc, J., Sala, A., Bergamaschi, S., Pecchioni, N.: A genotypic and phenotypic information source for marker-assisted selection of cereals: the cerealab database. Database (2011)
16. Nešić, S., Rizzoli, A.E., Athanasiadis, I.N.: Publishing and Linking Semantically Annotated Agro-Environmental resources to LOD with AGROPub. In: Proceedings of the 5th Metadata and Semantics Research Conference, MTSR, Izmir, Turkey (2011) (in press)
17. Nolin, M.A., Corbeil, J., Lamontagne, L., Dumontier, M.: Bio2rdf: Convert, provide and reuse. In: Nature Precedings Biocuration Meeting 2010, Tokyo, Japan, October 11 (2010)
18. Po, L., Sorrentino, S.: Automatic generation of probabilistic relationships for improving schema matching. Inf. Syst. 36(2), 192–208 (2011)
19. Rajbhandari, S., Keizer, J.: The agrovoc concept scheme - a walkthrough. Journal of Integrative Agriculture 11(5), 694–699 (2012),
    http://www.sciencedirect.com/science/article/pii/
    S2095311912600586
20. Sorrentino, S., Bergamaschi, S., Gawinecki, M.: NORMS: An automatic tool to perform schema label normalization. In: Abiteboul, S., Böhm, K., Koch, C., Tan, K.L. (eds.) ICDE, pp. 1344–1347. IEEE Computer Society (2011)
21. Walls, R.L., Athreya, B., Cooper, L., Elser, J., Gandolfo, M.A., Jaiswal, P., Mungall, C.J., Preece, J., Rensing, S., Smith, B., Stevenson, D.W.: Ontologies as integrative tools for plant science. American Journal of Botany 99(8), 1263–1275 (2012),
    http://www.amjbot.org/content/99/8/1263.abstract

Management of Multiple and Imperfect Sources in the Context of a Territorial Community Environmental System

Karima Zayrit[1], Eric Desjardin[1], and Herman Akdag[2]

[1] CReSTIC, Université de Reims Champagne-Ardenne
{Karima.zayrit,eric.desjardin}@univ-reims.fr
[2] LIASD, Université Paris 8
Akdag@ai.univ-paris8.fr

Abstract. The work presented in this paper is a part of Observox, a community environmental information system for the monitoring of agricultural practices and their pressure on water resources in the Vesle basin, Champagne-Ardenne, France. The construction of Observox is the result of several research projects, and it is based on a methodology involving the actors concerned by the issue of water quality. Furthermore such a system requires the use of information provided by multiple sources which are usually imperfect. To provide the most honest indicators to the system's users, we integrate the notion of information quality by a degree of confidence. Thus we present the use of two main frameworks for imperfect knowledge management in the environmental information system, the fuzzy logic for propagating imprecision and belief functions for merging classifications.

Keywords: Belief function, fuzzy sets, agronomy, imperfect data, data merging.

1 Introduction

The urban community of Reims is located on the watershed of the Vesle, a territory with a high agricultural and viticultural activity. It is supplied with drinking water by three major water catchment areas (Couraux, Fléchambault and Auménancourt / 15 million m3 per year). Since 1989, traces of phytosanitary molecules are detected in the water drawn from water catchment area of Couraux. This situation has for consequence a pressure on the water and thus has an impact on its quality. To better manage this risk, it was necessary to establish a community environmental information system for the monitoring of agricultural practices and their impact on water resources at the watershed of the Vesle.

Community environmental information systems for the monitoring of agricultural practices were developed in the last years in order to assess and solve some problems encountered in agricultural territories, with the help of the actors involved in these problems. Observox is based on a methodological approach inspired by several research works, and aims to engage, participate and mobilize different actors in the studied area

B. Murgante et al. (Eds.): ICCSA 2013, Part I, LNCS 7971, pp. 204–215, 2013.

(farmers, growers, technicians, water manager, citizens, politicians) around the protection and quality of water resources based on the expectations and needs.

Observox has the objective of informing territory's stakeholders concerned on the development of agricultural practices by providing a sustainable archiving of information and helping them in their efforts to improve their practices (choice of molecules, grassing ...). It aims to organize and deploy information on phytosanitary practices, the nature of the products and their link with changes in cropping systems.

In the absence of exhaustive knowledge on agricultural practices in the territory, Observox uses multiple sources of information, heterogeneous and subject to imperfection. The implementation of such systems therefore requires addressing the quality of information, in order to provide indicators as fair as possible to users. We have to manage information about uncertainty during the whole chain of treatments, from the first stage of data acquirement to their computation and visualization. To achieve these objectives, the mathematical theories of uncertainty [2] [4] provide a theoretical framework that allows to process and reason about imperfect information.

In this article, we focus in section 2 on the implementation methodology we have developed for the establishment of Observox. In section 3, we introduce the notion of imperfection and describe different sources of information used in the Community environmental information system. In section 4, we focus on the management of uncertain and imprecise information in a multi-sources environment using fuzzy sets and belief functions. This approach is illustrated by case studies under simulated agronomic data.

2   Conception Methodology of Observox

The methodology used for the co-construction of Observox comes from several research projects [13]. The works of [8] showed that it was difficult to follow this general methodology because of the reality of the ground and the specificity of every couple stake-territory. Furthermore, every stage of the methodology can be interpreted by different manners according to contexts.

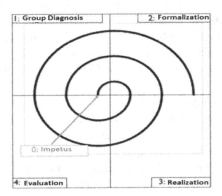

Fig. 1. Spiral of Implementation of Observox

Taking into account the specific features of the project (human and financial characteristic) and the results of these projects, we have developed a suitable method based on previous work: the Spiral of implementation of Observox.   Thus we focused on ways to ensure the mobilization of actors involved in the issue of water quality.

1. Group Diagnosis: To meet the needs of the territory and to be used by the actors concerned by the stake, a community must be co-constructed with them. For that purpose every spiral producing a version more succeeded in the device, must start with an analysis phase including a group diagnosis.

2. Formalization phase aims to put on paper the different ways of constructing the Observatory, including the drafting of two specifications: The website that defines all the elements allowing to create a functional device in line with requirement and exceptions of the actors; It specifies, through various descriptive fields, all indicators and dashboard that will be developed and viewed on the website. The collective elaboration made upstream of writing this specification, is a lever that allows mobilizing actors because it allows implying them in the conception of indicators adapted to their needs and integrates their idea of development.

3. Realization: In this phase begins data capitalization and legal tasks that relay on them.

4. Evaluation phase: This last phase of the conception of the spiral of OBSERVOX aims to face up what was produced to the reality on the ground and determine what elements to be revised, eliminated or to be added. It precedes a second phase of diagnostic group, starting a new turn of the conception spiral.

3   Multi-source and Imperfection

In order to understand the agricultural practices in the absence of shared and exhaustive knowledge, the project requires the exploitation of information resulting from multiple and varied sources. Moreover the information on the agricultural practices has the peculiarity to be multi-scale. They are collected at various levels of scales both for the spatial aspect and for the temporal aspect. Furthermore multi-varied knowledge can be of different natures: qualitative and quantitative, spatial and temporal. The context is further complicated by the fact information is partial, evolutionary and subject to imperfection. Imperfection can be due to imprecision, uncertainty, ambiguity, lack... Imprecision concerns the difficulty in formally and precisely expressing a state of reality. Uncertainty refers to the quality of the information; it expresses a doubt on the validity of knowledge and describes a partial knowledge of the reality. The incompleteness describes the lack of information or missing information provided by an information source. The ambiguity involves two or several interpretations for the same information which can be in conflict. A study of the information sources in the context of Observox helps to identify a number of such imperfections. We will now describe some of those we can use in France.

3.1   Graphic Parcel Register

The Graphic Parcels Register (RPG) serves us to acquire a basis of the truth on the studied ground. For needs and to discriminate between the various cultures of the territory, it was chosen to concentrate at first on five cultures: the wheat, the colza, the Lucerne, the potato and the sugar beet. These cultures are the most important in terms of surface of exploitation and in applied phytosanitary treatments. RPG's nomenclature used as actual knowledge on the studied territory, during the classification of satellite images suffers from indistinctness and from ambiguity. For example Wheat and Colza have separate and well informed classes that can be used to create a learning base. However this is not the case of three other cultures of interest within Observox area. In order to fill this imperfection, further studied were carried out in the territory and with farmers.

3.2   The Survey of Viticultural Practices on the Experimental Area

These surveys are conducted within the framework of the building of a recovery basin for water run-off in the studied territory; we interviewed the manager of each vineyard. This information sources allow us to describe the viticultural practices on this territory and to know the technical routes for the dimension of soil management and plant protection via the results of field survey and the database on the phytosanitary practices of the cooperative of Nogent-l'Abbesse/Cernay-les-Reims.

However some types of imperfection were identified as uncertain and imprecise due for example to declarative bias, expressed by the gap between what the farmer really makes and what he states. Similarly, sprayer type, model, and types of nozzles used, are factors that can influence the quality of the spray and therefore the amount of product or active substance actually applied. On the other hand, the estimation of the really treated surface is source of imprecision and uncertainty depending on the potential vegetation between rows of vineyard.

3.3   « Pratiques Culturales » (PK) Survey

Conducted every 5 years by the Department of Statistics and Forecasting of the Ministry of Agriculture, these surveys provide information about the quantities of phytopharmaceutical products applied on a sample of more than 20000 plots surveyed. They provide precise information on agricultural practices, crop management, and the use of phytosanitary products on a large sample of agricultural parcels for the most present cultures on the national territory. They concern the main crops of the French territory, namely: industrial beet, durum wheat, common wheat, colza, corn grain (bead), corn feed, barley, protéagineux pea, potato, sunflower and, since 2006, the vineyard. Due to a sampling chosen for a representation at national scale, the number of plots surveyed across the Vesle watershed is very small. The use of this source of information is almost managed with a great margin of uncertainty.

We can also note the absence of this survey on certain crops occupying a large area of the watershed as Lucerne. Other crops such as potato sunflower and corn are surveyed, but they are not sampled on the territory. In that case, we are facing uncertainty due to lack of information.

3.4   E-phy (Phy2X) Database

This is the catalog of all authorized plant protection products and their use in the French market. It associates to each commercial product its approved dose. This information source is used in our context to calculate or estimate indicator as TFI (treatment frequency index). But they are some mistakes and update is not frequent. Naturally, these errors induce a greater risk of providing wrong information if we do not manage a reliable index expressed with its imperfection.

4   Management of Imperfection

In order to build more accurate decisions such systems need to handle the quality of the information and especially to pay attention to the consideration of imperfect knowledge upon acquisition and during use. Geographical data must take into consideration the quality of the stored information and its analyses [7]. The study of the data quality is essential in order to obtain results we can trust. As says Goodchild[12] "Quality could be viewed as a measure of the difference between the data and the reality they represent, and becomes poorer as the data and the corresponding reality diverge". Therefore, the poorer the quality of data, the less they tell us about reality.

Handling imperfect data often leads to data quality issues i.e. the management of data imprecision, ambiguity, incompleteness and/or uncertainty. Modeling data imperfection in the information system allows to reduce decision uncertainty and also improves the data quality.

Taking into account data quality should cover the different process related to the acquisition, management, diffusion and use of information [7]. For this purpose, mathematical theories of uncertainty provide a theoretical framework for reasoning about processes and imperfect information. Two of them will be presented in this part.

4.1   Fuzzy Sets + Case Study

Observox exploits data coming from heterogeneous sources; the construction of a unique set of entities implies the combination of information coming from all sources. The built entities induce some imprecision in the definition of spatial features and quantitative attribute [1]. Fuzzy sets introduced by Zadeh in [2], can be considered as a natural tool to manage uncertainty [9], [10], [11].

According to this, agronomical entities could be modeled as fuzzy geographical entities. Those entities have a label, a fuzzy spatial shape and a set of fuzzy quantities (each quantity corresponds to a specific attribute such as population or a specific chemical). The definition of a geographical entity may be defined as follows.

Let Ω be the set of studied geographical entities $\{A_1,...,A_n\}$. Let be Ω the set of monitoring quantitative information ($Q_1,...,Q_m$) if one supervises m different information $(P_1,...,P_m)$ as for instance m different molecules or products . Let us define a fuzzy geographical entity A_i in Ω as an object described by:

- A label or concept LA_i member of an ontology.
- A fuzzy set FSA_i describing its spatial representation. The membership function μSA_i of FSA_i is defined on $\mathbb{R}^2$.
- A fuzzy quantity FQ_jA_i for each quantity Q_j(of P_j) in Q. The member-ship function μQ_jA_i of FQ_jA_i is defined on R^+.

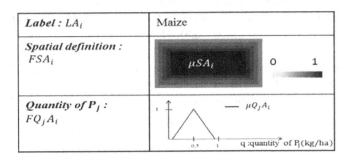

Label : LA_i	Maize
Spatial definition : FSA_i	μSA_i 0 1
Quantity of P_j : FQ_jA_i	

Fig. 2. Example of a fuzzy agronomical entity A_i

The imprecision, conceptualized using a classical fuzzy number for quantities and by fuzzy area for spatial feature, is propagated in the consideration of a fuzzy quantities at a specific location. Let consider x a location, we consider that the confidence in FQ_jA_i should be put into perspective with the membership degree $\mu SA_i, (x)$ in order to define FQ_jA_i, x with its membership function $\mu Q_jA_i, x$ as:

$$\mu Q_jA_i, x(q) = T\left(\mu SA_i(x), \mu Q_jA_i(q)\right) \tag{1}$$

with q in $\mathbb{R}^+$ and T an aggregation function usually a t-norm such as the multiplication or the minimum.

As our goal is to consider all the quantities of a specific product at each location of the space, an aggregation operator is now needed for obtaining the combined information. Then, we use the Zadeh's extension principle that allows to extend usual operation in the fuzzy set context such as the sum in our context (due to the additive aspect of product diffusion).

Thus, if we deal with an additive information P_j, using this hypothesis and Zadeh's extension principle, we define FQ_j, x the overall quantity at the position x by following the equation (2) for the definition of its membership function Q_j, x .

$$\mu Q_j, x(q) = sup_{q=z+t}\left(min_{A_i,A_k\ in\ \Omega^2, i\neq k}\left(\mu Q_jA_i, x(z), \mu Q_jA_k, x(t)\right)\right) \tag{2}$$

In order to view the usability of the previous approaches, we will work on simulated data. The study will be carried out on three plots of land: the first one P_1 is a lucerne field; the second one P_2 and the third one P_3 produce maize. At the position x the membership degrees are: $\mu SP_1(x) = 0.3$, $\mu SP_2(x) = 0.3$, $\mu SP_3(x) = 0.1$.

The study will be done for Bentazone, a molecule presented in chemical products. This corresponds to a major issue in water quality impact. The three plots are treated each year as follows. For the maize and lucerne fields, farmers use Basagran (a commercial chemical product) which has Bentazone content of 87%. In P1, they use approximately 1.3kg/ha of Basagran, i.e. a quantity estimated between 1.2 to 1.4 kg/ha and thus a triangular number (1.2,1.3,1.4). In P2, the farmers spread approximately 0.7 kg/ha of Basagran (between 0.6 to 0.8 kg/ha). In P3, they use roughly 0.6 kg/ha of Basagran (between 0.5 to 0.7 kg/ha).

Therefore, the quantities of Bentazone, expressed in kg/ha, could be represented by the following triangular fuzzy numbers: $\mu Q_j A_1(q) = (1.0, 1.1, 1.2)$, $\mu Q_j A_2(q) = (0.5, 0.6, 0.7)$, and $\mu Q_j A_3(q) = (0.4, 0.5, 0.6)$. They correspond to 87% of the Basagran quantities.

Using the Zadeh t-norm we produce three intermediary fuzzy sets $\mu Q_j A_1, x(q)$, $\mu Q_j A_2, x(q)$, $\mu Q_j A_3, x(q)$. Then we use them to compute the fuzzy quantity defined by $\mu Q_j, x(q)$ presented in (figure 3).

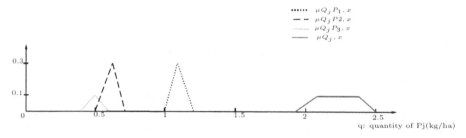

Fig. 3. The membership function $\mu Q_j, x$ of the fuzzy Bentazone quantity at x

All those results have to be stored in the information system with their genealogy.

4.2   Belief Function + Case Study

Now we focus on the exploitation of distinct ways to access to uncertain information in a multi-sources environment. For this purpose, we have chosen to work within the framework of the theory of belief functions. In terms of information fusion, this formalism provides tools, which allow us to combine information from diverse sources, manage conflict between different knowledge, and to weight information sources according to their reliability, and so to help in the decision making. Our goal is to use these tools for the management of imperfect data and establishing the reliability of the system.

In this section, we briefly review the main concepts underlying the theory of belief functions. The Belief function theory, introduced by Dempster [5] and formalized by Shafer [4], provides a framework for processing uncertain and imprecise data. An illustrative case study is given to represent the processes of classifiers fusion using the Dempster-Shafer theory, and shows the effectiveness of the proposed method.

Representing Evidence

Let be Ω a finite set of mutually exclusive hypotheses denoted by $\{h_1, h_2, .., h_n\}$, called the frame of discernment. In terms of land cover identification, Ω is the set of hypotheses about pixel class. The set of all subsets of Ω is denoted by 2^Ω. The mass function, called basic belief assignment (bba), is defined by the mapping of each subset of the space of discernment Ω onto $[0,1]$, such that:

$$\sum_{A \in 2^\Omega} m(A) = 1 \tag{3}$$

Subsets of Ω such that $m(A) > 0$ are called focal elements of m. The Dempster-Shafer theory provides a representation of both imprecision and uncertainty through the definition of two functions: plausibility *(Pl)* and belief *(Bel)*. The belief and the plausibility function, derived from m, are respectively defined from 2^Ω to $[0,1]$. The belief function gives the amount of evidence which implies the observation of A: $Bel(A) = \sum_{B \subseteq A, B \neq 0} m(B)$.

The plausibility function can be seen as the upper bound on the degree of support that should be assigned to A: $Pl(A) = \sum_{B \cap A \neq 0} m(B)$.

Thus, uncertainty about A is represented by the value of the val $[Bel(A), Pl(A)]$, which is called the belief interval and the length of this interval gives a measurement of the imprecision about the uncertainty value.

Evidence Combination

Given two mass functions, Dempster-Shafer theory provides a combination rule for combining them, which is defined in its normalized form, as follow:

$$m_{1 \oplus 2}(A) = m_1 \oplus m_2 = \frac{\sum_{A_1 \cap A_2 = A} m_1(A_1) * m_2(A_2)}{1 - K} \tag{4}$$

where $K = \sum_{A_1 \cap A_2 = \emptyset} m_1(A_1) * m_2(A_2)$ is the inconsistence of the combination.

The obtained mass is the combination of the mass function of each different source. The denominator in Dempster's rule is a normalization factor. K represents the mass which would be assigned to the empty set after combination and K is interpreted as a measure of conflict between the different sources. Conflict can occur if the sources are not reliable or if they give information about different phenomena. When K is equal to 1, the sources are in total conflict and cannot be combined. When K is equal to 0, the sources are in complete agreement.

Decision Making

Having computed the mass, belief and plausibility value for each simple and compound hypothesis, Dempster-Shafer theory provides decision rules, in order to decide which hypothesis is the more relevant depending on context.

The maximum of the plausibility (5)/credibility (6), a rather optimistic/pessimistic rule, selects between all the singletons, the one having the maximum of plausibility/credibility :

$$Dec(\Omega) = argmax_{A\in 2^{\Omega},|A|=1}Pl(A) \tag{5}$$

$$Dec(\Omega) = argmax_{A\in 2^{\Omega},|A|=1}Bel(A) \tag{6}$$

The maximum of pignistic probability [3], is generally considered as a compromise between the maximum of plausibility and the maximum of credibility. It is based on an equitable distribution of belief masses from non-singleton elements to singletons. It is given for all $A \in 2^{\Omega}$ by :

$$BetP(A) = \sum_{B\in 2^{\Omega},A\in B} \frac{m(B)}{|B| * \left(1 - m(\emptyset)\right)} \tag{7}$$

And $$Dec(\Omega) = argmax_{A\in\Omega}BetP(A) \tag{8}$$

Case Study: Application to Multisource Classification
The purpose of using the theory of belief functions in the structure of our information system is the management of multi-sources taking their quality into consideration. In fact, we want that the system provides information about the type of grown crop with its associated confidence and reliability for each location of the territory. The belief mass associated to the type of culture (using the pignistic probability for example) gives us information about the quality of the system response.

We consider the context of a simulated but realistic example for which we have results of credibilistic classifications in the scale of the pixel, from a processing satellite images S1 and a classification provided by an expert in a part of the study area according to the list of culture. Information on the type of crops is taken from the "Graphic Parcel Register". We thus try to identify the type of culture in this location. The belief masses (m_1 and m_2 respectively) provided by these classifiers are shown in Table 1.

Table 1. Masses m_1 and m_2 on Ω={$Wheat,Barley,Maize$}

Hypothesis	m_1	Hypothesis	m_2
{$Wheat$}	0,1	{$Wheat,Barley$}	0,3
{$Maize$}	0,5	{$Wheat, Maize$}	0,6
{$Wheat,Barley$}	0,4	Ω	0,1

Thus, we obtain the results of combination (table 2) for which at least one of the masses stemming from the classifiers is not null. We can observe that the conflict (K) is 0.15, which is not meaningful.

Table 2. Results of combination of masses m_1 and m_2 on $\Omega=\{$ *Wheat,Barley,Maize* $\}$

	$m_2(\{Wheat, Barley\})$ = 0.3	$m_2(\{Wheat, Maize\}$ = 0.6	$m_3(\Omega) = 0.1$
$m_1(\{Wheat\}) = 0.1$	{ *Wheat* }=0.03	{ *Wheat* }=0.06	{ *Wheat* }=0.01
$m_1(\{Maize\}) = 0.5$	K=0.15	{ *Maize* }=0.24	{Maize}=0.05
$m_1(\{Wheat, Barley\})$ = 0.4	{ *Wheat,Barley* }=0.12	{ *Wheat* }=0.24	{ *Wheat, Barley*} =0.04

By applying the combination of Dempster, we obtain the masses presented in the table 3.

Table 3. Results of combination of classifiers S_1 and S_2 on $\Omega=\{$ *Wheat,Barley,Maize* $\}$

Hypothesis	$m_{1\oplus2}$	Bel	Pl	BetP
{ *Wheat* }	0.4	0,4	**0,59**	**0,495**
{ *Maize* }	0.41	**0.41**	0.41	0.41
{*Barley*}	0.19	0	0.19	0.095

The study of the results gives: a very pessimistic reading would like that we choose the maize (maximum of faith), whereas a less pessimistic reading (maximum of probability pignistique) or optimist (maximum of admissibility) would give us the wheat. It is however necessary to qualify the possible choices by trying to reduce the conflict by balancing information sources and by privileging an approach negotiated with experts statements. This stage of knowledge merging allows then to supply to the users a result qualified by a degree of confidence. All of those results have to be stored in the information system with the merged data and their lineage (mathematical operators, algorithms, sources weighting...).

5   Conclusion

In this paper we present the context of the establishment of a community environmental information system, on the Vesle watershed. In the absence of exhaustive comprehensive knowledge on agricultural practices, the construction approach presented is based on a methodology that requires involving different actors of the territory concerned by the issue of water quality. The first work undertaken within the framework of Observox allowed us to understand the imperfection of information from its acquisition to its use. We address in this work the aspects of imperfect, uncertain, incomplete, partially wrong data coming from our diverse sources. Thus to model and manage these imperfection inside the information system, we have developed two approaches based respectively on fuzzy sets and on belief functions by integrating the quality notion of the information by a degree of confidence. In order to evaluate the feasibility of our approach, we considered two case studies simulated but realistic. The first results provide information integrating the quality notion ready to be stored in the agro-environmental information system of Observox.

To enrich our information system, we envisage in our further work to characterize the uncertainty representation with fuzzy numbers and belief functions. We are interesting also by the consideration of the imperfection in other types of indicators such as the treatment frequency index (TFI), we would like to focus other types of imperfect data representation as imprecise probabilities and establish a virtual platform of the environmental information system in order to integrate the different indicators with their confidence.

Acknowledgements. We would like to thank the Seine-Normandy Water Agency, Champagne-Ardenne Region Council, S.I.A.B.A.V.E, France and European Union, through the FEDER, for their funding of the AQUAL State-Region Planning Project.

References

1. Shi, W.: Principles of Modeling Uncertainties in Spatial Data and Spatial Analyses. CRC Press (2010)
2. Zadeh, L.A.: Fuzzy sets. Information and Control 8, 338–353 (1965)
3. Smets, P., Kennes, R.: The transferable belief model. Artificial Intelligence 66, 191–234 (1994)
4. Shafer, G.: A mathematical theory of evidence. Princeton University Press, Princeton (1976)
5. Dempster, A.: Upper and Lower Probabilities Induced by a Multivalued Mapping. In: Yager, R., Liu, L. (eds.) Classic Works of the Dempster-Shafer Theory of Belief Functions. STUDFUZZ, vol. 219, pp. 57–72. Springer, Heidelberg (2008)
6. Denœux, T., Younes, Z., Abdallah, F.: Representing uncertainty on set-valued variables using belief functions. Artificial Intelligence 174, 479–499 (2010)
7. Devillers, R.: Qualité de l'information géographique: Traité Igat. Hermes Science Publications (2005)
8. Le Ber, F., Nogry, S., Brassac, C., Benoît, M.: Capitalisation d'expériences pour la mise en place d'observatoires de pratiques agricoles. Revue internationale de Géomatique 21, 99–118 (2011)
9. Klir, G.J., Clair, U., Yuan, B.: Fuzzy set theory: foundations and applications. Prentice-Hall, Inc., Upper Saddle River (1997)
10. Fisher, P., Comber, A., Wadsworth, R.: Approaches to Uncertainty in Spatial Data. In: Devillers, R., Jeansoulin, R. (eds.) Fundamentals of Spatial Data Quality, pp. 43–59. ISTE (2010)

11. Bouchon-Meunier, B.: Logique floue et ses applications. Addison Wesley, Paris (1995)
12. Goodchild, M.: Foreword. In: Devillers, R., Jeansoulin, R. (eds.) Fundamentals of Spatial Data Quality (2006)
13. Passouant, M., Caron, P., Loyat, J., Tonneau, J.-P., Barzman, M.: Observatoire des agricultures et des territoires: mise à l'épreuve d'une méthode de conception. In: SAGEO 2007 (2007)

Error Correction for Fire Growth Modeling

Kathryn Leonard and Derek DeSantis

CSU Channel Islands, Camarillo, CA, USA
kleonard.ci@gmail.com

Abstract. We construct predictions of fire boundary growth using level set methods. We generate a correction for predictions at the subsequent time step based on current error. The current error is captured by a thin-plate spline deformation from the initial predicted boundary to the observed boundary, which is then applied to the initial prediction at the subsequent time step. We apply these methods to data from the 1996 Bee Fire and 2002 Troy Fire. We also compare our results to earlier predictions for the Bee Fire using the FARSITE method. Error is measured using the Hausdorff distance. We determine conditions under which error correction improves prediction performance.

1 Introduction

Developing accurate models for the growth of forest fires is a vexing problem with life-and-death implications. The physical interactions between variables in a fire are too complex to be captured by any solvable mathematical formulation. For example, humidity plays an important role in rate of fire spread (ROS), but the fire itself alters the humidity of the surrounding air. Additionally, collecting reliable measurements of these variables is often impossible during a fire event.

The model currently used by US fire departments, FARSITE [3], propagates fires locally along ellipses normal to the fire boundary via functions based on topography, atmosphere, vegetation, and elevation above ground level of the fire [6]. Implementation often relies on coarse approximations to real-time input parameters, or none at all. Recently, level set methods have been applied to model ROS [5]. Level set methods develop a global model of fire growth that depends on the geometry of the fire front as well as an external vector field that can encompass the aforementioned external variables. Again, implementations of the the level set model rarely account for the complications of the physical reality of the fire.

Not surprisingly, none of these models produces very accurate predictions. As a result, attention has turned to error correction [4], [7], whereby correcting error in a current prediction relies on errors at earlier times. The hope is that we can account for the barriers to sophistication in our implementations by learning from their failures.

In one such approach, Fujioka defines a notion of bias in [4] based on a polar representation of the fire boundary, and generates a correction based on removing that bias from the prediction estimates. His work concludes that the uncorrected estimates are more accurate. We compare our results with his in Section 3.2. In another approach, Rochoux, et. al, develop a probabilistic framework using simulations and controlled

B. Murgante et al. (Eds.): ICCSA 2013, Part I, LNCS 7971, pp. 216–227, 2013.

burns to generate a best linear unbiased estimator (BLUE) of the correction using techniques based on the Kalman Filter [7]. To date, the methodology has not been applied to real-world fire prediction.

Our approach uses fire intensity data from the California Troy and Bee fires to explore the idea that past fire behavior can realistically inform future fire prediction on the time scales for which data is available during an actual fire event. We also explore the relative accuracy of level set and FARSITE methods for modeling fire spread for the Bee fire, data that captures the first 105 minutes of fire growth.

We apply the level set method as implemented by Sumengen [9] to an initial fire boundary. We determine a mismatch between the predicted fire boundary and the observed fire boundary using thin plate spline (TPS) point matching as implemented by Chui and Rangarajan [2]. The level set prediction at the following time step is then adjusted according to the TPS deformation. Accuracy is measured using the Hausdorff distance between the observed and predicted boundaries. We present accuracy results for the Troy and Bee fires, and compare our results for the Bee fire with those found in Fujioka [4].

2   Methods

Troy fire data consists of 13 heat intensity aerial images at approximately 10 minute intervals beginning in the afternoon of June 19, 2002. The 1996 Bee fire data consists of three images at $15, 45$, and 105 minutes after ignition. All data was obtained from the USDA Forest Service. In addition, our data includes predicted boundaries generated by Fujioka's method described in [4] for the Bee Fire. Fujioka's method uses FARSITE, a Rothermel-based method, with unusually detailed wind data to generate predictions.

2.1   Preprocessing

Bee fire data consists of UTM coordinates of the fire boundary points, which we scaled down. To extract boundaries from the heat intensity images comprising the Troy fire data, we segment the fire area using k-means clustering with $k = 2$, then extract the coordinates of the boundary contour. Given the boundary coordinates, we compute the signed distance function of the boundary for input to the level set method.

2.2   Level Set Method

The level set method models contours evolving in time as the zero level sets of a function $\phi_t(x, y)$. The level set function satisfies the level set equation:

$$\frac{d\phi}{dt} + \boldsymbol{V} \cdot \nabla \phi = 0$$

where $\boldsymbol{V}(x, y)$ is a continuous vector field encoding both external forces and intrinsic geometry of the curve [8]. For the Troy fire, $\boldsymbol{V}$ contains coarse wind information and constant rate of spread normal to the curve. For the Bee fire, $\boldsymbol{V}$ is just the constant normal rate of spread (wind data was not available). We compute a first-order Godunov numerical solution as implemented in [9] with square mesh width $dx = dy = 0.5, 1.5$ iterations per minute of prediction, and $\alpha = 0.5$ in determining dt.

2.3 Thin-Plate Spline Matching

Thin-plate splines (TPS) approximate smooth planar deformations mapping one con-tour onto another by defining a function $f(v) = \sum_{i=1}^{n} a_i \phi(v - x_i)$ based on pairings of control points $\{(x_i, y_i)\}_{i=1}^{n}$ that minimizes the energy functional [1]:

$$\sum_{i=1}^{n} \|f(x_i) - y_i\| + \lambda \iint \left[\left(\frac{\partial^2 f}{\partial x^2} \right)^2 + 2 \left(\frac{\partial^2 f}{\partial x \partial y} \right) + \left(\frac{\partial^2 f}{\partial y^2} \right)^2 \right].$$

Given two sets of boundary points $\{x_i\}$ and $\{y_j\}$, the energy minimization is difficult to compute. As implemented in Chui and Rangarajan [2], an approximate minimization is found using deterministic annealing. At high temperatures, point sets are matched based on geometric features. As the temperature decreases, points are matched based on proximity. The output of the implementation is a point matching between the two sets of boundary points, and the weights and coefficients for computing the resulting transformation for any new input points. In our implementation, initial temperatures range from 400 to 7500 and final temperatures range from 12 to 1000, based on magni-tude of the coordinates.

2.4 Error Correction

At time $t = t_0$, we input points on the initial fire boundary B_0 to the level set method, producing an estimate P_1 for the observed boundary B_1 at $t = t_1$. In this first stage, there is no history available to construct a corrected prediction, so the corrected pre-diction $Q_1 = P_1$. We then compute the TPS matching between the level set prediction P_1 and the observed boundary B_1, producing a function $f_1 : \mathbb{R}^2 \to \mathbb{R}^2$. We begin the iteration at $t = t_i$, $i > 0$, with the observed boundary B_i and the TPS mapping f_i cap-turing the error between the level set prediction P_i and B_i. We then generate the level set prediction P_{i+1} of the observed boundary B_{i+1} at $t = t_{i+1}$ and correct it according to f_i, generating a revised prediction Q_{i+1} of B_{i+1}, where $Q_{i+1} = f_i(P_i)$.

2.5 Error Measurement

We use the Hausdorff distance to measure the error between predicted and observed fire boundaries. The Hausdorff distance captures the maximum Euclidean distance between two boundaries. Given two boundaries B_1, B_2,

$$d_H(B_1, B_2) = \max_{i,j=1,2} \max_{p \in B_i} \min_{q \in B_j, i \neq j} d(p, q)$$

where $d(p, q)$ is the standard Euclidean distance between points p, q in the plane.

3 Results

We find that neither the level set predictions $\{P_i\}$ nor the corrected predictions $\{Q_i\}$ satisfactorily capture the fire behavior. For the Troy fire data, both methods provide adequate predictions, with better accuracy sometimes with correction and sometimes without. For the Bee fire, neither corrected nor uncorrected method is satisfactory. We judge Fujioka's predicted boundaries to be superior. In other words, correction after the fact does not adequately compensate for the inability of the original implementation to adequately model the physical complexities of the fire.

3.1   Troy Fire Results

The Hausdorff distances between the observed boundary and the predicted boundaries are given in Table 1. We also show a sampling of the level set predictions without correction (cyan), with correction (green), and the observed boundaries (yellow) in Figures 1 - 9. Note that the corrected and uncorrected predictions for the boundary at time $t = 2$ are the same because there is not yet a history of error. Note also in Figures 7 and 9 how significant new growth areas emerge but are not captured at all by either corrected or uncorrected predictions.

Table 1. Hausdorff distance between level set predictions of fire boundaries and the observed boundaries, with and without TPS correction based on error at previous time step

Time-step	Level Set	Corrected
2	51.99	51.99
3	13.92	51.30
4	19.10	11.64
5	14.14	14.79
6	22.36	16.21
7	15.13	19.50
8	28.07	22.63
9	24.18	15.10
10	16.12	15.83
11	26.00	19.45
12	13.03	15.72

In certain cases, the correction contributes to a substantially more accurate prediction (Figures 3, 8), but often the two predictions are roughly the same distance from the boundary. As the images in Figures 1 - 10 show, the level set method errors tend to underestimate growth while the TPS-corrected predictions tend to overestimate growth.

3.2   Bee Fire Results

For the Bee fire, only three time steps of data are useable (Figure 11) giving a very small sample. We include results nonetheless because we are able to directly compare our predictions with FARSITE predictions. The Hausdorff distances between the observed boundary and the predicted boundaries are given in Table 2 for level set predictions,

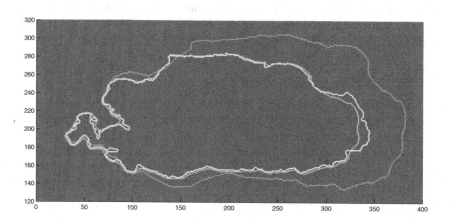

Fig. 1. Troy fire $t = 2$: level set prediction (cyan), TPS-corrected level set prediction (green), and observed boundary (yellow)

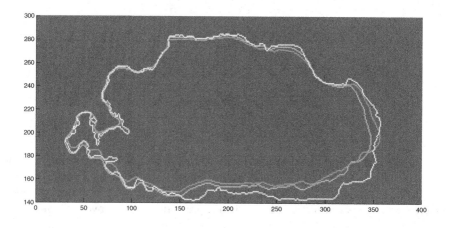

Fig. 2. Troy fire $t = 3$: level set prediction (cyan), TPS-corrected level set prediction (green), and observed boundary (yellow)

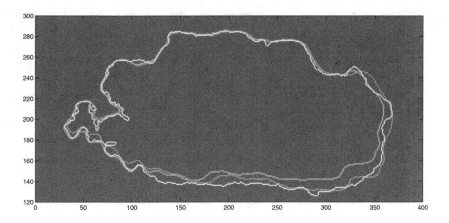

Fig. 3. Troy fire $t = 4$: level set prediction (cyan), TPS-corrected level set prediction (green), and observed boundary (yellow)

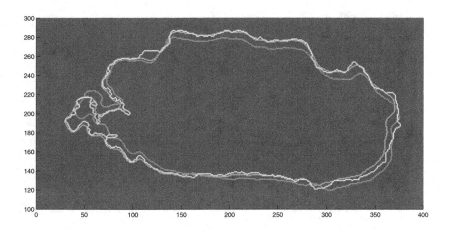

Fig. 4. Troy fire $t = 5$: level set prediction (cyan), TPS-corrected level set prediction (green), and observed boundary (yellow)

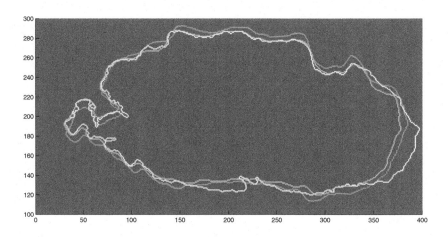

Fig. 5. Troy fire $t = 6$: level set prediction (cyan), TPS-corrected level set prediction (green), and observed boundary (yellow)

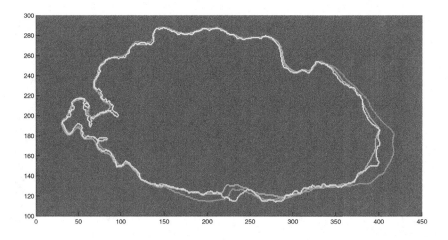

Fig. 6. Troy fire $t = 6$: level set prediction (cyan), TPS-corrected level set prediction (green), and observed boundary (yellow)

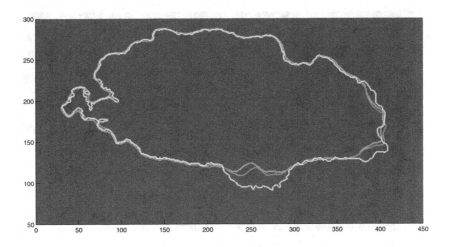

Fig. 7. Troy fire $t = 7$: level set prediction (cyan), TPS-corrected level set prediction (green), and observed boundary (yellow)

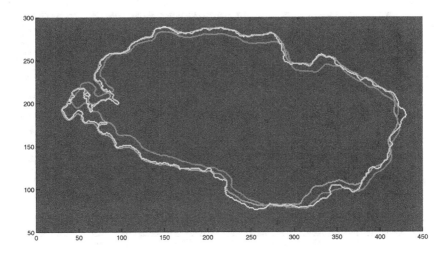

Fig. 8. Troy fire $t = 10$: level set prediction (cyan), TPS-corrected level set prediction (green), and observed boundary (yellow)

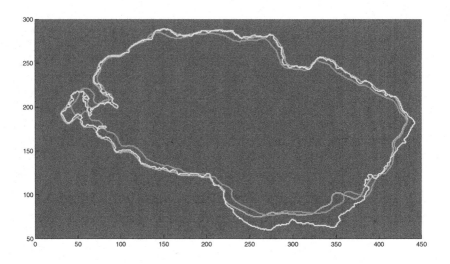

Fig. 9. Troy fire $t = 11$: level set prediction (cyan), TPS-corrected level set prediction (green), and observed boundary (yellow)

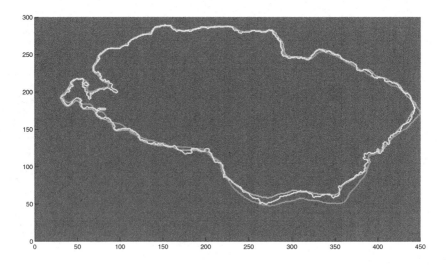

Fig. 10. Troy fire $t = 12$: level set prediction (cyan), TPS-corrected level set prediction (green), and observed boundary (yellow)

corrected level set predictions, and FARSITE predictions. We applied TPS correction to the FARSITE predictions but found no improvement. We do not include those results here. Again, recall that at $t = 60$ minutes, no history of error exists and so the corrected prediction is equal to the original prediction. Note also that the scale of these errors is different than the results for the Troy fire, as the coordinate systems are different.

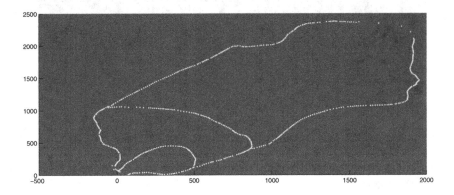

Fig. 11. Bee fire after 30, 60 and 105 minutes

Table 2. Hausdorff distance between FARSITE or level set predictions of fire boundaries and the observed boundaries, with and without TPS correction based on error at previous time step

Time	Level Set	Corrected	FARSITE
60	591.7	591.7	790.5
105	1907.7	1876.5	954.2

At the early stage, the two methods are comparable. At the later stage, however, FARSITE error is half the error for either of the level set predictions.

We show the FARSITE predictions (white), level set predictions without correction (green), and the observed boundaries (yellow) in Figures 12 and 13.

4 Discussion

We have shown that the level set method with and without TPS error correction models the Troy fire reasonably well. For time steps where the fire growth remains stable, error correction does improve estimates meaningfully. An adaptive scheme where the decision to correct or not is based on degree of change in atmospheric, terrain, or temperature factors may be desirable.

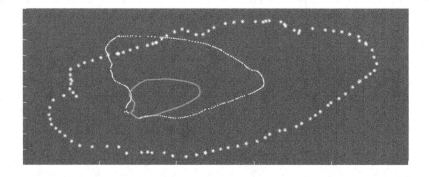

Fig. 12. Bee fire $t = 60$ minutes: FARSITE prediction (white), level set prediction (cyan), and observed boundary (yellow)

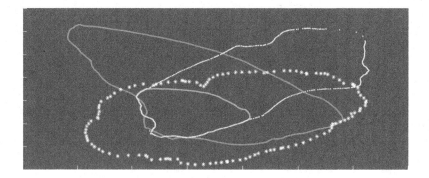

Fig. 13. Bee fire $t = 105$ minutes: FARSITE prediction (white), level set prediction (cyan), TPS-corrected level set prediction (green), and observed boundary (yellow)

We also show that the more accurate approximation for the Bee fire is the Rothermel-based model. We believe this is largely due to the superiority of the FARSITE method with detailed wind data to our implementation of the level set method with no wind data.

Some of the most unpredictable behavior arises from newly formed bulges in the fire, so-called "fire fingers," which are among the most dangerous of fire behaviors. Current work is underway to better model these localized events. Predicting when and where fire fingers are likely to arise will be helpful, even if precise modeling of their boundary evolution eludes us.

Our work demonstrates the challenges of applying historical error to correct current prediction of fire boundaries. Particularly during the early stages of a fire, or for a fire with unusual physical constraints, the orientation and magnitude of boundary evolution is rapidly changing. Sophistication equal to that desired in the original model of boundary evolution is likely necessary to produce useful correction. Past error alone, at least as globally measured, is not reliably informative. Future work will consider local error measures.

Acknowledgements. The authors gratefully acknowledge Francis Fujioka for introducing us to the fire modeling problem, sharing data and predictions from the Bee fire, and predictions for the Troy fire, Robert Tissell for sharing data from the Troy fire, and the National Science Foundation IIS-0954256 for funding our work.

References

1. Bookstein, F.: Principal Warps: Thin-plate Splines and the Decomposition of Deformations. IEEE Transactions in Pattern Analysis and Machine Intelligence 14(2), 239–256 (1992)
2. Chui, H., Rangarajan, A.: A New Algorithm for Non-rigid Point Matching. In: IEEE Conference on Computer Vision and Pattern Recognition (2000)
3. Finney, M.: FARSITE: Fire Area Simulator-Model, Development and Evaluation. Report RMRS-RP-4, US Department of Agriculture Forest Service, Rocky Mountain Research Station Paper (1998)
4. Fujioka, F.: A New Method for the Analysis of Fire Spread Modeling Errors. International Journal of Wildland Fire 11, 193–203 (2002)
5. Mallet, V., Keyes, D., Fendell, F.: Modeling Wildland Fire Propagation with Level Set Methods. Journal of Computers & Mathematics with Applications 57(7), 1089–1101 (2009)
6. Rothermel, R.: A Mathematical Model for Predicting Fire Spread in 17 Wildland Fuels. Research Paper INT-115, US Department of Agriculture Forest Service (1972)
7. Rochoux, M., Delmottea, B., Cuenot, B., Riccia, S., Trouvé, A.: Regional-scale simulations of wildland fire spread informed by real-time flame front observations. Proceedings of the Combustion Institute 34(2), 2641–2647 (2013)
8. Sethian, J.: Level Set Methods and Fast Marching Methods: Evolving Interfaces in Computational Geometry, Fluid Mechanics, Computer Vision, and Materials Science. Cambridge University Press (1999)
9. Sumengen, B.: A Matlab Toolbox Implementing Level Set Methods,
   http://barissumengen.com/level_set_methods

An Ontological Approach to Meet Information Needs of Farmers in Sri Lanka

Anusha Indika Walisadeera[1,2], Gihan N. Wikramanayake[2], and Athula Ginige[1]

[1] School of Computing, Engineering & Mathematics, University of Western Sydney,
Penrith, NSW 2751, Australia
waindika@cc.ruh.ac.lk, a.ginige@uws.edu.au
[2] University of Colombo School of Computing, Colombo 07, Sri Lanka
gnw@ucsc.cmb.ac.lk

Abstract. Farmers in Sri Lanka are badly affected by not being able to get vital information required to support their farming activities in a timely manner. Some of the required information can be found in government websites, agriculture department leaflets, and through radio and television programs on agriculture. Due to its unstructured and varied format, and lack of targeted delivery methods, this knowledge is not reaching the farmers. Therefore, this knowledge needs to be provided not only in a structured way, but also in a context-specific manner. To address this shortcoming an international collaborative research project was launched to develop a Social Life Network to provide necessary information to farmers using mobile devices. Agricultural information has strong local characteristics in relation to climate, culture, history, languages, and local plant varieties. These local characteristics as well as the need to provide information in a context-specific manner made us to develop an ontology for agriculture. In this paper we present the approach we used to derive contextual information related to the farmers and the ontological approach that we developed to meet information needs of the farmers at various stages of the farming life cycle.

Keywords: agricultural information, contextual information, knowledge representation, ontology, ontology development.

1   Introduction

In many developing countries, agriculture plays a major role in the country's economy; Sri Lanka is no exception. The agriculture sector in Sri Lanka is the main source of livelihood for the rural population, which accounts for 70% of the total population. Often we hear news about farmers in Sri Lanka not being able to sell their harvest due to oversupply or not getting the planned harvest, selecting the wrong seed types, or lack of necessary information at the right time such as information about market prices [1]. Therefore, flow of information in the agriculture sector must be strengthened to attain higher growth rates and to contribute to the overall economic development of the country. There are many issues to be investigated to achieve a successful delivery of information from agricultural experts to rural farmers. Some important issues are: what information is required, what should be the delivery

B. Murgante et al. (Eds.): ICCSA 2013, Part I, LNCS 7971, pp. 228–240, 2013.

methods, and how to customize the information to meet the needs of farmers in different regions. From time to time farmers need information such as accurate market prices, current supply and demand, seasonal weather, best cultivars and seeds, fertilizers and pesticides, information on pest and diseases and their control methods, harvesting and post harvesting methods, and information on farming machinery and practices, to make informed decisions at various stages of the farming cycle [2], [3]. Some of this information is available from government websites [4], [5], leaflets, and mass media in several different formats; text, audio, video. Sometimes different terminologies to express the same concept have been used. Due to its unstructured, varied formats, general nature of information, and lack of appropriate delivery methods this knowledge is not reaching the farmers.

Glendenning, Babu, and Asenso-Okyere [6] discussed clearly the importance of contextualized information and knowledge for the farmers in India. They further explained how effective this knowledge on their productivity and income since this information is more relevant to their farm enterprises and better reflects needs of the farmers. They therefore recommend that the existence of context-specific and relevant information should be considered when developing approaches for farmers as an agricultural extension.

According to the above analysis, we have identified that, farmers need information relevant to their context rather than generic information. For instance, farmers need agricultural information relevant to their situation such as the location of their farm land, their economical condition, their interest and belief, need and available equipments and so on. Then, this information would be more relevant and appropriate to farmers' needs and also could have a greater impact on their decision-making process.

Since context-specific information is more important to farmers for successful farming, we need a novel way to deliver agricultural information to farmers in a context specific manner. Social Life Networks for the Middle of the Pyramid (www.sln4mop.org) is an International Collaborative research project aiming to develop mobile based information system to support livelihood activities of people in developing countries [7]. Our research work is part of the Social Life Network project, aiming to provide information to farmers based on their context.

The idea of term *context* is treated in different ways in the literature [8], [9], [10]. One such definition is as follows [10]:

"Context is any information that can be used to characterize the situation of an entity. An entity is a person, place or object that is considered relevant to the interaction between a user and an application, including the user and applications themselves".

This definition describes context clearly and generally as it can be used to describe the situation of a participant in interactive way.

In this paper, we describe *context* specific to the farmers in Sri Lanka and the approach we developed to design the ontology to provide context-specific information and knowledge to farmers. It further discusses methodologies that were used for designing, technology selection for implementation as well as validation and evaluation techniques. The remainder of the paper is organized as follows. Section 2 describes the modeling of farmer context for this application. Section 3 presents the need for an agricultural ontology and related research in this field. Section 4 introduces design approach of the ontology for farmers in Sri Lanka. Finally, section 5 concludes the paper and describes the future direction.

2   Contextual Information

Our starting point was to gain a better understanding of information contexts specific to the farmers in Sri Lanka. Our main target group is farmers and people associated with agriculture in Sri Lanka such as researches, agriculture officers and instructors, and information specialists. To identify farmer's context clearly, we have extracted domain specific knowledge using the following reliable knowledge sources:

- Subject matter experts from universities in Sri Lanka and Australia, agriculture offices and farmers in Sri Lanka (by structured and unstructured interviews, group discussions, and questionnaires);
- Research articles and books [2], [3], [11], [12], [13];
- Published authoritative online data sources (the Department of Agriculture, Sri Lanka [4], the Ministry of Agriculture, Sri Lanka [5], and NAVAGOVIYA [14]);
- Mass media (newspapers, radio, and television) and meteorological data.

By analyzing information gathered from various sources, we have identified what information is really required by the farmers at various stages to make better decisions. The required information can be broadly divided into two types; static that changes very slowly over time and dynamic that rapidly changes; sometime on a daily or hourly basis. The required static information includes seed types and corresponding properties, fertilizers, pesticides, weather patterns, soil factor, disease management, and post harvest issues and management. The dynamic information consist of market prices, consumer behavior and demand, information related to places to buy and sell products and services, and information about what other farmers grow in different regions. Information important to farmers was identified in the form of questions in this study. Some examples are given in Table 1.

Table 1. Farmers' Problems

What are the suitable crops to grow?
What are the best cultivars?
What are the best fertilizers for selected crops and in what quantities?
When is the appropriate time to apply fertilizer?
What are the types of pests or crop diseases?
How to protect crops from disease?
Which are the most suitable control methods to a particular disease?
What are the symptoms of a specific disease?
What are the best techniques for harvesting?
What are the important factors to maintain quality of harvested crops?
Which post-harvest method is best for a particular crop?
What are the crops cultivated by other farmers and in what amounts?

When responding to these questions following factors need to be considered. Thus response can vary from farmer to farmer.

- **Farm environment:** information about environment based on location of farm such as elevation, rainfall, climate zone, temperature, humidity, sunlight, wind, soil, etc.
- **Types of farmers:** farmers are classified based on size of the cultivated farm land and estimated budget for cultivation. There are two main categories; garden farmers and commercial farmers. Commercial farmers can be further categorized as small-scale farmers, medium-scale farmers, and large-scale farmers. For example, when applying the fertilizers, information varies based on size of the farm land, budget, and number of employees.
- **Preferences of farmer:** farmers have their own preferences such as high yielding crops/varieties, preferred control methods (e.g. chemical, cultural, and biological) and fertilizers (e.g. organic, chemical, or bio), low labor cost crops, high disease and insect resistance crops/varieties, desired farming systems (e.g. shifting cultivation and bush cultivation) and techniques, etc.
- **Farming stages:** required information varies based on different stages of the farming life cycle. We therefore have reviewed existing farming stages to identify suitable farming stages for our application.

Lokanathan and Kapugama [2] categorized the information needs of farmers based on six stages of a crop cycle identified in De Silva and Rathnadiwakara [15]. It includes *Deciding stage, Seeding stage, Preparing and Planting stage, Growing stage, Harvesting, packing and storing stage*, and *Selling stage*.

Information needs of farmers are grouped by Narula and Nainwal [11] based on four stages from sowing the seeds to selling such as: *Pre-sowing, Pre-harvest, Post-harvest*, and *Market Information*.

By analyzing each of the farming stages of the above studies, we have recognized that, according to the farmers' needs identified in this section, the above two classifications of farming stages are not totally fit for our application. Selecting proper crops and cultivars is paramount for successful farming. Therefore in our application, crop selection is selected as the first farming stage. According to the farmers needs, post harvesting is an important factor. In our application we therefore have included Post-harvesting as a separate stage because this information can help farmers to reduce their post harvest losses. To improve overall decision making in farming, we specially have identified other farming stages according to following basis using the information needs identified earlier:

- by covering all required information needed by farmers;
- by placing right information at the right stage.

Fig. 1 shows the identified farming stages from crop selection to selling stage.

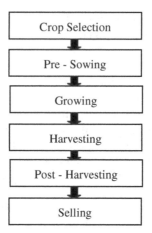

Fig. 1. Farming Stages

- *Crop Selection*: most important decision is *deciding which crops to grow*. Crop selection is a complex process because it depends on many factors. The environmental factors mostly affected this selection. Features of a crop (e.g. color, size, shape, flavor, and hardiness), farmer preferences, available resources, and market demand are other key determinant for this decision.
- *Pre-Sowing*: refers to *planning different activities related to growing* the selected crop(s). At this stage farmer needs information on quality agricultural inputs such as seed rate (i.e. needed seed quantity based on size of the farm land or number of plants according to the land size), required plant nutrients (e.g. growth regulators) and fertilizers, which types of irrigation facilities needed for selected crop, and information on best planting methods and new techniques for field preparation.
- *Growing*: includes information related to *planting and managing the crop during the growth phase*. Information on good agriculture practices (traditional practices and new technologies), common growing problems (diseases, weeds, insects, and nematode pests), and their symptoms, methods for management of growing problems (e.g. chemical, cultural, and biological methods) are required at this stage.
- *Harvesting*: at this stage, farmer needs information related to *harvesting* such as maturity time, methods and techniques of harvesting, expected average yield, labor cost, and total production cost for cultivation.
- *Post-Harvesting*: refers to proper *handling after harvest*. Required information includes post harvesting issues and management, packaging, grading, storing, standardization, transportation, and value added products.
- *Selling*: refers to *preparation for selling*. Mainly includes information related to the market such as market prices, consumer behavior and demand, and alternative marketing channels.

The nature of information required by farmers very much depends on what stage they are in, in the farming life cycle. For instance, when selecting crops, farmers are more interested to receive extra information about "diseases and its control methods". In this stage, we can recommend to *choose suitable cultivars (varieties)* as a control method for particular disease. In crop selection stage, farmers can select this cultivar to avoid the particular disease. If the farmers have actually faced particular disease then farmers have gone beyond *choose cultivars* since it is in the growing stage. In this stage we want to provide suitable disease control methods to manage this attack, for example apply suitable fungicide sprays.

This analysis suggests that, *farm environment, types of farmers, farmers' preferences,* and *farming stages* are the important factors needed to be considered when delivering agricultural information and knowledge to farmers. Therefore, this forms the context model related to the farmers for Social Life Networks. If we provide agricultural information according to identified context model, then this information is more relevant and will satisfy farmer's information needs. We therefore need to represent agricultural information and knowledge for farmers based on identified context model.

3 Need for a New Agricultural Ontology

Presently, many agricultural information systems have been developed to assist farmers' in decision-making and problem-solving by providing required information or knowledge through computers and mobile phones. Some systems include information about weather forecast, pest and diseases, and systems like *Fertilizer Expert System* which provides the tailor-made fertilizer to farmers [16]. We can see that, these systems address only a particular problem and focus is not holistic.

Making information context-specific way is more resource intensive as it requires procedures, methods, staff, and professional expertise to provide this information which is required at the farm level [17].

Ontology provides a structured view of domain knowledge and act as a repository of concepts in the domain. This structured view is essential to facilitate knowledge sharing, knowledge aggregation, information retrieval, and question answering [18]. Furthermore, ontology represents a better data model (richer knowledge) than a normal data model because ontology provides precise and well-defined relationships, strong semantic capabilities, inference mechanism, and reasoning support [19]. Therefore, ontology can be used to find a response to queries within a specified context in the domain of agriculture. The most quoted definition of ontology was proposed by Thomas Gruber as "*an ontology is an explicit specification of a conceptualization*" [20].

At present, there are few ontologies for the domain of agriculture, for instance, Thai Rice ontology [21] and Soil Science ontology [22].

Thai Rice ontology covers the domain of rice production from cultivation to harvesting in Thailand [21]. This is a prototype ontology developed for plant production using Thai rice. As existing, research knowledge repositories for plant production are not well organized; research policy administrators and researchers face many problems in finding relevant previous studies for research and development.

Therefore, Thai Rice ontology has been designed in a manner to facilitate the process of knowledge acquisition and information retrieval for research purposes.

In agriculture domain, there exist many well-established and authoritative controlled vocabularies. Thesaurus can be interpreted as a simple ontology and it is useful to build domain ontology. One of the most well-established and authoritative controlled vocabularies in agriculture is the AGROVOC Thesaurus of the Food and Agriculture Organization (FAO) [23]. However, there are several limitations and drawbacks with current vocabularies such as semantic ambiguity in definitions and usage of vocabularies; lack of high-level cross-domain concepts; and meaning of their relationships is not precisely defined [24].

Bansal and Malik are proposing an agricultural ontology for the crop production cycle to provide relevant information for farmers based on AGROVOC vocabulary in Semantic Web [19]. The point of this article seems to help individual farmers to get relevant and contextual agriculture information searching concepts on the user interfaces. However, it cannot be seen presenting information in context. Also there is no clear design approach presented to describe how to represent information in context.

The existing ontologies in the domain of agriculture are crop-specific thus too general and not specific enough to satisfy the farmers' need for timely information in context. We did not come across any agricultural ontology that has been developed to represent the agriculture domain knowledge in the context of farmer information needs. Having discovered this research gap we have focused on our attention to develop an agricultural ontology for farmers (i.e. farmer centric ontology in the domain of agriculture) to represent information needs according to farmer context.

4   Our Design Approach

As an initial step, this study focuses only on the static information to represent the knowledge which is important for the farmers to manage their farming activities. We took this decision because the static information represents data that rarely change over time while dynamic information such as market prices changes frequently and hard to obtain without an elaborate network to gather market data in real time.

There are several methodologies and techniques for building ontologies reported in the literature [25]. We reviewed these to identify suitable ontology development methodology to represent information in context. Main characteristics of existing methodologies which are used to build ontologies have been compared. The Grüninger and Fox's methodology [26] publish a formal approach to design ontology as well as providing a framework for evaluating the adequacy of the developed ontology. Its main strength is high degree of formality and focuses on building ontology based on first-order logic (FOL) by providing strong semantics. Being a formal ontology it is structurally and functionally rich enough for the description of the domain knowledge in context. It also provides a mechanism to address the drawbacks of terminological ambiguity in domain by defining rigorous, scientific, and meaningful terms. We therefore selected Grüninger and Fox's methodology, a FOL based approach to develop agricultural ontology for farmers.

It is important to highlight that ontologies inherently have a social nature, i.e. concrete artifact represents a model of consensus within a community and a universe of discourse [27]. Therefore, our ontology creation begins with the definition of a set of farmers' problems identified in section 2. For example, one of the most important questions made by the vegetable farmers is *what are the suitable crops to grow?* We take these real farmers' problems as a main motivation scenario of our application to provide information in context.

To represent information within the farmers' context, we need to formulate well-defined competency questions. Competency questions determine the scope of the ontology and use to identify the content of the ontology. The ontology should be able to represent the competency questions using its terminologies, axioms, and definitions. A knowledge-base created based on such ontology should be able to provide complete answers to the scenario questions [26]. In this application the completeness is the ability obtain answers to the scenario questions in context. To achieve this we had to derive a new way to formulate the competency questions in a systematic way. The following is our basis for formulating competency questions:

- by covering all information of the farming stages, and its constraints (restrictions);
- by introducing farmers' conditions based on the farmers' context.

For example, we have formulated following competency questions related to the crop selection stage:

- *Which crops are suitable to grow in the 'LowCountryDryZone' agro-ecology region?*
- *What are the crops involving in high labor cost?*
- *What Brinjal's cultivars are good for the 'Bacterial Wilt' disease?*
- *Which crop's cultivars are the most appropriate for 'WetZone' and 'Maha' season?*
- *What are the suitable vegetable crops for 'UpCountry', applicable to the 'Well-drained Loamy' soil, and average rainfall > 2000 mm?*

We have developed a novel approach to derive the competency questions shown above incorporating user context. This serves as a basis for formulating the competency questions in a user context because it satisfies the expressiveness and reasoning requirements of the ontology. The details of modeling competency questions and the design of the ontology are outside the scope of this paper and are explained in [28].

In order to answer these competency questions, we need to identify basic ontology components such as main concepts (e.g. Farmer, Crop, Zone, Season, Cultivar, Fertilizer), their properties (e.g. Zone has properties: maximum rainfall and minimum rainfall), and relationships between concepts (e.g. Crop *dependsOn* Zone, Crop *hasCultivar* Cultivar).

We also define the sub classes by considering instances of the concepts and their properties. For example, AgroZone is a subclass of Zone, because there are several additional properties specific to the AgroZone such as maximum and minimum temperature, and maximum and minimum elevation. Then the properties of concept Zone can be inherited by the AgroZone concept because of the taxonomic hierarchy (*is_a* relation).

Definitions of terms in the ontology and constrains in their interpretation are specified using set of axioms in FOL. For example, the following axioms express that the main climate zones in Sri Lanka (Dry zone, Intermediate zone, and Wet zone) which are categorized by annual rainfall (in mm):

$\forall$x (Zone(x) $\wedge$ ($\exists$y Integer(y) $\wedge$ avg_annual_rainfall(x,y) $\wedge$ (y<1750)) $\leftrightarrow$ Dry-zone(x))

$\forall$x (Zone(x) $\wedge$ ($\exists$y Integer(y) $\wedge$ avg_annual_rainfall(x,y) $\wedge$ (1750$\leq$y$\leq$2500))$\leftrightarrow$ Intermediate-zone(x))

$\forall$x (Zone(x) $\wedge$ ($\exists$y Integer(y) $\wedge$ avg_annual_rainfall(x,y) $\wedge$ (2500<y)) $\leftrightarrow$ Wet-zone(x))

These axioms are required to represent and answer a set of competency questions. If the defined axioms are insufficient to represent the formal competency questions and solutions to the questions, then, additional concepts, relationships, or axioms must be added to the ontology.

Next, we need to define the informal competency questions in a formal way using formal terminology. For example, the following query would model the answer to competency question: *Which crops are suitable to grow in the LowCountryDryZone region?*

($\exists$x)   (Crop(x)) $\wedge$ (AgroZone(LowCountryDryZone)) $\wedge$ dependsOn(x, LowCountryDryZone); *it generates a list of crops suitable for the 'LowCountryDryZone' region.*

The competency questions drive the development of the ontology. The answers to the competency questions can be retrieved from the knowledge base (i.e. ontology populated with instances). In this ontology, inference capability is to be represented by using inheritance and the first-order logic based axioms, i.e. it refers to the implicit knowledge derived from the ontology when reasoning procedures are applied to the ontology. We can use this ontology to organize domain knowledge by combining the farmers' context and can also deduce answers to queries based on context.

It is also important to check the validity of the ontology. Two aspects need to be validated; correctness of the contents and correctness of the construction. The content correctness depends on definitions of concepts, relationships between concepts, hierarchical structures, concept properties, and information constraints of the ontology.

The Delphi Method is a research technique that is used to obtain the responses to a problem from a group of domain experts [29]. We selected the Delphi technique to receive expert advice and responses to validate the content of the ontology. The validation process is done by agricultural experts by examining the correctness, relevancy, and inconsistency of the ontology components. In this application, structured paper-based questionnaires will be used for validation of the content correctness according to the Delphi method. The contents will be refined based on domain experts' feedbacks and comments.

However, creating ontology manually is tedious and time-consuming task. There are many ontology development tools and languages available to build a new ontology. By doing comparative study of ontology languages and tools, we have selected protégé [30] as ontology development environment and Web Ontology Language (OWL) [31] as ontology language. These are rich enough to represent our ontology since OWL is based on Description Logics (DLs) which is decidable fragment of first-order logic [32]. Fig. 2 shows the part of implemented ontology for farmers in Sri Lanka.

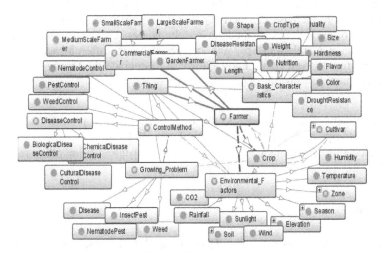

Fig. 2. Part of the Agricultural Ontology

We also used the competency questions to evaluate the ontological commitments to see whether the ontology meets the farmers' requirements. The ontology that we developed for crop selection phase was tested with a group of 32 farmers in Sri Lanka. We used a mobile based application for the evaluation. In initial version of the mobile based system, the farmer can receive the information about a particular crop, select suitable crops according to the region of the farm location and their own preferences. There was universal agreement among the farmers participated in the field trial to varying degree (strongly agree, agree, moderately agree) to the question "All information for the crop selection stage is provided". This information was provided to the farmers using farmer context via low cost Smart phones running Android operating system [33].

5   Conclusions

Farmers in Sri Lanka need necessary and relevant information to make optimal decisions for successful farming. Currently, not having agricultural knowledge repositories that can be easily accessed by farmers within their own context is a major problem. In this research project we have addressed this need.

In this study, we have clearly identified the context related to the farmers in Sri Lanka to be *farm environment, types of farmers, farmers' preferences*, and *farming stages*. Next we developed an ontological approach to meet information needs within the farmers' context. To represent information in context, we have selected first-order logic based approach because its features support us to response to farmers' queries in context. We have achieved this through modifying how competency questions are formulated in a well established methodology.

In the paper we presented a novel ontological approach to organize domain knowledge using context effectively. Knowledge organized in this manner can better assist the decision making process. In the case of farmers this can result in high quality production, increase productivity, improve post harvest handling, and better identification of agricultural methods and technologies. Using this approach we can add more concepts, relationships, and constraints with different scenarios to the ontology to provide even richer knowledgebase to upgrade the farming industry and to contribute more towards the economic growth of the country.

Acknowledgment. We acknowledge the financial assistance provided to carry out this research work by the HRD Program of the HETC project of the Ministry of Higher Education, Sri Lanka and the valuable assistance from other researchers working on the Social Life Network project.

References

1. Hettiarachchi, S.: Leeks Cultivators Desperate as Price Drops to Record Low. Sunday Times, Sri Lanka (2011)
2. Lokanathan, S., Kapugama, N.: Smallholders and Micro-enterprises in Agriculture: Information Needs and Communication Patterns. LIRNEasia, Colombo, Sri Lanka (2012)
3. De Silva, L.N.C., Goonetillake, J.S., Wikramanayake, G.N., Ginige, A.: Towards using ICT to Enhance Flow of Information to aid Farmer Sustainability in Sri Lanka. In: 23rd Australasian Conference on Information Systems (ACIS), Geelong, Australia, pp. 1–10 (2012)
4. Department of Agriculture, Sri Lanka, http://www.agridept.gov.lk
5. Ministry of Agriculture, Sri Lanka, http://mimrd.gov.lk
6. Glendenning, C.J., Babu, S., Asenso-Okyere, K.: Review of Agriculture Extension in India Are Farmers' Information Needs Being Met? International Food Policy Research Institute (2010)
7. Ginige, A.: Social Life Networks for the Middle of the Pyramid, http://www.sln4mop.org//index.php/sln/articles/index/1/3
8. Brown, P.J., Bovey, J.D., Chen, X.: Context-aware Applications: From the Laboratory to the Marketplace. IEEE Personal Communications 4(5), 58–64 (1997)
9. Dey, A.K.: Context-aware Computing: The CyberDesk Project. In: AAAI Spring Symposium on Intelligent Environments, pp. 51–54. AAAI Press, Menlo Park (1988)

10. Dey, A.K., Abowd, G.D.: Towards a Better Understanding of Context and Context-Awareness. In: Workshop on the What, Who, Where, When and How of Context-Awareness, affiliated with the CHI 2000 Conf. on Human Factors in Computer Systems, New York (2000)
11. Narula, S.A., Nainwal, N.: ICTs and Agricultural Supply Chains - Opportunities and Strategies for Successful Implementation Information Technology in Developing Countries. A Newsletter of the International Federation for Information Processing (IFIP) Working Group 9.4 Centre for Electronic Governance, Indian Institute of Management, Ahmedabad, vol. 20, pp. 24–28 (2010)
12. Decoteau, D.R.: Vegetable Crops. Prentice- Hall, Upper Saddle River (2000)
13. Kawtrakul, A.: Ontology Engineering and Knowledge Services for Agriculture Domain. Journal of Integrative Agriculture 11(5), 741–751 (2012)
14. NAVAGOVIYA, CIC (Private Sector), http://www.navagoviya.org
15. De Silva, H., Rathnadiwakara, D.: ICT Policy for Agriculture Based on a Transaction Cost Approach: Some Lessons from Sri Lanka. International Journal of ICT Research and Development in Africa (IJICTRDA) 1(1), 51–64 (2010)
16. Sriswasdi, W., Luengsrisagoon, S., Lorsuwansiri, N., Wuttilerdcharoenwong, S., Khunthong, V., Suksaengsri, T., Kawtrakul, A., Seebungkerd, N., Tananon, U., Narkwiboonwong, W., Pusittigul, A.: A Smart Mobilized Fertilizing Expert System: 1-2-3 Personalized Fertilizer. In: World Conference on Agricultural Information and IT, Tokyo, Japan, pp. 397–404 (2008)
17. Garforth, C., Angell, B., Archer, J., Green, K.: Improving Farmers' Access to Advice on Land Management: Lessons from Case Studies in Developed Countries. ODI Agricultural Research and Extension Network Paper, vol. 125. Overseas Development Institute, London (2003)
18. Gruber, T.R.: Toward Principles for the Design of Ontologies Used for Knowledge Sharing. International Journal of Human-Computer Studies 43, 907–928 (1995)
19. Bansal, N., Malik, S.K.: A Framework for Agriculture Ontology Development in Semantic Web. In: International Conference on Communication Systems and Network Technologies, pp. 283–286. IEEE Computer Society (2011)
20. Gruber, T.R.: A Translation Approach to Portable Ontology Specifications. Technical Report KSL 92-71, Knowledge System Laboratory, Stanford University, California (1993)
21. Thunkijjanukij, A., Kawtrakul, A., Panichsakpatana, S., Veesommai, U.: Rice Production Knowledge Management: Criteria for Ontology Development. Thai Journal of Agricultural Science 42(2), 115–124 (2009)
22. Heeptaisong, T., Srivihok, A.: Ontology Development for Searching Soil Knowledge. In: The 9th International Conference on e-Business (iNCEB2010), Bangkok, Thailand (2010)
23. Food and Agricultural Organization of United Nations, http://www.fao.org/aims
24. Fisseha, F.: Towards better Semantic Standards for Information Management AGROVOC and the Agricultural Ontology Service (AOS). UN FAO, Rome, Italy (2002)
25. Fernández-López, M., Gómez-Pérez, A.: Overview and analysis of methodologies for building ontologies. The Knowledge Engineering Review 17(2), 129–156 (2002)
26. Grüninger, M., Fox, M.S.: Methodology for the Design and Evaluation of Ontologies. In: Workshop on Basic Ontological Issues in Knowledge Sharing, Montreal, Canada (1995)
27. Fernandes, P.C.B., Guizzardi, R.S.S., Guizzzardi, G.: Using Goal Modeling to Capture Competency Questions in Ontology-based Systems. Journal of Information and Data Management (JIDM) 2, 527–540 (2011)

28. Walisadeera, A.I., Wikramanayake, G.N., Ginige, A.: Designing a Farmer Centred Ontology for Social Life Network. In: 2nd International Conference on Data Management Technologies and Applications (DATA 2013), Reykjavík, Iceland (2013)
29. Mattingley-Scott, M.: Delphi Method,
    `http://www.12manage.com/methods_helmer_delphi_method.html`
    (retrieved)
30. Knublauch, H., Fergerson, R.W., Noy, N.F., Musen, M.A.: The Protégé OWL plugin: An Open Development Environment for Semantic Web Applications. In: McIlraith, S.A., Plexousakis, D., van Harmelen, F. (eds.) ISWC 2004. LNCS, vol. 3298, pp. 229–243. Springer, Heidelberg (2004)
31. Patel-Schneider, P.F., Hayes, P., Horrocks, I.: OWL Web Ontology Language Semantics and Abstract Syntax. W3C Recommendation,
    `http://www.w3.org/TR/owl-semantics`
32. Baader, F., Calvanese, D., McGuinness, D.L., Nardi, D., Patel-Schneider, P.F. (eds.): The Description Logics Handbook – Theory and Applications, 2nd edn. Cambridge University Press (2008)
33. De Silva, L.N.C., Goonetillake, J.S., Wikramanayake, G.N., Ginige, A.: Farmer Response towards the Initial Agriculture Information Dissemination Mobile Prototype. In: 13th International Conference on Computational Science and Its Applications (ICCSA 2013), Ho Chi Minh City, Vietnam (2013)

Querying Spatial and Temporal Data by Condition Tree: Two Examples Based on Environmental Issues

Vincenzo Del Fatto[2], Luca Paolino[1], Monica Sebillo[1],
Giuliana Vitiello[1], and Genoveffa Tortora[1]

[1] University of Salerno, 84084 Fisciano (SA) - Italy
{lpaolino,msebillo,gvitiello,gtortora}@unisa.it
[2] Free University of Bolzano-Bozen, 39100 Bolzano-Bozen
vincenzo.delfatto@unibz.it

Abstract. The need to perform complex analysis and decision making tasks has motivated growing interest in Geographic Information Systems (GIS) as a means to compare different scenarios and simulate the evolution of phenomena. However, data and function complexity may critically affect human interaction and system performances during planning and prevention activities. This is especially true when the scenarios of interest involve space or time in imprecise contexts.

In this paper we propose a visual language which drives users to perform queries involving discrete objects by considering their temporal component. Moreover, in order to allow queries closer to the user mental model we add a specific hint for relaxing constraints and allowing fuzzy conditions. The visual language will be tested on two specific contexts concerning with the fire risk and the air pollution.

Keywords: Visual language, GIS, Spatial Database, Temporal Conditions.

1 Introduction

Computers and their applications changed the face of many traditional activities related to territory including agriculture, environment, industry and so on. In particular, Geographic Information Systems thanks to their specific ability to enhance the interconnection between information and space, have strongly encouraged the development of new decision models able to maximize productivity and effectiveness. Unfortunately, the gap between who should use those systems and who develops them is far to be filled. For this reason, the development of new techniques able to support the interaction of new unskilled users with these technologies seems very desirable. In this context, visual languages may provide users with the necessary abilities for filling the gap and make them as autonomous as possible.

In this paper, we present a new visual notation for querying spatio-temporal databases, which can be integrated for managing spatio-temporal data warehouse and can be profitably used for extracting useful information about data in a large variety of domains. The novelty of this research contribution consists in a uniform approach to

B. Murgante et al. (Eds.): ICCSA 2013, Part I, LNCS 7971, pp. 241–252, 2013.

query at the same time spatial and temporal information which has been integrated with a hint that allows to relax the conditions in order to "fuzzify" queries.

The paper can be structured as follows. Section 2 provides readers with an overview of the past approaches proposed for querying spatial database systems. Section 3 introduces the running example used to describe the visual notation introduced in Section 4. Finally, Section 5 summarizes results and proposes the next steps of the research as well.

2   Related Work

In this Section we present the state of the art of the research concerning visual approaches for querying and exploring structured data organized as Spatial Databases, GIS and Spatial Data Warehouses. These approaches are all related in some aspects with our work.

Some previous papers have been the basis of the current visual representation: in [6] we presented a visual language for querying discrete objects in a 3D domain; in [16], [18], [21] visual languages able to compose queries concerning with discrete objects and continuous fields were defined. More generally, in [11], [19], [20] we faced the issue of querying spatial databases by means of query languages.

Other works related to our research deal with visual languages for querying GIS or spatial databases. An excellent survey about visual query languages can be found in [5], where significant work has been analyzed and relevant issues have been outlined for the design of next generation visual query systems. A significant visual approach for GIS querying is represented by sketch-based visual languages which adopt a query-by-example approach where particular spatial elements configurations performed by users are interpreted by the system. Sketch! is one of the first languages which adopted that approach for composing spatial queries [12]. In Spatial-Query-By-Sketch [7] users interact with a touch sensitive screen to sketch the example spatial configurations. Even if some approaches offer support to the user during the drawing phase, exact queries can be generally ambiguous due to the several possible interpretations of the visual configurations [8]. Another approach is followed by the Spatial Exploration Environment (SEE) [10], which is an integrated framework that adopts the visual paradigm for spatial query specification and result visualization. It relies on a visual query interface for two-dimensional spatial data and an underlying visual query system, SVIQUEL, which allows the specification of topological and directional relationships between objects through direct manipulation.

As for the visual composition of conditions, other work related to our proposal deal with visual search systems for databases, such as Filter/flow [13], Kaleidoquery [14] and FindFlow [9]. In Filter/flow, users adopt the pipe metaphor to describe the Boolean logic. Each condition is considered as a filter for the water flow. If two conditions are satisfied at the same time (AND) then they would be located as a sequence of faucets, whereas if at least just one is satisfied, then the flow would be divided into two minor flows which may be interrupted by faucets, namely the filter conditions. Downstream the faucets, flows are newly connected. Filter/flow is similar from the graphical point of view but it has been defined for use with structured data.

Kaleidoquery specifies AND and OR connections by using a representation similar to the one used to represent the AND and OR gates in electronics. With the AND gate a bulb will only light when both switches are closed (conditions are true) but with the or gate a bulb will light when either or both switches are closed. FindFlow creates a tree representation for searching data where each node contains the partial result of the query specified by using filters on arcs necessary to reach the node from the root.

As for Data Warehouses, in [4] and [23] authors define OLAP queries by using a graphical representation of DWs. Users interact with graphical elements of the language by setting dimensions levels queries. Subsequently, the visual sentence is translated into a classical OLAP query. The same approach is used in [25], where authors use sequence diagrams to visually define sequence of OLAP queries. The authors associate one diagram for each dimension where a sequence is a set of Roll-Up/Drill-Down operations, represented by arrows, and a state which corresponds to a Slice operation visualized as a class object. Other works addressing in particular SOLAP queries define a correspondence between geographic layers of the visual language and the spatial dimensions levels, allowing to define spatial predicates on spatial levels (spatial slice) and spatial Roll-Up and Drill-Down operations on the spatial dimension [22]. However, the visual language does not provide a visual representation of all SOLAP concepts: aggregation functions, classical dimensions, (spatial) measures. Then, Drill and Slice operators on non-spatial dimensions cannot be defined by means of the visual language, as well as the choice of measures and aggregation functions. In[24] authors propose a visual query language for Spatial Data Warehouses where all the multidimensional elements and the spatial predicates are represented by Unified Modeling Language (UML) stereotypes. Then, they implement the spatial slice operator by means of Object Constraint Language (OCL) constraints on these stereotypes. However, this model does not allow users to visually define Drill operators. Moreover, using OCL to define topological and distance operators is a quite complex task for non-expert computer science users. In previous works, no visual language for spatial DWs allowed defining spatial DW data structures and drilling and slicing SOLAP operators using simple visual icons arrangements or sketch. In [2] the authors attempted to fill this gap. Differently from other approaches based on browsing, they described a method based on the query composition, by describing the conceptual structure for including the relationships with dimensions and measures, by defining the Nested Rectangle and the Condition Tree visual metaphors for aggregating measures and complex conditions, and by defining the Grouping Metaphor, a visual metaphor able to easily describe grouping operations. This particular approach has been implemented in Phenomena [16], a visual environment for querying heterogeneous spatial data, in order to test the interface.

3   The Running Examples

In this Section we illustrate two use scenarios which will be used throughout the paper, in order to simplify the comprehension of our proposal and illustrate how it works. Based on these scenarios, different spatio-temporal queries are formulated and then used in the following sections.

3.1   NASA LANCE - FIRMS, 2011.MODIS Hotspot / Active Fire Detections

The first scenario deals with active fire data from NASA/LANCE – FIRMS [15], a dataset which contains real time data about active fire hotspot in the real world (see Fig. 1). The Fire Information for Resource Management System (FIRMS) integrates remote sensing and GIS technologies to deliver global MODIS fire locations and burned area information to natural resource managers and other stakeholders around the world. The data are collected by the MODIS (MODerate Resolution Imaging Spectroradiometer) instrument, which is on board NASA's Earth Observing System (EOS) Terra (EOS AM) and Aqua (EOS PM) satellites. The MODIS fire locations are good for determining the location of active fires, providing information on the spatial and temporal distribution of fires and comparing data between years. The attributes in the active fire data are the followings:

   • Latitude and Longitude: The center point location of the 1km (approx.) pixel flagged as containing one or more fires/hotspots (fire size is not 1km, but variable).
   • Brightness: The brightness temperature, measured (in Kelvin).
   • Scan and Track: The actual spatial resolution of the scanned pixel.
   • Date: Acquisition date of the hotspot/active fire pixel.
   • Time: Time of the overpass of the satellite (in UTC).

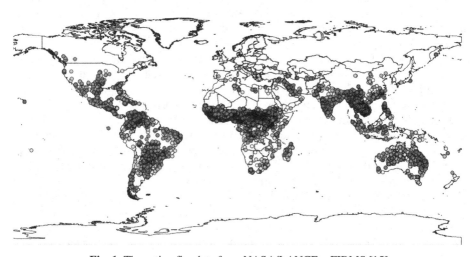

Fig. 1. The active fire data from NASA/LANCE – FIRMS [15]

Based on the described data, the following two spatio-temporal queries are formulated:

   • Query 1 *"What are the fires that occurred near the Italy country?"*
   • Query 2 *"What are the fires that occurred in the south and in the north of Brazil?"*

3.2   BRACE – Monitoring of Air Quality

The BRACE database [3], is an Italian database (see Fig. 2) that contains information about networks, stations and measuring sensors used for monitoring of air quality and related data of concentration of pollutants (measured in $\mu g/m^3$).

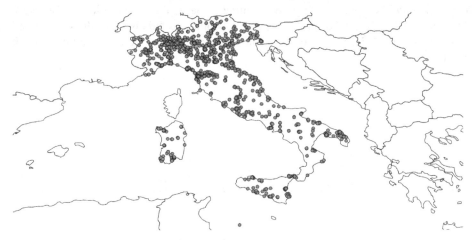

Fig. 2. Stations and measuring sensors location from BRACE

Based on the described data, the following two spatio-temporal queries are formulated:

- Query 3 *"What are the pollutant concentration measured near Salerno during the last two August?"*
- Query 4 *"What are the pollutant concentration measured in the north of Naples in the last July?"*

In the next Section, the main concepts of the visual query language are described. The language is meant to allow the visual formulation of queries and their subsequent translation into the underlying textual query language.

4   The Visual Slice Notation

In this section, we present the visual notation of the proposed visual language. Basically, it is composed by two main concepts. The Condition Tree which is the visual structure allowing to compose nodes representing the conditions.

4.1   The Condition Tree

Starting from the idea of the Predicate Tree proposed in a previous paper [17], in this paper we propose the visual language named Condition Tree (CT), able to hide the complexities of query composition by means of a simple and intuitive structure.

As depicted in Fig. 3, a Condition Tree C is defined as a Root Node R whose value is always set to the Boolean value True. Such a node may be connected with one or more Tree T. In turn, a Tree T is either an empty tree Ø or a Condition Node N, whose value is a predicate which may result either true or false. The Condition Node N may be also connected with zero or more Tree T. From a graphical point of view, the Condition Tree supports users in defining visual complex conditions through its structure. In this structure, nodes represent simple conditions and edges represent AND connectors. Moreover, edges starting from the same node are ORed connected to each other.

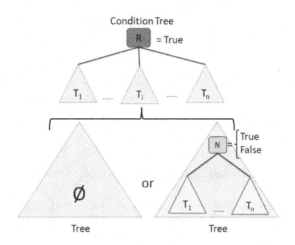

Fig. 3. Visual definition of a Condition Tree

4.2 The Spatial Fuzzy Condition Nodes

When spatial database querying is necessary, it is firstly required to understand which granularity level and which requests designers want to satisfy, namely it is necessary to define the Calculus which they refer to. In this section, we have to refer to two specific calculus: the first concerning the topological relationships while the second the Cardinal directions. The method proposed in this paper to represent fuzzy conditions is based on the Region Connection Calculus 8, RCC8 for short:

- 1-*disjoint*,
- 2-*meet*,
- 3-*partially overlap*,
- 4-*tangent proper part*,
- 5-*nontangent proper part*,
- 6-*tangent proper part inverse*,
- 7-*nontangent proper part inverse*, and
- 8-*equal*.

and the Cardinal Directional Calculus, CDC4 for short:

- ⋏ 1-*east,*
- ⋏ 2-*north,*
- ⋏ 3-*west,*
- ⋏ 4-*south.*

Relations are expressed in numeric values where each value represents the percentage area of two objects under either a specific topological or Cardinal relation. In order to represent this kind of relationships we propose a visual metaphor which allows to represent spatial combined conditions bounded by means of fuzzy values. Two objects represented by two ovals are located inside the rectangle. It is possible to move the objects within the rectangle as we want in order to figure out the desired condition. On the basis of the visual arrangement is determined which topological condition the user wants to represent and the fuzziness desired. As an example, in Fig. 4 we show two example of a disjoint condition representation. In both cases, ovals are disjoint but the strength of the condition derives from the distance of the two oval. In order to visually represent this strength of the relationship, a bar located under the rectangle gives the idea of the topological strength. As an example, the brown bar gives the idea of the topological strength. Clearly, in the second image the disjoint condition is stronger.

Fig. 4. Two examples of Disjoint fuzzy representations with two different fuzzy values

In the same way, it is possible to represent conditions for the Cardinal relationships. Differently from the previous representation, a new bar representing the Cardinal condition strength and two lines dividing the rectangle into four parts for helping users to understand where locating objects. In Fig. 5, we give an example of possible representations. As for the topological relationship, on the left side, the green oval is partly located over the red oval. In this case, the topological bar is almost in half. On the other hand, on the right side of the image, the green is much more overlapping the red. In this case, the brown bar is higher. As for the Cardinal direction, on the left side just a part of the green is in the East side whereas on the right side the green is totally in the East side. As we can see, the Cardinal left bar is lower than the right one.

The first query "*What are the fires that occurred near the Italy country?*" requires just one fuzzy topological node in the tree. Indeed, as shown in Fig. 6(a), there is a node which locates "fires" (green) near the "Italy" country (red). The concept of near

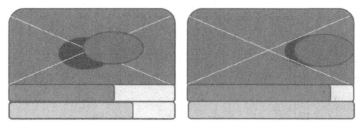

Fig. 5. Two examples of Overlap-East condition fuzzy representations with two different fuzzy values

is managed by the system moving the green oval closer or farther. The fuzzy value is indicated by the brown bar located under the node. The second query *"What are the fires that occurred in the south and in the north of the Brazil?"* is a little bit more complicated. In this case, as shown in Fig. 6(b), there are two nodes because we are looking for both the fires in North and in South of Brazil. For this situation, it is necessary to define a branch under the root indicating both the conditions. Each node is primarily arranged by overlapping the Brazil (red) and the fires (green) to each other, then we move the composition in the top for the North selection, while for the South selection we move the composition in the bottom. By activating the Cardinal option a new bar indicating the direction appears in the bottom of each node which indicates the fuzzy value of the Cardinal direction.

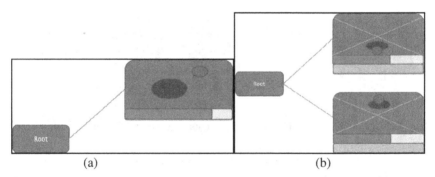

(a)   (b)

Fig. 6. The Running example queries concerning with the NASA/LANCE – FIRMS dataset (a) Query 1, (b) Query 2

4.3   The Temporal Fuzzy Condition Nodes

Following the principles of the Allen's theory proposed in [1], we propose a set of temporal conditions representations which may be used to compose complex queries. Basically, each node contains two temporal bars which represent a generic time interval. The temporal bars are differently colored in order to distinguish the left temporal side (green) from the right temporal side (red) in the condition (i.e. T1 before T2, T1 is the left temporal side and temporal T2 is the right side). In Fig. 7, a small set of examples of temporal nodes is shown.

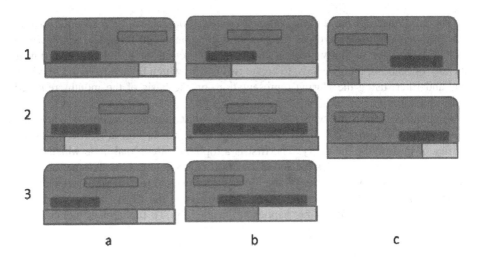

Fig. 7. A set of temporal nodes representing After

In order to move from crisp to fuzzy, we relaxed the concept of temporal relationship. Basically, in fuzzy terms a visual representation may represent more than one condition (i.e., Fig. 7 (3a) may represent time overlap, time finish and, in some way, time after). To solve this ambiguity we select the condition having the higher percentage, so in this specific case it represents a finish. As the green bar is moved to the left side, the predominant condition became overlap, and so on). Similarly to the spatial condition, each representation is provided with a brown bar which indicates the fuzziness of the relationship.

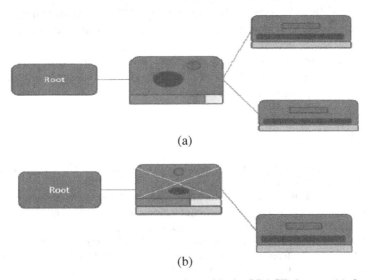

Fig. 8. The Running example queries concerning with the BRACE dataset. (a) Query 3, (b) Query 4.

The third query *"What are the pollutant concentration measured near Salerno during July and August?"* involves the temporal component. In this case, as shown in Fig. 8(a), we have to firstly create a node which select all the pollutants (green) near the city of Salerno, then we have to create two nodes, which branch from the previous one and filter, using the During representation, on the basis of the months required (July and August).

Finally, the fourth query is *"What are the pollutant concentration measured at north of Naples in the last July?"*. In this case, as shown in Fig. 8(b), two nodes are sufficient to describe the query. The first is a topological-cardinal node which select the fires (green) near Naples located at North. It is possible to note that the topological bar indicates a medium fuzzy value. In this case, we are relaxing the query involving also some fires located inside Naples but always in the North side. To complete the query we have to filter fires happened during July. To this aim we connect one temporal node with the During condition.

5   Conclusion

In this paper, we presented a visual notation for querying spatio-temporal database or spatio-temporal data warehouse which can be profitably used for extracting useful information concerning with data in a large variety of domains. Specifically, in order to prove its effectiveness, we applied this visual notation on the NASA/LANCE – FIRMS a dataset which contains real time data about active fire hotspot in the world, and BRACE an Italian database that contains information about networks, stations and measuring sensors used for monitoring of air quality and related data of concentration of pollutants. In both cases, the paper clearly highlights the simplicity of query building. The novelty of this paper may be focused on two specific aspects: a uniform approach which allows to query at the same time spatial and temporal information and an hint which gives the opportunity to relax the conditions in order to "fuzzify" the results.

At the moment, this is a preliminary result and it needs to be implemented and successively further investigated in terms of efficacy and effectiveness in order to understand the real support it can provide to users.

References

1. Allen, F.J.: Maintaining Knowledge about Temporal Intervals. Communications of the ACM 26(11), 832–843 (1983)
2. Bimonte, S., Del Fatto, V., Paolino, L., Sebillo, M.: A Visual Query Language for Spatial Data Warehouses. Lecture Notes in Geoinformation and Cartography, pp. 43–60 (2010)
3. BRACE database, http://www.brace.sinanet.apat.it
4. Cabibbo, L., Torlone, R.: From a Procedural to a Visual Query Language for OLAP. In: Procs of the Intl. Conference on Scientific and Statistical Database Management, Capri, Italy, pp. 74–83. IEEE Computer Society (1998)

5. Catarci, T., Costabile, M.F., Levialdi, S., Batini, C.: Visual Query Systems for Databases: A Survey. Journal of Visual Languages and Computing 8(2), 215–260 (1997)
6. Del Fatto, V., Paolino, L., Pittarello, F.: A Usability-Driven Approach to the Development of a 3D Web-GIS Environment. Journal of Visual Languages and Computing 18(3), 280–314 (2007)
7. Egenhofer, M.: Query Processing in Spatial Query by Sketch. Journal of Visual Languages and Computing 8(4), 403–424 (1997)
8. Ferri, F., Pourabbas, E., Rafanelli, M.: The syntactic and semantic correctness of pictorial configuration query geographic databases by PQL. In: Proceedings of the 17th ACM Annual Symposium on Applied Computing (ACM SAC 2002), Madrid, Spain, pp. 432–437 (2002)
9. Hansaki, T., Shizuki, B., Misue, K., Tanaka, J.: FindFlow: A visual interface for information search based on intermediate results. In: Asia-Pacific Symposium on Information Visualization, Tokyo, Japan, vol. 60 (2006)
10. Kaushik, S., Rundensteiner, E.: SEE: A Spatial Exploration Environment Based on a Direct-Manipulation Paradigm. IEEE Transactions on Knowledge and Data Engineering 13(4), 654–670 (2001)
11. Laurini, R., Paolino, L., Sebillo, M., Tortora, G., Vitiello, G.: A spatial SQL extension for continuous field querying. In: Proceedings - International Computer Software and Applications Conference, vol. 2, pp. 78–81 (2004)
12. Meyer, B.: Beyond Icons: Towards new metaphors for visual query languages for spatial information systems. In: Proceedings of International Workshop on Interfaces to Database Systems (IDS 1992), Glasgow, UK, pp. 113–135 (1992)
13. Morris, A.J., Abdelmoty, A.I., El-Geresy, B.A., Jones, C.B.: A Filter Flow Visual Querying Language and Interface for Spatial Databases. GeoInformatica 8(2), 107–141 (2004)
14. Murray, N., Paton, N., Goble, C.: Kaleidoquery: A visual query language for object databases. In: ACM Working Conference on Advanced Visual Interfaces, L'Aquila, Italy, pp. 247–257 (1998)
15. NASA LANCE - FIRMS, MODIS Hotspot / Active Fire Detections. Data set (2011), http://earthdata.nasa.gov/data/nrt-data/firms
16. Paolino, L., Laurini, R., Sebillo, M., Tortora, G., Vitiello, G.: Phenomena - A Visual Environment for Querying Heterogenous Spatial Data. Journal of Visual Languages and Computing 20(6), 420–436 (2009)
17. Paolino, L., Sebillo, M., Tortora, G., Vitiello, G.: The Predicate Tree – A Metaphor for Visually Describing Complex Boolean Queries. In: Qiu, G., Leung, C., Xue, X.-Y., Laurini, R. (eds.) VISUAL 2007. LNCS, vol. 4781, pp. 524–536. Springer, Heidelberg (2007)
18. Paolino, L., Tortora, G., Sebillo, M., Vitiello, G., Laurini, R.: Phenomena - A Visual Query Language for Continuous Fields. In: Proceedings of the 11th ACM International Symposium on Advances in Geographic Information Systems (GIS 2003), pp. 147–153. ACM, New York (2003)
19. Paolino, L., Sebillo, M., Tortora, G., Vitiello, G.: Integrating Discrete and Continuous Data in an OpenGeospatial-Compliant Specification. Transactions in GIS 14(6), 731–753 (2010)
20. Paolino, L., Sebillo, M., Tortora, G., Vitiello, G.: An OpenGIS®-based approach to define continuous field data within a visual environment. In: Bres, S., Laurini, R. (eds.) VISUAL 2005. LNCS, vol. 3736, pp. 83–93. Springer, Heidelberg (2006)
21. Paolino, L., Sebillo, M., Tortora, G., Vitiello, G.: Towards a new approach to query search engines: The Search Tree visual language. Software - Practice and Experience 40, 735–750 (2010)

22. Pourabbas, E., Rafanelli, M.: A Pictorial Query Language for Querying Geographic Databases using Positional and OLAP Operators. SIGMOD Record 31(2), 22–27 (2002)
23. Ravat, F., Teste, O., Tournier, R., Zurfluh, G.: Graphical Querying of Multidimensional Databases. In: Ioannidis, Y., Novikov, B., Rachev, B. (eds.) ADBIS 2007. LNCS, vol. 4690, pp. 298–313. Springer, Heidelberg (2007)
24. Glorio, O., Trujillo, J.: Designing Data Warehouses for Geographic OLAP Querying by Using MDA. In: Gervasi, O., Taniar, D., Murgante, B., Laganà, A., Mun, Y., Gavrilova, M.L. (eds.) ICCSA 2009, Part I. LNCS, vol. 5592, pp. 505–519. Springer, Heidelberg (2009)
25. Trujillo, J., Palomar, M., Gomez, J., Song, I.-Y.: Designing Data Warehouses With OO Conceptual Models. Computer 34(12), 6 (2001)

Contribution of Model-Driven Engineering to Crop Modeling

Guillaume Barbier[1,2], Véronique Cucchi[1], and David R.C. Hill[2]

[1] ITK, 5 rue de la cavalerie, Montpellier, 34000
{guillaume.barbier,veronique.cucchi}@itkweb.com
[2] ISIMA/LIMOS, UMR 6158 CNRS, Blaise Pascal University, BP 10125, Aubiere 63177
david.hill@univ-bpclermont.fr

Abstract. This work was initiated to tackle issues met by the ITK Company in developing and designing new crop models for decision support systems. At the crossroads of two disciplines, Computer Science and Agronomy, we propose the use of Model-Driven Engineering which has the potential to be the future of software engineering. This paper presents the model-driven approach retained to achieve a full-fledge crop modeling and simulation environment. The meta-model and graphical concrete syntax designed are overcoming the lack of formal tool for conceptual modeling. The presented prototype permits to improve ITK production process through the use of code generation techniques.

Keywords: Crop Models Design, Model-Driven Engineering, Domain Specific Language, Visual Modeling, Code Generation.

1 Introduction

The ITK firm develops web-based Decision Support Systems (DSS) for wine growers and croppers. These DSSs are based on mechanistic models of plant growth and development. Historically, crop modeling has followed the development pattern of software engineering practice and technologies. First, with the use of procedural languages like Fortran [1] then the onset of object-oriented programming has led to the adoption of this paradigm and the resulting research on genericity [2]. Concurrently different initiatives have achieved modularity like APSIM [3] or DSSAT [4]. Modularity interest is still advocated for the most recent plant-modeling tools based on components [5]. Model-Driven Engineering (MDE) is thought to be the next major step in software engineering and should in turn bring its advances to modeling and simulation (M&S) in agronomy as it does already in other M&S fields [6,7].

The work presented in this paper has been initiated to provide an answer to ITK specific production problem through the use of MDE techniques. Indeed, in the crop model development process adopted at ITK, agronomists first design an informal conceptual model. This conceptual model describes the main characteristics of the intended crop model. It depicts the model hierarchy, the ordering of the biophysical processes and flows of information using flowchart representations like in [8,9]. Once

B. Murgante et al. (Eds.): ICCSA 2013, Part I, LNCS 7971, pp. 253–263, 2013.
© Springer-Verlag Berlin Heidelberg 2013

designed, the conceptual model is used as a basis to achieve a prototype in the Matlab® environment. Due to the lack of formal tool for conceptual modeling [10], this prototype implementation soon becomes the *de facto* reference for the crop model. Afterwards, software engineers use this reference to implement the model in the Java language to achieve its integration into the DSS running on a Java Enterprise Edition platform. This development process is suboptimal: managing both codes evolution causes losses in production costs. Moreover it presents disjunction risks which cannot be overlooked.

MDE offers possibilities which could overcome such software production problem and in particular it can address ITK specific needs. In addition, such possibilities could also prove to be of interest for the crop modeling community in general. First of all, MDE makes it possible to use domain-specific graphical languages [11] which could lead to the definition and design of conceptual models editors. These editors may be further enhanced with mathematical and logical expressions design to achieve the complete design of a crop model. The obtained set of editors may be improved by enabling the crop model executability [12]. And lastly, MDE offers code generation techniques [13] to obtain the corresponding Java implementation. Our work is aimed at producing an environment providing all these functionalities; we named it CMF (Crop Model Factory) as an analogy to software factories [14].

To better understand the presented work, an overview of MDE concepts is followed by the description of the retained MDE approach (Section 2). In line with this approach, the analysis of the crop modeling domain has been achieved (Section 3). This analysis enables the design of a metamodel which is the core of CMF (Section 4). This metamodel has then been coupled with a graphical concrete syntax to provide conceptual model editors to CMF (Section 5). CMF has been enhanced with code generation capabilities to provide the Java code corresponding to the conceptual model (Section 6).

2   Steps of a Model-Driven Approach

This section does not aim at providing an extensive presentation of Model-Driven Engineering (MDE), interested readers may refer to [11,15-17], however basic knowledge and information on the approach applied to achieve our crop modeling factory should improve the understanding of the following sections. Basically, MDE main goal is to achieve industrialization of software production through the use of models. Modeling is nothing new in software engineering, but rather than using contemplative models as before, MDE permits to obtain productive model [18]: models from which code is generated. MDE makes it possible through the design of Domain-Specific Languages (DSLs). A given DSL is intended to enclose the concepts and rules of a given application domain. With a narrow-enough domain, the number of concepts handled by the DSL are limited. This results in the DSL being more concise than a general-purpose language; therefore, it enables a productivity increase. Moreover it may be tailored so that domain experts may use it without having to develop any software engineering skill.

In MDE, the DSL concepts and their association rules are defined by a metamodel. Any model expressed using the DSL is said to conform to the metamodel. The conformity relationship is denoted as χ. The metamodel is the abstract syntax of the DSL; it is not intended to be used by end-users in their modeling task. The model design is achieved by handling the concrete syntax of the DSL. The concrete syntax is a set of textual and/or graphical notations. As a model designed with the DSL uses the restricted set of concepts defined by the metamodel, it is possible to apply transformations to the model to obtain code expressed in a general purpose language. The transformation relationship is denoted as τ. A complete taxonomy of model transformations can be found in [19].

Figure 1 gives an overview of the approach we have retained to produce the crop model factory. From a set of legacy applications $(S_1...S_n)$ reverse engineering is applied to achieve the domain analysis. The models $(M_1...M_n)$ recovered from the reverse engineering enable the identification of the domain essential concepts and thus, the metamodel design. In parallel, the legacy applications reengineering leads to a framework design used by the reengineered applications $(S'_1...S'_n)$. The main goal of designing this framework is to limit the amount of code that has to be generated through transformation. The $M_1...M_n$ models are then reengineered using the DSL to obtain the $M'_1...M'_n$ models conforming to the metamodel. These latter are used to design the transformation τ which is validated with the reengineered applications $S'_1...S'_n$.

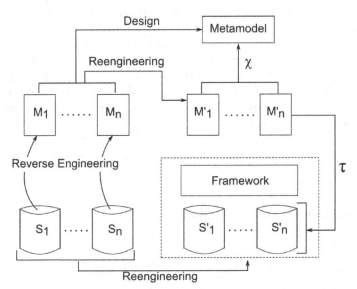

Fig. 1. The model-driven approach retained to design the crop model factory

Following this model-driven approach, the first steps of our work was dedicated to the analysis of the crop modeling domain. This analysis was led by using ITK legacy crop models and also by conducting an extensive bibliographic review of the domain. The results of this analysis are exposed in the next section.

3   Mechanistic Modeling in Agronomy: A Domain Analysis

Three legacy crop models were used to conduct the analysis. Those are combinations of different published formalisms for wheat [20,21], vine [22,23] and cotton [24]. The complementary literature review led to considering the case of Functional Structural Plant Models (FSPMs) [25]. FSPMs consider explicitly the plant architecture and interactions between organs at fine-grained scale. Though powerful research tools they have been ruled out of the analysis since their level of detail seems unfit for prediction at the field scale. However an explicit topology is considered in our study as it is used in the vine and cotton model.

Two different sets of concepts have been identified in the crop models: on one hand the structure describing the soil-plant system and on the other hand the biophysical processes (e.g. light interception, potential evapotranspiration) interacting with this structure. The former may be very basic like in the big-leaf models [26], have a higher complexity like in the leaf-layered model Sirius [21] or detail each organ [24]. The last two imply that the plant structure is dynamic: new layers or organs are created during simulation.

In any way, the soil-plant structure contains state variables which are read and/or updated by the biophysical processes. These processes are characterized by their inputs, outputs, parameters and the logical and mathematical expressions ruling their behavior. This behavior is defined at a specific time step; different processes pertaining to the same crop model may be configured with different time steps. In most cases, the processes are considered sequentially during simulation and always are in the context of ITK crop models. Modelers naturally adopt a logical decomposition of their models. They consider large biophysical concepts and divide them into sub-models until they obtain the simplest processes. Lastly, an ever-present process seems to require specific care: the process associated to the plant development commonly called the phenological model. Behaving like a state machine, this model impacts the other processes at simulation time.

This analysis, more detailed in [27], enabled the definition of a set of concepts and rules. This identification is the prerequisite to the design of the core element of CMF: its metamodel.

4   A Metamodel for Crop Models

This metamodel has been defined and refined thanks to the reverse engineering and refactoring of the legacy models used in the ITK DSS as advised by [28]. The evolution of our previous work can be seen in the two following publications [27] and [29]. We present here the metamodel in use in the current operational prototype of our crop model factory. In Figure 2, is shown an excerpt of the metamodel oriented towards the representation of the different processes and their logical decomposition. The hierarchy of biophysical processes is handled through a Composite design pattern [30]. A *CompositeModel* is used to define a crop model hierarchy and the terminal elements of this tree-like hierarchy are *ExecutableModel*s which describe the mathematical logic of the crop model. A model execution supposes a simulation sequence which gives the order by which *ExecutableModel*s are processed at simulation time, a timestep attribute of the *Model* class defines the models respective timescales. This sequence is formalized by the *firstInFlow* and *nextInFlow* relationships. An *ExecutableModel* can be either an *AtomicModel* or a *PretreatmentModel*. Finally *AlternateSequenceModel* can define alternative sequences depending on logical conditions.

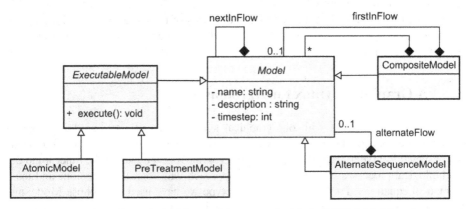

Fig. 2. Excerpt of our Crop Model Metamodel presented with the Unified Modeling Language

With regards to the data flow between modeling elements we have identified two types of behavior (Figure 3): entities providing data (*DataSource*) or consuming data (*DataSink*). Data is sourcing from model *Outputs* and *DataProviders*, which give access to external data sources (e.g., weather database), and from *Adapters*. The latter are our answer to the variety of model linking and data conversion possibilities.

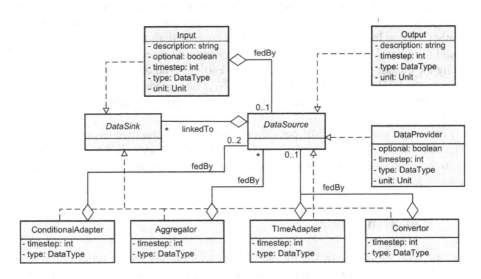

Fig. 3. Part of C3M showing the concepts associated to data flows

Adapters also behave as data sinks as do the model *Inputs*. Four types of adapters have been identified so far. The *Convertor* adapter ensures conversion between a data source expressed in units different from those of the sources linked to it. The *Time-Adapter* permits to link sources and sinks belonging to executable models having different time steps. The *Aggregator* adapter merges different sources into a single one. Lastly, the *ConditionalAdapter* permits to switch from a data source to another depending on certain conditions.

As stated in section 2, the metamodel is the basis of the expected DSL. Its graphical concrete syntax has been defined and the editors which permit conceptual modeling have been achieved.

5   A Graphical Syntax for Agronomists

The semantics associated with our graphical syntax is specific to our domain even though some concepts are common with other simulation domains. The visual aspects are designed to give a general description of crop models. To handle the inner logic of the biological processes we prefer a textual approach with logical expressions and mathematical equations. To produce a first prototype we have used the Eclipse Modeling Framework (EMF), the Graphical Modeling Framework (GMF) and EuGENia [31].

Figure 4.A exhibits elements of the graphical syntax we have retained to design crop models within CMF. It shows the main elements of the syntax needed to understand the screen caption of CMF while the vine model is being edited (Figure 4.B). The open diagram corresponds to a subpart of the vine model. This part is dedicated to computing the daily actual evapotranspiration, representing the actual loss of water by the soil-plant system. Effective computation is done by the atomic model *computeETR* (1). This requires the reference evapotranspiration ETP (2) - which represents the theoretical loss of water by soil-plant system should there be an unlimited amount of accessible water - as an input. However ETP data cannot always be provided by a weather station. In this case, the ETP must be computed, thus an *AlternateSequenceModel* (3) is used to access the alternative sequence. This sequence consists in computing the quantity of sunrays received by the earth (4) and then in using the result to compute the ETP (5). A ConditionalAdapter (6) is then needed to determine which of the computed ETP or the provided ETP should be used as an input for the *computeETR* model.

The current editors of CMF permit to design a conceptual model in an efficient way. The rather simple syntax will soon be tested with new models developed at ITK. From the models designed with these editors, CMF is already able to automatically generate Java code.

6   Code Generation

The approach we have retained for the code generation rely on the serialization features of the GMF generated editors. GMF editors produce XML Metadata Interchange

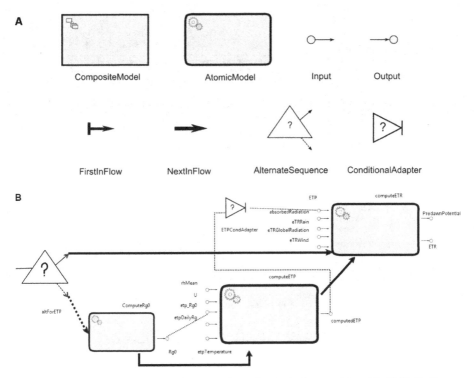

Fig. 4. A. Excerpt of the graphical syntax defined for designing models with CMF. B. Screenshot of CMF while editing part of the vine model.

files that can be processed by the Acceleo Model to Text (M2T) software (http://www.eclipse.org/acceleo/). Acceleo generates from this file the final Java code as expected from our crop model factory. The code generation relies on the use of templates and the Object Constraint Language (OCL). The use of templates permits to obtain readable code and thus quality code (see as an example figure 5). This code is correctly indented and presents comments linked to the information given by the modeler at design time. In our case, the templates have been designed to produce code which relies on a software framework we have designed and obtained by reengineering our legacy models in accordance with the approach explained in Section 2. An excerpt of this framework is presented in figure 6.

In this framework some elements stem from concepts enclosed in our metamodel, whereas some others are specific constructs dedicated to a better simulation execution. For instance the *Workflow* class handles the simulation sequence depending on the current time step; whereas the time flow is delegated by the *Simulator* to the *Clockwork* class which implements the *IClockworkAccess* interface. This interface enables an easy access to various time representations (e.g. Julian date, hour of day, date comparisons). The data flow handling is achieved by the *DataLinker* class which links the various data sinks and sources thanks to a common reference to the same data representation.

```
DispeauWorkflow.java ⊠
    import com.itk.dispeau.simulator.main_model.reservoirmodel.ReservoirModel;
    import com.itk.dispeau.simulator.main_model.plantmodel.dispeauphenologicalmodel.D:
    import com.itk.dispeau.simulator.main_model.plantmodel.lineargrowthmodel.LinearGro

    //providers with a below day frequency imports
    import com.itk.dispeau.simulator.dataproviders.HourlyGlobalRadiationProvider;

    //Conditional Flows
    import com.itk.dispeau.simulator.main_model.vatbrissonfeddes.AltForETP;

    public class DispeauWorkflow extends Workflow {
        //executable model references
        private SunCoordinates sunCoordinates;
        private Interception interception;
        private ComputeRg0 computeRg0;
        private ComputeETP computeETP;
        private ComputeETR computeETR;
        private ReservoirModel reservoirModel;
        private DispeauPhenologicalModel phenoModel;
        private LinearGrowthModel linearGrowthModel;

        //Conditional Flows
        private AltForETP altForETP;

        //Provider with a below day frequency
        private HourlyGlobalRadiationProvider hourlyGlobalRadiationProvider;

        public DispeauWorkflow(){
            providers = new ArrayList<DataProvider>(9);
        }

        @Override
        public void execute(int timestepNumber) throws DataAccessException{

            hourlyGlobalRadiationProvider.readNextValue();
```

Fig. 5. Excerpt of the Java workflow class generated with correct indents and comments

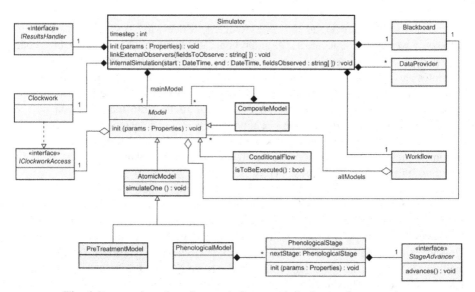

Fig. 6. Presentation of our Java code framework for Java code generation

The use of the Observer design pattern eased the handling of simulation outputs. A class registers a list of observers on the corresponding data sources and as the outputs of a simulation are entirely configurable, we can easily adapt the simulation outputs to the targeted DSS needs. This can be done without modifying the source code of the model and/or the simulator and it has been found a valuable asset for model validation. A first empirical validation of CMF generative capabilities has been achieved by comparing generated code for the vine and wheat models with the refactored legacy ones.

The tree-like structure of our models enabled an easy parsing of the hierarchical model without having to ponder over possible circular cross-references and thus it facilitated the code generation of the essential parts of the code (structure, models, blackboard, data sources and data sinks). As an example, from 60 visual elements to conceptualize the vine model, 1500 code lines have been generated with a mere 300 lines left to hand-code corresponding to the behavior of the executable models. The biggest advantage is that each executable model matches a single Matlab® function. Thus, to complete the code with the appropriate behavior, the software engineers only have to translate the content of this function into the generated executable model. This prevents them from having to ponder over the relationships (execution sequence and dataflow) between the different models which were the biggest source of error during the translation process. This improves the development process and limits the disjunction risk between the Matlab and Java implementations. That way, a significant step has been achieved in providing an environment for improving the software productivity for DSS at ITK.

7   Conclusion

We have presented in this paper the key elements leading to a prototype of Crop Model Factory (CMF) in the field of agronomy. Thanks to a reverse engineering and a reengineering of legacy models at ITK, we have proposed, with a model-driven approach, a metamodel for mechanistic models in agronomy. Relying on MDE Tools we have diminished the risks of code disjunction that were observed at ITK between the model prototypes written in Matlab® and industrial models written in Java. This is particularly the case since we have tackled automatic Java code generation using Acceleo, a part of the Eclipse Model to Text (M2T) project which implements the Object Management Group Model to Text Language specification.

In future work we will complete the Java code generation by enabling the definition of the processes (with mathematics and logics). The formalization of a textual syntax has already started and this inner logic is directly available in functions written in the Matlab® prototypes and its integration in CMF will be the next step of our developments. Another point that we would like to handle is the improvement in consistency checking of the designed models. Avoidance of loops in the simulation flow or preventing model inputs to be unlinked are part of the rules expected to be integrated in CMF. We plan to use the Object Constraint Language (OCL) which is well-integrated into EMF and GMF to handle this part.

References

1. Bouman, B.A.M., Van Keulen, H., van Laar, H.H., Rabbinge, R.: The 'School of de Wit' crop growth simulation models: a pedigree and historical review. Agricultural Systems 52(2/3), 171–198 (1996)
2. Gauthier, L., Gary, C., Zekki, H.: GPSF: A Generic and Object-Oriented Framework for Crop Simulation. Ecological Modelling 116(2-3), 253–268 (1999)
3. McCown, R.L., Hammer, G.L., Hargreaves, J.N.G., Holzworth, D.P., Freebairn, D.M.: APSIM: A Novel Software System for Model Development, Model Testing and Simulation in Agricultural Systems Research. Agricultural Systems 50, 255–271 (1996)
4. Jones, J.W., Hoogenboom, G., Porter, C.H., Boote, K., Batchelor, W.D., Hunt, L.A., Wilkens, P.W., Singh, U., Gijsman, A.J., Ritchie, J.T.: The DSSAT Cropping System Model. European Journal of Agronomy 18, 235–265 (2003)
5. Fournier, C., Pradal, C., Louarn, G., Combes, D., Soulié, J.-C., Luquet, D., Boudon, F., Chelle, M.: Building modular FSPM under OpenAlea: concepts and applications. In: 6th International Workshop on Functional-Structural Plant Models, pp. 109–112. University of California, Davis (2010)
6. de Lara, J., Vangheluwe, H.: AToM^3: A Tool for Multi-formalism and Meta-modelling. In: Kutsche, R.-D., Weber, H. (eds.) FASE 2002. LNCS, vol. 2306, pp. 174–188. Springer, Heidelberg (2002)
7. Touraille, L., Traoré, M.K., Hill, D.R.C.: A Model-driven Software Environment for Modeling, Simulation and Analysis of Complex Systems. In: Spring Simulation Multiconference - Symposium on Theory of Modeling and Simulation (TMS/DEVS), Boston, USA, pp. 229–237 (2011)
8. Lawless, C., Semenov, M.A., Jamieson, P.D.: A wheat canopy model linking leaf area and phenology. European Journal of Agronomy 22, 19–32 (2005)
9. Hakojarvi, M., Hautala, M., Ahokas, J., Oksanen, T., Maksimow, T., Aspiala, A., Visala, A.: Platform for simulation of automated crop production. Agronomy Research 8(1), 797–806 (2010)
10. Robinson, S.: Editorial. The Future's Bright the Future's... Conceptual Modelling for Simulation! Journal of Simulation 1, 149–152 (2007)
11. Kelly, S., Tolvanen, J.-P.: Domain-Specific Modeling: Enabling Full Code Generation, 448 p. Wiley - IEEE Computer Society Press (2008)
12. Muller, P.-A., Fleurey, F., Jézéquel, J.-M.: Executability into Object-Oriented Meta-Languages. In: Briand, L.C., Williams, C. (eds.) MoDELS 2005. LNCS, vol. 3713, pp. 264–278. Springer, Heidelberg (2005)
13. Hemel, Z., Kats, L.C.L., Visser, E.: Code generation by model transformation: a case study in transformation modularity. Software and Systems Modeling 9, 183–198 (2008)
14. Greenfield, J., Short, K.: Software Factories Assembling Applications with Patterns, Models, Frameworks and Tools. In: OOPSLA 2003, Anaheim, Canada, pp. 16–27 (2003)
15. Bézivin, J.: On the Unification Power of Models. Software and Systems Modeling 4(2), 171–188 (2005)
16. Favre, J.M.: Foundations of Model (Driven) (Reverse) Engineering: Models – Episode I: Stories of the Fidus Papyrus and of the Solarius. In: Proceedings of Dagsthul Seminar on Model-Driven Reverse Engineering, pp. 1–30 (2004)
17. Favre, J.M.: Foundations of Meta-Pyramids: Languages and Metamodels - Episode II: Story of Thotus the Baboon. Post-Proceedings of Dagsthul Seminar on Model Driven Reverse Engineering, pp. 1–28 (2004)

18. Selic, B.: The Pragmatics of Model-Driven Development. IEEE Software 20(5), 19–25 (2003)
19. Mens, T., Van Gorp, P.: A Taxonomy of Model Transformation. Electronic Notes in Theoretical Computer Science 152, 125–142 (2006)
20. Gate, P.: Ecophysiologie du blé. Lavoisier, 429 p. (1995)
21. Jamieson, P.D., Semenov, M.A., Brooking, I.R., Francis, G.S.: *Sirius*: A Mechanistic Model of Wheat Response to Environmental Variation. European Journal of Agronomy 8, 161–179 (1998)
22. Brisson, N., Perrier, A.: A semiempirical model of bare soil evaporation for crop simulation models. Water Resources Research 27(5), 719–727 (1991)
23. Louarn, G.: Analyse et Modélisation de l'Organogénèse et de l'Architecture d'un Rameau de Vigne (Vitis vinifiera L.). Montpellier, Ecole nationale supérieure agronomique de Montpellier. PhD, 198 p. (2009)
24. Jallas, E.: Improved Model-Based Decision Support by Modeling Cotton Variability and using Evolutionary Algorithms. Mississippi, Mississippi State University. PhD, 239 p. (1998)
25. Vos, J., Evers, J.B., Buck-Sorlin, G.H., Andrieu, B., Chelle, M., de Visser, P.H.B.: Functional-Structural Plant Modelling: A New Versatile Tool in Crop Science. Journal of Experimental Botany 61(8), 2101–2115 (2010)
26. Brisson, N., Ripoche, D., Jeuffroy, M.H., Ruget, F., Nicoullaud, B., Gate, P., Devienne-Barret, F., Antoniotelli, R., Durr, C., Richard, G., Beaudoin, N., Recous, S., Tayot, X., Plenet, D., Cellier, P., Machet, J.M., Meynard, J.M., Delecolle, R.: STICS: A Generic Model for the Simulation of Crops and their Water and Nitrogen Balance. Agronomie 18, 311–346 (1998)
27. Barbier, G., Pinet, F., Hill, D.R.C.: MDE in Action: First Steps towards a Crop Model Factory. In: Proceedings of the ESM 2011 European Simulation and Modeling Conference, Guimarães, Portugal, pp. 130–137 (2011)
28. Favre, J.M., Bézivin, J., Bull, I.: Evolution, rétro-ingénierie et IDM: du code aux modèles". In: L'ingénierie dirigée par les modèles - Au-delà du MDA, Hermes science, Lavoisier édition, pp. 185–215 (2006)
29. Barbier, G., Flusin, J., Cucchi, V., Pinet, F., Hill, D.R.C.: Vine Model Design using a Domain-Specific Modeling Language. Prototype and Proof of Concept. In: Proceedings of the ESM 2012 European Simulation and Modelling Conference, Essen, Germany, M. Klumpp, pp. 100–106 (2012)
30. Gamma, E., Helm, R., Vlissides, J.: Design Patterns: Elements of Reusable Object-Oriented Software, 403 p. Addison-Wesley (1995)
31. Kolovos, D.S., Rose, L.M., Abid, S.B., Paige, R.F., Polack, F.A.C., Botterweck, G.: Taming EMF and GMF Using Model Transformation. In: Petriu, D.C., Rouquette, N., Haugen, Ø. (eds.) MODELS 2010, Part I. LNCS, vol. 6394, pp. 211–225. Springer, Heidelberg (2010)
32. Barbier, G., Cucchi, V., Pinet, F., Hill, D.R.C.: CMF: A Crop Model Factory to Improve Scientific Models Development Process. In: Progressions and Innovations in Model-Driven Software Engineering, 16 p. IGI Global (in press)

Farmer Response towards the Initial Agriculture Information Dissemination Mobile Prototype

Lasanthi N.C. De Silva[1], Jeevani S. Goonetillake[1], Gihan N. Wikramanayake[1], and Athula Ginige[2]

[1] The University of Colombo School of Computing, Sri Lanka
{lnc,jsg,gnw}@ucsc.lk
[2] The University of Western Sydney, Australia
A.Ginige@uws.edu.au

Abstract. Timely and relevant agriculture information is essential for farmers to make effective decisions. Finding the right approach to provide this information to empower farmers is vital due to the high failure rate in current agricultural information systems. As most farmers now have mobile phones we developed a mobile based information system. We used participatory action research methodology to enable high farmer participation to ensure sustainability of the solution. The initial version of the application based on the preliminary studies focused on the crop choosing stage of the farming life cycle. This initial prototype was evaluated with a sample of farmers to check their willingness in adapting such technology, usefulness of provided information and usability of the application in order to support their day to day decision making process. The sample group strongly endorsed the various aspects of the prototype application and provided valuable insights for improvement.

Keywords: Farming Life Cycle, Action Research, Crop Choosing Stage, Mobile Technology, Agriculture Information Systems.

1   Introduction

Today we are in an era where we have access to almost any information regardless of our current location. The evolving technology is the prime reason for this availability and accessibility of information. Information is vital to make optimal decisions at the right time. However, making this information available at the right time via correct technology targeting the right population is the challenge unrelieved among the research community.

Even though the technology is in place, today there are people in developing counties who are unable to access information that are crucial for their livelihood activities due to lack of information visibility. Farming community in Sri Lanka is such a group of people. As a result farmers face many challenges within their entire farming life cycle. Main reason behind this issues is the lack of information visibility at the time of decision making [1]. This results in farmers not being able to make optimum

B. Murgante et al. (Eds.): ICCSA 2013, Part I, LNCS 7971, pp. 264–278, 2013.

decisions at different stages of the farming life cycle making a huge impact on the farmer's revenue.

Not being able to meet the expected yield quality, supply and the demand at the market level are some of the main issues faced by these people. As a result, a huge increase in suicide rate has been reported among the farming community [2]. According to reports this is mainly due to the frustration that they undergo being unable to pay their debts. As such poverty has risen creating a huge impact on the sustainability of the agriculture sector. Consequently, this has generated an impact on the younger generation as they tend to go away from the farming industry threatening the near future of the country. Further information related to the issues faced by the farming community and the need for better farmer centric information flow model can be found in [1].

In our preliminary studies it was identified that farmers need information at the right stage of the farming life cycle to make better informed decisions [1, 3]. The information need varies mainly depending on the stage of the farming life cycle [4]. Further, it was identified that the way this information should reach the farmer should be made more efficient due to the inefficacy of the existing information dissemination methods such as face to face communication via agriculture officers, web sites and other applications implemented targeting the farming community. The preliminary studies [1, 3] and other surveys carried out by different researchers [5-7] highlighted the need for a systematic approach to address the information gap among the farming community. Moreover, providing information along is not sufficient to empower the farming community [8]. Thus, it is essential to identify how the farmers can be motivated to use this information in decision making. This motivated us to conduct in depth research to explore how this can be made more efficient using evolving technology in a systematic way.

This work is a part of an international Collaborative research project which aims in developing an artefact based on mobile technologies for people in developing countries. Due to the prevailing need Sri Lanka is chosen as the test bed for this research. Various researchers based on their expertise provide their valuable research insights and findings [1, 3, 8-15] to this national project to make it a success. More information regarding the collaborative project and the members associated with it can be found on www.sln4mop.org web site. The work presented in this paper describes the evaluation study carried out with our first application to assist farmers with crop selection which was developed aiming to address the problems farmers in Sri Lanka face due to lack of information visibility. .

The rest of the paper is comprised of the following sections. Following this introduction, section 2 discusses the methodology that was adapted in this study. The main reason behind choosing this methodology is highlighted under this section. It also describes the systematic process which is adapted in this study. Section 3 provides a detail description of the evaluations. Objectives, evaluation planning and the actual evaluation setup were discussed in this section. Results that were obtained followed by the discussions will be found in Section 4. The way forward is illustrated in section 5 and finally the concluding remarks in section 6.

2   Research Methodology

The study adapted participatory action research to ensure active stakeholder collaboration. This methodology has been extensively used in healthcare and education to successively improve and enhance their current practices [16-19]. Action research is a systematic investigation where collaboration between the researchers and stakeholders are highly anticipated [20-22].   Due to this systematic behaviour it looks into the practical problem in a scientific and a holistic manner.

Specific characteristics of action research; the practical nature, change and professional development, cyclical process and high user participation [23] made us use this approach to make an intervention to the current practical situation faced by the farming community. The key point in applying action research to a problem of this nature is the ability to integrate the research and the actions rather than applying them separately. Further, action research addresses the utility of the system from the stakeholder point of view [24], which is lacking in the existing systems developed for the farming community. As stated by Denzin and Lincoin "it sees human beings as co-creating their reality through participation, experience and action" [25] thus adding more benefits while enhancing relationships, communication and participation between all stakeholders [26].

Kurt Lewin who is known as the farther of Action research described it as a cycle of planning, executing and fact finding [27]. Gradually, with more research, researchers started to define the action research life cycle more precisely [21, 22, 28]. As stated by Susman and Everd action research life cycle consists of five main phases; Diagnosing, Action planning, Action taking, Evaluating and Specifying learning [22].

1. Problem Diagnosis: At this phase the specific problem and the broader application domain context relevant to the problem is identified. In order to understand the application domain we have conducted several preliminary field visits. These included meetings with Department of Agriculture officers and a group of farmers at Dambulla where high population of people are engaged in cultivation. Based on the information collected we carried out a causal analysis [1] and identified the information need by the farmers at the different stages of the farming life cycle to make informed decisions. Further analysis carried out with aid of other agriculture experts such as agriculture extension officers and officials involved in preparing reports to higher authorities enabled us to identify gaps in the current information flow model [1]. Having identified these ground level issues we have come up with a conceptual model to meet the information gap of the farmers [3]. This led us to design the farmer centric information flow model via a mobile phone to intervene the current situation faced by the farming community at different stages of the farming life cycle.

2. Action Planning: We adopted mobile technologies having identified the increase use of mobile phones even among the farming community irrespective of the education level. Another aspect observed by the researchers is the facts behind Smart phones. Due to the rapid growth in mobile technologies now Smartphone comes

with different types of sensors and other multimedia capabilities. Thus, by using these mobile phones one can deliver valuable information to the farming community while capturing context specific information such as geo coordinates using these capabilities. Moreover, most significantly the prices of the Smartphone are rapidly decreasing. It was further observed that now low end Smartphone are within the affordable range of Sri Lankan farmers. Thus, feasibility in providing information via a Smartphone is identified to be the best solution in order to address the lack of information visibility among the farming community. Another reason behind choosing this technology is the high failure rate of the current computer based applications developed for the farming community [3]. Hence we have planned our solution to address the information need of the farmers at different stages of the farming life cycle, using especially low cost Smartphones.

Farming life cycle undergoes several stages. Mainly, crop choosing, crop growing and crop harvesting. Crop choosing stage where farmers decide what to grow in the coming season is the most crucial stage of the farming life cycle. The selected crop(s) will go through different stages of the farming life cycle. At each stage farmer has to invest labour, time and money to obtain good revenue from the harvest. Thus, if they choose a wrong crop at this stage their entire expectations might go in vain. As such we decided to select this phase of the farming life cycle to plan the first set of actions. Actual development of the first mobile prototype targeting the crop choosing stage came about only after few iterations of user studies carried out with aid of mobile paper prototypes. The information need at this stage of the farming life cycle made us to prepare the paper prototypes of the actual mobile interfaces to obtain real user feedback on the proposed design. This is a technique which can be used to get real user feedback quickly and easily to refine the design based on the user need [29]. This design was evaluated using a sample set of farmers and some agriculture extension officers at the Dambulla region. Evaluation results gave us promising insights to developing the actual model to address the information needs of the farming community.

3. Action Taking: Based on the findings we designed the first mobile prototype targeting the crop choosing stage of the farming life cycle. Issues related to the human computer interaction were addressed by the Italian Research team; one of the partners of the overall project. Thus further research on suitable mobile interfaces [12, 13] were carried out by this team of researchers. Ontology approach [14, 15] to designing the agriculture knowledge base was researched by some other researchers from Sri Lanka and Australia. The backend and the aggregation module were developed by the researchers in Australia and USA. Through this collaboration research effort the initial mobile information system mainly targeting the crop choosing stage was developed. The first working prototype was then evaluated with the aid of 32 farmers.

4. Evaluation and Reflective Learning: Next two sections of this paper will describe the evaluation process of the initial mobile information system and the results

derived from these user studies. This will be further supported using a discussion based on the analysis of the statistics.

3   Mobile Prototype Evaluation

3.1   Objectives

The main objectives behind this mobile prototype evaluation was to

- assess the farmer reaction towards the mobile technology.
- identify the usability issues in using the application.
- find out detail information to facilitate their decision making in relation to crop selection.

3.2   Design of Evaluation Study and Main Instruments

Evaluation study activities were designed in order to assess the above mentioned objectives. Foremost, farmers' ability to adapt to the new technology was assessed while letting them few minutes to play with the mobile phone. As the application was designed for a low cost Smartphone, this activity was identified as an important aspect to make them familiar in using a Smartphone. The farmer reaction was further captured using a questionnaire at the end of this activity.

The initial prototype and the questionnaire are the main instruments used in this evaluation study. The questionnaire included both multiple choice questions and open ended questions to encourage and capture wide range of answers based on the participants' knowledge. This gave us the capability to capture farmers' ideas freely. One such open ended question was to identify the factors / functionality which attracted the farmers towards using this application. The prototype that we used for the trials was in Sinhala language which is the native language in Sri Lanka. However, in this paper we are showing the English language version except for two interfaces.

The prototype included a basic login facility to identify the farmer and directed to an interface where the 6 main stages of the farming life cycle is included as shown in Fig. 1(a). As mentioned earlier the prototype targeted mainly the crop choosing stage. Thus, only the crop planning functionalities were available in the initial version used for the evaluation. Crop planner function directed the user to the next screen, illustrated in Fig. 1(b). It included 3 main categories of crop namely vegetables, fruits and other. These categories were identified based on the preliminary field trials carried out with a sample of farmers and other agriculture experts. Suitable vegetables and varieties were listed based on the region and the season.

A colour coding scheme was used to visually represent the current production level of a crop as shown in Fig. 1(c). Specific colours were used to represent different thresholds and when it reaches a specific threshold farmers were warned of the danger (highlighted using Red) of selecting the same crop as it may create an oversupply at the market level. Once the farmer selects a specific crop variety it shows the variety

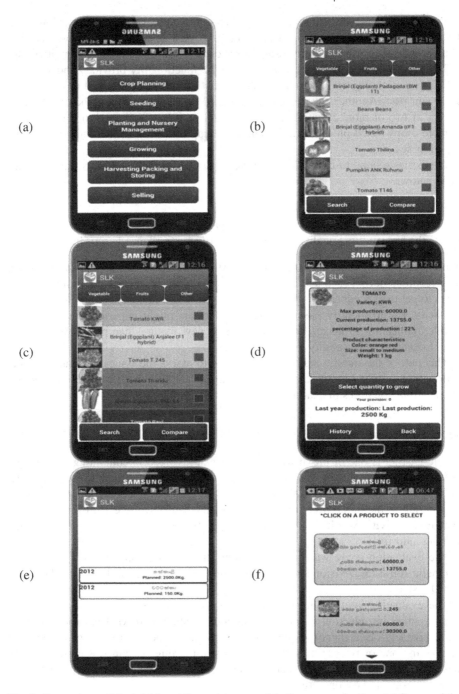

Fig. 1. Screenshots of the initial mobile prototype used in the user evaluation. (a) Stages of the farming Life Cycle, (b) Vegetables types and varieties (c) Colour code scheme (d) Special information related to a vegetable type (e) History (f) Comparison facility

specific special characteristics such as yield colour, weight, length/size etc. Moreover, it also illustrates special statistics (refer Fig 1(d)) such as current production and last year production to make farmers aware of the current as well as the last year situation.

The image illustrated in Fig 1(e) relates to the functionality provided under History. This functionality was included to show what farmer has cultivated or planned in the recent past. Another special feature included in this prototype is the comparison facility of two or more crops. As shown in Fig 1(f), it has facilitated to compare two or more crops based on one's desire and to decide on what to select minimizing the risk of losses.

The issues related to usability were investigated by assigning 3 individual tasks as listed below.

- Select a crop and a quantity to be cultivated.
- Compare two or more crops using the comparison facility.
- Use the history functionality to check the cultivated or planned crops in recent past.

In order to gain an understanding of the effort required, the starting and the end time were recorded during each task. After performing the tasks separate questionnaire was given to get their feedback on the initial prototype. Main intension behind this part of the study was to identify possible problems farmers might experience in using the mobile application to make decisions and what new features can be incorporated in order to enhance the usability.

This part of the questionnaire was also used to assess the issues in relation to the information provided for the crop choosing stage of the farming life cycle and to identify new functionalities that are needed.

3.3   Evaluation Setup

Matale District is one of the districts in Sri Lanka involved in cultivating wide range of vegetables. Farmers in this area engage in producing vegetables in large scale mainly due to the rich weather and soil conditions. Thus, as a test bed we have chosen this district for the initial investigation. Within this district we selected two main agrarian service divisions where a high percentage of the population is engaged in the farming industry.

The evaluation study was conducted with 18 farmers during the first day at the Dambulla Agrarian Division and 14 farmers on the second day at the Galewela Agrarian Division.

The evaluation study comprised of a demonstration session where farmers were given a small introduction to what the research is about and what is expected from them. Five key researchers took part in the evaluation process in the two consecutive days taking one or two farmers at a time. First a training session was carried out to make farmers familiar with the touch screen technology. Their demographic data was collected using a questionnaire.

Next the crop planner prototype was demonstrated while illustrating the key features incorporated in to the application. Then the activities mentioned on section 3.2 were given in order to evaluate the prototype.

4   Evaluation Results and Discussion

4.1   Demography

Basic characteristics of the sample population mentioned in the evaluation process are listed in Table 1. As shown in this table 75% of the sample population is of age less than 50 years. When considering the education level of the farmers, it is evident that today most of them are well educated. Based on the percentages listed in Table 1 around 91% have at least completed their secondary education to the level of taking Ordinary Level examination.

Table 1. Demographic Data of the Sample Population

Age		%	Education		%
Galewela	14	44%	Primary	2	6%
Dambulla	18	56%	Secondary (O/L)	22	69%
Age		%	Secondary (A/L)	5	16%
21 - 30	4	13%	Diploma	0	0%
31 - 40	9	28%	Graduate	2	6%
41 - 50	11	34%	Post Graduate	0	0%
> 51	8	25%	No Proper Education	1	3%

4.2   Farmer Reaction towards Technology

While analysing these statistics it was clear that today's farming community represent a portion of well educated population when compared to latter days. Thus, their willingness to get expose to new technology is high.

Computer vs. Internet

Basically, we had questions to find out related technology usage among farmers. This is to find out how many use a computer or the Internet facility in their daily activities. In general the computer usage among farmers was reported to be low. This was consistent with findings during our earlier field trials. In this sample population it was reported to be around 25%. Also the Internet usage among the farming community is relatively low. It was around 3% for the sample population. There was one farmer even not heard about the Internet. Rest had a basic idea of what Internet is and 9% of them were frequent Internet users. Out of the frequent Internet users two of them accessed Internet at least once a day and the other on average once in 2 weeks.

We have also checked their awareness of the availability of different services provided via Internet through a mobile or a computer. As illustrated on the pie chart in Fig. 2, 56% of the sample population was aware of at least one service provided via the Internet while unawareness reported to be around 44%. The statistics of the awareness of different services provided via Internet is depicted on the bar chart in

Fig. 2. As such out of these 56% of the farmers more than 50% were aware of the services such as money transfers, market price information, news and weather related information that can be accessed via Internet.

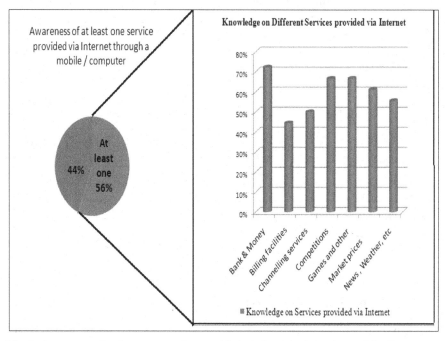

Fig. 2. Awareness of at least one service provided via Internet through a mobile /a computer

Computer vs. Mobile

The most significant fact is that, when compared to computer usage, the mobile phone usage is very high among the farming community. For the sample population it was recorded to be around 81%. A preliminary study was also conducted using a sample of farmers during the problem diagnosis stage to identify the mobile phone availability among the farmers. Compared with this study the sample characteristics demonstrated similar statistics for mobile phone availability irrespective of the education level. However, their main intension behind using a mobile phone was basically to make and receive calls. This percentage was recorded to be around 96%. Compared to this, percentage accessibility of other services such as short message service (SMS), Multimedia Message Service (MMS), internet etc are found to be relatively low.

Though Smartphone usage among farmers were found to be very low, all most all the farmers to whom the prototype was demonstrated got used to the technology within about 5 to 10 minutes. During this time period, they were been capable of taking a call, type a short message and to take a picture using the camera. According to the observations in general they have valued this type of technology and showed their eagerness in adoption even though, most of the farmers are new to the touch screen technology.

4.3   Usability Evaluation

The usability issues were monitored with the aid of the activities listed in section 3.2. According to the observations in general the initial version was a success as most of the farmers used the mobile device in few minutes time and were able to do the activities as instructed by the researchers. However, in some instances it was identified that some farmers had issues in navigation due to poor wordings and onscreen instructions. Further, it was also observed that some farmers tend to select wrong options as the button areas were too small.

According to the farmer response around 56% was attracted to the idea presented using the colour coding scheme. Farmers were bit concern on the accuracy of yield information through this method and the ability to make a correct decision based on the colour code. A percentage around 47% farmers found the information provided with respect to crop types and different varieties very useful for them to continue in using the application. As our initial prototype contains all possible vegetables and their different varieties on the same screen without having a better classification, farmers found it difficult to select or search for the crop varieties that they were looking for. 34% responses were received in favour of the comparison facility and around 25% for the information provided using the functionality history. However, some have also mentioned the importance of showing more information such as the price sold and the issues faced with respect to the selected crops in the previous seasons will add more value to the history functionality. Some mentioned that they were attracted as information was presented in their native language "Sinhala" and the presentation of information which is clear for any novice users to learn and understand. Similarly, some have liked the application as it provides more valuable information which can be accessed in lesser time and cost. Around 81% of the correspondents mentioned that there is nothing they can identify as an unwanted feature. However, some have mentioned that it is difficult to find the next action due to the lack of clarity, making it harder to use the application.

Further, based on the questionnaire which included questions as listed in Table 2 was evaluated on a liker scale; strongly agree (SA), agree (A), neutral (N), disagree (D) and strongly disagree (SD). To better visualize the responses the percentages recorded under SA and A, D and SD were aggregated and based on Fig. 3 the above claims were further verified.

As such we were been success in delivering a message using a colour code schema. According to the bar chart illustrated in Fig. 3, 100% of the participants fully agreed to the fact that the message trying to deliver using such a colour coding scheme is clear to them. They were also fascinated by the added functionalities such as comparison facility and history. According to the observations almost all got familiar with the application during the evaluation process. Further, according to the statistics a total of

93% fully agreed that the prototype evaluated is easy to use and can be learnt easily by anyone. However, 7% stressed the need for providing a better training before using the application to gain most out of it.

Table 2. Interview Questions based a 5 liker scale

1.	All information for the crop choosing stage is provided
2.	Information is sufficient for decision making
3.	Provide knowledge on different crop varieties
4.	Knowledge on history is important
5.	Market prices of the previous year are important in deciding a crop
6.	Crop comparison facility is essential in deciding a crop
7.	Information provided using the Colour code is clear
8.	Colour code usage is important in deciding a crop
9.	Functionality provided in this system can be easily learnt by anyone

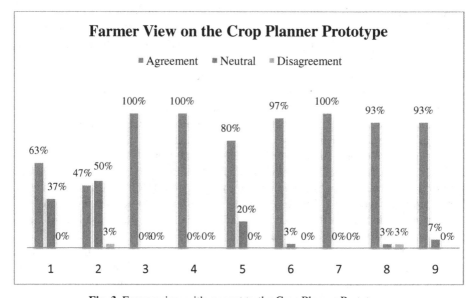

Fig. 3. Farmer view with respect to the Crop Planner Prototype

4.4 Detail Information to Facilitate Decision Making in Relation to Crop Selection

It was identified that farmers do consider different factors when choosing a crop during this stage of cultivation. Despite crop knowledge presented in the prototype is not complete; almost all were willing to use the system. This was visible mainly due to the ability of making better decisions that they have started seeing within the information provided via the application. They have also mentioned that this will add new knowledge to their limited experience to enhance their farming activities. At the same time, they have started seeing the ability in making decisions with help of this information to enhance their cultivation practices. Ability to acquire knowledge at any time from any place was the benefit they have seen from this type of technology when compared to other existing approaches.

In total as shown in Fig. 3, 63% of the farmers agreed that the initial prototype has covered the basic information needs at the crop choosing stage. Rest expected more information related to crop variety and seeds. They also agreed that this information is essential knowledge during this stage of farming life cycle which they lack in current practices.

5   Way Forward: Specifying Learning

As mentioned earlier in the paper the main intension of this evaluation process is to check the farmer motivation and willingness to use the technology to obtain required information to make optimal decisions on time. In order to facilitate this we have only added some basic functionality to the first prototype. Thus, limited information relevant to the crop selection stage was made available in the initial prototype.

After analysing the evaluation results it is evident that there is a need for more features or functionalities in addition to what was provided within the initial prototype. Information with respect to quality seeds, new cultivation techniques, possible pest and disease attacks were among the farmer feedback. Most of these responses were also similar to what was gathered from the preliminary studies [3]. Due to the time limitation though these were not incorporated in the initial version we would look into these features before going on with further development to aid the crop choosing stage of the farming life cycle. In addition to that, we would also look into the new requirements identified such as water availability, cost prediction per crop, market condition etc and will further analyse how these can be incorporated in order to facilitate decision-making.

Now we are in the process of developing the next version of the mobile prototype addressing the missing needs for the crop choosing stage. Further, we are planning few more user evaluations based on this initial version with other stakeholders of the farming community such as agriculture extension officers, agriculture experts and vendors. This will aid us further verification as it is essential to review other viewpoints to support better decision making. Further, it will open up opportunities in resource sharing strengthening sustainability and making the application a success.

The initial version will be enhanced based on the farmer feedback as well as based on other studies that will be conducted among other stake holders in the near future. The second version will be deployed in the near future to record their usage pattern. Thereby, we will get more feedback to make it better.

6   Conclusion

As a whole farmers wanted to use the application with more features included to acquire new knowledge for their farming practices. At the same time they have witnessed the importance of such technology to bridge the information gap and thereby to make optimal decisions at the right phase of the farming life cycle to enhance their livelihood. Most notably their readiness in moving towards the technology to enhance their current practices was extremely astounding. Thus, they have shown a positive reaction towards using the mobile application. Moreover, they have highlighted the importance of using such techniques to help educate the inexperience younger generation to encourage them to retain within the farming industry. It was also mentioned the importance of providing highly accurate, recent information if they are to rely on the application. The importance of making linkages between other agriculture research institutes all over the world to obtain new knowledge is also pointed out by some farmers to make this application a success. Moreover, some have brought up the idea of providing the information for free and to give a thorough training for the farmers to teach them how to use it and what can be get out of the application to increase the popularity towards using the application.

Thus, it is obvious that the possibility of presenting the information using right technology in a user friendly manner will aid the farming community to make better decisions. Further, this will add new opportunities while ensuring sustainability both in terms of farmer and the solution.

Acknowledgements. Authors of this paper would like to provide our sincere thanks to all the farmers in two regions who took part in the evaluation process regardless of their valuable time and bad weather condition. We would also like to thank the officers at the agrarian service centre for facilitating these meetings and for gathering farmers to make this a success. Further, we value their comments and suggestions provided from the beginning of the project to enhance the application. Last but not least we would appreciate the commitment received in developing the mobile information system from the members of the international collaborative research team. Their invaluable comments and suggestions made the evaluation process a success. Some farmer evaluation expenses were funded by the National Science Foundation, Sri Lanka.

References

1. De Silva, L., et al.: Towards using ICT to Enhance Flow of Information to aid Farmer Sustainability in Sri Lanka. In: Australasian Conference on Information Systems (ACIS), Geelong, Australia (2012)
2. Amaranath, S.: Sri Lanka suicide rate one of the world's highest. World Socialist Web Site (2012), http://www.wsws.org/en/articles/2012/09/sril-s28.html
3. De Silva, L., et al.: A Holistic Mobile Based Information System to Enhance Farming Activities in Sri Lanka. In: The IASTED International Conference on Engineering and Applied Science (EAS), Colombo, Sri Lanka (2012)
4. Lokanathan, S., Kapugama, N.: Smallholders and Micro-enterprises in Agriculture: Information needs & communication patterns, Colombo, Sri Lanka. LIRNE asia, pp. 1–48 (2012)
5. Punchihewa, D.J., Wimalaratne, P.: Towards an ICT enabled farming Community. In: E-Governance in Practice, India (2010)

6. Pavitrani, A.D.S., et al.: The effectiveness of existing ICT modules in addressing issues of Farming community in Sri Lanka: Empirical Study. In: National Information Technology Conference, Colombo, Sri Lanka (2011)

7. Parikh, T.S., Patel, N., Schwartzman, Y.: A Survey of Information Systems Reaching Small Producers in Global Agricultural Value Chains. In: International Conference on Information and Communication Technologies for Development (ICTD), Bangalore, India (2007)

8. Ginige, T., Richards, D.: A model for Enhancing Empowerment in Farming using Mobile based Information System. In: 23rd Australasian Conference on Information Systems (ACIS), Geelong, Australia (2012)

9. Ginige, A.: Social Life Networks for the Middle of the Pyramid (2011) (cited January 31, 2012)

10. Ginige, T., Ginige, A.: Towards next generation mobile applications for MOPS: investigating emerging patterns to derive future requirements. In: International Conference on Advances in ICT for Emerging Regions (ICTer). IEEE, Sri Lanka (2011)

11. Ginige, A., Ginige, T., Richards, D.: Architecture for Social Life Network to Empower People at the Middle of the Pyramid. In: 4th International United Information Systems Conference (UNISCON), Yalta, Ukraine (2012)

12. Giovanni, P.D., et al.: Building Social Life Networks through Mobile Interfaces– the Case Study of Sri Lanka Farmers. In: IX Conference of the Italian Chapter of AIS (ITAIS), Rome, Italy (2012)

13. Giovanni, P.D., et al.: User Centered Scenario based Approach for Developing Mobile Interfaces for Social Life Networks. In: 34th International Conference on Software Engineering (ICSE) - UsARE Workshops, Zurich, Switzerland (2012)

14. Walisadeera, A.I., Wikramanayake, G.N., Ginige, A.: An Ontological approach to meet Information needs of farmers in Sri Lanka. In: 13th International Conference on Computational Science and its Applications, Vietnam (2013)

15. Walisadeera, A.I., Wikramanayake, G.N., Ginige, A.: Designing a farmer centred Ontology for Social Life Network. In: 2nd International Conference on Data Management Technologies and Applications (DATA), Reykjavík, Iceland (2013)

16. Ferrance, E.: Action Research. The Education Alliance (2000)

17. Marshall, M., et al.: Development of an information source for patients and the public about general practice services: an action research study. Health Expectations: An International Journal of Public Participation in Health Care and Health Policy 9(3), 265–274 (2006)

18. Dickinson, A., et al.: Hospital mealtimes: action research for change? The Proceedings of the Nutrition Society 64(3), 269–275 (2005)

19. Sax, C., Fisher, D.: Using Qualitative Action Research to Effect Change: Implications for Professional Education. Teacher Education Quarterley 28 (Spring 2001)

20. O'Brien, R.: An Overview of the Methodological Approach of Action Research. In: Richardson, R. (ed.) Theory and Practice of Action Research, Toronto (2001)

21. Baskerville, R.L.: Investigating Information Systems with Action Research. Communication of the Association for Information Systems 2 (1999)

22. Susman, G.I., Evered, R.D.: An Assessment of the Scientific Merits of Action Research. Administrative Science Quarterly 23, 582–603 (1978)

23. Denscombe, M.: The Good Research Guide for small-scale social research projects, pp. 122–131. McGraw-Hill Open University Press, Berkshire (2007)

24. Jarvinen, P.: Action research as an approach in design science. In: European Academy of Managment (EURAM) Conference. Net Publications, Munich (2005)
25. Denzin, N.K., Lincoln, Y.S.: The SAGE Handbook of Qualitative Research, 3rd edn. Sage, Thousand Oaks (2005)
26. Stringer, E.T.: Action Research: A handbook for Practitioners. Sage, Thousand Oaks (1996)
27. Lewin, K.: Action Research & Minority Problems. J. of Social Issues 4(2), 34–46 (1946)
28. Elliott, J.: Action Research for Educational Change. Open University Press, Philadelphia (1991)
29. Snyder, C.: Paper Prototyping: The Fast and Easy Way to Design and Refine User Interfaces. Elsevier, San Francisco (2003)

On the Use of a Priori Knowledge in Pattern Search Methods: Application to Beam Angle Optimization for Intensity-Modulated Radiation Therapy

Humberto Rocha[1], Joana M. Dias[1,2],
Brigida C. Ferreira[3,4], and Maria do Carmo Lopes[3,4]

[1] INESC-Coimbra, Rua Antero de Quental, 199
3000-033 Coimbra, Portugal
[2] Faculdade de Economia, Universidade de Coimbra,
3004-512 Coimbra, Portugal
[3] I3N, Departamento de Física, Universidade de Aveiro,
3810-193 Aveiro, Portugal
[4] Serviço de Física Médica, IPOC-FG, EPE,
3000-075 Coimbra, Portugal
hrocha@mat.uc.pt, joana@fe.uc.pt, brigida@ua.pt,
mclopes@ipocoimbra.min-saude.pt

Abstract. Pattern search methods are widely used for the minimization of non-convex functions without the use of derivatives. One of the main features of pattern search methods is the flexibility to incorporate different search strategies taking advantage of the imported global optimization techniques without jeopardizing their convergence properties. Pattern search methods can also be adapted to problem contexts where the user can provide points incorporating a priori knowledge of the problem that can lead to an objective function improvement. Here, an automated incorporation of a priori knowledge in pattern search methods is implemented instead of an algorithm that requires the user's contribution. Moreover, a priori knowledge can also play a role on the choice of the initial point(s), an important aspect in the success of a global optimization process. Our pattern search approach is tailored for addressing the beam angle optimization (BAO) problem in intensity-modulated radiation therapy (IMRT) treatment planning that consists of selecting appropriate radiation incidence directions and may influence the quality of the IMRT plans, both to enhance better organs sparing and to improve tumor coverage. Beam's-eye-view dose ray tracing metrics are used as a priori knowledge of the problem both to decide the initial point(s) and to be incorporated within a pattern search methods framework. A couple of retrospective treated cases of head-and-neck tumors at the Portuguese Institute of Oncology of Coimbra is used to discuss the benefits of incorporating a priori dosimetric knowledge in pattern search methods for the optimization of the BAO problem.

Keywords: Pattern Search Methods, Beam's-Eye-View Dose Metrics, IMRT, Beam Angle Optimization.

B. Murgante et al. (Eds.): ICCSA 2013, Part I, LNCS 7971, pp. 279–292, 2013.

1   Introduction

Pattern search methods are widely used for the minimization of non-convex functions without the use of derivatives or approximations to derivatives. Pattern search methods are organized around two steps at every iteration: the poll step and the search step. The poll step guarantees global convergence to stationary points by performing a local search in a neighborhood around the current iterate using the concepts of positive bases. The search step consists of a search away from the current iterate, free of rules as long as the search is finite. One of the main features of pattern search methods is the flexibility to incorporate different search strategies in the search step, taking advantage of the imported global optimization techniques, without jeopardizing their convergence properties. Different techniques for global optimization have been successfully incorporated on the search step, including surrogate optimization, e.g., radial basis functions [15], or global optimization, e.g., particle swarm optimization [19]. Pattern search methods can also be adapted to problem contexts where the user can provide points incorporating physical or a priori knowledge of the problem that can lead to an objective function decrease [1]. Here, an automated incorporation of a priori knowledge in pattern search methods is implemented instead of an algorithm that requires the user's contribution. Moreover, a priori knowledge can also play a role on the choice of the initial point(s), an important aspect in the success of a global optimization process.

The pattern search approach presented in this work is tailored for addressing the beam angle optimization (BAO) problem in intensity-modulated radiation therapy treatment planning. The intensity-modulated radiation therapy (IMRT) is a modern type of radiation therapy, whose planning leads to complex optimization problems, including the BAO problem - the problem of deciding which incidence radiation beam angles should be used. The pattern search methods framework has been used to address the BAO problem successfully due to its ability to avoid local entrapment and its need for few function value evaluations to converge [14,15]. Here, a priori knowledge of the problem is incorporated in pattern search methods using beam's-eye-view dose ray tracing metrics. The beam's-eye-view concept considers topographic and/or dosimetric criteria to rank the candidate beam directions.

The goal of the paper is twofold: to discuss the influence of a priori dose metric knowledge on the choice of the initial point(s) and to discuss the benefits of incorporating a priori knowledge in pattern search methods. A couple of retrospective treated cases of head-and-neck tumors at the Portuguese Institute of Oncology of Coimbra is used to discuss these benefits for the optimization of the BAO problem. Our approach is tailored to address this particular problem but it can be easily extended for other general problems. The paper is organized as follows. In the next Section we describe the BAO problem. Beam's-eye-view dose ray tracing metrics and its use within the pattern search methods framework is presented in Section 3. Computational tests using clinical examples of head-and-neck cases are presented in Section 4. In the last Section we have the conclusions.

2   Beam Angle Optimization in IMRT Treatment Planning

The BAO problem consists on the selection of appropriate radiation incidence directions in radiation therapy treatment planning and may be decisive for the quality of the treatment plan, both for appropriate tumor coverage and for enhance better organs sparing. Many attempts to address the BAO problem can be found in the literature including simulated annealing [3], genetic algorithms [9], particle swarm optimization [11] or other heuristics incorporating a priori knowledge of the problem [13]. The BAO problem is quite difficult, and yet to be solved in a satisfactory way, since it is a highly non-convex optimization problem with many local minima [4].

In IMRT, the radiation beam is modulated by a multileaf collimator, transforming the beam into a grid of smaller beamlets of independent intensities, allowing the irradiation of the patient using non-uniform radiation fields from selected angles aiming to deliver a dose of radiation to the tumor minimizing the damages on the surrounding healthy organs and tissues. The IMRT treatment planning is usually a sequential process where initially a given number of beam directions are selected followed by the fluence map optimization (FMO) at those beam directions. Obtaining the optimal fluences for a given beam angle set is time consuming due to the dosimetric calculations required. For that reason, many of the previous BAO studies are based on a variety of scoring methods or approximations to the FMO to gauge the quality of the beam angle set. However, when the BAO problem is not based on the optimal FMO solutions, the resulting beam angle set has no guarantee of optimality and has questionable reliability. Therefore, our approach for modeling the BAO problem uses the optimal solution value of the FMO problem as the measure of the quality for a given beam angle set. Thus, we will present the formulation of the BAO problem followed by the formulation of the FMO problem we used. Here, we will assume that the number of beam angles is defined a priori by the treatment planner and that all the radiation directions lie on the same plane.

2.1   BAO Model

Let us consider n to be the fixed number of (coplanar) beam directions, i.e., n beam angles are chosen on a circle around the CT-slice of the body that contains the isocenter (usually the center of mass of the tumor). In our formulation, instead of a discretized sample, all continuous $[0°, 360°]$ gantry angles will be considered. Since the angle $-10°$ is equivalent to the angle $350°$ and the angle $370°$ is the same as the angle $10°$, we can avoid a bounded formulation. A simple formulation for the BAO problem is obtained by selecting an objective function such that the best set of beam angles is obtained for the function's minimum:

$$\min f(\theta_1, \ldots, \theta_n)$$
$$s.t. \ (\theta_1, \ldots, \theta_n) \in \mathbb{R}^n. \tag{1}$$

Here, for the reasons stated before, the objective $f(\theta_1, \ldots, \theta_n)$ that measures the quality of the set of beam directions $\theta_1, \ldots, \theta_n$ is the optimal value of the FMO problem for each fixed set of beam directions. The FMO model used is presented next.

2.2   FMO Model

In order to solve the FMO problem, i.e., to determine optimal fluence maps, the radiation dose distribution deposited in the patient needs to be assessed accurately. Each structure's volume is discretized into small volume elements (voxels) and the dose is computed for each voxel considering the contribution of each beamlet. Typically, a dose matrix D is constructed from the collection of all beamlet weights, by indexing the rows of D to each voxel and the columns to each beamlet, i.e., the number of rows of matrix D equals the number of voxels (N_v) and the number of columns equals the number of beamlets (N_b) from all beam directions considered. Therefore, using matrix format, we can say that the total dose received by the voxel i is given by $\sum_{j=1}^{N_b} D_{ij} w_j$, with w_j the weight of beamlet j. Usually, the total number of voxels is large, reaching the tens of thousands, which originates large-scale problems. This is one of the main reasons for the difficulty of solving the FMO problem.

For a given beam angle set, an optimal IMRT plan is obtained by solving the FMO problem - the problem of determining the optimal beamlet weights for the fixed beam angles. Many mathematical optimization models and algorithms have been proposed for the FMO problem, including linear models [18], mixed integer linear models [10] and nonlinear models [2]. Here, we will use this later approach that penalizes each voxel according to the square difference of the amount of dose received by the voxel and the amount of dose desired/allowed for the voxel. This formulation yields a quadratic programming problem with only linear non-negativity constraints on the fluence values [18]:

$$\min_w \sum_{i=1}^{N_v} \frac{1}{v_S} \left[\underline{\lambda}_i \left(T_i - \sum_{j=1}^{N_b} D_{ij} w_j \right)_+^2 + \overline{\lambda}_i \left(\sum_{j=1}^{N_b} D_{ij} w_j - T_i \right)_+^2 \right]$$

$$s.t. \quad w_j \geq 0, \ j = 1, \ldots, N_b,$$

where T_i is the desired dose for voxel i of the structure v_S, $\underline{\lambda}_i$ and $\overline{\lambda}_i$ are the penalty weights of underdose and overdose of voxel i, and $(\cdot)_+ = \max\{0, \cdot\}$. This nonlinear formulation implies that a very small amount of underdose or overdose may be accepted in clinical decision making, but larger deviations from the desired/allowed doses are decreasingly tolerated [2].

The FMO model is used as a black-box function and the conclusions drawn regarding BAO coupled with this nonlinear model are valid also if different FMO formulations are considered.

3 Pattern Search Methods Incorporating a Priori Knowledge

The incorporation of a priori knowledge in pattern search methods is done using beam's-eye-view dose ray tracing metrics. We will briefly describe the concept of beam's-eye-view and the strategy used to take advantage of the incorporation of the resulting metrics in the pattern search method framework applied to the BAO problem.

3.1 Beam's-eye-view Dose Metrics

Conventional beam's-eye-view (BEV) tools consider only geometric criteria, i.e., topographic localization of tumor volume(s) versus surrounding healthy structures, to evaluate each candidate beam direction. The use of beam's-eye-view dose metrics (BEVD) was introduced by Pugachev and Xing [12] to evaluate and rank the irradiation beam directions using a score function that accounts for beam modulation unlike the traditional BEV. An intensity-modulated beam can intercept a large volume of an organ at risk (OAR) or normal tissue and may not be necessarily a bad beam direction, which makes the geometrical criteria used by BEV limited. The computation of a metric to gauge the quality of incidence radiation directions should also consider the dose tolerances of the involved structures. Thus, in IMRT, a score function based on dosimetric criteria is more appropriate to measure the quality of a radiation beam direction.

The measure of the quality of a radiation beam direction adopted is based on sensitive structures tolerance dose as a determinant factor for deliverable target dose. A given incidence radiation direction is preferred if it can deliver more dose to the target without exceeding the tolerance dose of the OARs or normal tissue located on the path of the beam [12,13]. The BEVD score for a given beam angle corresponds to the computation of the maximum achievable intensity for each beamlet involved, which depends on the locations and tolerances of the OARs along the path of the beamlet.

The BEVD score calculation of a beam requires the assumption of a single incident beam. Initially, all beamlets are assigned with an intensity that assures the delivery of a dose that fulfills the prescription to every target voxel. Beamlet intensities are then iteratively updated until tolerance dose for every structure's voxel crossed by the all the beamlets is not exceeded. The intensities obtained for each beamlet correspond to the maximum usable intensity of the beamlet without exceeding the tolerance of the sensitive structures. Finally, a forward dose calculation using the maximum usable beamlet intensities is performed and the score of a given beam direction is computed [12,13]:

$$S_k = \frac{1}{N_T} \sum_{i \in Target} \left(\frac{d_{ik}}{D_P^T} \right)^2,$$

where N_T is the number of voxels in the target, D_P^T is the target prescription dose and d_{ik} is the "maximum" dose delivered to the target voxel i by the radiation beam direction k.

The BEVD score is based on an intuitive consideration of the deliverable dose capability to the target of a single beam direction. This information can be used to construct initial point(s) whose neighborhood may be worth of being thoroughly explored. However, the optimal beam configuration for an IMRT treatment should balance the BEVD score and the beam interplay as a result of the overlap of radiation fields. Thus, BEVD scores are used as a priori knowledge to construct an insightful algorithm for beam angle optimization. This a priori knowledge of the problem is used by a pattern search methods framework.

3.2   Pattern Search Methods Incorporating BEVD

Pattern search methods are derivative-free optimization methods that use the directions of positive bases to explore the search space, such that iterate progression is solely based on a finite number of function evaluations in each iteration, without explicit or implicit use of derivatives. We will briefly describe pattern search methods for unconstrained optimization problems such as the beam angle problem formulated in (1).

Pattern search methods use the concept of positive bases (or positive spanning sets) to move towards a direction that would produce a function decrease. A positive basis for $\mathbb{R}^n$ can be defined as a set of nonzero vectors of $\mathbb{R}^n$ whose positive combinations span $\mathbb{R}^n$ (positive spanning set), but no proper set does. The motivation for directional direct search methods such as pattern search methods is given by one of the main features of positive basis (or positive spanning sets) [7]: there is always a vector $\mathbf{v}^i$ in a positive basis (or positive spanning set) that is a descent direction unless the current iterate is at a stationary point, i.e., there is an $\alpha > 0$ such that $f(x^k + \alpha \mathbf{v}^i) < f(x^k)$. This is the core of directional direct search methods and in particular of pattern search methods.

Pattern search methods are iterative methods generating a sequence of non-increasing iterates $\{x_k\}$. Given the current iterate x^k, at each iteration k, the next point x^{k+1}, aiming to provide a decrease of the objective function, is chosen from a finite number of candidates on a given mesh M_k defined as

$$M_k = \{x^k + \alpha_k \mathbf{V} \mathbf{z} : \mathbf{z} \in \mathbb{Z}_+^p\},$$

where α_k is the mesh-size (or step-size) parameter, $\mathbb{Z}_+$ is the set of nonnegative integers and $\mathbf{V}$ denote the $n \times p$ matrix whose columns correspond to the p ($\geq n+1$) vectors forming a positive spanning set.

Pattern search methods consider two steps at every iteration. The first step consists of a finite search on the mesh, free of rules, with the goal of finding a new iterate that decreases the value of the objective function at the current iterate. This step, called the search step, has the flexibility to use any strategy, method or heuristic, or take advantage of a priori knowledge of the problem at hand, as long as it searches only a finite number of points in the mesh. The search step provides the flexibility for a global search since it allows searches away from the neighborhood of the current iterate, and influences the quality of the local minimizer or stationary point found by the method. If the search step fails to produce a decrease in the objective function, a second step, called the poll step, is performed around the current iterate. The poll step follows stricter rules and, using the concepts of positive bases, attempts to perform a local search in a mesh neighborhood around x^k, $\mathcal{N}(x^k) = \{x^k + \alpha_k \mathbf{v} : \text{for all } \mathbf{v} \in P_k\} \subset M_k$, where P_k is a positive basis chosen from the finite positive spanning set $\mathbf{V}$. For a sufficiently small mesh-size parameter α_k, the poll step is guaranteed to provide a function reduction, unless the current iterate is at a stationary point [1]. So, if the poll step also fails to produce a function reduction, the mesh-size parameter α_k must be decreased. On the other hand, if the search or the poll steps obtain an improved value for the objective function, the mesh-size parameter is increased or held constant.

The efficiency of pattern search methods improved significantly by reordering the poll directions according to descent indicators built from simplex gradients [6]. Here, the poll directions are reordered according to the BEVD scores meaning that directions with higher dosimetric value are tested first. Adding to the efficiency provided by an insightful reordering of the poll directions, the search step was recently provided with the use of minimum Frobenius norm quadratic models to be minimized within a trust region, which can lead to a significant improvement of direct search for smooth, piecewise smooth, and noisy problems [5]. The prior knowledge of the problem is also included in this step to take advantage of BEVD scores. A trial point is tested by considering the current best beam angle configuration and replacing the beam direction with smallest BEVD score by a beam direction with larger score that is not in the close neighborhood of the remaining beam directions. Last, but not least, the prior knowledge of the problem is used on the choice of initial point(s) by considering initial beam angle sets whose beam directions correspond to the largest BEVD scores. The strategy sketched is tailored for addressing the BAO problem taking advantage of prior knowledge of the problem:

Algorithm 1 (PSM framework incorporating BEVD).

0. Initialization Set $k = 0$. Compute BEVD scores for each beam angle. Choose a positive spanning set $\mathbf{V}$, $\alpha_0 > 0$, and $x^0 \in \mathbb{R}^n$ considering the beam directions with largest BEVD scores.

1. Search step Evaluate f at a finite number of points in M_k with the goal of decreasing the objective function value at x^k. If $x^{k+1} \in M_k$ is found satisfying $f(x^{k+1}) < f(x^k)$, go to step 4. Both search step and iteration are declared successful. Otherwise, go to step 2 and search step is declared unsuccessful.

2. Poll step This step is only performed if the search step is unsuccessful. Reorder the poll directions according to the BEVD scores. If $f(x^k) \leq f(x)$ for every x in the mesh neighborhood $\mathcal{N}(x^k)$, then go to step 3 and shrink M_k. Both poll step and iteration are declared unsuccessful. Otherwise, choose a point $x^{k+1} \in \mathcal{N}(x^k)$ such that $f(x^{k+1}) < f(x^k)$ and go to step 4. Both poll step and iteration are declared successful.

3. Mesh reduction Let $\alpha_{k+1} = \frac{1}{2} \times \alpha_k$. Set $k = k + 1$ and return to step 1.

4. Mesh expansion Let $\alpha_{k+1} = \alpha_k$. Set $k = k + 1$ and return to step 1.

4   Computational Results for Head-and-neck Clinical Examples

Our tests were performed on a 2.66Ghz Intel Core Duo PC with 3 GB RAM. In order to facilitate convenient access, visualization and analysis of patient treatment planning data, as well as dosimetric data input for treatment plan optimization research, the computational tools developed within MATLAB and CERR – computational environment for radiotherapy research [8] are used widely for IMRT treatment planning research.

The incorporation of BEVD into the pattern search methods framework was tested using two clinical examples of retrospective treated cases of head-and-neck tumors at the Portuguese Institute of Oncology of Coimbra (IPOC). In general, the head-and-neck region is a complex area to treat with radiotherapy due to the large number of sensitive organs in this region (e.g., eyes, mandible, larynx, oral cavity, etc.). For simplicity, in this study, the OARs used for treatment optimization were limited to the spinal cord, the brainstem and the parotid glands. The spinal cord and the brainstem are some of the most critical organs at risk (OARs) in the head-and-neck tumor cases. These are serial organs, i.e., organs such that if only one subunit is damaged, the whole organ functionality is compromised. Therefore, if the tolerance dose is exceeded, it may result in functional damage to the whole organ. Thus, it is extremely important not to exceed the tolerance dose prescribed for these type of organs. Other than the spinal cord and the brainstem, the parotid glands are also important OARs. The parotid gland is the largest of the three salivary glands. A common complication due to parotid glands irradiation is xerostomia (the medical term for dry mouth due to lack of saliva). This decreases the quality of life of patients undergoing radiation therapy of head-and-neck, causing difficulties to swallow. The parotids are parallel organs, i.e., if a small volume of the organ is damaged, the rest of the organ functionality may not be affected. Their tolerance dose depends strongly on the fraction of the volume irradiated. Hence, if only a small fraction of the organ is irradiated the tolerance dose is much higher than if a larger fraction is irradiated. Thus, for these parallel structures, the organ mean dose is generally used instead of the maximum dose as an objective for inverse planning optimization. The tumor to be treated plus some safety margins is called planning target volume (PTV). For the head-and-neck cases in study it was separated in two parts with different prescribed doses: PTV1 and PTV2. The prescription dose for the target volumes and tolerance doses for the OARs considered in the optimization are presented in Table 1.

The patients' CT sets and delineated structures are exported via Dicom RT to a freeware computational environment for radiotherapy research – CERR. We used CERR 3.2.2 version and MATLAB 7.4.0 (R2007a). An automatized procedure for dose computation for each given beam angle set was developed, instead of the traditional dose computation available from IMRTP module accessible from CERR's menubar. This automatization of the dose computation was essential for integration in our BAO algorithm. To address the convex

Table 1. Prescribed doses for all the structures considered for IMRT optimization

Structure	Mean dose	Max dose	Prescribed dose
Spinal cord	–	45 Gy	–
Brainstem	–	54 Gy	–
Left parotid	26 Gy	–	–
Right parotid	26 Gy	–	–
PTV1	–	–	70.0 Gy
PTV2	–	–	59.4 Gy
Body	–	80 Gy	–

nonlinear formulation of the FMO problem we used a trust-region-reflective algorithm (*fmincon*) of MATLAB 7.4.0 (R2007a) Optimization Toolbox.

We choose to implement the incorporation of BEVD scores into the pattern search methods framework taking advantage of the availability of an existing pattern search methods framework implementation used successfully by us to tackle the BAO problem [15,16,17] – the last version of SID-PSM [5,6]. The spanning set used was the positive spanning set ($[e \; -e \; I \; -I]$. Each of these directions corresponds to, respectively, the rotation of all incidence directions clockwise, the rotation of all incidence directions counter-clockwise, the rotation of each individual incidence direction clockwise, and the rotation of each individual incidence direction counter-clockwise.

Treatment plans with five to nine equispaced coplanar beams are used at IPOC and are commonly used in practice to treat head-and-neck cases [2]. Therefore, treatment plans of five coplanar orientations were obtained using *SID-PSM* and using an algorithm that incorporates a priori knowledge in pattern search, denoted *BEVD-PSM*. These plans were compared with the typical 5-beam equispaced coplanar treatment plans denoted *equi*. Since we want to improve the quality of the typical equispaced treatment plans, a starting point considered is the equispaced coplanar beam angle set. The choice of this initial point and the non-increasing property of the sequence of iterates generated by pattern search methods imply that each successful iteration correspond to an effective improvement with respect to the usual equispaced beam configuration. A different initial point is considered using the BEVD scores. Beforehand, we need to compute the BEVD scores that will be the prior knowledge of the problem to be incorporated in the BAO optimization algorithm. For each patient, the scores for every beam angle are computed as described in Section 3.1. The initial point using the BEVD scores is obtained considering the peaks of the BEVD score curve that are not too close and correspond the solution obtained directly using BEVD criteria [12]. The obtained scores for the two patients and the corresponding initial points considered are displayed in Fig. 1.

The results of BAO optimization concerning the improvement of the objective function value for the two clinical cases of head-and-neck tumors using *SID-PSM* and *BEVD-PSM* considering the equispaced configuration (equi) and

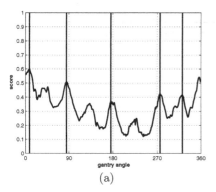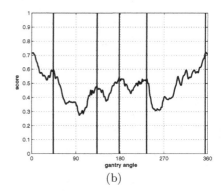

(a)   (b)

Fig. 1. BEVD scores as a function of the gantry angle and the BEVD initial point for cases 1 and 2, 1(a) and 1(b) respectively

the beam's-eye-view configuration (bevd) as starting points are presented in Table 2. Overall, the comparison of the best beam angle configurations obtained by both approaches with the equispaced beam angle configuration in terms of objective function value is clearly favorable to the pattern search approaches, regardless of the initial point used. It is important to emphasize that the use of a priori knowledge through BEVD scores has a positive influence both in the choice of the initial point and also incorporated in the pattern search methods algorithm. That conclusion can be also withdrawn from the simple inspection of Fig. 2 where the performances of *SID-PSM* and *BEVD-PSM* are compared with respect to the objective function value decrease versus the number of function evaluations. The benefits of using an initial point that takes into account the dosimetric characteristics of the case at hand are highlighted by the comparison between *SID-PSM* starting with the equispaced configuration (equi) and the the beam's-eye-view configuration (bevd), favorable to the later. Even starting from higher function values, when the bevd configuration does not correspond to better function values compared with the equi configuration, as in case 2, it seems to be advantageous to start from search regions where the neighbor beam directions of the initial configuration also have high BEVD scores. The advantage of incorporating BEVD scores in pattern search methods is clear since the best results are obtained by *BEVD-PSM*.

Despite the improvement in FMO value, the quality of the results can be perceived considering a variety of metrics. Typically, results are judged by their cumulative dose-volume histogram (DVH). The DVH displays the fraction of a structure's volume that receives at least a given dose. Another metric usually used for plan evaluation is the volume of PTV that receives 95% of the prescribed dose. Typically, 95% of the PTV volume is required. Using only 5 beam directions makes harder to obtain a satisfactory target coverage. DVH results for the two cases are displayed in Fig. 3. For clarity, the DVHs are split in *PTV*1 and *PTV*2

Table 2. FMO value improvement obtained by *SID-PSM* and *BEVD-PSM* compared with the typical equispaced coplanar treatment plans, *equi*, considering the equispaced configuration (equi) and the beam's-eye-view configuration (bevd) as starting points.

	equi	equi + *SID-PSM*		bevd + *SID-PSM*		bevd + *BEVD-PSM*	
Case	Fvalue	Fvalue	% decrease	Fvalue	% decrease	Fvalue	% decrease
1	165.8	144.1	13.1%	139.9	15.6%	136.5	17.7%
2	228.9	180.6	21.1%	179.2	21.7%	177.4	22.6%

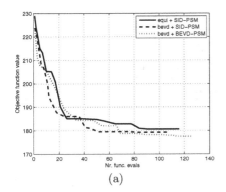
(a)

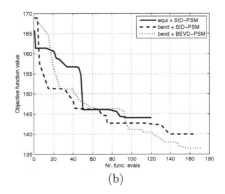
(b)

Fig. 2. History of the 5-beam angle optimization process using *SID-PSM* and *BEVD-PSM*, considering the equispaced configuration (equi) and the beam's-eye-view configuration (bevd) as starting points, for cases 1 and 2, 2(a) and 2(b) respectively

and the remaining structures distributed as an attempt to better visualize the results. The asterisks indicate 95% of PTV volumes versus 95% of the prescribed doses. The results displayed in Fig. 3 confirm the benefits of using the optimized beam directions obtained and used in *BEVD-PSM* treatment plans, with an improved target coverage and generally better organ sparing compared to the equispaced beam angle configuration, typically used in clinical practice.

5 Conclusions

The BAO problem is a continuous global highly non-convex optimization problem known to be extremely challenging and yet to be solved satisfactorily. This paper proposes an alternative approach to the BAO problem which is yet another step on the quest that may take us closer to find the global or near global optimum in a clinical acceptable time. The pattern search methods framework had already proved to be a suitable approach for the resolution of the non-convex

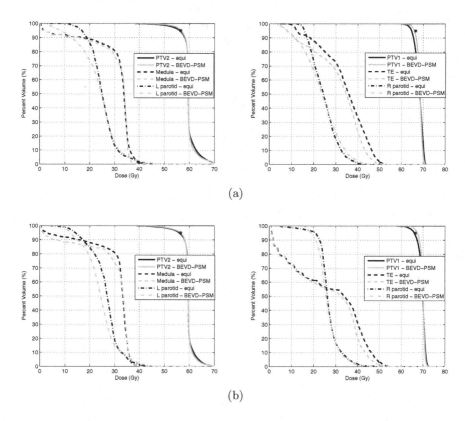

Fig. 3. Cumulative dose volume histogram comparing the results obtained by *equi* and *BEVD-PSM* for cases 1 and 2, 3(a) and 3(b) respectively

BAO problem due to their structure, organized around two phases at every iteration. The poll step, where convergence to a local minima is assured, and the search step, where flexibility is conferred to the method since any strategy can be applied. The search step is provided with the use of minimum Frobenius norm quadratic models to be minimized within a trust region, which can lead to a significant improvement of direct search for the type of problems at hand. A novel approach for the resolution of the BAO problem, incorporating prior knowledge in a pattern search methods framework, was proposed and tested using a set of clinical head-and-neck cases. For the clinical cases retrospectively tested, the use of prior knowledge of the patient in our tailored approach showed a positive influence on the quality of the local minimizer found. The improvement of the solutions in terms of objective function value corresponded, for the head-and-neck cases tested, to high quality treatment plans with better target coverage and with improved organ sparing. Moreover, the choice of the initial point also benefits from the prior knowledge of the problem leading to better solutions.

Acknowledgements. This work was supported by QREN under Mais Centro (CENTRO-07-0224-FEDER-002003) and FEDER funds through the COMPETE program and Portuguese funds through FCT under project grant PTDC/EIA-CCO/121450/2010. This work has also been partially supported by FCT under project grant PEst-C/EEI/UI0308/2011. The work of H. Rocha was supported by the European social fund and Portuguese funds from MCTES.

References

1. Alberto, P., Nogueira, F., Rocha, H., Vicente, L.N.: Pattern search methods for user-provided points: Application to molecular geometry problems. SIAM J. Optim. 14, 1216–1236 (2004)
2. Aleman, D.M., Kumar, A., Ahuja, R.K., Romeijn, H.E., Dempsey, J.F.: Neighborhood search approaches to beam orientation optimization in intensity modulated radiation therapy treatment planning. J. Global Optim. 42, 587–607 (2008)
3. Bortfeld, T., Schlegel, W.: Optimization of beam orientations in radiation therapy: some theoretical considerations. Phys. Med. Biol. 38, 291–304 (1993)
4. Craft, D.: Local beam angle optimization with linear programming and gradient search. Phys. Med. Biol. 52, 127–135 (2007)
5. Custódio, A.L., Rocha, H., Vicente, L.N.: Incorporating minimum Frobenius norm models in direct search. Comput. Optim. Appl. 46, 265–278 (2010)
6. Custódio, A.L., Vicente, L.N.: Using sampling and simplex derivatives in pattern search methods. SIAM J. Optim. 18, 537–555 (2007)
7. Davis, C.: Theory of positive linear dependence. Am. J. Math. 76, 733–746 (1954)
8. Deasy, J.O., Blanco, A.I., Clark, V.H.: CERR: A Computational Environment for Radiotherapy Research. Med. Phys. 30, 979–985 (2003)
9. Dias, J., Rocha, H., Ferreira, B.C., Lopes, M.C.: A genetic algorithm with neural network fitness function evaluation for IMRT beam angle optimization. Cent. Eur. J. Oper. Res. (at press), doi:10.1007/s10100-013-0289-4
10. Lee, E.K., Fox, T., Crocker, I.: Integer programming applied to intensity-modulated radiation therapy treatment planning. Ann. Oper. Res. 119, 165–181 (2003)
11. Li, Y., Yao, D., Yao, J., Chen, W.: A particle swarm optimization algorithm for beam angle selection in intensity modulated radiotherapy planning. Phys. Med. Biol. 50, 3491–3514 (2005)
12. Pugachev, A., Xing, L.: Pseudo beam's-eye-view as applied to beam orientation selection in intensity-modulated radiation therapy. Int. J. Radiat. Oncol. Biol. Phys. 51, 1361–1370 (2001)
13. Pugachev, A., Xing, L.: Incorporating prior knowledge into beam orientation optimization in IMRT. Int. J. Radiat. Oncol. Biol. Phys. 54, 1565–1574 (2002)
14. Rocha, H., Dias, J.M., Ferreira, B.C., Lopes, M.C.: Beam angle optimization using pattern search methods: initial mesh-size considerations. In: Proceedings of the 1st International Conference on Operations Research and Enterprise Systems (2012)
15. Rocha, H., Dias, J.M., Ferreira, B.C., do Carmo Lopes, M.: Incorporating Radial Basis Functions in Pattern Search Methods: Application to Beam Angle Optimization in Radiotherapy Treatment Planning. In: Murgante, B., Gervasi, O., Misra, S., Nedjah, N., Rocha, A.M.A.C., Taniar, D., Apduhan, B.O. (eds.) ICCSA 2012, Part III. LNCS, vol. 7335, pp. 1–16. Springer, Heidelberg (2012)

16. Rocha, H., Dias, J.M., Ferreira, B.C., Lopes, M.C.: Beam angle optimization for intensity-modulated radiation therapy using a guided pattern search method. Phys. Med. Biol. 58, 2939–2953 (2013)
17. Rocha, H., Dias, J.M., Ferreira, B.C., Lopes, M.C.: Selection of intensity modulated radiation therapy treatment beam directions using radial basis functions within a pattern search methods framework. J. Glob. Optim. (at press), doi:10.1007/s10898-012-0002-5
18. Romeijn, H.E., Ahuja, R.K., Dempsey, J.F., Kumar, A., Li, J.: A novel linear programming approach to fluence map optimization for intensity modulated radiation therapy treatment planing. Phys. Med. Biol. 48, 3521–3542 (2003)
19. Vaz, A.I.F., Vicente, L.N.: A particle swarm pattern search method for bound constrained global optimization. J. Global Optim. 39, 197–219 (2007)

A Note on Totally Regular Variables and Appell Sequences in Hypercomplex Function Theory

Carla Cruz[1], M. Irene Falcão[1,2], and Helmuth R. Malonek[1,3]

[1] Centro de Investigação e Desenvolvimento em Matemática e Aplicações,
Universidade de Aveiro
carla.cruz@ua.pt
[2] Departamento de Matemática e Aplicações, Universidade do Minho
mif@math.uminho.pt
[3] Departamento de Matemática, Universidade de Aveiro
hrmalon@ua.pt

Abstract. The aim of our contribution is to call attention to the relationship between totally regular variables, introduced by R. Delanghe in 1970, and Appell sequences with respect to the hypercomplex derivative. Under some natural normalization condition the set of all paravector valued totally regular variables defined in the three dimensional Euclidean space will be completely characterized. Together with their integer powers they constitute automatically Appell sequences, since they are isomorphic to the complex variables.

Keywords: totally regular variables, Appell sequences, hypercomplex differentiable functions.

1 Introduction

Some years ago, authors of this note (see [6]) introduced for the first time monogenic power-like functions (i.e. Appell sequences with respect to the hypercomplex derivative) as examples for the generation of *monogenic* (cf. [3]), or *Clifford holomorphic* (cf. [10]) functions by special polynomials given in terms of a paravector variable and its conjugate. Meanwhile Appell sequences have been subject of investigations by different authors with different methods and in various contexts (cf. [2]). The concept of a *totally regular variable*, introduced by R. Delanghe in [5] and later also studied by Gürlebeck ([7], [9]) for the special case of quaternions, has some obvious relationship with the latter. It describes the set of linear monogenic functions whose integer powers are also monogenic (without demanding to form an Appell sequence as it is the case for the integer powers of the complex variable $z = x + iy$). Indeed, the simple example of the totally regular Fueter-polynomials (cf. [10], [12]) shows, that not every totally regular variable and its integer powers form an Appell sequence with respect to the hypercomplex derivative. From the other side, the Appell sequence constructed in [6] is not constituted by a totally regular variable and its integer powers.

B. Murgante et al. (Eds.): ICCSA 2013, Part I, LNCS 7971, pp. 293–303, 2013.
© Springer-Verlag Berlin Heidelberg 2013

These facts motivated us to ask for the relationship between totally regular variables and Appell sequences with respect to the hypercomplex derivative in the case of a paravector valued variable in $\mathbb{R}^3$. Therefore we characterize completely the set of all paravector valued totally regular variables. The higher dimensional case can be treated in the same way. In view of our aim to connect totally regular variables with Appell sequences, we are using a natural normalization condition for the set of all paravector valued totally regular variables. We prove that under that normalization condition all totally regular variable constitute automatically Appell sequences, since they are isomorphic to the complex variables. We finish with some remarks on the role of polynomials in terms of the totally regular Fueter-polynomials (which are not normalized in the aforementioned way) as well as their use in the construction of Appell sequences with respect to the hypercomplex derivative.

2   Basic Notations

As usual, let $\{e_1, e_2, \ldots, e_n\}$ be an orthonormal basis of the Euclidean vector space $\mathbb{R}^n$ with a non-commutative product according to the multiplication rules

$$e_k e_l + e_l e_k = -2\delta_{kl}, \quad k, l = 1, \ldots, n,$$

where δ_{kl} is the Kronecker symbol. The set $\{e_A : A \subseteq \{1, \ldots, n\}\}$ with

$$e_A = e_{h_1} e_{h_2} \cdots e_{h_r}, \quad 1 \leq h_1 < \cdots < h_r \leq n, \quad e_\emptyset = e_0 = 1,$$

forms a basis of the 2^n-dimensional Clifford algebra $C\ell_{0,n}$ over $\mathbb{R}$. Let $\mathbb{R}^{n+1}$ be embedded in $C\ell_{0,n}$ by identifying $(x_0, x_1, \ldots, x_n) \in \mathbb{R}^{n+1}$ with

$$x = x_0 + \underline{x} \in \mathcal{A}_n := \text{span}_\mathbb{R}\{1, e_1, \ldots, e_n\} \subset C\ell_{0,n}.$$

Here, $x_0 = \text{Sc}(x)$ and $\underline{x} = \text{Vec}(x) = e_1 x_1 + \cdots + e_n x_n$ are, the so-called, scalar and vector parts of the paravector $x \in \mathcal{A}_n$. The conjugate of x is given by $\bar{x} = x_0 - \underline{x}$ and its norm by $|x| = (x\bar{x})^{\frac{1}{2}} = (x_0^2 + x_1^2 + \cdots + x_n^2)^{\frac{1}{2}}$.

To call attention to its relation to the complex Wirtinger derivatives, we use the following notation for a generalized Cauchy-Riemann operator in $\mathbb{R}^{n+1}$, $n \geq 1$:

$$\bar{\partial} := \frac{1}{2}(\partial_0 + \partial_{\underline{x}}), \quad \partial_0 := \frac{\partial}{\partial x_0}, \quad \partial_{\underline{x}} := e_1 \frac{\partial}{\partial x_1} + \cdots + e_n \frac{\partial}{\partial x_n}.$$

Definition 1 (Monogenic function).
$\mathcal{C}^1$-functions f satisfying the equation $\bar{\partial} f = 0$ (resp. $f\bar{\partial} = 0$) are called left monogenic (resp. right monogenic).

We suppose that f is hypercomplex-differentiable in Ω in the sense of [8,12], that is, it has a uniquely defined areolar derivative f' in each point of Ω (see also [13]). Then, f is real-differentiable and f' can be expressed by real partial derivatives as $f' = \partial f$ where, analogously to the generalized Cauchy-Riemann operator, we use $\partial := \frac{1}{2}(\partial_0 - \partial_{\underline{x}})$ for the conjugate Cauchy-Riemann operator. Since a hypercomplex differentiable function belongs to the kernel of $\overline{\partial}$, it follows that, in fact, $f' = \partial_0 f = -\partial_{\underline{x}} f$ which is similar to the complex case.

In general, $C\ell_{0,n}$-valued functions defined in some open subset $\Omega \subset \mathbb{R}^{n+1}$ are of the form $f(z) = \sum_A f_A(z) e_A$ with real valued $f_A(z)$. However, in several applied problems it is very useful to construct $\mathcal{A}_n$-valued monogenic functions as functions of a paravector with special properties. In this case we have

$$f(x_0, \underline{x}) = \sum_{j=0}^{n} f_j(x_0, \underline{x}) e_j \tag{1}$$

and left monogenic functions are also right monogenic functions, a fact which follows easily by direct inspection of the corresponding real system of first order partial differential equations (*generalized Riesz system*).

Example 1.

1. Consider the $\mathcal{A}_3$-valued function

$$f(x) = f(x_0, x_1, x_2, x_3) = x_1 x_2 x_3 - x_0 x_2 x_3 e_1 - x_0 x_1 x_3 e_2 - x_0 x_1 x_2 e_3.$$

Simple calculations allow to conclude that $\overline{\partial} f = 0$ which means that f is left monogenic. Since f is of the form (1), it follows that f is also right monogenic. Moreover, $f'(x) = \partial_0 f(x) = -x_2 x_3 e_1 - x_1 x_3 e_2 - x_1 x_2 e_3$.

2. Consider now the $\mathcal{A}_2$-valued functions

$$f_k(x_0, x_1, x_2) = (x_0 + x_1 e_1 + x_2 e_2)^k, \ k = 1, 2, \ldots.$$

It follows easily that

$$\begin{aligned}
\overline{\partial} f_1 &= -1; \\
\overline{\partial} f_2 &= -2x_0; \\
\overline{\partial} f_3 &= -3x_0^2 + (x_1^2 + x_2^2); \\
\overline{\partial} f_4 &= \left(-4x_0^2 + 4(x_1^2 + x_2^2)\right) x_0.
\end{aligned}$$

In fact, by induction, on can prove that

$$\overline{\partial} f_n = \begin{cases} \displaystyle\sum_{k=0}^{r-1} (-1)^{1+k} \binom{2r}{2k+1} x_0^{2r-1-2k} (x_1^2 + x_2^2)^k, & \text{if } n = 2r; \\[4mm] \displaystyle\sum_{k=0}^{r} (-1)^{1+k} \binom{2r+1}{2k+1} x_0^{2r-2k} (x_1^2 + x_2^2)^k, & \text{if } n = 2r+1. \end{cases}$$

Therefore, neither $z := f_1(x)$ nor any of its nonnegative integer powers are left or right monogenic functions.

We use also the classical definition of sequences of Appell polynomials [1] adapted to the hypercomplex case.

Definition 2 (Generalized Appell sequence).
A sequence of monogenic polynomials $(\mathcal{P}_k)_{k \geq 0}$ *of exact degree* k *is called a generalized Appell sequence with respect to the hypercomplex derivative if*

1. $\mathcal{P}_0(x) \equiv 1$,
2. $\mathcal{P}'_k = k\,\mathcal{P}_{k-1}, \ k = 1, 2, \ldots$.

The second condition is the essential one while the first condition is the usually applied normalization condition which can be changed to any constant different from zero.

3   Totally Regular Variables

Underlining the fact that, in general, an integer power of a hypercomplex variable is not monogenic, Delanghe introduced the following concept (see [5])

Definition 3 (Totally regular variable).
A totally regular variable is a linear monogenic function of the form

$$z = x_0 e_{A_0} + x_1 e_{A_1} + \ldots + x_n e_{A_n} \in C\ell_{0,n} \tag{2}$$

whose integer powers are monogenic.

Depending on the choice of e_{A_k}, Delanghe obtained for the general Clifford Algebra valued case, where $e_i^2 = \varepsilon_i e_0$, for real ε_i, $(i = 1, \ldots, n)$, necessary and sufficient conditions for a hypercomplex variable z to be totally regular [5, Theorem 4]. For our purpose here, we would like to call attention to the following weaker result, involving a much simpler condition.

Theorem 1. *[5, Corollary 1 of Theorem 4] Any monogenic variable* z *of the form* (2) *for which*

$$e_{A_k} e_{A_l} = e_{A_l} e_{A_k}, \quad k, l = 0, \ldots, n, \tag{3}$$

is totally regular.

Additionally, Delanghe showed that (3) is only sufficient, by referring to the special case of the totally regular variable $z = x_2 e_1 e_2 + x_3 e_1 e_3$, with $e_1^2 = \varepsilon_1 \neq 0$, $e_2^2 = e_3^2 = 0$, for which clearly $e_1 e_2 \cdot e_1 e_3 \neq e_1 e_3 \cdot e_1 e_2$.

Later on Gürlebeck [7] studied the case of quaternion valued ($\mathbb{H}$ - valued) variables in the form of

$$z = x_0 d_0 + x_1 d_1 + x_2 d_2 + x_3 d_3 \in \mathbb{H}, \tag{4}$$

with $d_k = a_{k0}e_0 + a_{k1}e_1 + a_{k2}e_2 + a_{k3}e_1e_2$ not necessarily linearly independent (see also [9]). In order to obtain $\mathbb{H}$-totally regular variables he found a necessary and sufficient condition, expressed by the rank of the matrix

$$A = \begin{pmatrix} a_{01} & a_{02} & a_{03} \\ a_{11} & a_{12} & a_{13} \\ a_{21} & a_{22} & a_{23} \\ a_{31} & a_{32} & a_{33} \end{pmatrix}, \tag{5}$$

which can be rewritten as follows:

Theorem 2. *Let z be a quaternionic holomorphic variable of the form (4). The following statements are equivalent:*

  I.   *z is a a totally regular variable;*

  II.   *$d_k d_l = d_l d_k$, $l, k = 0, 1, 2, 3$;*

  III. *The rank of the matrix (5) is less than 2.*

We note that the general form of a totally regular variable has not been explicitly determined, neither in the general case (2) nor in the quaternionic case (4). The aim of the present work is to characterize totally regular variables defined in $\mathbb{R}^3$.

Following this idea we study here the case of linear paravector valued functions of three real variables, subject to a normalization condition with respect to the real variable x_0. This normalization condition is given in terms of the hypercomplex derivative by demanding that

$$\partial z = z\partial = 1. \tag{6}$$

This is motivated by the fact that at the same time we are looking for the characterization of all totally regular variables whose integer powers form an Appell sequence in the sense of Definition 2 as we know it from the complex case for $z = x + iy$.

We note that not every totally regular variable (TRV) and its powers form an Appell sequence. In addition the first degree polynomial of an Appell sequence is not necessarily a TRV. The following examples illustrate these situations.

Example 2.

1. The variables
$$z_s := x_s - x_0 e_s, \ s = 1, 2, \tag{7}$$

   are TRV, which are not Appell sequences in the sense of Definition 2, because

$$\overline{\partial} z_s^k = 0, \ k = 1, 2, \ldots \quad \text{but} \quad \partial z_s = \partial_0 z_s = -e_s \neq 1.$$

2. A sequence of the form considered in [6]

$$\mathcal{P}_k(x) = \sum_{s=0}^{k} \frac{1}{2^s} \binom{k}{s} \binom{s}{\lfloor \frac{s}{2} \rfloor} x_0^{k-s} \underline{x}^s, \tag{8}$$

is an Appell sequence which does not consist of a TRV and its powers, since besides the fact that

$$\tilde{z} := \mathcal{P}_1(x) = x_0 + \tfrac{1}{2}(x_1 e_1 + x_2 e_2)$$

is not a TRV, we also have $\tilde{z}^k \neq \mathcal{P}_k$, $k > 1$.

3. The variables

$$\hat{z}_s := x_0 + x_s e_s, \ \ s = 1, 2, \tag{9}$$

are TRV and their powers form an Appell sequence, because

$$\overline{\partial}\hat{z}_s^k = 0 \quad \text{and} \quad \partial \hat{z}_s^k = \partial_0 \hat{z}_s^k = k \hat{z}_s^{k-1}, \ s = 1, 2, \ldots.$$

4 The Explicit Form of Paravector Valued Totally Regular Variables

As mentioned before, for reasons of applications and simplicity we concentrate on the computation of the explicit form of TRV given by

$$z = x_0 d_0 + x_1 d_1 + x_2 d_2 \in \mathcal{A}_2 \subset Cl_{0,2}. \tag{10}$$

We first note that from $\overline{\partial}z = 0$ it follows that

$$d_0 + e_1 d_1 + e_2 d_2 = 0. \tag{11}$$

In addition, the application of the normalization condition (6) implies immediately that

$$d_0 = 1 \tag{12}$$

and therefore, combining (11) and (12) we obtain as a first condition on the d_k's the following relation

$$1 + e_1 d_1 + e_2 d_2 = 0. \tag{13}$$

For z to be TRV we also need that the square of z and all other integer powers of z are monogenic. We will see, that the case of z^2 implies conditions which guarantee the same property for all integer powers. Since

$$\begin{aligned}
\overline{\partial}z^2 &= x_0(1 + e_1 d_1 + e_2 d_2) + (1 + e_1 d_1)x_1 d_1 + +(1 + e_2 d_2)x_2 d_2 \\
&\quad + \tfrac{1}{2}(x_2 e_1 + x_1 e_2)(d_1 d_2 + d_2 d_1) \\
&= x_0(1 + e_1 d_1 + e_2 d_2) + (1 + e_1 d_1 + e_2 d_2)x_1 d_1 + (1 + e_1 d_1 + e_2 d_2)x_2 d_2 \\
&\quad + \tfrac{1}{2}(x_1 e_2 - x_2 e_1)(d_1 d_2 - d_2 d_1)
\end{aligned}$$

and taking into account condition (13), we obtain a second condition on the d_k's, namely

$$d_1 d_2 - d_2 d_1 = 0. \tag{14}$$

Notice that (14) is identical to the necessary and sufficient conditions, mentioned in Theorem 2. For a detailed analysis of the consequences of (13) and (14) we use the notation of [7] and write

$$d_1 = a_{10} + a_{11}e_1 + a_{12}e_2,$$
$$d_2 = a_{20} + a_{21}e_1 + a_{22}e_2,$$

with $a_{lm} \in \mathbb{R}$, $l, m = 0, 1, 2$. Therefore, from (13) it follows easily that

$$a_{11} + a_{22} = 1, \tag{15}$$
$$a_{12} = a_{21} \tag{16}$$

and

$$a_{10} = a_{20} = 0, \tag{17}$$

while condition (14) implies

$$a_{11}a_{22} - a_{12}a_{21} = 0. \tag{18}$$

We note that, based on (10) and (12), the matrix (5) has the form

$$A = \begin{pmatrix} 0 & 0 & 0 \\ a_{11} & a_{12} & 0 \\ a_{21} & a_{22} & 0 \end{pmatrix},$$

which has obviously rank less than 2, due to (18).

Relation (16) together with (18) gives

$$a_{11}a_{22} = \lambda^2, \qquad \text{for some real } \lambda.$$

Let us now consider the two possible cases, for the values of the parameter λ.

Case A: $\lambda \neq 0$.

In this first case, a_{11} and a_{22} have the same sign and as a consequence of (15), both coefficients are positive. Therefore we can define

$$i_1^2 := a_{11}; \quad i_2^2 := a_{22} \quad \text{with} \quad i_1^2 + i_2^2 = 1,$$

in order to write

$$\lambda^2 = (i_1 i_2)^2.$$

Remark: Because of

$$i_1^2 + i_2^2 = 1,$$

we can choose, for instance,

$$i_1 = t, \quad i_2 = \sqrt{1 - t^2}, \quad (\text{with } |t| = |i_1| \leq 1),$$

or

$$i_1 = \cos\alpha, \quad i_2 = \sin\alpha, \text{ (for some angle } \alpha).$$

The relation with the roots of unity is obvious and permits interesting applications (see [4]).

The consequences of case **A** for the general form of the TRV z are the following:

$$z = x_0 + x_1(i_1^2 e_1 + i_1 i_2 e_2) + x_2(i_1 i_2 e_1 + i_2^2 e_2)$$
$$= x_0 + i_1 x_1(i_1 e_1 + i_2 e_2) + i_2 x_2(i_1 e_1 + i_2 e_2)$$
$$= x_0 + (i_1 x_1 + i_2 x_2)(i_1 e_1 + i_2 e_2),$$

where the constant "imaginary unit"

$$\hat{i} := i_1 e_1 + i_2 e_2$$

is such that $\hat{i}^2 = -i_1^2 - i_2^2 = -1$. Writing

$$x_{\hat{i}} := i_1 x_1 + i_2 x_2,$$

we recognize the isomorphism with $z = x + yi \in \mathbb{C}$:

$$x \rightarrow x_0; \quad y \rightarrow x_{\hat{i}}; \quad i \rightarrow \hat{i}.$$

Moreover, under the conditions of case A, z is a TRV whose integer powers

$$z^k = [x_0 + (i_1 x_1 + i_2 x_2)(i_1 e_1 + i_2 e_2)]^k = (x_0 + x_{\hat{i}}\hat{i})^k$$

form an Appel sequence, because obviously $(z^k)' = k z^{k-1}$ and $z^0 = 1$.

Consider now the second case:

Case B: $\lambda = 0$

If

$$a_{11} \neq 0 \quad \text{and} \quad a_{22} = 0,$$

then $a_{11} = 1$ and $z = x_0 + x_1 e_1$ (trivial case). On the other hand, if

$$a_{11} = 0 \quad \text{and} \quad a_{22} \neq 0,$$

then $a_{22} = 1$ and $z = x_0 + x_2 e_2$ (also a trivial case).

The above considerations can be summarized as follows:

Theorem 3. *The set of all linear monogenic variables of the form*

$$z = x_0 + x_1 d_1 + x_2 d_2 \in \mathcal{A}_2 \subset Cl_{0,2},$$

which are TRV explicitly consists of pseudo-complex variables of the form

$$z_{\hat{\imath}} = x_0 + (i_1 x_1 + i_2 x_2)(i_1 e_1 + i_2 e_2) = x_0 + x_{\hat{\imath}} \hat{\imath},$$

with $(i_1, i_2) \in \mathbb{R}^2$ *and* $i_1^2 + i_2^2 = 1$.

Moreover, due to their isomorphism with the complex variable $z = x + yi$ *these pseudo-complex variables together with their integer powers* $z_{\hat{\imath}}^k$ *form automatically an Appell sequence with respect to the hypercomplex derivative.*

5   Concluding Remarks

Even the consideration of homogeneous polynomials of degree k, with a "relaxed" binomial expansion (characteristic property of Appell sequences) of the form

$$\mathcal{P}_k(z) = \sum_{0}^{k} m_s \binom{k}{s} x_0^{k-s} [X_1(x_1, x_2) e_1 + X_2(x_1, x_2) e_2]^s, \tag{19}$$

where $X_i(x_1, x_2)$, $i = 1, 2$, are real valued functions in x_1 and x_2, leads only to the cases **A** e **B** of TRV with $m_s \equiv 1$ or to the case where

$$m_s = \frac{1}{2^s} \binom{s}{\lfloor \frac{s}{2} \rfloor}, \quad s = 0, 1, \ldots, k, \tag{20}$$

with $X_1(x_1, x_2) = x_1$ and $X_2(x_1, x_2) = x_2$ (not covered by **A** or **B** and not based on the integer powers of a TRV, since $\mathcal{P}_1^2 \neq \mathcal{P}_2$).

Polynomials of the form (19) as elements of generalized Appell sequences of paravector valued monogenic polynomials in $\mathcal{A}_2$ have been studied in [14]. It was proved that both mentioned cases, i.e. where $m_s \equiv 1$ or m_s given by (20), are the only one examples of Appell sequences with respect to the hypercomplex derivative and normalized as in Definition 2. This means that with the exception of polynomials (19) in the special form ·

$$\mathcal{P}_k(z) = \sum_{0}^{k} \frac{1}{2^s} \binom{s}{\lfloor \frac{s}{2} \rfloor} \binom{k}{s} x_0^{k-s} (x_1 e_1 + x_2 e_2)^s,$$

all other Appell sequences with respect to the hypercomplex derivative and normalized as in Definition 2, consist of totally regular variables (TRV) and its integer powers in the form

$$z_{\hat{\imath}} = x_0 + (i_1 x_1 + i_2 x_2)(i_1 e_1 + i_2 e_2) = x_0 + x_{\hat{\imath}} \hat{\imath}.$$

Further, let us mention the following. If we admit that the usually used normalization condition $\mathcal{P}_0 = 1$ (or initial value of the polynomial of degree 0) in Definition 2 is changed to $\mathcal{P}_0 = -e_1$, resp. $\mathcal{P}_0 = -e_2$, (a possibility that we mentioned), then also the TRV in the examples of Section 4

$$z_s = x_s - x_0 e_s = -e_s(x_0 + x_s e_s), \quad s = 1, 2, \tag{21}$$

form together with its integer powers Appell sequences, which can be verified by straightforward calculations. The initial value appears as the constant factor $-e_1$ resp. $-e_2$ of the considered $z_{\hat{\imath}}^k$ with $(i_1, i_2) = (1,0)$, resp. $(i_1, i_2) = (0,1)$. Of course, the same is true for other choices of initial values of the polynomial of degree 0 and constant factors of the "natural" two copies of the complex variable $z = x + yi$, i.e. for the first degree polynomials

$$z_r = x_0 + x_r e_r, \; r = 1, 2.$$

But since both TRV z_s of the form (21) are the first degree Fueter polynomials (see [10]), we mention finally a remark of Habetha in [11, p. 233], on the use of those "natural" copies of several complex variables, i.e. $x_0 + x_s e_s = e_s z_s$, with $s = 1, 2$, instead of Fueter polynomials for the power series representation of any monogenic function. Theorem 3 shows (here only for the case of $\mathbb{R}^3$), that also the more general pseudo-complex variables of the form $z_{\hat{\imath}} = x_0 + (i_1 x_1 + i_2 x_2)(i_1 e_1 + i_2 e_2) = x_0 + x_{\hat{\imath}} \hat{\imath}$ can play a decisive role in the power series representation of any monogenic function. Of course, this is also true in the general case for $\mathbb{R}^{n+1}$ where one has to work analogously with a parameter set $(i_1, i_2, \ldots, i_n)$.

Acknowledgements. This work was supported by *FEDER* founds through *COMPETE*–Operational Programme Factors of Competitiveness ("Programa Operacional Factores de Competitividade") and by Portuguese funds through the *Center for Research and Development in Mathematics and Applications* (University of Aveiro) and the Portuguese Foundation for Science and Technology ("FCT–Fundação para a Ciência e a Tecnologia"), within project PEst-C/MAT/UI4106/2011 with COMPETE number FCOMP-01-0124-FEDER-022690. The research of the first author was also supported by FCT under the fellowship SFRH/BD/44999/2008.

References

1. Appell, P.: Sur une classe de polynômes. Ann. Sci. École Norm. Sup. 9(2), 119–144 (1880)
2. Bock, S., Gürlebeck, K.: On a generalized Appell system and monogenic power series. Math. Methods Appl. Sci. 33(4), 394–411 (2010)
3. Brackx, F., Delanghe, R., Sommen, F.: Clifford analysis. Pitman, Boston-London-Melbourne (1982)
4. Cruz, C., Falcão, M.I., Malonek, H.R.: On pseudo-complex bases for monogenic polynomials. In: Sivasundaram, S. (ed.) 9th International Conference on Mathematical Problems in Engineering, Aerospace and Sciences (ICNPAA 2012). AIP Conference Proceedings, vol. 1493, pp. 350–355 (2012)
5. Delanghe, R.: On regular-analytic functions with values in a Clifford algebra. Math. Ann. 185, 91–111 (1970)
6. Falcão, M.I., Cruz, J., Malonek, H.R.: Remarks on the generation of monogenic functions. 17th Inter. Conf. on the Appl. of Computer Science and Mathematics on Architecture and Civil Engineering, Weimar (2006)

7. Gürlebeck, K.: Über interpolation und approximation verallgemeinert analytischer funktionen. Wiss. Inf. 34, 21 S.(1982)
8. Gürlebeck, K., Malonek, H.: A hypercomplex derivative of monogenic functions in $\mathbb{R}^{n+1}$ and its applications. Complex Variables Theory Appl. 39, 199–228 (1999)
9. Gürlebeck, K., Sprößig, W.: Quaternionic analysis and elliptic boundary value problems. International Series of Numerical Mathematics, vol. 89. Birkhäuser Verlag, Basel (1990)
10. Gürlebeck, K., Habetha, K., Sprößig, W.: Holomorphic functions in the plane and n-dimensional space. Birkhäuser Verlag, Basel (2008)
11. Habetha, K.: Function theory in algebras. In: Complex Analysis, Methods, Trends, and Applications. Math. Lehrb., vol. 61, pp. 225–237. Akademie-Verlag, Berlin (1983)
12. Malonek, H.: A new hypercomplex structure of the euclidean space $\mathbb{R}^{m+1}$ and the concept of hypercomplex differentiability. Complex Variables 14, 25–33 (1990)
13. Malonek, H.: Selected topics in hypercomplex function theory. In: Eriksson, S.L. (ed.) Clifford Algebras and Potential Theory, pp. 111–150. 7. University of Joensuu (2004)
14. Malonek, H.R., Falcão, M.I.: On paravector valued homogeneous monogenic polynomials with binomial expansion. Advances in Applied Clifford Algebras 22(3), 789–801 (2012)

A Genetic Algorithm for the TOPdTW at Operating Rooms

Gabriel Mota, Mário Abreu, Artur Quintas, João Ferreira, Luis S. Dias,
Guilherme A.B. Pereira, and José A. Oliveira

Centre ALGORITMI, Escola de Engenharia, Universidade do Minho,
Campus Azurém, 4800-058 Guimarães, Portugal

Abstract. This paper presents a genetic algorithm for the Team Orienteering Problem with double Time Windows (TOPdTW). The aim is to study TOPdTW to model a real problem that arises within the operating rooms in a hospital. The Genetic Algorithm uses a peculiar way to construct solutions that only generates valid solutions, which improves the global performance. This constructive algorithm reads the chromosome and decides which operation is scheduled next in the route. The algorithm was tested using some public instances of the TOPTW and instances generated for TOPdTW. The computational results are presented.

1 Introduction

Health care services are in increasing demand in modern societies, because the access of people to the heath care has been democratized, and also due to the aging of populations. However, due to budgetary constraints the available resources in the health care facilities cannot be increased infinitely. Given this situation, the list of patients waiting for health care is also increasing. So, the units providing the health care have to rationalize their activities in order to increase the use of their resources. Nowadays the message that the managers send to their employees is that one must seek to increase the level of service jointly with reducing the operating costs.

The rationalization of health care services is a subject that has received attention from academic researchers over time and this interest has increased in recent years. Joint work of Operations Research practitioners with health care professionals promoted the publication of several studies in this subject. Examples can be pointed out from some studies such as: designing ambulance interiors layout [1], improving hospital logistics [2], and also the research related with shift scheduling for personnel [3,4,5,6].

As van Essen et al. [7] stated, the operating room (OR) is the most expensive resource in a hospital. Also, Su et al. [8] refer to the level of expenses in the ORs, which account for approximately 40% of a hospital's total expenses, according to reporting data of 2005. Attending the importance of this level of expenses the work in the ORs should be well planned in order to fully occupy their availability. Several studies concerned with the ORs [7,8,9,10,11] presented different

B. Murgante et al. (Eds.): ICCSA 2013, Part I, LNCS 7971, pp. 304–317, 2013.

methodologies that can be integrated in decision support systems. Cardoen et al. [12] provide an important review of recent operational research on operating room planning and scheduling.

However, planning and scheduling the operating room is a hard problem due to the high level of constraints to be considered. There are several works that deal with this problem trying to provide the optimal solution using (mixed) integer linear programming (MILP) [7,8,9,10,13,14,15,16]. Particularly, these formulations attempt to solve different aspects of the planning problems of the operating room, but all of them share the conclusion that it is not an easy task solving large instances optimally. Often, to deal with large instances of these problems one should use heuristic procedures to obtain solutions in an acceptable computation time.

In this paper we intend to study the possibility of model planning and scheduling problems of operating rooms in terms of choice elective patients, aiming for throughput maximization. Indeed, several simplifications were made to the real problem, however the model allows to verify if the selection of elective surgeries and its scheduling in the operating rooms could be done using a new variant of the Team Orienteering Problem with Time Windows (TOPTW) as a basis model. For this new variant of TOPTW a MILP and also a Genetic Algorithm are presented to solve the problem.

The paper is structured into six sections. Section 2 discusses some known versions of Team Orienteering Problem (TOP) and TOPTW, their application to the real problems, and also the developed methodology. Section 3 describes various topics related with the real problem of Operating Rooms Scheduling. Also, Section 3 proposes a mixed integer linear programming (MILP) model for Operating Rooms Scheduling and discusses the model. Section 4 describes the Genetic Algorithm (GA) for TOPTW and its adaption for the Operating Rooms Scheduling. Section 5 presents preliminary computational results with practical-sized instances. Section 6 discusses the main conclusions of this study.

2   TOP and TOPTW Variants

Team Orienteering Problem (TOP) is a fairly recent concept, first suggested by Butt and Cavalier [17] under the name of Multiple Tour Maximum Collection Problem. Later, Chao et al. [18] formally introduced the problem and designed one of the most frequently used sets of benchmark instances. In 2006, Archetti et al. [19] achieved many of the currently best-known solutions for the TOP instances by presenting two versions of Tabu Search, along with two metaheuristics based on Variable Neighbourhood Search (VNS). Also, Vieira et al. [20] and Ferreira et al. [21] presented recently evolutionary procedures to the TOP. In the Team Orienteering Problem it is necessary to choose a subset of nodes in a graph to visit and collect a prize, using a given set of paths. Each node could be visited at most by one path. If there are time constraints to visit a node, we have the Team Orienteering Problem with Time Windows (TOPTW) and can be considered a generalization of the TOP.

There were several published studies about TOPTW. Lin and Yu [22] presented two versions of simulated annealing based heuristic approach for TOPTW. The computational study indicates that both versions of the SA presented in this paper are capable of producing high quality TOPTW solutions. Labadie et al. [23] proposed a Variable Neighborhood Search (VNS) procedure based on the idea of exploring, most of the time, granular instead of complete neighborhoods in order to improve the algorithm's efficiency without losing effectiveness. Computational results have shown that a proposed VNS approach is already an effective algorithm, whereas the introduction of the granularity can improve the algorithm's efficiency while maintaining effectiveness. Garcia et al. [24] model the tourist planning problem, integrating public transportation, as the time-dependent team orienteering problem with time windows (TD-TOPTW) in order to allow personalized electronic tourist guides to create personalized tourist routes in real-time. They develop and compare two different approaches to solve the TD-TOPTW. Both algorithms are applied on a set of test instances based on real data for the city of San Sebastian. The main contribution of this research is that the authors develop two approaches that solve the TD-TOPTW in real-time. Also, Vansteenwegen et al.[25] deal with the planning problem of a personalized electronic tourist guide that was modeled as a TOPTW. The main contribution of this paper is a simple, fast and effective iterated local search meta-heuristic to solve the TOPTW. On a large set of test instances, the average gap with the best-known solutions is only 1.8% and the computation time is decreased with a factor of several hundreds compared to other algorithms. Even when the computation time is limited to 1 s, high quality results are obtained. This is achieved by speeding up the evaluation of possible improvements and the specific implementation of the shake step to better explore the whole solution space. All this makes the algorithm appropriate for the personalized tourist guide application. Tricoire et al. [26] presents the multi-period orienteering problem with multiple time windows (MuPOPTW), a new routing problem combining objective and constraints of the orienteering problem (OP) and team orienteering problem (TOP), constraints from standard vehicle routing problems, and original constraints from a real industrial case. Another study from Garcial et al. [27] deals with a problem that can be directly related to the Multi Constrained Team Orienteering Problem with Time Windows (MCTOPTW). It introduces an Iterated Local Search (ILS) based algorithm to solve the MCTOPTW fast and effectively. The algorithm performs well on a published set of Selective Vehicle Routing Problems with Time Windows (SVRPTW) and a large set of new MCTOPTW instances. Dohn et al. [28] consider the manpower allocation problem with time windows, job-teaming constraints and a limited number of teams (m-MAPTWTC). Given a set of teams and a set of tasks, the problem is to assign to each team a sequential order of tasks to maximize the total number of assigned tasks. An integer-programming model is presented for the problem, which is decomposed using Dantzig-Wolfe decomposition. The problem is solved by column generation in a branch-and-price framework. Simultaneous execution of tasks is enforced by the branching scheme.

3   The TOPdTW

In this paper a model is presented to deal with a planning and scheduling problem that arises in Operating Rooms in a hospital. The bibliography about Operating Rooms is huge, and it is not possible to discuss all the different ways that the main problem and its variants were modelled. The model that this study presents is a variant of TOPTW. In this variant not only the vertex has a time window to be visited, but also the path will have a time window to be fulfilled. This variant is called Team Orienteering Problem with double Time Windows (TOPdTW). To the best of our knowledge, TOPdTW was never presented before, and no other studies using this model were applied to the planning of ORs. For this new TOP variant a MILP and also a GA are presented to solve the problem. This problem is a generalization of TOPTW, because in TOPTW all paths could be started at time zero, and the ending time may occur in a time defined by the last node visited in the path. In TOPdTW a path must be started at a time that respects the available time of a vehicle, or facility, or even an Operating Room. This strategy allows using a facility (vehicle / Operating Room / etc.) several times during a given planning horizon.

3.1   The Real Problem

In a hospital there are several Operating Rooms that could be used in different shifts along the different days of a week. In this paper, the number of paths is given by the number of rooms times the number of shifts times the number of days. Given an available list of possible surgeries, that has different levels of importance, time windows to be considered, and a duration to be processed, the objective is to select the best set of elective surgeries to be scheduled in the given horizon planning. This is a very elementary approach to the real problem, since it does not consider directly other important resources, just as medical staff, rostering personnel constraints, specific equipment, and other relevant issues. However it is also possible to extend this model to make it more representative of a real problem. Furthermore, it is believed that this model could be used at a planning level of decisions that could be then enhanced at an operational level of decisions, promoting several decisions to be considered in a decision support system (DSS).

3.2   The Mixed Integer Linear Programming

The aim of the present study is to solve the TOPdTW, which means to develop a method that determines v paths, which start in the same location and have the same destination, in order to maximize the total profit made in each path, while respecting the time windows constraints. A determined number of vehicles is given at the instance formulation and a path is generated for each vehicle. Following the MILP model presented by Labadie et al. [23] and the model presented by Vansteenwegen et al. [25] the objective function of the TOPdTW is presented next in Equation 1, where v represents the number of vehicles; x_{ijd} is a binary variable set to 1 if the vehicle d goes from node i to j and to 0 otherwise; p_i represents a positive integer score (profit) associated with the node i. The objective function consists of finding v feasible routes that maximize the total reward or profit.

$$max \sum_{d=1}^{v} \sum_{i=1}^{n-1} \sum_{j=2}^{n} p_i x_{ijd}, \quad i \neq j \tag{1}$$

Additionally to the TOPTW formulation presented in [25] the constraints (2) and (3), that model the time window of each path are added, as well as the parameters R_d and F_d that represent the ready time for path d and the finished time for path d. This means that the path d has a time window $[R_d; F_d]$ to be performed.

$$R_d + c_{1jd} x_{1jd} - s_{jd} \leq M(1 - x_{1jd}), \quad (i = 2, \dots, n; d = 1, \dots, v;) \tag{2}$$

$$s_{jd} + T_j + c_{jnd} x_{jnd} \leq F_d + M(1 - x_{jnd}), \quad (j = 2, \dots, n-1; d = 1, \dots, v;) \tag{3}$$

, where c_{ijd} is the time needed to travel from location i to j by path d; T_j is a service or visiting time at node i; s_{jd} is the time instant to start the service at node i; and M is a constant with a large value.

As with the problems TOP and TOPTW also TOPdTW is NP-hard, and then the use of heuristics is of great importance to quickly generate solutions to be available in a DSS. In this study, a genetic algorithm was chosen because it is easy to implement.

4   The Genetic Algorithm

The Genetic Algorithm (GA) is a search heuristic that imitates the natural process of evolution as it is believed to happen to all the species of living beings. This method uses nature-inspired techniques such as mutation, crossover, inheritance and selection, to generate solutions for optimization problems. The success of a GA depends on the type of problem to which the algorithm is applied and its complexity.

In a GA, the chromosomes or individuals are represented as strings, which encode candidate solutions for an optimization problem, that later evolve towards better solutions. Designing a GA requires a genetic representation of the solution domain, as well as a fitness function to evaluate the solutions produced. The GA evolutionary process starts off by initializing a population of solutions (usually randomly), which will evolve and improve during three main steps:

- Selection: A portion of each successive generation is selected, based on their fitness, in order to breed the new and probably better fit generation.
- Reproduction: The selected solutions produce the next generation through mutation and/or crossover, propagating the most crucial changes to the future generations by inheritance.
- Termination: Once a stopping condition is met, the evolutionary process ends.

The developed genetic algorithm consists on three components. The most elementary one is the chromosome, which represents a set of vehicles and their respective routes. The next component is the evolutionary process, responsible for executing crossovers and mutations within the population of chromosomes. The third and last component of the algorithm controls the evolutionary process, ensures the validation of chromosomes according to the limitations imposed by the TOPdTW and its instances, and also carries out the evaluation of the chromosomes based on a fitness function.

In the GA presented here, a possible solution for the TOPdTW is represented in the form of a chromosome. At each new generation, a renewed population of chromosomes is obtained; each vehicle available and each route includes a sequence of customers to be visited under a given time limit and within their respective time windows. An example of a valid solution is showed in Fig. 1. There are six customers denoted as numbered vertices and two vehicles available. There is also a starting point (S) and an ending point (E). In this case, vertex 6 is not included in any route because otherwise it would turn the solution invalid by violating the time constraints. While addressing the team orienteering

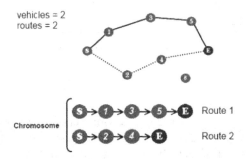

Fig. 1. Representation of a valid solution example, for an instance with six vertices and with two available vehicles

problem with double time windows, the fitness function of the algorithm was set to correspond to the sum of all the collected rewards in each customer visited during the identified routes. In order to produce solutions for the TOPdTW, the algorithm starts by generating an initial population of valid chromosomes. Then, some genetic operators are applied to the population in order to promote their evolution towards better fitness levels. This evolution process is repeated until a stopping criterion is met. As explained before, the chromosomes in the GA contain routes to be assigned to the available vehicles in a certain TOPdTW instance. Consequently, the creation of a new random chromosome is in fact the creation of a group of random valid routes. All these routes include the required starting and ending points. In order to assemble a route, customers are added in from an availability list that is common to all routes. The attempt to add a customer to a route is only successful if that addition keeps that route feasible while not exceeding the given time budget and the respective time windows constraints for each customers are met. In other words, the new customer insertion will be tried between two other customers. To ensure that the given time budget and the time windows constraints are met, the time budget, from the beginning of the route, until the previous customer to where the new one is attempting to be inserted, is calculated. Then the new customer respective travel distance and operation time are inserted. The time budget and time windows constraints are again validated and if feasible, the time for the rest of the route is calculated and it is verified if the given time budget was not exceeded. Once added to a route, a customer is then removed from the list and marked as checked. Each and every customer in the list is tested for an insertion in the current route. A chromosome must contain as much routes as the number of vehicles available in a chosen TOPdTW instance. Figure 2 presents a simple scheme of how a random route is created. The algorithm uses parallel processing in order to assemble all the routes in a chromosome. In respect to the evolutionary process, it includes

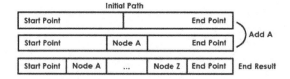

Fig. 2. Creation of a random route

two genetic operations: crossover and mutation. The crossover procedure is done by exchanging routes between two chromosomes, resulting in the creation of two new chromosomes. The routes to be exchanged are randomly selected, yet entire blocks of consecutive routes are copied to the new chromosomes, as it can be observed in Fig. 3. The chromosomes used for crossover are chosen based on

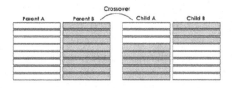

Fig. 3. Crossover procedure between two parents, resulting in two children chromosomes (each bar represents the route of a vehicle)

a roulette-wheel selection, also known as fitness proportionate selection, which assesses the probability of a chromosome being used in combinatorial methods. Therefore, a chromosome with high fitness level is more likely to be selected as a parent for the next generation. The other genetic operation used is mutation and it consists on the removal of a random customer from a randomly chosen route within a chromosome. Then, an attempt is made in order to insert one or more customers from the availability list of chromosome. The customers in this list are checked one by one in a random order, and when the current customer represents a valid option, it is added to the route. During this process, an attempt is made in order to add as many new customers as possible to the route. In Fig. 4, a simple mutation is presented, where only one customer is removed from a route. There is also the possibility to perform more complex mutations by removing more than one customer from a route, and the process is executed in a similar way as in the simple mutation. There are two special classes of

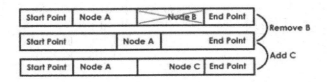

Fig. 4. Crossover procedure between two parents, resulting in two children chromosomes (each bar represents the route of a vehicle)

chromosomes within the population, the elite and the sub-elite groups, to which different rules are applied. The elite class is the group of the highest fitness chromosomes within the population in a certain generation, and these chromosomes are immune to mutation until they are replaced by new chromosomes in further generations. As for the sub-elite class, it includes the fittest chromosomes immediately after the elite ones. This group is kept intact during the crossover phase but suffers mutation right after. During the evolutionary process, the resultant chromosomes from both crossover and mutation processes are kept, even if their fitness is lower than the chromosomes that originated them.

5   Computational Test Results and Discussion

In order to evaluate this model several computational experiments were performed. As a first step the experiments were conducted with a set of TOPTW instances whose optimal values are known. Five instances of Solomon and Derosiers [29] of the first set were chosen (c101, c102, c104, c108 and c109) and three different numbers of paths were considered.

5.1   MILP Model

Using the *NEOS Server with AMPL/Gurobi* [30,31,32] the MILP could not find a solution for all vertices. Several experiments with subsets of vertices of different sizes were performed to find out the maximum number of vertices that it is possible to solve optimally. Tables 1-3 present these results. It is possible to verify that only instance c101 could be solved for all vertices. For the others it was not possible to obtain results considering more than the first 30 vertices. PK means "Process Killed" after 180 minutes. When it is available, the time (in seconds) to reach the solution is indicated in parentheses. As the number of paths increase, the number of vertices in the instance is also increased, because it becomes easier to obtain the optimal solution. To evaluate the MILP and the *NEOS*

Table 1. NEOS Server results for the test problems of Solomon (paths = 2)

Instance/Nodes	10	15	20	25	30	35	40	45	50	100
c101	-	-	-	-	-	-	-	-	-	590(280s)
c102	150	260	360(173s)	450(14585s)	PK	PK	PK	PK	PK	PK
c104	150	260	360	PK	PK	PK	PK	PK	PK	PK
c108	150	260	PK	PK	PK	PK	PK	PK	PK	PK
c109	150	260	360(15s)	PK	PK	PK	PK	PK	PK	PK

Table 2. NEOS Server results for the test problems of Solomon (paths = 3)

Instance/Nodes	10	15	20	25	30	35	40	45	50	100
c101	-	-	-	-	-	-	-	-	-	810(12105s)
c102	150	260	360	460(2s)	PK	PK	PK	PK	PK	PK
c104	150	260	360(1s)	460(24s)	520(7s)	PK	PK	PK	PK	PK
c108	150	260	360	460(2s)	PK	PK	PK	PK	PK	PK
c109	150	260	360(1s)	460(6s)	520(14224s)	PK	PK	PK	PK	PK

Table 3. NEOS Server results for the test problems of Solomon (paths = 4)

Instance/Nodes	10	15	20	25	30	35	. 40	45	50	100
c101	150	260	360	460	520	640(12s)	720(22s)	720(45s)	770(65s)	PK
c102	150	260	360	460(1s)	520(566s)	PK	PK	PK	PK	PK
c104	150	260	360(1s)	460(20s)	520(27s)	PK	PK	PK	PK	PK
c108	150	260	360	460	520(19s)	PK	PK	PK	PK	PK
c109	150	260	360	460(2s)	520(7s)	PK	PK	PK	PK	PK

Server with AMPL/Gurobi [30,31,32] solving the TOPdTW, five instances of 50 surgeries were generated randomly, varying the time windows and processing time for each elective surgery. The MILP could not find a solution for all vertices. Several experiments with subsets of vertices of different sizes were performed to find out the maximum number of vertices that it is possible to solve optimally. Two operating rooms and four shifts were considered, resulting in 8 paths. Table 4 presents results of the computational experimentation. Two operating rooms and nine shifts were considered, resulting in 18 paths. Table 5 presents results of the computational experimentation.

Table 4. NEOS Server results for the test problems of Operating Rooms (paths = 8)

Instance/Nodes	10	15	20	25	30	50
OR01	24	36(3s)	PK	69(255s)	PK	PK
OR02	31	44(6s)	59(698s)	PK	PK	PK
OR03	35(1s)	52(1s)	69(27s)	83(274s)	PK	PK
OR04	27(1s)	45(6s)	62(19s)	PK	PK	PK
OR05	27(2s)	41(3s)	55(92s)	PK	PK	PK

Table 5. NEOS Server results for the test problems of Operating Rooms (paths = 18)

Instance/Nodes	10	15	20	25	30	50
OR01	24	36(19s)	55(19s)	69(83s)	PK	PK
OR02	31	44(3s)	71(285s)	PK	PK	PK
OR03	35	52	69	PK	99(360s)	PK
OR04	27	45(11s)	PK	75(117s)	94(3643s)	PK
OR05	27(4s)	41(10s)	55(233s)	PK	PK	PK

5.2 Genetic Algorithm

In order to evaluate the GA, several experiments with the same set of instances were performed, using two different population sizes. For each instance 10 runs were performed. Table 6 presents the results of these runs and some statistics are also presented such as the Best Solution, Worst Solution, and Average Value. Also the Best Known Solution (BKS) for each instance is presented. The presented algorithm was implemented using the JAVA Swing Framework, and the resulting software tool incorporates a Graphical User Interface that allows adjustment of various parameters. The tests were run on a desktop computer with an Intel Core i7 quad-core at 2.7 GHz processor and 16GB of RAM at 1600 MHz. During the tests, the maximum number of generations to be produced for each instance was set to be the stopping criterion, with its value limited to 3000. The CPU time to perform 1000 iterations is around 2 minutes using a total population of 520 chromosomes, with 10 being in the Elite and another 10 in the Sub-Elite. Most of the times, the best solution was found in the first 500 iterations.

To evaluate the GA solving the TOPdTW, the same five instances of 50 surgeries were considered, with two operating rooms and nine shifts (18 paths). Table 7 presents results of the computational experimentation. The UB column shows the total value of prizes of all nodes. This value can be seen as a coarse Upper Bound, because in this type of problems (TOP, TOPTW, TOPdTW) normally it is not possible to visit all vertices. The GA performs reasonably and presents deviations below 25% of the "Upper Bound". These are preliminary results, and therefore may be considered promising.

Table 6. Results for the test problems of Solomon (paths = 4)

Instance	PopSize	Run1	Run2	Run 3	Run 4	Run 5	Run 6	Run 7	Run 8	Run 9	Run 10
c101	Pop520	960	980	970	950	950	950	970	970	960	950
	Pop120	970	980	990	940	940	960	930	940	950	980
c102	Pop520	1080	1080	1060	1100	1050	1120	1070	1080	1050	1100
	Pop120	1080	1030	1050	1060	1080	1030	1010	1070	1090	1090
c104	Pop520	1120	1140	1130	1140	1130	1040	1090	1090	1100	1130
	Pop120	1130	1090	1110	1130	1130	1130	1090	1120	1120	1120
c108	Pop520	1070	1050	1080	1060	1080	1070	1040	1060	1070	1070
	Pop120	1030	1050	1020	1070	1070	1050	1080	1070	1070	1030
c109	Pop520	1100	1100	1110	1080	1160	1100	1110	1090	1070	1110
	Pop120	1080	1090	1090	1090	1130	1120	1130	1100	1070	1130

Instance	PopSize	BestSolution	WorstSolution	Avg	BKS	Avg/BKS
c101	Pop520	980	950	961	1020	0,942
	Pop120	990	930	958	1020	0,939
c102	Pop520	1120	1050	1079	1150	0,938
	Pop120	1090	1010	1059	1150	0,921
c104	Pop520	1140	1090	1121	1260	0,89
	Pop120	1130	1090	1117	1260	0,887
c108	Pop520	1080	1040	1065	1130	0,942
	Pop120	1080	1020	1054	1130	0,933
c109	Pop520	1160	1070	1103	1190	0,927
	Pop120	1130	1070	1103	1190	0,927
Global Average						0,925

Table 7. Results for the test problems of Operating Rooms (paths = 18)

Instance	PopSize	Run1	Run2	Run 3	Run 4	Run 5	Run 6	Run 7	Run 8	Run 9	Run 10
OR01	Pop520	109	112	114	112	113	109	111	113	112	111
	Pop120	111	111	112	116	108	110	113	114	113	114
OR02	Pop520	98	99	95	99	99	99	99	97	97	100
	Pop120	98	99	99	100	96	100	95	98	99	94
OR03	Pop520	139	143	141	142	144	141	142	142	144	142
	Pop120	141	144	140	140	144	142	144	143	142	142
OR04	Pop520	99	98	99	99	98	99	100	97	100	98
	Pop120	97	98	98	98	99	97	100	100	97	97
OR05	Pop520	123	123	125	124	121	122	124	122	124	123
	Pop120	119	122	122	123	121	122	127	123	122	121

Instance	PopSize	BestSolution	WorstSolution	Avg	UB	Avg/UB
OR01	Pop520	114	109	111,6	140	0,797
	Pop120	116	108	112,2	140	0,801
OR02	Pop520	100	95	98,2	140	0,701
	Pop120	100	94	97,8	140	0,699
OR03	Pop520	144	139	142	156	0,91
	Pop120	144	140	142,2	156	0,912
OR04	Pop520	100	97	98,7	152	0,649
	Pop120	100	97	98,1	152	0,645
OR05	Pop520	125	121	123,1	142	0,867
	Pop120	127	119	122,2	142	0,861
Global Average						0,784

6   Conclusions and Future Work

This paper presented a genetic algorithm for the Team Orientering Problem with double Time Windows that is a new variant of TOP. The aim to study TOPdTW it to model a real problem that arises within the operating rooms in a hospital. The Genetic Algorithm was tested using some public instances of the TOPTW and instances generated for TOPdTW. The computational results presented are preliminary, however they show the potential of this methodology to be applied for these types of problems. The solution obtained in TOPdTW reveals a level of temporal occupation of the operating rooms, which seem promising.

There are still some enhancements that can be done to improve the results, like modifying the crossover and mutations procedures. Improvements can also be achieved by finding a better balance between parameters such as the total population size and the number of elite and sub-elite chromosomes. A possible way of doing this is to use dynamic parameters to set the behaviour of the evolution process within the genetic algorithm. This could be achieved by implementing a Machine Learning algorithm that would tune the parameters of the genetic algorithm by evaluating its performance during the tests and would apply the best parameter configuration to overcome adversities while aiming for better results. The assessment of the software tool developed for the presented study was important in order to identify its functionalities, advantages and limitations. Future experimentations will focus on the usage of the C++ programming language, which might perform faster than JAVA.

Acknowledgments. This work was funded by the "Programa Operacional Fatores de Competitividade - COMPETE" and by the FCT - Fundação para a Ciência e Tecnologia in the scope of the project: FCOMP-01-0124-FEDER-022674. The authors would like to thank the anonymous reviewers for their valuable comments and suggestions to improve the paper.

References

1. Alejo, J.S., Martín, M.G., Ortega-Mier, M., García-Sánchez, A.: Mixed integer programming model for optimizing the layout of an ICU vehicle. BMC Health Services Research 9, 224 (2009)
2. Lapierre, S., Ruiz, A.: Scheduling logistic activities to improve hospital supply systems. Computers & Operations Research 34, 624–641 (2007)
3. Topaloglu, S.: A shift scheduling model for employees with different seniority levels and an application in healthcare. European Journal of Operational Research 198, 943–957 (2009)
4. Jaumard, B., Semet, F., Vovor, T.: A generalized linear programming model for nurse scheduling. European Journal of Operational Research 107, 1–18 (1998)
5. Cheang, B., Li, H., Lim, A., Rodrigues, B.: Nurse rostering problems - a bibliographic survey. European Journal of Operational Research 151, 447–460 (2003)
6. Burke, E., De Causmaecker, P., Berghe, G.V., Van Landeghem, H.: The state of the art of nurse rostering. Journal of Scheduling 7, 441–499 (2004)
7. van Essen, J., Hans, E., Hurink, J., Oversberg, A.: Minimizing the waiting time for emergency surgery. Operations Research for Health Care 1, 34–44 (2012)
8. Su, M.C., Lai, S.C., Wang, P.C., Hsieh, Y.Z., Lin, S.C.: A SOMO-based approach to the operating room scheduling problem. Expert Systems with Applications 38, 15447–15454 (2011)
9. Augusto, V., Xie, X., Perdomo, V.: Operating theatre scheduling with patient recovery in both operating rooms and recovery beds. Computers & Industrial Engineering 58, 231–238 (2010)
10. van Essen, J.T., Hurink, J.L., Hartholt, W., van den Akker, B.J.: Decision support system for the operating room rescheduling problem. Health Care Management Science 15, 355–372 (2012)
11. Hovlid, E., Bukve, O., Haug, K., Aslaksen, A.B., von Plessen, C.: A new pathway for elective surgery to reduce cancellation rates. BMC Health Services Research 12, 154 (2012)
12. Cardoen, B., Demeulemeester, E., Belien, J.: Operating room planning and scheduling: A literature review. European Journal of Operational Research 201, 921–932 (2010)
13. Ben Bachouch, R., Guinet, A., Hajri-Gabou, S.: An integer linear model for hospital bed planning. International Journal of Production Economics 143, 833–843 (2012)
14. Hulshof, P., Boucherie, R., Hans, E., Hurink, J.: Tactical resource allocation and elective patient admission planning in care processes. Health care management science (2013)

15. Pham, D.N., Klinkert, A.: Surgical case scheduling as a generalized job shop scheduling problem. European Journal of Operational Research 185, 1011–1025 (2008)
16. Jebali, A., Hadj Alouane, A.B., Ladet, P.: Operating rooms scheduling. International Journal of Production Economics 99, 52–62 (2006)
17. Butt, S.E., Cavalier, T.M.: A heuristic for the multiple tour maximum collection problem. Computers & Operations Research 21, 101–111 (1994)
18. Chao, I.M., Golden, B., Wasil, E.A.: The team orienteering problem. European Journal of Operational Research 88, 464–474 (1996)
19. Archetti, C., Hertz, A., Speranza, M.G.: Metaheuristics for the team orienteering problem. Journal of Heuristics 13, 49–76 (2007)
20. Vieira, F., Macedo, J., Carcao, T., Leite, T., Murta, D., Ferreira, J., Dias, L., Pereira, G., Oliveira, J.A.: Developing tools for the team orienteering problem - A simple genetic algorithm. In: Proceedings of 2nd International Conference on Operations Research and Enterprise Systems (ICORES 2013), Barcelona, Spain, February 16-18 (2013)
21. Ferreira, J., Quintas, A., Oliveira, J.A., Pereira, G., Dias, L.: Solving the team orienteering problem - Developing a solution tool using a genetic algorithm approach. In: Snasel, V., Kromer, P. (eds.) Proceedings of 17th Online World Conference on Soft Computing in Industrial Applications (WSC17), December 21-31 (2012)
22. Lin, S.W., Yu, V.F.: A simulated annealing heuristic for the team orienteering problem with time windows. European Journal of Operational Research 217, 94–107 (2012)
23. Labadie, N., Mansini, R., Melechovsky, J., Wolfler Calvo, R.: The Team Orienteering Problem with Time Windows: An LP-based Granular Variable Neighborhood Search. European Journal of Operational Research 220, 15–27 (2012)
24. Garcia, A., Vansteenwegen, P., Arbelaitz, O., Souffriau, W., Linaza, M.T.: Integrating public transportation in personalised electronic tourist guides. Computers & Operations Research 40, 758–774 (2013)
25. Vansteenwegen, P., Souffriau, W., Vanden Berghe, G., Van Oudheusden, D.: Iterated local search for the team orienteering problem with time windows. Computers & Operations Research 36, 3281–3290 (2009)
26. Tricoire, F., Romauch, M., Doerner, K.F., Hartl, R.F.: Heuristics for the multi-period orienteering problem with multiple time windows. Computers & Operations Research 37, 351–367 (2010)
27. Garcia, A., Vansteenwegen, P., Souffriau, W., Arbelaitz, O., Linaza, M.: Solving Multi Constrained Team orienteering Problems to Generate Tourist Routes. Working paper (2009)
28. Dohn, A., Kolind, E., Clausen, J.: The manpower allocation problem with time windows and job-teaming constraints: A branch-and-price approach. Computers & Operations Research 36, 1145–1157 (2009)
29. Solomon, M., Desrosiers, J.: Time windows constrained routing and scheduling problems. Transportation Science 22, 1–22 (1998)
30. Czyzyk, J., Mesnier, M.P., Moré, J.J.: The NEOS server. IEEE J. Computational Sci. Engrg. 5, 68–75 (1998)
31. Gropp, W., Moré, J.J.: Optimization environments and the NEOS server. In: Buhmann, M.D., Iserles, A. (eds.) Approximation Theory and Optimization: Tributes to M. J. D. Powell, pp. 167–182. Cambridge University Press, Cambridge (1997)
32. Dolan, E.D.: NEOS server 4.0 administrative guide. Technical Memorandum ANL/MCS-TM-250, Mathematics and Computer Science Division, Argonne National Laboratory, Argonne, IL (2001)

A Computational Study on Different Penalty Functions with DIRECT Algorithm

Ana Maria A.C. Rocha[1,2] and Rita Vilaça[2]

[1] Department of Production and Systems, School of Engineering
arocha@dps.uminho.pt
[2] Algoritmi R&D Centre
University of Minho, 4710-057 Braga, Portugal
rita.pinto.vilaca@gmail.com

Abstract. The most common approach for solving constrained optimization problems is based on penalty functions, where the constrained problem is transformed into an unconstrained problem by penalizing the objective function when constraints are violated. In this paper, we analyze the implementation of penalty functions, within the DIRECT algorithm. In order to assess the applicability and performance of the proposed approaches, some benchmark problems from engineering design optimization are considered.

Keywords: Global optimization, constrained optimization, DIRECT algorithm, penalty function.

1 Introduction

This paper aims to illustrate the behavior of some penalty functions combined with the DIRECT algorithm for solving nonlinear global optimization problems of the form:

$$\begin{aligned}
\text{minimize} \quad & f(x) \\
\text{subject to} \quad & g_i(x) \leq 0, \ i = 1, \ldots, m \\
& x \in \Omega,
\end{aligned} \tag{1}$$

where $f : \mathbb{R}^n \to \mathbb{R}$, $g : \mathbb{R}^n \to \mathbb{R}^m$ are nonlinear continuous functions, $\Omega = \{x \in \mathbb{R}^n : l \leq x \leq u\}$ and x corresponds to the main variables, decision or controllable through which will optimize $f(x)$. We do not assume that functions f and g are convex. There may exist many local minima in the feasible region. This class of global optimization problems arises frequently in engineering applications. Specially for large scale problems of type (1), derivative-free and stochastic methods are the most well-known and used methods.

DIRECT is a deterministic global optimization algorithm developed by Jones, Perttunen and Stuckman [19] and the name "DIviding RECTangles" describes the way the algorithm moves towards the optimum. The DIRECT method starts the iterative process by dividing the solution space into hyperrectangles, using criteria (size and value of the function center) well defined. Hence, the algorithm,

B. Murgante et al. (Eds.): ICCSA 2013, Part I, LNCS 7971, pp. 318–332, 2013.

based on a space-partitioning scheme, performs both global exploration and local exploitation [9].

DIRECT was initially developed to solve difficult global optimization problems with simple bound constraints [9,10]. However, over the years the method has been improved and some modifications were done in order to solve other kind of problems [11,12].

There is a set of strategies for solving nonlinear constrained problems consisting of transforming the constrained problem into a sequence of unconstrained subproblems, whose solutions are related in some way to the solution of the original problem. Solutions of the successive unconstrained subproblems will eventually converge to the solution of the original constrained problem. The subproblems created by penalty methods involve a penalty function that incorporates the objective function and the constraints of the problem. The penalty function may include one or more penalty parameters which determine the relative importance of each constraint, or set of constraints. When these parameters are suitably modified, the effects of constraints become increasingly evident in the sequence of generated problems.

Penalty methods are simple to implement, are applicable to a broad class of problems and take advantage of powerful unconstrained minimization methods. Thus, they have been widely accepted in practice as an effective class of methods for constrained optimization. The appropriate selection of penalty parameter values is vital for a fast convergence. A function that determines its value at the beginning of each round of optimization may be defined. Different penalty functions may emerge depending on the way penalties vary throughout the minimization process.

In this paper we intend to analyze the performance of different penalty approaches embedded in the DIRECT algorithm, namely, the l_1 penalty function, Quadratic Penalty function, Dynamic Penalty function, Hyperbolic Penalty function and Augmented Lagrangian Penalty function. Some preliminary results are presented when solving a benchmark of engineering design optimization problems with the proposed methods.

The remainder of this paper is as follows. We briefly describe the DIRECT algorithm in Section 2. In Section 3 we briefly introduce the penalty function technique and in Section 4 the proposed penalized DIRECT algorithm approaches are outlined. Section 5 describes the experimental results and finally we draw the conclusions of this study in Section 6.

2 DIRECT Algorithm

The DIRECT optimization algorithm was first presented in [19], and emerged by a modification to Lipschitzian optimization, that has proven to be effective in a wide range of application domains. The motivation for the DIRECT algorithm comes from a different way of looking at the Lipschitz constant. In particular, the Lipschitz constant is viewed as a weighting parameter that indicates how much emphasis is to be placed on global versus local search.

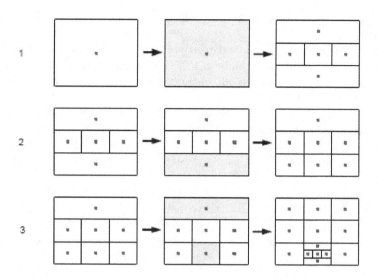

Fig. 1. First three iterations of DIRECT on a sample problem

DIRECT is a sampling algorithm and requires no knowledge of the objective function gradient. Furthermore, the algorithm samples points in the search space, and uses the information it has obtained to decide where to search next. DIRECT can be very useful when the objective function is a "black box" function or simulation. An example of using DIRECT to solve a large industrial problem can be found in [4].

DIRECT begins the optimization by transforming the domain of the problem into the unit hypercube and its center point, c, is evaluated. Then the hypercube is divided in three hyperrectangles along the coordinate with best objective function value. That is, the evaluation of the function at points $c \pm \delta e_i$ for $i = 1, 2, ..., n$ where c is the central point of the hypercube, δ is a third side of the hypercube and e_i is the ith unit vector. Then, the new points evaluated in this coordinate will be the central points of the new hyperrectangles. At each iteration, a set of *potentially optimal hyperrectangles* are defined for further divisions [10]. Subsequently, the central hyperrectangle that contains c must be partitioned into another three hyperrectangles along the coordinate with the second best value of the objective function and so on. Figure 1 shows an example of DIRECT algorithm.

DIRECT searches locally and globally by dividing all hyperrectangles that meet the following criteria. Thus, for a hyperrectangle i, the value of the objective function at c_i, the center of the hyperrectangle, and d_i, the distance from the center point to the vertices are required. A hyperrectangle j is said to be potentially optimal if there exists some $K > 0$, thought as the Lipschitz approximation value, such that,

$$f(c_j) - Kd_j \leq f(c_i) - Kd_i, \text{ for all } i \neq j \tag{2}$$

$$f(c_j) - Kd_j \leq f_{min} - \epsilon |f_{min}| \tag{3}$$

where f_{min} is the current best function value and $\epsilon > 0$ is a positive constant [11]. The parameter $\epsilon > 0$ was introduced in order to avoid an exhaustive local search. Tests performed with classic problems of global optimization showed that values between 10^{-7} and 10^{-2} did not degrade the performance of DIRECT. The value of 10^{-4} was suggested in [19] presenting results more robust, in practice.

In the last years many modifications have been suggested by some authors with the aim of improving the DIRECT [11,12,14]. For global optimization of noisy functions, [12] presented a new approach, designated by, noisy DIRECT algorithm. This algorithm can be divided into two phases. In the first phase, the algorithm must find multiple promising regions of interest. In the second phase, local search algorithms must be applied around those promising regions aiming to refine solutions. A method involving a non-traditional penalty function and a heuristic for determining the penalty parameters, was proposed by [14]. The method only solves inequality constrained problems. For the same type of problems, a traditional exact l_1 penalty function approach was implemented with DIRECT in [9], where the penalty parameter is maintained constant along the optimization process.

3   Penalty Function Technique

The most common approach to solve constrained optimization problems is based on penalty functions. In this approach, penalty terms are added to the objective function to penalize the objective function value of any solution that violates the constraints.

Penalty methods were originally proposed by Courant in 1940s [7] and later expanded by Carroll [3] and Fiacco & McCormick [8]. The idea of this method is to transform a constrained optimization problem into a sequence of unconstrained subproblems by adding a certain value to the objective function based on the amount of constraint violation present in a certain solution. The solutions of the successive unconstrained subproblems will eventually converge to the solution of the original constrained problem.

Although penalty functions are very simple and easy to implement they often require one or more penalty parameters that are usually problem dependent and chosen with a priori knowledge by users. The basic penalty approach defines a penalty function for each point x, herein denoted by $\Phi(x)$, by adding to the objective function value a penalty term, $P(x, \mu)$, that aims to penalize infeasible solutions

$$\Phi(x) = f(x) + P(x, \mu), \tag{4}$$

where $f(x)$ is the objective function of the constrained problem and μ is a positive real number denoted by penalty parameter. The goal of the penalty method is to solve a sequence of unconstrained subproblems, minimizing the penalty function.

Ideally the penalty parameter should be kept as low as possible, just above the limit below which infeasible solutions are near-optimal (this is called the minimum penalty rule) [5]. This is due to the fact that if the penalty is too high or too low, then the problem may be difficult to solve. A large penalty discourages the exploration of the infeasible region since the very beginning of the search process. On the other hand, if the penalty is too low, a lot of the search time will be spent exploring the infeasible region because the penalty will be negligible with respect to the objective function. Thus, the selection of appropriate penalty parameter values is vital for a fast convergence and accuracy. Further, the initial penalty parameter value also has a key role in the convergence behavior of the method.

According to (1) the level of constraints violation is measured by the vector

$$v_j(x) = \max\{0, g_j(x)\}, \ j = 1, \ldots, m \, . \tag{5}$$

and a measure of constraints violation for a point x, is given by

$$\text{viol} = \sum_{j=1}^{m} v_j(x).$$

If viol $= 0$ then x is a feasible point, otherwise is infeasible.

4   The Proposed Penalized DIRECT Algorithm

Different penalty functions have been presented in the past which can be classified according to the way the penalties are added [1,5,13,23]. In this study, the l_1 penalty, the quadratic, the dynamic, the hyperbolic and the augmented Lagrangian penalty functions are implemented within the DIRECT algorithm. Different penalties may emerge depending on the way penalties vary throughout the minimization process. A brief description of these penalties follows.

4.1 l_1 Penalty

A traditional exact l_1 penalty function approach was already implemented in DIRECT [9]. In the linear penalty method the penalty term is the l_1 norm of the constraint violation. Although continuous, the penalty function l_1 is not differentiable at all points. This is the major disadvantage of the l_1 penalty method [18].

In this method the penalty term [9] is given by

$$P(x, \mu) = \mu \sum_{i=1}^{m} \max(0, g_i(x)) \tag{6}$$

where μ is the penalty parameter. Finally, the sequence of subproblems, parameterized by the penalty parameter μ, is solved by DIRECT algorithm. We remark that in [9] the penalty parameter is maintained constant during the optimization process.

4.2   Quadratic Penalty

In the quadratic penalty, the penalty term is based in the square of the constraint violation [18]. The penalty term of quadratic penalty method can be defined as,

$$P(x,\mu) = 2\mu \sum_{i=1}^{m} \max(0, g_i(x))^2 \tag{7}$$

where μ is the penalty parameter that tends to infinite. The update of the penalty parameter, in a k iteration, is given by

$$\mu^{k+1} = \min(\mu^{max}, \alpha\mu^k)$$

where $\alpha > 1$ and μ^{max} is an upper bound to the penalty parameter μ.

4.3   Dynamic Penalty

Joines and Houck [13] proposed that the penalty parameters should vary dynamically along the search according to exogenous schedule, as

$$P(x,\mu) = \mu \sum_{i=1}^{m} \max(0, g_i(x))^{\gamma(max(0,g_i(x)))} \tag{8}$$

where μ is a dynamically modified penalty parameter. This penalty parameter is updated by

$$\mu^{k+1} = \begin{cases} (C(k+1))^{\alpha} & \text{if viol} > 0 \\ \mu^k & \text{otherwise} \end{cases} \tag{9}$$

where k represents the iteration counter, and the constants C and α are set as 0.5 and 2, respectively. The power of the constraint violation, $\gamma(.)$, is a violation dependent constant: $\gamma(z) = 1$ if $z \leq 1$, and $\gamma(z) = 2$, otherwise. See, examples of dynamic penalties in [17,20].

Another interesting and quite efficient rule for the penalty update, found in the literature [20], is given by $\mu^{k+1} = k\sqrt{k}$. Note that, the penalty parameter does not depend on the number of constraints although the pressure on infeasible solutions increases as k increases.

4.4   Hyperbolic Penalty

In the hyperbolic penalty method the sequence of subproblems is obtained by controlling two parameters in two different phases of the optimization process. In the first phase, the initial parameter λ increases, thus causing an increase in the penalty, P, to the points outside the feasible region and directing the search to the feasible region. This phase continues until a feasible point is obtained. From this point on, λ remains constant and the values of τ decrease sequentially.

In this context, [26] proposed the hyperbolic penalty function below, where only problems with inequality constraints are considered

$$P(x, \lambda, \tau) = \sum_{i=1}^{m} (\lambda_i g_i(x) + \sqrt{(\lambda_i)^2 g_i(x)^2 + (\tau_i)^2}) \tag{10}$$

with $\lambda \geq 0$, $\tau \geq 0$ are penalty parameters and $\lambda \to \infty$, $\tau \to 0$.

The penalty parameters are updated by

$$\begin{cases} \lambda_i^{k+1} = \min(\lambda^{\max}, \gamma_\lambda \lambda_i^k) \text{ and } \tau_i^{k+1} = \tau_i^k \text{ if } \max(0, g_i(x^{k+1})) > 0, \\ \tau_i^{k+1} = \max(\tau^{\min}, \gamma_\tau \tau_i^k) \text{ and } \lambda_i^{k+1} = \lambda_i^k, \text{ otherwise} \end{cases}$$

for $i = 1, \ldots, m$, where $\lambda^{\max}$ and $\tau^{\min}$ are upper and lower bounds respectively to the penalty parameters λ and τ. The goal is to define safeguards to prevent the subproblems from becoming ill-conditioned and more difficult to solve as λ increases or τ decreases.

We remark, the hyperbolic penalty function allows the use of optimization methods which use derivatives information, such as quasi-Newton method, to obtain the solution of the problem, since this is a continuously differentiable function.

4.5   Augmented Lagrangian Penalty

The augmented Lagrangian method uses the Lagrangian multipliers vector which reduces the possibility of generating ill-conditioned subproblems [2].

Combining the quadratic penalty function and the Lagrangian function, it is possible to obtain the penalty term of the augmented Lagrangian function [16],

$$P(x, \delta, \mu) = \frac{1}{2\mu} \sum_{i=1}^{m} (\max(0, \delta_i + \mu g_i(x))^2 - \delta_i^2) \tag{11}$$

where μ is the penalty parameter and δ is the Lagrange multipliers vector associated with the inequality constraints. The initial values set to the Lagrange multipliers and penalty parameter need not be large to occur a good approximation to the solution of the first subproblem of the sequence of subproblems to be solved [18].

An augmented Lagrangian penalty method is a multiplier-based method requiring [2]:

- The sequence $\{\delta^k\}$ must be bounded;
- The sequence of penalty values $\{\mu^k\}$ must satisfy $0 < \mu^k \leq \mu^{k+1}$, for all k and $\mu^k \to \infty$.

The updating scheme for the Lagrange multipliers δ_i associated with the constraints $g_i(x) \leq 0$, $i = 1, \ldots, m$ is given by

$$\delta_i^{k+1} = \max(0, \delta_i^k + \mu^k g_i(x^{k+1}))$$

and the penalty parameter is updated by

$$\mu^{k+1} = \min(\mu^{\max}, \alpha\mu^k)$$

where $\alpha > 1$ and $\mu^{\max}$ is an upper bound to the penalty parameter.

4.6   The Penalized DIRECT Algorithm

The general steps of the penalized DIRECT algorithm is described in Algorithm 1. At the end of the algorithm ϕ_{min} is the global optimal solution.

Algorithm 1. Penalized DIRECT algorithm

1: normalize the original domain Ω to the unit hypercube in $\mathbb{R}^n$ with center c_1.
2: evaluate $\phi(c_1)$; $\phi_{min} = \phi(c_1)$; set $k = 1$
3: evaluate $\phi(c_1 \pm \frac{1}{3}e_i)$, $1 \leq i \leq n$ and divide hypercube
4: **while** stopping criterion is not reached **do**
5:    identify the set S of all potentially optimal hyperrectangles/cubes
6:    **for all** $j \in S$ **do**
7:       identify the longest side(s) of hyperrectangle j
8:       update the penalty parameter
9:       evaluate ϕ at centers of new hyperrectangles and divide hyperrectangle j into smaller hyperrectangles
10:      update ϕ_{min}
11:   **end for**
12:   set $k = k + 1$
13: **end while**

The stopping criterion could be based on the maximum number of iterations and a maximum number of function evaluations.

5   Numerical Experiments

In this section, we report the numerical results obtained by running the Algorithm 1 based on the linear, quadratic, dynamic, hyperbolic and augmented Lagrangian penalty functions using six benchmark engineering design problems.

The implementation details of the proposed approach could be found in [25], as well as its application on a well-known and hard optimization problem from the chemical and bio-process engineering area.

Problems of practical interest are important for assessing the effectiveness of any algorithm. Thus, Table 1 contains a summary of the characteristics of the selected problems, where all of them have simple bounds and inequality constraints [6,15,21,22] and where n is the number of variables of the problem, m is the number of inequality constraints and $f(x^*)$ is the optimal solution known in the literature.

Table 1. Characteristics of the engineering design problems

Problem	n	Type of Objective Function	m	$f(x^*)$
Spring	3	quadratic	4	0.0126
Speed	7	nonlinear	11	2994.4991
Brake	4	quadratic	6	0.1274
Tubular	2	linear	2	26.5313
3-Bar	2	linear	3	263.896
4-Bar	4	linear	1	1400

5.1   Parameter Sensitivity Analysis

The choice of penalty parameters is a hard and very important task since the performance of DIRECT depends on the magnitude of the penalty parameters. On some problems, an extremely large penalty parameter is necessary for the algorithm to converge to a feasible point. However, a large penalty parameter is a critical issue for DIRECT, since could bias the algorithm away from hyper-rectangles near the boundary of feasibility [9].

Regarding this, it turns out to be very important to perform a sensitivity analysis in order to find the best values to provide to each problem. Depending on the penalty function and the problem, the values of the parameters could be different.

Below, it is graphically shown the influence of the tolerance ϵ, see (3), when solving different problems, for the local performance assessment of the algorithm. Figure 2 shows results for the l_1 penalty function when solving the 3-Bar problem. The bars of the right plot represent the number of function evaluations using the l_1 penalty and the circles represents the value of the objective

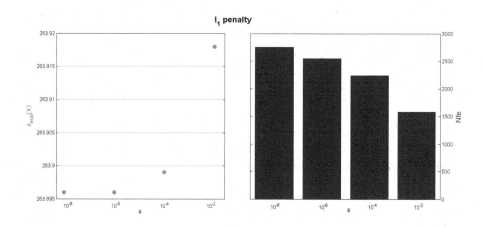

Fig. 2. Sensitivity analysis of parameter ϵ for the 3-Bar problem when using l_1 penalty function

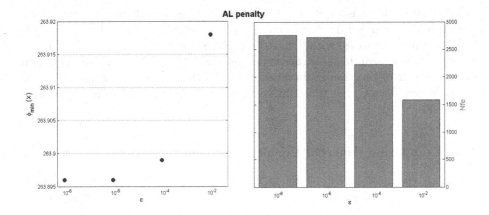

Fig. 3. Sensitivity analysis of parameter ϵ for the 3-Bar problem when using the augmented Lagrangian function

function attained for each ϵ value. Similar quantities obtained by the augmented Lagrangian function are shown in Fig. 3.

Figures 2-3 show that for lower values of ϵ, the value of the achieved penalty function value is, also, lower, although spending higher number of function evaluations. Thus, it is concluded that a reasonable value of ϵ, $\epsilon = 10^{-6}$, would be used for both penalties.

In addition to the parameter ϵ, which balances between local and global search, other parameters are associated with the penalty algorithm that affect its performance. For example, with the augmented Lagrangian penalty, initial values for the penalty parameter μ, for the multipliers vector δ, and for the penalty update α are required.

Table 2 shows the influence of these parameters in the augmented Lagrangian penalty algorithm, for 3-Bar problem. In the table, δ^1, refers to the initial value for all coordinates of the multipliers vector, 'Fopt' is the obtained solution and

Table 2. Penalty and multiplier values of augmented Lagrangian function for 3-Bar problem

μ^1	δ^1	α	Fopt	Nfe
1	1	2	141.9291*	1487
10	10	2	195.4431*	7659
100	100	2	231.6233*	5137
1000	1	1.5	263.8959	3567
1000	1	2	263.8959	2609
1000	10	2	263.8959	2985
1000	1000	2	263.8959	5011
* infeasible solution				

'Nfe' denotes the number of function evaluations required to achieve a solution with a tolerance of $\epsilon = 10^{-6}$.

The solutions marked with * violate the constraints. By analyzing the Table 2 it appears that with initial higher values of μ, a better solution is obtained with fewer number of function evaluations. On the other hand, the algorithm seems to be more efficient with small initial values of Lagrange multipliers than with large ones.

5.2   Comparing Different Penalty Functions

To assess the performance of the penalty functions, the previously referred engineering design problems are used. The results are reported in Table 3 and include the number of iterations (Nit), the number of objective function evaluations (Nfe) and the optimal solution found (Fopt) with the l_1, quadratic, hyperbolic, dynamic and augmented Lagrangian penalty functions. The stopping criterion was based on three conditions: a relative tolerance to the known optimal solution of 0.001, a maximum of 250000 function evaluations or a maximum of 3000 iterations. We remark that we also stopped the algorithm when the penalty parameters stabilized and the solution could not improve, since DIRECT is a deterministic algorithm.

Table 3. Comparison results of the penalty functions for the engineering problems

		Spring	Speed	Brake	Tubular	3-Bar	4-Bar
l_1	Fopt	0.0144	2995.1358	0.1303	26.5313	263.8958	1400.0013
	Nfe	38659	249917	164389	113713	50191	2679
	Nit	890	2599	35	2785	1181	50
Quadratic	Fopt	0.0144	2995.1358	0.1274	26.5313	263.8958	1400.0001
	Nfe	157199	249979	14319	3841	48721	3497
	Nit	3000	2616	158	108	1251	57
Dynamic	Fopt	0.0225	2995.6809	0.1362	26.5425	263.8969	1400.0003
	Nfe	106443	224611	295167	179545	142283	3167
	Nit	3000	2074	39	3000	3000	54
Hyperbolic	Fopt	0.0225	2994.5967	0.1303	26.5316	263.8962	1400.0003
	Nfe	114339	200599	101887	53035	5327	1943
	Nit	3000	3000	3000	1172	199	55
Aug. Lagrangian	Fopt	0.0201	2995.7613	0.1362	26.5313	263.8958	1400.0001
	Nfe	89387	289381	450969	6449	2609	3615
	Nit	3000	3000	42	177	110	58

The results reported in Table 3 show that, in general, all functions show good results in all problems, *i.e.*, there is no evidence of a penalty function obtaining the best results for all problems. However, the quadratic penalty function achieved the best results for Brake, Tubular and 4-Bar problems. The hyperbolic, dynamic and augmented Lagrangian penalty functions only achieved good results for one problem. A study based on the adaptive penalty function when solving these engineering problems could be found in [24].

During the experiments, we noticed, for some problems, the need to use a rather low penalty parameter, *i.e.*, the penalty function need not be so penalizing. For example, in the dynamic penalty the update of the μ parameter is based on the number of iterations and it was realized that after some iterations it was obtained a large penalty parameter, which does not benefit the output solution of the algorithm.

5.3   Behavior in Search Space

For a better understanding of the algorithm's performance, the search space visited by DIRECT combined with the penalty functions is presented below.

Since the 3-Bar problem is a bidimensional problem it is possible to represent the points in a Cartesian plane. Figure 4 contains a graphical representation of the search points of the algorithm in the space of the 3-Bar problem. The feasible points are marked with blue stars and the infeasible ones are marked with red points for all tested penalties. Figure 4a) plots the feasible and infeasible points visited by the DIRECT algorithm when combined with the augmented Lagrangian during the 110 iterations; Fig. 4b) shows the visited points by DIRECT with quadratic penalty function (1251 iterations); and Fig. 4c) depicts the points generated by DIRECT with the dynamic penalty (3000 iterations). The contours of the 3-Bar problem's functions (the objective and three constraints) are exhibited in Figure 4d) where the red star locates the global optimal solution.

We may conclude that DIRECT with the tested penalties performed an efficient and effective search around the area of the global optimal solution. However, the augmented Lagrangian function is able to achieve a better performance, visiting fewer points, since the plot presents a less dense cloud when compared with the quadratic and dynamic penalty functions.

6   Conclusion

This paper presents the performance of the DIRECT algorithm combined with quadratic, dynamic, hyperbolic and augmented Lagrangian penalty functions when solving six constrained engineering design problems.

In order to achieve the best solutions found by each algorithm, a sensitivity analysis to some parameters of the algorithm, namely the ϵ tolerance of DIRECT and penalty parameters of the tested penalty functions is carried out. We conclude that a consistent value for the penalty parameters appropriate for all tested penalty functions and for all problems is difficult to be found.

Generally, we may conclude that the obtained results with the different penalty functions showed competitive results when compared with the results from l_1 penalty function implemented in the DIRECT [9].

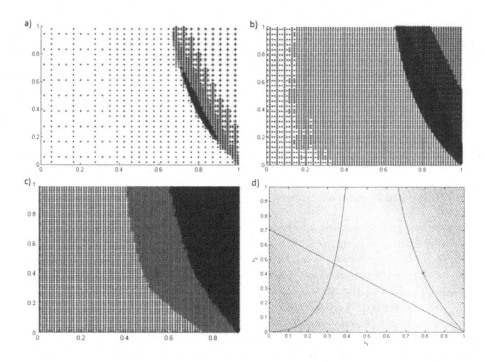

Fig. 4. Graphical representation of the points visited by DIRECT, for the 3-Bar problem. a) Augmented Lagrangian penalty function. b) Quadratic penalty. c) Dynamic penalty. d) Contours of the 3-Bar problem's functions.

Acknowledgment. The authors would like to thank the financial support from FEDER COMPETE (Operational Programme Thematic Factors of Competitiveness) and FCT (Portuguese Foundation for Science and Technology) Project FCOMP-01-0124-FEDER-022674.

References

1. Barbosa, H.J.C., Lemonge, A.C.C.: An adaptive penalty method for genetic algorithms in constrained optimization problems. In: Iba, H. (ed.) Frontiers in Evolutionary Robotics, pp. 9–34. I-Tech Education Publ., Austria (2008)
2. Bertsekas, D.P.: Constrained Optimization and Lagrange Multipliers Methods. Academic Press, New York (1982)
3. Carroll, C.W.: The created response surface technique for optimizing nonlinear restrained systems. Operations research 184, 9–169 (1961)
4. Carter, R.G., Gablonsky, J.M., Patrick, A., Kelley, C.T., Eslinger, O.J.: Algorithms for noisy problems in gas transmission pipeline optimization. Optimization and Engineering 2, 139–157 (2002)

5. Coello Coello, C.A.: Theoretical and numerical constraint-handling techniques used with evolutionary algorithms: a survey of the state of the art. Computer Methods in Applied Mechanics and Engineering 191, 1245–1287 (2002)
6. Costa, M.F.P., Fernandes, E.M.G.P.: Efficient solving of engineering design problems by an interior point 3-D filter line search method. In: Simos, T.E., Psihoyios, G., Tsitouras, C. (eds.) AIP Conference Proceedings, vol. 1048, pp. 197–200 (2008)
7. Courant, R.: Variational methods for the solution of problems of equilibrium and vibrations. Bulletin of the American Mathematical Society 49, 1–23 (1943)
8. Fiacco, A.V., McCormick, G.P.: Extensions of SUMT for nonlinear programming: equality constraints and extrapolation. Management Science 12(11), 816–828 (1966)
9. Finkel, D.E.: Global optimization with the DIRECT algorithm. PhD thesis, North Carolina state University (2005)
10. Finkel, D.E., Kelley, C.T.: Convergence Analysis of the DIRECT Algorithm. North Carolina State University: Center for Research in Scientific Computing, Raleigh (2004)
11. Gablonsky, J.M.: Modifications of the DIRECT algorithm. PhD Thesis, North Carolina State University, Raleigh, North Carolina (2001)
12. Henderson, S.G., Biller, B., Hsieh, M.-H., Shortle, J., Tew, J.D., Barton, R.R.: Extension of the direct optimization algorithm for noisy functions. In: Proceedings of the 2007 Winter Simulation Conference (2007)
13. Joines, J., Houck, C.: On the use of nonstationary penalty functions to solve nonlinear constrained optimization problems with GAs. In: Proceedings of the First IEEE Congress on Evolutionary Computation, Orlando, FL, pp. 579–584 (1994)
14. Jones, D.R.: The DIRECT Global Optimization Algorithm. The Encyclopedia of Optimization. Kluwer Academic (1999)
15. Lee, K.S., Geem, Z.W.: A new meta-heuristic algorithm for continuous engineering optimization: Harmony search theory and practice. Computer Methods in Applied Mechanics and Engineering 194, 3902–3933 (2005)
16. Lewis, R., Torczon, V.: A Globally Convergent Augmented Lagrangian Pattern Search Algorithm for Optimization with General Constraints and Simple Bounds. SIAM Journal on Optimization 4(4), 1075–1089 (2012)
17. Liu, J.-L., Lin, J.-H.: Evolutionary computation of unconstrained and constrained problems using a novel momentum-type particle swarm optimization. Engineering Optimization 39, 287–305 (2007)
18. Nocedal, J., Wright, S.: Numerical Optimization. Springer Series in Operations Research (1999)
19. Jones, D.R., Perttunen, C.D., Stuckman, B.E.: Lipschitzian optimization without the lipschitz constant. Journal of Optimization Theory and Application 79(1), 157–181 (1993)
20. Petalas, Y.G., Parsopoulos, K.E., Vrahatis, M.N.: Memetic particle swarm optimization. Annals of Operations Research 156, 99–127 (2007)
21. Ray, T., Liew, K.M.: A swarm metaphor for multiobjective design optimization. Engineering Optimization 34(2), 141–153 (2002)
22. Rocha, A.M.A.C., Fernandes, E.M.G.P.: Hybridizing the electromagnetism-like algorithm with descent search for solving engineering design problems. International Journal of Computer Mathematics 86, 1932–1946 (2009)
23. Tessema, B., Yen, G.G.: A Self Adaptive Penalty Function Based Algorithm for Constrained Optimization. In: IEEE Congress on Evolutionary Computation, pp. 246–253 (2006)

24. Vilaça, R., Rocha, A.M.A.C.: An Adaptive Penalty Method for DIRECT Algorithm in Engineering Optimization. In: Simos, T.E., Psihoyios, G., Tsitouras, C., Anastassi, Z. (eds.) AIP Conference Proceedings, vol. 1479, pp. 826–829 (2012)
25. Vilaça, R.: Sofia Pinto: Extending the DIRECT algorithm to solve constrained nonlinear optimization problems: a case study. MSc Thesis, University of Minho (2012)
26. Xavier, A.E.: Hyperbolic penalty: a new method for nonlinear programming with inequalities. International Transactions in Operational Research 8, 659–671 (2001)

Multilocal Programming:
A Derivative-Free Filter Multistart Algorithm

Florbela P. Fernandes[1,3], M. Fernanda P. Costa[2,3],
and Edite M.G.P. Fernandes[4]

[1] Polytechnic Institute of Bragança, ESTiG, 5301-857 Bragança, Portugal
fflor@ipb.pt
[2] Department of Mathematics and Applications, University of Minho,
4800-058 Guimarães, Portugal
mfc@math.uminho.pt
[3] Centre of Mathematics
[4] Algoritmi R&D Centre,
University of Minho,
4710-057 Braga, Portugal
emgpf@dps.uminho.pt

Abstract. Multilocal programming aims to locate all the local solutions
of an optimization problem. A stochastic method based on a multistart
strategy and a derivative-free filter local search for solving general constrained optimization problems is presented. The filter methodology is
integrated into a coordinate search paradigm in order to generate a set
of trial approximations that might be acceptable if they improve the
constraint violation or the objective function value relative to the current one. Preliminary numerical experiments with a benchmark set of
problems show the effectiveness of the proposed method.

Keywords: Multilocal programming, multistart, derivative-free, coordinate search, filter method.

1 Introduction

Multilocal programming has a wide range of applications in the engineering field
[8,10,18,19] and aims to compute all the global and local/non-global solutions of
constrained nonlinear optimization problems. The goal of most multistart methods presented in the literature is to locate multiple solutions of bound constrained
optimization problems [1,21,23,24] (see also [15] and the references therein included). Multistart may also be used to explore the search space and converge
to a global solution of nonlinear optimization problems [6]. When a multistart
strategy is implemented, a local search procedure is applied to randomly generated (sampled) points of the search space aiming to converge to the multiple
solutions of the problem. However, the same solutions may be found over and
over again. To avoid convergence to an already computed solution, some multistart methods use clustering techniques to define prohibited regions based on the

B. Murgante et al. (Eds.): ICCSA 2013, Part I, LNCS 7971, pp. 333–346, 2013.
© Springer-Verlag Berlin Heidelberg 2013

closeness to the previously located solutions. Sampled points from these prohibited regions are discarded since the local search procedure would converge most certainly to an already located solution. MinFinder is an example of a clustering algorithm that competes with multistart when global and some local minimizers are required [21,22]. Alternatively, niching, deflecting and stretching techniques may be combined with global optimization methods, like the simulated annealing, evolutionary algorithm and the particle swarm optimization, to discover the global and some specific local minimizers of a problem [17,18,20]. A glowworm swarm optimization approach has been proposed to converge to multiple optima of multimodal functions [13].

The purpose of this paper is to present a method based on a multistart technique and a derivative-free deterministic local search procedure to obtain multiple solutions of an optimization problem. The novelty here is that a direct search method and the filter methodology, as outlined in [3,7], are combined to construct a local search procedure that does not require any derivative information. The filter methodology is implemented to handle the constraints by forcing the local search towards the feasible region. Bound, as well as linear and nonlinear inequality and equality constraints may be treated by the proposed local search procedure.

Direct search methods are popular because they are straightforward to implement and do not use or approximate derivatives. Like the gradient-based methods, direct search methods also have their niche. For example, the maturation of simulation-based optimization has led to optimization problems in which derivative-free methods are mandatory. There are also optimization problems where derivative-based methods cannot be used since the objective function is not numerical in nature [11].

The problem to be addressed is of the following type:

$$\begin{aligned}
\min \ & f(x) \\
\text{subject to } & g_j(x) \leq 0, \quad j = 1, ..., m \\
& l_i \leq x_i \leq u_i, \ i = 1, ..., n
\end{aligned} \tag{1}$$

where, at least one of the functions $f, g_j : \mathbb{R}^n \longrightarrow \mathbb{R}$ is nonlinear and $F = \{x \in \mathbb{R}^n : g(x) \leq 0, \ l \leq x \leq u\}$ is the feasible region. Problems with general equality constraints can be reformulated in the above form by introducing $h(x) = 0$ as an inequality constraint $|h(x)| - \tau \leq 0$, where τ is a small positive relaxation parameter. This kind of problems may have many global and local optimal solutions and so, it is important to develop a methodology that is able to explore the entire search space and find all the minimizers guaranteeing, in some way, that convergence to a previously found minimizer is avoided.

This paper is organized as follows. In Section 2, the algorithm based on the multistart strategy and on the filter methodology is presented. In Section 3, we report the results of our numerical experiments with a set of benchmark problems. In the last section, conclusions are summarized and recommendations for future work are given.

2   Multistart Coordinate Search Filter Method

The methodology used to compute all the optimal solutions of problem (1), hereafter called MCSFilter method, is a multistart algorithm coupled with a clustering technique to avoid the convergence to previously detected solutions. The exploration feature of the method is carried out by a multistart strategy that aims at generating points randomly spread all over the search space. Exploitation of promising regions is made by a simple local search approach. In contrast to the line search BFGS method presented in [24], the local search proposal, a crucial procedure inside a multistart paradigm, relies on a direct search method, known as coordinate search (CS) method [11], that does not use any analytical or numerical derivative information.

Since the goal of the local search is to converge to a solution of a constrained optimization problem, started from a sampled approximation, progress towards an optimal solution is measured by a filter set methodology, as outlined in [7], which is integrated into the local search procedure. The filter methodology appears naturally from the observation that an optimal solution of the problem (1) minimizes both constraint violation and objective function [3,4,7,9]. Thus, the proposed CS method is combined with a (line search) filter method that aims at generating trial iterates that might be acceptable if they improve the constraint violation or the objective function relative to the current iterate.

2.1   The Multistart Strategy

Multistart is a stochastic algorithm that repeatedly applies a local search to sampled points (randomly generated inside $[l, u]$) $x_i = l_i + \lambda(u_i - l_i)$ for $i = 1, \ldots, n$, where λ is a random number uniformly distributed in $[0, 1]$, aiming to converge to the solutions of a multimodal problem. When a multistart strategy is applied to converge to the multiple solutions, some or all of the minimizers may be found over and over again. To avoid convergence to a previously computed solution, a clustering technique based on computing the regions of attraction of previously identified minimizers is to be integrated in the algorithm. The region of attraction of a local minimizer, y_i, associated with a local search procedure **L**, is defined as:

$$A_i \equiv \{x \in [l, u] : \mathbf{L}(x) = y_i\}, \tag{2}$$

where $\mathbf{L}(x)$ is the minimizer obtained when the local search procedure **L** starts at point x. The ultimate goal of a multistart algorithm is to invoke the local search procedure N times, where N is the number of solutions of (1). If a sampled point $x \in [l, u]$ belongs to a region of attraction A_j then the minimizer y_j would be obtained when **L** is started from x. Ideally, the local search procedure is to be applied only to a sampled point that does not belong to any of the regions of attraction of already computed minimizers, or equivalently to the union of those regions of attraction, since they do not overlap. However, computing the region of attraction A_i of a minimizer y_i is not an easy task. Alternatively, a stochastic procedure may be used to estimate the probability, p, that a sampled point will not belong to the union of the regions of attraction of already computed minimizers, i.e.,

$$p = Prob[x \notin \cup_{i=1}^{k} A_i] = \prod_{i=1}^{k} Prob[x \notin A_i] \approx Prob[x \notin A_n]$$

where A_n is the region of attraction of the nearest to x minimizer y_n (see details in [24]). The value of p may be approximated using $Prob[x \notin B(y_n, R_n)]$, where $B(y, R)$ represents a sphere centered at y with radius R.

Let the maximum attractive radius of the minimizer y_i be defined by:

$$R_i = \max_j \left\{ \left\| x_i^{(j)} - y_i \right\| \right\}, \tag{3}$$

where $x_i^{(j)}$ is one of the sampled points that led to the minimizer y_i. Given x, let $d_i = \|x - y_i\|$ be the distance of x to y_i. If $d_i < R_i$ then x is likely to be inside the region of attraction of y_i. However, if the direction from x to y_i is ascent then x is likely to be outside the region of attraction of y_i and the local search procedure is to be implemented started from x, since a new minimum could be computed with high probability. Thus, using these arguments, similarly to [24], the probability that $x \notin A_i$ is herein estimated by:

$$Prob(x \notin A_i) = \begin{cases} 1, & \text{if } z \geq 1 \text{ or the direction from } x \text{ to } y_i \text{ is ascent} \\ \varrho\, \phi(z,r), & \text{otherwise} \end{cases}$$
$$\tag{4}$$

where $\varrho \in [0, 1]$ is a fixed parameter, $z = d_i/R_i \in (0, 1)$, r is the number of times y_i has been recovered so far and $\phi(z, r) \in (0, 1)$ is taken as

$$\phi(z, r) = z \exp(-r^2(z - 1)^2), \text{ where } \lim_{z \to 0} \phi(z, r) \to 0, \lim_{z \to 1} \phi(z, r) \to 1,$$
$$\lim_{r \to \infty} \phi(z, r) \to 0.$$

In this derivative-free approach, the direction from x to y_i is considered ascent when $f(x + \beta(y_i - x)) - f(x) > 0$ for a small $\beta > 0$. In the gradient-based approach [24], ϱ is a function that depends on the directional derivative of f along the direction from x to y_i.

Figure 1 illustrates the behavior of the multistart method when converging to four minimizers. The point represented by a ⋆ (in black) is the first sampled point that converges to a minimizer. The points represented by ◇ (in full blue) lie outside the region of attraction and thus the local search procedure is applied to be able to converge to a minimizer. For example, the global minimizer (leftmost and bottom of the figure) has been recovered seven times. The other points (represented by ◦) are sampled points inside the region of attraction of a minimizer. They have been discarded, i.e., the local search has not been applied to them.

Algorithm 1 shows the multistart algorithm. Although this type of methods is simple, they would not be effective if a bad stopping rule is used. The main goal of a stopping rule is to make the algorithm to stop when all minimizers have

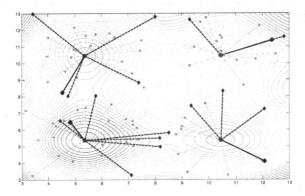

Fig. 1. Illustration of the multistart method

been located with certainty. Further, it should not require a large number of local searches to decide that all minimizers have been found (see [14]). A simple stopping rule uses the estimate of the fraction of uncovered space $P(k) = \frac{k(k+1)}{t(t-1)}$, where k is the number of recovered minimizers after having performed t local search procedures. The multistart algorithm then stops if $P(k) \leq \epsilon$, for a small $\epsilon > 0$.

2.2   The Coordinate Search Filter Procedure

The coordinate search filter (CSFilter) method, combining a derivative-free paradigm with the filter methodology, is proposed as the local search procedure **L**, to compute a minimizer y starting from a sampled point $x \in [l, u]$. Briefly, a minimizer y of the constrained optimization problem (1) is to be computed, starting from x. The basic idea behind this approach is to interpret (1) as a bi-objective optimization problem aiming to minimize both the objective function $f(x)$ and a nonnegative continuous aggregate constraint violation function $\theta(x)$ defined by

$$\theta(x) = \|g(x)_+\|^2 + \|(l - x)_+\|^2 + \|(x - u)_+\|^2 \qquad (5)$$

where $v_+ = \max\{0, v\}$. Therefore, the proposed CSFilter approach computes an approximate minimizer, y, to the bi-objective optimization problem

$$\min_x \left(\theta(x), f(x)\right). \qquad (6)$$

The filter technique incorporates the concept of nondominance, present in the field of multi-objective optimization, to build a filter that is able to accept trial approximations if they improve the constraint violation or objective function value. A filter $\mathcal{F}$ is a finite set of points y, corresponding to pairs $(\theta(y), f(y))$, none of which is dominated by any of the others. A point y is said to dominate a point y' if only if $\theta(y) \leq \theta(y')$ and $f(y) \leq f(y')$.

Algorithm 1. Multistart algorithm

Require: Parameter values; set $Y^* = \emptyset^\dagger$, $k = 1$, $t = 1$;
1: Randomly generate $x \in [l, u]$; compute $A_{\min} = \min_{i=1,\ldots,n}\{u_i - l_i\}$;
2: Compute $y_1 = \mathbf{L}(x)$, $R_1 = \|x - y_1\|$; set $r_1 = 1$, $Y^* = Y^* \cup y_1$;
3: **repeat**
4:     Randomly generate $x \in [l, u]$;
5:     Set $o = \arg\min_{j=1,\ldots,k} d_j \equiv \|x - y_j\|$;
6:     **if** $d_o < R_o$ **then**
7:         **if** the direction from x to y_o is ascent **then**
8:             Set $p = 1$;
9:         **else**
10:             Compute $p = \varrho\,\phi(\frac{d_o}{R_o}, r_o)$;
11:         **end if**
12:     **else**
13:         Set $p = 1$;
14:     **end if**
15:     **if** $\zeta^\ddagger < p$ **then**
16:         Compute $y = \mathbf{L}(x)$; set $t = t + 1$;
17:         **if** $\|y - y_j\| > \gamma^* A_{\min}$, for all $j = 1, \ldots, k^\S$ **then**
18:             Set $k = k + 1$, $y_k = y$, $r_k = 1$, $Y^* = Y^* \cup y_k$; compute $R_k = \|x - y_k\|$;
19:         **else**
20:             Set $R_l = \max\{R_l, \|x - y_l\|\}^\natural$; $r_l = r_l + 1$;
21:         **end if**
22:     **else**
23:         Set $R_o = \max\{R_o, \|x - y_o\|\}$; $r_o = r_o + 1$;
24:     **end if**
25: **until** the stopping rule is satisfied

---

† - Y^* is the set containing the computed minimizers.
‡ - ζ is a uniformly distributed number in $(0, 1)$.
§ - $y \notin Y^*$.
$^\natural$ - $\|y - y_l\| \leq \gamma^* A_{\min}$.

A rough outline of a coordinate search filter is as follows. At the beginning of the optimization, the filter is initialized to $\mathcal{F} = \{(\theta, f) : \theta \geq \theta_{\max}\}$, where $\theta_{\max} > 0$ is an upper bound on the acceptable constraint violation.

Let $\mathcal{D}_\oplus$ denote de set of $2n$ coordinate directions, defined as the positive and negative unit coordinate vectors, $\mathcal{D}_\oplus = \{e_1, e_2, \ldots, e_n, -e_1, -e_2, \ldots, -e_n\}$. The search begins with a central point, the current approximation $\tilde{x}$, as well as $2n$ trial approximations $y_c^i = \tilde{x} + \alpha d_i$, for $d_i \in \mathcal{D}_\oplus$, where $\alpha > 0$ is a step size. The constraint violation value and the objective function value of all $2n + 1$ points are computed. If some trial approximations improve over $\tilde{x}$, reducing θ or f by a certain amount (see (7) and (8) below), and are acceptable by the filter, then the best of these non-dominated trial approximations, y_c^{best}, is selected, and the filter is updated (adding the corresponding entries to the filter and removing any dominated entries). Then, this best approximation becomes the new central point in the next iteration, $\tilde{x} \leftarrow y_c^{best}$. If, on the other hand, all trial approximations y_c^i are dominated by the current filter, then all y_c^i are rejected, and a restoration phase is invoked.

To avoid the acceptance of a point y_c^i, or the corresponding pair $\left(\theta(y_c^i), f(y_c^i)\right)$, that is arbitrary close to the boundary of $\mathcal{F}$, the trial y_c^i is considered to improve over $\tilde{x}$ if one of the conditions

$$\theta(y_c^i) \leq (1 - \gamma_\theta)\,\theta(\tilde{x}) \text{ or } f(y_c^i) \leq f(\tilde{x}) - \gamma_f\,\theta(\tilde{x}) \tag{7}$$

holds, for fixed constants $\gamma_\theta, \gamma_f \in (0,1)$.

However, the filter alone cannot ensure convergence to optimal points. For example, if a sequence of trial points satisfies $\theta(y_c^i) \leq (1 - \gamma_\theta)\,\theta(\tilde{x})$ then it could converge to an arbitrary feasible point. Therefore, when $\tilde{x}$ is nearly feasible, $\theta(\tilde{x}) \leq \theta_{\min}$ for a small positive $\theta_{\min}$, the trial approximation y_c^i has to satisfy only the condition

$$f(y_c^i) \leq f(\tilde{x}) - \gamma_f\,\theta(\tilde{x}) \tag{8}$$

instead of (7), in order to be acceptable.

The best non-dominated trial approximation is selected as follows. The best point y_c^{best} of a set $Y = \{y_c^i : y_c^i = \tilde{x} + \alpha d_i, d_i \in \mathcal{D}_\oplus\}$ is the point that satisfies one of two following conditions:

- if there are some feasible points in Y, y_c^{best} is the point that has the less objective function value among the feasible points:

$$\theta\left(y_c^{best}\right) = 0 \text{ and } f\left(y_c^{best}\right) < f\left(y_c^i\right) \text{ for all } y_c^i \in Y \text{ such that } \theta\left(y_c^i\right) = 0; \tag{9}$$

- otherwise, y_c^{best} is the point that has less constraint violation among the non-dominated infeasible points

$$0 < \theta\left(y_c^{best}\right) < \theta\left(y_c^i\right) \text{ and } y_c^i \notin \mathcal{F}. \tag{10}$$

We remark that the filter is updated whenever the trial approximations y_c^i verify conditions (7) or (8) and are non-dominated.

When it is not possible to find a non-dominated best trial approximation, and before declaring the iteration unsuccessful, a restoration phase is invoked. In this phase, the most nearly feasible point in the filter, $x_{\mathcal{F}}^{inf}$, is recuperated and the search along the $2n$ coordinate directions is carried out about it to find the set $Y = \{y_c^i : y_c^i = x_{\mathcal{F}}^{inf} + \alpha d_i, d_i \in \mathcal{D}_\oplus\}$. If a non-dominated best trial approximation is found, this point becomes the central point of the next iteration and the iteration is successful. Otherwise, the iteration is unsuccessful, the search returns back to the current $\tilde{x}$, the step size is reduced, for instance $\alpha = \alpha/2$, and new $2n$ trial approximations y_c^i are generated about it. If a best non-dominated trial approximation is still not found, the step size reduction is repeated since another unsuccessful iteration has occurred. When α falls below $\alpha_{\min}$, a small positive tolerance, the search terminates since first-order convergence has been attained [11]. At each unsuccessful iteration, the CSFilter algorithm reduces the step size and tries again the coordinate search about the current point $\tilde{x}$.

Thus, to judge the success of the CSFilter algorithm, the below presented conditions are applied

$$\begin{cases} \alpha \leq \alpha_{\min} \\ \theta(y_c^{best}) < 0.01\,\theta_{\min} \\ \left|f(y_c^{best}) - f(y)\right| < 10^{-6}\left|f(y)\right| + 10^{-8}, \end{cases} \tag{11}$$

where $0 < \alpha_{\min} << 1$ and y is the previous current point. The proposed algorithm for the local procedure is presented in Algorithm 2.

Algorithm 2. CSFilter algorithm

Require: x (sampled in the Multistart algorithm) and parameter values; set $\tilde{x} = x$, $x_{\mathcal{F}}^{inf} = x$, $y = \tilde{x}$;

1: Initialize the filter;
2: Set $\alpha = \min\{1, 0.05\frac{\sum_{i=1}^{n} u_i - l_i}{n}\}$;
3: **repeat**
4:    Compute the trial approximations $y_c^i = \tilde{x} + \alpha d_i$, for all $d_i \in \mathcal{D}_\oplus$;
5:    **repeat**
6:       Check acceptability of trial points y_c^i using (7) and (8);
7:       **if** there are some y_c^i acceptable by the filter **then**
8:          Update the filter;
9:          Choose y_c^{best} using (9) or (10);
10:         Set $y = \tilde{x}$, $\tilde{x} = y_c^{best}$; update $x_{\mathcal{F}}^{inf}$ if appropriate;
11:      **else**
12:         Compute the trial approximations $y_c^i = x_{\mathcal{F}}^{inf} + \alpha d_i$, for all $d_i \in \mathcal{D}_\oplus$;
13:         Check acceptability of trial points y_c^i using (7) and (8);
14:         **if** there are some y_c^i acceptable by the filter **then**
15:            Update the filter;
16:            Choose y_c^{best} using (9) or (10);
17:            Set $y = \tilde{x}$, $\tilde{x} = y_c^{best}$; update $x_{\mathcal{F}}^{inf}$ if appropriate;
18:         **else**
19:            Set $\alpha = \alpha/2$;
20:         **end if**
21:      **end if**
22:   **until** new trial y_c^{best} is acceptable
23: **until** the conditions (11) are satisfied

---

3   Numerical Results

To analyze the performance of the MCSFilter algorithm, a set of 30 test problems is used (see Table 1). The set contains bound constrained problems, inequality and equality constrained problems, multimodal objective functions, with one global and some local, more than one global, and a unimodal optimization problem. Table 1 reports the acronym of the tested problems, under 'Prob.', references with details of the models and the known number of solutions, 'Min'. Five minimization problems, 2Dt+1, 2Dt+2, MMO+1, CB6+1 and BR+1 are defined from well-known problems by adding constraints: 2Dt+1 comes from 2Dt by adding the constraint $(x_1 + 5)^2 + (x_2 - 5)^2 - 100 \leq 0$; 2Dt+2 is 2Dt+1 with an additional linear constraint $-x_1 - x_2 - 3 \leq 0$; MMO+1 comes from MMO by adding the linear constraint $-2x_1 - 3x_2 + 27 \leq 0$; CB6+1 comes from CB6 with the constraint $(x_1 + 1)^2 + (x_2 - 1)^2 \leq 2.25$; BR+1 comes from BR with the additional constraint $(x_1 - 5)^2 + 2(x_2 - 10)^2 \leq 100$. Problem g8 is a maximization problem that was rewritten as a minimization problem. g11 has an equality constraint which was transformed into an inequality constraint using $\tau = 10^{-5}$.

Table 1. Problem, reference and known number of solutions

Prob.		Min	Prob.		Min	Prob.		Min
ADJ	[8]	3	SHK10	[2,14,16]	10	MMO+1	[5]	4
CB6	[2,8,14]	6	2Dt	[14,18]	4	CB6+1		4
BR	[2,14,18]	3	3Dt	[14,18]	8	BR+1		3
GP	[2,8,14]	4	4Dt	[14,18]	16	g8	[9,12,18]	2
H3	[2,8,14]	3	5Dt	[14,18]	32	g9	[9,18]	1
H6	[2,8,14]	2	6Dt	[14,18]	64	g11	[9]	2
MMO	[18]	4	8Dt	[14,18]	256	EX. 2.2	[10]	2
SBT	[16]	760	10Dt	[14,18]	1024	EX. 3.3	[10]	2
SHK5	[14,16]	5	2Dt+1	[5]	4	EX. 6.17	[10]	4
SHK7	[14,16]	7	2Dt+2	[5]	5	EX. 1	[19]	2

The MCSFilter method was coded in MatLab and the results were obtained in a PC with an Intel Core i7-2600 CPU (3.4GHz) and 8 GB of memory. In the CSFilter method, we set after an empirical study: $\gamma_\theta = \gamma_f = 10^{-5}$, $\alpha_{min} = 10^{-5}$, $\theta_{min} = 10^{-3}$, $\theta_{max} = 10^3 \max\{1, 1.25\theta(x_{in})\}$, where x_{in} is the initial point in the local search. We also set $\varrho = 0.5$, $\beta = 0.001$, $\gamma^* = 0.1$ and $\epsilon = 0.1$ in the stopping rule of the multistart algorithm. Each problem was solved 10 times and the average values are reported.

Table 2 contains the results obtained when solving the bound constrained problems, where the columns show:

- the average number of computed minimizers, 'Min$_{av}$';
- the average number of function evaluations, 'nfe$_{av}$';
- the average time (in seconds) 'T$_{av}$'.

For comparative purposes, Table 2 also reports:

(i) in columns 5–7, the results presented in [18], relative to the problems BR and nDt with $n = 2, 4, 6, 8, 10$;

(ii) in columns 8–9, the results presented in [14], relative to the problems CB6, BR, GP, H3, H6, SHK5, SHK7, SHK10 and nDt with $n = 4, 5, 6$.

Table 2. Numerical results obtained with bound constrained problems

Prob. (n)	MCSFilter algorithm			results in [18]			results in [14]	
	Min_{av}	nfe_{av}	T_{av}	Min_{av}	nfe_{av}	T_{av}	Min_{av}	nfe_{av}
ADJ (2)	2.5	5768	27.970	-	-	-	-	-
CB6 (2)	6[†]	1869.1	0.262	-	-	-	6	5642
BR (2)	3[†]	1571.1	0.221	3	2442	0.45	3	2173
GP (2)	4[†]	13374.9	2.021	-	-	-	4	5906
H3 (3)	2.9	2104.3	0.313	-	-	-	3	3348
H6 (6)	2[†]	6559.2	0.840	-	-	-	2	3919
MMO (2)	4[†]	1328.3	0.200	-	-	-	-	-
SBT (2)	25.2	9276.2	5.738	-	-	-	-	-
SHK5 (4)	4.6	6240.3	0.782	-	-	-	5	8720
SHK7 (4)	6.4	8335.2	1.036	-	-	-	7	11742
SHK10 (4)	8.6	10312.6	1.293	-	-	-	10	16020
2Dt (2)	4[†]	1372.6	0.193	2	1067	0.17	-	-
3Dt (3)	8[†]	3984.4	0.521	-	-	-	-	-
4Dt (4)	16[†]	11718.5	1.471	2	3159	0.29	16	17373
5Dt (5)	31.9	32881.7	4.373	-	-	-	32	37639
6Dt (6)	63.8	102490.3	16.105	2	10900	0.75	64	81893
8Dt (8)	254.3	659571.6	368.674	1	36326	2.28	-	-
10Dt (10)	1016	3863756	18563	1	58838	3.71	-	-

[†] - All the minimizers were computed in all runs.

- Not available.

The algorithm presented in [18] implements a function stretching technique combined with a simulated annealing approach, known as SSA method. We observe that the MCSFilter algorithm has a good performance and is able to find almost all minimizers of the problems with acceptable number of function evaluations. The algorithm finds all minimizers in all runs when solving eight of the 18 tested bound constrained problems and an average of 85% of the minimizers in the remaining ones (94% when the problem SBT is excluded from these statistics). In terms of time needed to find all the solutions, the worst cases are observed with the problems 8Dt and 10Dt. An average of 1.45 seconds and 2593.7 function evaluations are required to compute each solution of problem 8Dt, and an average of 18.3 seconds, with an average of 3802.9 function evaluations, to compute each solution in 10Dt. We remark that the global minimum has been always identified in all runs and in all problems of Table 2. We observe from the comparison with the results of [14] that MCSFilter is slightly more efficient although the multistart method implemented in [14] seems to retrieve a greater number of minimizers.

When a comparison is made with the number of solutions reported in [18], we find that for problems 2Dt and 4Dt, our method finds all the minimizers while SSA identifies only two in each problem. The average number of function evaluations and time to locate each minimizer required by the MCSFilter algorithm for problems 6Dt and 8Dt are smaller than those of the SSA method. Furthermore, MCSFilter finds almost all the minimizers, while SSA finds only two and one respectively. For problem BR, both methods obtained the same number of minimizers, although MCSFilter requires a smaller number of function evaluations and time than SSA. Results for the problem MMO, with 50 and 100 variables are presented in [18], with 4 and 6 found minimizers respectively. Their results were obtained after 1000000 function evaluations. For comparative purposes, we implemented this condition to stop the MCSFilter algorithm and obtain the following results: $Min_{av}=48$, when $n = 50$, and $Min_{av}=14$, when $n = 100$. In both cases MCSFilter method finds more minimizers than SSA. A total of 100000 function evaluations were used by SSA (in [18]) to find one minimizer of problem 10Dt. Using the same condition to stop MCSFilter, we obtain $Min_{av}=85.3$. We note that the problems of the set nDt have 2^n minimizers, where n is the number of variables. We may conclude that the MCSFilter method consistently finds more minimizers.

Table 3. Results obtained with inequality and equality constrained problems

Prob. (n, m)	MCSFilter algorithm				other results		
	Min_{av}	nfe_{av}	T_{av}		Min_{av}	nfe_{av}	T_{av}
2Dt+1 (2,1)	3.9	10127.6	7.986	[5]	3.8	9752.5	51.6
2Dt+2 (2,2)	4.6	28065.4	24.242	[5]	3.1	13417.4	69.9
MMO+1 (2,1)	$4^\dagger$	1858.5	0.527	[5]	4	7630.1	8.8
CB6+1 (2,1)	3.4	12319.1	40.626				
BR+1 (2,1)	$3^\dagger$	4128.9	2.525				
g8 (2,2)	1	1930	3.087	[6]	$1^\S$	4999	-
				[9]	$1^\S$	56476	-
				[18]	5	67753	-
g9 (7,4)	$1.4^\ddagger$	5767.7	0.737	[6]	$1^\S$	38099	-
				[9]	$1^\S$	324569	-
				[18]	1	183806	-
g11 (2,1)	2	84983.3	191.629	[6]	$1^\S$	139622	-
				[9]	$1^\S$	23722	-
EX. 2.2 (2,2)	$2^\dagger$	2469.2	2.315				
EX. 3.3 (2,1)	1.7	6435.9	9.403				
EX. 6.17 (2,3)	3.7	53292.5	141.114				
EX. 1 (2,1)	$2.1^\ddagger$	65133.0	593.220				

$\dagger$ - All the minimizers were computed in all runs.

$\ddagger$ Some cases of premature convergence have been observed, thus identifying a new minimizer.

$\S$ Only one global minimizer was required to be found.

- Not available.

Table 3 lists the results obtained with inequality and equality constrained problems. A comparison is made with the results reported:

(i) in [5,6], where a multistart method coupled with a stochastic approach to derive approximate descent directions and a filter technique is used;
(ii) in [9], which implements a filter simulated annealing method;
(iii) in [18], which implements the SSA method with a penalty function technique.

We may conclude that the presented MCSFilter has a good performance since the average number of function evaluations required to locate each minimizer is smaller than those of the other methods in comparison. We also observe that the algorithm finds all minimizers in all runs when solving MMO+1, BR+1 and EX. 2.2 and an average of 94% of the minimizers in the remaining nine problems. Thus, we may conclude that MCSFilter is able to consistently find almost all minimizers within a reduced time.

4   Conclusions

We present a multistart algorithm based on a derivative-free filter method to solve multilocal programming problems. The method is based on a multistart strategy which relies on the concept of regions of attraction in order to avoid convergence to previously found local minimizers. The proposal for the local procedure is a coordinate search combined with a filter method to generate a sequence of approximate solutions that improve either the constraint violation or the objective function relative to the previous approximation. A set of benchmark problems was used to test the algorithm and the results are very promising. One problematic issue of the proposed MCSFilter method is related to the large number of search directions in the set $\mathcal{D}_{\oplus}$. For large dimensional problems, the computational effort in terms of number of function evaluations and consequently CPU time greatly increases with n. We have observed that the proposed method consistently locates all or almost all minimizers of a problem. To improve the effectiveness of the algorithm, a stopping rule that balances the number of estimated minimizers with local search calls is to be devised in the future.

Acknowledgments. This work was financed by FEDER funds through COM-PETE (Operational Programme Thematic Factors of Competitiveness) and by portuguese funds through FCT (Foundation for Science and Technology) within the projects FCOMP-01-0124-FEDER-022674 and PEst-C/MAT/UI0013/2011.

References

1. Ali, M.M., Gabere, M.N.: A simulated annealing driven multi-start algorithm for bound constrained global optimization. J. Comput. Appl. Math. 233, 2661–2674 (2010)

2. Ali, M.M., Khompatraporn, C., Zabinsky, Z.B.: A numerical evaluation of several stochastic algorithms on selected continuous global optimization test problems. J. Glob. Optim. 31, 635–672 (2005)
3. Audet, C., Dennis Jr., J.E.: A pattern search filter method for nonlinear programming without derivatives. SIAM J. Optimiz. 14(4), 980–1010 (2004)
4. Costa, M.F.P., Fernandes, E.M.G.P.: Assessing the potential of interior point barrier filter line search methods: nonmonotone versus monotone approach. Optimization 60(10-11), 1251–1268 (2011)
5. Fernandes, F.P., Costa, M.F.P., Fernandes, E.M.G.P.: Stopping rules effect on a derivative-free filter multistart algorithm for multilocal programmnig. In: ICACM 2012, 6 p. (2012), file:131-1395-1-PB.pdf,
   http://icacm.iam.metu.edu.tr/all-talks
6. Fernandes, F.P., Costa, M.F.P., Fernandes, E.M.G.P.: A derivative-free filter driven multistart technique for global optimization. In: Murgante, B., Gervasi, O., Misra, S., Nedjah, N., Rocha, A.M.A.C., Taniar, D., Apduhan, B.O. (eds.) ICCSA 2012, Part III. LNCS, vol. 7335, pp. 103–118. Springer, Heidelberg (2012)
7. Fletcher, R., Leyffer, S.: Nonlinear programming without a penalty function. Math. Program. 91, 239–269 (2002)
8. Floudas, C.A., Pardalos, P.M., Adjiman, C.S., Esposito, W.R., Gumus, Z.H., Harding, S.T., Klepeis, J.L., Meyer, C.A., Schweiger, C.A.: Handbook of Test Problems in Local and Global Optimization. Kluwer Academic Publishers (1999)
9. Hedar, A.R., Fukushima, M.: Derivative-Free Filter Simulated Annealing Method for Constrained Continuous Global Optimization. J. Glob. Optim. 35, 521–549 (2006)
10. Hendrix, E.M.T., G.-Tóth, B.: Introduction to Nonlinear and Global Optimization. Springer Optimization and Its Applications, vol. 37 (2010)
11. Kolda, T.G., Lewis, R.M., Torczon, V.: Optimization by Direct Search: New Perspectives on Some Classical and Moddern Methods. SIAM Rev. 45(3), 385–482 (2003)
12. Koziel, S., Michalewicz, Z.: Evolutionary algorithms, homomorphous mappings, and constrained parameter optimization. Evol. Comput. 7(1), 19–44 (1999)
13. Krishnanand, K.N., Ghose, D.: Glowworm swarm optimization for simultaneous capture of multiple local optima of multimodal functions. Swarm Intell. 3, 87–124 (2009)
14. Lagaris, I.E., Tsoulos, I.G.: Stopping rules for box-constrained stochastic global optimization. Appl. Math. Comput. 197, 622–632 (2008)
15. Marti, R.: Multi-start methods. In: Glover, F., Kochenberger, G. (eds.) Handbook of Metaheuristics, pp. 355–368. Kluwer Academic Publishers (2003)
16. Ozdamar, L., Demirhan, M.: Experiments with new stochastic global optimization search techniques. Comput. Oper. Res. 27, 841–865 (2000)
17. Parsopoulos, K.E., Vrahatis, M.N.: On the computation of all global minimizers through particle swarm optimization. IEEE T. Evolut. Comput. 8(3), 211–224 (2004)
18. Pereira, A., Ferreira, O., Pinho, S.P., Fernandes, E.M.G.P.: Multilocal Programming and Applications. In: Zelinka, I., et al. (eds.) Handbook of Optimization. Intelligent Systems Series, pp. 157–186. Springer (2013)
19. Ryoo, H.S., Sahinidis, N.V.: Global optimization of nonconvex NLPs and MINLPs with applications in process design. Comput. Chem. Eng. 19(5), 551–566 (1995)
20. Singh, G., Deb, K.: Comparison of multi-modal optimization algorithms based on evolutionary algorithms. In: GECCO 2006, pp. 1305–1312. ACM Press (2006)

21. Tsoulos, I.G., Lagaris, I.E.: MinFinder: Locating all the local minima of a function. Computer Phys. Com. 174, 166–179 (2006)
22. Tsoulos, I.G., Stavrakoudis, A.: On locating all roots of systems of nonlinear equations inside bounded domain using global optimization methods. Nonlinear Anal. Real 11, 2465–2471 (2010)
23. Tu, W., Mayne, R.W.: Studies of multi-start clustering for global optimization. Int. J. Numer. Meth. Eng. 53(9), 2239–2252 (2002)
24. Voglis, C., Lagaris, I.E.: Towards "Ideal Multistart". A stochastic approach for locating the minima of a continuous function inside a bounded domain. Appl. Math. Comput. 213, 1404–1415 (2009)

Performance Evaluation of Flooding Schemes on Duty-Cycled Sensor Networks: Conventional, 1HI, and 2HBI Floodings

Boram Hwang[1], Minhan Shon[1], Mihui Kim[2], Dongsoo S. Kim[1],
and Hyunseung Choo[3,*]

[1] College of Information and Communication Engineering, Sungkyunkwan University, Korea
{boramhw,minari95,dskim61}@skku.edu
[2] Department of Computer Engineering, Hankyong National University, Korea
mhkim@hknu.ac.kr
[3] Department of Interaction Science Sungkyunkwan University, Korea
choo@ece.skku.ac.kr

Abstract. The 1-Hop Information flooding (1HI) and 2-Hop Backward Information flooding (2HBI) schemes for choosing the retransmission nodes have been studied to solve the broadcast storm Problem for data flooding in wireless sensor networks (WSNs). These schemes do not consider duty cycles to save the energy of sensor node. Recently duty cycle approach is very commonly used as it performs well to save the energy of sensor nodes We analyze the performance of Conventional flooding, 1HI and 2BHI with duty cycles.In the well-known test environment, the result with 30% duty cycle shows that 2HBI scheme can reduce energy consumption up to 65%, increase the number of floodings up to 2 times, and increase the flooding duration up to 6 times. The comparison shows that 2HBI scheme takes the longest flooding time with duty cycle approach, where as it takes the shortest flooding time in non-duty cycled networks. Through various experiments, this paper suggests that proper schemes for various deployment conditions.

Keywords: Wireless Sensor Networks, Flooding, Duty Cycle, Delay Time, Energy Consumption.

1 Introduction

In Wireless Sensor Networks (WSNs), redundancy and collision are the problems of fundamental flooding operation [1]. 1HI scheme and 2HBI scheme are proposed to solve those problems and these two schemes also manage the energy consumption efficiently [2-3]. The efficient energy consumption is a important issue because sensor node has a limited energy [4]. Unfortunately, 1HI and 2HBI scheme have not been studied on duty cycled WSNs. At recently applying duty cycle research is progressing

---

* Corresponding author.

B. Murgante et al. (Eds.): ICCSA 2013, Part I, LNCS 7971, pp. 347–357, 2013.
© Springer-Verlag Berlin Heidelberg 2013

activity, duty cycled flooding scheme study is needed. This paper analyses the impact of duty cycle on several flooding schemes, and then suggests a suitable flooding scheme and a reasonable duty cycled value for each particular environment.

To analyses the impact of duty cycle on the conventional flooding, 1HI flooding, 2HBI flooding schemes, the performance of these schemes in 30%, 50%, and 80% duty cycled scenarios are compared with those of the schemes in non-duty-cycled environment, i.e. all sensor nodes always active. Trivially, 100% duty-cycled is equivalent with non-duty-cycled. All sensor nodes have the same transmission range. We deploy the sensor nodes randomly and use the Mica2 energy mode in the analysis environment. The data arrive every 60 time unit and the experiments are finished until a sensor node depletes its energy. The number of floodings, the energy consumption, and the flooding time is measured to analysis. The number of floodings represents how many flooding are from the time when the first data arriving in the network to the time when one sensor node uses its all energy. It is used to measure the workload of the sensor nodes. The energy consumption is measured to check how much energy consumed by the sensor nodes in duty-cycled environment. Flooding time is used to measure the delay time to flood a message in the network, i.e. all sensor nodes receive the broadcast message.

According to the result, three flooding schemes cause the reduction of energy consumption, the increment of number of flooding and flooding time. According to result of analysis, 64.8% reduction in energy consumption, 1.9% increment in the number of floodings, the flooding time of 2HBI is increased 6 times in 30% duty-cycled environment. In Conventional schemes, 1.4% increment in the number of flooding, 45.1% reduction in energy consumption, and 2.8 times increment in flooding time in 30% duty-cycled scenario. In 1HI schemes, 1.7% increment in the number of flooding, 60.2% reduction in energy consumption, and 2.9 times increment in flooding time in 30% duty-cycled scenario. In 2HBI schemes, 1.9% increment in the number of flooding, 64.79% reduction in energy consumption, and 6 times increment in flooding time in 30% duty-cycled environment. Especially, 2HBI scheme takes the longest flooding time with duty cycle approach, and it takes the shortest flooding time in non-duty cycled networks. The suitable flooding scheme with duty cycle for each environment in WSNs is follow. The energy consumption, network lifetime, and the number of floodings are the important factors for the environment. 2HBI scheme is suitable for the non-duty-cycled networks, in which the important factor of the environment is the flooding time.

In chapter 2, related flooding schemes are explained. Chapter 3 introduces the performance evaluation method and chapter 4 analyses the results. Finally, a conclusion is presented in chapter 5.

2   Related Work

2.1   Conventional Flooding

Conventional flooding is a traditional scheme and a simple protocol [2, 5]. In this scheme, a sensor node broadcasts data to all neighbor nodes in its transmission range. After receiving the data, the node repeats the data until the destination of the data is the node itself or the data is reached the maximum hop [1]. If the sensor node receives the data which it sent, it stops the repeating. Conventional flooding does not require maintain costly topology and complex algorithms.

The flooding scheme has implosion and overlap problems. Implosion means that the duplicate messages are sent because sensor node broadcasts the data to all neighbor nodes Overlap means that more than two sensor nodes sensing the same data from same area [1, 5]. Sensor node cannot use their energy well because of above problems [6].

2.2   1HI Flooding

The 1HI scheme was proposed to reduce the number of retransmission nodes. When a source node generates the data in the network, it attaches a forwarding set that includes retransmission nodes to the data. The sensor node receives the data and then checks whether or not the first. If the sensor node has already received the data, the data is discarded. Nevertheless if the sensor node is the first to receive the data, it checks the node list in forwarding set. If there is a sensor node ID in the forwarding set, the sensor node makes a forwarding set. For the optimization step, sensor node ID and geographical information of one hop neighbors are used. The receiver node optimizes its forwarding set by removing the nodes covered by the sender node and the node which its transmission range is covered by other nodes. This scheme is a simple protocol and incurs little overhead. Moreover, it uses energy efficiently, since it reduces the number of retransmission and collisions more effectively than the conventional flooding scheme. However, the number of retransmission nodes is still high, because this scheme uses only 1-hop information to optimize the forwarding set.

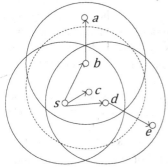

Consider the example in Fig. 1. The source node s makes its own forwarding set, and then broadcasts the data with the forwarding set. Nodes b, c, and d exist in the transmission range of source node. However, node c does not send the data, since the transmission coverage is covered by nodes s, b, and d. Nodes b and d are in the forwarding set of source node s. The data is received for the first time by nodes b, c, and d. These nodes check the forwarding set in data. If there is its own ID, the node optimize their own forwarding sets, and broadcast the data with the forwarding sets. However, node c dose not broadcast, since the ID is not in the forwarding set. In conventional flooding, nodes b, c, and d broadcast the data, but the 1HI scheme reduces energy consumption by choosing the retransmission node.

Fig. 1. An example of 1HI Flooding

2.3   2HBI Flooding

The 2HBI scheme uses 2-hop backward information to reduce the number of retransmission nodes, which allows nodes to save more energy than 1HI. In this scheme, it is assumed that all sensor nodes have the same transmission range, a unique ID, and know the two-hop backward information regarding sensor node ID, geographical information, and the forwarding set. The 2-hop backward information is this. For example, let a node i send data to a node j, which is a 1-hop backward node. If a node k sent the data to node i, node k is a 2-hop backward node of node j. The 2HBI scheme optimizes the forwarding set, and sends the data and forwarding set as in the 1HI scheme. All sensor nodes optimize the forwarding set based on the following rules. Rule 1: the retransmission node removes some nodes from its forwarding set that have already been covered by its sender node. Rule 2: if a node is in an overlapping zone, the retransmission node that is close to the sender removes the nodes from its forwarding set. Rule 3: based on the forwarding set information of the 2-hop backward node, the retransmission node removes the nodes that are in the neighbor coverage area of the 2 hop backward node from the forwarding set.

An example is shown in Fig. 2, sensor node s generates the data and broadcasts it and its forwarding set. Nodes a, b, and c broadcast the data after optimizing their own forwarding set and incorporating it into the data. The step of node f receiving the optimization forwarding set from node b is as follows. The initial forwarding set of node f is $F(f) = \{b, e, g, j\}$. Node f removes the node e, which received data from a sender node based on rule 1. Then, the forwarding set $F(f) = \{g, j\}$. Node j is in an overlapping zone between nodes f and d, and it is removed from the forwarding set, since it is in the coverage area of node d, which is farther from the sender node b. Based on rule 2. The forwarding set of node f is $F(f) = \{g\}$. Node g is in the coverage area of the previous-hop retransmission node s (which is a next-hop retransmission node of 2-hop backward node of node f). It is removed from the forwarding set based on rule 3. Then, the forwarding set of node f contains nothing. Finally, the forwarding set is $F(f) = \{\}$, and this node does not the retransmission. This scheme efficiently optimizes the forwarding set, which allows sensor nodes to save energy by reducing the number of transmissions, receiving, and collisions.

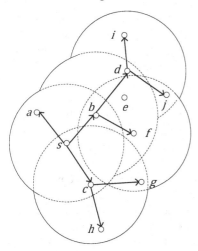

Fig. 2. An example of 2HBI Flooding

2.4   Duty Cycle

In WSNs, the main factors of wasted energy consumption are collision, overhearing, control overhead, and idle listening. The most energy is wasted during the idle listening state [7-9]. Idle listening is the state when sensor node just waits without any sending or receiving. Generally, this phenomenon occurs when sensor node waits the receiver node to send the data. Duty cycle is used to improve energy efficiency by reducing the time in the idle listening state.

Duty cycle changes the sensor node communication module to sleep and wake up states periodically [10]. The sensor node can save the energy at sleep state. However, in sleep state data cannot be received and it makes the data delay. Data transmission can occur at sleep state.

3   Evaluation Method

This chapter introduces the environment to analyze the impact of duty cycle on flooding scheme. The object flooding schemes are Conventional, 1HI and 2HBI schemes. The performance evaluation metrics are the number of floodings, energy consumption, flooding time, and number of transmissions and Table 1 defines the metrics.

The number of floodings is an important metric to measure the amount of sensors working. Energy consumption is used to measure how much energy a sensor node is using. The flooding time is used to measure the delay, since the flooding schemes incur flooding delay when duty cycled. To examine the amount of working, the number of transmissions is used.

Table 1. Performance metrics and their definitions

Performance Metrics	Definition
Energy consumption	The total amount of energy consumed by all nodes in data inter arrival time (T)
Flooding time	The time spent for all nodes to receive data since the data was generated
Number of transmissions	The total number of transmissions in a flooding, including retransmissions

The 30%, 50%, 80%, and 100% duty cycles are used. A period comprises ten time slots, and all sensor nodes are synchronized [11-13]. A sensor node has just one state, which can be wake up, sleep, transmission, and receive on a time unit states. In the sleep state, data can be transmitted, but not received. The duty cycle % setting is described as follows. If the 30% duty cycled, wake up process initiating slots are selected randomly from ten time slots. And then, wake up slots and three slots in a row are assigned to be active.

The simulation environment is same as the 2HBI paper to verify the experiment and set the following environment to reduce the time of simulation. In a simulation, with a topology size of 100x100, every sensor node has the same 20m transmission range, and a Mica 2 energy model is used, with sensor node locations randomly chosen and data occurring every 60 time units. The average results are derived from 60 simulations. In each simulation, the sensor nodes have different wake up times. A sensor node makes a forwarding set before flooding. It is assumed that every sensor node knows the wake up time of the neighboring nodes. In our case sensor node is fixed and is connected to a power supply. When a node except the source node uses its all energy, the simulation is finished.

Table 2. Mica2 energy model [14]

State	Transmission	Receive	Wake up	Sleep
Consumed Energy	27mJ	10mJ	8mJ	15μJ

4   Performance Results on Various Duty Cycles

In chapter 4, we evaluate the result. For simulation collision environment and collision free environment are considered. Nevertheless, in cases where result from both environment show similar trend, only duty cycle environment results are presented. In this paper, we evaluate the number of floodings, energy consumption, and flooding time on various duty cycles.

4.1   The Number of Floodings

Fig 3 shows the flooding result in an environment with collisions, and Fig 4 shows the result for the collision-free environment. According to the Fig 3 and 4, the schemes increase floodings are 2HBI, 1HI, and the conventional schemes. The number of floodings are increaser 1.4 times in Conventional, 1.7 time in 1HI, 1.9 times in 2HBI than on 100% duty cycle with collisions environment. First, the conventional scheme on 30% duty cycle use 45.1% energy less than that of 100% duty cycled, up to 60.3% in 1HI and up to 64.8% in 2HBI. The sensor nodes waste energy for the wake up state with high duty cycle. The network time is expanded, and the number of transmissions increased, since the wake up state is increased. As a result, the floodings are increased, regardless of the schemes having low duty cycle. In increasing order of floodings is due to the number of transmission gaps. The network time is expanded, and floodings are increased through the reduction of the energy consumed, because the transmission state needs a large amount of energy. This result is same as the collision free environment.

The differences between the collision and collision-free cases are as follows. Retransmission due to collision does not exist in a conflict-free environment, and most of the transmissions are successful. Therefore, the number of floodings of the three flooding schemes is increased. The other characteristic is that among the schemes, there is a little gap between the

30, 50, and 80% duty cycles. The schemes in order of increasing gaps of floodings are the conventional, 1HI, and 2HBI schemes in the collision environment. However, floodings of the three schemes are decreased in the collision-free environment. The collision and collision-free environments affect the floodings.

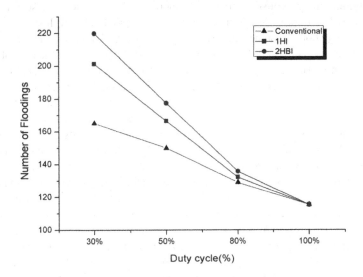

Fig. 3. The Number of Floodings with Collisions

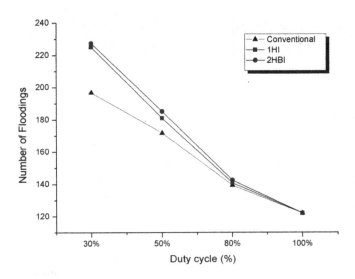

Fig. 4. The Number of Floodings without Collisions

4.2   Energy Consumption

Three characteristics can be observed about the energy consumption. The energy consumption on high duty cycle increaser than low duty cycle, and the schemes listed in increasing order of consumption are the conventional, 1HI, and 2HBI schemes. Additionally, the energy consumption gap is increased among the three schemes with low duty cycle. The implications of the three characteristics are as follows.

First, the conventional scheme on 30% duty cycle use 45.1% energy less than that of 100% duty cycle, up to 60.3% in 1HI and up to 64.8% in 2HBI. Because the time of a node spent in the wake up state is increased by increasing the duty cycle. The amount of energy consumption of the entire sensor node is increased by increasing the energy consumption in the wake up mode. Second, The gap of energy consumption between conventional, and 1HI, 2HBI on high duty cycle is about 1.06J and 5J on low duty cycle. Because the number of transmission gap increase on low duty cycle. A sensor node broadcasts the data right away after receiving the data without duty cycle. However, the wake up time of the sensor node is reduced by decreasing the duty cycle. The sensor node waits for the wake up time, and many sensor node data transmissions occur during the wake up time. Consequently, the number of collisions and retransmissions is increased. Also, the 1HI and 2HBI schemes have lower numbers of retransmission nodes than the conventional scheme, and consequently, the numbers of collisions and retransmissions are low as well. Third, the schemes listed in order of increasing energy consumption are the conventional, 1HI, and 2HBI schemes over the entire duty cycle. The main factor is the gap among the three schemes. Nevertheless, the amount of energy consumed for transmission is important. Since the transmission requires the most amount of energy comparing to other states.

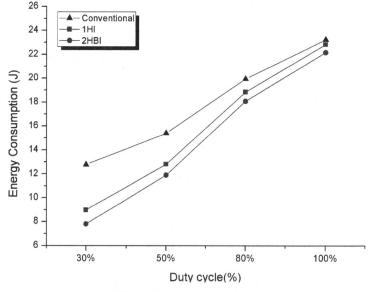

Fig. 5. Energy consumption

4.3   Flooding Time

The flooding time is longer in low duty cycle environment because of transmission delay, the number of collisions increased, the big forwarding set, longer waiting time of sensor nodes. The collision is increased because sensor nodes transmit the data at same active time slot of receiver. The smaller the forwarding set is, the higher probability the waiting time of receiver increases. If there are many retransmission nodes, receiving opportunity is improved and the data delay is reduced. However, in 2 HBI, there are a few retransmission nodes; the opportunity of transmission thus is low. This leads to the longer flooding time.

Figure 6 illustrates that the 2HBI flooding time is the shortest one among the three schemes in 100% duty cycle with collision environment. However, it incurs the poor performance compared with others on 30% duty cycle. The sensor node transmits the data immediately on 100% duty cycle, thus the data delay is decreased because of small forwarding set. Nevertheless, in 30% duty cycle environment the sensor node should wait longer time and the number of collision is higher than those on 100% duty cycle.

In the flooding time results, there are many cases as follows: (Conventional > 1HI, 2HBI), (Conventional > 1HI = 2HBI), (Conventional = 1HI = 2HBI), (1HI > Conventional > 2HBI), (1HI > 2HBI > Conventional). The flooding time depends on the wake up time of sensor node, which leads to a variety of results. However, the flooding time cannot significantly affect the energy consumption.

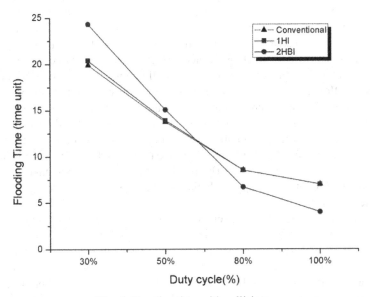

Fig. 6. Flooding time with collisions

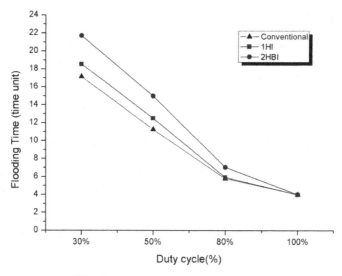

Fig. 7. Flooding time without collisions

5   Conclusion

This paper analysis the impact of duty cycle on Conventional, 1HI, and 2HBI flooding schemes. As the results, above mentioned flooding schemes reduce the energy consumption and increase the number of floodings in network lifetime but at the downside flooding time is also increased. In Conventional flooding with the introduction of 30% duty cycle, the number of floodings in network lifetime is increased up to 1.4 times, energy consumption is reduced by 45.1% and flooding time is increased up to 2.8 times. In 1HI flooding with the introduction of 30% duty cycle, the number of floodings in network lifetime is increased up to 1.7 times, energy consumption is reduced by 60.3%, and flooding time is increased up to 2.9 times. In 2HBI flooding with the introduction of 30% duty cycle, the number of floodings in network lifetime is increased up to 1.9 times, energy consumption is reduced by 64.79%, and flooding time is increased up to 6 times. The above results clearly mentions that, 2HBI scheme takes the longest flooding time with duty cycle approach, and it takes the shortest flooding time in non-duty cycled networks.

Characteristics of each flooding scheme in terms of energy consumption and flooding time with duty cycling are summarized below. Conventional flooding scheme use less energy than other two schemes but shows the fastest flooding time. 1HI flooding scheme use the energy efficiently comparing to Conventional flooding and shows similar energy consumption as 2HBI. Whereas it's flooding time is shorter than 2HBI and similar to Conventional flooding. 2HBI flooding reduces the energy consumption more than the other two schemes but the flooding time is the highest. For example, 2HBI on low duty cycle is suitable if the energy consumption is an important factor, whereas 2HBI scheme on 100% duty cycle is suitable when important factor is flooding time.

Based on the discovered features of this paper, efficient flooding schemes of the duty cycle when applied to environments are to be suggested in next research. To lessen the delay of time, a scheme of shortened flooding time using features of 2HBI in the low duty cycle or an idea of using active time of sensor nodes can be suggested. There can be more suggestions regarding not only flooding time but also energy efficiency, network lifetime, floodings, et cetera for flooding schemes in a duty cycle.

Acknowledgement. This research was supported in part by MKE and MEST, Korean government, under ITRC NIPA-2013-(H0301-13-3001), WCU NRF (No. R31-2010-000-10062-0) and PRCP(2012-0005861) through NRF, respectively.

References

1. Akyildiz, I.F., Su, W., Sankarasubramaniam, Y., Cayirci, E.: A Survey on Sensor Networks. IEEE Communications Magazine 40, 102–114 (2002)
2. Liu, H., Jia, X., Wan, P.-J., Liu, X., Yao, F.F.: A Distributed and Efficient Flooding Scheme Using 1-Hop Information in Mobile Ad Hoc Networks. IEEE Transactions on Parallel and Distributed System 18, 658–671 (2007)
3. Le, T.D., Choo, H.: Towards an Efficient Flooding Scheme Exploiting 2-Hop Backward Information in MANETs. IEICE Transactions on Communications E92-B (April 2009)
4. Ganesan, D., Cerpa, A., Ye, W., Yu, Y., Zhao, J., Estrin, D.: Networking Issues in Wireless Sensor Networks. Journal of Parallel and Distributed Computting (2004)
5. Ho, C., Obraczka, K., Tsudik, G., Viswanath, K.: Flooding for Reliable Multicast in Multi-Hop Ad Hoc Networks. Wireless Network 7(6), 627–634 (2001)
6. Chiang, T.-C., Chang, J.-L., Lin, S.-W.: A Distributed Multicast Protocol with Location-Aware for Mobile Ad-Hoc Networks. In: Jin, D., Lin, S. (eds.) MSEC 2011. AISC, vol. 129, pp. 691–697. Springer, Heidelberg (2011)
7. Gu, Y., He, T.: Data Forwarding in Extremely Low Duty-Cycle Sensor Networks with Unreliable Communication Links. In: Proc. ACM SenSys (2007)
8. Zhao, Y.Z., Miao, C., Ma, M., Zhang, J.B., Leung, C.: A Survey and Projection on Medium Access Control Protocols for Wireless Sensor Networks. ACM Computing Surveys 45 (November 2012)
9. Lu, G., Krishnamachari, B., Raghavendra, C.: An Adaptive Energy Efficient and Low-Latency Mac for Data Gathering in Wireless Sensor Networks. IEEE International Parallel & Distributed Processing Symposium(IPDPS) (2004)
10. Tang, L., Sun, Y., Gurewitz, O., Johnson, D.B.: PW-MAC: An Energy-Efficient Predictive-Wakeup MAC Protocol for Wireless Sensor Networks. In: Proc. IEEE International Conference on Computer Communications, pp. 1305–1313 (April 2011)
11. Marot, M., Kusy, B., Ledeczi, A.: The Flooding Time Synchronization Protocol. In: Proc. ACM SenSys (2004)
12. Koo, J., Panta, R.K., Bagchi, S., Montestruque, L.: A Tale of Two Synchronizing Clocks. In: Proc. ACM SenSys (2009)
13. Lenzen, C., Sommer, P., Wattenhofer, R.: Optimal Clock Synchronization in Networks. In: Proc. ACM SenSys (2009)
14. Rev A, Crossbow Technology, Inc., Document Part Number: 6020-0042-04

Formal Verification of Cyber-Physical Systems: Coping with Continuous Elements

Muhammad Usman Sanwal[1] and Osman Hasan[2]

[1] Research Center for Modeling and Simulation (RCMS)
[2] School of Electrical Engineering and Computer Science (SEECS)
National University of Sciences and Technology (NUST),
Islamabad, Pakistan
{muhammad.usman1,osman.hasan}@seecs.nust.edu.pk

Abstract. The formal verification of cyber-physical systems is a challenging task mainly because of the involvement of various factors of continuous nature, such as the analog components or the surrounding environment. Traditional verification methods, such as model checking or automated theorem proving, usually deal with these continuous aspects by using abstracted discrete models. This fact makes cyber-physical system designs error prone, which may lead to disastrous consequences given the safety and financial critical nature of their applications. Leveraging upon the high expressiveness of higher-order logic, we propose to use higher-order-logic theorem proving to analyze continuous models of cyber-physical systems. To facilitate this process, this paper presents the formalization of the solutions of second-order homogeneous linear differential equations. To illustrate the usefulness of our foundational cyber-physical system analysis formalization, we present the formal analysis of a damped harmonic oscillator and a second-order op-amp circuit using the HOL4 theorem prover.

1 Introduction

Cyber-physical systems (CPS) [25] are characterized as computational systems, with software and digital and/or analog hardware components, that closely interact with their continuously changing physical surroundings. These days, CPS are widely being used and advocated to be used in a variety of applications ranging from ubiquitous consumer electronic devices, such as tele-operated health-care units and autonomous vehicles, to not so commonly used but safety-critical domains, such as tele-surgical robotics, space-travel and smart disaster response and evacuation. Due to the tight market windows or safety-critical nature of their applications, it has become a dire need to design error-free CPS and thus a significant amount of time is spent on ensuring the correctness of CPS designs.

Traditionally, physical and continuous aspects of a CPS are analyzed by capturing their behaviors by appropriate differential equations [31] and then solving these differential equations to obtain the required design constraints. This kind of analysis can be done using paper-and-pencil proof methods or computer based

B. Murgante et al. (Eds.): ICCSA 2013, Part I, LNCS 7971, pp. 358–371, 2013.
© Springer-Verlag Berlin Heidelberg 2013

numerical techniques. Whereas, the software and digital hardware components of a CPS are usually analyzed using computer based testing or simulation methods, where the main idea is to deduce the validity of a property by observing its behavior for some test cases. However, all the above mentioned analysis techniques, i.e., paper-and-pencil proof methods, numerical methods and simulation, cannot ascertain the absence of design flaws in a design. For example, paper-and-pencil proof methods are error prone due to the human error factor. Moreover, it is quite often the case that many key assumptions of the results obtained using paper-and-pencil proof methods are in the mind of the mathematician and are not documented. Such missing assumptions may also lead to erroneous CPS designs. Similarly, computer based numerical methods cannot attain 100% accuracy as well due to the memory and computation limitations and round-off errors introduced by the usage of computer arithmetics. Thus, given the above mentioned inaccuracies, these traditional techniques should not be relied upon for the analysis of CPS, especially when they are used in safety-critical areas, such as medicine and transportation, where inaccuracies in the analysis could result in system design bugs that in turn may even lead to the loss of human life.

In the past couple of decades, formal methods [3] have been successfully used for the precise analysis of a variety of software, hardware and physical systems. The main principle behind formal analysis of a system is to construct a computer based mathematical model of the given system and formally verify, within a computer, that this model meets rigorous specifications of intended behavior. Two of the most commonly used formal verification methods are model checking [2] and higher-order-logic theorem proving [18]. Model checking is an automatic verification approach for systems that can be expressed as a finite-state machine. Higher-order-logic theorem proving, on the other hand, is an interactive approach but is more flexible in terms of tackling a variety of systems. The rigorous exercise of developing a mathematical model for the given system and analyzing this model using mathematical reasoning usually increases the chances for catching subtle but critical design errors that are often ignored by traditional techniques like paper-and-pencil based proofs, numerical methods or simulation.

Given the extensive usage of CPS in safety-critical applications, there is a dire need of using formal methods for their analysis. However, the frequent involvement of ordinary differential equations (ODEs) in their analysis is a main limiting factor in this direction. For example, ODEs are essential for modeling the motion of mechanical parts, analog circuits and control systems, which are some of the most common elements of any CPS. Thus, automatic state-based formal methods, like model checking, and automatic theorem provers cannot be used to model and analyze the true CPS models due to their inability to model continuous systems. This is the main reason why most of the formal verification work about CPS utilizes their abstracted discrete models (e.g., [28]). These limitations can be overcome by using higher-order-logic theorem proving [13] for conducting the formal analysis of CPS since the high expressiveness of higher-order logic can be leveraged upon to model elements of continuous nature. However, the main challenge in this direction is the enormous human guidance required in the formal verification of CPS due to the non-decidable nature of higher-order logic. In order to minimize this effort, we propose to develop a formal library of foundational results that can be built upon along with the automatic simplifiers to automatically reason about the correctness of CPS in a theorem prover.

In this paper, as a first step towards the proposed direction, we present the formal reasoning support for the solutions of second-order homogeneous linear differential equations [14], i.e., a simple yet the most widely used class of differential equations. In particular, we present a formal definition that can be used to specify arbitrary order homogeneous linear differential equations and the formal verification of some mathematical facts, like a couple of general solutions of second-order homogeneous linear differential equations and the quadratic formula, that allow us to reason about the correctness of their solutions in a very straightforward way. The prime advantage of these results is that they greatly minimize the user intervention for formal reasoning about differential equations and thus facilitate the usage of higher-order-logic theorem proving for verifying the solutions of differential equations for CPS. In order to demonstrate the practical effectiveness and utilization of the reported formalization, we utilize it to analyze a damped harmonic oscillator and a second-order op-amp, in this paper.

Our formalization primarily builds upon the higher-order-logic formalization of the derivative function and its associated properties. This formalization is available in a number of theorem provers like HOL4 [16], PVS [7] and CoQ [10]. Our work is based on Harrison's formalization [16] that is available in the HOL4 theorem prover [27]. The main motivations behind this choice include the past familiarity with HOL4, the availability of formalized transcendental functions, which play a key role in the reported work, and the general richness of Harrison's real analysis related theories. Though, it is important to note here that the ideas presented in this paper are not specific to the HOL4 theorem prover and can be adapted to any other LCF style higher-order-logic theorem prover as well.

The rest of the paper is organized as follows: Section 2 presents an overview of the related work. Section 3 presents a brief introduction to theorem proving and the HOL4 theorem prover. This is followed by a brief introduction to the HOL4 formalization of the derivative and integration functions in Section 4. In Section 5, we present the formalization of the solutions of the second-order homogeneous linear differential equation. The illustrative examples are presented in Section 6. Finally, Section 7 concludes the paper.

2   Related Work

Formal methods have been extensively used these days for analyzing CPS due to their ever increasing usage in various safety and financial-critical domains. Zhang et. al [21] proposed to use formal specification for CPS in order to reduce the infinite set of test parameters in a finite set. Similarly, the aspect-oriented programming based on the UML and formal methods is utilized for QoS modeling of CPS in [20]. Moreover, in order to formally specify CPS along with their continuous aspects, a combination of formal methods Timed-CSP, ZimOO and differential (algebraic) equations is used in [30]. Even though such rigorous formal specifications allow us to catch bugs in the early stages of the design but they do not guarantee error-free analysis since the analysis or verification is not based on formal methods.

For formal verification of CPS, model-checking has been frequently explored. For example, Akella [1] proposed to dicretize the events causing the change of flow and thus model the CPS as a deterministic state model with discrete values of flow within its physical components. This model is then used to formally verify insecure interactions between all possible behaviors of the given CPS using model checking. Similarly, Bu et. al [6] used hybrid model checking for verifying CPS. However, this verification is also not based on true continuous models of the system and instead a short-run behavior of the model is observed by providing numerical values of various parameters in order to reduce the state-space. A statistical model checker has been recently utilized to analyze some aspects of CPS [9]. However, this approach also suffers from the classical model checking issues, like the state-space explosion and inability to reason about generic mathematical relations. Thus the model checking approach, even though is capable of providing exact solutions, is quite limited in terms of handling true continuous models of CPS and thus various abstractions [28] have to be used for attaining meaningful results. The accuracy of the analysis is thus compromised, which is undesirable in the case of analyzing safety-critical applications of CPS.

Higher-order-logic theorem proving is capable of overcoming all the above mentioned problems. Atif et. al [22] used the HOL4 theorem prover for the probabilistic analysis of cyber-physical transportation systems. However, their focus was only on the formal verification of probabilistic aspects of CPS and they did not tackle the continuous aspects, especially the ones that require to be modeled by ODEs, which is the main focus of the current paper.

3   Preliminaries

In this section, we give a brief introduction to theorem proving in general and the HOL theorem prover in particular. The intent is to introduce the main ideas behind this technique to facilitate the understanding of this paper for the CPS research community.

3.1   Theorem Proving

Theorem proving [13] is a widely used formal verification technique. The system that needs to be analyzed is mathematically modeled in an appropriate logic and the properties of interest are verified using computer based formal tools. The use of formal logics as a modeling medium makes theorem proving a very flexible verification technique as it is possible to formally verify any system that can be described mathematically. The core of theorem provers usually consists of some well-known axioms and primitive inference rules. Soundness is assured as every new theorem must be created from these basic or already proved axioms and primitive inference rules.

The verification effort of a theorem in a theorem prover varies from trivial to complex depending on the underlying logic [15]. For instance, first-order logic [12] is restricted to propositional calculus and terms (constants, function names and free variables) and is semi-decidable. A number of sound and complete first-order logic automated reasoners are available that enable completely automated proofs. More expressive logics, such as higher-order logic [5], can be used to model a wider range of problems than first-order logic, but theorem proving for these logics cannot be fully automated and thus involves user interaction to guide the proof tools. For analyzing continuous aspects of systems, we have to use higher-order logic as first-order logic is not expressive enough to represent real numbers and calculus fundamentals.

3.2 HOL Theorem Prover

HOL is an interactive theorem prover developed by Mike Gordon at the University of Cambridge for conducting proofs in higher-order logic. It utilizes the simple type theory of Church [8] along with Hindley-Milner polymorphism [23] to implement higher-order logic. HOL has been successfully used as a verification framework for both software and hardware as well as a platform for the formalization of pure mathematics.

Secure Theorem Proving. In order to ensure secure theorem proving, the logic in the HOL system is represented in the strongly-typed functional programming language ML [24]. An ML abstract data type is used to represent higher-order logic theorems and the only way to interact with the theorem prover is by executing ML procedures that operate on values of these data types. The HOL core consists of only 5 basic axioms and 8 primitive inference rules, which are implemented as ML functions. Soundness is assured as every new theorem must be verified by applying these basic axioms and primitive inference rules or any other previously verified theorems/inference rules.

Terms. There are four types of HOL terms: constants, variables, function applications, and lambda-terms (denoted function abstractions). Polymorphism, types containing type variables, is a special feature of higher-order logic and is thus supported by HOL. Semantically, types denote sets and terms denote members of these sets. Formulas, sequences, axioms, and theorems are represented by using terms of Boolean types.

Theories. A HOL theory is a collection of valid HOL types, constants, axioms and theorems, and is usually stored as a file in computers. Users can reload a HOL theory in the HOL system and utilize the corresponding definitions and theorems right away. The concept of HOL theory allows us to build upon existing results in an efficient way without going through the tedious process of regenerating these results using the basic axioms and primitive inference rules.

HOL theories are organized in a hierarchical fashion. Any theory may inherit types, definitions and theorems from other available HOL theories. The HOL system prevents loops in this hierarchy and no theory is allowed to be an ancestor and descendant of a same theory. Various mathematical concepts have been formalized and saved as HOL theories by the HOL users. These theories are available to a user when he first starts a HOL session. We utilized the HOL theories of Booleans, lists, sets, positive integers, *real* numbers and calculus in our work. In fact, one of the primary motivations of selecting the HOL theorem prover for our work was to benefit from these built-in mathematical theories.

Writing Proofs. HOL supports two types of interactive proof methods: forward and backward. In forward proof, the user starts with previously proved theorems and applies inference rules to reach the desired theorem. In most cases, the forward proof method is not the easiest solution as it requires the exact details of a proof in advance. A backward or a goal directed proof method is the reverse of the forward proof method. It is based on the concept of a *tactic*; which is an ML function that breaks goals into simple sub-goals. In the backward proof method, the user starts with the desired theorem or the main goal and specifies tactics to reduce it to simpler intermediate sub-goals. Some of these intermediate sub-goals can be discharged by matching axioms or assumptions or by applying built-in decision procedures. The above steps are repeated for the remaining intermediate goals until we are left with no further sub-goals and this concludes the proof for the desired theorem.

The HOL theorem prover includes many proof assistants and automatic proof procedures [15] to assist the user in directing the proof. The user interacts with a proof editor and provides it with the necessary tactics to prove goals while some of the proof steps are solved automatically by the automatic proof procedures.

4   Derivatives and Integrals in HOL4

Harrison [16] formalized the *real number theory* along with the fundamentals of calculus, such as limits of a function, derivatives and integrals and verified most of their classical properties in HOL4. The derivative of a function f, of data type (`real` $\rightarrow$ `real`), is defined as follows [16]:

Definition 1. *Derivative of a Function (Relational Form)*
$\vdash \forall$ f l x. (f diffl l) x = ((λ h.(f (x + h) - f x) / h) $\rightarrow$ l) (0)

where (f $\rightarrow$ y0)(x0) represents the HOL4 definition of limit of a function f $lim_{(x \rightarrow x0)} f(x) = y0$. Definition 1 provides the derivative of a function f at point x as the limit value of $\frac{f(x+h)-f(x)}{h}$ when h approaches 0, which is the standard mathematical definition of the derivative function. Now, the differentiability of a function f is defined as the existence of its derivative [16].

Definition 2. *Differentiability of a Function*
⊢ ∀ f x. f differentiable x = ∃l. (f diffl l) (x)

A functional form of the derivative, which can be used as a binder, is also defined using the Hilbert choice operator @ as follows [16]:

Definition 3. *Derivative of a Function (Functional Form)*
⊢ ∀ f x. deriv f x = @l. (f diffl l) x

The function `deriv` accepts two parameters f and x and returns the derivative of function f at point x.

The above mentioned definitions associated with the derivative function have been accompanied by the formal verification of most of their classical properties, such as uniqueness, linearity and composition [16]. Moreover, the derivatives of some commonly used transcendental functions have also been verified. For example, the derivative of the Exponential function has been verified as follows:

Theorem 1. *Differential of the Exponential Function*
⊢ ∀ g m x. ((g diffl m) x ⇒
    ((λ.x. exp (g x)) diffl (exp (g x) * m)) x)

where `exp x` represents the exponential function e^x and $(\lambda x.f(x))$ represents the lambda abstraction function which accepts a variable x and returns $f(x)$.

Similarly, the Gauge integral has also been formalized as a function Dint (a, b) f k [16], which mathematically describes $\int_a^b f(x) \, dx = k$. The corresponding functional form is given as follows:

Definition 4. *Integral of a Function (Functional Form)*
⊢ ∀ a b f . integral (a,b) f = @k. Dint (a,b) f k

Many interesting properties of integration have been formally verified in [16]; one of them being the second fundamental theorem of calculus.

Theorem 2. *Second Fundamental Theorem of Calculus*
⊢ ∀ f f' a b. (a ≤ b) ∧
    (∀x. a ≤ x ∧ x ≤ b ⇒ (f diffl f'(x))(x))
        ⇒ Dint(a,b) f' (f(b) - f(a))

We build upon the above mentioned formalization to develop formal reasoning support for second-order homogeneous linear differential equations in the next section.

5   Second-Order Homogeneous Differential Equations

A second-order homogeneous linear differential equation can be mathematically expressed as follows:

$$p_2(x)\frac{d^2 y(x)}{dx} + p_1(x)\frac{dy(x)}{dx} + p_0(x)y(x) = 0 \tag{1}$$

where terms p_i represent the coefficients of the differential equation defined over a function y. The equation is linear because (i) the function y and its derivatives appear only in their first power and (ii) the products of y with its derivatives are also not present in the equation. By finding the solution of the above equation, we mean to find functions that can be used to replace the function y in Equation (1) and satisfy it.

We proceed to formally represent Equation (1) by first formalizing an nth-order derivative function as follows:

Definition 5. *N^{th}-order Derivative of a Function*
⊢ (∀ f x. n_order_deriv 0 f x = f x) ∧
  (∀f x n.n_order_deriv (n+1) f x = n_order_deriv n (deriv f x) x)

The function n_order_deriv accepts an integer n that represents the order of the derivative, the function f that represents the function that needs to be differentiated, and the variable x that is the variable with respect to which we want to differentiate the function f. It returns the nth-order derivative of f with respect to x. Now, based on this definition, we can formalize the left-hand-side (LHS) of an nth-order differential equation in HOL4 as the following definition.

Definition 6. *LHS of a Nth-order Differential Equation*
⊢ ∀ P y x. diff_eq_lhs P y n x =
  sum(0,n)(λn.(EL n L x) * (n_order_deriv n y x))

The function diff_eq_lhs accepts a list P of coefficient functions corresponding to the p_i's of Equation (1), the differentiable function y, the order of differentiation n and the differentiation variable x. It utilizes the HOL4 functions sum (0,n) f and EL n L, which correspond to the summation ($\sum_{i=0}^{n-1} f_i$) and the nth element of a list L_n, respectively. It generates the LHS of a differential equation of nth order with coefficient list P. The second-order differential equation of Equation (1) can now be formally modeled by instantiating variable n of Definition 6 by number 3.

If the coefficients p_i's of Equation (1) are constants in terms of the differentiation variable x then, using the fact that the derivative of the exponential function $y = e^{rx}$ (with a constant r) is a constant multiple of itself $dy/dx = re^{rx}$, we can obtain the following solution of Equation (1):

$$Y(x) = c_1 e^{r_1 x} + c_2 e^{r_2 x} \tag{2}$$

where c_1 and c_2 are arbitrary constants and r_1 and r_2 are the roots of the auxiliary equation $p_2 r^2 + p_1 r^1 + p_0 = 0$ [31]. In this paper, we formally verify this result which plays a key role in formal reasoning about the solutions of second-order homogeneous linear differential equations in a higher-order-logic theorem prover.

Theorem 3. *General Solution of a Homogeneous Linear Differential Equation*
⊢ ∀ a b c c1 c2 r1 r2 x.
  (c + (b * r1) + (a * (r1 pow 2)) = 0) ∧

```
(c + (b * r2) + (a * (r2 pow 2)) = 0) ⇒
(diff_eq_lhs (const_list [c; b; a])
   (λx. c1 * (exp (r1 * x) + c2 * (exp (r2 * x)) 3 x = 0)
```

where $[c; b; a]$ represents the list of constants corresponding to the coefficients p_0, p_1 and p_2 of Equation (1), $r1$ and $r2$ represent the roots of the corresponding auxiliary equation as given in the assumptions, $c1$ and $c2$ are the arbitrary constants and x is the variable of differentiation. The function const_fn_list used in the above theorem transforms a constant list to the corresponding constant function list recursively as follows:

Definition 7. *Constant Function List*
```
⊢ (const_fn_list [] = []) ∧
(∀ h t. const_fn_list (h::t) = (λ(x:real). h) :: (const_fn_list t))
```

The function diff_eq_lhs permits coefficients that are functions of the variable of differentiation but Theorem 3 is valid only for constant coefficients. Thus, using const_fn_list we provide the required type for the coefficient list of the function diff_eq while fulfilling the requirement of Theorem 3. The formal reasoning about Theorem 3 is primarily based on Theorem 1 and the linearity property of higher-order derivatives, which has been verified in our work for *class C^n* functions, i.e., the functions for which the first n derivatives exist for all x as the following higher-order-logic theorem:

Theorem 4. *Linearity of nth-order Derivative*
```
⊢ ∀ f g x a b.
   (∀m x. m ≤ n ⇒ (λx. n_order_deriv m f x) differentiable x) ∧
   (∀m x. m ≤ n ⇒ (λx. n_order_deriv m g x) differentiable x) ⇒
     (n_order_deriv n (λx. a * f x + b * g x) x =
       a * n_order_deriv n f x + b * n_order_deriv n g x)
```

where variables a and b represent constants with respect to variable x. The formal reasoning about Theorem 4 involves induction on variable n, which represents the order of differentiation, and is primarily based on the linearity property of the first order derivative function [16].

If the roots of Equation (1) are real and repeated then we can obtain the following solution:

$$Y(x) = c_1 e^{rx} + c_2 x e^{rx} \tag{3}$$

where c_1 and c_2 are arbitrary constants and r is the real and repeated root of the auxiliary equation $p_2 r^2 + p_1 r^1 + p_0 = 0$ [31]. Just like Theorem 3, we also formally verified that this solution satisfies Equation (1).

Theorem 5. *Differential Equation with repeated roots*
```
⊢ ∀ a b c c1 c2 r1 r2 x.
   (c + (b * r) + (a * r²) = 0) ∧ (b² - (4 * a * c) = 0) ∧ b ≠ 0 ⇒
     (diff_eq_lhs (const_list [c; b; a])
       (λy. c1 * (exp (r * y)) + c2 * (y * exp (r * y))) 3 x = 0)
```

where r represents the sole root of the corresponding auxiliary equation and the rest of the variables are the same as Theorem 3. The formal reasoning about Theorem 5 is also mainly based on Theorems 1 and 4 and the well-known quadratic formula which was also formally verified in our development as the following theorem in our development.

Theorem 6. *Quadratic Formula*
```
⊢ ∀ a b c x. (a ≠ 0) ∧ (4 * a * c < b pow 2) ⇒
aux_eq_roots_list [((-b + sqrt (b pow 2 - 4 * a * c)) / (2 * a));
   ((-b - sqrt (b pow 2 - 4 * a * c)) / (2 * a))]
      (const_fn_list [a; b; c]) x
```

where the functions `sqrt` and `pow` represent the square-root and square of a real number, respectively. The theorem essentially says that the roots of the auxiliary equation $ax^2 + bx + c$ are given by the first list argument of the function `aux_eq_roots_list`. The assumption (4 * a * c < b pow 2) guarantees that the roots are always real.

We will build upon the results presented in this section to formally reason about the applications in Section 6 of this paper.

6 Applications

In this section, we formally analyze two widely used continuous components of CPS, i.e., a damped harmonic oscillator and a second-order op-amp circuit. The first one being a widely used mechanical model for CPS whereas the later is a widely used analog component of CPS.

6.1 Damped Harmonic Oscillator

Behaviors of many CPS are mathematically equivalent to harmonic oscillators, that is, these systems can be described by an ODE that is similar to the ODE of the damped harmonic oscillator. Some prominent examples include the spring mass system and the classical RLC circuit [11]. Therefore, the analysis of the damped harmonic oscillator, depicted in Figure 1, has gained interest from the CPS community [4,26]. The ODE of a damped harmonic oscillator can be expressed as

$$\frac{d^2x}{dt^2} + z\frac{dx}{dt} + w^2x = 0 \tag{4}$$

Where x is the extension produced, $w = \sqrt{(k/m)}$ represents the angular frequency of the oscillator in terms of spring constant k and the mass of spring m and $z = \sqrt{(b/m)}$ represents the damping ratio in terms of the damping coefficient b and the mass of spring m.

Based on the formalization of the last section, the solution of the above differential equation can be formally verified as the following theorem:

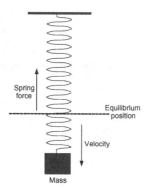

Fig. 1. Damped Harmonic Oscillator

Theorem 7. *Damped Harmonic Oscillator*
```
⊢ ∀ w z c1 c2 x. (z² = w²) ⇒
(diff_eq_lhs (const_list [w²; z; 1])
  (λx.c1 * exp((-z + sqrt (z² - w²)) * x) +
      c2 * exp((-z - sqrt (z² - w²)) * x)) x = 0)
```

This theorem is formally verified primarily using Theorem 3. The assumptions of Theorem 7 declare the relationships between the various parameters that are required for the solution to hold. This is one of strengths of the proposed theorem proving based verification as all the assumptions have to be explicitly stated besides the theorem for its formal verification. Thus, there is no chance of missing a critical assumption which often occurs in paper-and-pencil proof methods. It is also important to note the generic and continuous nature of Theorem 7 as all the variables are of type **real** and they are universally quantified. Such results cannot be obtained via state-based formal methods tools.

6.2   Second-Order Op-Amp

As a second example, consider a second-order op-amp circuit (Figure 2), which has a wide range of applications in oscillators, filters, audio buffers and line drivers [19,29]. Thus, it is a widely used continuous component of many CPS.

The behavior of this circuit can be modeled as the following differential equation [19]:

$$\frac{d^2v}{dt^2} - (\frac{1}{R^2C^2})v = 0 \tag{5}$$

We formally verified its solution, i.e., an expression for the voltage v, as the following theorem.

Theorem 8. *Second-order Op-Amp Circuit*
```
⊢ ∀ R L c1 c2. (R * C ≠ 0) ⇒
  (diff_eq_lhs (const_list [-1/(R² * C²);0;1])
    (λx. c1 * exp ((1/R*C) * x) + c2 * (x * exp ((1/R*C) * x)) x = 0))
```

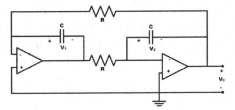

Fig. 2. Second-order Op-Amp Circuit

The proof of the above theorem is based on Theorem 5, which presents the formally verified solution of the homogeneous linear differential equation with real and repeated roots, along with some arithmetic reasoning, which can be done in an automatic manner using the HOL4 arithmetic simplifiers.

The proof script for all the theorems, presented in this section, is composed of just 300 lines approximately. This is far less than the proof script for the formalization, presented in the previous section, which is more than 3500 lines of HOL code. This fact clearly indicates the usefulness of our foundational formalization presented in Sections 5 of this paper. Just like the case studies, presented in this section, our formalization results can be utilized to automatically verify interesting properties of a wide variety of CPS in a straight-forward manner and the results would be guaranteed to be correct due to the inherent soundness of theorem proving.

7   Conclusions

In this paper, we propose to use higher-order-logic theorem proving to analyze continuous aspects of CPS. Due to the high expressiveness of the underlying logic, we can formally model the continuous components of CPS while capturing their true behavior and the soundness of theorem proving guarantees correctness of results. To the best of our knowledge, these features are not shared by any other existing CPS analysis technique. The main challenge in the proposed approach is the enormous amount of user intervention required due to the undecidable nature of the logic. We propose to overcome this limitation by formalizing the foundational mathematical theories so that these available results can be built upon to minimize user interaction. As a first step towards this direction, we presented the formalization of the solutions of second-order homogenous linear differential equations in this paper. Based on this work, we are able to formally analyze the damped harmonic oscillator and the second order op-amp circuit, which are both quite frequently encountered in CPS analysis, in a very straightforward way.

The proposed approach opens the doors to many new directions of research. We are working on developing reasoning support for non-homogeneous linear differential equations. Moreover, the calculus theories available in HOL-Light

[17] are based on multivariate real numbers and thus can model complex numbers. Our formalization can be ported in a very straight-forward manner to this formalization of complex numbers in HOL-Light, which would enable handling the formal analysis of CPS that can be modeled in the complex plane only.

Acknowledgment. This work was supported by the National Research Program for Universities grant (number 1543) of Higher Education Commission (HEC), Pakistan.

References

1. Akella, R., McMillin, B.M.: Model-Checking BNDC Properties in Cyber-Physical Systems. In: Computer Software and Applications Conference, pp. 660–663 (2009)
2. Baier, C., Katoen, J.: Principles of Model Checking. MIT Press (2008)
3. Boca, P.P., Bowen, J.P., Siddiqi, J.I.: Formal Methods: State of the Art and New Directions. Springer (2009)
4. Broman, D., Lee, E.A., Tripakis, S., Toerngren, M.: Viewpoints, Formalisms, Languages, and Tools for Cyber-Physical Systems. In: 6th International Workshop on Multi-Paradigm Modeling (2012)
5. Brown, C.E.: Automated Reasoning in Higher-order Logic. College Publications (2007)
6. Bu, L., Wang, Q., Chen, X., Wang, L., Zhang, T., Zhao, J., Li, X.: Towards Online Hybrid Systems Model Checking of Cyber-Physical Systems' Time-Bounded Short-Run Behavior. SIGBED (2), 7–10 (2011)
7. Butler, R.W.: Formalization of the Integral Calculus in the PVS Theorem Prover. Journal of Formalized Reasoning 2(1), 1–26 (2009)
8. Church, A.: A Formulation of the Simple Theory of Types. Journal of Symbolic Logic 5, 56–68 (1940)
9. Clarke, E.M., Zuliani, P.: Statistical model checking for cyber-physical systems. In: Bultan, T., Hsiung, P.-A. (eds.) ATVA 2011. LNCS, vol. 6996, pp. 1–12. Springer, Heidelberg (2011)
10. Cruz-Filipe, L.: Constructive Real Analysis: a Type-Theoretical Formalization and Applications. PhD thesis, University of Nijmegen (April 2004)
11. Daneshbod, Y., Latulippe, J.: A Look at Damped Harmonic Oscillators through the Phase Plane. Teaching Mathematics and its Applications 30(2)
12. Fitting, M.: First-Order Logic and Automated Theorem Proving. Springer (1996)
13. Gordon, M.J.C.: Mechanizing Programming Logics in Higher-Order Logic. In: Current Trends in Hardware Verification and Automated Theorem Proving, pp. 387–439. Springer (1989)
14. Strang, G.: Calculus, 2nd edn. Wellesley College (2009)
15. Harrison, J.: Formalized Mathematics. Technical Report 36, Turku Centre for Computer Science (1996)
16. Harrison, J.: Theorem Proving with the Real Numbers. Springer (1998)
17. Harrison, J.: A HOL theory of Euclidean space. In: Hurd, J., Melham, T. (eds.) TPHOLs 2005. LNCS, vol. 3603, pp. 114–129. Springer, Heidelberg (2005)
18. Harrison, J.: Handbook of Practical Logic and Automated Reasoning. Cambridge University Press (2009)

19. Alexander, K., Sadiku, M.N.O.: Fundamentals of Electric Circuits. McGraw-Hill (2008)
20. Liu, J., Zhang, L.: QoS Modeling for Cyber-Physical Systems using Aspect-Oriented Approach. In: 2011 Second International Conference on Networking and Distributed Computing (ICNDC), pp. 154–158 (2011)
21. Zhang, L., Hu, J., Yu, W.: Generating Test Cases for Cyber Physical Systems from Formal Specification, pp. 97–103. Springer (2011)
22. Mashkoor, A., Hasan, O.: Formal Probabilistic Analysis of Cyber-Physical Transportation Systems. In: Murgante, B., Gervasi, O., Misra, S., Nedjah, N., Rocha, A.M.A.C., Taniar, D., Apduhan, B.O. (eds.) ICCSA 2012, Part III. LNCS, vol. 7335, pp. 419–434. Springer, Heidelberg (2012)
23. Milner, R.: A Theory of Type Polymorphism in Programming. Journal of Computer and System Sciences 17, 348–375 (1977)
24. Paulson, L.C.: ML for the Working Programmer. Cambridge University Press (1996)
25. Rajkumar, R., Lee, I., Sha, L., Stankovic, J.J.: Cyber-Physical Systems: The next Computing Revolution. In: 2010 47th ACM/IEEE Design Automation Conference (DAC), pp. 731–736 (2010)
26. Shi, J., Wan, J., Yan, H., Suo, H.: A survey of Cyber-Physical Systems. In: Wireless Communications and Signal Processing (WCSP), pp. 1–6 (2011)
27. Slind, K., Norrish, M.: A Brief Overview of HOL4. In: Mohamed, O.A., Muñoz, C., Tahar, S. (eds.) TPHOLs 2008. LNCS, vol. 5170, pp. 28–32. Springer, Heidelberg (2008)
28. Thacker, R.A., Jones, K.R., Myers, C.J., Zheng, H.: Automatic Abstraction for Verification of Cyber-Physical Systems. In: Proceedings of the 1st ACM/IEEE International Conference on Cyber-Physical Systems, pp. 12–21. ACM (2010)
29. Jung, W.: Op Amp Applications Handbook. Newnes (2004)
30. Zhang, L.: Aspect Oriented Formal Techniques for Cyber Physical Systems. Journal of Software 7(4), 823–834 (2012)
31. Zill, D.G., Wright, W.S., Cullen, M.R.: Advanced Engineering Mathematics, 4th edn. Jones and Bartlett Learning (2009)

Implementation of Enhanced Android SPICE Protocol for Mobile Cloud

Jun-Kwon Jung[1], Sung-Min Jung[1], Tae-Kyung Kim[2],
and Tai-Myoung Chung[3,*]

[1] Department of Electrical and Computer Engineering
Sungkyunkwan University, Suwon, Korea
{jkjung,smjung}@imtl.skku.ac.kr
[2] Department of liberal art,
Seoul theological university, Bucheon, Korea
tkkim@stu.ac.kr
[3] College of Information and Communication Engineering
Sungkyunkwan University, Suwon, Korea
tmchung@ece.skku.ac.kr

Abstract. Cloud computing is the top issue of IT industry. In addition, the mobile platform has become a major keyword of IT technologies. Thus, we introduce the mobile cloud computing that consist of the union of two concepts. This conflated model can provide a high performance computing in mobile platform. To provide the mobile cloud computing service, it needs various IT technologies. Among these technologies, remote access protocol is one of major issues because it is directly related in a performance of mobile cloud computing. There are two major standard protocols to provide remote access. They are Remote Frame Buffer protocol(RFB protocol) and Remote Desktop Protocol(RDP protocol). RFB protocol relays the view image from a server to a client. It has a simple architecture, but it makes many network traffics structurally. On the other hand, RDP protocol sends the message of a server event to client and the client creates and shows the view image to users. It has smaller traffic than the RFB protocol, but it is only used in Windows platform due to the license problem. The other protocol called SPICE can solve problems from two standard protocols. This protocol has similar architecture to RDP protocol and it is an open-source program. However, SPICE protocol cannot support Android platform now. In this paper, we introduce the way to convert SPICE protocol to support Android platform. However, the modified protocol has some additional delay. Therefore, we also propose the way to enhance the performance of this protocol. It can provide a high performance and be suitable for the mobile cloud.

Keywords: Android, SPICE, Mobile Cloud, Remote access protocol.

---

* Corresponding Author.

B. Murgante et al. (Eds.): ICCSA 2013, Part I, LNCS 7971, pp. 372–381, 2013.

1   Introduction

Cloud computing is one of major issues in IT technologies. Recently, various investigation agencies such as Gartner and IDC have submitted the forecast reports related with increasing of cloud computing. For this trend, many people are interested in the cloud computing. It is divided to a public cloud and a private cloud. Between two types, the public cloud has more important elements than the other because the public cloud is used in wide area and has more limitation than the private cloud. Among many organizations, the government agencies in many countries invest in public cloud services. In Gartner research, Cloud computing market will be growing up rapidly. In particular, the part related with business process services is the best growth. Figure 1 shows the public cloud services forecast overview.

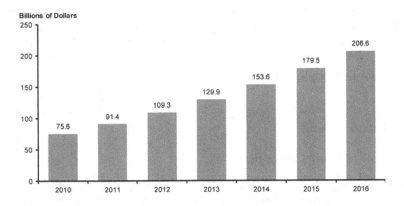

Fig. 1. Public cloud services market size[1]

In addition, people who use mobile devices such as smart phones or tablet PC are increasing more and more. Due to a user explosion of a mobile device, people began to pay attention to the mobile cloud computing. It is a model of high performance computing with accessing to a virtual machine at a mobile device. It requires the remote access protocol to serve it due to providing the connection between the mobile terminal and a virtual machine in cloud server. Because the performance of a remote access protocol is directed related with a mobile cloud computing, this is a major issue in mobile cloud computing.

To provide the remote control, there are two standard remote access protocols[2]. They are the RFB protocol and the RDP protocol. The RFB protocol has a simple architecture. It just sends the output display image from a remote server to a client. The client receives this image and shows directly. The RDP protocol sends the event message and the client creates the remote view from received message. In a mobile cloud computing, RDP protocol is better than RFB protocol because the RDP protocol needs smaller traffic than RFB protocol. Thus, the RDP protocol is good for mobile cloud computing. However, it cannot use at mobile cloud computing due to Microsoft license problem. Therefore, the alternative instead of two standard protocols is needed. The SPICE protocol is the alternative of two protocols. The architecture of this is similar to the RDP protocol and it is an open-source protocol. Thus, it is able to be used in mobile platform and it can be expected small traffics. However, it cannot be supported in mobile platform yet.

In this paper, we introduce the way to provide the SPICE protocol to support in Android platform. The Android SPICE protocol is a name of modified SPICE protocol to run in Android platform. However, it has some delays. These delays are occurred by additional encoding process in itself. Therefore, we also propose the solution of these delays. This protocol can be changed the encoding module and removed unnecessary image processing. As a result, enhanced Android SPICE protocol gets about half of delay reducing from the image encoding. Organization of this paper is as follows. In section 2, we explain the concept of mobile cloud computing and the typical remote access protocols. Then we introduce the possibility of Android SPICE protocol and the way to change SPICE protocol to support Android platform in section 3. Also, we delineate the problem about performance of Android SPICE protocol and propose the solution of this problem in section 4. Finally, in section 5, we present the conclusion and future works.

2   Related Work

2.1   Mobile Cloud Computing

The attentions about cloud computing is increased. For this trend, various services of cloud computing are being developed and offered. In addition, the number of mobile device users is also increased rapidly. Thus, mobile device users require the mobile cloud computing service. In this paper, the mobile cloud computing service provides high performance computing to users from a mobile Virtual Machine(VM) in the cloud server[3]. In order to connect the mobile VM, the service can use the remote access protocol such as RFB, RDP or SPICE. The remote access protocol is the important element of performance of the service. If the protocol has low performance, users experience some control delay. Figure 2 shows the structure of mobile cloud computing.

2.2   Remote Access Protocols

There are many protocols to provide a connection with mobile virtual machine. Among them, the RFB protocol and the RDP protocol are standard protocols. The RFB protocol captures the screen view from a remote server and shows this view to client[4]. This is simple structure. Also, many open source programs are based on this protocol. Thus, many users can freely to use this program based on this protocol. Typical program using RFB protocol is Virtual Network Computing(VNC)[4]. VNC also has free version. Figure 3 shows the overview of the RFB protocol.

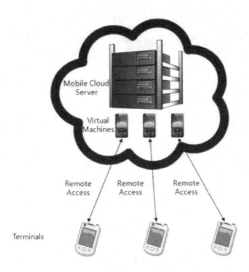

Fig. 2. Mobile cloud architecture

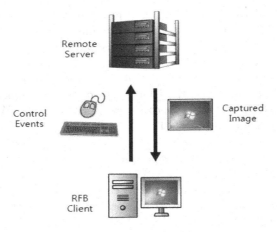

Fig. 3. RFB protocol

The RDP protocol sends the message of events from a remote server to a client[5]. A client creates the remote view image from this message and shows the remote view to users. Since the RDP protocol sends a message data instead of a remote server image, this protocol can save a network traffic. However, a client device has to guarantee a minimum performance to make remote view from the message. In addition, it is used to Windows server only but a client services can support most OS because Microsoft has a license of the RDP protocol. Figure 4 shows the overview of RDP protocol.

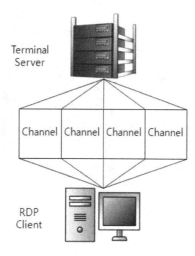

Fig. 4. RDP protocol

2.3 SPICE Protocol

In mobile cloud environment, the service provider needs the effective remote access protocol. Due to the characteristic of mobile environment, reducing network traffic is major consideration. From this perspective, the RFB protocol sends the capture image of remote server. Thus, it occurs considerable network traffic by using the remote view image. Therefore, the RDP protocol is superior to the RFB protocol. However, the RDP protocol cannot support mobile platforms. Also, the RDP protocol supports server platform only Windows due to Microsoft licenses of this protocol. Thus, mobile cloud service provider cannot select it.

The SPICE protocol is a solution of above problems. It is one of open-source protocol among remote access protocols and it deals with various types of messages such as accessing, controlling, and receiving inputs. The structure of this is similar to the RDP protocol. The SPICE protocol manages the message which is created by remote server. This protocol consists of many channels[7]. These channels can communicate separately.

Due to the specifications like the RDP protocol, the SPICE protocol can serve with small network traffic. Also, this protocol support working in virtualization environment. This protocol is developed to support Virtual Desktop Infrastructure(VDI). Therefore, the SPICE protocol is best suitable for mobile cloud service than other remote access protocols. Figure 5 shows the architecture of the the SPICE protocol.

Although the SPICE protocol has some advantages, it cannot support mobile platform yet. It supports just PC platform such as Windows and Linux. To use SPICE protocol in Android, SPICE protocol need to be converted. Because Android is based on Linux kernel, SPICE protocol for Linux can be converted to

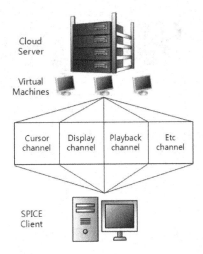

Fig. 5. SPICE protocol

Android version. Therefore, next section describes usefulness of SPICE protocol in Android platform and how to convert from Linux to Android.

3   SPICE for Android

The SPICE protocol in Linux platform has a possibility of conversion to use other platforms. In particular, if it can be changed to Android version, the Android SPICE protocol will be effective remote access protocol. The SPICE protocol in Linux platform can be changed to Android version, because Android platform is based on Linux kernel[6]. To convert this, some modules are needed to be cross-compiled to mobile platform earlier. These modules such as iconv, glib, openssl, and jpeg help to run the SPICE protocol. Also, the Android Native Development Kit(NDK) helps to create the library file from cross-compiled SPICE protocol object with other essential modules. Created library file name is 'libspicec.so'. This supports the Android platform to use the SPICE protocol.

Figure 6 shows the structure of the Android SPICE protocol. It works on a Java Native Interface(JNI) and created remote view by Android SPICE protocol is shown at an activity. Due to the structure of Android, the remote view image is not directly shown to user. It is created as a native library but it cannot show this view directly. The Android SPICE protocol must relay the remote view image to an activity. For this process, this protocol fulfills additional image encoding. First, this protocol creates BGRA raw image when client receive the message then it converts this image to RGB raw image. The converted image is encoded to jpeg image by libjpeg module. The encoded image is sent to main activity and then it represents remote view finally. In this activity, the received remote view image is additionally resized and rotated for convenience of end user. As a result, the Android SPICE protocol needs to such additional process unlike Linux SPICE protocol which is shown the remote view through using x11.

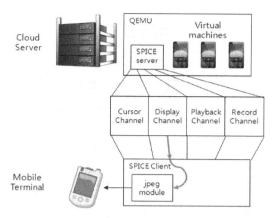

Fig. 6. Android SPICE protocol

This processes make a delay to represent the remote view. It consumes about 400ms per each frame. Thus, this processing delay occurs inconvenient use of this protocol. Therefore, The next section introduces the cause of the delay and how to solve this problem.

4   Improvement of SPICE for Android

Simply converted SPICE protocol is insufficient to support the Android platform. This protocol need to be improved by removing some delay points in it. If the defect of this protocol is cleared, it changes to best remote access protocol for the Android platform. There are two major delay points in this protocol. One is a jpeg encoding. This protocol encodes a RGB raw image to a jpeg image before relaying to an activity. If it can show raw image directly, the protocol can relay a raw image to an activity without encoding process. The createBitmap() method can support a raw image view. Thus, the test of representing dummy raw image is successful. However, it is very slow. In a Galaxy Note with a resolution of 1280×800, white dummy raw image is shown for about $200 \sim 250ms$ each frame. The size of it is about 3 Mbytes($1280 \times 800 \times 3$). The device cannot handle this processing load which consists of dozens of frames in each second yet. Therefore, Android SPICE protocol cannot remove a jpeg encoding process. Because this process is not removed, the other way that it is changing faster encoding module is selected. The jpeg-turbo library in Linux is an alternative of existing encoding module. If this protocol can replace a jpeg module with a jpeg-turbo, it has slightly reduced the encoding delay. The jpeg-turbo module can be cross-compiled similar to an iconv module. In addition, the activity has an extra bitmap creating process. This process works the changing a remote view to user suitably. It is also additional delay point. If this process may be removed, the Android SPICE protocol will be fast more. Figure 7 shows the delay points and enhanced points in it.

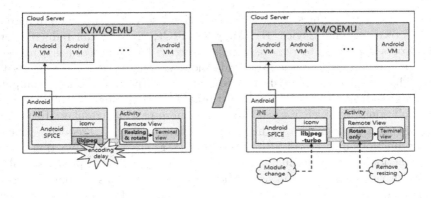

Fig. 7. Encoding delay points and modified points in Android SPICE protocol

The test result with a jpeg-turbo module has better than we expected. The replaced module brings reduced response time. Table 1 and Figure 8 present the results of improving the SPICE protocol.

Table 1. Processing delay time of Android SPICE protocol(ms)

	libjpeg	libjpeg-turbo	time-difference
trial 1	221	106	115
trial 2	228	125	103
trial 3	222	109	113
trial 4	336	123	213
trial 5	238	152	86
average	249	123	126

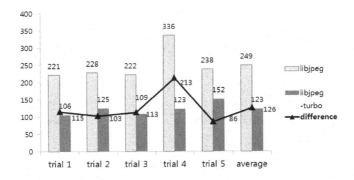

Fig. 8. Time difference between two SPICE protocols(ms)

Moreover, if it assumes that users want cloud services with same environment as own terminal device, the Android SPICE protocol can reduce a few more processing delays to remove resizing process. This protocol has a process of resizing and rotating view to provide more various server environments to users. This characteristic satisfies a requirement of the mobile cloud computing. Thus, Android SPICE protocol must provide the expandability of platforms and resolutions. However, we want to reduce more processing delay. If the Android SPICE protocol improves this process, it brings additional reducing delays. Among functions of Android SPICE protocol, there are a rotate function and a zoom function in optional functions. Therefore, Android SPICE protocol is able to provide multiple resolution using these optional functions. In our test, this manner gets about 160ms delay reducing. Table 2 presents the results of reducing time of this manner. This result is better than Table 1, but this way limits to support various resolution of mobile virtual machine. For the expandability of mobile cloud services, we do not recommend unconditional removal of resizing process to service provider.

Table 2. Processing delay time of removing resizing process(ms)

	before	after	delay-difference
trial 1	222	62	160
trial 2	246	43	203
trial 3	200	49	151
trial 4	193	62	131
trial 5	225	58	197
average	223.2	54.8	168.4

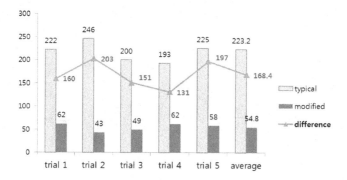

Fig. 9. Delay difference by removing resizing process(ms)

5   Conclusion

The cloud computing services grows up and up, and many people use the smart devices. According to these trends, users want the mobile cloud computing. This provides higher mobile computing resource. To offer better mobile cloud computing, the mobile cloud is required an effective remote access protocol. This protocol has many types, but the SPICE protocol is the best one from two perspectives - performance and platform compatibility. However, due to the mobile platform support problem, we introduce the Android SPICE protocol that is modified SPICE protocol to use in Android platform. In addition, because performance problems of Android SPICE protocol, we also find the points of delay of the Android SPICE protocol and propose the enhanced Android SPICE protocol. New encoding module and removal of additional image object decreases of process delay in Android SPICE protocol. The improved protocol makes a decrease of processing delay about 50%. The Android SPICE protocol will be improved more and more for providing high speed mobile cloud services. Moreover, another mobile platform needs SPICE protocol for mobile cloud. For future work, we study about SPICE protocol for iOS and the other mobile platform.

Acknowledgement. This work was supported by Priority Research Centers Program through the National Research Foundation of Korea(NRF) funded by the Ministry of Education, Science and Technology(2012-0005861).

References

1. Gartner, "Forecast Overview: Public Cloud Services, Worldwide, 2011-2016" (August 2012)
2. Jang, S.-J.: Design of the Remote Management System in the Windows Operating System. In: IJCSNS 2011 (2011)
3. Richardson, T.: The RFB Protocol. RealVNC Ltd. (2005)
4. MSDN (February 2013),
   http://msdn.microsoft.com/en-us/library/aa383015(VS.85).aspx
5. Red hat, "Spice remote computing protocol definition v1.0". Red hat, Inc. (2009)
6. Porting snappy/libspicec.so on Android-ARM (September 2012),
   http://blog.csdn.net/rozenix/article/details/6277647
7. spice-client-android (October 2012),
   http://code.google.com/p/spice-client-android/

The Permission-Based Malicious Behaviors Monitoring Model for the Android OS

Min-Woo Park[1], Young-Hyun Choi[1],
Jung-Ho Eom[2], and Tai-Myoung Chung[1,*]

[1] Internet Management Technology Laboratory,
Electrical and Computer Engineering,
Sungkyunkwan University, Republic of Korea
[2] Daejeon University, Republic of Korea
{mwpark,yhchoi}@imtl.skku.ac.kr
eomhun@gmail.com,tmchung@ece.skku.ac.kr

Abstract. Nowadays, the smartphone has become the critical issue of users information security. The smartphone services not only principal functions of a cellular phone but also personalized functions such as mobile banking and GPS location tracking. It means that an attacker will be able to obtain a variety of personal information when an attacker compromises the smartphone. For this reason, malwares target environments have been changed to the smartphone platforms from computers. The android OS has the permission mechanism that was designed to prevent malicious behaviors of installed applications. However most of the users are not concerned about permissions of application to install. Thus, the android OS needs an advanced security mechanism that will be able to warn malicious behaviors of installed applications to the user. In this paper, we proposed permission-based application monitoring model, which is performed at the Android OS-level.

Keywords: Mobile Security, Android Malware Analysis, Malicious Behavior Detecting, Android Permission Mechanism.

1 Introduction

The Android phone is composed of various devices, which are network interfaces (e.g., cellular network interface, WiFi interface, Bluetooth network interface, and near field communication interface) and sensors (e.g., GPS and gyroscope). The Google opens the Android Software Development Kit [1] which is able to access and control these devices to the 3rd party application developer for developing user applications and provides an official market, called GoolgePlay, for distributing 3rd party developed applications. As a result, a wide variety of applications have been developed and distributed. According to the announcement of the Android Official Blog, it had reached 25 billion downloads and 675,000 total apps in the GooglePlay [2,8]. The Android phone users can personalize

---

* Corresponding author.

B. Murgante et al. (Eds.): ICCSA 2013, Part I, LNCS 7971, pp. 382–395, 2013.
© Springer-Verlag Berlin Heidelberg 2013

their phone through downloading and installing freely these 3rd party developed applications from official markets or unofficial markets. Nowadays, people can be provided with various convenient services by using personalized Android phone.

However, the smartphone quickly has become the critical point of users information security as it started to take large portion of our lives. It is able to perform various new tasks by using mobility and network connectability besides the existing tasks of the PC. As a result, a small hand-held smartphone has a lot of complex and sensitive information on users pictures, contacts, interests, social relationship, business information and etc. The smartphone targeted malware has increased significantly with the rising use of smartphone [3]. In particularly, the Android OS recently has become the main target of the malware.

The Android OS has three security weaknesses: (i) policies of the Android OS are opened, so user can install an untrustworthy application which is downloaded from unofficial markets; (ii) official market of the Android OS does not check whether 3rd party developed application is malicious or not in application registration process. Thus, it is difficult to block registration of application that contains malicious code in advance. Actually, lots of malwares have been distributed through official market such as Android.Adsms, Android.DroidDream, and Android.Zeahache [4]; (iii) every applications are authorized their permissions at installing time and these authorized permissions are unalterable until application is updated or deleted. The Android OS does not report about permissions of application except installing time. Thus, if user does not find mis-authorized permissions at installing time, finding these mis-authorized permissions is very difficult; (iv) permission authorization process is very inflexible. User cannot give permissions to application flexibly. User just determine whether installing application or not. In many cases user doesnt care about permissions when user installs application [5].

Advanced security model is necessary to protect the Android devices from increasing threats of malwares [13]. The Android OS is primitively designed to restrict every behavior, escape from sandbox, of user applications. It means user application cannot communicate with other applications and access to databases or devices [6]. This policy increases the overall security of the Android OS but bring restricting functionality of applications. To relieve restrictions of functionality of applications, the Android OS provides permission mechanism. A permission of the Android OS is composed of the component that is accessible by the application and operation like ACCESS, READ, WRITE, SET, GET, and so on.

In this paper, we propose the permission-based malicious behavior monitoring model. Our model traces permission validation check methods for detecting malicious behavior of user applications. Permission validation check methods are called by system processes when user application access to the component that is protected by permission mechanism. If user application contains malicious codes, it must accesses the component such as ContentProviders.

In the following Sections 2 we describe trend, types, and researches for detecting of malwares on the Android platform. In Section 3 we discuss permission mechanism of the Android OS. While in Section 4, we propose security model for detecting and reporting malicious behaviors. Lastly, we conclude this paper in Section 5.

Table 1. The number of discovered malwares for the Android OS

Discovered date	The number of discovered malwares
2010	5
2011	42
2012	82
total	127

Table 2. Classification of Android malwares that are discovered over the period 2010-2012

Malware type	The number of malwares	Rate
Trojan	130	84.4%
Spyware	19	12.3%
Potentally Unwanted App	2	1.3%
Security Assessment Tool Trojan	1	0.65%
total	157	100%

2   Android Malware

2.1   Trend of Android Malware

After the first-ever malicious program, AndroidOS.FakePlayer, for the Android OS was detected in August 2010, the number of malicious program for the Android OS has been steadily increasing [7]. Table 1 shows the number of discovered malwares for the Android OS from Aug 2010 to Dec 2012. This information was extracted from the public database of Symantec [4]. The total number of Android malwares that are registered on the database of Symantec is 153 in period 2010-2012. Among malwares that are registered, only 127 of them have specific date of detection. According to this information, android malwares has increased annually and this trend is expected to continue: 5 malwares were discovered in 2010, 42 malwares were discovered in 2011, and 82 malwares were discovered in 2012.

2.2   Types of Android Malware

Great feature of the smartphone is able to be personalized to the needs of the individual by using user applications. Malware authors are generally using this feature to propagate malicious code. To attract users attention, malware authors pass their malware off as useful application on the application market or create and distribute malware that is repackaged legitimate application with malicious codes. Type of these malwares is classified as a Trojan. The Trojan is a large part of malwares in Android malwares. Android malwares are classified as Trojan, Spyware, Potentially Unwanted Application, or Security Assessment Trojan and the classification results are shown in Table 1.

- **Trojan:** A Trojan is a type of malware which appears to provide useful functions that are whatever user wants. A Trojan malware performs complex and various operations to accomplish its purpose. A Trojan is downloaded (and installed) by user.
- **Spyware:** A Spyware stealthily collects information (e.g., SMS, E-mail, and GPS location) and reports to malware authors. Sometimes user installs spyware, such as *Android.Beaglespy* [9], himself on their smartphone in case it gets lost.
- **Potentially Unwanted Application:** Applications of this group are actually not malware but these applications can be used maliciously. For instance, *Android.Ecardgrabber* [10] reads credit card information through NFC module and prints this information on screen. *Android.Ecardgrabber* can be misused for credit card information leakage.

2.3   Android Malware Detection

Misused detection is a signature-based malware detection mechanism. The signature is a kind of malicious patters that are extracted through malware analysis. Misused detection is mainly used mechanism for anti-virus tools in the PC platform. It has some problems in applying misuse detection methods to the Android platform: (i) mobile devices are always suffering from a lack of resources. Detection overhead of misused detection methods depends on the amount of malicious signatures. As the number of android malware increases, more resources will be consumed for misused detection. From a long-term perspective, application-based misused detection methods are not suitable in mobile environments; (ii) the Android platform allows downloading applications from unofficial markets and installing these applications. Thus, a fledgling developer who is a lack of security knowledge can easily distribute his applications through unofficial markets. As a result, signatures for misused detection are increased as potentially malicious applications are on the rise; (iii) it is difficult to determine the policies about applications that contains malicious code but it is not downloaded due to low recognition. It is inefficient to create a signature for these applications. However, if misused detection cannot find known malwares, misused detection loses its reliability.

Anomaly detection is a behavior-based malware detection mechanism. Anomaly detection is usually using machine learning to determine whether a series of operation is malicious behavior or not. Anomaly detection can detect unknown malwares, but a lot of discussion about its accuracy. Bose et al [11] proposes behavioral detection mechanism by training Support Vector Machines (SVMs).

3   Permission of the Android OS

In this section, we describe the permission architecture of the Android OS and the enforcing mechanism of permission.

3.1   Architecture of the Permission

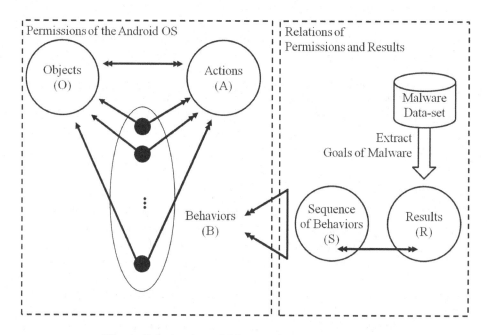

Fig. 1. Relationship of Objects, Actions, and Results

Fig. 1 shows relationship of Object, Action, Behavior, Sequence of Behavior, and Result. The upper case denotes a set (e.g., each O, A, B, S, and R mean each set of object, action, behavior, sequence of behavior, and result). The arrows denote relationship between entities. In case that both sides are double-headed arrows, it means many-to-many relationship. In other case that only one side is double-headed arrow, it means one-to-many relationship. In this figure, the left-side indicates original architecture of permission in the Android OS and the right-side indicates relation of sequence of behaviors and purpose or goal of malware.

The permission of the Android OS is composed of the Object, which is a set of protected components, and the Action, which is a set of restricted operations.

The Object and the Action are arranged Table 3 and Table 4. The Behavior means individual permissions such as to send SMS messages. Each Behavior maps one Object to relate many Actions. The Sequence of Behaviors and The Results are additional concept for anomaly detection. The Sequence of Behaviors is a set of sequential operations that are analyzed to potentially malicious operations. The Results is purpose or goal of malwares. It is extracted from data-set of Android malwares.

Table 3. Objects

Object Type	Objects
User Information	Contact, E-mail, SMS, MMS, Phone, Call_log, Web history and bookmark, account, calendar, and etc.
Device Information	IMEI, IMSI, Phone number, APN, Device manufacture and model, Version of OS and SDK, WiFi state, BLUETOOTH, CAMERA, status bar, External storage, system setting, and etc.

Table 4. Actions

Actions	Descriptions
SEND	Process sends Message by using Internet or cellular network
READ	Process accesses to OBJECT and gets DATA from OBJECT
WRITE	Process accesses to OBJECT and sets DATA to OBJECT
DELETE	Process accesses to OBJECT and delete DATA from OBJECT
BLOCK	Process deletes or hides received Message from user
ACCESS	Process takes reading or writing permission about OBJECT
DISCOVER	Process discovers paired bluetooth devices
PAIR	Process pairs smartphone with bluetooth device
BROADCAST	Process received broadcast messages such as packet incoming
CALL	Process dials some numbers without user's cognitive
INSTALL	Process setups additional application including malicious codes
GET	Process takes information
KILL	Process stops the other process
REBOOT	Process restart compromised device

3.2 Enforcing mechanism of the Permission

The Android OS is primitively designed to restrict every behavior, escape from sandbox, of user applications. It means user application is restricted to call some APIs that are related to sensitive services such as to control devices (e.g., CAMERA, BLUETOOTH, WiFi, gyroscope and etc) or to access internal databases. The Android OS checks permissions of user application when this application tries to access components that can affect other applications and global phone state. Permissions of user process are specified on the AndroidManifest.xml. Every user process has its AndroidManifest.xml file which has information for running user application such as unique identifier for user application, application components, permissions, and etc. Fig. 2 is an example code, which is declaring part of user applications permissions on the AndroidManifest.xml.

The Android OS is composed of the user space and the kernel space. Every user application is running in the user space. User application is written in the Android API libraries. The Android API libraries have no ability to directly access other applications or kernel. The libraries just provide interfaces to access other applications or kernel through the internal APIs. The internal APIs are running in the kernel space and these APIs can access user applications or change kernel states. The system process checks validation of permissions when application, which is running in the user space, uses the Android API libraries [12]. Generally, permission check is done using following methods in the Context class:

- checkPermission();
- checkCallingPermission();
- checkCallingOrSelfPermission();
- enforcePermission();
- enforceCallingPermission();
- enforceCallingOrSelfPermission().

```
<!-- Allows an application to send SMS messages. -->

<permission android:name="android.permission.SEND_SMS"

        android:permissionGroup="android.permission-group.COST_MONEY"

        android:protectionLevel="dangerous"

        android:label="@string/permlab_sendSms"

        android:description="@string/permdesc_sendSms" />
```

Fig. 2. Declaring part of user applications permissions on the AndroidManifest.xml

Enforcing permissions is different depending on the components [6]. Generally, permissions are checked at any call into a service. Fig. 3 shows source code that is part of internal class IccSmsInterfaceManager.java. Activity and service permissions is checked during start methods like Context.startActivity() and Context.startService(). ContentProvider, which is kind of internal database, permission is composed of two detail permissions: read permission and write permission. Read permission is checked during ContentResolver.query() and Write permission is checked during ContentResolver.insert(), ContentResolver.update(), and ContentResolver.delete().

```
public void sendText(String destAddr, String scAddr,

    String text, PendingIntent sentIntent, PendingIntent deliveryIntent) {

  mPhone.getContext().enforceCallingOrSelfPermission(

     "android.permission.SEND_SMS",

     "Sending SMS message");

  if (Log.isLoggable("SMS", Log.VERBOSE)) {

    log("sendText: destAddr=" + destAddr + " scAddr=" + scAddr +

       " text='"+ text + "' sentIntent=" +

      sentIntent + " deliveryIntent=" + deliveryIntent);

    }

  mDispatcher.sendText(destAddr, scAddr, text, sentIntent, deliveryIntent);

}
```

Fig. 3. Source code of sendText() methods in IccSmsInterfceManager class

4   The Permission-Based Malicious Behaviors Monitoring Model

The permission-based malicious behavior monitoring model detects and reports malicious behavior of user applications through tracing permission validation check processes.

4.1   Concept of the Monitoring Model

In the Android platform, most of infection is occurred by user due to download and install Trojan malware. Some people do not want to pay for downloading applications. Thus, these people download paid application from unofficial markets. Malware authors use these habits for malware propagation. Malware authors repackage paid application with malicious code and distribute it through unofficial markets for free. Sometime malware is disguised as a useful application. If user can confirm behaviors of downloaded applications, he will be able to judge whether downloaded applications are malicious or not. In fact, permission mechanism is responsible for warning to about potential malicious of applications. However, user generally doesnt pay attention to permission list of application [5]. Moreover, permission list is notified only during installing time, not running time. As a result, the warning role of permission mechanism has become the titular role. Because of this reason, the Android OS needs the monitoring model that can trace behaviors of user application and report potential malicious of application during running time.

4.2   Users Role in Our Model

The normal application: The Trojan application:

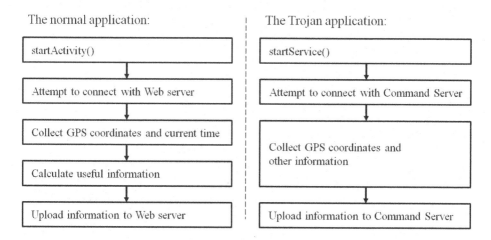

Fig. 4. Assumed high-level behaviours of the two example applications

Distinguishing malicious behavior from normal behavior is a very difficult task.
So, the monitoring model assigns responsibility of decision, whether malicious
or not, to user. For example, we compare two applications. One is the normal
behavior application that tracks GPS coordinates for recording jogging history.
Another one is the Trojan malware that leaking GPS coordinates. Fig. 4 shows
assumed behaviors of two applications. The normal application needs user interfaces. So, this application is started by startActivity(). When the normal
application is launched, it attempt to connect with its web server that provides
useful information such as jogging history, amount of exercise, calorie consumption, and so on. In the running, the normal application collects GPS coordinates
and creates useful information. If jogging is over, this application uploads jogging history to the web server. The Trojan malware is leaking GPS coordinates
stealthily. So, the Trojan malware does not need user interfaces. If the Trojan malware has user interfaces, it will be fake user interfaces for hiding itself.
The Trojan malware is executing at the background, so the Trojan malware is
started by startService(). The Trojan malware is launched successfully, it trying to communicate with Command Server. Then, the Trojan malware performs
malicious behaviors, which are pre-defined or received from Command Server. In
this example, malicious behaviors are collecting GPS coordinates and uploading
information to Command Server. Behaviors of the normal application and the
Trojan malware are very similar, thus it is impossible to detecting malicious
behavior just by using difference of behaviors. For distinguishing the Trojan
malware, additional patterns are necessary like IP address of Command Server,
package name, and so on. However, if detecting mechanism depends on these

additional patterns for detecting, it will not be able to detect unknown malicious codes.

We define the Android malwares as that application performs unexpected behaviors. So, user determines whether running application is malicious or not.

4.3 Monitoring System Architecture

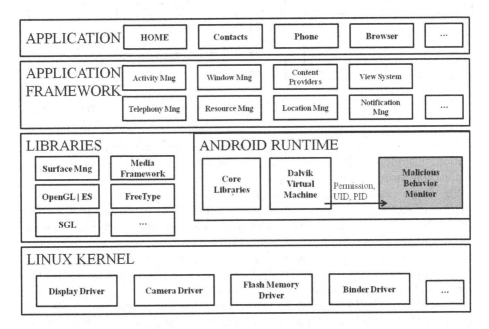

Fig. 5. Location of the Malicious Behavior Monitor in System Architecture

The malicious behavior monitor is core component, which is responsible for detecting and reporting malicious behavior of user application. Fig. 5 shows location of the malicious behavior monitor in the Android OS system architecture. This monitor is located in the android runtime layer for communication with the dalvik virtual machine (DVM). The malicious behavior monitor receives validation results of permission from the DVM for tracing sensitive behaviors of user application in real-time. The malicious behavior monitor records sequential behaviors of each application and reports warning when it catches suspicious behaviors.

The malicious behavior monitor uses the permission mechanism of the Android OS for detecting malicious behaviors of each application. The Android OS restricts operations that will be able to affect other applications or change global phone state by using application permissions. Thus, if the malware invokes one or more restricted API libraries that escape the sandbox of user application to

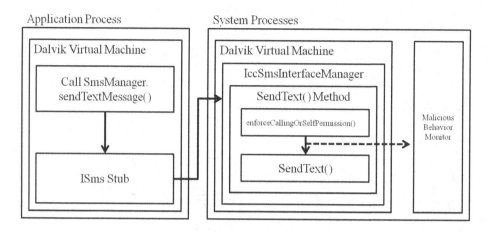

Fig. 6. An example of permission validation in IccSmsInterfaceManager class

accomplish its purpose, the malware is many times experienced permission validation. Therefore, the malicious behavior monitor can trace behaviors of each
application through validation results of permission.

Fig. 6 is a detail example of permission' validation check. It figures sequential flow of method calls abstractly when sendTextMessage(), which is defined in
the SmsManager class, is invoked in user application. If sendTextMessage() is invoked, the control of user application is changed to system process and then internally implemented method is invoked. The internally implemented method of the
SmsManager.sendTextMessage() is the IccSmsInterfaceManager.sendText(). The
permission checking is operated in this internal method. In the sendText() case,
enforceCallingOrSelfPermission() is invoked for permission validation and it veri-
fies that SEND_SMS permission is specified in its the AndroidManifest.xml file of
the user. If SEND_SMS permission is not specified in the AndroidManifest.xml,
enforceCallingOrSelfPermission() returns the SecurityException. Therefore, it is
necessary to internal method modification for receiving validation results.

Fig. 7 shows the malicious behavior monitors architecture. The malicious behavior monitor is composed of three modules and two data. First module is a
communication module, which is responsibility for communication with the other
components such as the DVM. This module delivers validation results from permission validation checking methods (e.g., enforceCallingOrSelfPermission) to
the table handler module or reports analyzing result to user through popup
message or status bar. Second module is a table handler module, which handles
the application behavior state table. The application behavior state table has information of running applications behaviors. The columns of this table are index
number, insert timestamp, last update timestamp, and the head pointer of each
applications behavior data that is a linked-list data structure. The table handler
module is notified validation results through the communication module. If this
module is received report about new activity or service, it adds new entry on the
application behavior state table. If not, this module finds the entry on the table

and inserts new behavior information on the end of the behavior pointer. This module delivers table update notification to an analyzer module. The analyzer module is core module in the malicious behavior monitor. This module compares a series of applications behaviors with the data-set of malicious behaviors. In this data-set, a set of potentially malicious behaviors is stored. If updated a set of behaviors are same with a set of potentially malicious behaviors, this module requests to send report to the communication module.

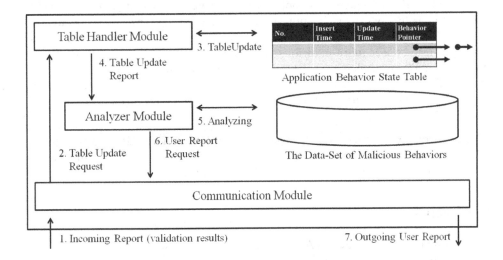

Fig. 7. The malicious behavior monitor's architecture

An operation order of the malicious behavior monitor is described in fig. 7: (i) a validation result is incoming to the communication module; (ii) the communication module requests table update to the table handler module; (iii) the table handler module updates the application behavior state table; (iv) the table handler module reports result of table update to the analyzer module; (v) the analyzer module compares applications behaviors with stored potentially malicious behaviors; (vi) according to analyzing results, the analyzer requests user report to the communication module; (vii) the communication reports malicious behaviors of application through popup message or status bar.

5 Conclusion

Recently, security for Android platform is highly researched as risks of the Android malware are sharply increased. In particular, research about malware detecting mechanism has become very important research field. In the Android OS, the permission mechanism has responsibility to notify user of applications risk as basic security feature. However, this role, actually, is not accomplished. In many cases, most users do not care about notified permission list of application. As a result, advanced security feature is necessary for the Android OS. In this paper, we propose the permission-based malicious behavior monitoring model. This monitoring model is located in android runtime layer, and it traces operation sequences for detecting malicious application. The monitoring model reports presumed malicious behaviors to user in real-time, and users determine whether these behaviors are malicious or not. We must consider following features for the future works: (i) it is necessary to design detection mechanism about two or more malicious applications cooperation for their purpose. Our model traces sequential behaviors of each application and decides whether these behaviors are malicious or not through tracing result. However, if a series of behaviors are split into two or more malicious applications, our model cannot trace these whole of malicious behaviors; (ii) it is necessary to improve detection accuracy for reliability about security reports of out model. If our model lost reliability, it has become same state with the Android OSs permission mechanism.

Acknowledgments. This work was supported by Priority Research Centers Program through the National Research Foundation of Korea(NRF) funded by the Ministry of Education, Science and Technology(2012-0005861).

References

1. Android Developers, `http://developer.android.com/sdk/index.html`
2. Android Official Blog. Google Play hits 25 billion downloads,
   `http://officialandroid.blogspot.kr/2012/09/`
   `google-play-hits-25-billion-downloads.html`
3. Felt, A.P., Finifter, M., Chin, E., Hanna, S., Wagner, D.: A Survey of Mobile Malware in the Wild. In: The 1st ACM Workshop on Security and Privacy in Smartphones and Mobile Devices, pp. 3–14. ACM, New York (2011)
4. Symantec, A-Z listing of threats & risks,
   `http://www.symantec.com/security_respo-nse/landing/azlisting.jsp?`
   `inid=us_sr_flyout_reporting_allthreats`
5. Felt, A.P., Ha, E., Egelman, S., Haney, A., Chin, E., Wanger, D.: Android Permissions: User Attention, Comprehension, and Behavior. In: The Eighth Symposium on Usable Privacy and Security, Article No. 3. ACM, New York (2012)
6. Android Developers, Permissions,
   `http://developer.android.com/guide/topics/secur-ity/permissions.html`
7. Symantec, AndroidOS.FakePlayer,
   `http://www.symantec.com/security_response/w-riteup.jsp?`
   `docid=2010-081100-1646-99`
8. Polla, M.L., Martinelli, F., Sgandurra, D.: A survey on Security for Mobile devices. IEEE Communications Surveys & Tutorials 15(1), 446–471
9. Symantec, Android.Beaglespy,
   `http://www.symantec.com/security_response/write-up.jsp?`
   `docid=2012-091010-0627-99`

10. Symantec, Android.Ecardgrabber,
    http://www.symantec.com/security_response/-writeup.jsp?
    docid=2012-062215-0939-99
11. Bose, A.: Hu. X., Shin, K.G., Park, T.: Behavioral Detection of Malware on Mobile
    Handsets. In: The 6th International Conference on Mobile Systems, Applications,
    and Services, pp. 225–238. ACM, New York (2008)
12. Felt, A.P., Chin, E., Hanna, S., Song, D., Wanger, D.: Android Permissions Demys-
    tified. In: The 18th ACM Conference on Computer and Communications Security,
    pp. 627–638. ACM, New York (2011)
13. Enck, W., Ongtang, M., McDaniel, P.: Understanding Android Security. IEEE
    Security and Privacy 7(1), 50–57 (2009)

A Method for Cricket Bowling Action Classification and Analysis Using a System of Inertial Sensors

Saad Qaisar[1], Sahar Imtiaz[1], Paul Glazier[2], Fatima Farooq[1], Amna Jamal[1],
Wafa Iqbal[1], and Sungyoung Lee[3]

[1] School of Electrical Engineering & Computer Science
National University of Science & Technology
Islamabad, Pakistan
[2] Sheffield Hallam University, United Kingdom
[3] Kyung Hee University, Korea
{11mseesimtiaz,saad.qaisar}@seecs.edu.pk, paul@paulglazier.info,
{ffaruq,wafa.iqbal1}@gmail.com, A0090875@nus.edu.sg,
sylee@oslab.khu.ac.kr

Abstract. A number of similar structured wireless sensors, constituting a Wireless Sensor Network (WSN) are used for activity recognition— particularly for coaching of a bowler to practice correct bowling action in the game of cricket. Several experiments are conducted for training certain algorithms, like K-means and Hidden Markov Model, etc., and the real-time data acquired by a subject under study, or a cricket bowler, is tested for statistical characteristics' comparison. This paper explains the whole implemented system and the prime application in which it can assist in the field of cricket.

Keywords: components: WSN, Bluetooth, K-means, MM, HMM, Cricket, Sports biomechanics.

1 Introduction and Motivation

A Wireless Sensor Network (WSN) comprises a number of devices known as wireless sensors that are designed specifically for acquiring data and processing it according to requirements [1]. In context of research, WSN is a major field in which extensive research work is being done nowadays. However, as the technological evolution continues, many unexplored areas are being discovered and thus, present an excellent opportunity for the researchers to expand their knowledge by innovation in the area of wireless sensors and wireless communications.

The latest technology has enabled us to design such compact and versatile devices that can greatly assist us in daily-life activities [2]. Weiser investigated how the pre-existing technologies can be used to integrate the latest information technology in daily-life activities of the people [3]. WSNs are deployed for a variety of applications, most of them related to daily life activities. Extensive research has been conducted in the past decades, and is still going on, particularly in the area of activity recognition.

B. Murgante et al. (Eds.): ICCSA 2013, Part I, LNCS 7971, pp. 396–412, 2013.
© Springer-Verlag Berlin Heidelberg 2013

Activity recognition is a process of identifying certain activities of a subject under study that are being carried out in real-time, by comparing them with the information readily stored in the system. Human Activity Recognition is another perspective of activity recognition in which human activities are recognized by deploying external sensors such as motion, acceleration and video sensors [4]. Usually, the activity recognition process is used for medicine or biomechanics related applications. A WSN for activity recognition of a patient suffering from Parkinson's disease has been discussed by Akay [5] and thus, the movement of various body parts is monitored which eventually helps in the recovery process. Activity recognition can also be used for security purposes or for monitoring the data traffic in a particular space at a particular instant of time as done by Weaver et. al [6]. Recently, inertial sensors are integrated with antenna for developing a compact device capable of performing the body motion analysis [7].

Activity recognition can be used for certain applications in sports technology. Previously, Aerts [8] has developed a wearable device for monitoring different factors like velocity, acceleration, body temperature, etc., for sports like jogging, swimming, cycling and similar sports. Recently, inertial sensors have been extensively deployed for various applications, and activity recognition is one of them. The use of inertial sensors for activity recognition purpose has been further extended for deployment in particular fields of sports and training; swimming, golf, sprint-running and tennis to name a few. Sports regulatory bodies have also shown keen interest in developing compact devices that can be easily used for training and monitoring of developing players and athletes [9 -10]. Cricket is one major sport in which researchers are carrying out advance level studies in order to develop an efficient device that can be specifically used to monitor and detect legality of a move during a bowling action [9, 11]. Researchers like Portus and Wixted [9, 11] are working in collaboration with some testing facilities to achieve a solution for monitoring and detecting an illegal bowling action performed by a player on-field, instead of doing the same at a dedicated facility and under unnatural circumstances.

Wixted et. al. [11] have proposed a method for elbow axis alignment to compare the output of sensors for known data sets for legal and illegal deliveries, and the system was vali-dated with the help of a video based motion capture system. Another approach of detecting the illegal bowling action has been discussed by Wixted et. al. [9] by monitoring the angle of bowling arm during a delivery, by comparing the data collect-ed from MEMS devices with that obtained using Vicon markers. Research is also being carried out in determining the accuracy of wearable sensor device data in comparison with the data obtained using the testing facilities available today.

Although the problem of designing an efficient and compact device to detect the legality of bowling action in the game of cricket has been addressed by many researchers [9-10], however, the coaching technique using miniature sensors has not been given much attention. Further, the equipment cost made the solution unviable for use by the developing players or for personal training purposes. The use of video based system in well-established laboratories was also of concern for personnel due to privacy issues. Besides, the use of cameras limited the application in terms of viewing angle possibility with a fixed number of camera devices.

As mentioned earlier, coaching is also necessary to properly train the bowlers to comply with the rules of cricket. Most of the time, the coaching is done by showing the videos of experienced bowlers to the developing players. Some players also practice in the vicinity of biomechanics laboratory specialized for monitoring the bowling action. Nowadays, inertial sensor based devices are used for coaching in sports like sprint running [12], tenpin bowling [13], baseball [14], etc. Similar coaching strategy can also be applied for coaching the cricket bowlers using inertial sensors.

As mentioned in the work by Eric Guetenberg et. al. [15], parameter extraction is also possible without direct calculation of angle and position to identify an activity. Our system is different from the one proposed by Eric Guetenberg et. al. [15] as the acquired data from inertial sensors is used to train the HMM to generate the corresponding template for checking the validity of a test data set against it. The identification of key events is not necessary since the training data set is acquired only for one particular action, signifying only a specific bowling style. This paper is based on an activity recognition system specifically designed for coaching a cricket bowler. The Micro-Electro-Mechanical Systems (MEMS) based Inertial Motion Units (IMUs) namely gyroscope and accelerometer, are used in this system. The data is collected from the sensors wirelessly and is passed on to a computational unit where it is compared with another data set that complies with the standards specified by the regulatory body (International Cricket Council (ICC), in this case). Care is taken that the rule of arm extension angle not exceeding 15 degrees [16] is followed in the standard data set, based on which the test data is classified as incorrect if it is not closely correlated to the standard data set. Section 2 of the paper discusses the design challenges that are faced at the system level, followed by a description of implementation methodology of the proposed system in section 3. The test scenarios and experimental setup for testing the implemented system is discussed in section 4, whereas the results and a discussion on critique are provided in section 5. The conclusion of the proposed system and the results is covered under section 6.

2   System Design Challenges

Proper coaching and training of a player in a game requires very precise details about parameters associated with movement so as to correctly follow the rules of the game. The technical challenges that are faced at every stage, from component level design to the system level design, while designing a system for training a bowler for a particular bowling action can be divided into two major categories; device level and, data processing and classification inference level, both of which are discussed as follows:

2.1   A. Device Level Structure

Keeping in view the work reported in literature [11], the designing of inertial sensor was accomplished by thorough investigation of each related parameter. MEMS based IMUs namely gyroscope and accelerometer are preferable to use because of their compactness in size, minimal power consumption and motion data capturing characteristics [17, 18].

Sensor data resolution is the next important parameter, as it specifies the number of samples of acceleration and angular velocity obtained to correctly model the action. However, care must be taken that higher data resolution may not lead to extra power consumption.

2.2   B. Choice of Communication Technology

Once the components are properly calibrated, a suitable technique needs to be chosen for communicating the data collected by the device to the processing module. The chosen wireless communication technology should meet the requirements for adequate data rate, proper networking, data aggregation process, low power consumption, greater communication range and most importantly the accuracy. Greater the accuracy of data acquired, the more reliable is the system functioning. Keeping in view all these constraints, the best suitable communication technology is chosen to be Bluetooth Class 2. Its transmission range is between 10 meters to 33 feet, power consumption is very low (about 25mW), and most importantly its operating frequency lies in the unlicensed spectrum [19]. Also, Bluetooth is feasible for use in most WSNs [20] and also because high data rate communication is required over a short range as it is an efficient technology for reducing end-to-end data transmission time [21]. Furthermore, the Adaptive Frequency Hopping (AFH) capability in Bluetooth technology makes it feasible for usage with other technologies operating in the 2.4 GHz spectrum, without interference [19]. The prime advantage of using inertial sensors coupled with wireless technology is the ease of mobility. Such devices can be easily used by coaches to train the developing players as well as by the players themselves while playing on field. Another advantage of such device design is that privacy of a player is not affected, since only the data associated with the bowling action is captured, instead of the player profile as done in the video based activity recognition processes. Moreover, the security of the player is not invaded since no information about physical features of the player is used for checking the bowling action.

2.3   C. Processing of Device Data for Testing the Correct Action

After successful acquisition of accurate data, the main task is to define sequence of steps to process the data and finally, to determine the correctness of action associated with it. The acquired data is first calibrated by passing it through a filter, which removes the noise in the data as well. The calibrated data is then refined using data estimation techniques commonly employed for mathematical analysis.

After noise removal and data refinement, the preprocessed data will contain two types of values; one for acceleration in the x, y and z axes, and the second for angular velocity obtained through gyroscope. This preprocessed data is used for feature extraction on the basis of which the classification engine will work. The classification engine comprises certain algorithms that have been deployed in the previous literature for machine learning. The choice of classification algorithm depends on the accuracy of detection required as well as the features extracted from the data. High level accuracy is required for this system to correctly classify the bowling action as correct or incorrect, therefore such algorithm is to be chosen that ensures an accuracy of nearly 100 percent.

3   Methodology

Based on the design challenges, discussed in the previous section, the proposed methodology can be divided into the following sub-categories:

3.1   A. Device Level Design

The initial WSN design is limited to three sensors to be placed on the arm. The sensors are similarly structured and are capable of acquiring data. The IMU is calibrated for a data rate of 150Hz so that data samples can be collected accurately without significant power consumption, however the calibration can be increased till 400Hz at the cost of greater power consumption. The data transmission is accomplished using Bluetooth class2. A microprocessor is used for interfacing the IMU and the Bluetooth transmission module to complete the device level design.

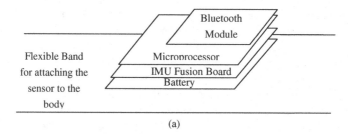

(a)

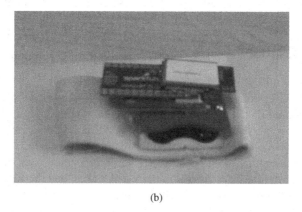

(b)

Fig. 1. The sensor module (a) schematic and (b) final form

The schematic of the sensor is shown in figure 1 (a). The sensor is made up of four separate parts; and all of them are connected in such a way that the sensor, as a whole, is compact and comfortable for use in lab environment or on-field. The microprocessor, Arduino Pro-Mini, is connected to the IMU 3000 Fusion Board. Transceiver operation is completed by the use of Bluetooth class2 compatible module. The whole assembly is connected to a 3.6 Volt rechargeable Lithium battery, placed at the bottom of the wearable sensor device and all the boards are connected in overlapping fashion to achieve compactness. The combined assembly is attached to a flexible band, so as to support its fixture on the desired body part. Figure 1(b) shows the sensor module used for the experiment discussed in this paper.

The placement of these sensors on the arm has to be carefully decided so as to correctly determine the anomaly in action, if any. Since the objective is to test the arm extension angle with respect to a standard data set, the three sensors are to be placed on the upper arm, the elbow joint and the wrist of the subject under test. This placement of three sensors will ensure maximum accuracy in determination of correctness of the delivery. Figure 2 shows the sensors placed at the upper arm, elbow joint and the wrist of a test subject.

The implemented system works according to a procedure outlining several steps. The first among them is to acquire real-time data from the sensor using Bluetooth. This is achieved by interfacing the Bluetooth module attached to the sensor, to the computation machine where either it is stored to be used later, or is processed on real-time basis. This data is pre-processed through calibration by a filter using signal processing techniques and then interpolated to smooth out the filtered data [22].

The output of preprocessing block gives a six tuple data; i.e., triple axis (X, Y, Z) acceleration and angular velocity. From this data, useful information has to be obtained for further processing. After detailed research and extensive literature review [14] following features were decided on

- Mean
- Mode
- Standard Deviation
- Peak to Peak Value
- Minimum
- Maximum
- First and Second derivative

These features are supposed to be evaluated at each data point and therefore take a shape of a multi-axis waveform. The resulting wave is then fed to the algorithmic block comprising the algorithms as elaborated in the following sub-section.

3.2   B. Classification Algorithms

The training of a bowler for a specific bowling action is accomplished by adopting the technique of unsupervised machine learning, in which the algorithms are first trained according to the available set of data known to be accurate and the sample data

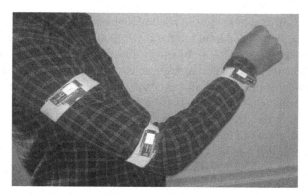

Fig. 2. Placement of sensors on the test subject's arm

is matched with the trained data set in order to check any anomalies. Unsupervised machine learning was opted over supervised learning due to its advantage in not binding the user or developers to manually train the data. The implemented system uses the following three algorithms for this purpose; however, certain other algorithms can also be used:

3.3   K-Means Clustering

The k-means clustering algorithm has been used in two ways in the presented work; one as a classification algorithm, the other as a means for generating observations for other classification algorithms (i.e. MM and HMM).

K-means is an unsupervised method of clustering where a set of data is clustered into k clusters. The algorithm has no prior knowledge of which cluster the data may belong to. The only inputs to the algorithm are the data and the desired number of clusters denoted by 'k'.

Typically, for k-means the data is not 1 dimensional. Each dimension represents a feature. A simple flow graph for a 1 dimensional k-means is illustrated in Figure 3. Note that for 2-dimensional data, each row represents a data record and each column represents a feature.

The number of clusters 'k' for k-means clustering was decided based on the number of 'events'. The event extraction is accomplished by using the 'min-max' algorithm. This algorithm is based on the concept that any local minima or maxima in the data represent a noticeable change in movement. Using this information the data is divided into 'events'. Each event is then fed into the clustering algorithm.

As a classification algorithm, the three-dimensional data was clustered and a template was generated in the form of k-means string for comparison. The data was clustered and a string of the form 'xxxx......xxx' was formed where 'x' can be any number ranging from 1 to k.

The k-means clustering is also used for generating the observations for MM and HMM. In order to cluster the events, a variation of the basic k-means algorithm

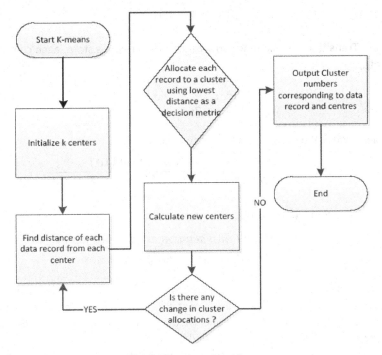

Fig. 3. K-means algorithm

was coded which could cluster three dimensional data. Each data record is k-means clustered and hence quantized. This reduces the number of values for the dataset as well as makes it easier to generate observations for the MM and HMM algorithms.

3.4   Markov Models

A MM is a probabilistic model that consists of a sequence of states. It is defined by the following parameters:

P_i: The probability of starting in a particular state
A: The state transition probability matrix which describes the probability of being in state i after state j

These are used to generate a model for MMs. The training data is first clustered according to the events using the special three dimensional k-means algorithm. The clustered data serves as training data for the sequence of states and hence both the state transition matrix and initial probability matrix. These parameters are then stored in a file to be used later.

For a three state model the matrices are as follows

Initial Probability Matrix:

$$[\pi_1 \quad \pi_2 \quad \pi_3]$$

State Transition Matrix: (each row represents previous state, each column next state)

$$\begin{bmatrix} a_{11} & a_{12} & a_{13} \\ a_{21} & a_{22} & a_{23} \\ a_{31} & a_{32} & a_{33} \end{bmatrix}$$

For each state S_i the transition probability to state S_j is a_{ij}, where

$$a_{ij} = \frac{No.\, of\, transitions\, from\, i\, to\, j}{Total\, No.\, of\, transitions\, from\, i\, to\, all\, states}$$

For each state S_i the initial probability is π_i, where

$$\pi_i = \frac{No.\, of\, training\, sequences\, starting\, with\, i}{Total\, No.\, of\, training\, sequences}$$

Here, the states 'S' are the average numbers of extremes in the whole data set. For clustering purposes, cluster all the events or states, identified together, individually for the entire data set and store their centers. Then k-means clustering is applied, keeping time as the feature, and each state is allotted a number.

The probability P of a sequence O of length n, with $\{a, \pi\}$ given, is found by

$$P = \pi_{O(1)} \prod_{i=1}^{n-1} a_{O(i)O(i+1)}$$

Using the probability P, the algorithm is trained for the given data set initially, and afterwards, the new data set, i.e., the one acquired on real-time basis, is compared with the initially given data set. For clustering purpose, the distance metric is decided as logarithm of the forward probability P and the new data set's distance metric is compared with a threshold value. If majority of the data set clusters are greater than or equal than the threshold, the activity is recognized as correct. Else, an anomalous action is detected.

3.5   Hidden Markov Models

In most real world scenarios, one cannot determine all possible states for a particular occurrence. To model the hidden states, Hidden Markov Models are mostly used. An HMM is a generative probabilistic model used for generating hidden states from observed data [23, 24]. HMMs imply that there is a set of hidden elements between the known and the unknown states that connect them together. These hidden elements are known as hidden states; whereas the known elements are known as observations. In order to simplify the problem the known states are considered to be hidden states.

This forms the basis of the HMM which can then be improved by adding or removing states and modifying corresponding probabilities.

An HMM models probabilistically the following parameters:

Pi: The probability of starting in a particular state
A: The state transition matrix
B: The confusion matrix (the probability of a certain observation occurring given a particular state)

The HMM parameters are very similar to MM parameters except the confusion matrix. State information is derived in the same way as done in MM. Observation information is derived by extracting features and quantizing them into k levels, resulting in k observations for each state.

For a three state model with three observations the matrices are as follows:

Initial Probability Matrix:

$$[\pi_1 \quad \pi_2 \quad \pi_3] \tag{A}$$

State Transition Matrix: (each row represents previous state, each column next state)

$$\begin{bmatrix} a_{11} & a_{12} & a_{13} \\ a_{21} & a_{22} & a_{23} \\ a_{31} & a_{32} & a_{33} \end{bmatrix} \tag{B}$$

Confusion Matrix: (each row represents state, each column observation)

$$\begin{bmatrix} b_{11} & b_{12} & b_{13} \\ b_{21} & b_{22} & b_{23} \\ b_{31} & b_{32} & b_{33} \end{bmatrix} \tag{C}$$

For each state S_i the transition probability of observation O_j is b_{ij}, where

$$b_{ij} = \frac{No.\,of\ observations\ j\ for\ state\ i}{Total\ No.\,of\ observations\ for\ state\ i} \tag{1}$$

For each state S_i the transition probability to state S_j is a_{ij}, where

$$a_{ij} = \frac{No.\,of\ transitions\ from\ i\ to\ j}{Total\ No.\,of\ transitions\ from\ i\ to\ all\ states} \tag{2}$$

For each state S_i the initial probability is π_i, where

$$\pi_i = \frac{No.\,of\ training\ sequences\ starting\ with\ i}{Total\ No.\,of\ training\ sequences} \tag{3}$$

Here, the states 'S' are the average numbers of extremes in the whole data set.

For running the forward algorithm, given the observation sequence O, and parameters $\{\pi, a, b\}$, we can calculate the probability or likelihood of O as:

$$a_1(i) = \pi_i b_{iO(1)} \tag{4}$$

for $i = 1:N$, where N is the number of states, and

$$a_{t+1}(i) = \sum_{j=1}^{N} a_t(j) a_{ij} b_{iO(t+1)} \tag{5}$$

for $j = 1:N$ and $t = 1:T$, where T is the length of the observation. The probabilities are found by:

$$P_a = \sum_{j=1}^{N} a_T(j) \tag{6}$$

For expectation maximization, backward algorithm was implemented on HMMs. Given the observation sequence O, and parameters $\{\pi, a, b\}$, we can calculate the probability or likelihood of O as:

$$\beta_T(i) = 1$$

for $i = 1:N$, where N is the number of states, and

$$\beta_t(i) = \sum_{j=1}^{N} \beta_{t+1}(j) a_{ji} b_{jO(t+1)}$$

for $j = 1:N$ and $t = 1:T$, where T is the length of the observation. The probabilities are found by:

$$P_\beta = \sum_{j=1}^{N} \beta_1(j) b_{jO(1)} \pi_j$$

For calculating the best path in HMMs, the Viterbi algorithm has been used. Given the observation sequence O, and parameters $\{\pi, a, b\}$, we can calculate the best sequence W, of O as:

$$a_1(i) = \pi_i b_{iO(1)}$$

for $i = 1:N$, where N is the number of states, and

$$a_{t+1}(i) = max_j [a_t(j) a_{ij}] b_{iO(t+1)}$$

for $j = 1:N$ and $t = 1:T$, where T is the length of the observation.

$$\omega_{t+1}(j) = argmax_j\big[a_t(j)a_{ij}\big]$$

$$p* = max_j\big[a_T(j)\big]$$

$$\omega_T(j) = argmax_j\big[a_T(j)\big]$$
$$W_t = argmax_i\big[w_t(i)\big]$$

W_t is the best path state sequence. The number of hidden states is decided on the basis of local extremes present in the data set and feature extraction is done which is later quantized feasibly. The distance metric for k-means clustering comprises two factors; the logarithm of forward probability, and the log probability obtained from Viterbi algorithm (the best path sequence). The decision metric is the threshold, the same method as done in MMs. If the resulting clusters of data set lie above or equal to the threshold, the data set for the subject under study is recognized as correct action, otherwise it is detected as incorrect.

Based on the above analysis, the states are determined using the average number of extremes in the whole data set, whereas the observations are derived using the k-means clustering algorithm on the extracted events for HMM. For the three sensors placed on the arm of the test subject, the mathematical notations are detailed in table 1.

4   Test Scenarios and Experimental Analysis

To test the functional and computational efficiency of the system, different scenarios can be implemented. For example the proposed system can be tested for fast bowling, spin bowling, medium pace bowling, etc. However, this paper discusses the test case for medium pace bowling only, due to ease of analysis.

The experimentation is performed using three sensors placed on right arm of the subject at the wrist, elbow joint and the upper arm. The placement of the sensors is decided by keeping in view the objective of comparing the arm extension angle with a standard data set. The comparison results will assist the test subject in training himself for that specific bowling action complying with the rules set forth by ICC [16]. A total

Table 1. Meanings of Notations Used

Symbol	Meaning
k	Number of clusters into which the data is divided
pi	Probability of a state being an initial state
a	State transition probability
b	Probability of being in a state given an observation
P	Probability of an observed sequence given an HMM

of 40 samples have been collected, from which 23 correct samples are used for training the algorithm, while the remaining 17 samples have been used for testing upon the trained model. The sampling rate of the sensor data is 150 Hz and all the samples are transmitted via Bluetooth serial communication and are recorded in a file stored in PC. In this way, the data stored in the file can be utilized for on-spot processing as well as for comparison with other test data sets to check for improvement after the training process.

As mentioned in previous sections, the real-time data is filtered, then processed using k-means algorithm. The resulting data after applying k-means is processed through HMM and finally, the likelihood is calculated using Viterbi Algorithm. The whole procedure is designed as a GUI (graphical user interface) on MATLAB and the results are presented as a graph showing the test data, accompanied by the likelihoods obtained after processing. Table 2 enlists the different parameter settings for the experimental setup.

Table 2. Parameter Settings For Experimental Setup

Parameter	Value
k	5
No. of observations	280
Order of filter	100
Normalized frequency of filter	0.005 Hz

5   Results and Critique

The data was initially classified based upon the trained model obtained through k-means algorithm. The classification accuracy obtained using k-means model was about 66%, keeping edit distance as the distance metric and a threshold value as the decision metric. To improve the data classification accuracy, the same data set was used for training a Markov Chain Model and the data samples were tested for classification accuracy. 68.09% accuracy was achieved, considering logarithm of calculated forward probability P as the distance metric and a threshold value as a decision metric.

To further improve the classification accuracy, Hidden Markov Model was used and a model template was generated by training the HMM using the same training data set as used in previous two algorithms. At present, the training data set is treated separately for the three locations, i.e., wrist, elbow joint and upper arm, rather than using a concatenated data set comprising the data for the three locations altogether. The accuracy of correct data classification greatly depends on correct placement of sensors. Therefore, the sensor placement must be checked very carefully while collecting either the training data set or the test data set.

Figure 4 shows the plots of accelerometer and gyroscope's readings for one correct data set of wrist joint, three for triple axes gyroscope and three for x-, y-, z-axes accelerometer. The same set of readings is plotted for an incorrect data set for wrist joint, as shown in figure 5. The irregularity in plot for incorrect data set is evident, which shows the deviation of the bowling action from the correct bowling action.

Figure 6 shows the classification graph for wrist joint for 17 data sets. As seen from the figure, the data is classified 100% accurately. Particularly, in this case, medium paced bowling action has been performed which does not involve variable wrist joint movement; that is why the wrist joint movement is classified as either correct or incorrect with 100% accuracy.

Similar tests were performed on the data samples for elbow joint and upper arm and the classification results achieved using HMM trained model are shown in table 3, whereas graphical representation is given in figure 7. Overall, 17 data sets were used, out of which 14 were for correct action. The wrist data was classified correctly with 100% accuracy, elbow joint data with 88.24%, whereas the upper arm data was 82.35% correctly classified. The overall classification accuracy, for all three joints is thus evaluated as 90.2%.

These results are presented using bowling action for only medium paced bowling. The results may vary if trials are performed using other bowling actions. Also note that since the presented model is trained for only a particular bowling action, other bowling actions, though classified as legal bowling actions by ICC, will be placed under 'incorrect' data if tested against the presented model. Thus, depending on the different bowling actions, a trained model can be developed against each action and the test data can be classified using appropriate model.

The accuracy of classification is heavily dependent on the training data set. It is necessary that the training data set should be collected from the bowler who correctly follows the ICC rules for arm extension during a delivery. The more accurate the training data set, higher is the probability of correct classification of a test data set as correct or incorrect bowling action. Highly accurate results can be obtained by using the data sets taken from bowling actions of professional bowlers complying strictly with the rules of ICC.

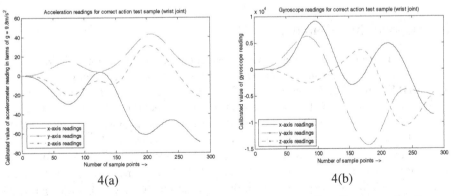

4(a)   4(b)

Fig. 4. Plot of (a) triple axis accelerometer and (b)gyroscope readings for correct bowling action's data set for wrist joint

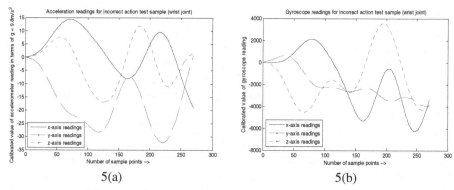

5(a)   5(b)

Fig. 5. Plot of triple axis accelerometer (a) and gyroscope (b) readings for incorrect bowling action's data set for wrist joint

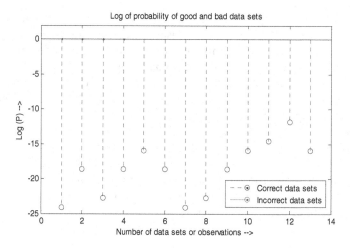

Fig. 6. Log of probability of the test data set given a trained Hidden Markov Model (HMM)

6   Concluding Remarks

A system for coaching of cricket bowlers has been proposed by using the HMM model. A compact design for inertial sensors has been proposed to acquire data for training the HMM model and consequently, for test data acquisition, to test the bowling action. The system is tested for medium paced bowling action and overall accuracy of 90.2% has been found while validating the test data set against the trained HMM model. To achieve better coaching results, similar strategy can be extended to the sensor placement on all key body parts in order to correctly train a bowler for a specific bowling action. This system can be extended for implementing an algorithm for efficient calculation of angle and position so as to correctly detect the illegality in bowling action in case of non-compliance of the rules of the game.

(Note: Since the probability of each bad data set was computed as zero and log (0) gives the value '-Inf', therefore it is not visible in Fig. 6)

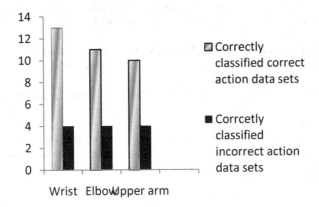

Fig. 7. Classification results for correct and incorrect action data sets for wrist, elbow joint and upper arm

References

1. Casas, R., Gracia, H.J., Marco, A., Falco, J.L.: Synchronization in Wireless Sensor Networks Using Bluetooth. In: The Third International Workshop on Intelligent Solutions in Embedded Systems Hamburg University of Technology, Hamburg, May 20 (2005)
2. Lu, C., Fu, L.: Robust Location-Aware Activity Recognition Using Wireless Sensor Network in an Attentive Home. IEEE Transactions on Automation Science and Engineering 6(4), 598–609 (2009)
3. Rennie, J., Press, M.: The computer in the 21st century, Scientific Amer. Special Issue (The Computer in the 21st Century), pp. 4–9 (1995)
4. Zia Uddin, M., et al.: Human Activity Recognition via 3-D joint angle features and Hidden Markov models. In: 17th IEEE International Conference on Image Processing, September 26-29 (2010)
5. Akay, M.: Intelligent Wearable Monitor Systems and Methods. US Patent number 2005/0240086 (October 27, 2005)
6. Weaver, et al.: Apparatus and Method for Processing Data Collected via Wireless Network Sensors. US Patent number 2011/0035271 (February 10, 2011)
7. Kan, Y.-C., Chen, C.-K.: A Wearable Inertial Sensor Node for Body Motion Analysis. IEEE Sensors Journal 12(3), 651–657 (2012)
8. Aerts, S.G.E.: Wearable Device. US Patent number 2007/0161874 (July 12, 2007)
9. Wixted, A., Spratford, W., Davis, M., Portus, M., James, D.: Wearable Sensors for on Field near Real Time Detection of Illegal Bowling Actions. In: Conference Proceedings for Conference of Science, Medicine & Coaching in Cricket, Sheraton Mirage Gold Coast, Queensland, Australia, June 1-3, pp. 165–168 (2010)
10. Gaffney, M., O'Flynn, B., Mathewson, A., Buckley, J., Barton, J., Angove, P., Vcelak, J.Ó., Conaire, C., Healy, G., Moran, K., O'Connor, N.E., Coyle, S., Kelly, P., Caulfield, B., Conroy, L.: Wearable wireless inertial measurement for sports applications. In: Proc. IMAPS-CPMT Poland 2009, Gliwice – Pszczyna, Poland, September 22-24 (2009)
11. Wixted, A., James, D., Portus, M.: Inertial Sensor Orientation for Cricket Bowling Monitoring. IEEE Sensors, 1835–1838 (October 28-31, 2011)

12. Cheng, L., Hailes, S.: Analysis of Wireless Inertial Sensing for Athlete Coaching Support. In: IEEE Global Telecommunications Conference 'IEEE GLOBECOM' 2008, November 30-December 4 (2008)

13. Hon, T.M., Senanayake, S.M.N.A., Flyger, N.: Biomechanical Analysis of 10-Pin Bowling Using Wireless Inertial Sensor. In: IEEE/ASME International Conference on Advanced Intelligent Mechatronics, Singapore, July 14-17 (2009)

14. Ghasemzadeh, H., Jafari, R.: Coordination Analysis of Human Movements with Body Sensor Networks: A Signal Processing Model to Evaluate Baseball Swings. IEEE Sensors Journal 11(3), 603–610 (2011)

15. Guenterberg, E., Yang, A.Y., Ghasemzadeh, H., Jafari, R., Bajcsy, R., Sastry, S.S.: A Method for Extracting Temporal Parameters Based on Hidden Markov Models in Body Sensor Networks With Inertial Sensors. IEEE Transactions on Information Technology In Biomedicine 13(6), 1019–1030 (2009)

16. ICC Regulations For The Review of Bolwers Reports With Suspected Illegal Bowling Actions, accessed from ICC website link:
    http://static.icc-cricket.yahoo.net/ugc/documents/
    DOC_C26C9D9E63C44CBA392505B49890B5AF_1285831722391_859.pdf

17. Guenterberg, E., Ghasemzadeh, H., Jafari, R.: A Distributed Hidden Markov Model for Fine-grained Annotation in Body Sensor Networks. In: The Sixth International Workshop on Wearable and Implantable Body Sensor Networks, BSN 2009, June 3-5 (2009)

18. Aminian, K., Najafi, B.: Capturing human motion using bodyfixed sensors: outdoor measurement and clinical applications. Computer Animation and Virtual Worlds 15(2), 79–94 (2004)

19. http://www.bluetooth.com/Pages/Basics.aspx
    (accessed on May 10, 2012)

20. Casas, R., et al.: Synchronization in Wireless Sensor Networks Using Bluetooth. In: The Third International Workshop on Intelligent Solutions in Embedded Systems, May 20. Hamburg University of Technology, Hamburg (2005)

21. Wang, D., et al.: A Wireless Sensor Network Based on Bluetooth for Telemedicine Monitoring System. In: The Proceedings of IEEE International Symposium on Microwave, Antenna, Propagation and EMC Technologies for Wireless Communications (2005)

22. Ghasemzadeh, H., Loseu, V., Jafari, R.: Collaborative signal processing for action recognition in body sensor networks: A distributed classification algorithm using motion transcripts. In: International Conference on Information Processing in Sensor Networks, IPSN 2010 (2010)

23. Sutton, C., McCallum, A.: An Introduction to Conditional Random Fields for Relational Learning. In: Getoor, L., Taskar, B. (eds.) Introduction to Statistical Relational Learning. MIT Press (2006)

24. Kim, E., Helal, S., Cook, D.: Human Activity Recognition and Pattern Discovery. IEEE Pervasive Computing 9(1), 48–53 (2010)

A New Back-Propagation Neural Network Optimized with Cuckoo Search Algorithm

Nazri Mohd. Nawi, Abdullah Khan, and Mohammad Zubair Rehman

Faculty of Computer Science and Information Technology,
Universiti Tun Hussein Onn Malaysia (UTHM)
P.O. Box 101, 86400 Parit Raja, Batu Pahat, Johor Darul Takzim, Malaysia
nazri@uthm.edu.my, hi100010@siswa.uthm.edu.my,
zrehman862060@gmail.com

Abstract. Back-propagation Neural Network (BPNN) algorithm is one of the most widely used and a popular technique to optimize the feed forward neural network training. Traditional BP algorithm has some drawbacks, such as getting stuck easily in local minima and slow speed of convergence. Nature inspired meta-heuristic algorithms provide derivative-free solution to optimize complex problems. This paper proposed a new meta-heuristic search algorithm, called cuckoo search (CS), based on cuckoo bird's behavior to train BP in achieving fast convergence rate and to avoid local minima problem. The performance of the proposed Cuckoo Search Back-Propagation (CSBP) is compared with artificial bee colony using BP algorithm, and other hybrid variants. Specifically OR and XOR datasets are used. The simulation results show that the computational efficiency of BP training process is highly enhanced when coupled with the proposed hybrid method.

Keywords: Back propagation neural network, cuckoo search algorithm, local minima, and artificial bee colony algorithm.

1   Introduction

Artificial Neural Networks (ANN) provide main features, such as: flexibility, competence, and capability to simplify and solve problems in pattern classification, function approximation, pattern matching and associative memories [1-3]. Among many different ANN models, the multilayer feed forward neural networks (MLFF) have been mainly used due to their well-known universal approximation capabilities [4]. The success of ANN mostly depends on their design, the training algorithm used, and the choice of structures used in training. ANN are being applied for different optimization and mathematical problems such as classification, object and image recognition, Signal processing, seismic events prediction, temperature and weather forecasting, bankruptcy, tsunami intensity, earthquake, and sea level etc. [5-9]. ANN has the aptitude for random nonlinear function approximation and information processing which other methods does not have [10]. Different techniques are used in the past for optimal network performance for training ANNs such as back propagation neural network

B. Murgante et al. (Eds.): ICCSA 2013, Part I, LNCS 7971, pp. 413–426, 2013.

(BPNN) algorithm [11]. However, the BPNN algorithm suffers from two major drawbacks: low convergence rate and instability. The drawbacks are caused by a risk of being trapped in a local minimum [12], [13] and possibility of overshooting the minimum of the error surface [14]. Over the last years, many numerical optimization techniques have been employed to improve the efficiency of the back propagation algorithm including the conjugate gradient descent [15] [16]. However, one limitation of this procedure, which is a gradient-descent technique, is that it requires a differentiable neuron transfer function. Also, as neural networks generate complex error surfaces with multiple local minima, the BPNN fall into local minima instead of a global minimum [17], [18]. Evolutionary computation is often used to train the weights and parameters of neural networks. In recent years, many improved learning algorithms have been proposed to overcome the weakness of gradient-based techniques. These algorithms include global search technique such as hybrid PSO-BP [19], artificial bee colony algorithm [20-22], evolutionary algorithms (EA) [23], particle swarm optimization (PSO) [24], differential evolution (DE) [25], ant colony, back propagation algorithm [26], and genetic algorithms (GA) [27].

In order to overcome the weaknesses of the conventional BP, this paper proposed a new meta-heuristic search algorithm, called cuckoo search back propagation (CSBP). Cuckoo search (CS) is developed by Yang and Deb [28] which imitates animal behavior and is useful for global optimization [29], [30], [31]. The CS algorithm has been applied independently to solve several engineering design optimization problems, such as the design of springs and welded beam structures [32], and forecasting [33]. For these problems, Yang and Deb showed that the optimal solutions obtained by CS are far better than the best solutions obtained by an efficient particle swarm optimizer or genetic algorithms. In particular, CS can be modified to provide a relatively high convergence rate to the true global minimum [34].

In this paper, the convergence behavior and performance of the proposed Cuckoo Search Back-propagation (CSBP) on XOR and OR datasets is analyzed. The results are compared with artificial bee colony using BPNN algorithm, and similar hybrid variants. The main goals are to decrease the computational cost and to accelerate the learning process using a hybridization method.

The remaining paper is organized as follows: Section 2 gives literature review on BPNN. Section 3, explains Cuckoo Search via levy flight. In section 4, the proposed CSBP algorithm, and simulation results are discussed in Section 5. Finally, the paper is concluded in the Section 6.

2   Back-Propagation Neural Network (BPNN)

The Back-Propagation Neural Network (BPNN) is one of the most novel supervised learning ANN algorithm proposed by Rumelhart, Hinton and Williams in 1986 [35]. BPNN learns by calculating the errors of the output layer to find the errors in the hidden layers. Due to this ability of back-propagating, it is highly suitable for problems in which no relationship is found between the output and inputs. The gradient descent method is utilized to calculate the weights and adjustments are made to the network to minimize the output error. The BPNN algorithm has become the standard algorithm used for training multilayer perceptron. It is a generalized least mean squared (LMS) algorithm that minimizes a criterion equals to the sum of the squares of the errors between the actual and the desired outputs. This principle is;

$$E_p = \Sigma_{i=1}^{j}(e_i)^2 \tag{1}$$

Where the nonlinear error signal is

$$e_i = d_i - y_i \tag{2}$$

d_i and y_i are respectively, the desired and the current outputs for the ith unit. P denotes in (1) the pth pattern; j is the number of the output units. The gradient descent method is given by;

$$w_{ki} = -\mu \frac{\partial E_p}{\partial w_{ki}} \tag{3}$$

Where w_{ki} is the weight of the ith unit in the (n-1)[th] layer to the kth unit in the nth layer. The BP calculates errors in the output layer ∂_l, and the hidden layer, ∂_j are using the formulas in Equation (4) and Equation (5) respectively [36] [37]:

$$\partial_l = \mu(d_i - y_i)f'(y_i) \tag{4}$$

$$\partial_j = \mu \Sigma_i \partial_l w_{lj} f'(y_i) \tag{5}$$

Here d_i is the desired output of the ith   output neuron, y_i is the actual output in the output layer, y_i is the actual output value in the hidden layer, and   k is the adjustable variable in the activation function. The back propagation error is used to update the weights and biases in both the output and hidden layers. The weights, w_{ij} and biases, b_i, are then adjusted using the following formulae;

$$w_{ij}(k + 1) = w_{ij}(k) + \mu \partial_j y_i \tag{6}$$

$$w_{lj}(k + 1) = w_{lj}(k) + \mu \partial_j x_l \tag{7}$$

$$b_i(k + 1) = b_i(k) + \mu \partial_j \tag{8}$$

Here k is the number of the epoch and μ is the learning rate.

3   Cuckoo Search Viva Levy Flight

Levy Flights have been used in many search algorithms [38]. In Cuckoo Search algorithm levy flight is an important component for local and global searching [34]. In sub sections, levy flight and then cuckoo search algorithm based on levy flights is explained.

a.   Levy Flight

In nature, the flight movement of many animals and insects is recognized as a random manner. The foraging path of an animal commonly has the next move based on the current state and the variation probability to the next state. Which way it chooses be determined by indirectly on a probability that can be modeled mathematically. Levy Flights is a random walk that is characterized by a series of straight jumps chosen from a heavy-tailed probability density function [34]. In statistical term, it is a stochastic algorithm for global optimization that finds a global minimum [38]. A levy flight process step length can be calculated by using Mantegna algorithm using equation (9).

$$S = \frac{u}{|v|^{\frac{1}{\beta}}} \tag{9}$$

Note that u and v are drawn from normal distribution with respect to these two random variables;

$$u \sim N(0. \sigma_u{}^2), \qquad v \sim N(0. \sigma_v{}^2) \tag{10}$$

The symbol $\sim$ in (10) means the random variable obeys the distribution on right hand side; that is, samples should draw from the distribution. The $\sigma_u{}^2$ and $\sigma_v{}^2$ present in equation (10) are the variance of distributions which come from equation (11);

$$\sigma_u = \left\{ \frac{\Gamma(1+\beta)\sin(\tau\beta/2)}{\Gamma[(1+\beta)/2]\beta 2^{(\beta-1)^2}} \right\}^{\frac{1}{\beta}}, \qquad \sigma_v = 1 \tag{11}$$

b.   Cuckoo Search (CS) Algorithm

Cuckoo Search (CS) algorithm is a novel meta-heuristic technique proposed by Xin-Shen Yang [28]. This algorithm was stimulated by the obligate brood parasitism of some cuckoo species by laying their eggs in the nests of other host birds. Some host nest can keep direct difference. If an egg is discovered by the host bird as not its own, it will either throw the unknown egg away or simply abandon its nest and build a new nest elsewhere. Some other species have evolved in such a way that female parasitic cuckoos are often very specialized in the mimic in color and pattern of the eggs of a few chosen host species. This reduces the probability of their eggs being abandoned and thus increases their population. The CS algorithm follows three idealized rules:

a.   Each cuckoo lays one egg at a time, and put its egg in randomly chosen nest;
b.   The best nests with high quality of eggs will carry over to the next generations;
c.   The number of available host nests is fixed, and the egg laid by a cuckoo is discovered by the host bird with a probability $pa \in [0, 1]$.

In this case, the host bird can either throw the egg away or abandon the nest, and build a completely new nest. The rule-c defined above can be approximated by the fraction pa $\in [0, 1]$ of the n nests that are replaced by new nests (with new random solutions). For a maximization problem, the quality or fitness of a solution can simply be proportional to the value of the objective function. In this algorithm, each egg in a nest represents a solution, and a cuckoo egg represents a new solution, the aim is to use a new and a potentially better solution (cuckoo) to replace a not so good solution in the nests. Based on these three rules, the basic steps of the Cuckoo Search (CS) can be summarized as the following pseudo code:

When generating new solutions x^{t+1} for a cuckoo i, a Levy flight is performed

$$x_i^{t+1} = x_i^t + \alpha \oplus levy(\lambda) \tag{12}$$

Where $\alpha > 0$ is the step size, which should be related to the scales of the problem of interest. The product $\oplus$ means entry wise multiplications. The random walk via Levy flight is more efficient in exploring the search space as its step length is much longer in the long run. The Levy flight essentially provides a random walk while the random step length is drawn from a Levy distribution as shown in the Equation 12:

$$Lavy \sim u = t^{-\lambda}, 1 < \lambda \leq 3 \tag{13}$$

This has an infinite variance with an infinite mean. Here the steps essentially construct a random walk process with a power-law step-length distribution and a heavy tail. Some of the new solutions should be generated by Levy walk around the best solution obtained so far, this will speed up the local search. However, a substantial fraction of the new solutions should be generated by far field randomization whose locations should be far enough from the current best solution. This will make sure the system will not be trapped in local minima.

The pseudo code for the CS algorithm is given below:

Step 1: Generate initial population of N host nest   i= 1... N
Step 2: while (fmin<MaxGeneration) or (stop criterion)
Do
Step 3: Get a cuckoo randomly by Levy flights and evaluate its fitness F_i
Step 4: Choose randomly a nest j among N.
Step 5: if $F_i > F_j$ Then
Step 6: Replace j by the new solution.
End if
Step 7: A fraction pa of worse nest are abandoned and new ones are built.
Step 8: Keep the best solutions (or nest with quality solutions).
Step 9: Rank the solutions and find the current best.
End while

---

4   The Proposed CSBP Algorithm

The CS is a population based optimization algorithm, and like other meta-heuristic algorithms, it starts with a random initial population, The CS algorithm basically works in three steps: selection of the best source by keeping the best nests or solutions, replacement of host eggs with respect to the quality of the new solutions or cuckoo eggs produced based randomization via Levy flights globally (exploration) and discovering of some cuckoo eggs by the host birds and replacing according to the quality of the local random walks (exploitation) [29]. Each cycle of the search consists of several steps initialization of the best nest or solution, the number of available host nests is fixed, and the egg laid by a cuckoo is discovered by the host bird with a probability $pa\ \varepsilon\ [0,1]$.

In the proposed CSBP algorithm, each best nest represents a possible solution (i.e., the weight space and the corresponding biases for BPNN optimization in this paper). The weight optimization problem and the size of population represent the quality of the solution. In the first epoch, the best weights and biases are initialized with CS and then these weights are passed to the BPNN. The weights in BPNN are calculated and compared with best solution in the backward direction. In the next cycle CS will updated the weights with the best possible solution and CS will continue searching the best weights until the last cycle/ epoch of the network is reached or either the MSE is achieved.

The pseudo code of the proposed CSBP algorithm is:

Step 1: CS is initializes and passes the best weights to BPNN
Step 2: Load the training data
Step 3**: While**   MSE<stopping criteria
Step 4: Initialize all cuckoo nests
Step 5: Pass the cuckoo nests as weights to network
Step 6: Feed forward neural network runs using the weights initialized with CS
Step 7: Calculate the error backward
Step8: CS keeps on calculating the best possible weight at each epoch until the network is converged.
End While

5   Experiments and Results

To test the performance of the proposed CSBP, Boolean datasets of 4-bit OR, 2-bit and 3-bit XOR were used. The simulation experiments were performed on a 1.66 GHz AMD Processor with 2GB of RAM. For performing simulations Matlab 2009b software was used. The proposed (CSBP) algorithm is compared with artificial bee colony Levenberg Marquardt algorithm (ABC-LM), artificial bee colony back propagtion (ABC-BP) algorithm and standard BPNN based on mean squarer error, epochs and CPU time. The three layer feed forward neural network are used for each problem.

I.e. input layer, one hidden layer, and output layers. The number of hidden nodes is design of five and ten neurons. In the network structure the biases nodes are also used and the log sigmoid activation function is placed as the activation function for the hidden and output layers nodes. For each problem, trial is limited to 1000 epochs. And minimum error is keeping 0. A total of 20 trials are run for each case. The network results are stored in the result file for each trial.

The first test problem is the Exclusive OR (XOR) Boolean function of two   binary input to a single binary output as (0 0; 0 1; 1 0 ; 1   1) -to -(0; 1; 1 ;0). From the Table 1, we can see that the proposed CSBP method performs well on 2-bit XOR dataset. The CSBP converges to global minima in 21.23 second of CPU time with an average accuracy of 100% and achieves a MSE of 0. While other algorithms fall behind in-terms of MSE, CPU time and accuracy. The Figure 1 shows that the CSBP performs well and converges to global minima within 134 epochs for the 2-5-1 network structure.

Table 1. CPU time, Epochs and MSE error for 2- bit XOR dataset with 2-5-1 ANN architecture

Algorithm	ABC-BP	ABC-LM	BPNN	CSBP
CPU TIME	172.3388	123.9488	42.64347	21.23
EPOCHS	1000	1000	1000	149
MSE	2.39E-04	0.125	0.220664	0
Accuracy (%)	96.47231	71.69041	54.6137	100

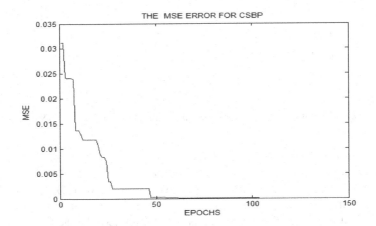

Fig. 1. The CSBP convergence performance on 2 bit XOR with 2-5-1 ANN architecture

Table 2 give an idea about the CPU time, number of epochs and the mean square error for the 2 bit XOR data sets with ten hidden neurons. From the table, we can identify that the proposed CSBP method has better result than the ABC-BP, ABC-LM and BPNN algorithm. The CSBP achieves a MSE of 0 in 100 epochs and in 28.9 second of CPU time. Meanwhile, the other algorithms have short performed with large MSE's and CPU times. The Figure 2 shows the convergence performance of CSBP algorithm for the 2-10-1 network architecture.

Table 2. CPU time, Epochs and MSE error for 2- bit XOR dataset with 2-10-1 ANN architecture

Algorithm	ABC-BP	ABC-LM	BPNN	CSBP
CPU TIME	197.34	138.96	77.63	46.99
EPOCHS	1000	1000	1000	203
MSE	8.39E-04	0.12578	0.120664	0
Accuracy (%)	96.8	71.876	54.6137	100

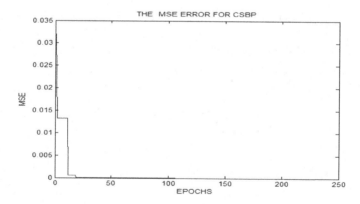

Fig. 2. The CSBP convergence performance on 2 bit XOR with 2-10-1 ANN architecture

In the second case we used 3- bit Exclusive OR test problem of three input to a single binary output as (1 1 1; 1   1 0 ; 1 0 1; 1 0 0; 0 1 1; 0   1 0; 0 0 1; 0 0 0) –to-(1; 0; 0; 1; 0; 1; 1; 0). In the simulation, we selected a 3- bit exclusive OR dataset for 3-5-1 architecture.   In three bit XOR input if the number of binary inputs is odd, the output is 1, otherwise the output is 0. Tables 3 describe the CPU time, number of epochs and the MSE of the XOR dataset with five hidden neurons. We can see in the Table 3 that the proposed algorithm has an MSE of 5.4E-04 within 938 epochs and in 149.53 of CPU time. In 3-bit XOR dataset has better performances in MSE and accuracy. But it has take long CPU time for the processing then the BPNN, and ABC

Table 3. CPU time, Epochs and MSE error for 3- bit XOR dataset with 3-5-1 ANN architecture

Algorithm	ABC-BP	ABC-LM	BPNN	CSBP
CPU TIME	172.3388	123.7925	50.03059	149.53
EPOCHS	1000	1000	1000	938
MSE	0.08	0.01063	0.25077	5.4E-04
Accuracy (%)	86.47231	78.83231	47.626	98.7351

Table 4 provides an idea about the CPU time, number of epochs and the mean square error for the 3 bit XOR test problem with ten hidden neurons. The proposed CSBP method with 3-10-1 network structure also outperforms the other algorithm in term of CPU time and MSE errors. The CSBP gets an MSE of 7.4E-04 in 161.53 CPU time. While for the same network structure the ABC-BP achieves a less MSE of 0.0845. The conventional back propagation has 0.256 MSE achieved in 75.13748 CPU cycles.

Table 4. CPU time, Epochs and MSE error for 3- bit XOR dataset with 3-10-1 ANN architecture

Algorithm	ABC-BP	ABC-LM	BPNN	CSBP
CPU TIME	185.3388	153.7925	75.13748	161.53
EPOCHS	1000	1000	1000	1000
MSE	0.0845	0.01054	0.256	7.4E-04
Accuracy (%)	86.8971	79.83231	49.626	98.88

In the third case we used four bit OR test problem of four input to a single binary output as (1 1 1 1;1 1 1 0;1 1 0 1;1 1 0 0; 1 0 1 1;1 0 0 1; 1 0 0 0;0 1 1 1;0 1 1 0; 0 1 0 1; 0 1 0 0; 0 0 1 1 ; 0 0 1 0; 0 0 0 1;1 0 1 0; 0 0 0 0) –to –(1; 1; 1; 1; 1; 1; 1; 1; 1; 1; 1; 1; 1; 1; 1; 0). In four bit OR dataset, if all inputs are 0, the output is 0, otherwise the output will be 1. For the 4-5-1 network architecture, it has twenty five connection weights and six biases. Tables 5, confirms the CPU time, number of epochs, the mean square error, and accuracy for the 4 bit OR test problem with five hidden neurons. The proposed CSBP converged to a MSE of 0 within 51 epochs. While the ABC-LM algorithm has an MSE of 1.82E-10 and the ABC-BP has the MSE of 1.91E-05.

Table 5. CPU time, Epochs and MSE error for 4- bit OR dataset with 4-5-1 ANN architecture

Algorithm	ABC-BP	ABC-LM	BPNN	CSBP
CPU TIME	162.4945	118.7274	63.28089	8.486525
EPOCHS	1000	1000	1000	51
MSE	1.91E-10	1.82E-10	0.052778	0
Accuracy (%)	99.97	99.99572	89.83499	100

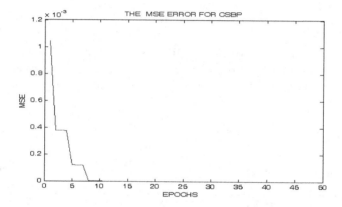

Fig. 3. The CSBP convergence performance on 4 bit OR with 4-5-1 ANN architecture

Table 6 illustrates the CPU time, number of epochs and the MSE for the 4 bit OR test problem with ten hidden neurons. In four bit OR datasets the proposed algorithm has outperformed other algorithms with 99 percent accuracy. Figure 4 shows the convergence performance of the proposed CSBP for the 4-10-1 network architecture.

Table 6. CPU time, Epochs and MSE error for 4- bit OR dataset with 4-10-1 ANN architecture

Algorithm	ABC-BP	ABC-LM	BPNN	CSBP
CPU TIME	180.4945	129.7274	67.28089	8.678
EPOCHS	1000	1000	1000	36

Table 6. *(Continued)*

MSE	1.67E-10	1.76E-10	0.05346	0
Accuracy(%)	99.47	99.78	89.83499	100

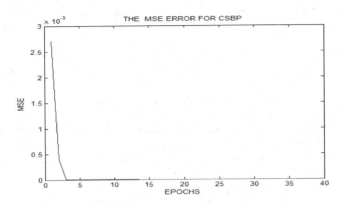

Fig. 4. The CSBP convergence performance on 4 bit OR with 4-10-1 ANN architecture

6   Conclusion

BPNN algorithm is one of the most widely used and a popular procedure to optimize the feed forward neural network training. Conventional BPNN algorithm has some drawbacks, such as getting stuck in local minima and slow speed of convergence. Nature inspired meta-heuristic algorithms provide derivative-free solution to optimize complex problems. A new meta-heuristic search algorithm, called cuckoo search (CS) is proposed to train BPNN to achieve fast convergence rate and to minimize the training error. The performance of the proposed CSBP algorithm is compared with the ABC-LM, ABC-BP and BPNN algorithms by means of simulation on three datasets such as 2-bit, 3-bit XOR and 4-bit OR. The simulation results show that the proposed CSBP is far better than the previous methods in terms of simplicity, convergence rate and accuracy. In future the proposed model will be used on the benchmarks classification datasets collected from UCI machine learning repository.

Acknowledgment. The Authors would like to thank Office of Research, Innovation, Commercialization and Consultancy Office (ORICC), Universiti Tun Hussein Onn Malaysia (UTHM) and Ministry of Higher Education (MOHE) Malaysia for financially supporting this Research under Fundamental Research Grant Scheme (FRGS) vote no. 1236.

References

1. Dayhoff, J.E.: Neural-Network Architectures: An Introduction, 1st edn. Van Nostrand Reinhold Publishers, New York (1990)
2. Rehman, M.Z., Nazri, M.N.: Studying the Effect of adaptive momentum in improving the accuracy of gradient descent back propagation algorithm on classification problems. International Journal of Modern Physics (IJMPCS) 9(1), 432–439 (2012)
3. Ozturk, C., Karaboga, D.: Hybrid Artificial Bee Colony algorithm for neural network training. In: IEEE Congress of Evolutionary Computation (CEC), pp. 84–88 (2011)
4. Haykin, S.: Neural Networks, A Comprehensive Foundation. Prentice Hall, New Jersey (1999)
5. Du, K.L.: Clustering: A neural network approach. Neural Networks 23(1), 89–107 (2010)
6. Guojin, C., Miaofen, Z., et al.: Application of Neural Networks in Image Definition Recognition, Signal Processing and Communications. In: ICSPC, pp. 1207–1210 (2007)
7. Romano, M., Liong, S., et al.: Artificial neural network for tsunami forecasting. Asian Earth Sciences 36, 29–37 (2009)
8. Hayati, M., Mohebi, Z.: Application of Artificial Neural Networks for Temperature forecasting. World Academy of Science, Engineering and Technology 28(2), 275–279 (2007)
9. Perez, M.: Artificial neural networks and bankruptcy forecasting: a state of the art. Neural Computing & Application 15, 154–163 (2006)
10. Contreras, J., Rosario, E., Nogales, F.J., Conejos, A.J.: ARIMA Models to Predict Next-Day Electricity Prices. IEEE Transactions on Power Systems 18(3), 1014–1020 (2003)
11. Leung, C., Member, C.: A Hybrid Global Learning Algorithm Based on Global Search and Least Squares Techniques for back propagation neural network Networks. In: International Conference on Neural Networks, pp. 1890–1895 (1994)
12. Ahmed, W.A.M., Saad, E.S.M., Aziz, E.S.A.: Modified Back Propagation Algorithm for Learning Artificial Neural Networks. In: Eighteenth National Radio Science Conference (NRSC), pp. 345–352 (2001)
13. Leigh, W., Hightower, R., Modena, N.: Forecasting the New York Stock exchange composite index with past price and invest rate on condition of volume spike. Expert System with Applications 28(1), 1–8 (2005)
14. Wen, J., Zhao, J.L., Luo, S.W., Han, Z.: The Improvements of BP Neural Network Learning Algorithm. In: 5th Int. Conf. on Signal Processing WCCC-ICSP, pp. 1647–1649 (2000)
15. Lahmiri, S.: Wavelet transform, neural networks and the prediction of s & p price index: a comparativepaper of back propagation numerical algorithms. Business Intelligence Journal 5(2), 235–244 (2012)
16. Nawi, N.M., Ransing, R.S., Salleh, M.N.M., Ghazali, R., Hamid, N.A.: An improved back propagation neural network algorithm on classification problems. In: Zhang, Y., Cuzzocrea, A., Ma, J., Chung, K.-I., Arslan, T., Song, X. (eds.) DTA and BSBT 2010. CCIS, vol. 118, pp. 177–188. Springer, Heidelberg (2010)
17. Gupta, J.N.D., Sexton, R.S.: Comparing backpropagation with a genetic algorithm for neural network training. The International Journal of Management Science 27, 679–684 (1999)
18. Nawi, N.M., Ghazali, R., Salleh, M.N.M.: The development of improved back-propagation neural networks algorithm for predicting patients with heart disease. In: Zhu, R., Zhang, Y., Liu, B., Liu, C. (eds.) ICICA 2010. LNCS, vol. 6377, pp. 317–324. Springer, Heidelberg (2010)

19. Zhang, J., Lok, T., Lyu, M.: A hybrid particle swarm optimization back propagation algorithm for feed forward neural network training. Applied Mathematics and Computation 185, 1026–1037 (2007)
20. Shah, H., Ghazali, R., Nawi, N.M., Deris, M.M.: Global hybrid ant bee colony algorithm for training artificial neural networks. In: Murgante, B., Gervasi, O., Misra, S., Nedjah, N., Rocha, A.M.A.C., Taniar, D., Apduhan, B.O. (eds.) ICCSA 2012, Part I. LNCS, vol. 7333, pp. 87–100. Springer, Heidelberg (2012)
21. Shah, H., Ghazali, R., Nawi, N.M.: Hybrid ant bee colony algorithm for volcano temperature prediction. In: Chowdhry, B.S., Shaikh, F.K., Hussain, D.M.A., Uqaili, M.A. (eds.) IMTIC 2012. CCIS, vol. 281, pp. 453–465. Springer, Heidelberg (2012)
22. Karaboga, D., Akay, B., Ozturk, C.: Artificial bee colony (ABC) optimization algorithm for training feed-forward neural networks. In: Torra, V., Narukawa, Y., Yoshida, Y. (eds.) MDAI 2007. LNCS (LNAI), vol. 4617, pp. 318–329. Springer, Heidelberg (2007)
23. Yao, X.: Evolutionary artificial neural networks. International Journal of Neural Systems 4(3), 203–222 (1993)
24. Mendes, R., Cortez, P., Rocha, M., Neves, J.: Particle swarm for feedforward neural network training. In: Proceedings of the International Joint Conference on Neural Networks, vol. 2, pp. 1895–1899 (2002)
25. Ilonen, J., Kamarainen, J.I., Lampinen, J.: Differential Evolution Training Algorithm for Feed-Forward Neural Networks. Neural Processing Letters 17(1), 93–105 (2003)
26. Liu, Y.-P., Wu, M.-G., Qian, J.-X.: Evolving Neural Networks Using the Hybrid of Ant Colony Optimization and BP Algorithms. In: Wang, J., Yi, Z., Żurada, J.M., Lu, B.-L., Yin, H. (eds.) ISNN 2006. LNCS, vol. 3971, pp. 714–722. Springer, Heidelberg (2006)
27. Khan, A.U., Bandopadhyaya, T.K., Sharma, S.: Comparisons of Stock Rates Prediction Accuracy using Different Technical Indicators with Backpropagation Neural Network and Genetic Algorithm Based Backpropagation Neural Network. In: Proceedings of the First International Conference on Emerging Trends in Engineering and Technology. IEEE Computer Society, Nagpur (2008)
28. Yang, X.S., Deb, S.: Cuckoo search via Lévy flights. In: Proceedings of World Congress on Nature & Biologically Inspired Computing, India, pp. 210–214 (2009)
29. Yang, X.S., Deb, S.: Engineering Optimisation by Cuckoo Search. International Journal of Mathematical Modelling and Numerical Optimisation 1(4), 330–343 (2010)
30. Tuba, M., Subotic, M., Stanarevic, N.: Modified cuckoo search algorithm for unconstrained optimization problems. In: Proceedings of the European Computing Conference (ECC 2011), Paris, France, pp. 263–268 (2011)
31. Tuba, M., Subotic, M., Stanarevic, N.: Performance of a Modified Cuckoo Search Algorithm for Unconstrained Optimization Problems. WSEAS Transactions on Systems 11(2), 62–74 (2012)
32. Yang, X.S., Deb, S.: Engineering optimisation by cuckoo search. International Journal of Mathematical Modelling and Numerical Optimisation 1(4), 330–343 (2010)
33. Chaowanawate, K., Heednacram, A.: Implementation of Cuckoo Search in RBF Neural Network for Flood Forecasting. In: Fourth International Conference on Computational Intelligence, Communication Systems and Networks, pp. 22–26 (2012)
34. Walton, S., Hassan, O., Morgan, K., Brown, M.: Modified cuckoo search: A new gradient free optimisation algorithm. Chaos, Solitons & Fractals 44(9), 710–718 (2011)
35. Rumelhart, D.E., Hinton, G.E., Williams, R.J.: Learning Internal Representations by error Propagation. In: Parallel Distributed Processing: Explorations in the Microstructure of Cognition, vol. 1 (1986)

36. Cichocki, A., Unbehauen, R.: Neural Network for Optimization and Signal Processing. Wiley, Chichester (1993)
37. Lippman, R.P.: An introduction to computing with neural networks. IEEE ASSP. Mag. 4(2) (April 1987)
38. Pavlyukevich, I.: Levy flights, non-local search and simulated annealing. Journal of Computational Physics 226(2), 1830–1844 (2007)

Functional Link Neural Network – Artificial Bee Colony for Time Series Temperature Prediction

Yana Mazwin Mohmad Hassim and Rozaida Ghazali

Faculty of Computer Science and Information Technology
Universiti Tun Hussein Onn Malaysia (UTHM),
86400 Batu Pahat, Johor, Malaysia
{yana,rozaida}@uthm.edu.my

Abstract. Higher Order Neural Networks (HONNs) have emerged as an important tool for time series prediction and have been successfully applied in many engineering and scientific problems. One of the models in HONNs is a Functional Link Neural Network (FLNN) known to be conveniently used for function approximation and can be extended for pattern recognition with faster convergence rate and lesser computational load compared to ordinary feedforward network like the Multilayer Perceptron (MLP). In training the FLNN, the mostly used algorithm is the Backpropagation (BP) learning algorithm. However, one of the crucial problems with BP learning algorithm is that it can be easily gets trapped on local minima. This paper proposed an alternative learning scheme for the FLNN to be applied on temperature forecasting by using Artificial Bee Colony (ABC) optimization algorithm. The ABC adopted in this work is known to have good exploration and exploitation capabilities in searching optimal weight especially in numerical optimization problems. The result of the prediction made by FLNN-ABC is compared with the original FLNN architecture and toward the end we found that FLNN-ABC gives better result in predicting the next-day ahead prediction.

Keywords: Temperature prediction, Functional Link Neural Network, Artificial Bee Colony Algorithm.

1 Introduction

Artificial Neural Networks (ANNs) have been known to be successfully applied in a variety of real world tasks includes prediction, classification, signal processing, image recognition and especially in industry, business and science [1, 2]. The most common architecture of ANNs is the Multi-layer feed forward network known as Multilayer perceptron (MLP). Since the MLP has multilayered structure, the network requires excessive training time for learning [3]. This is because, the number of weight and the training time will increase as the number of layers and the nodes in layer increases [3, 4]. In order to overcome the drawback of MLP, another type of network known as Higher Order Neural Networks (HONNs) have been introduced [5]. HONNs are a type of feed forward neural network which have single layer trainable weights that can help

B. Murgante et al. (Eds.): ICCSA 2013, Part I, LNCS 7971, pp. 427–437, 2013.
© Springer-Verlag Berlin Heidelberg 2013

in speeding up the training process. They are simple in their architecture and also able to reduce the number of required training parameters in the network. One of the models in HONNs is a Functional Link Neural Network (FLNN) known to be conveniently used for function approximation and can be extended for pattern recognition with faster convergence rate and lesser computational load [6].

FLNN is a class of higher order neural network (HONNs) that utilized higher combination of it inputs[6, 7]. FLNN can capture non-linear input-output mapping, provided that, they are fed with an adequate set of functional inputs[8]. Pao [6] pointed out that FLNN may be conveniently used for function approximation and can be extended for pattern recognition with faster convergence rate and lesser computational load. In the training of FLNN the mostly used algorithm is the Backpropagation (BP) learning algorithm. However, one of the crucial problems with BP-learning algorithm is that it can easily get trapped in local optima [8]. To improve this, the Artificial Bee Colony (ABC) optimization algorithm is proposed in this work to be used to optimize the weight in FLNN instead of the BP-learning algorithm.

In this paper, we described an overview of FLNN for time series prediction task particularly on the training of the network and the proposed Artificial Bee Colony (ABC) optimization as learning algorithm in order to achieve better learning for the network. The rest of this paper is organized as follows: A model description regarding the FLNN and Artificial Bee Colony optimization technique are given in section 2. The proposed FLNN-ABC for the learning scheme is detailed in section 3. The implementation of FLNN-ABC on time series temperature prediction is presented in section 4. Finally, the paper is concluded in section 5.

2   Functional Link Neural Network-Artificial Bee Colony

In this section, the properties and learning scheme of FLNN and ABC optimization are briefly discussed.

2.1   Functional Link Neural Network

Functional Link Neural Network is a class of HONNs created by Pao [7] and has been successfully used in many applications such as system identification [9-14], channel equalization [3], classification [15-18], pattern recognition [19, 20] and prediction [21, 22]. In this paper, we would discuss on the FLNN for the prediction task. FLNN is much more modest than MLP since it has a single-layer network compared to the MLP whilst able to handle a non-linear separable classification and functions approximation tasks. The FLNN architecture is basically a flat network without any hidden layer which has make the learning algorithm used in the network less complicated [23]. In FLNN, the input vector is extended with a suitably enhanced representation of the input nodes, thereby artificially increasing the dimension of the input space [6, 7].

In this work we focused on Functional link neural networks with generic basis architecture. This model uses a tensor representation. Pao [7], Patra [10], Namatamee [24] has demonstrated that this architecture is very effective for classification task.

Fig. 1 depicts the FLNN structure up to second order with 3 inputs. The first order consist of the 2 inputs x_1, and x_2 while the second order of the network is the extended input based on the product unit of x_1x_2. The learning part of this architecture on the other hand, consists of a standard Backpropagation as the training scheme.

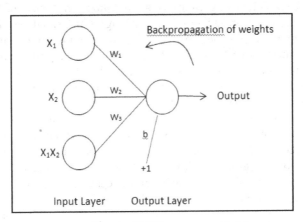

Fig. 1. The 2nd order FLNN structure with 2 inputs

2.2   FLNN Learning Scheme

In most previous researches, the learning algorithm used for training the FLNN is the Backpropagation (BP) [8, 16, 22, 23, 25-27]. BP learning is developed by Rumelhart [28] in which the network is provided with examples of the inputs and desired outputs to be computed, and then the error (difference between actual and expected results) will be calculated. The idea of the backpropagation algorithm is to reduced error, until the networks learned the training data. The training began with initializing random weights for the FLNN network with the goal to adjust these weights set in order to achieve the minimal error through the learning phase.

Even though BP is the mostly used algorithm in training the FLNN, the algorithm however, has several limitations which affect the performance of FLNN-BP network. The FLNN-BP network is prone to get trapped in local minima especially for the training on non-linearly separable classification problems. This is caused by the gradient descent method used by BP-learning algorithm in which the algorithm itself is strictly depends on the shape of the error surface. Since a common error surface on non-linearly separable classification problems may have many local minima and are multimodal, this has typically makes the algorithm susceptible to get stuck in some local minima when moving along the error surface during the training phase.

Another limitation of BP-learning algorithm inherit by FLNN-BP is that, the network is very dependent on the choices of initial values of the weights set as well as the parameters in the algorithm such as the learning rate and momentum [8] which make it not very easy to meet the desired convergence criterion during the training.

For these reasons, a further investigation to improve a learning algorithm used in tuning the learnable weights in FLNN is desired.

2.3   Artificial Bee Colony Optimization

The Artificial Bee Colony algorithm is an optimization tool, which provides a population-based search procedure [29]. The algorithm simulates the intelligent foraging behaviour of a honey bee swarm for solving multidimensional and multimodal optimization problem [30]. In population-based search procedure, each individual population called foods positions are modified by the artificial bees while the bee's aim is to discover the places of food sources with high nectar amount and finally the one with the highest nectar.

In this model, the colony of artificial bees consists of three groups, which are employed bees, onlookers and scouts [31]. For each food source there is only one artificial employed bee. The number of employed bees in the colony is equal to the number of food sources around the hive. Employed bees go to their food source and come back to the hive with three information regarding the food source; 1) the direction 2) its distance from the hive and 3) the fitness. The employed bees then perform waggle dance to let the colony evaluate the information. Onlookers watch the dances of employed bees and choose food sources depending on the dances. After waggle dancing on the dance floor, the dancer goes back to the food source with follower bees that were waiting inside the hive. This foraging process is called local search method as the method of choosing the food source is depending on the experience of the employed bees and their nest mates [30]. The employed bee whose food source has been abandoned becomes a scout and starts to search for finding a new food source randomly without using experience. If the nectar amount of a new source is higher than that of the previous one in their memory, they memorize the new position and forget the previous one [30]. This exploration managed by scout bees is called global search methods.

Several studies done by [30-32] has described that the Artificial Bee Colony algorithm is very simple, flexible and robust as compared to the existing swarm based algorithms: Genetic Algorithm (GA) , Differential Evolution (DE) and Particle Swarm Optimization (PSO) in solving numerical optimization problem. As in classification task in data mining, ABC algorithm also provide a good performance in gathering data into classes [33]. Hence motivated by these studies, the ABC algorithm is utilized in this work as an optimization tool to optimize FLNN learning for a prediction task.

3   FLNN-ABC Learning Scheme

Inspired by the robustness and flexibility offered by the population-based optimization algorithm, we proposed the implementation of the ABC algorithm as the learning scheme to overcome the disadvantages caused by backpropagation in the FLNN training. The proposed flowchart is presented in Fig. 2. In the initial process, the FLNN architecture (weight and bias) is transformed into objective function along with the training dataset. This objective function will then fed to the ABC algorithm in order to search for the optimal weight parameters. The weight changes are then tuned by the ABC algorithm based on the error calculation (difference between actual and expected results).

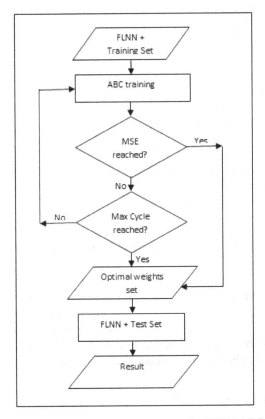

Fig. 2. The Proposed training scheme for FLNN-ABC

Based on the ABC algorithm, each bee represents the solutions with a particular set of weight vector. The ABC algorithm for training the FLNN is summarized as follow:

1) Cycle 0:
2) Initialize optimization parameter of FLNN, $j = 1,2, .. D$.
   where D are the number of weights and biases in FLNN (D= weights + biases)
3) Initialize a population of scout bee with random solution x_i, $i = 1,2, .. SN$.
   where SN denotes the size of population (Solution Numbers).
4) evaluate fitness of the population
5) cycle 1:
   i.   form new population v_i for the employed bees using:

$$v_{ij} = x_{ij} + \phi_{ij}(x_{ij} - x_{kj})$$

where k is a solution in the neighbourhood of i, Φ is a random
number in the range [-1, 1] and evaluate them.

ii.   Apply the greedy selection process between x_{ij} and v_{ij}
iii.   Calculate the probability values p_i for the solutions x_i using:

$$p_i = \frac{fit_i}{\sum\limits_{n=1}^{SN} fit_n}$$

iv.   Produce the new solutions v_i for the onlookers from the solutions x_i
       elected depending on p_i and evaluate them:
v.   Apply the greedy selection process for onlookers
vi.   Determine the abandoned solution for the scout, if exists, and replace
       it with a new randomly produced solution x_i using:

$$x_i^j = x_{min}^j + \text{rand}(0,1)(x_{max}^j - x_{min}^j)$$

vii.   Memorize the best solution
6) cycle=cycle+1
7) Stop when cycle = Maximum cycle.

4   FLNN-ABC for Time series Temperature Prediction

Weather forecasting is the application of science and technology on predicting the
state of the atmosphere of a location for monitoring seasonal changes in the weather.
In meteorology domain, weather is measureable in terms of temperature, atmospheric
pressure, humidity, wind speed and direction and also cloudiness [34]. Temperature
forecast is very important in ensuring the successful of weather forecast and also has
significant impact especially in the agriculture activities and water resources [35].

4.1   Experiment Setting

In this study, we used a univariate time series data of daily temperature forecasting for
the location of Batu Pahat, Johor ranging from the year 2005 to 2009. The data was
obtained from Malaysian Meteorological Department, Malaysia. The simulation
experiments were carried out on a 1.66 GHz Core 2 Duo Intel Workstation with 1GB
RAM. The comparison of standard FLNN-BP training and FLNN-ABC algorithms is
discussed based on the simulation results implemented in Matlab 2010a software for
both FLNN-BP and FLNN-ABC network models. Both FLNN-BP and FLNN-ABC
will be compared with MLP network trained with BP learning (MLP-BP) as to
evaluate in term of network complexity.

The temperature data series is partition into three parts; 50% for training set, and
25% for each validation set and Test set. Both network models were trained for 1000
epochs/cycles on the training set. Training is stopped when the minimum error of
0.0001 reached; or when the maximum of 1000 epochs/cycles reached. We also
implemented an early stopping measurement to avoid overfitting. For early stopping

measurement, the error on the validation set is measure after 15 epochs/cycles of training. The training is stopped when the validation phase detected an increase error in the validation set for both FLNN-BP and FLNN-ABC. Meanwhile, the test set is used for evaluating the network performance on the unseen data.

Table 1. Parameters setting

	MLP-BP	FLNN-BP	FLNN-ABC
Network structure	5-4-1	15-1	15-1
Optimization parameters/dimensions	29	16	16
Learning Rate	0.5	0.5	-
Momentum	0.5	0.5	-
Colony Size	-	-	50

The number of input node for FLNN was set to 5 up to 2nd order of input enhancement. This was done through trial-and-error procedure between 4 and 8 number of nodes. The 5 inputs with 2nd order inputs enhancement was selected as it gave better output result with less number of trainable parameters (weights + bias) for the FLNN network. The same trial-and-error procedure also performed on the selection of MLP network and the best MLP network structure selected was 5 input nodes with single hidden layer of 4 nodes. The parameters setting for the experiment are presented as Table 1.

As for forecasting horizon, we have chosen a one-step-ahead prediction since the main target is to predict the upcoming measure of daily temperature. The learning rate for MLP-BP and FLNN-BP was set to 0.05 with momentum value of 0.5, while the colony size of 50 bees with weight range of [-2, 2] was set for the FLNN-ABC architecture. The average results of 10 simulations were determined for both FLNN-BP and FLNN-ABC. The Mean Squared Error (MSE), Mean Absolute Error (MAE) and Normalized Mean Squared Error (NMSE) were used to evaluate each network performance.

5   Simulation Results

The comparison of simulation results for MLP-BP and FLNN-BP and FLNN-ABC on the time series temperature prediction data is presented in Table 2 below. Comparison of MLP and FLNN network in term of network complexity showed that the FLNN network required less numbers of parameters (trainable weights + biases) than MLP. The less numbers of parameters indicate that the network required less computational load as there are small numbers of weight and bias to be updated at every epoch or cycle. It can be seen from Table 2 that, training by FLNN-ABC resulted the lowest MSE which is 0.0063 as compared to FLNN-BP and MLP-BP with both are 0.0069 and 0.0075 respectively. When performing on the unseen data, FLNN-ABC also gives better MSE result which is 0.0066 thus outperform both FLNN-BP and MLP-BP with the difference of 0.0002 between FLNN-BP and FLNN-ABC and 0.0004 between

MLP-BP and FLNN-ABC on testing set. FLNN-ABC also gained lower MAE rather than FLNN-BP and MLP-BP which is 0.0633 for FLNN-ABC, 0.0641 for FLNN-BP and 0.0663 for MLP-BP. The lower MAE value indicated that FLNN-ABC was able to produce close forecast to the actual temperature data by outperforming the standard FLNN-BP with the ratio of 1.2×10^{-2}. Results from Table 2, also shows that the FLNN-ABC gives lower NMSE compared to both traditional FLNN-BP and standard MLP-BP which shows that the predicted and the actual values obtained by the FLNN-ABC are better in term of measuring the overall deviations of scatter between the prediction and the actual values. On the whole, the performance of training the FLNN network with ABC gives a better prediction result on unseen data when compare to FLNN-BP model and also with less network complexity as compared to MLP-BP.

Table 2. Performance Evaluations on Test Set

	MLP-BP	FLNN-BP	FLNN-ABC
Number of trainable nodes	29	16	16
MAE	0.0663	0.0641	0.0633
NMSE	0.8343	0.8026	0.6431
MSE Training	0.0075	0.0069	0.0063
MSE Testing	0.0070	0.0068	0.0066

The temperature forecast made by FLNN-ABC and standard FLNN-BP on test sets are graphically presented as in Fig. 3 and Fig. 4. The blue line represents the actual values while the black line refers to the predicted values. From both figures, it is shown that the FLNN-ABC has the ability to follow the actual trend as compared to standard FLNN-BP with minimum error forecast. Hence it can be seen that training scheme by ABC algorithm has facilitate the FLNN with better learning by providing a good exploration and exploitation capabilities in searching optimal weights set in the FLNN weights space as compared to BP learning [31, 36].

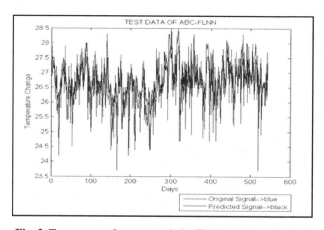

Fig. 3. Temperature forecast made by FLNN-ABC on Test set

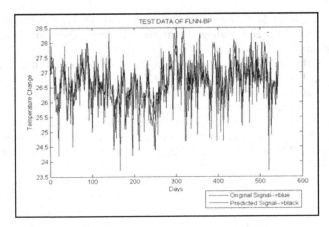

Fig. 4. Temperature forecast made by FLNN-BP on Test set

6   Conclusion and Future Work

In this work, the experiment has demonstrated that the FLNN-ABC performs the temperature prediction task quite well. Implementing the ABC algorithm as a learning scheme for FLNN has shown a significant higher results than backpropagation during experiment in terms of the lowest MSE, NMSE, RMSE and MAE. Since ABC algorithm combine the exploration and exploitation process in it search strategy, it can successfully avoid local minima trapping and provide the FLNN network with better ability in searching for optimal weights set during the training phase. Thus, ABC algorithm can be considered as an alternative learning scheme for training the Functional Link Neural Network instead of standard BP learning algorithm with better scheme in finding minimal error. As for future works, we are considering investigating the use of multivariate data to expand the FLNN-ABC ability for weather forecasting.

Acknowledgement. The authors wish to thank the Ministry of Higher Education Malaysia and Universiti Tun Hussein Onn Malaysia for the scholarship given in conducting these research activities.

References

1. Zhang, G.P.: Neural networks for classification: a survey. IEEE Transactions on Systems, Man, and Cybernetics, Part C: Applications and Reviews 30(4), 451–462 (2000)
2. Liao, S.-H., Wen, C.-H.: Artificial neural networks classification and clustering of methodologies and applications – literature analysis from 1995 to 2005. Expert Systems with Applications 32(1), 1–11 (2007)
3. Patra, J.C., Pal, R.N.: A functional link artificial neural network for adaptive channel equalization. Signal Processing 43(2), 181–195 (1995)

4. Chen, A.-S., Leung, M.T.: Regression neural network for error correction in foreign exchange forecasting and trading. Computers & Amp; Operations Research 31(7), 1049–1068 (2004)
5. Giles, C.L., Maxwell, T.: Learning, invariance, and generalization in high-order neural networks. Applied Optics 26(23), 4972–4978 (1987)
6. Pao, Y.H., Takefuji, Y.: Functional-link net computing: theory, system architecture, and functionalities. Computer 25(5), 76–79 (1992)
7. Pao, Y.H.: Adaptive pattern recognition and neural networks (1989)
8. Dehuri, S., Cho, S.-B.: A comprehensive survey on functional link neural networks and an adaptive PSO–BP learning for CFLNN. Neural Computing & Applications 19(2), 187–205 (2010)
9. Patra, J.C., Bornand, C.: Nonlinear dynamic system identification using Legendre neural network. In: The 2010 International Joint Conference on Neural Networks, IJCNN (2010)
10. Patra, J.C., Kot, A.C.: Nonlinear dynamic system identification using Chebyshev functional link artificial neural networks. IEEE Transactions on Systems, Man, and Cybernetics, Part B: Cybernetics 32(4), 505–511 (2002)
11. Abbas, H.M.: System Identification Using Optimally Designed Functional Link Networks via a Fast Orthogonal Search Technique. Journal of Computers 4(2) (2009)
12. Emrani, S., et al.: Individual particle optimized functional link neural network for real time identification of nonlinear dynamic systems. In: 2010 The 5th IEEE Conference on Industrial Electronics and Applications (ICIEA), (2010)
13. Nanda, S.J., et al.: Improved Identification of Nonlinear MIMO Plants using New Hybrid FLANN-AIS Model. In: IEEE International Advance Computing Conference, IACC 2009 (2009)
14. Teeter, J., Mo-Yuen, C.: Application of functional link neural network to HVAC thermal dynamic system identification. IEEE Transactions on Industrial Electronics 45(1), 170–176 (1998)
15. Raghu, P.P., Poongodi, R., Yegnanarayana, B.: A combined neural network approach for texture classification. Neural Networks 8(6), 975–987 (1995)
16. Abu-Mahfouz, I.-A.: A comparative study of three artificial neural networks for the detection and classification of gear faults. International Journal of General Systems 34(3), 261–277 (2005)
17. Liu, L.M., et al.: Image classification in remote sensing using functional link neural networks. In: Proceedings of the IEEE Southwest Symposium on Image Analysis and Interpretation (1994)
18. Dehuri, S., Cho, S.-B.: Evolutionarily optimized features in functional link neural network for classification. Expert Systems with Applications 37(6), 4379–4391 (2010)
19. Klaseen, M., Pao, Y.H.: The functional link net in structural pattern recognition. In: 1990 IEEE Region 10 Conference on Computer and Communication Systems, TENCON 1990 (1990)
20. Park, G.H., Pao, Y.H.: Unconstrained word-based approach for off-line script recognition using density-based random-vector functional-link net. Neurocomputing 31(1-4), 45–65 (2000)
21. Majhi, R., Panda, G., Sahoo, G.: Development and performance evaluation of FLANN based model for forecasting of stock markets. Expert Systems with Applications 36(3, pt. 2), 6800–6808 (2009)
22. Ghazali, R., Hussain, A.J., Liatsis, P.: Dynamic Ridge Polynomial Neural Network: Forecasting the univariate non-stationary and stationary trading signals. Expert Systems with Applications 38(4), 3765–3776 (2011)

23. Misra, B.B., Dehuri, S.: Functional Link Artificial Neural Network for Classification Task in Data Mining. Journal of Computer Science 3(12), 948–955 (2007)
24. Namatame, A., Veda, N.: Pattern classification with Chebyshev neural network. International Jounal of Neural Network 3, 23–31 (1992)
25. Haring, S., Kok, J.: Finding functional links for neural networks by evolutionary computation. In: Van de Merckt, T., et al. (eds.) Proceedings of the Fifth Belgian–Dutch Conference on Machine Learning, BENELEARN 1995, Brussels, Belgium, pp. 71–78 (1995)
26. Dehuri, S., Mishra, B.B., Cho, S.-B.: Genetic Feature Selection for Optimal Functional Link Artificial Neural Network in Classification. In: Fyfe, C., Kim, D., Lee, S.-Y., Yin, H. (eds.) IDEAL 2008. LNCS, vol. 5326, pp. 156–163. Springer, Heidelberg (2008)
27. Sierra, A., Macias, J.A., Corbacho, F.: Evolution of functional link networks. IEEE Transactions on Evolutionary Computation 5(1), 54–65 (2001)
28. Widrow, B., Rumelhart, D.E., Lehr, M.A.: Neural networks: applications in industry, business and science. Commun. ACM 37(3), 93–105 (1994)
29. Pham, D., et al.: The Bees Algorithm, in Technical Note, Manufacturing Engineering Centre, Cardiff University, UK (2005)
30. Karaboga, D.: An Idea Based on Honey Bee Swarm for Numerical Optimization, Erciyes University, Engineering Faculty, Computer Science Department, Kayseri/Turkiye (2005)
31. Karaboga, D., Basturk, B.: On the performance of artificial bee colony (ABC) algorithm. Applied Soft Computing 8, 687–697 (2007)
32. Akay, B., Karaboga, D.: A modified Artificial Bee Colony algorithm for real-parameter optimization. Information Sciences (2010) (in press, corrected proof)
33. Karaboga, D., Ozturk, C.: A novel clustering approach: Artificial Bee Colony (ABC) algorithm. Applied Soft Computing 11(1), 652–657 (2011)
34. Chapter 2. Weather and Climate, http://www.nasa.gov
35. Husaini, N.A., Ghazali, R., Mohd Nawi, N., Ismail, L.H.: Jordan pi-sigma neural network for temperature prediction. In: Kim, T.-h., Adeli, H., Robles, R.J., Balitanas, M. (eds.) UCMA 2011, Part II. CCIS, vol. 151, pp. 547–558. Springer, Heidelberg (2011)
36. Mohmad Hassim, Y.M., Ghazali, R.: Using Artificial Bee Colony to Improve Functional Link Neural Network Training. Applied Mechanics and Materials 263, 2102–2108 (2013)

A New Cuckoo Search Based Levenberg-Marquardt (CSLM) Algorithm

Nazri Mohd. Nawi, Abdullah Khan, and Mohammad Zubair Rehman

Faculty of Computer Science and Information Technology,
Universiti Tun Hussein Onn Malaysia (UTHM)
P.O. Box 101, 86400 Parit Raja, Batu Pahat, Johor Darul Takzim, Malaysia
`nazri@uthm.edu.my, hi100010@siswa.uthm.edu.my,`
`zrehman862060@gmail.com`

Abstract. Back propagation neural network (BPNN) algorithm is a widely used technique in training artificial neural networks. It is also a very popular optimization procedure applied to find optimal weights in a training process. However, traditional back propagation optimized with Levenberg marquardt training algorithm has some drawbacks such as getting stuck in local minima, and network stagnancy. This paper proposed an improved Levenberg-Marquardt back propagation (LMBP) algorithm integrated and trained with Cuckoo Search (CS) algorithm to avoided local minima problem and achieves fast convergence. The performance of the proposed Cuckoo Search Levenberg-Marquardt (CSLM) algorithm is compared with Artificial Bee Colony (ABC) and similar hybrid variants. The simulation results show that the proposed CSLM algorithm performs better than other algorithm used in this study in term of convergence rate and accuracy.

Keywords: Artificial neural network, back propagation, local minima, Levenberg-Marquardt, cuckoo search algorithm.

1 Introduction

Artificial Neural Networks (ANNs) is known for its competence in providing main features, such as: flexibility, ability of learning by examples, and capability to solve problems in pattern classification, function approximation, optimization, pattern matching and associative memories [1],[2]. Due to their powerful capability and functionality, ANN provides an alternative approach for many engineering problems that are difficult to solve by conventional approaches. ANNs are widely used in many areas such as signal processing, control, speech production, speech recognition and business [1]. Among many different neural network models, the multilayer feed- forward neural networks (MLFF) have been mainly used due to their well-known universal approximation capabilities [3]. The mostly popular MLFF training algorithms are the back-propagation (BP) algorithm and Levenberg Marquardt (LM), which are gradient-based methods [4]. Different techniques have been used in finding an optimal network performance for training ANNs such as evolutionary algorithms (EA),

B. Murgante et al. (Eds.): ICCSA 2013, Part I, LNCS 7971, pp. 438–451, 2013.
© Springer-Verlag Berlin Heidelberg 2013

genetic algorithms (GA), particle swarm optimization (PSO), differential evolution (DE), and back propagation algorithm [5-8]. Therefore, a variety of NN models have been proposed. The most commonly used method to train the NN is based on back propagation [9-10]. The back-propagation (BP) learning has become the most standard method and process in adjusting weight and biases for training an ANNs in many domains [11]. Unfortunately, the most commonly used Error Back Propagation (EBP) algorithm [12-13] is neither powerful nor fast. It is also not easy to find the proper neural network architectures. Moreover another limitation of gradient-descent based technique, is that it requires a differentiable neuron transfer function. Also, as ANN generate complex error surfaces with multiple local minima, the BP fall prey to local minima instead of a global minima [14].

In recent years, many improved learning algorithms have been proposed to overcome the weakness of gradient-based techniques. These algorithms include a direct optimization method using a poly tope algorithm [14], a global search technique such as evolutionary programming [15], and genetic algorithm (GA) [16]. The standard gradient-descent BP is not trajectory driven, but population driven. However, the improved learning algorithms have explorative search features. Consequently, these methods are expected to avoid local minima frequently by promoting exploration of the search space. The Stuttgart Neural Network Simulator (SNNS) which was developed a decade ago and is constantly improving [17]. The SNNS uses many different algorithms including Error Back Propagation [13], Quick prop algorithm [18], Resilient Error Back Propagation [19], Back percolation, Delta-bar-Delta, Cascade Correlation [20] etc. Unfortunately, all these algorithms are derivatives of steepest gradient search and training is relatively slow. For fast training, second order learning algorithms have to be used. The most effective method is Levenberg Marquardt algorithm (LM) [21], which is a derivative of the Newton method. This is a relatively complex algorithm since not only the gradient but also the Jacobian Matrix must be calculated. The LM algorithm was developed only for layer-by layer architectures, which is far from optimum [22]. LM algorithm is ranked as one of the most efficient training algorithms for small and medium sized patterns. LM algorithm was successfully implemented for neural network training in [23], but only for multilayer perceptron (MLP) architectures known as LMBP. LM has proved its mettle in improving the convergence speed of the network. It is the due to the good collaboration of Newton's method and steepest descent [24]. Not only it has the speed advantage of Newton's method, but also has the convergence character of the steepest descent method. Even though the LM algorithm is frequently used for fast convergence [26] but still, it is not devoid of local minima problem [26-27], [34].

In order to overcome the issues of slow convergence and network stagnancy, this paper propose a new algorithm that combines Cuckoo Search (CS) developed in 2009 by Yang and Deb [28-29] and Levenberg-Marquardt back propagation (LMBP) algorithm to train ANN for Exclusive-OR (XOR) datasets. The proposed Cuckoo Search Levenberg-Marquardt (CSLM) algorithm helps in reducing the error and avoids local minima. The remaining of the paper is organized as follows.

The remaining paper is organized as follows: Section 2 gives literature review on ANN. Section 3, explains Cuckoo Search via levy flight. In section 4, explain the proposed CSLM algorithm and the simulation results are discussed in section 5 respectively. Finally, the paper is concluded in the Section 6.

2   Artificial Neural Networks

Artificial Neural Networks (ANNs) imitates the learning processes of human cognitive system and the neurological functions of the brain. ANN works by processing information like Biological neurons in the brain and consists of small processing units known as Artificial Neurons, which can be trained to perform complex calculations [30]. An Artificial Neural Network (ANN) consists of an input layer, one or more hidden layers and an output layer of neurons. In ANN, every node in a layer is connected to every other node in the adjacent layer. An Artificial Neuron can be trained to store, recognize, estimate and adapt to new patterns without having the prior information of the function it receives. This ability of learning and adaption has made ANN superior to the conventional methods. Due to its capability to solve complex time critical problems, it has been widely used in the engineering fields such as biological modelling, financial forecasting, weather forecasting, decision modelling, control systems, manufacturing, medicine, ocean and space exploration etc [31-32]. ANN are usually classified into several categories on the basis of supervised and unsupervised learning methods and feed-forward and feed backward architectures [30].

3   Cuckoo Search (CS) Algorithm

Cuckoo Search (CS) algorithm is a novel meta-heuristic technique proposed by Xin-Shen Yang [28]. This algorithm was stimulated by the obligate brood parasitism of some cuckoo species by laying their eggs in the nests of other host birds. If an egg is discovered by the host bird as not their own then they will either throw the unknown egg away or simply abandon its nest and build a new nest somewhere else. Some other species have evolved in such a way that female parasitic cuckoos are often very specialized in the mimicking the color and pattern of the eggs of a few chosen host species. This reduces the probability of their eggs being abandoned and thus increases their reproductively. The CS algorithm follows the three idealized rules:

- Each cuckoo lays one egg at a time, and put its egg in randomly chosen nest;
- The best nests with high quality of eggs will carry over to the next generations;
- The number of available host nests is fixed, and the egg laid by a cuckoo is discovered by the host bird with a probability pa [0, 1].

In this case, the host bird can either throw the egg away or abandon the nest, and build a completely new nest. The last assumption can be approximated by the fraction pa of the n nests that are replaced by new nests (with new random solutions). For a maximization problem, the quality or fitness of a solution can simply be proportional to the value of the objective function. In this algorithm, each egg in a nest represents a solution, and a cuckoo egg represents a new solution, the aim is to use the new and potentially better solutions (cuckoos) to replace a not so good solution in the nests. Based on these three rules, the basic steps of the Cuckoo Search (CS) can be summarized as;

Step 1: Generate initial population of N host nest   i= 1... N
Step 2: *while* (f$_{min}$< MaxGeneration) or (stop criterion)
Do
Step 3: Get a cuckoo randomly by Levy flights and evaluate its fitness F_i.
Step 4: Choose randomly a nest j among N.
Step 5: *if   Fi >Fj*
Then
Step 6: Replace j by the new solution.
Step 7: *end if*
Step 8: A fraction (pa) of worse nest are abandoned and new ones are built.
Step 9: Keep the best solutions (or nest with quality solutions).
Step 10: Rank the solutions and find the current best.
Step 11: *end while*

When generating new solutions x^{t+1} for a cuckoo *i*, a Levy flight is performed;

$$x_i^{t+1} = x_i^t + \alpha \oplus levy(\lambda) \tag{1}$$

Where, $\alpha > 0$ is the step size which should be related to the scales of the problem of interest. The product $\oplus$ means entry wise multiplications. The random walk via Levy flight is more efficient in exploring the search space as its step length is much longer in the long run.

$$Lavy \sim u = t^{-\lambda}, 1 < \lambda \leq 3 \tag{2}$$

This has an infinite variance with an infinite mean. Here the steps essentially construct a random walk process, a power-law step-length distribution with a heavy tail. Some of the new solutions should be generated by Levy walk around the best solution obtained so far, this will speed up the local search. However, a substantial fraction of the new solutions should be generated by far field randomization whose locations should be far enough from the current best solution. This will make sure the system will not be trapped in local optimum.

4   The Proposed CSLM Algorithm

The CS is a population based optimization algorithm, and similar to many others, meta-heuristic algorithms start with a random initial population, The CS algorithm essentially works with three components: selection of the best source   by keeping the best nests or solutions, replacement of host eggs with respect to the quality of the new solutions or cuckoo eggs produced based randomization via Levy flights globally (exploration) and discovering of some cuckoo eggs by the host birds and replacing according to the quality of the local random walks (exploitation) [33]. In Figure 1, each cycle of the search consists of several steps initialization of the best nest or solution, the number of available host nests is fixed, and the egg laid by a cuckoo is discovered by the host bird with a probability *pa* ε [0, 1].

In the proposed Cuckoo Search Levenberg-Marquardt (CSLM) algorithm, each best nest or solution represents a possible solution (i.e., the weight space and the corresponding biases for NN optimization in this study) to the considered problem and the size of a population represents the quality of the solution. The initialization of weights is compared with output and the best weight cycle is selected by cuckoo. The cuckoo will continue searching until the last cycle to find the best weights for the network. The solution that is neglected by the cuckoo is replaced with a new best nest. The main idea of this combined algorithm is that CS algorithm is used at the beginning stage of searching for the optimum to select the best weights. Then, the training process is continued with the LM algorithm using the best weights of CS algorithm. The LM algorithm interpolate between the Newton method and gradient descent method. The LM algorithm is the most widely used optimization algorithm. It outperforms simple gradient conjugate descent and gradient methods in a wide other variety of problems [35]. The flow diagram of proposed CSLM algorithm is shown in

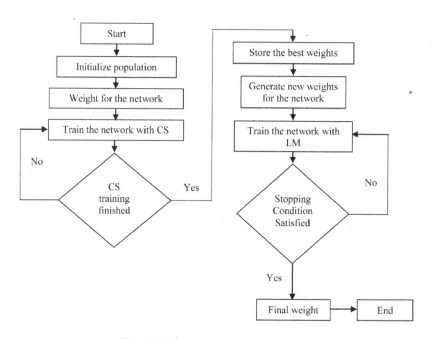

Fig. 1. The Proposed CSLM Algorithm

Figure 1. In the first stage CS algorithm finished its training, then LM Algorithm start training with the weights from CS algorithm and the LM continue to train the network until the stopping criteria or Mean Square Error (MSE) is achieved.

5   Experiments and Results

In ordered to illustrate the performance of the proposed CSLM algorithm in training ANN. CSLM algorithm is tested on 2-bit, 3-bit XOR, and 4-bit OR parity problems. The simulation experiment are performed on an AMD Athlon 1.66 GHz CPU with a 2-GB RAM. The software used for simulation process is MATLAB 2009b.

The proposed CSLM algorithm is compared with Artificial Bee Colony Levenberg- Marquardt (ABCLM), Artificial Bee Colony Back-Propagation (ABCBP) and conventional Back-Propagation Neural Network (BPNN) algorithms respectively. The performance measure for each algorithm is based on the Mean Square Error (MSE). The three layers feed forward neural network architecture (i.e. input layer, one hidden layer, and output layers.) is used for each problem. The number of hidden nodes is varied to 5 and 10 neurons. In the network structure, the bias nodes are also used and the log-sigmoid activation function is applied. . For each problem, trial is limited to 1000 epochs. And MSE criteria is kept to 0. A total of 20 trials are run for each case. The network results are stored in the result file for each trial.

5.1   The 2 Bit XOR Problem

The first test problem is the Exclusive OR (XOR) Boolean function of two binary input to a single binary output as (0 0; 0 1; 1 0 ; 1 1) to (0; 1; 1 ;0). In the simulations, we used 2-5-1, 2-10-1 MLFF network for two bit XOR problem. The parameters range for the upper and lower band is used [5,-5], [5,-5], [5,-5], [1,-1] respectively. For the CSLM, ABCLM, ABCBP and BPNN, Table 1 and Table 2

Table 1. CPU Time, Epochs and MSE for **2-5-1** ANN Structure

Algorithm	ABCBP	ABCLM	BPNN	CSLM
CPUTIME	172.3388	123.9488	42.64347	14.41
EPOCHS	1000	1000	1000	126
MSE	2.39E-04	0.125	0.220664	0
Accuracy (%)	96.47231	71.69041	54.6137	100

Table 2. CPU Time, Epochs and MSE for 2-10-1 ANN Structure

Algorithm	ABCBP	ABCLM	BPNN	CSLM
CPUTIME	197.34	138.96	77.63	18.61
EPOCHS	1000	1000	1000	153
MSE	8.39E-04	0.12578	0.120664	0
Accuracy (%)	96.8	71.876	54.6137	100

shows the CPU time, number of epochs and the MSE for the 2 bit XOR test problem with 5 and 10 hidden neurons. Figure 2 and Figure 3 shows the mean square error of CSLM algorithm and ABCBP algorithm for the 2-5-1 network structure.

The CSLM algorithm avoids the local minima and trained the network successfully within 145, 153 epochs as seen in the Table 1 and Table 2. Both Tables, show that CSLM can converge successfully for almost every kind of network structure. In Figure 2, the CSLM algorithm can be seen to converge within 153 epochs which is quite superior convergence rate as compared to the other algorithms. The ABCBP algorithm is showing a lot of oscillations in the trajectory path and not converging within 100 epochs as shown in the Figure 3. BPNN shows many failures in convergence to the global solution. The average CPU time and MSE of CSLM is also found to be less than BPNN, ABCBP, and ABCLM algorithms.

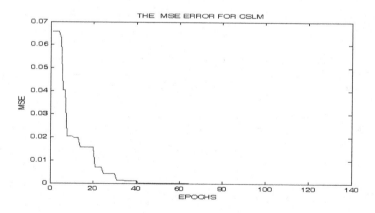

Fig. 2. MSE for CSLM using 2-5-1 ANN Structure

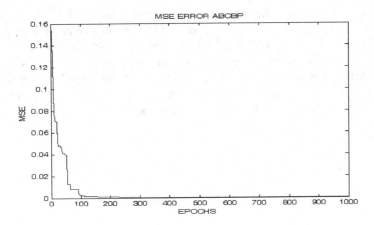

Fig. 3. MSE for ABCBP using 2-5-1 ANN Structure

5.2   The 3 Bit XOR Problem

In the second case we used three bit Exclusive OR test problem of three input to a single binary output as (1 1 1; 1   1 0 ; 1 0 1; 1 0 0; 0 1 1; 0 1 0 ; 0 0 1; 0 0 0) –to-(1; 0; 0; 1; 0; 1; 1; 0). In three bit XOR   input if the number of binary inputs is odd, the output is 1, otherwise the output is 0. Also for the three bit input we apply 3-5-1, and 3-10-1, feed forward neural network structure. In Figure 4 and Figure 5 we can see the simulation result of 3 bit XOR problem. The parameter range is same as used for two bit XOR problem. For the 3-5-1 network structure it has twenty connection weights and six biases, and for the 3-10-1 the network has forty connection weights and eleven biases.

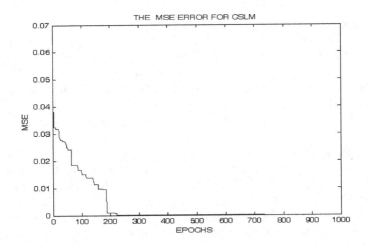

Fig. 4. MSE for CSLM using 3-5-1 ANN Structure

In Figure 4 and Figure 5 we can see the simulation results of 3 bit XOR problem. Figure 4 illustrates the 'MSE vs. Epochs' of CSLM algorithm. While Figure 5 shows the 'MSE vs. Epochs' of the BPNN algorithm. CSLM algorithm can be seen converging with 0 MSE within 80.53 seconds CPU time and 671 epochs in Figure 4. While in Figure 5, ABCLM is seen as converging with 0.0056 MSE and 125.5 seconds CPU time. In three bit XOR dataset, CSLM has fulfilled all criterion during convergence; such as less MSE, less CPU cycles, high Accuracy, and less no of epochs.

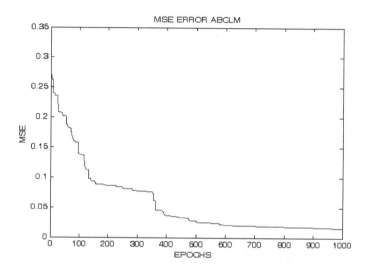

Fig. 5. MSE for BPNN using 3-5-1 ANN Structure

5.3   The 4 Bit OR Problem

The third problem is 4 bit OR. In case we used four bit OR test problem of four input to a single binary output as (1 1 1 1;1 1 1 0;1 1 0 1;1 1 0 0; 1 0 1 1;1 0 0 1; 1 0 0 0;0 1 1 1;0 1 1 0; 0 1 0 1; 0 1 0 0; 0 0 1 1 ; 0 0 1 0; 0 0 0 1;1 0 1 0; 0 0 0 0) –to –(1; 1; 1; 1; 1; 1; 1; 1; 1; 1; 1; 1; 1; 1; 1; 0). The network structure is same as the 2 and 3 bit XOR problem. In 4 bit OR input if the number of inputs all is 0, the output is 0, otherwise the output is 1. Again for the four bit input we apply 4-5-1, 4-10-1, feed forward neural network structure. For the 4-5-1 network structure it has twenty five connection weights and six biases, and for the 4-10-1 the network has fifty connection weights and eleven biases. Table 3 and Table 4 illustrates the CPU time, epochs, and MSE performance of the proposed CSLM algorithm, ABCBP, ABCLM and BPNN algorithms respectively. Figure 6, and Figure 7, shows the 'MSE performance vs. Epochs' for the 4-5-1, and 4-10-1 network structure of the proposed CSLM algorithm. While in Figure 8, we see the 'MSE performance vs. Epochs' for the network structure 4-5-1 of ABCBP algorithm, and Figure 9 shows the 'MSE performance vs. Epochs' of ABCLM algorithm for 4-10-1 network structure.

Table 3. CPU time, Epochs and MSE error for 4-5-1 ANN structure

Algorithm	ABCBP	ABCLM	BPNN	CSLM
CPU TIME	162.4945	118.7274	63.28089	6.16091
EPOCHS	1000	1000	1000	43
MSE	1.91E-10	1.82E-10	0.052778	0
Accuracy (%)	99.97	99.99572	89.83499	100

Table 4. CPU time, Epochs and MSE error for 4-10-1 ANN structure

Algorithm	ABCBP	ABCLM	BPNN	CSLM
CPU TIME	180.4945	129.7274	67.28089	8.65
EPOCHS	1000	1000	1000	46
MSE	1.67E-10	1.76E-10	0.05346	0
Accuracy (%)	99.47	99.78	89.83499	100

Table 3 illustrates the CPU time, MSE and epochs for the 4-5-1 network structure. While, Table 4 shows the CPU time MSE error and epochs with 4-10-1 network structure. From both Tables, we can see that CSLM algorithm outperforms ABCBP, ABCLM, and BPNN in-terms of CPU time, epochs, MSE, and accuracy. Figure 6

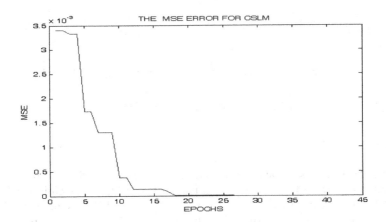

Fig. 6. MSE for CSLM using 4-5-1 ANN Structure

and Figure 7 shows the 'MSE performance vs. Epochs' for the 4-5-1, and 4-10-1 network structure for CSLM algorithm. In both Figures, CSLM is converging within 43, 46 epochs. For the 4-5-1 and 4-10-1 network structure CSLM again has 0 MSE. While in Figure 8 we see that for the network structure 4-5-1, ABCBP is converging within 1000 epochs which is quite a large number if compared with CSLM. Also, ABCLM algorithm for 4-10-1 network structure shows a large MSE and more CPU time then the proposed CSLM algorithm.

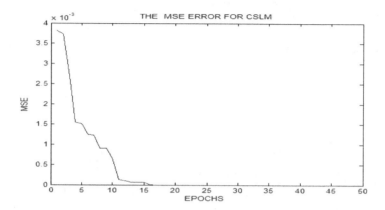

Fig. 7. MSE for CSLM using 4-10-1 ANN Structure

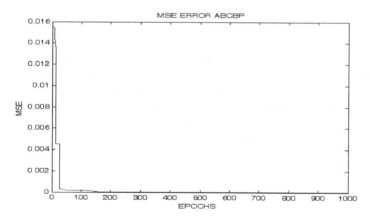

Fig. 8. MSE for ABCBP using 4-10-1 ANN Structure

5.4   Overall Result

The overall MSE results of the algorithms for 2 bit XOR, 4 bit OR problem, are given in above Tables. From the above Tables, it is clear that the CSLM algorithm obtained the best results as compared to the ABCBP, ABCLM, and BPNN in term of CPU time, MSE, and accuracy. the over all results show that CSLM perform better result on 4-bit OR dataset then 2 and 3 bit XOR dataset in term of CPU time and epochs.

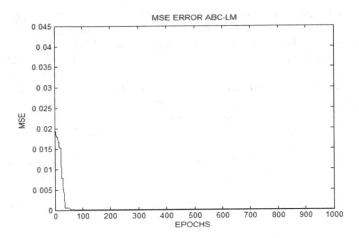

Fig. 9. MSE for ABCLM using 4-10-1 ANN Structure

6   Conclusion

Traditional back propagation optimized with Levenberg marquardt training algorithm has some drawbacks such as getting stuck in local minima, and network stagnancy [4]. In this paper, an improved Levenberg-Marquardt back propagation (LMBP) algorithm integrated and trained with Cuckoo Search (CS) algorithm. In the proposed CSLM algorithm, first CS algorithm trains the network and LM continues training by taking the best weight-set from CS and tries to minimize the training error avoided local minima problem and achieve fast conversances. The proposed CSLM algorithm is used to train MLFF on the 2-bit XOR, 3- Bit XOR and 4-Bit OR benchmark problems. The results show that the CSLM is simple and generic for optimization problems and has better convergence rate and accuracy than the ABCLM, ABCBP and BPNN algorithms. In future this proposed model will be used benchmarks data classification.

Acknowledgements. The Authors would like to thank Office of Research, Innovation, Commercialization and Consultancy Office (ORICC), Universiti Tun Hussein Onn Malaysia (UTHM) and Ministry of Higher Education (MOHE) Malaysia for financially supporting this Research under Fundamental Research Grant Scheme (FRGS) vote no. 1236.

References

1. Dayhoff, J.E.: Neural-Network Architectures: An Introduction, 1st edn. Van Nostrand Reinhold Publishers, New York (1990)
2. Mehrotra, K., Mohan, C., Ranka, S.: Elements of Artificial Neural Networks. MIT Press, Cambridge (1997)

3. Rehman, M.Z., Nazri, M.N.: Studying the Effect of adaptive momentum in improving the accuracy of gradient descent back propagation algorithm on classification problems. International Journal of Modern Physics (IJMPCS) 9(1), 432–439 (2012)

4. Ozturk, C., Karaboga, D.: Hybrid Artificial Bee Colony algorithm for neural network training. In: IEEE Congress of Evolutionary Computation (CEC), pp. 84–88 (2011)

5. Leung, C., Member, C.: A Hybrid Global Learning Algorithm Based on Global Search and Least Squares Techniques for back propagation neural network Networks. In: International Conference on Neural Networks, pp. 1890–1895 (1994)

6. Yao, X.: Evolutionary artificial neural networks. International Journal of Neural Systems 4(3), 203–222 (1993)

7. Mendes, R., Cortez, P., Rocha, M., Neves, J.: Particle swarm for feed forward neural network training. In: Proceedings of the International Joint Conference on Neural Networks, vol. 2, pp. 1895–1899 (2002)

8. Nawi, N.M., Ghazali, R., Salleh, M.N.M.: The development of improved back-propagation neural networks algorithm for predicting patients with heart disease. In: Zhu, R., Zhang, Y., Liu, B., Liu, C. (eds.) ICICA 2010. LNCS, vol. 6377, pp. 317–324. Springer, Heidelberg (2010)

9. Abid, S., Fnaiech, F., Najim, M.: A fast feedforward training algorithm using a modified form of the standard backpropagation algorithm. IEEE Transactions on NeuralNetworks 12, 424–430 (2001)

10. Yu, X., OnderEfe, M., Kaynak, O.: A general backpropagation algorithm for feedforward neural networks learning. IEEE Transactions on Neural Networks 13, 251–259 (2002)

11. Chronopoulos, A.T., Sarangapani, J.: A distributed discrete time neural network architecture forpattern allocation and control. In: Proceedings of the International Parallel and Distributed Processing Symposium (IPDPS 2002), Florida, USA, pp. 204–211 (2002)

12. Wilamowski, B.M.: Neural Networks and Fuzzy Systems. CRC Press (2002)

13. Rumelhart, D.E., Hinton, G.E., Williams, R.J.: Learning Internal Representations by error Propagation. In: Parallel Distributed Processing: Explorations in the Microstructure of Cognition, vol. 1 (1986)

14. Gupta, J.N.D., Sexton, R.S.: Comparing backpropagation with a genetic algorithm for neural network training. The International Journal of Management Science 27, 679–684 (1999)

15. Salchenberger, L.M., Cinar, E.M., Lash, N.A.: Neural Networks: A New Tool for Predicting Thrift Failures. Decision Sciences 23(4), 899–916 (1992)

16. Sexton, R.S., Dorsey, R.E., Johnson, J.D.: Toward Global Optimization of Neural Networks: A Comparison of the Genetic Algorithm and Back propagation. Decision Support Systems 22, 171–186 (1998)

17. SNNS (Stuttgart Neural Network Simulator) (2013), http://wwwra.informatik.unituebingen.de/SNNS/ (Accessed February 2, 2013)

18. Fahlman, S.E.: Faster-Learning Variations on Back-Propagation: An Empirical Study. In: Proceedings of the 1988 Connectionist Models Summer School. Morgan-Kaufmann, Los Altos (1988)

19. Riedmiller, M., Braun, H.: A direct adaptive method for faster backpropagation learning: The RPROP algorithm. In: Proceedings of the IEEE International Conference on Neural Networks, ICNN 1993 (1993)

20. Fahlman, S.E., Lebiere, C.: The Cascade-Correlation Learning Architecture. In: Touretzky, D.S. (ed.) Advances in Neural Information Processing Systems 2. Morgan-Kaufmann, Los Altos (1990)

21. Hagan, M.T., Menhaj, M.B.: Training feedforward networks with the Marquardt algorithm. IEEE Transactions on Neural Networks 23, 899–916 (1994)
22. Wilamowski, B.M., Cotton, N., Kaynak, O.: Neural Network Trainer with Second Order Learning Algorithms. In: 11th International Conference on Intelligent Engineering Systems, Budapest, Hungary (2007)
23. Hagan, M.T., Menhaj, M.B.: Training feedforward networks with the Marquardt algorithm. IEEE Trans. Neural Networks 5(6), 989–993 (1994)
24. Cao, X.P., Hu, C.H., Zheng, Z.Q., Lv, Y.J.: Fault Prediction for Inertial Device Based on LMBP Neural Network. Electronics Optics & Control 12(6), 38–41 (2005)
25. Haykin, S.: Neural Networks: A Comprehensive Foundation. China Machine Press, Beijing (2004)
26. Xue, Q., et al.: Improved LMBP Algorithm in the Analysis and Application of Simulation Data. In: 2010 International Conference on Computer Application and System Modeling (2010)
27. Yan, J., Cao, H., Wang, J., Liu, Y., Zhao, H.: Levenberg-Marquardt algorithm applied to forecast the ice conditions in Ningmeng Reach of the Yellow River. In: Fifth International Conference on Natural Computation (2009)
28. Yang, X.S., Deb, S.: Cuckoo search via Lévy flights. In: Proceedings of World Congress on Nature & Biologically Inspired Computing, India, pp. 210–214 (2009)
29. Yang, X.S., Deb, S.: Cuckoo search via levy flights. In: Nature Biologically Inspired Computing (NaBIC), pp. 210–214 (2009)
30. Deng, W.J., Chen, W.C., Pei, W.: Back-propagation neural network based importance-performance analysis for determining critical service attributes. Expert Systems with Applications 4(2) (2008)
31. Kosko, B.: Neural Network and Fuzzy Systems, 1st edn. Prentice Hall of India (1994)
32. Lee, T.: Back-propagation neural network for the prediction of the short-term storm surge in Taichung harbor, Taiwan. Engineering Applications of Artificial Intelligence 21(1) (2008)
33. Yang, X.S., Deb, S.: Engineering optimisation by Cuckoo Search. International Journal of Mathematical Modelling and Numerical Optimisation 1, 330–343 (2010)
34. Kumar, M.P.: Backpropagation Learning Algorithm based on Levenberg Marquardt Algorithm. In: Proceedings of CSCP, pp. 393–398 (2012)
35. Nourani, E., Rahmani, A.M., Navin, A.H.: Forecasting Stock Prices using a hybrid Artificial Bee Colony based Neural Network. In: International Conference on Innovation, Management and Technology Research (ICIMTR 2012), Malacca, Malaysia (2012)
36. Chaowanawatee, K., Heednacram, H.: Implementation of Cuckoo Search in RBF Neural Network for Flood Forecasting. In: Fourth International Conference on Computational Intelligence, Communication Systems and Networks, pp. 22–26 (2012)

Dynamic Context for Document Search and Recovery

José Rodríguez[1], Manuel Romero[1], and Maricela Bravo[2]

[1] Computing Department, CINVESTAV-IPN
Distrito Federal, México, CP 07300
rodriguez@cs.cinvestav.mx, jmromero@computacion.cs.cinvestav.mx
[2] Systems Department, UAM-Azcapotzalco
Distrito Federal, México, CP 02200
mcbc@correo.azc.uam.mx

Abstract. From an Information Retrieval perspective there are many works which have been proposed to deal with the problem of retrieving and searching relevant documents. One of the main drawbacks of traditional approaches is related with their static context similarity evaluation, this issue has been addressed by manually refining the query . In this paper we describe a context-based method to dynamically improve the query during document search. A set of experiments were conducted to evaluate the precision and recall of the proposed method, evaluation of results show the benefits of this novel method.

Keywords: document context, document search, dynamic context.

1 Introduction

Information Retrieval (IR) is the research area concerned with the development of methods and tools for supporting document search aiming at recovering relevant documents. Document search methods depend mainly on the document format, which may fall in one of the following classifications: *structured* documents, which are documents coded with a structured language, most of them based on XML notations; *unstructured,* mainly free text documents, and *semi-structured.* In particular, in this work we address the problem of searching and retrieving relevant *unstructured* documents obtained from a collection of public available documents.

One of the most important tasks performed by a researcher, when he is writing a paper or report, is the search of papers that are closely related to the topic being addressed. One way to do this search is through the use of a popular Web search tool to find information related with the search criteria provided by the user; however this approach does not get always the most suitable results. In a second way, sometimes the researcher have access to a database of some scientific publisher, where he can gets the publications related with the topics he he is interested in, even if this approach may give better results, they are not always the most suitable neither.

B. Murgante et al. (Eds.): ICCSA 2013, Part I, LNCS 7971, pp. 452–463, 2013.

In the both cases the task for searching information is a very time-consuming and error prone task. One way to solve this problem is through the development of intelligent search tools with lower computational costs that facilitate the automation of retrieval of documents.

In this paper we present a proposal for the search of documents based in his context. Our approach considers the existence of a given document search space and requires the specification of an initial document, which is used to construct an initial context. This initial context is composed of m terms extracted from the document with their calculated frequencies. The context of search is carried out by comparing the initial context with each of the rest of the documents from the document search space. At each comparison stage the context is improved as better terms are found in the current document being compared. In this way the documents more related to the context are preserved and the non-or less related are rejected.

This document is structured as follows: section 2 presents the theoretical basis for this proposal and the related works, section 3 presents the architecture and theoretical concepts of our proposal, in section 4 the experimental results are presented and analyzed, finally section 5 contains the conclusions.

2   Related Work

2.1   Relevance FeedBack the Rocchio Algorithm

When working with collections of documents, the presence of synonyms gives as result that the same concept can be referred by different words. In consequence it may be difficult to construct a good query when the collection is not well known. Nevertheless some particular promising documents could be used to engage an interactive query refinement process.

The presence of synonym has an impact on the recall of information retrieval systems. One way to address this issue is by using relevance feedback and manually refining the query; however this is a slow and error prone process. This make necessary to count with an automatically query refinement mechanism i.e. a method that adjust the query relative to the documents that appear to match the initial query. Relevance Feedback is implemented mainly as [1]: 1)User issues a query (normally short and simple). 2) The system returns an initial set or results. 3) User marks some returned documents as relevant or nonrelevant. 4) The system computes a new representation of the information based on the users feedback. 5)A new set of documents is returned by the system.

The Rocchio algorithm [2] is one of the most well known algorithms for implementing relevance feedback. In this algorithm users have a partial knowledge, an original query vector, and two sets of relevant and nonrelevant documents. In every interaction, a new query is generated by displacing the original query towards the centroid of relevant documents and far from the nonrelevant centroid documents. Weight parameters are attached to both sets to control the balance between the analyzed documents and the original query.

The success of relevance feedback depends on: first having enough knowledge in order to be capable of making the initial query. Second, it is necessary to have relevant documents similar to each other i.e. relevant documents clustered around a centroid prototype. However, because clusters of documents are not easy to find in the Web relevance feedback has not been widely used in Web search.

2.2 Cluster Based Approaches

From the Information Retrieval perspective there are many works which have been proposed to deal with the problem of retrieving relevant documents. One popular approach is using a clustering approach. A cluster-based approach consists of calculating similarity distances between all documents in order to execute an appropriate clustering algorithm, such as: hierarchical agglomerative clustering (HAC), partition clustering (K means) or self-organizing maps (SOM). One of the main drawbacks of this approach is its performance when there is the need to execute clustering on a large collection of documents.

One of the most representative works of a cluster-based approach is the adaptive document clustering approach proposed by Yu, Wang, & Chen [3]. In this proposal authors state that relevant documents tend to belong to the same cluster. Their approach starts by assigning a random position to each document and, as queries are continuously processed, relevant documents are shifted closer to each other. In [4] Na et al. proposed a query-based similarity for adaptive online document clustering. The main advantages of their approach are: *collection adaptability*, when the collection of documents changes the system only needs to relearn the similarities; and *user adaptability*, the system calculates similarities considering the user interests.

Most of clustering algorithms use the Vector Space Model (VSM) proposed by Salton, Wong and Yang in [5], where each document is represented by a vector of weights corresponding to text features. A very important component in the VSM-based approach is the Bag of Words (BOW), which considers as features the different terms in the documents (we use term as a synonim of word in the bag). The BOW model is a popular method which represents different terms of a document in a vector of term frequencies where the order between terms is not considered.

Another efficient VSM-based method is the Generalized Vector Space Model (GVSM) [6]. GVSM represents each document as a vector containing the similarities with the rest of the documents in the collection. The GVSM introduces the concept of document context, where a collection of similar documents describe and share a document context, which is a main issue in this paper. The Context Vector Model [7] is a VSM-based method used to describe a term context vector for storing similarities between the terms in the context vector. A new promising trend in VSM representation is the incorporation of ontologies. Two related works are the Hotho, Maedche and Staab proposal of an ontology-based text document clustering [8]; and the Jing et.al [9] ontology-based distance measure for text clustering.

Current advances on VSM show improvements on discovering and representing correlations between terms in a vector. Kalogeratos and Likas [10] describe the Global Text Context Vector (GTCV-VSM) for text document representation. Their main contribution is a method that 1) captures local contextual information for each term occurrence in the term sequences of documents, 2) combines local contexts to define a global context of that term, 3) constructs a semantic matrix with all global contexts, and 4) uses this matrix to represent a feature space where document search and clustering can be solved.

The main differences between related work and the current solution reported in this paper are evaluated in accordance with the characteristics showed in Table 1:

Table 1. Evaluation characteristics of related work

Characteristics	Yu, Wang, & Chen [3]	Seung-Hoon Na [4]	GVSM [6]	Hotho, Maedche and Staab [8]	Jing et.al [9]	GTCV-VSM [10]	Our Proposal
VSM representation	Yes	Yes	Yes	No	No	Yes	Yes
BOW (term order is not relevant)	Yes	Yes	Yes			Yes	Yes
Document contexts	No	Yes	Yes	Yes	Yes	Yes	Yes
Document vector as a query formulation	No	No	No	No	No	Yes	Yes
Dynamic change of document context	No	Yes	No	No	No	No	Yes

3   Document Recovery Using Dynamic Context

In this section we describe a novel VSM method for document recovery based on dynamic context measuring. Our method considers a scenario where the user has an initial document, this document is used to build the initial context represented as a vector of term-frequencies. This initial context will improved at comparison stage by adding new terms extracted from the document being analysed.

The documents are represented in the vector space as a set of terms, the vector is composed of a term to represent a dimension and a related frequency for determining the value for that dimension. In this proposal the bag of words is considered the context of the document.

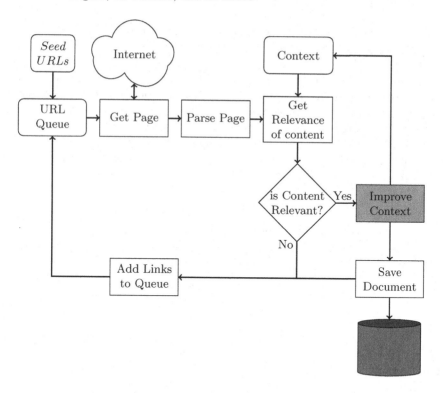

Fig. 1. Architecture of our proposal

Figure 1 shows the general architecture of our proposal. The main component in this architecture is the mechanism for the improvement of the context. This mechanism takes a bag of words from the new relevant document and makes improvement over the original base context.

3.1   Document Context Representation

In order to obtain the document context representation, the set of documents must be preprocessed and represented as vector spaces. The weight for every term is determinated according to the frequency of that term inside the document. The preprocessing of documents consists of three steps: tokenization of the text contained into each document, elimination of StopWords to avoid redundancies and noisy terms, and stemming terms form the remaining text.

Let $D = \{d_1, d_2, \ldots, d_n\}$ be a set of n documents, where each document is represented as a vector, composed of a set of m different pairs where each pair is formed of the term w_j and its frequency f_j.

$$d_i = \{(w_1, f_1), (w_2, f_2), \ldots, (w_j, f_j)\}, \tag{1}$$

with $0 < i <= n$, and $0 < j <= m$

For each document the context is obtained by its term-frequency vector representation. Where the frequency of every term is normalized to compose the bag of words, in this way a vector of dimension n is constructed and every term of the document represents a dimension of the vector.

3.2   Dynamic Document Context Calculation

Given an initial document, his context q is represented as a vector we call it the initial context consisting of the set of m different terms w_j and their frequencies f_j respectively so $q = \{(w_1, f_1), (w_2, f_2), \ldots, (w_j, f_j)\}$, the process of dynamically calculating the context to recover relevant documents starts by measuring the similarity between the initial context q with the context d_i of each documents using the Cosine Similarity [11] (2).

The relevance of links is evaluated using the approach of Cheng et. al. [12] where the context of links is taken into account.

$$simC_{ij} = \frac{\sum_k x_{ik} x_{jk}}{(\sum_k x_{ik}^2 \sum_k x_{jk}^2)^{\frac{1}{2}}}, \tag{2}$$

Where $simC_{ij}$ is the evaluated rate of similarity and x_{ij} is the frequency of term k inside of document i. The dynamic context is generated by mixing every bag of words obtained from every relevant document with the actual bag of words. In this way the new context is refined with a best domain of search.

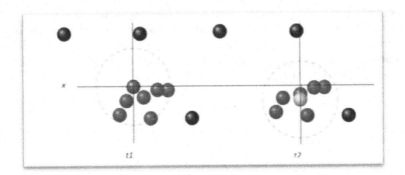

Fig. 2. Dynamic context

Figure 2 shows our dynamic document context proposal. In the first iteration identified by $t1$, the dot located at $(x, t1)$ represents the context initial surrounded by a threshold (represented by the dotted circle), this graphical representation allows immediate identification of the set of relevant documents (inside the dotted circle). In $t2$ the new improved context is located at $(x',$ $t2)$, representing a x-x displacement from its original place, surrounded by its threshold to include the new relevant documents. This dynamic context calculation allows the threshold to adapt to new documents and improve the domain. In both cases the dots outside the dotted circles (threshold) represent non-relevant documents and therefore are not considered for the dynamic context calculation.

The new context is a mixture of the terms from the actual context and the set of terms found in the new relevant document. This improved context is generated as follows:

1. A new relevant text for the actual context (B) is obtained.
2. A bag of words is obtained from the new relevant text, by applying (1).
3. For every element i in the context of the last relevant text recovered D, if $D[i]$ is not inside of B, this element is added to B.
4. If B has already $D[i]$, the average of $B[i]$ and $D[i]$ is evaluated and $B[i]$ is actualized with the calculated average.
5. The least meaning elements of the context are eliminated.

The steps explained above can be represented as:

$$B' = \begin{cases} \{(w_i, f_i)|(w_i, f_i) = (w_{iB}, f_{iB}) \text{ if } \forall \ w_i \in B \wedge w_i \notin D\} \\ \{(w_i, f_i)|(w_i, f_i) = (w_{iD}, f_{iD}) \text{ if } \forall \ w_i \in D \wedge w_i \notin B\} \\ \{(w_i, f_i)|w_i = w_{iB} \text{ y } f_i = (f_{iB} + f_{iD})/2 \text{ if } \forall \ w_i \in B \cap D\}, \end{cases} \quad (3)$$

Where B' is the new bag of words, (w_i, f_i) are the pairs composing the new context, according to (1), w_{iB} and w_{iD} are terms of the set B and D respectively and f_{iB} and f_{iD} are de frequencies for the i-th term inside of the documents B and D. In our proposal two thresholds are used: α to evaluate if the document could be downloaded and β to evaluate the evolution of the context.

In this way the mechanism tries to improve the search and recovery of relevant sources for a particular domain of knowledge. Moreover the least relevant documents are eliminated from the set after the new context is defined, keeping the context in the core of the set of relevant documents.

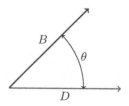

Fig. 3. Vector representation

The improvement of context through the evaluation of the average of every element in the vectors is represented graphically in figure 3. Where B represents the actual context, in figure 4 B' represents the new evaluated context and θ represents the angle between them.

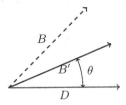

Fig. 4. Vector representation after context merge

If $\cos(\theta)$ gives a value greater than the defined threshold the context will be improved with the average of both vectors. In figure 4 the generated context (B') is closer to the context of the relevant document (D).

4   Experiments

Our proposal has been tested on our local repositories. The test has been implemented by the execution of a crawler opening and analyzing every document and evaluating his relevance for the context. By using our local repositories of documents we can know if the system is recovering documents that are relevant for the context.

To test our proposal we have used three sets of documents, where each set is composed of 125 documents. Before tests were executed, the documents were compared against the initial document base to calculate the similarity matrix. For every set of documents a different document base was used. Each document set represents a given area of knowledge, but they don't have necessarily the same topics. Additionally, some documents from different knowledge areas have been added to the sets in order to evaluate where they are placed by the crawler with regard to the threshold.

The tests were executed using different values for the thresholds α and β, in order to determine the best combination according to the measures of precision and recall. The mean F-measure has been evaluated too for every test. Table 2 shows the values obtained for every test using a threshold $\beta = 0.70$. In this tests a high value for the harmonic mean is searched, keeping the balance between the values of precision an recall. As it can be seen in table 2 the best results for every set are: $\alpha = 0.6$ for the set 1, $\alpha = 0.65$ for the set 2, and $\alpha = 0.55$ for set 3.

Table 3 shows the results for the second test, whit a threshold $\beta = 0.75$. In this case the best values for α are: 0.55 for test 1, 0.65 for set 2, and 0.55 for

Table 2. Results for $\beta = 0.70$

$\beta = 0.70$	set 1			set 2			set 3		
$\alpha =$	**0.55**	**0.60**	**0.65**	**0.55**	**0.60**	**0.65**	**0.55**	**0.60**	**0.65**
Precision	0.6511	0.8333	1.0	0.6585	0.7222	0.913	0.9	0.9166	1.0
Recall	0.9032	0.8064	0.5483	0.9310	0.8965	0.7241	0.7823	0.4782	0.2173
F-measure	0.7567	0.8196	0.7083	0.7714	0.8	0.8076	0.8372	0.6285	0.3571

Table 3. Results for $\beta = 0.75$

$\beta = 0.75$	set 1			set 2			set 3		
$\alpha =$	**0.55**	**0.60**	**0.65**	**0.55**	**0.60**	**0.65**	**0.55**	**0.60**	**0.65**
Precision	0.8055	0.8846	1.0	0.5918	0.6923	0.8666	0.9545	1.0	1.0
Recall	0.9354	0.7419	0.4838	1.0	0.9310	0.8965	0.913	0.7826	0.3478
F-measure	0.8656	0.8070	0.6521	0.7435	0.7941	0.8813	0.9333	0.878	0.5161

set 3. The value of α in both tests indicates that a tradeoff between precision and recall i.e. when the value of α raises the precision raises too and the recall decreases.

Table 4. Results without improving the context

	set 1	set 2	set 3
$\alpha = 0.6$			
Precision	0.8846	0.6585	1.0
Recall	0.7419	0.931	0.7826
F-measure	0.8070	0.7714	0.878

Table 4 shows the results without the evolution on the context, as in the previous tests the threshold for α was set to 0.6, as we can see compared against table 3 our proposal has better results for set 2 however for sets 1 and 3 we get similar results, the reason is that with a greater β we have less documents taking part in the definition of the context nevertheless with a greater β we are assuring that the context is influenced mainly by the relevant documents.

Comparing the results of test 3 against results in table 2, our proposal gives a better balance between precision and recall for sets 1 and 2 but not for set 3. The reason is that with a more small value for β the context is influenced by less (or non) relevant documents that could make the lost of relevant documents

As we can see it is very important to define a higher value for β avoiding in this way a negative influence over the definition of the context. Better results can be obtained with our proposal as the context is improved with the information from new documents with a high coefficient of similarity.

Figures 5 a) to f) show the results for the first 40 documents for every set. Figures a) and b) show the results for set 1, figure c) and d) for set 2 and figure d) and f) for set 3. In these figures we can see:

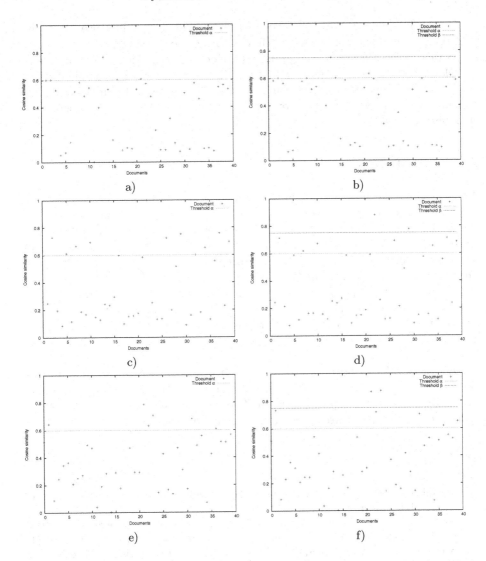

Fig. 5. Document similarity without context improve (left side) and with improved contex (right side)

For set 1: in figure a) (without the improvement of context) we get 6 relevant docu-ments, in figure b (with the improvement of context) we get 7 relevant documents, while discarding at the same time non-relevant documents recovered in a).

For set 2: in figure c) we get 11 relevant documents, in d) improving the context we get 9 relevant documents, in this case some recovered documents in c) have been discarded.

For set 3: in figure e) we get 6 relevant documents, while in f) we get 7, the additional document was recovered thanks to the improved context.

As we can see, through the improvement of context some documents are found as relevant, these documents were not considered when a non-dynamic context was used. Moreover we can see that our proposal can discard non-relevant documents. We can say that the context move towards the relevant documents avoiding the rest.

5   Conclusions

In this paper we have described a novel approach for document search and recovery based on the generation of a dynamic context at runtime. We have compared our results against the traditional method of the VSM and we got better results. A collateral results we have found is that the manipulation of thresholds must be make carefully, because they can get us closer or far from the relevant sources of information.

We think that our proposal gives good results on search and recovery of documents, the analysis of results shows us that the recovered information is improved by using the dynamic context. Our proposal is capable of refine the search and recovering information as exploring documents with a threshold greater than the defined β.

Our proposal can be improved in several aspects, by example: it could be necessary to find the right combination of contexts where the bag of words could be the optimum for a particular search. The use of semantic information for the definition of the bag of words could be considered too.

References

1. Manning, C.D., Raghavan, P., Schutze, H.: Introduction to information retrieval. Cambridge University Press (2008)
2. Rocchio, J.J.: Relevance feedback in information retrieval. In: The Smart System-experiments in Automatic Document Processing, pp. 313–323. Prentice Hall (1971)
3. Yu, C.T., Wang, Y.T., Chen, C.H.: Adaptive document clustering. In: Proceedings of the 8th Annual International ACM Conference on Research and Development in Information Retrieval, Montreal, Canada, pp. 197–203 (1985)
4. Na, S.H., Kang, I.S., Lee, J.H.: Adaptive document clustering based on query-based similarity. Inf. Process. Manage. 43(4), 887–901 (2007)
5. Salton, G., Wong, A., Yang, C.S.: A vector space model for automatic indexing. Commun. ACM 18(11), 613–620 (1975)
6. Wong, S.K.M., Ziarko, W., Wong, P.C.N.: Generalized vector space model in information retrieval. In: Proceedings of the 8th Annual International ACM Conference on Research and Development in Information Retrieval, Montreal, Canada, pp. 18–25 (1985)
7. Billhardt, H., Borrajo, D., Maojo, V.: A context vector model for information retrieval. Journal of the American Society for Information Science and Technology 53(3), 236–249 (2002)

8. Hotho, A., Maedche, A., Staab, S.: Ontology-based text document clustering. Journal of Knstliche Intelligence 16(4), 48–54 (2002)
9. Jing, L., Zhou, L., Ng, M., Huang, J.: Ontology-based distance measure for text clustering. In: Proceedings of the Text Mining Workshop, SIAM International Conference on Data Mining, Boston, Massachusetts (2006)
10. Kalogeratos, A., Likas, A.: Text document clustering using global term context vectors. Knowledge and Information Systems 31, 455–474 (2012)
11. Lee, M.D., Pincombe, B.M., Welsh, M.B.: An empirical evaluation of models of text document similarity. In: Bara, B.G., Barsalou, L., Bucciarelli, M. (eds.) Proceedings of the XXVII Annual Conference of the Cognitive Science Society, Stresa, Italy, pp. 1254–1259 (2005)
12. Cheng, Q., Beizhan, W., Pianpian, W.: Efficient focused crawling strategy using combination of link structure and content similarity. In: IEEE International Symposium on IT in Medicine and Education, Bangalore, India, pp. 1045–1048 (2008)

Multiscale Discriminant Saliency for Visual Attention

Anh Cat Le Ngo[1], Kenneth Li-Minn Ang[2],
Guoping Qiu[3], and Jasmine Seng Kah-Phooi[4]

[1] School of Engineering, The University of Nottingham, Malaysia Campus
[2] Centre for Communications Engineering Research, Edith Cowan University
[3] School of Computer Science, The University of Nottingham, UK Campus
[4] Department of Computer Science & Networked System, Sunway University

Abstract. The bottom-up saliency, an early stage of humans' visual attention, can be considered as a binary classification problem between center and surround classes. Discriminant power of features for the classification is measured as mutual information between features and two classes distribution. The estimated discrepancy of two feature classes very much depends on considered scale levels; then, multi-scale structure and discriminant power are integrated by employing discrete wavelet features and Hidden markov tree (HMT). With wavelet coefficients and Hidden Markov Tree parameters, quad-tree like label structures are constructed and utilized in maximum a posterior probability (MAP) of hidden class variables at corresponding dyadic sub-squares. Then, saliency value for each dyadic square at each scale level is computed with discriminant power principle and the MAP. Finally, across multiple scales is integrated the final saliency map by an information maximization rule. Both standard quantitative tools such as NSS, LCC, AUC and qualitative assessments are used for evaluating the proposed multiscale discriminant saliency method (MDIS) against the well-know information-based saliency method AIM on its Bruce Database wity eye-tracking data. Simulation results are presented and analyzed to verify the validity of MDIS as well as point out its disadvantages for further research direction.

1 Visual Attention - Computational Approach

Visual attention is a psychological phenomenon in which human visual systems are optimized for capturing scenic information. Robustness and efficiency of biological devices, the eyes and their control systems, visual paths in the brain have amazed scientists and engineers for centuries. From Neisser [26] to Marr [25], researchers have put intesive effort in discovering attention principles and engineering artificial systems with equivalent capability. For last two decades, this research field is dominated by visual saliency principles which proposes an existence of a saliency map for attention guidance. The idea is further promoted in Feature Integration Theory (FIT) [34] which elaborates computational principles of saliency map generation with center-surround operators and basic image features such as intensity, orientation and colors. Then, Itti et al. [21] implemented and released the first complete computer algorithms of FIT theory [1]. Feature Integration Theory are widely accepted as principles behind visual

---

[1] http://ilab.usc.edu/toolkit/

B. Murgante et al. (Eds.): ICCSA 2013, Part I, LNCS 7971, pp. 464–484, 2013.
© Springer-Verlag Berlin Heidelberg 2013

attention partly due to its utilization of basic image features. Moreover, this hypothesis is supported by several evidences from psychological experiments. However, it only defines theoretical aspects of visual attention with saliency maps, but does not investigate how such principles would be implemented algorithmically. It leaves research field open for many later saliency algorithms [21],[16],[32],[19], etc. Saliency might be computed as a linear contrast between features of central and surrounding environments across multiple scales in the center-surround operation. Saliency is also modeled as phase difference in Fourier Transform Domain [20], or saliency at each location depends on statistical modeling of the local feature distribution [32]. Though many approaches are mentioned in long and rich literature of visual saliency, only a few are built on a solid theory or linked to other well-established computational theory. Among the approaches, Neil Bruce's work [6] nicely established a bridge between visual saliency and information theory. It put a first step for bridging two alien fields; moreover, visual attention for first time could be modeled as information system. Then, information-based visual saliency has continuously been investigated and developed in several works [23],[22], [29], [14]. The distinguish points between these works are computational approaches for information retrieval from features. The process attracts much interest due to difficulty in estimating information of high dimension data like 2-D image patches. It usually runs into computational problems which can not be efficiently solved due to the curse-of-dimensionality; moreover, central and surrounding contexts are usually defined in ad-hoc manners without much theoretical supports. To encounter the difficulties, Danash Gao et al. has simplified the information extraction step as a binary classification problem with decision theory. Two classes are identified as center and surround contexts then discriminant power or mutual information between features and classes are estimated as saliency values for each location. This formulation of visual saliency approach is named as Discriminant Saliency (DIS) of which underlying principles are carefully elaborated by Gao et al. [18]. Its significant point is estimating information from class distributions given input features rather than from the input features themselves. Therefore, computational load is significantly reduced since only simple class distribution need estimating rather than complex feature distribution.

Spatial features have dominated influence on saliency values; however, scale-space features do have a decisive role in visual saliency computation since center or surround environments are simply processing windows with different sizes. In signal processing, scale-space and spectral space are two sides of a coin; therefore, there is a strong relation between scale-frequency-saliency in visual attention problem. Several researchers [3,30,27,33] outlined that fixated regions have high spatial contrast or showed that high-frequency edges allow stronger discrimination between fixateed over non-fixated points. In brief, they all come up with one conclusion increased predictability at high frequencies. Though these studies emphasizes a greater visual attraction to high frequencies (edges, ridges, other structures of images), there are other works focusing on medium frequency. Bruce et al. [7] found that fixation points tend to prefere horizontal and vertical frequency content rather than random position, and these oriented content have more noticeable difference in medium frequencies. More interestingly, choices of frequency range for visual processing may depend on encountering visual context [2]. For example, luminance contrast explained fixation locations better in natural image category and slightly worse in urban scenes category provided that all images are applied low-pass filters as preprocessing steps. Perhaps, that attention system might include different range of frequencies in generating optimal eye-movements. Diversity in spectral space usage means utilization of several different scales in scale-space theory. It can be assumed that both high frequency (small scale) and medium frequency (medium scale) constitutes an ecological relevance and compromise between information requirement and available attentional capacity in the early stage of visual attention when observers are not driven by performing any specific tasks.

Though multi-scale nature have been emphasized as implicit element of human visual attention, it is often ignored in several visual saliency algorithm. For example, DIS approach [18] considers only one fixed-size window and it may lead to inconsideration of significant attentive features in a scene. Therefor, DIS approach needs constituting under the multi-scale framework to form multiscale discriminant saliency (MDIS) approach. This is the main motivation as well as contribution of this paper which are organized as follows. Section 2 reviews principles behind DIS [14] and focuses on its important assumption and limitation. After that, MDIS approach is carefully elaborated in section 3 with several relating contents such as multiple dyadic windows for binary classification problem in subsection 3.1, multiscale statistical model of wavelet coefficients in subsection 3.2, maximum likehood (MLL) and maximum a posterior probability (MAP) computation of dyadic subsquares in subsections 3.3, 3.4. Then, all steps of MDIS are combined for final saliency map generation in subsection 3.5. Quantitative and qualitative analysis of the proposed method with different simulation modes are discussed in section 4; moreover, simulation data of MDIS in comparisons with the well-known information-based saliency method AIM [6] are presented with a number of interesting conclusions. Finally, main contributions of this paper as well as further research direction are stated in the conclusion section 5.

2   Visual Attention - Discriminant Saliency

Saliency mechanism plays a key role in perceptual organization; therefore, recently several researchers attempt to generalize principles for visual saliency. In the decision theoretic point of view, saliency is regarded as power for distinguishing salient and non-salient classes; moreover, discriminant saliency combines classical center-surround hypothesis with derived optimal saliency architecture. In other word, saliency of each image location is identified by the discriminant power of a feature set with respect to the binary classification problem between center and surround classes. Based on decision theory, this discriminant saliency detector can work with variety of stimulus modalities, including intensity, color, orientation and motion. Moreover, various psychophysic property for both static and motion stimuli are shown to be accurately satisfied quantitatively by DIS saliency maps.

Perceptual systems evolve for producing optimal decisions about the state of surrounding environments in a decision-theoretic sense with minimum probability of error. Beside accurate decisions, the perceptual mechanisms should be as efficient as possible. Mathematically, the problem needs defining as (1) a binary classification of interest stimuli (salient features) against the null hypothesis (non-salient features) and (2) measurement of discriminant power from extracted visual features as saliency at each location in the visual field. The discriminant power is estimated in classification process with respect to two classes of stimuli: stimuli of interest and null stimuli of all uninterested features. Each location of visual field can be classified whether it includes stimuli of interest optimally with lowest expected probability of error. From pure computational standpoint, the binary classification for discriminant features are widely studied and well-defined as tractable problem in the literature. Moreover, the discriminant saliency concept and the decision theory appear in both top-down and bottom-up problems with different specifications of stimuli of interest [16],[14].

The early stages of biological vision are dominated by the ubiquity of "center-surround" operator; therefore, bottom-up saliency is commonly defined as how certain the stimuli at each location of central visual field can be determined against other stimuli in its surround. In other words, "center-surround" hypothesis is a natural binary classification problem which can be solved by well-established decision theory. In this problem, classes can be defined as follows.

- Center class: observations within a central neighborhood W_l^1 of visual fields location l.
- Surround class: observations within a surrounding window W_l^0 of the above central region.

At each location, likelihood of either hypothesis depends on the visual stimulus, of a predefined set of features X. The saliency at location l should be measured as discriminating power of features X in W_l^1 against features X in W_l^0. In other words, discriminant saliency value is proportional to distance between feature distributions of center and surrounding classes.

Feature responses within the windows are drawn from the predefined feature sets X in a process. Since there are many possible combinations and orders of how such responses are assembled, the observations of features can be considered as a random process, $X(l) = (X_1(l), \ldots, X_d(l))$ of dimension d. This random process is drawn conditionally on the states of hidden variable $Y(l)$, which is either center or surround state. Feature vector $x(j)$ such that $j \in W_l^c, c \in \{0, 1\}$ are drawn from classes c according to the conditional probability density $P_{X(l)|Y(l)}(x|c)$ where $Y(l) = 0$ for surround or $Y(l) = 1$ for center. The saliency of location l, $S(l)$ is equal to the discriminant power of X for the classification of the observed feature vectors. That discriminant concept is quantified by the mutual information between feature, X and class label, Y.

$$S(l) = I_l(X;Y)$$
$$= \sum_c \int p_{X,Y}(x,c) log \frac{p_{X,Y}(x,c)}{p_X(x)p_y(c)} dx$$

Though binary classification and decision theory makes discriminant saliency computationally feasible, it is only true for low-dimensional data. Computer vision and visual attention need to deal with high-dimensional input images especially when it involves statistics and information theory. As mentioned previously, observations of feature responses $X(l)$ are considered as a random process in d-dimensional space.

Mutual information estimation in such high-dimensional space encounters serious obstacles due to the curse of dimensionality. As these problems persist, saliency algorithms would never be biologically plausible and computationally feasible. Therefore, discriminant saliency algorithms have to be approximated by taking into account statistical characteristics of natural images as well as mathematical simplification. Dashan Gao and Nuno Vasconcelos have proposed a feasible algorithm for mutual estimation. Mathematically, it can be formulated as follows.

$$I_l(X;Y) = H(Y) - H(Y|X) \tag{1}$$

$$= E_X(H(Y) + E_{Y|X}[log P_{Y|X}(c|x)]) \tag{2}$$

$$= E_X \left[H(Y) + \sum_{c=0}^{1} P_{Y|X}(c|x) log P_{Y|X}(c|x) \right] \tag{3}$$

$$= \frac{1}{|W_l|} \sum_{j \in W_l} \left[H(Y) + \sum_{c=0}^{1} P_{Y|X}(c|x(j)) log P_{Y|X}(c|x(j)) \right] \tag{4}$$

where $H(Y) = -\sum_{c=0}^{1} P_Y(c) log P_Y(c)$ is the entropy of classes Y and $-E_{Y|X}$ $[log P_{Y|X}(c|x)]$ is the conditional entropy of Y given X. Given a location 1, there are corresponding center W_l^1 and surround W_l^0 windows along with a set of associated feature responses $x(j), j \in W_l = W_l^0 \cup W_l^1$. The mutual information can be estimated by replacing expectations with means of all samples inside the join windows W_l. The conditional entropy $H(Y|X)$ can be computed by analytically deriving MAP $P(Y|X)$ given that transformed features are modelled by Generalized Gaussian Distribution (GGD) and only binary classification is considered. Lets name Gao's proposal for discriminant saliency computation as DIS; more details about DIS can be found in their publications [18][14][17][15].

While DIS successfully defines discriminant saliency in information-theoretic senses, its implementation has certain limits. Feature responses are randomly sampled in a single fixed-size window; therefore, it is obviously biased toward objects with distinctive features fitted in that window size. As previous visual attention has confirmed involvement of multi-scales factor in visual attention, DIS needs extension from a fixed-scale process to a multi-scale process. In theory, DIS can be carried out with different size of windows, and this approach certainly produces image responses and saliency values at multiple scales. However, such an approach are not recommended for both computational and biological efficiency. Moreover, it causes high redundancy in saliency values across multiple scales. In order to solve the multi-scale problems systematically, DIS need integration with multiple scale processing techniques such as wavelet .

3   Multiscale Discriminant Saliency

Expansion from a fixed window-size to multiscale processing is a common problem of algorithms development for computer vision applications. Therefore, there are several framework with multi-scale processing capability, which can be used to develop a so called Multiscale Discriminant Saliency (MDIS). A chosen framework has to include both binary classification and multi-scale processing. In other words, it needs classifying a single image point into two or more separate classes with prior knowledge from other scales. With respect to these requirements, a multiscale image segmentation framework should be a great starting point for MDIS as DIS can be considered as simplification of image segmentation in a sense that it only needs to classify a data point into two classes (center-or-surround). DIS only uses this binary classification as intermediate step to measure discriminant power of center-surrounding features, and classification step of DIS does not emphasize on accuracy of segmentation results.

Typical algorithms employ a classification window of some size in a vague hope that all included pixels are belong to the same class. Obviously, these algorithms and DIS have similar problems with choices of processing window sizes. Clearly, the size of classification is crucial to balance between the classification reliability and accuracy. A large window usually provides rich statistical information and enhance reliability of the algorithm. However, it simultaneously risks including heterogeneous elements in the window and inevitably reduces segmentation accuracy. Appropriate window sizes are equivalently vital for DIS to avoid local maxima in discriminant power as well. If sizes of classification windows are too large or too small, the algorithm risks losing useful discriminative features or being too susceptible to noise. In brief, sizes of processing windows and a number of involved data points have vital impact on both DIS and a segmentation problem.

3.1 Multiple Classification Windows

Multiscale segmentation employs many classification window of different sizes and classifying results are later combined to obtain more accurate segmentation at fine scales. MDIS adapts a similar approach to increase classification efficiency between features of center and surrounding classes. In this paper, dyadic squares (or blocks) are implemented as classification windows with different sizes. Lets assume an initial square image s with $2^J \mathrm{x} 2^J$ of $n := 2^{2J}$ pixels, the dyadic square structures can be generated by recursively dividing x into four square sub-images equally. As a result, it has the popular quad-tree structure, commonly employed in computer vision and image processing problems. In this tree structure, each node is related to a direct above parent node while it plays a role of parent node itself for four direct below nodes 1. Each node of the quad-tree corresponds to a dyadic square, and let denote a tree-node in scale j by d_i^j whereof i is a spatial index of the dyadic square node. Given a random field image X, the dyadic squares are also random fields which are formulated as D_i^j mathematically. In following sections, we sometime use D_i (dropping scale factor j) as general randomly-generated dyadic square regardless of scales.

With this predefined structures of classification windows, we will classify each nodes d_i or dyadic squares into either of binary classes center-surround. Using the likelihood of such classification, we can estimate mutual information between features and corresponding labels. Such a mutual information has been proved to quantify discriminant power of central against surrounding features at each location or corresponding discriminant saliency value. This estimation requires multiple pixel pdf models for each class in multiple scales, and these distributions can be learned through wavelet-based statistical models.

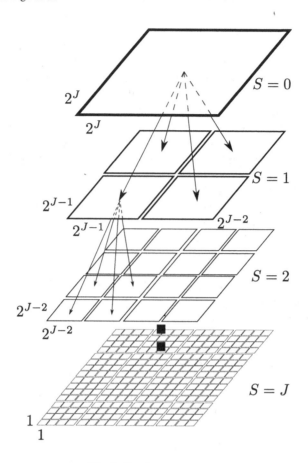

Fig. 1. Quad-tree structure

3.2   Multiscale Statistical Model

The complete joint pixel pdf is typically overcomplicated and difficult to model due to their high dimension nature. Unavailability of simple distribution model in practice motivates statistical modeling on transform domain which is often less complex and easier to be estimated. Obviously, joint pixel pdf could be well approximated as marginal pdf of transformed coefficients. Since wavelet transform well-characterizes semantic singularity of natural images (general image category considered in this study), it provides a suitable domain for modeling statistical property of singularity-rich images. Natural images are full of edges, ridges other highly structural features as well as textures; therefore, wavelet transforms as multi-scale edge detectors well represents that singularity-rich content in at multiple scales and three different directions. Noted that, only normal discrete wavelet transform (DWT) is considered in this study for the sake of simplicity though concepts can be adapted into other wavelet transforms as well. Henceforth, whenever wavelet transform is mentioned, it should refer to DWT instead of stating otherwise. As multi-scale edge detectors, responses of wavelet transform overlying a singularity are large wavelet coefficients while wavelets overlying a smooth region yield small coefficients. Deploying thresholds over wavelet coefficients leads to binary classification of singularity against non-singularity features. Moreover, statistical model of images can be simplified under "'restructure'" singularity representation in multiscale manner. Quad-tree structure of dyadic squares in pixel domain can be mirrored in wavelet decomposition since four wavelet coefficients at a given scale nest inside one at the next coarser level. For example, Haar wavelet coefficients at each quad-tree nodes are generated by Harr wavelet transform of the corresponding dyadic image square. The marriage between singularity detector and multi-scale quad-tree structure implies the singularity property at each spatial location persists through scales along the branches of the quad-tree. The singularity characterization along scales makes the wavelet domain well-suited for modeling natural images. In fact, statistical modeling of wavelet coefficients have quite comprehensive literature; here we concentrate on the hidden Markov tree (HMT) of Crouse, Nowak and Baraniuk [13]. In consideration of both marginal and joint wavelet coefficient statistics, the HMT model introduces a (hidden) state variable of either "'large'" or "'small'" to each wavelet coefficients. Then, the marginal density of wavelet coefficients is modeled as a two-density Gaussian mixture in which a "'large'" state refers to a large variance Gaussian and a "'small'" state represents small variance Gaussian distribution. This Gaussian mixture closely matches marginal statistics observed in natural images [28],[1], [10]. With the HMT, persistence of large or small coefficients are captured across scales using Markov-1 chain. It models dependencies between hidden states across scale in a tree structure, parallel to that of wavelet coefficients and dyadic squares. Combining both the Gaussian Mixture Model (GMM) and Markov State Transition into a vector $\mathcal{M}$, the wavelet HMT is able to approximate the overall joint pdf of the wavelet coefficients $\mathbf{W}$ by a high-dimensional but highly structured Gaussian mixture models $f(\mathbf{w}|\mathcal{M})$. Highly structural nature of wavelet coefficients allows efficient implementation of HMT-based processing. The HMT model parameters $\mathcal{M}$ can be learned through the iterative expectation and maximization (EM) algorithm with cost $\mathcal{O}(n)$ per iteration [13] or predefined for a particular image category [31]. After the parameters $\mathcal{M}$ are estimated by EM algorithm, we need to compute statistical characteristics of wavelet coefficients given the wavelet transform $\tilde{\mathbf{w}}$ of a test image $\tilde{\mathbf{x}}$ and a set of HMT parameters $\mathcal{M}$. It is a realization of the HMT model in which computation of the likelihood $f(\tilde{\omega}|\mathcal{M})$ that $\tilde{\omega}$ requires only a simple $\mathcal{O}(n)$ unsweep through the HMT tree from leaves to root [13]. Conveniently, HMT models wavelet coefficients in a tree parallel to quad-tree structures of wavelet coefficients and dyadic squares. Therefore, statistical behavior of each dyadic square $\tilde{\mathbf{d}}_i$ can also be approximately modeled by a HMT branch rooted at node i. As mentioned earlier, maximum likelihood of the sub-tree $\mathcal{T}_i$ is computed simply by upsweeping from corresponding leave-nodes at scale $S = J$ to the root node at scale $S = j$. If the "upsweeping" operation is done from tips to toes, from leave nodes to root node at scale "S = 0", we could find out likelihood probability of the whole image. The partial likelihood calculations at intermediate scales j of the HMT tree is denoted as $f(\tilde{\mathbf{d}}_i|\mathcal{M})$, and it is also statistical behavior of the corresponding dyadic sub-square under the HMT model as well.

The above model leads to a simple multi-scale image classification algorithm. Supposed that there are two main classes: center and surround environment $c \in \{1, 0\}$, we have specified or trained an HMT tree for each class with parameters $\mathcal{M}_c$. When the above likelihood calculation is deployed on each node of the HMT quad-tree given the wavelet transform $\tilde{\omega}$ of an image $\tilde{\mathbf{x}}$. For each node of the tree, HMT yields the likelihood $f(\mathbf{d_i} | \tilde{\mathcal{M}}_c), c \in \{1, 0\}$ for each dyadic sub-image d_i. With the multiscale likelihoods at hand, we can choose the most suitable class c for a dyadic sub-square $\tilde{\mathbf{d_i}}$ as follows.

$$\hat{c}_i^{ML} := argmax_{c \in \{1, 2\}} f(\tilde{\mathbf{d_i}} | \mathcal{M}_c)$$

The most likely label $\hat{c}_i^{ML}$ for each dyadic sub-square $\tilde{\mathbf{d_i}}$ can be found by simple comparison of likelihood among available classes. Moreover, likelihood of the whole tree just needs $\mathcal{O}(n)$ operations for an n-pixel image.

3.3 Multiscale Likelihood Computation

For a given set of HMT model parameters $\mathcal{M}$, it is straight forward to compute the likelihood $f(\tilde{\mathbf{w}} | \mathcal{M})$ by upsweeping from leaves nodes to current node in a single sub-band branch [13]. Moreover, likelihoods of all dyadic squares of the image can be obtained simultaneously in upsweep operations along the tree as well due to compatibility between HMT , wavelet decompositions and dyadic square quad-trees.

To obtain the likelihood of a sub-tree $\mathcal{T}_i$ of wavelet coefficients rooted at w_i, we have deployed wavelet HMT trees and learn parameters Θ for multi-scale analysis [13]. The conditional likelihood $\beta_i(m) := f(\mathcal{T}_i | S_i = m, \Theta)$ can be retrieved by sweeping up to node i (see [13]); then, the likelihood of the coefficients in $\mathcal{T}_i$ can be computed as.

$$f(\mathcal{T} | \Theta) = \sum_{m=S,L} \beta_i(c) p(S_i = c | \Theta) \tag{5}$$

with $p(S_i = c | \Theta)$ state probabilities can be predefined or obtained during traning [31].

From the previous discussion about relations between wavelet coefficients and dyadic squares, it is obvious that each dyadic square $\mathbf{d_i}$ is well represented by three sub-bands $\{\mathcal{T}_i^{LH}, \mathcal{T}_i^{HL}, \mathcal{T}^{HH}\}$. All above likelihood can be estimated by upsweeping operations in their corresponding trees independently. In DIS approach , Gao [18] have also assumed that correlation between feature channels would not affect discriminant powers; moreover, there is de-correlation characteristics of DWT. Therefore, the likelihood of a dyadic square is formulated as product of three independent likelihoods of three wavelet sub-bands at each scale.

$$f(d_i | \mathcal{M}) = f(\mathcal{T}_i^{LH} | \Theta^{LH}) f(\mathcal{T}_i^{HL} | \Theta^{HL}) f(\mathcal{T}_i^{HH} | \Theta^{HH}) \tag{6}$$

Using the above equation, the likelihood can be computed for each dyadic square down to 2x2 block scale. Noteworthy, sub-band HH of the wavelet transform is not utilized in our computation since it is low-passed approximation of original images. Therefore, it is vulnerable to pixel brightness and lightning conditions of scenes. Since dealing with shades, lightning changes is out of this paper's scope, the final HH sub-band is discarded. The above simple formulation of likelihood is usually employed in block-by-block classification or "'raw'" classification since it does not exploit any possible relationship at different scales. Therefore, classification decisions between classes (center and surround) are lack of inheritance across dyadic scales since likelihood values are estimated independently at each scale. There still remains a problem of integrating classification processes across all scales or at least the direct coarser scale.

3.4 Multiscale Maximum a Posterior

In the previous section, binary classification between between two states have been realized under the wavelet Hidden Markov Tree model [13]. It bases on comparison of likelihood given a system parameters. However, realization of DIS and MDIS, the equation 4, needs a posterior probability $p(c_i^j|d^j)$ given c_i^j and $d^j = d_i^j$ are class labels and features of an image at a dyadic scale j and location i.

In order to estimate the MAP $p(c_i^j|d^j)$, we need to employ Bayesian approach and capture dependencies between dyadic squares at different scales. Though many approximation techniques [8],[9],[24],[5] are derived for a practical MAP, the Hidden Markov Tree (HMT) by Choi [11] is proven to be a feasible solution. Choi introduces hidden label tree modeling instead of estimating joint probability of high-dimensional dyadic squares. Due to strong correlation between the considered square and its parents as well as their neighbors, the class labels of these adjacent squares should affect its class label decision. For example, if parent squares belongs to the center class, their sub-square most likely belongs to the same class as well; neighbor squares holds similar influences.

A possible solution for modeling these relations between squares is using a general probabilistic graph [13]; however, the complexity exponentially increases with number of neighborhood nodes. Choi [12] proposes alternative simpler solution based on context-based Bayesian approach. For the sake of simplicity, causal contexts are only defined by states of the direct parent node and its 8 intermediate neighbors. Lets denote the context for D_i as $\mathbf{v_i} \equiv [v_{i,0}, v_{i,1}, \ldots, v_{i,8}]$ where $v_{i,0}$ refers to context from a direct parent node, the other contexts from neighboring sub-squares. The triple $\mathbf{v_i} \rightarrow C_i \rightarrow \mathbf{D_i}$ forms a Markov-1 chain, relating prior context $\mathbf{v_i}$ and node features $\mathbf{D_i}$, to classify labels C_i. Moreover, class labels of parent nodes and its neighbors $\mathbf{v_i}$ are chosen as discrete values, then it simplifies the modeling considerably. Given prior context, independence can be assumed for label classification at each node; therefore, it is allowed to write.

$$p(\mathbf{c^j}|\mathbf{v^j}) = \prod_i p(c_i^j)|\mathbf{v_i^j} \tag{7}$$

The property of Markov-1 chain assumes that $\mathbf{D_i}$ is independent from $\mathbf{v_i}$ given C_i; therefore, the posterior probability of classifying $\mathbf{c^j}$ given $\mathbf{d^j}$, $\mathbf{v^j}$ is written as follows.

$$p(\mathbf{c^j}|\mathbf{d^j}, \mathbf{v^j}) = \frac{f(\mathbf{d^j}|\mathbf{c^j})p(\mathbf{c^j}|\mathbf{v^j})}{f(\mathbf{d^j}|\mathbf{v^j})} \tag{8}$$

As independence is assumed for label decision from classifying processes, it yields.

$$p(\mathbf{c^j}|\mathbf{d^j}, \mathbf{v^j}) = \frac{1}{f(\mathbf{d^j}|\mathbf{v^j})} \prod_i f(\mathbf{d^j}|c^j)p(c^j|\mathbf{v^j}) \tag{9}$$

and the marginalized context-based posterior

$$f(c_i^j|\mathbf{d^j}, \mathbf{v^j}) \propto f(\mathbf{d_i^j}\, c_i^j)p(c_i^j|\mathbf{v_i^j}) \tag{10}$$

It greatly simplifies MAP posterior estimation since it no longer needs to deal with joint prior conditions of features and contexts. Rather than that, it only need to obtain two separated likelihood of the dyadic square given the class value C_i, $f(\mathbf{d_i}^j|c_i^j)$ and prior context provided through $\mathbf{v_i}$, $p(c_i^j|\mathbf{v_i^j})$.

While it is straightforward to retrieve the likelihood $f(\mathbf{d_i}^j|c_i^j)$ by up-sweeping operations with given HMT model parameters at each scale, the complexity of prior context estimation greatly depends on the context choice. Though general context may give better prior information for classification, it also greatly complicates modeling and it is difficult to summarize information conveyed by $\mathbf{v_i^j}$ as well. In other words, we run on the risk of context dilution, especially with insufficient training data [8],[9],[13].

To balance simplicity and generalization of prior information, we will employ a simple context structure inspired by the hybrid tree model [5] for context-labeling tree. Instead of including all neighboring sub-squares, the simplified context only involves labels from the parent square $C_{\rho(i)}$ and majority vote of the class labels from neighboring squares $C_{\mathcal{N}_i}$. Since there are only two class labels $N_c := 0, 1$, the prior context $\mathbf{v_i} := \{C_{\rho(i)}, C_{\mathcal{N}_i}\}$ can only been drawn from $N_c^2 = 4$ different values $\{0, 0\}, \{0, 1\}, \{1, 0\}, \{1, 1\}$. While the choice of such simple context is rather ad-hoc, it provides sufficient statistic for demonstrating the effectiveness of multi-scale decision fusion. Another advantage of the context structure simplification is that only few training data are required for probability estimation.

Any decision about labels at a scale j depends on prior information of labels on a scale $j - 1$; therefore, we can maximize MAP, in the equation 11, in multi-scale coarse-to-fine manner by fusing the HMT likelihoods $f(\mathbf{d_i}|c_i)$ given the label tree prior $p(c_i^j|\mathbf{v_i})$. That fusion will pass down MAP classification decisions through scales to enhance across-scale coherency. Moreover, posterior probability of a class label c_i given features and context are computed and maximized coherently across multiple scale.

$$\hat{c}_i^{MAP} = argmax_{c_i^j \in 0,1} f(c_i^j|\mathbf{d^j}, \mathbf{v^j}) \tag{11}$$

3.5 Multiscale Discriminant Saliency

The core idea of discriminant saliency is discriminant power between two classes center and surrounds. Though the discriminant power is measured by sample means of mutual information, the underlying mechanism is making use of difference between generalized gaussian distributions (GGD) given either center label or surrounding label. Since GGD of wavelet coefficients usually have zero-mean, it is well-characterized with only variance parameters. Dashan Gao [18] tries to estimate the scale parameter or variance of GGD (see section 2.4 [18] for more details) by the maximum a probability process.

$$\hat{\alpha}^{MAP} = \left[\frac{1}{\mathcal{K}} \left(\sum_{j=1}^{n} |x(j)|^\beta + \nu \right) \right]^{\frac{1}{\beta}} \tag{12}$$

The above MAP is later used for deciding whether a sample point or a image data point belongs to the center or surround class (see [18] for a detailed proof and explanation). Therefore, the more difference between MAP estimation of the center's variance parameter α_1 and the surround's variance parameter α_0, the more discriminant power is for classifying interest versus null hypothesis.

The idea of modeling wavelet distributions with multiple classes' variances can be realized by Gaussian Mixture Model (GSM) [28],[10] as well. In binary classification problem with only two classes, there are mixtures of two states with Gaussian Distribution (GD) of distinguishing variances. It is reasonable to name "'large'" and "'small'" states according to their comparison in terms of variance values. Now the only difference between GSM models and Gao's proposal [14] are whether GD or GGD should be used. Though GGD is more sophisticated with distribution shape parameter β, several factors support validity of simple GD modeling given the class labels as hidden variables. Empirical results from estimation have shown that the mixture model is simple yet effective [28],[5]. Modeling wavelet coefficients with hidden classes of "'large'" and "'small'" variance state are basic data models in Wavelet-based Hidden Markov Model (HMT) [13]. With wavelet HMT, image data are processed in coarse-to-fine multi-scale manner; therefore, MAP of a state C_i^j given input features from a sub-square D_i^j can be inherently estimated across scales $j = 0, 1, \ldots, J$. More details about this multi-scale MAP estimation by wavelet HMT are discussed in the previous sections 3.4. Combination MAP estimation, the equation 10, and mutual information estimation, the 4, the equation 4 yields a formulation for multiscale DIS.

$$I_i^j(C^j; \mathbf{D^j}) = H(C^j) + \sum_{c=0}^{1} P_{C^j|\mathbf{D^j}}(c_i^j|\mathbf{d^j}) log P_{C^j|\mathbf{D^j}}(c_i^j|\mathbf{d^j}) \qquad (13)$$

where $H(C^j) = -p(C^j)log(p(C^j))$ is entropy estimation of classes across the scale j, and the posterior probability can be estimated by modeling wavelet coefficients in HMT frameworks. This matter has been discussed in previous sections; therefore, it is not repeated here. The equation 13 yields discriminant power across multiple scales; meanwhile a strategy is needed for combining them across scales. In this paper, a simple maximum rule is applied for selecting discriminant values from multiple scales into a singular discriminant saliency map at a sub-square d_i.

$$I_i(C|\mathbf{D}) = max\left(I_i^j(C^j; \mathbf{D^j})\right) \qquad (14)$$

4   Experiments and Discussion

In the light of decision theory, saliency maps are considered as binary filters applied at each image locations. According to a certain saliency threshold, each spot can be labeled as interesting or uninteresting. If a binary classification map is considered as saliency map which leads to visual performance of human beings, it would be a significant undervaluation of human visual attention system. Psychology experiments shows much better capacity of biologically plausible visual attention system than that of binary classification maps. Therefore, our proposed method just use binary classification between center-surround environment as an intermediate step to develop information-based saliency map. At each location, mutual information between distributions of classes and features shows strength of discriminant power between interesting vs non-interesting classes given input dyadic sub-squares. From discrete binary values, the saliency representation has continuous ranges of discriminant power. Inevitably, the generated saliency maps become more correlated to the results of human visual attention maps.

Besides appropriate saliency representation, we also need reliable ground-truth to compare with those maps. As our research purpose is deepening knowledge about relationship between multi-scale discriminant saliency approaches and human visual attention, the ground truth data must be gotten from psychological experiments in which human subjects are tested with different natural scenes. Moreover, the research scope only focuses at bottom-up visual saliency, the early stage of visual attention without interference of subjects' prior knowledge and experiences. Human participants should be naive about purposes of experiments and should not know contents of displaying scenes in advance. After these prerequisites are satisfied, human responses on each scene can be accurately collected through eye-tracking equipments. It records collection of eye-fixations for each scene, and these raw data are basic form of ground truths for evaluating efficiency of saliency methods.

Assumed that ground-truth data are available, quantitative methods can be applied for evaluation of MDIS Saliency results. In an effort of standardizing evaluating process of evaluating saliency methods, we only utilize one of the most common and accessible database and evaluation tools in visual attention fields. In regards of available database and ground-truths, Niel-Bruce database [6] is certainly the most popular dataset used in information-based saliency studies. While proposing his An InfoMax (AIM) saliency approach, the first information-based visual saliency, Bruce simultaneously releases his testing database as well. The reasonably small database with 120 different color images which are tested by 20 different subjects. Each object observes displayed image in random order on a 12 inch CRT monitor positioned 0.75 from their location for 4 seconds with a mask between each pair of images. Importantly, no particular instructions are given except observing the image. Above brief description clarified validity of this database. When setting up our simulations on this database, AIM method, of which codes are released along, is included as referenced method. We compare our proposed saliency method MDIS against AIM in terms of performance, computational load, etc. Though DIS [14] is the closest approach to our proposed MDIS, implementation from the author is not available for comparison. Meanwhile, AIM also derives saliency value from information theory with slightly different computation, self-information instead of mutual-information in MDIS or DIS. Therefore, it would be considered the second best as referenced method for our later evaluation of MDIS.

As valid database is set, proper numerical tools are necessary for simulation data analysis. In regards of fairness and accuracy of the evaluation, we employ a set of three measurements LCC, NSS, and AUC recommended by Ali Borji et al. [4] since evaluation codes can be retrieved freely from the authors' website [2]. Three evaluation scores are used for analyzing a method to ensure the reliability of qualitative con-

---

[2] https://sites.google.com/site/saliencyevaluation/

clusion and any conclusion is free from the choice of metric. First linear correlation coefficient (LCC) measures the strenght of linear relationship between two variables $CC(G, S) = cov(G, S)$, where G and S are the standard deviation of ground-truths and corresponding saliency maps. LCC values variates from -1 to +1 while the correlation changes from total inverse to perfect linear relation between G and S maps. While LCC measures matching between saliency maps and eye-fixation maps as a whole, normalized scanpath saliency (NSS) treats eye-fixations as random variables which are classified by proposed saliency maps. The measurement is an average of saliency values at human eye positions according to saliency approaches. NSS values are in between 0 and 1. $NSS = 1$ indicates saliency values at eye-fixation locations are one standard deviation above average, while $NSS = 0$ indicates no better performance of saliency maps than randomly generated maps. The previous two measurements deals with saliency maps directly while the last quantitative tool AUC utilized saliency maps as binary classification filters with various thresholds. By regularly increasing/decreasing threshold values over the range of saliency values, we can have a number of binary classifying filters. Then, deploying these filters on eye-fixation maps produce several true positive rates (TPR) and false positive rate (FPR) as vertical axis and horizontal axis values of Receiver Operating Characteristics (ROC) curve. Area under curve (AUC) is a simple quantitative measurement to compare ROC of different saliency approaches. Perfect AUC prediction means a score of 1 while 0.5 indicates by chance-performance level.

As mentioned previously, AIM is chosen as the referenced information-based saliency method. It is chosen due to the well-established reputation as well as freely accessible code and experiment database [3]. Due to multiscale natures of the proposed MDIS, saliency maps for each dyadic-scale levels can be extracted as well as the final MDIS saliency, integrated across scales by the equation 14. The availability of saliency maps at multiple scales as well as combined one allows evaluation of discriminant power concept for saliency in scale-by-scale manner or on the whole. We denote integrated MDIS as HMT0; while separated saliency maps are named as HMT1 to HMT5 in accordance with coarse-to-fine order. By examining MDIS in different aspects, we would observe its effectiveness in predictions of eye-fixation points and how selection of classifying window sizes might affect its performance. In addition, comparisons against AIM would contribute a general view how MDIS performs against a well-known information-based saliency method.

In our proposed saliency techniques, Hidden Markov Tree (HMT) plays a key role of modeling statistical properties of images. It extracts model parameters at each scale by considering distribution of feature given hidden variables (center-surround labels in our method). Therefore, training is a necessary step before model parameters can be approximated in trained Hidden Markov Model (THMT). However, training steps are not necessarily needed if data are restricted in a specific category; for example, natural images in this paper. Romberg et al. [31] have studied this issue and proposed a Universal Hidden Markov Tree (UHMT) for the natural image class. In other words, model parameters can be fixed without any training efforts; the approach would greatly reduce computational load for MDIS. However, a necessary evaluation step needs carrying out to compare performances of UHMT and THMT in terms of LCC, NSS, AUC and TIME

---

[3] http://www-sop.inria.fr/members/Neil.Bruce/

(the computational time). Noteworthy that, we will use UHMT or THMT instead of HMT alone when representing experimental data in order to signify which tree-building method is employed.

In quantitative method, general ideas can be drawn about how the proposed algorithms perform in average. However, such evaluation method lacks of specific details about successful and failure cases. The simulation result has been averaged out in quantitative methods. In an effort of looking for pros and cons of the algorithm, we perform qualitative evaluation for saliency maps generated by MDIS in multiple scales. Furthermore, AIM saliency maps are generated and compared with MDIS maps to discover advantages and disadvantages of each method.

4.1   Quantitative Evaluation

After general review of how simulations are setup and evaluated in the previous section, following are data representation and analysis of the conducted experiments. In this paper, five dyadic scales are deployed for any HMT training and evaluation; therefore, we have simulation modes from (U)THMT1 to (U)THMT5 of MDIS depending on whether training stages is deployed (THMT) or universal parameters are used (UHMT). Saliency maps could be combined according to the maximization of mutual information rule, the equation 14, to form two (U)THMT0 modes for saliency maps. AIM is involved in the simulations as benchmarking approach, and LCC, NSS, AUC and TIME are chosen as numerical evaluation tools. Below are two tables of simulation results. Table 1 shows simulation data of all universal HMT modes while table 2 summarizes data of all trained HMT modes.

Table 1. UHMT - MDIS - Experiment Data

Observations	UHMT0	UHMT1	UHMT2	UHMT3	UHMT4	UHMT5	AIM
LCC	0.01434	-0.00269	0.01294	0.01349	**0.01604**	0.00548	0.01576
NSS	0.21811	0.19772	0.27819	0.32868	**0.42419**	0.13273	0.12378
AUC	**0.89392**	0.53862	0.60520	0.69065	0.83615	**0.89234**	0.72275
TIME(s)	0.39617	0.39617	0.39617	0.39617	0.39617	0.39706	50.41714

Table 2. THMT - MDIS - Experiment Data

Observations	THMT0	THMT1	THMT2	THMT3	THMT4	THMT5	AIM
LCC	**0.02382**	**0.02582**	0.01156	0.01604	0.01143	0.00512	0.01576
NSS	**0.48019**	0.38096	0.31855	0.32491	0.29662	0.36932	0.12378
AUC	**0.88357**	0.60922	0.64633	0.71972	0.81192	**0.89532**	0.72353
TIME(s)	2.32734	2.32734	2.32726	2.32726	2.32726	2.32726	50.41714

Looking at TIME rows of both tables 1 and 2, we present necessary processing time for each method or each mode. Generally, computational loads, proportional to processing time, of all modes in either UHMT or THMT rows are almost similar since the parameters of full-depth Hidden Markov Tree need estimating before computation of saliency values for each scale. In comparison between UHMT and THMT in terms of processing time, UHMT is faster than THMT as UHMT uses predefined parameters for HMT and MDIS computation instead of adapting HMT to each image. When comparing both UHMT and THMT modes of MDIS with AIM, our proposed methods are much faster than AIM. The well-known AIM directly estimate self-information from high-dimensional by ICA algorithm while MDIS statistically models two hidden states: "'large'" state "'small'" states in sparse and structural features. Computational load or processing time of the mentioned AIM and proposed MDIS with different modes can be seen in the figure 2(c).

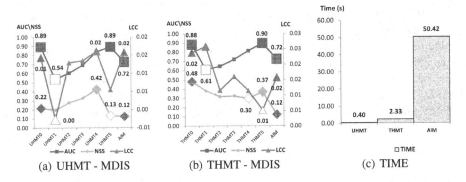

(a) UHMT - MDIS   (b) THMT - MDIS   (c) TIME

Fig. 2. Performance of UHMT-MDIS, THMT-MDIS in AUC,NSS,LCC and TIME

Though HMT-MDIS significantly reduces computational load for computation of information-based saliency, more verifications are necessary for their performances in terms of accuracy. We begin with evaluating two modes UHMT and THMT separately against AIM in terms of three numerical tools LCC, NSS, AUC together, Figures 2(a), 2(b). Then all modes of HMTs are summarized in three plots (the top row of Figure 3) corresponding to three evaluating schemes NSS, LCC, AUC from left to right. Especially in the figure 3, simulation modes of the same scale level are placed next to each other for example UHMT0 is next to THMT0, UHMT1 is next to THMT1 and etc. It is intentionally arranged in that way to compare performances of different simulation modes in the same scale level. Noteworthy that, in both tables 1 and 2 for each row is identified **maximum** values and minimum values by using corresponding text styles. Identification of extreme values only involves derivatives (UHMT and THMT) of MDIS modes except referenced method AIM. In the figures 2(a), 2(b), extreme values are also specially marked. For example, maximum values have big solid markers while big but empty markers represent minimum points. Especially, AIM have big markers with distinguishing big cross-board texture while integrated saliency modes (U)THMT0 have small cross-board textures. These special markers help to highlight interesting facts in comparison between simulation modes of MDIS or MDIS and AIM. The same marking policy is applied for data representation in the figure 3. Meanwhile, each line in this figure has an arrow head for showing a trending direction (increasing/decreasing) when simulation mode is changed from UHMT to THMT for each scale level.

Firstly, the MDIS approach with universal parameters for each level of hidden Markov tree is analyzed in terms of accurate performance since it requires very little effort in saliency computation. Obviously it is raising a question about their accuracy in classification between center and surround class as well as generated saliency maps. According to simulation data in the table 1 with highlighted extremum, UHMT performs pretty well against AIM in all three measurements LCC, NSS and AUC. For example, MDIS with UHMT4 mode (4x4 square blocks) surpasses AIM in all measurements. It confirms validity and efficiency of our proposed methods in the information-based saliency map research field. When performances of different UHMT-MDIS modes are considered, UHMT4 with 4x4 squares have the most consistent evaluation among all dyadic scales with maximum LCC and NSS and the second best AUC value. However, UHMT0-MDIS, integration of saliency values across scales, does not have better performance than other UHMTs except for AUC level. It shows inconsistent side of deploying HMT with predefined universal parameters while no traning effort is done for adapting the tree into image statistical mutliscale structures

Secondly, training stage is included in the simulation of MDIS with THMT mode (Trained Hidden Markov Tree). With additional adaptivity, THMT might improves the saliency evaluation and produce more consistent results than UHMT might. This subjection is solidified by simulation data in the table 2 and they are also plotted in the

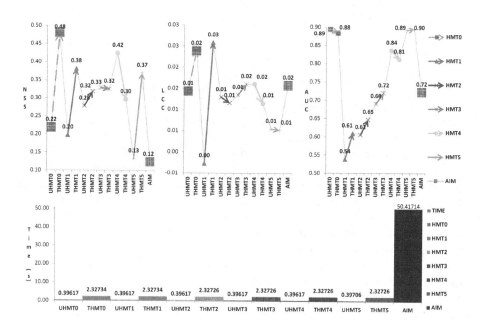

Fig. 3. Summary of all MDIS against AIM

figure 2(b). As observed in the table, all **maximum** values locate at the THMT0 column, THMT0-MDIS over-performs AIM in all evaluating schemes. Again, the rationale of MDIS is confirmed and partly proved. Furthermore, effectiveness of traning stages are clearly shown when comparing THMT0 against UHMT0. Though AUC of THMT0 is smaller than that of UHMT0, THMT0 evaluation is better than their counterparts in both NSS and LCC scheme. This confirms usefulness of trained Hidden Markov Tree models for each sample image. In addition, the figure 2(b) shows supremacy of THMT0 mode, the across-scale integration mode of MDIS over other singular saliency maps at different dyadic scales in any measurement. Noteworthy, that LCC of THMT0 mode is a little bit smaller LCC of THMT1 mode; however, this small difference can be safely ignored. Comparison of UHMT-MDIS and THMT-MDIS mode-by-mode or a-dyadic-scale by a-dyadic-square between data in Table 2(a) and Table 2(b) are shown in the figure 3. According to the figure, there are slight improvement of THMT1, THMT2 over UHMT1, UHMT2, equivalence of THMT3, UHMT3, and a reverse trend that UHMT4, UHMT5 are comparable or slightly better than THMT4, THMT5. It seems that training processes are more important when big classification windows are used. Meanwhile universal approaches of HMT work pretty well if window sizes get smaller. Two possible reasons for this observation are statistical natures of dyadic squares and characteristics of trained processes. A bigger square has richer joint-distribution of features; therefore, UHMT with fixed parameters can not marginally model that distribution well. However, parameters of THMT can be learn from analyzing images; then, significant improvement is achieved. While smaller sub-squares are less statistically distinguishing, they are successfully modeled by universal parameters of HMT. Then training processes might become redundant since UHMT would perform as well as THMT would do.

4.2   Qualitative Evaluation

In this section, saliency maps are analyzed qualitatively or visually. From this analysis, we want to identify (i) on which image contexts UHMT-MDIS, THMT-MDIS work well, (ii) how scale parameters affect formation of saliency maps, and (iii) how MDIS in general is compared with AIM.

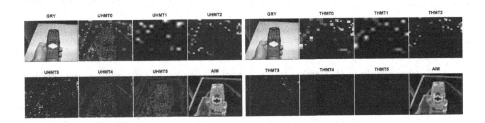

Fig. 4. Saliency Maps 1

In figure 4, the first example with central objects shows an example of good UHMT performance but bad THMT performance. All scale levels of THMT suppress features of the most obvious objects in the image center. Meanwhile, UHMT4 and UHMT5 capture significant features points of that objects; therefore, UHMT0 has much better saliency map thant THMT0. In this case, the best saliency map of MDIS approach, UHMT0, is reasonably competitive against AIM saliency map.

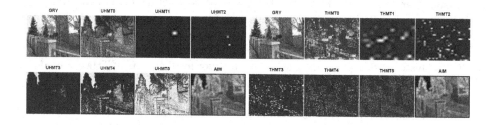

Fig. 5. Saliency Maps 2

In contrast with the first example, the figure 5, a general outdoor scene, shows an opposite situation where THMT generates more reasonable saliency maps. While UHMT0 map covers whole region of sky despite no interesting features, THMT0 correctly select available objects on the scene. Similarly, THMT does extract more meaningful features than UHMT does (see UHMT1-5 and THMT1-5 in the figure 5. In addition, the best saliency map THMT0 or THMT5 highlights more discriminant features than AIM saliency map.

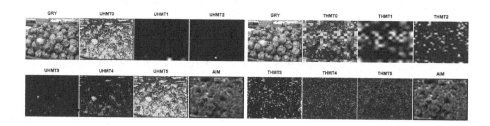

Fig. 6. Saliency Maps 3

The third example is chosen such that complex scenes are presented to the saliency methods. In the figure 6, there are several fruits on the shelf; it is considerably complicated due to richness of edges, textures, as well as color. In general, both MDIS and AIM only partially succeed in detecting saliency regions in this image since none of them successfully high-light the fruit with different color on the shelf (the fruit inside a red circle, Figure 6). Though most MDIS with different modes and scale levels do not explicitly detect that fruit, UHMT3 and UHMT4 salency maps are able to highlight the the location of that fruit (see UHMT3 an UHMT4 saliency maps, Figure 6). The sample matches with the fact that UHMT4 data in the table 1 has extremely good performance in all evaluation schemes. Surprisingly, there are some cases, like the figure

6 when appropriate choices of scales and parameters of predefined HMT models can over-perform all HMT models. The interesting example of the figure 6 opens another research direction about how HMT model can be learned and optimized; however, it is the question for another research paper.

5   Conclusion

In conclusion, multiple discriminant saliency (MDIS), an extension of DIS [17] under dyadic scale framework, has strong theoretical foundation as it is quantified by information-theory and adapted to multiple dyadic-scale structures. The performance of MDIS against AIM is simulated on the standard database with well-established numerical tools; furthermore, simulation data proves competitiveness of MDIS over AIM in both accuracy and speed. However, MDIS fails to capture salient regions in some complex scenes; therefore, the next research step is improving MDIS accuracy in such cases. In addition, implementation of MDIS algorithm in embedded systems are also considered as a possible extension.

References

1. Abramovich, F., Sapatinas, T., Silverman, B.W.: Wavelet thresholding via a bayesian approach. Journal of the Royal Statistical Society: Series B (Statistical Methodology) 60(4), 725–749 (1998)
2. A\cc\ik, A., Onat, S., Schumann, F., Einhäuser, W., König, P.: Effects of luminance contrast and its modifications on fixation behavior during free viewing of images from different categories. Vision Research 49(12), 1541 (2009)
3. Baddeley, R.J., Tatler, B.W.: High frequency edges (but not contrast) predict where we fixate: A bayesian system identification analysis. Vision Research 46(18), 2824–2833 (2006)
4. Borji, A., Sihite, D.N., Itti, L.: Quantitative analysis of Human-Model agreement in visual saliency modeling: A comparative study. IEEE Transactions on Image Processing (2012)
5. Bouman, C., Shapiro, M.: A multiscale random field model for bayesian image segmentation. IEEE Transactions on Image Processing 3(2), 162–177 (1994)
6. Bruce, N., Tsotsos, J.: Saliency based on information maximization. Advances in Neural Information Processing Systems 18, 155 (2006)
7. Bruce, N.D.B., Loach, D.P., Tsotsos, J.K.: Visual correlates of fixation selection: a look at the spatial frequency domain. In: IEEE International Conference on Image Processing, ICIP 2007, vol. 3, p. III–289 (2007)
8. Cheng, H., Bouman, C.A., Allebach, J.P.: Multiscale document segmentation 1. In: IS and T Annual Conference, pp. 417–425 (1997)
9. Cheng, H., Bouman, C.: Trainable context model for multiscale segmentation. In: Proceedings of the 1998 International Conference on Image Processing, ICIP 1998, vol. 1, pp. 610–614 (October 1998)
10. Chipman, H.A., Kolaczyk, E.D., McCulloch, R.E.: Adaptive bayesian wavelet shrinkage. Journal of the American Statistical Association 92(440), 1413–1421 (1997)
11. Choi, H., Baraniuk, R.: Multiscale image segmentation using wavelet-domain hidden markov models. IEEE Transactions on Image Processing 10(9), 1309–1321 (2001)
12. Choi, H., Baraniuk, R.: Multiscale texture segmentation using wavelet-domain hidden markov models. In: Conference Record of the Thirty-Second Asilomar Conference on Signals, Systems & Computers, vol. 2, pp. 1692–1697 (November 1998)

13. Crouse, M., Baraniuk, R.: Simplified wavelet-domain hidden markov models using contexts. In: Proceedings of the 1998 IEEE International Conference on Acoustics, Speech and Signal Processing, vol. 4, pp. 2277–2280 (May 1998)
14. Gao, D., Mahadevan, V., Vasconcelos, N.: The discriminant center-surround hypothesis for bottom-up saliency. Advances in Neural Information Processing Systems 20, 1–8 (2007)
15. Gao, D., Mahadevan, V., Vasconcelos, N.: On the plausibility of the discriminant center-surround hypothesis for visual saliency. Journal of Vision 8(7) (2008)
16. Gao, D., Vasconcelos, N.: Discriminant saliency for visual recognition from cluttered scenes. Advances in Neural Information Processing Systems 17, 481–488 (2004)
17. Gao, D., Vasconcelos, N.: Discriminant interest points are stable. In: IEEE Conference on Computer Vision and Pattern Recognition, CVPR 2007, pp. 1–6 (2007)
18. Gao, D., Vasconcelos, N.: Decision-theoretic saliency: computational principles, biological plausibility, and implications for neurophysiology and psychophysics. Neural Computation 21(1), 239–271 (2009)
19. Harel, J., Koch, C., Perona, P.: Graph-based visual saliency. In: NIPS (2007)
20. Hou, X., Zhang, L.: Saliency detection: A spectral residual approach. In: IEEE Conference on Computer Vision and Pattern Recognition, CVPR 2007, pp. 1–8 (2007)
21. Itti, L., Koch, C., Niebur, E.: A model of saliency-based visual attention for rapid scene analysis. IEEE Transactions on Pattern Analysis and Machine Intelligence 20(11), 1254–1259 (1998)
22. Le Ngo, A.C., Ang, L.M., Seng, K.P., Qiu, G.: Improvement and evaluation of visual saliency based on information theory. In: 2010 International Computer Symposium (ICS), pp. 500–505 (2010)
23. Le Ngo, A.C., Qiu, G., Underwood, G., Ang, L.M., Seng, K.P.: Visual saliency based on fast nonparametric multidimensional entropy estimation. In: 2012 IEEE International Conference on Acoustics, Speech and Signal Processing (ICASSP), pp. 1305–1308 (2012)
24. Li, J., Gray, R., Olshen, R.: Multiresolution image classification by hierarchical modeling with two-dimensional hidden markov models. IEEE Transactions on Information Theory 46(5), 1826–1841 (2000)
25. Marr, D., Marr, D.: Early processing of visual information. Philosophical Transactions of the Royal Society of London. B, Biological Sciences 275(942), 483–519 (1976)
26. Neisser, U.: Cognitive psychology. APA (1967)
27. Parkhurst, D., Law, K., Niebur, E.: Modeling the role of salience in the allocation of overt visual attention. Vision Research 42(1), 107–124 (2002)
28. Pesquet, J., Krim, H., Leporini, D., Hamman, E.: Bayesian approach to best basis selection. In: 1996 IEEE International Conference on Acoustics, Speech, and Signal Processing, ICASSP 1996. Conference Proceedings, vol. 5, pp. 2634–2637 (May 1996)
29. Qiu, G., Gu, X., Chen, Z., Chen, Q., Wang, C.: An information theoretic model of spatiotemporal visual saliency. In: 2007 IEEE International Conference on Multimedia and Expo., pp. 1806–1809 (July 2007)
30. Reinagel, P., Zador, A.M.: Natural scene statistics at the centre of gaze. Network: Computation in Neural Systems 10(4), 341–350 (1999)
31. Romberg, J., Choi, H., Baraniuk, R.: Bayesian tree-structured image modeling using wavelet-domain hidden markov models. IEEE Transactions on Image Processing 10(7), 1056–1068 (2001)
32. Sun, Y., Fisher, R.: Object-based visual attention for computer vision. Artificial Intelligence 146(1), 77–123 (2003)
33. Tatler, B.W., Baddeley, R.J., Gilchrist, I.D.: Visual correlates of fixation selection: Effects of scale and time. Vision Research 45(5), 643–659 (2005)
34. Treisman, A.M., Gelade, G.: A feature-integration theory of attention. Cognitive Psychology 12(1), 97–136 (1980)

A Memetic Algorithm
for Waste Collection Vehicle Routing Problem
with Time Windows and Conflicts

Thai Tieu Minh[1], Tran Van Hoai[2], and Tran Thi Nhu Nguyet[3]

[1] Ho Chi Minh City Institute of Physics,
Vietnam Academy of Science and Technology
[2] Faculty of Computer Science & Engineering,
Ho Chi Minh City University of Technology
[3] Faculty of Computer Engineering,
Ho Chi Minh City University of Information Technology

Abstract. We address an application of vehicle routing problem (VRP) in the real life, namely waste collection problem. Constraints are considered including conflicts between waste properties, time windows of the waste, and multiple landfills. A combination of flow and set partitioning formulation is suggested to model the problem in case of multi-objective optimization. To minimize the total traveling time and number of vehicles of solution, we propose using a memetic algorithm (MA) with λ-interchange mechanism. The λ-interchange operator is modified to be compatible with new sub-routes construction for the multiple landfills.

In experiments, we compare the result of proposed MA method with some good results published as well as other meta-heuristic algorithms. The density of conflict matrix is also considered to understand its influence on the quality of the solution. Experimental results show that our approach can be competitive to other results in the VRP with time windows. Furthermore, the algorithm outperforms others in the VRP with time windows and conflict.

Keywords: waste collection, vehicle routing problems, time windows, conflicts, meta-heuristics.

1   Introduction

Vehicle Routing Problem with Time Windows and Conflicts (VRPTWC) is a new variant of VRP in which constraints are a combination of VRP with Time Windows (VRPTW) and VRP with Conflicts (VRPC). The time window constraints limit the time when vehicles visit the stop and the conflict between kinds of the waste will not allow them to be loaded together in one vehicle. Even though the weight of waste in a certain vehicle is nearly empty, the vehicle still cannot collect more if the conflict happens in between the waste groups. These constraints are often encountered in the hazardous materials transportation problem or the industrial waste collection. In the practice, the problems are

B. Murgante et al. (Eds.): ICCSA 2013, Part I, LNCS 7971, pp. 485–499, 2013.
© Springer-Verlag Berlin Heidelberg 2013

Table 1. Properties of waste and conflict matrix between properties

Group	Characteristic property	conflict matrix						
		1	2	3	4	5	6	7
1	Flammability		x					
2	Explosive	x						
3	Oxidation				x			
4	Corrosion			x				
5	Infection						x	x
6	Toxicity					x		
7	Biochemical toxicity					x		

more complex when many details are added such as multiple landfills and the lunch break at noon of the drivers. The waste collection activities in our problem are described below.

A certain company is in charge of collecting the hazardous industry waste in a city and the infrastructure of the company includes a depot where vehicles are stored in, and some landfills (treatment stations) where the waste can be disposed. Vehicles begin a working day at the depot and they must return to the depot in order to complete the working day and the waste is collected at factories (or stops). If the load in each vehicle reaches a threshold of capacity, the vehicle must visit a certain landfill to dispose its load. In the context of the paper, a sub-route is a path that vehicles start collecting the waste until the waste is removed completely.

Depending on the working time at each stop, we have a time window which the vehicle must visit on time. Vehicles must wait if they visit the stop before the earliest time of time window. However, vehicles cannot visit the stop after the latest time of time window. In addition, each vehicle has a constraint of total traveling distance and a constraint of total capacity load.

About waste conflict constraints, we base on information about hazardous waste properties of Decision No. 23/2006/QD-BTNMT of Ministry of Natural Resources and Environment Vietnam[1]. The hazardous waste properties are classified into seven groups and each group has one characteristic property, as presented in Table 1. Waste cannot be loaded together in the same route if conflict happens between groups. This constraint is released if only if vehicles return to the landfill and release the waste. The conflict matrix identifies the pairs of groups that are conflict each other and absolutely they must be contained separately. For example, the flammability group cannot be loaded together the explosive group.

In the mathematical viewpoint, the conflict constraints in the waste collection VRP are difficult to formulate, because these constraints are not exactly the same in each sub-route, route, as well solution in waste collection. Belong to the properties of the waste collected, the constraints will limit some other properties

---

[1] http://www.chatthainguyhai.net/documents/qd_23-2006-qd-btnmt.pdf

in sub-route until the waste is disposed on the landfill. These constraints are called partial constraints because they only appear and disappear in sub-routes. In addition, one landfill may be visited many times to dispose the waste. Thus we cannot represent the problem as a model that all vertices have just one in-degree and one out-degree.

The study proceeds as follows. In section 1, we introduce the VRPTWC and its constraints. In section 2, we review existing works relating to the problem and in section 3 we represent some definitions, the structure of the solution and the mathematical model. More details about the memetic algorithm are discussed in section 4. In section 5, we compare our experiments with the recent studies and in section 6, we provide conclusions.

2   Literature Review

Early research on VRPTW is started by Pullen and Webb [9], Savelsbergh [10], Knight and Hofer [6], Madsen [7]. While heuristics were found by Solomon and Desrochers [12] to be more effective and efficient, the works on optimal approaches in solving VRPTW of practical size had been very limited. Solomon [11] proposed benchmark problem sets for the VRPTW and heuristics such as saving heuristics, time-oriented and nearest-neighbor heuristics, insertion heuristics and time-oriented sweep heuristics. These heuristics continue to be used in works [13], [5], [1].

Desrochers et al. [2] introduced mathematical formulations for the VRPTW problem with $O(n^2)$ variables. Optimization algorithms as branch and bound, dynamic programming and set partitioning were used to compute the lower bound of the problem. Additionally, three types of approximation algorithms applied widely to unconstrained routing problems including construction methods, iterative improvement methods and incomplete optimization were reviewed too.

Tung and Pinnoi [13] applied techniques of VRPTW for waste collection activities in Hanoi, Vietnam. The problem is complicated by several time windows, inter-arrival time constraints at each customer point and multiple landfill locations. They formulated the problem into a mixed-integer program and proposed a heuristic procedure consisting of construction and improvement phases. Besides that, they modified the Solomon's I1 heuristic and improved phases by combining the power of Or-opt and 2-opt together.

Kim et al.[5] continued extending the VPRTW for Waste Management. Set of landfill locations are considered in the problem instead of one landfill as in [13]. Moreover, the problem has the lunch break constraints, vehicles must stop with an hour from 11 AM to 12 PM. For this problem, [5] presented a shape metric S_m in order to quantify the route (or cluster) compactness and developed a new clustering-based algorithm for waste collection VRPTW.

The paper of Hamdi-Dhaoui et al. [4] developed heuristic methods for Vehicle Routing Problem with Conflicts. This study presented a mathematical model, lower bound for the problem. ILS (Iterated Local Search), GRASP-ELS (Greedy

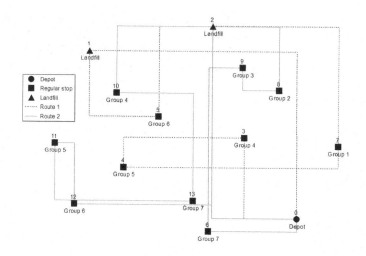

Fig. 1. An instance of the solution in waste collection VRPTWC

Randomized Adaptive Search Procedure - Evolutionary Local Search) heuristics are also introduced in their study.

As studies presented above, the VRP with time window and the VRP with conflict were considered independently. The problem that we are considering in this paper is more complicated, VRP in the combination of both time window and conflicts. A mathematical model of the problem will be expressed as well as a meta-heuristic algorithm will be proposed to solve the problem in the next part.

3   Definitions and Model

Considering a complete graph $G = (V, E)$ has $V = \{0, 1, ..., n + m - 1, n + m\}$ as the set of vertices including a depot, landfills, stops and E is the set of edges in V. We note the depot as vertex 0, the set N of stops are vertices from 1 to n, and the set M of landfills are vertices from $n + 1$ to $n + m$. Distance between vertices used in this study is Manhattan distance. Each stop or landfill, in the set of vertices V, has a time window frame $[e_i, l_i]$ to represent the earliest time and latest time that vehicles can visit the vertex. At the depot, time window frame $[e_0, l_0]$ represents the time that vehicles start collecting waste and the time that vehicles must return to.

We define $\mathbb{S}$ as a set of feasible solutions of the problem in the searching space. A solution $\sigma \in \mathbb{S}$ is a set of routes $\sigma = \{r^1, r^2, ..., r^{k-1}, r^k\}$, which route r^i is the i-th route of the solution and traveled exactly by one vehicle and $k = ||s||$ is the number of vehicles used in the solution. The routes start and finish at the depot. However they are not Hamiltonian cycles as same as the normal VRPTW because one landfill can be visited many times as showed in Fig 1.

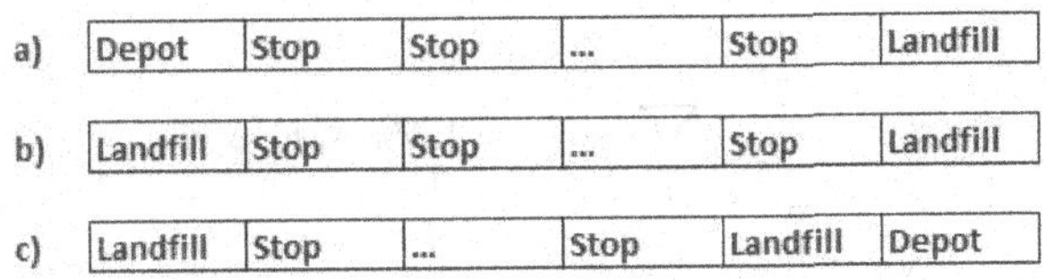

a)	Depot	Stop	Stop	...	Stop	Landfill

b)	Landfill	Stop	Stop	...	Stop	Landfill

c)	Landfill	Stop	...	Stop	Landfill	Depot

Fig. 2. Three types of a sub-route in waste collecting vehicle routing problem

A route is divided into several sub-routes $r^i = \{\gamma_1^i, \gamma_2^i, ..., \gamma_{p_i}^i\}$ in which γ_j^i is sub-route j-th and $n_s = ||r^i||$ is the number of sub-routes in the route i-th of the solution. A sub-route γ_j^i is a path that begins when the vehicle starts collecting the waste and ends when the vehicle releases all waste and returns to the depot. The structure of a sub-route is illustrated in Fig 2.

Let u_i denote the demand to collect at the vertex i and s_i denote the service duration at the vertex i. Depot and landfill have no the demand. The d_{ij} is the Manhattan distance and t_{ij} is the moving time from vertex i to j that corresponds to edge $(i, j) \in E$. K is the set of existing vehicles in depot, W is the amount of commodity limit of the vehicles and L is maximum distance which the vehicle can move in per day. Let a_{ij} be a binary coefficient that takes value 1 if waste property i conflicts with waste property j otherwise it takes value 0. We define some decision variables: $x_{ij}^k = 1$ if and only if the edge (i, j) is traveled by vehicle k, otherwise it is equal to 0; v_i^k is the starting service time at the stop i serviced by vehicle k; w_i^k is the total waste load of vehicle k after visiting stop i.

Objectives of the problem are to minimize the total length of routes, the waiting time, the total moving time and the number of vehicles, that means the problem has multi-objectives. There are many ways to solve this problem in which *weighted sum of objective functions* is the most popular approach. The method transforms the global objective function from the vector space to the scalar space thus the objective of the problem is:

$$min(\theta_1 C_d + \theta_2 C_m + \theta_3 C_w) \qquad (1)$$

$$C_d = \sum_{k \in K} \sum_{(i,j) \in E} t_{ij} x_{ij}^k \qquad (2)$$

$$C_m = \sum_{k \in K} \sum_{(i,j) \in E} d_{ij} x_{ij}^k \qquad (3)$$

$$C_w = \sum_{k \in K} \sum_{(i,j) \in E} (s_i^k - e_i) x_{ij}^k \qquad (4)$$

$\theta_1, \theta_2, \theta_3$: the weights assigned to each objective in order to express the relative importance of them.

Constraints:

$$\sum_{\forall k \in K} \sum_{\forall (i,j) \in E} x_{ij}^k = 1, \forall k \in K \tag{5}$$

$$\sum_{i=0}^{N+M} x_{ij}^k = \sum_{i=0}^{N+M} x_{ji}^k, \forall k \in K, \forall j \in V \tag{6}$$

$$\sum_{j=1}^{N} x_{0j}^k \leq 1, \forall k \in K \tag{7}$$

$$\sum_{j=N+1}^{N+M} x_{j0}^k \leq 1, \forall k \in K \tag{8}$$

$$\sum_{j=1}^{N} x_{j0}^k = 0, \forall k \in K \tag{9}$$

$$\sum_{k \in K} \sum_{(i,j) \in E} x_{ij}^k d_{ij} \leq L \tag{10}$$

$$w_i^k + u_j - w_j^k \leq (1 - x_{ij}^k)\mathbb{M}, \forall k \in K, \forall (i,j) \in E \tag{11}$$

$$0 < w_i^k \leq W, \forall k \in K, \forall i \in N \tag{12}$$

$$w_i^k = 0, \forall k \in K, \forall i \in \{0\} \cup M \tag{13}$$

$$E \leq v_i^k \leq L, \forall k \in K, \forall i \in \{0\} \cup M \tag{14}$$

$$v_i^k + s_i + t_{ij} - v_j^k \leq (1 - x_{ij}^k)\mathbb{M}, \forall k \in K, \forall (i,j) \in E \tag{15}$$

$$e_i \sum_{j=1}^{N+M} x_{ij}^k \leq v_i^k \leq l_i \sum_{j=1}^{N+M} x_{ij}^k \tag{16}$$

Constraint (5) and (6) require that each stop is serviced by at most one vehicle. Constraint (7) and (8) assure that each vehicle only starts and returns to depot once or not at all. Constraint (9) does not allow vehicles to return directly to the depot from stops, that assure vehicles must visit the landfill before.

Constraint (10) enforces each route must have the length less than value L. Waste volume constraint (11), (12) and (13) ensure that when the total load w_i^k reaches the vehicle capacity, the vehicle must go to the landfill to remove the waste. Time window constraint (13), (14) and (15) relate to the starting service time v_i^k. In order to service any stop, the starting service time is in time windows of that stop. In constraint (11) and (15), $\mathbb{M}$ is a large constant.

Let $\Gamma = \{\gamma_1, \gamma_2, ..., \gamma_q\}$ denotes a set of all feasible sub-routes in G, each element $\gamma_r \in \Gamma$ is a sub-route of G. We define that a binary variable y_{ij}^{rk} is equal to 1 if and only if edge $(i, j) \in \gamma_r$ and traveled by vehicle k. The conflict constraint:

$$\sum_{r=1}^{q} a_{ij} y_{ij}^{rk} = 0, \forall k \in K, \forall (i, j) \in E \tag{17}$$

4   Memetic Algorithm

MA implementation is divided into five major steps: initial population, selection, crossover, mutation and adaptation. The evolution of population is driven respectively by the selection, recombination, mutation while adaptation helps individuals be self-improvement.

At first, an initial population is created by algorithm in Algorithm A. Then population is improved in each generation by selection, crossover, mutation and adaptation. The process is continued until a stop condition is satisfied. The MA algorithm is implemented as follows.

Algorithm MA

(1) INITIAL population *pop*
(2) SET *cont* = stop condition
(3) **while** *cont* != true **do**
(4)   *pop* = naturalSelect(*pop*)
(5)   *pop* = recombine(*pop*)
(6)   *pop* = mutate(*pop*)
(7)   *pop* = adapt(*pop*)
(8)   *cont* = Update stop condition
(9) FINISH

4.1   Initial Population

In order to setup an initial population, we use the ideal proposed by [13]. These authors started from Solomon's I1 Insert heuristic [11] and modified it by some criteria as presented in (18), (19), (20) and (21).

$$C_1(i, u, j) = \alpha_1 C_{11}(i, u, j) + \alpha_2 C_{12}(i, u, j) \tag{18}$$

$$C_{11}(i, u, j) = d_{iu} + d_{uj} - \mu d_{ij} \tag{19}$$

$$C_{12}(i, u, j) = v_{ju} - v_j \tag{20}$$

$$C_2 = \rho d_{0u} C_{11} - C_1(i, u, j) \tag{21}$$

in which:

d_{ij}: distance from vertex i to vertex j

v_{ju} and v_j: starting service time at vertex j after and before inserting vertex u into the sub-route.

α_1: the weight of the distance increases when inserting a new stop into the current sub-route.

α_2: the weight of the waiting time increases when inserting a new stop into the current sub-route $\alpha_1 + \alpha_2 = 1$.

ρ: the weight expresses the distance between a depot and a inserted stop.

μ: the distance adjusted factor.

The stop selected to insert into the route had the largest value in expression (18) without violating any constrains. However, this method is only designed to create one feasible solution and support for optimizing local search algorithm. For the set of feasible solutions, these heuristics need to be improved.

Algorithm A

(1) UNROUTED all stops
(2) CREATE new solution S
(3) CREATE new route R
     SET ref = Depot
     SET $dist$ = Maximum travel distance
(4) **if** CAN create new sub-route **then**
         SR = createSubroute(ref, $dist$)
         Repeat
             SET $stop$ = The stop has value C_2 is max
             INSERT $stop$ into SR
             ROUTED $stop$
         Until CAN'T find the stop insert into SR
         DELETE Depot in the last of SR
         INSERT SR into the route R
         SET ref = The landfill at the end of SR
         UPDATE $dist$ = The remain travel distance
         GOTO (4)
     else
         if EXIST the sub-route in R **then**
             INSERT the Depot into the last of R
             INSERT R into the solution S
             GOTO (3)
         else
             GOTO (5)
(5) FINISH

In function *createSubroute* of Algorithm A, a random stop will be selected to insert into the first position of a new sub-route. It is not same the implementation of [13], the first stop of route has the best fitness value. Thus, the feasible solutions in population are created randomly.

4.2   Representation

Each individual is represented as a sequence of vertices. In chromosome representation, the depot and the landfills are removed from the sequence of vertices. That helps to advance in using crossover and mutation operator. In decode process, they are restored back. First, the depot will be inserted at the beginning and the ending position of the chromosome string. In the next step, the checking constraint process is executed, when a constraint is violated, a suitable landfill will be inserted into the chromosome string. The suitable landfill can be found by scanning all the landfills of the problem to get the landfill with the minimum distance.

4.3   Fitness Function

Based on objective function in expression (1), the fitness function can be expressed as:

$$C = \theta_1 C_d + \theta_2 C_m + \theta_3 C_w \tag{22}$$

in which:

C_d: The total moving distance of all vehicles.
C_m: The total moving time of all vehicles.
C_w: The total waiting time of all vehicles.

Depending on the importance of each objective, weights have great or small value correspondingly. In our study, we consider minimizing the total traveling distance, total move time and total wait time. Regarding optimizing the number of vehicles, this objective is not considered as main component in the global objective function. However, usually when the other objectives of the solution are enough good, the number of vehicles must be acceptable.

4.4   Operators and Heuristics

Next, we discuss about operators used in MA to solve waste collecting VRPTWC. Individuals in the initial population created by Algorithm A may be not dominant. In the searching process, they are evolved after each generation. MA tries

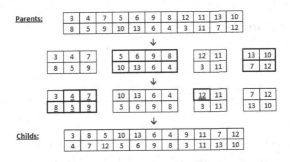

Fig. 3. Steps in crossover process implemented by PMX operator

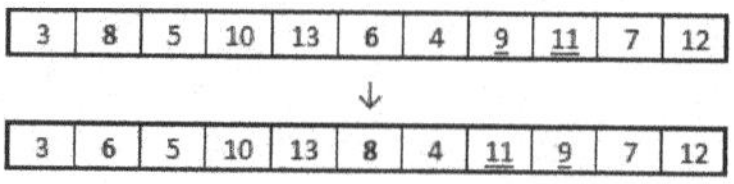

| 3 | 8 | 5 | 10 | 13 | 6 | 4 | 9 | 11 | 7 | 12 |

↓

| 3 | 6 | 5 | 10 | 13 | 8 | 4 | 11 | 9 | 7 | 12 |

Fig. 4. An example of random exchange mutation operator

to find dominant solutions in the set of solutions and its operators are essential tools that help the search explores the searching space.

Partial-mapped crossover operator (PMX) is the main component to reproduce next generation proposed by Goldberg [3]. With two selected individuals, PMX exchanges one or many genes (sequences of vertices) between two chromosomes and rearranges them when exchange phase finished. Fig. 3 shows performance of each step in crossover process of PMX operator. Two new permutation strings that are generated maybe do not satisfy the constraints of the problem. However, encoding process will re-corrected them.

Random exchange mutation is an operator used in the mutation. It exchanges two or more genes randomly in only one chromosome. Fig. 4 illustrates a example of the random exchange mutation operator with three crossover points. In the figure, the gene (vertex) at position 2 exchanges with the gene at position 6 and similarly for genes 8 and 9.

λ-*interchange operator* (λ-opt) is used in the adaptation. A local search, here is Hill Climbing, uses this operator to explore search space. The λ-interchange was introduced first in [8] for the capacitated clustering problem. However, the authors only implement for the routes that are Hamiltonian cycles. In this paper, we extend λ-interchange for using with the routes which are not simple cycles.

Assuming γ_p^i and γ_q^j are two sub-routes p-th and q-th of two routes r^i and r^j in the solution $\sigma \in \mathbb{S}$. We denote $\varsigma \subset \gamma_p^i$ and $\xi \subset \gamma_q^j$ are two paths in sub-routes, $||\varsigma|| \leq \lambda$ and $||\xi|| \leq \lambda$, they do not store the landfills and the depot. λ-interchange generates a new feasible solution by exchanging the subsets of two sub-routes of the solution. If γ'^i_p and γ'^j_q are denoted two new sub-routes, we have $\gamma'^i_p = (\gamma_p^i \backslash \varsigma) \bigcup \xi$ and $\gamma'^j_q = (\gamma_q^j \backslash \xi) \bigcup \varsigma$ and the new solution $\sigma' = \{r^1, ..., r'^i, ..., r'^j, ..., r^k\}$ with $r'^i = (r^i \backslash \gamma_p^i) \bigcup \gamma'^i_p$ and $r'^j = (r^j \backslash \gamma_q^j) \bigcup \gamma'^j_q$ that are two new routes.

4.5   Stop Condition

In order to estimate the capability of convergence of population after each generation, we define fitness function of population of each generation as:

$$C_i(pop) = \beta_1 avg_i(pop) + \beta_2 max_i(pop) \tag{23}$$

in which:

$C_i(pop)$: is used to evaluate the development of population in i-th generation

$avg(i)$: is the average fitness value of all individual in population at i-th generation.

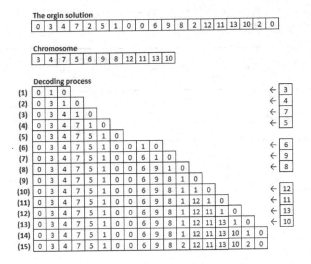

Fig. 5. Illustrate encoding and decoding process of a chromosome

$max(i)$: is the best fitness value of all individual in population at i-th generation.

The value $C_{i-1}(pop)$ of the previous generation is stored and if after i-th generation it is not developing than the previous generation and this happens more than n-times, the algorithm will be stopped. It is necessary for saving the computational time because in many cases, after a number of loops, the population set converged before it reaches the stop condition of the problem.

4.6 Encoding and Decoding

Encoding and decoding process are designed to support for optimizing the number of vehicles. Encoding is a simple process in which the landfills and the depot are removed from the solution. Decoding is different, the constraints are always checked in the process to form the solution. Base on Algorithm A, we change the next stop chosen with the best fitness value by the stop in the chromosome to create the solution again. When the next stop in chromosome is inserted into the solution and if it is violative the constraints, a suitable landfill will be inserted into the solution. Thus, the violated constraints are released and the next stop will continue to be inserted into the solution.

One interesting important detail in the processes is the solution that after decoding is not sure like as the original solution before encoding. The example in Fig 5 shows that clearly. The origin solution has total four sub-routes, however, the result of decoding process has only three sub-routes. In some cases, decoding process reduces not only the number of sub-routes but also the number of routes (the number of vehicles) in the solution. Thus, our encoding and decoding method can help MA to minimize the number of vehicles.

Table 2. Comparison of the solution of the waste collection VRPTW with no waste conflict by our experiments and [5]

	Value	Problem sets			
		102	277	335	444
B.-I. Kim	N_v	3	3	5	10
	TD	205	521	191	82
	T_c	3	37	28	372
MA	N_v	2	3	5	9
	TD	174	483	206	80
	T_c	4	111	647	248

5 Computational Results

Our experiments are implemented by GNU C++ language in a CentOS 6.3 machine of 2.2GHz Xeon processor. Algorithms include Hill Climbing (HC), Generic Algorithm (GA), and MA with λ-interchange mechanism in the adaptation.

The benchmark data sets for waste collection VRPTW used in this paper were proposed by [5]. However, they need the information about waste properties to use for the waste collection VRPTW with conflict. The waste property is a new column and attached into the origin benchmark and Waste properties are denoted from 1 to 7 as shown in Table 1. These values are inserted sequentially into the waste property column in which each row stores information about property groups. In case a certain row stores the information of a depot or landfill, the value 0 will be inserted into the column of this row.

We experiment with four benchmark sets [2] including: 102 stops, 277 stops, 355 stops and 444 stops . ID column is used to identify the index of nodes, X and Y columns (in feet) are coordinates of nodes in city. Early and late columns are the earliest service starting time and the latest service starting time in HHMM format. Service column is the needed time to load waste at stop in seconds. Load column is the volume of waste load in yards. Type column expresses types of vertex (0: depot, 1: regular stop, 2: landfill) and property column is properties of the waste. In the benchmarks, the difficulty of the time window constraints is descending while the number of stops is increasing.

First, MA is compared with experiments of [5] in case the conflict constraints are not considered. Table 2 presents computational result with four benchmark sets of [5] and MA. The computational results are shown in Table 2 including: computational time (T_c) in second, the number of vehicles (N_v), total travel distance (TD) in mile and waiting time (T_w) in minute. We gain better solutions with total traveling distance value and the number of vehicles are less than (our implementations do not consider break-time lunch).

---

[2] The benchmarks are uploaded in
http://www.cse.hcmut.edu.vn/~hoai/VRPTWC/benchmark/

Table 3. Comparison between the best solution results of our algorithms for the Waste collection VRPTWC benchmark problem sets

	Value	Problem sets			
		102	277	335	444
HC	F	-307	-511	-269	-86
	T_c'	6	365	1692	1769
	N_v	4	3	8	11
	TD	280	511	269	86
	T_w	139	0	0	0
GA	F	-294	-527	-263	-91
	T_c	1	9	44	14
	N_v	3	3	6	10
	TD	272	527	263	91
	T_w	107	0	0.	0
MA	F	-271	-521	-249	-56
	T_c	3256	9006	63471	74224
	N_v	3	3	7	7
	TD	230	521	249	56
	T_w	202	0	0	0

In next step, we compare MA with other algorithms in case of the waste collecting VRPTW and conflict. Compared with other algorithms, MA is better, however, its computational time is later. In Table 3, computational time shows time cost of algorithms, total travel distance is the travel distance of all routes of a solution and waiting time is the time vehicles took to wait of all routes of the solution. Fitness value (F) provides an overview of the quality of a solution while other values as total travel distance, waiting time and the number of vehicles provide the detail of information.

Furthermore, the influence of the density of conflict matrix is also considered in our experiment. Density of conflict matrix D_e is defined as:

$$D_e = \frac{\text{Number elements 1 in conflict matrix}}{\text{Total number elements in conflict matrix}} \quad (24)$$

Figure 6 shows the influence of the density of conflict matrix on computational time, total travel distance and the number of vehicles. Total travel distance, the number of vehicles and waiting time are proportional the density of the conflict matrix. The values will be less than if the conflict matrix is sparse and these will be greater than if the conflict matrix is dense.

However, that rule does not accommodate with the computational time value. Belong to each benchmark set, the extremum of computational time is different. To survey rule of the distribution of the extremum in each benchmark set, it needs to consider the mathematical viewpoint.

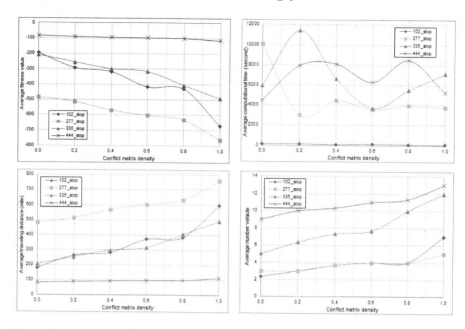

Fig. 6. The influence of the density of conflict matrix on fitness value, computational time, total travel distance and the number of vehicles

6   Conclusion

In this paper, we introduce a new type of VRP that is the waste collection VRPTW with conflict. The problem has multi-objectives and partial conflict constraints in sub-routes. Furthermore, we presented its mathematical model by combining the flow model and the set partitioning model. The memetic algorithm using global search technique and local search with the λ-interchange mechanism is proposed to solve this problem.

In case of the waste collection VRPTW with no conflict, our algorithm can improve the quality of solution. In the waste collection VRPTW with conflict, memetic algorithm with λ-interchange outperformed HC and GA. The experimental results show that the traveling distance and the number of vehicles are large if the conflict matrix has the large density while the computing time in each benchmark is not proportional the density of the conflict matrix. Further, the research also focuses on how the density of conflict matrix influences the result of proposed MA method.

Acknowledgements. Work funded by Contract No. 260/HD-SKHCN (2011-2013).

References

1. Bräysy, O., Gendreau, M.: Vehicle routing problem with time windows, part i: Route construction and local search algorithms. Transportation Science 39(1), 104–118 (2005)
2. Desrochers, M., Lenstra, J., Savelsbergh, M., Soumis, F.: Vehicle routing with time windows: Optimization and approximation. In: Golden, B.L., Assad, A.A. (eds.) Vehicle Routing: Methods and Studies, pp. 65–84. North-Holland (1988)
3. Goldberg, D., Lingle, R.: Alleles, loci and the traveling salesman problem. In: Proceedings of the Second International Conference on Genetic Algorithms. Lawrence Eribaum Associates, Mahwah (1985)
4. Hamdi-Dhaoui, K., Labadie, N., Zhu, A.Y.Q.: The vehicle routing problem with conflicts. In: Proceedings of the 18th IFAC World Congress (2011)
5. Kim, B.I., Kimb, S., Sahoo, S.: Waste collection vehicle routing problem with time windows. Computers & Operations Research 3, 3624–3642 (2005)
6. Knight, K., Hofer, J.: Vehicle scheduling with timed and connected calls: A case study. Operational Research Quarterly 19, 299–310 (1968)
7. Madsen, O.B.G.: Optimal scheduling of trucks - a routing problem with tight due times for delivery. In: Optimization Applied to Transportation Systems, pp. 126–136 (1976)
8. Osman, I.H., Christofides, N.: Capacitated clustering problems by hybrid simulated annealing and tabu search. International Transactions in Operational Research 1(3), 317–336 (1994)
9. Pullen, H.G.M., Webb, M.H.J.: A computer application to a transport scheduling problem. The Computer Journal 10, 10–13 (1967)
10. Savelsbergh, M.W.P.: Local search in routing problem with time windows. Annals of Operations Research 4, 285–305 (1985)
11. Solomon, M.M.: Algorithms for the vehicle routing and scheduling problems with time window constraints. Operational Research 35(2), 254–265 (1987)
12. Solomon, M.M., Desrochers, M.: Time windows con-strained routing and scheduling problem. Transportation Science 22, 1–13 (1988)
13. Tung, D.V., Pinnoi, A.: Vehicle routing-scheduling for waste collection in hanoi. European Journal of Operational Research 125, 449–468 (2000)

Efficient Distributed Algorithm of Dynamic Task Assignment for Swarm Robotics

Rafael Mathias de Mendonça[1,*], Nadia Nedjah[1],
and Luiza de Macedo Mourelle[2]

[1] Department of Electronics Engineering and Telecommunications
Faculty of Engineering
State University of Rio de Janeiro
rmathias.mendonca@gmail.com, nadia@eng.uerj.br
[2] Department of Systems Engineering and Computation
Faculty of Engineering
State University of Rio de Janeiro
ldmm@eng.uerj.br

Abstract. This paper proposes a distributed control algorithm to implement dynamic task allocation in a swarm robotics environment. In this context, each robot that integrates the swarm must run the algorithm periodically in order to control the underlying actions and decisions. The algorithm was implemented and extensively tested. The corresponding performance and effectiveness are promising.

Keywords: Dynamic task allocation, Swarm robotics, Swarm intelligence.

1 Introduction

Swarm intelligence was proposed in late 1980 after the observation of the social behavior of some species of insects and birds, as reported in [1]. These species perform collectively simple tasks in order to achieve complex goals, usually impossible to be carried out by only one individual of the swarm. The central idea of this collective behavior is to perform a complex task by dividing it into simpler tasks that are easily performed by members of the swarm. The approach of swarm robotics emerged from the application of the rules of swarm intelligence to groups of mobile robots with limited processing capacity.

Dynamic allocation of tasks to the robots is a necessary process for proper management of the swarm. This process allows the distribution of the tasks to be performed, among the swarm of robots, in such a way that a pre-defined proportion of execution of tasks is achieved. The ratio should be determined in order to permit the achievement of the overall objective of the swarm. Task allocation is a dynamic process because it needs to be continuously adjusted in response to

---

* The work of this author is supported by CAPES, the Coordination of Improvement of Higher Education Personnel of the Brazilian Federal Government.

B. Murgante et al. (Eds.): ICCSA 2013, Part I, LNCS 7971, pp. 500–510, 2013.

changes in the environment and/or in terms of performance of the swarm. An immediate solution to solve this problem is based on the centralized approach. Nonetheless, a distributed allocation represents a better imitation of the behavior of the social species in natural swarms, where there is no centralized control mechanism. Therefore, task allocation in swarms of robots must arise as a result of a distributed process. This decentralization increases the complexity of the problem, because the robot does not have a complete view of the environment. Each robot must take local control decisions without full knowledge about what the other robots have been doing in the past, are doing in the present or will do in the future.

Several applications that are automatized using swarm robotics, require dynamic task allocation. For instance, in situations which pose some risk or not viable for human presence to perform a given task, the swarm of robots would be able to self-organize to form groups, where each group perform a specific task with a main goal of performing together significant and complex action. Wit the aid of dynamic task allocation, the swarm would gain greater flexibility and reliability.

In this paper, we propose a distributed algorithm that aims at fulfilling dynamic allocation of tasks in a swarm of robots. The algorithm is simple and decentralized, allowing efficient implementation on robots with limited storage and processing power. The algorithm is tested in a swarm of Elisa III robots to demonstrate its effectiveness and efficiency. The results of the experiments are reported.

The remainder of this paper is organized into six sections. Initially in Section 2, we present formally the problem for dynamic allocation of tasks. Then, in Section 3, we introduce some related work. Subsequently, 4, we detail the main steps of the proposed algorithm to solve the problem at hand. Thereafter, in Section 5, we comment on some of the implementation issues on the used robots. Then, in Section 6, we discuss the results of the performed experiments. Last but not least, in Section 7, we draw some conclusion with respect to the proposed algorithm and point out some relevant directions for future work.

2 Dynamic Task Assignment

Generally, a swarm of robots needs to perform a complex task that is divided into other less complex tasks. Therefore, the assignment of these tasks to robots in the swarm is a requirement for many applications. To accomplish the task assignment, one first needs to correctly determine the distribution of the swarm in the environment and then allocate the tasks to the robots. There are several strategies for dynamic allocation of tasks. The allocation may be random, depending only on the robot's ability to effectively perform the selected task allocated, sequential, parallel or some combination of the two, which achieves a balance between time spent in communication and execution.

In order to provide a formal description of the dynamic task assignment problem, let $T = \{t_1, t_2, \ldots, t_\tau\}$ be the set of τ tasks to be performed by the η robots of the swarm. The tasks must be allocated to the robots in a way that a pre-scribed proportion P is always respected. Let $P = \{p_1, p_2, \ldots, p_\tau\}$, such that $0 \leq p_i \leq 1$ and $p_1 + p_2 + \ldots + p_\tau = 1$. In order to satisfy the imposed task proportion, we must, for all task $t \in T$, satisfy $S_t = p_t \times \eta$, wherein $S_t \leq \eta$ is the total number of robots assigned to task t.

3   Related Work

The analysis of collective behavior in Multi-Robot Systems is a relatively new field of study that encompasses a variety of sciences such as Mathematics, Physics and Biology. The first relevant studies in this area with the expressive intention to decentralize control in the process of allocating tasks are presented in [2] and [3]. In these works, studies with mixed groups of robots are reported, wherein the swarm is divided into robots that possess some kind of centralized communi-cation and those, that draw decisions only on local observations. In [4], general properties of aggregation in multi-robot systems using macroscopic phenomeno-logical models are studied.

New formulations for building phenomenological models aimed at a collective behavior of groups of robots are developed in [5] and [6]. These new formulations are tested in simple collaborative experiments as in [7], wherein the stick-pulling experiment is evaluated with a group of reactive robots.

Most of the approaches supra-cited are either implicitly or explicitly based on the theory of stochastic processes. The Microscopic Probabilistic model de-veloped in [8], [9] and [10] is an example of the stochastic approach explored in the study of the collective behavior of a group of robots. The impact of sev-eral stochastic events imposed in parallel, for each robot of the swarm, enabled the drawing of some properties about the behavior of robots in Multi-Robot Systems.

4   Proposed Algorithm

The proposed algorithm for dynamic task allocation is periodically executed by the η robots, active in the swarm. Note that the algorithm executions occur simultaneously, one in each robot as shown in Algorithm 1, wherein t_i identifies the task currently allocated to robot i. Let a be the address of this robot and T_i represents the set of task identifiers, allocated to the other robots of the swarm, excluding robot i, as perceived by that robot. The robot may choose one of the τ available tasks.

Initially, the *adjustment step* performs the necessary work to permit the se-lection of the task that the robot should perform to maintain the proportion balance imposed. This is done via several task adjustments, named *AdjsutTask*. When the prescribed proportion is achieved, considering the task assignments for all the swarm's robots, the robots enters the *execution step*. During this step, all robots execute their assigned task, via *ExecuteTask* until the occurrence of some event that disturbs the proportion balance. Thus, in order to re-establish

---

Algorithm 1. Distributed dynamic task allocation for one robot

---

Require: Number of tasks τ, Number of robots η, Robot addresses A, Proportion P;
Ensure: Proportion P is maintained at almost all time by the swarm;
1: Init all required data structures;
2: **while** *true* **do**
3:     **send** current robot task t_i to all other $(\eta - 1)$ robots;
4:     **receive** tasks $T_{i,j}$ from all other $j = 1 \ldots (\eta - 1)$ robots, $A[j] \neq a$;
5:     **if** proportion P achieved **then**
6:         $ExecuteTask(t_i)$ for a given period of time;
7:     **else**
8:         $AdjsutTask(T_i, t_i)$, considering the set of tasks $1 \ldots \tau$ and proportion P;
9:     **end if**
10: **end while**;

---

the required proportion, the robots go back and enter the adjustment step. The same process is iterated. Note that while executing the task assigned to it, the robot communicates with the swarm robots to by sending its task and receiving the other's and checks the achievement of the imposed proportion.

Algorithm 2 describes the proposed method for implementing the adjustment step and thus tuning the tasks of the robot to achieve the prescribed task proportion. In this algorithm, P is the desired proportion and T represents the tasks performed by each of the other $\eta - 1$ robots, i.e. excluding the robot i that is executing the adjustment process. Note that the address of this robot is a and those of the other's are stored in A.

In Algorithm 2, first, the tasks performed by the swarm are counted according to the task currently allocated. The count result is made available in C, which consists of τ counters, one for each permitted task. Note that at the end of the counting process, the counter $C[u]$ indicates how many robots are currently associated with task u.

After the counting process, an analysis, to determine whether robot i must change task, is initiated. The robot's task t is kept whenever the number of robots already allocated to that task is suitable or below the desired ratio as defined by P[t]. The only case in which the robot changes task is characterized by the fact that an excessive number of robots are also allocated to task t. Note that for a task u, $\delta[u]$ informs the difference between the number of robots associated with task u in the current task allocation T and the desired number as defined in $P[u]$.

During the decision process, the difference δ is evaluated considering the robot task t. If this difference is negative, which identifies the case of an excess of robots performing this task. In such a case, the process of choosing a new task for this robot begins. The selection process consists of finding the task t' with the largest positive difference δ, and hence requires more robots performing it.

Before taking the decision of changing the task of the robot, an assessment of what will be the other robots decision is done to avoid that simultaneous equal decision that invalidates the convergence of algorithm. For this purpose,

Algorithm 2. Adjusting the task of a robot

Require: Robot id i, robot task t_i, swarm tasks as perceived by robot T_i;
1: **for** $r := 1 \to \eta - 1$ **do**
2:     $C[T_i[r]] := C[T_i[r]] + 1$;
3: **end for**
4: $C[t_i] := C[t_i] + 1$;
5: **if** $P[t_i] - C[t_i] < 0$ **then**
6:     **for** $t := 1 \to \tau$ **do**
7:         **if** $P[t] - C[t] > 0$ **then**
8:             $\delta[t] := P[t] - C[t]$;
9:         **end if**
10:    **end for**
11:    $t' := t_k | \delta[t_k]$ is the largest, $k = 1 \ldots \tau$;
12: **end if**
13: $count := 0$;
14: **for** $r := 1 \to \eta - 1$ **do**
15:    let t be the task assigned to robot r;
16:    **if** $T_i[t] \neq t'$ and $P[t] - T_i[t] < 0$ and $A[r] < a$ **then**
17:        $count := count + 1$;
18:    **end if**
19: **end for**
20: **if** $P[t'] - C[t'] - count > 0$ **then**
21:    **return** t'
22: **else**
23:    **return** t_i
24: **end if**

the robot with the lowest address is given more priority to make the task change. The number of robots that more priority than the considered robot is computed, based on the robot task allocation and their respective address. This number, which coincides with the ideal number of task allocation shifts necessary to meet the required proportion for task t'. Whenever, this number does not affected the decision, the robot modifies its current task to the new task t'. Otherwise, it keeps the allocated task unchanged.

5   Implementation Issues

The proposed algorithm was implemented in a swarm of robots Elisa III [11], shown in Fig. 1. This robot is equipped with a centered RGB LED, that allows representing any color by combining the intensity of the colors Red, Green and Blue. The obtained color is intensified by a light diffuser located at the top of the robot. In this implementation, each task is represented by a distinct color in order to observe the progress of task adjustments done by the robots, visually. Thus, we used the colors Blue, Yellow, Pink, Purple and Green to represent the allowed five tasks.

Fig. 1. Elisa III Robots

Each robot is equipped with a transceiver module [11], suitable for wireless communication applications with ultra-low power consumption in the communication range of Radio Frequency (RF). The chip is designed to operate in the ISM radio band (Industrial, Scientific and Medical band basis) of 2.4 GHz.Thus, each robot communicates with each other using RF via a base station connected to a computer.

The base station is connected to a PC via USB and transfer data to and from the robots wirelessly. Likewise, a transceiver module, embedded in the robot, communicates via communication SPI (Serial Peripheral Interface) with the micro-controller of the robot and transfers data to and from the PC , wirelessly, as shown in Fig. 2.

Fig. 2. Communication from and to the robot via RF

Each robot is identified by a unique address in the swarm. this address is stored in a specific memory address of the robot EEPROM. Every message coming from the base station has a destination address, which should coincide with one of the addresses of the robot in the swarm. Upon reception of a message, the robot compares the destination address of the received message with its own address to assess whether it is the intended by this message. If so, the message is saved and interpreted by the robot.

The communication between the robots is done by an RF base. The composition, packing and unpacking of the messages are handled by a computer in-the-loop that continuously polls the station. The polling is performed once every millisecond, which is a restriction on the maximum speed of communication. To overcome this limitation, an optimized protocol was implemented, wherein the packet sent from the PC to the base station contains commands to four robots simultaneously. The base station is responsible of separating the received packet into four individual packets of 16 bytes each, before sending them to the indicated destination address. The same procedure is used during the reception form the robots. In this case, the base station is responsible for receiving the packets of 4 robots, assembles them into a single message of 64 bytes, and send it to the computer. This procedure allows higher throughput communication, making it 4x faster.

The packets generated by the robots to be transferred to the base station are composed of 16 bytes, as shown in Fig 3. The first byte is used to identify the validity of the packet, the following two bytes represent the sender address of the robot, the fourth byte represents the task performed by this robot and the remaining 12 bytes are not used.

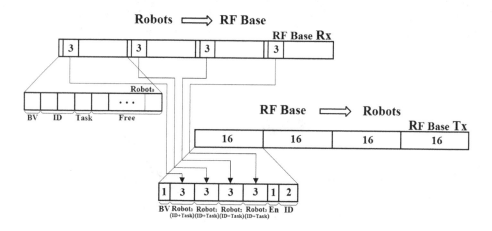

Fig. 3. Packet format of the messages from and to the robots

Considering each packet received from the base station, the PC takes the four addresses and task groups and composes a new package of 16 bytes. The first byte of this package is the validity flag of the package to be verified by the robot upon reception, the 12 following bytes represent the address and task information received, the fourteenth byte is a sending signal enabler and the remaining two bytes contain the address of the target robot. This package is replicated four times using four distinct robot addresses to compose the message of 64 bytes to be sent to the base station to be transmitted the destination addresses. Thus, the same information (robot ID and task) is transmitted to four robots at a time until all the robots of the swarm are encompassed.

The enabling signal generated and sent by the PC to robots of the swarm has the function of managing the process of packet transmission from the robots to the PC. This signal enables the transmission a group of four robots to transmit at a time. Note that the transmission time for a group of 4 robots depends on the total number of robots that compose the swarm. This synchronization trick is an alternative to reduce the packet traffic, thus decreasing the chances of packages being lost due to limited buffer size at the base station.

6 Performance Results

In order to analyze the performance of the proposed algorithm with respect to both execution time and communication time, several tests were performed. We varied the swarm configuration in terms of number of robots as well as the number of distinct tasks that these must perform respecting a prescribed task distribution. The number of robots in the swarm was chose between 5 and 15 robots, and the number of valid tasks, between 2 and 5 tasks. For each swarm configuration, we used a specific distribution of the tasks, as shown in Table 1. Thus, we handled a total of 25 tests performed for each swarm configuration. Aiming at reducing the external influence in robot communication, the obtained results were filtered, disregarding the largest 5 results for each group. Thus, a total of 400 valid tests were taken into account.

Table 1. Distribution the proportion among the tasks.

# Tasks	Proportion				
	Task 0	Task 1	Task 2	Task 3	Task 4
2	60%	40%	–	–	–
3	20%	30%	50%	–	–
4	10%	15%	30%	45%	–
5	5%	10%	20%	30%	35%

The tests were conducted initializing all robots swarm with the same Task, namely Task 0. After this step, the communication process was started counting the number of ticks and the number of messages sent and received by the RF base station until the swarm reaches and holds the desired task distribution. A tick is the duration of a clock cycle of the processor used to handle the work done by the RF base. The chart of Fig. 4 shows the yield results during the performed tests.

In order to obtain the number of messages exchanged between the robots, via the RF base, we count every valid message sent and received by the RF base, as well as the number of elapsed ticks so far. When the prescribed proportion is achieved as perceived by a central view, at the RF base, the state of the three counters is registered. Note that it is possible for a message to be lost in its way to destination and thus it does not reach some robots of the swarm.

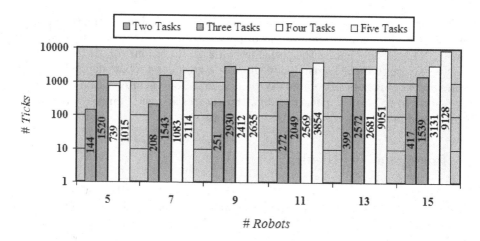

Fig. 4. The execution times in terms of of ticks yield during the performed tests

Thus, the distribution of tasks may oscillate around the prescribed proportion before becoming stable. Therefore, we keep the counters incrementing during all the time. Their states are registered every time the proportion is achieved. The experiment is halted when the task distribution among the robots keeps stable for at some time. The numbers of sent and received messages are presented in Fig. 5 and Fig. 6, respectively.

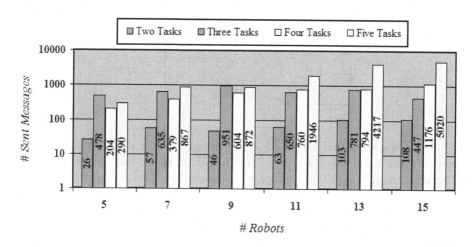

Fig. 5. The number of sent messages during the performed tests

We observe that the number of ticks of the system and the messages sent and received by the RF base increases with the number of robots in the swarm in most cases. We also see that the number of tasks has a direct influence on the

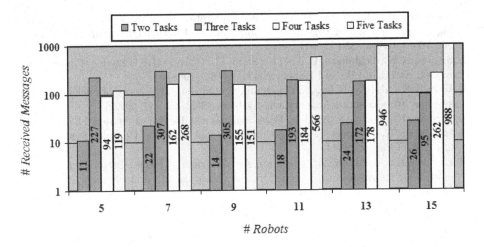

Fig. 6. The number of received messages during the performed tests

complexity of the decision and task assignment process, resulting in an increase in terms of the sent and received messages, and thus on the number of ticks.

7   Conclusions

In this paper we propose the use of an decentralized algorithm to manage the task assignment among a swarm of robots in order to achieve and maintain pre-defined proportion. The proposed algorithm is an efficient solution for task assignment in a swarm of robots. The decision process is conducted independently by each robot using a simple and straightforward process. The results of the tests show the execution time of the proposed algorithm as well as the statistics of the messages generated by the communications process.

In a future work, we intend to investigate the impact of different proportions in order to evaluate the influence of these proportions in the convergence process.

Acknowledgments. We are grateful to FAPERJ (*Fundação de Amparo à Pesquisa do Estado do Rio de Janeiro*, http://www.faperj.br) and CNPq (*Conselho Nacional de Desenvolvimento Científico e Tecnológico*, http://www.cnpq.br) and CAPES (*Coordenação de Aperfeiçoamento de Pessoal de Ensino Superior*, http://www.capes .gov.br) for their continuous financial support.

References

1. Bonabeau, E., Dorigo, M., Theraulaz, G.H.: Swarm Intelligence: From Natural to Artificial Systems. Oxford University Press INC, New York (1999)

2. Sugawara, K., Sano, M.: Cooperative acceleration of task performance: Foraging behavior of interacting multi-robots system. Physica D: Nonlinear Phenomena 100(3-4), 343–354 (1997)

3. Sugawara, K., Yoshihara, I., Abe, K., Sano, M.: Cooperative behavior of interacting robots. Artificial Life and Robotics 2(2), 62–67 (1998)

4. Kazadi, S., Abdul-Khaliq, A., Goodman, R.: On the convergence of puck clustering systems. Robotics and Autonomous Systems 38(2), 93–117 (2002)

5. Lerman, K., Galstyan, A.: Two Paradigms for the Design of Artificial Collectives. In: Tumer, K., Wolpert, D. (eds.) Collectives and Design of Complex Systems, pp. 231–256. Springer (2004)

6. Lerman, K., Martinoli, A., Galstyan, A.: A review of probabilistic macro- scopic models for swarm robotic systems. Swarm Robotics, 143–152 (2005)

7. Lerman, K., Galstyan, A., Martinoli, A., Ijspeert, A.: A macroscopic analytical model of collaboration in distributed robotic systems. Artificial Life 7(4), 375–393 (2001)

8. Martinoli, A.: Swarm intelligence in autonomous collective robotics: From tools to the analysis and synthesis of distributed control strategies. Ph. D. Thesis (1999)

9. Martinoli, A., Ijspeert, A., Gambardella, L.: A probabilistic model for un- derstanding and comparing collective aggregation mechanisms. Advances in Artificial Life, 575–584 (1999)

10. Ijspeert, A.J., Martinoli, A., Billard, A., Gambardella, L.M.: Collabora- tion through the exploitation of local interactions in autonomous collective robotics: The stick pulling experiment. Autonomous Robots 11(2), 149–171 (2001)

11. Nordic Semiconductor. Technical report, nRF24L01+ Single Chip 2.4GHz Transceiver (2008),
http://www.nordicsemi.com/kor/Products/2.4GHz-RF/nRF24L01P

Implementing an Interconnection Network Based on Crossbar Topology for Parallel Applications in MPSoC

Fábio Gonçalves Pessanha[1], Luiza de Macedo Mourelle[2],
Nadia Nedjah[1], and Luneque Del Rio de Souza e Silva Júnior[3]

[1] Department of Electronics Engineering and Telecommunications
Faculty of Engineering
State University of Rio de Janeiro
pessanha1078@gmail.com, nadia@eng.uerj.br
[2] Department of Systems Engineering and Computation
Faculty of Engineering
State University of Rio de Janeiro
ldmm@eng.uerj.br
[3] Pos-Graduation Program on Systems Engineering and Computation
Federal University of Rio de Janeiro
luneque@hotmail.com

Abstract. Multi-Processor System on Chip (MPSoC) offers a set of processors, embedded in one single chip. A parallel application can, then, be scheduled to each processor, in order to accelerate its execution. One problem in MPSoCs is the communication between processors, necessary to run the application. The shared memory provides the means to exchange data. In order to allow for non-blocking parallelism, we based the interconnection network in the crossbar topology. In this kind of interconnection, processors have full access to their own memory module simultaneously. On the other hand, processors can address the whole memory. One processor accesses the memory module of another processor only when it needs to retrieve data generated by the latter. This paper presents the specification and modeling of an interconnection network based on crossbar topology. The aim of this work is to investigate the performance characteristics of a parallel application running on this platform.

Keywords: Interconnection Network, Crossbar Topology, MPSoC, Shared Memory.

1 Introduction

During the 80's and 90's, engineers were trying to improve the processing capability of microprocessors by increasing clock frequency [10]. Afterwards, they tried to explore parallelism at the instruction level with the concept of pipeline [1] [2]. However, the speedup required by software applications was gradually

B. Murgante et al. (Eds.): ICCSA 2013, Part I, LNCS 7971, pp. 511–525, 2013.
© Springer-Verlag Berlin Heidelberg 2013

becoming higher than the speedup provided by these techniques. Besides this, the increase in clock frequency was leading to the increase in power required, to levels not acceptable. The search for smaller devices with high processing capability and with less energy consumption have turned solutions based on only one processor obsolete. This kind of solution has been restricted to low performance applications. On the other hand, there are few applications of this sort, for which microcontrollers are best employed.

In order to reach specific performance requirements, such as throughput, latency, energy consumed, power dissipated, silicon area, design complexity, response time, scalability, the concept of Multi-Processor System on Chip (MPSoC) was explored. In this concept, several processors are implemented in only one chip to provide the most of parallelism possible. MPSoCs require an interconnection network [6] to connect the processors, as shown in Fig. 1. Interconnection networks, beyond the context of MPSoCs, are implemented in different topologies, such as shared-medium, direct, indirect and hybrid [5] [3].

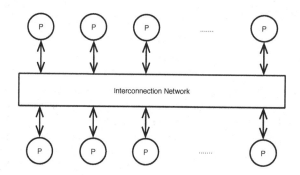

Fig. 1. Interconnection Network in a Multi-Processor System

2 The Crossbar Topology

The crossbar network allows for any processor to access any memory module simultaneously, as far as the memory module is free. Arbitration is required when at least two processors attempt to access the same memory module. However, contention is not an usual case, happening only when processors share the same memory resource, for example, in order to exchange information. In this work, we consider a distributed arbitration control, shared among the switches connected to the same memory module. In Fig. 2, the main components are introduced, labeled according to their relative position in the network, where i identifies the row and j identifies the column. For instance, component A(j) corresponds to the arbiter [4] for column j. For the sake of legibility, we consider 4 processors ($0 \leq i \leq 3$) and 4 memory modules ($0 \leq j \leq 3$).

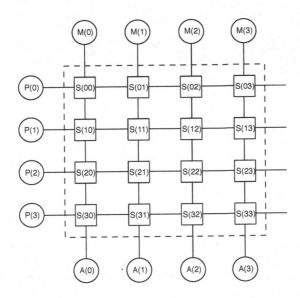

Fig. 2. Crossbar Components

The processor is based on the PLASMA CPU core, designated MLite_CPU (MIPS Lite Central Processor Unit) [7], shown in Fig. 3. In order to access a memory module M(j), processor P(i) must request the corresponding bus B(j) and wait for the response from the arbiter A(j). A bus access control is, then, implemented, as depicted in Fig. 4, which decodifies the two most significant bits of the current address (ADDRESS(i)<29:28>), generating the corresponding memory module bus request REQ_IN(i,j). If P(i) is requesting its primary bus B(j), for which $i = j$, the arbiter sets signal PAUSE_CONT(i), pausing the processor until arbitration is complete, when, then, the arbiter resets this signal. On the other hand, if P(i) is requesting a secondary bus B(j), for which $i \neq j$, signal BUS_REQ(i) is set, pausing the processor until arbitration is complete, when, then, the arbiter sets signal DIS_EN(i), activating the processor.

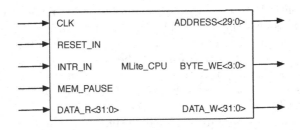

Fig. 3. MLite_CPU Interface Signals

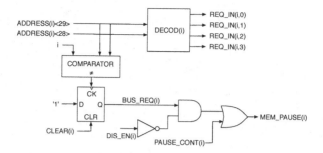

Fig. 4. Bus Access Control

The croosbar switch is basically a set of tri-state gates, controlled by the arbiter, as shown in Fig. 5. Signal COM_DISC(i,j) is set by the arbiter, after arbitration is complete, estabilishing the communication between processor P(i) and memory module M(j).

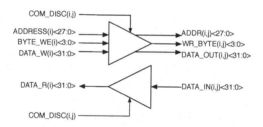

Fig. 5. Crossbar Switch

2.1   Network Controller

The network controller is composed of the arbiter A(j) and a set of controllers, one for each processor, implemented by state machines SM(j)<0:N-1>, as shown in Fig. 6. Upon receiving a bus request, through signals REQ_IN(i,j)<0:N-1>, the arbiter A(j) schedules a processor to be the next bus master, based on the round-robin algorithm, by activating the corresponding signal GRANT(i,j). State machines are used to control the necessary sequence of events to transition from the present bus master to the next one.

There are two types of state machines: primary and secondary. A primary state machine, designated SM_P(j), controls processor P(i) bus accesses to its primary bus B(j), for which $i = j$. A secondary state machine, designated SM_S(i,j),

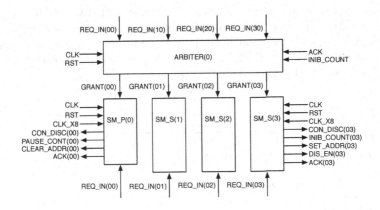

Fig. 6. Network Controller

controls processor P(i) bus accesses to a secondary bus B(j), for which $i \neq j$. Therefore, for each arbiter, there will be one primary state machine and $N - 1$ secondary state machines.

The primary state machine is described by Alg. 1. State **Reset** is entered whenever signal RESET goes to 1, setting signal PAUSE_CONT(i), which suspends P(i), and resetting all the others. As RESET goes to 0, SM_P(j) enters state **Con_1**, establishing the connection of P(i) to B(j), for $i = j$, by setting signal CON_DISC(i,j), according to Fig. 5. Once in state **Cont**, P(i) starts the bus access, as signal PAUSE_CONT(i) goes to 0. While there are no requests from other processors, so GRANT(i,j)=1 for $i = j$, and P(i) is not requesting any other B(j), so REQ(i,j)=1 for $i = j$, the primary state machine stays in state **Cont**. If another processor requests B(j), for $i = j$, then the arbiter resets signal GRANT(i,j), for $i = j$, and the primary state machine enters state **Pause**, in order to suspend P(i), by setting signal PAUSE_CONT(i). Next, SM_P(j) enters state **P_Disc**, in order to disconnect P(i) from B(j), by resetting signal CON_DISC(i,j). It stays in this state until the arbiter gives B(j) back to P(i), by setting signal GRANT(i,j), for $i = j$. SM_P(j), then, returns to state **Con_1**, where P(i) reestablishes its connection to B(j). On the other hand, from state **Cont**, the other possibility is that P(i) requests another B(j), for $i \neq j$, resetting signal REQ(i,j), for $i = j$. In this case, SM_P(j) enters state **Wait**, in order to check if the arbiter has already granted the secondary bus to P(i), in which case signal GRANT(i,j), for $i = j$, goes to 0, or not yet, in which case signal GRANT(i,j), for $i = j$, remains in 1. In the first situation, SM_P(j) enters state **Disc**, in order to disconnect P(i) from B(j), for $i = j$, by resetting signal CON_DISC(i,j). In the second situation, SM_P(j) enters state **Ack**, setting signal Ack(j), in order to force the arbiter to reset signal GRANT(i,j), as shown in Fig. 6. Once in state **Disc**, SM_P(j) stays in this state until the arbiter grants once again B(j) to P(i), for $i = j$, meaning that P(i) is now requesting access to its primary bus. As signal GRANT(i,j) goes to 1, SM_P(j) enters state **Con_2**, where P(i) is then connected to its primary bus, as signal CON_DISC(i,j) goes to 1. Then, SM_P(j) enters state **Clear**, in order to reset signal BUS_REQ, according to Fig. 4, which was set when P(i) was addressing a secondary memory module, for which $i \neq j$. This signal, when set, pauses P(i), until the secondary state machine sends the control for P(i) to access the secondary bus.

Algorithm 1. Primary State Machine

Reset:	$PAUSE\_CONT \leftarrow 1$; $CON\_DISC \leftarrow 0$; $ACK\_ME \leftarrow 0$; $CLEAR \leftarrow 0$;
	if $RESET = 1$ **then** goto **Reset**;
Con_1:	$PAUSE\_CONT \leftarrow 1$; $CON\_DISC \leftarrow 1$; $ACK\_ME \leftarrow 0$; $CLEAR \leftarrow 0$;
	if $CLK = 0$ **then** goto **Con_1**;
Cont:	$PAUSE\_CONT \leftarrow 0$; $CON\_DISC \leftarrow 1$; $ACK\_ME \leftarrow 0$; $CLEAR \leftarrow 0$;
	if $GRANT = 1$ **and** $REQ = 0$ **then** goto **Wait**;
	else if $GRANT = 0$ **and** $REQ = 1$ **then** goto **Pause**;
Wait:	$PAUSE\_CONT \leftarrow 0$; $CON\_DISC \leftarrow 1$; $ACK\_ME \leftarrow 0$; $CLEAR \leftarrow 0$;
	if $GRANT = 1$ **then** goto **ACK_ME else** goto **Disc**;
Ack:	$PAUSE\_CONT \leftarrow 0$; $CON\_DISC \leftarrow 1$; $ACK\_ME \leftarrow 1$; $CLEAR \leftarrow 0$;
Disc:	$PAUSE\_CONT \leftarrow 0$; $CON\_DISC \leftarrow 0$; $ACK\_ME \leftarrow 0$; $CLEAR \leftarrow 0$;
	if $GRANT = 0$ **then** goto **Disc**;
Con_2:	$PAUSE\_CONT \leftarrow 0$; $CON\_DISC \leftarrow 1$; $ACK\_ME \leftarrow 0$; $CLEAR \leftarrow 0$;
	if $CLK = 0$ **then** goto **Con_2**;
Clear:	$PAUSE\_CONT \leftarrow 0$; $CON\_DISC \leftarrow 1$; $ACK\_ME \leftarrow 1$; $CLEAR \leftarrow 1$;
	goto **Cont**;
Pause:	$PAUSE\_CONT \leftarrow 1$; $CON\_DISC \leftarrow 1$; $ACK\_ME \leftarrow 0$; $CLEAR \leftarrow 0$;
P_Disc:	$PAUSE\_CONT \leftarrow 1$; $CON\_DISC \leftarrow 0$; $ACK\_ME \leftarrow 0$; $CLEAR \leftarrow 0$;
	if $GRANT = 0$ **then** goto **P_Disc else** goto **CON_1**;

The secondary state machine is described by Alg. 2. During initialization, when signal Reset is 1, SM_S(i,j) stays in state **Reset** until the arbiter grants a secondary bus B(j) to processor P(i). When signal GRANT(i,j) goes to 1, SM_S(i,j) enters state **Wait_1**, followed by state **Wait_2**, in order to give time to the corresponding primary state machine to pause P(i) and disconnect it from B(j), for $i = j$. In this case, either P(i) is requesting a secondary bus or another processor is requesting B(j) as secondary bus. In the first situation, SM_P(i,j) enters state **Wait** and in the second situation SM_P(i,j) enters state **Pause**, as discribed above. Observe that only the corresponding signal GRANT(i,j) is set, according to P(i) and B(j) in question. Next, SM_S(i,j) enters state **Con**, where signal CON_DISC(i,j) is set, connecting P(i) to B(j). Once in state **Dis**, signal DIS_EN(i,j) goes to 1, activating P(i), as shown in Fig. 4. Recall that P(i), for $i = j$, was paused by the primary state machine, either because it requested a secondary bus or its primary bus is being requested by another processor. Once P(i) finishes using B(j), for which $i \neq j$, SM_S(i,j) enters state **En**, resetting signal DIS_EN(i,j) and pausing P(i). Next, SM_S(i,j) enters state **Disc**, resetting signal CON_DISC(i,j) and disconnecting P(i) from B(j). Then, SM_S(i,j) enters state **Ack**, in order to tell the arbiter it finished using B(j), by setting signal

ACK_ME, which makes the arbiter select the next bus master. Observe that signal INIB_COUNT goes to 1 as soon as SM_S(i,j) leaves state **Reset**, stopping the counter that controls the time limit for P(i) to use B(j), for $i = j$, since this processor is not using its primary bus.

Algorithm 2. Secondary State Machine

Reset: $CON\_DISC \leftarrow 0; DIS\_EN \leftarrow 0; INIB\_COUNT \leftarrow 0;$
$ACK\_ME \leftarrow 0;$if $GRANT = 0$ **then** goto **Reset**;
Wait_1: $CON\_DISC \leftarrow 0; DIS\_EN \leftarrow 0; INIB\_COUNT \leftarrow 1; ACK\_ME \leftarrow 0;$
Wait_2: $CON\_DISC \leftarrow 0; DIS\_EN \leftarrow 0; INIB\_COUNT \leftarrow; ACK\_ME \leftarrow 0;$
Con: $CON\_DISC \leftarrow 1; DIS\_EN \leftarrow 0; INIB\_COUNT \leftarrow 1; ACK\_ME \leftarrow 0;$
if $CLK = 0$ **then** goto **Con**;
Dis: $CON\_DISC \leftarrow 1; DIS\_EN \leftarrow 1; INIB\_COUNT \leftarrow 1; ACK\_ME \leftarrow 0;$
if $CLK = 0$ **then** goto **Dis**;
En: $CON\_DISC \leftarrow 1; DIS\_EN \leftarrow 0; INIB\_COUNT \leftarrow 1; ACK\_ME \leftarrow 0;$
Disc: $CON\_DISC \leftarrow 0; DIS\_EN \leftarrow 0; INIB\_COUNT \leftarrow 1; ACK\_ME \leftarrow 0;$
Ack: $CON\_DISC \leftarrow 0; DIS\_EN \leftarrow 0; INIB\_COUNT \leftarrow 1; ACK\_ME \leftarrow 1;$
goto **Reset**;

---

3   Experimental Results

In order to analyse the performance of the proposed architecture, we used the Particle Swarm Optimization (PSO) method [8][9] to optimize an objective function. This method was chosen due to its intensive computation, being a strong candidate for parallelization. In this method, particles of a swarm are distributed among the processors and, at the end of each iteration, a processor accesses the memory module of another one in order to obtain the best position found in the swarm. The communication between processors is based on three strategies: ring, neighbourhood and broadcast.

3.1   Particle Swarm Optimization

The PSO method keeps a swarm of particles, where each one represents a potential solution for a given problem. These particles transit in a search space, where solutions for the problem can be found. Each particle tends to be attracted to the search space, where the best solutions were found. The position of each particle is updated by the velocity factor $v_i(t)$, according to Eq. 1:

$$x_i(t + 1) = x_i(t) + v_i(t + 1) \tag{1}$$

Each particle has its own velocity, which drives the optimization process, leading the particle through the search space. This velocity depends on its performance, called cognitive component, and on the exchange of information with its neighbourhood, called social component. The cognitive component quantifies the performance of particle i, in relation to its performance in previous iterations. This component is proportional to the distance between the best position found by the particle, called $Pbest_i$, and its actual position. The social component quantifies the performance of particle i in relation to its neighbourhood. This component is proportional to the distance between the best position found by the swarm, called $Gbest_i$, and its actual position. In Eq. 2, we have the definition of the actual velocity in terms of the cognitive and social components of the particle:

$$v_i(t+1) = v_i(t) \times w(t) + c_1 \times r_1(Pbest_i - x_i(t)) + c_2 \times r_2(Gbest_i - x_i(t)) \quad (2)$$

Components r_1 and r_2 control the randomness of the algorithm. Components c_1 and c_2 are called the cognitive and social coeficients, controlling the trust of the cognitive and social components of the particle. Most of the applications use $c_1 = c_2$, making both components to coexist in harmony. If $c_1 \gg c_2$, then we have an excessive movement of the particle, making difficul the convergence. If $c_2 \gg c_1$, then we could have a premature convergence, making easy the convergence to a local minimum.

Component w is called the inertia coeficient and defines how the previous velocity of the particle will influence the actual one. The value of this factor is important for the convergence of the PSO. A low value of w promotes a local exploration of the particle. On the other side, a high value promotes a global exploration of the space. In general, we use values near to one, but not too close to 0. Values of w greater than 1 provide a high acceleration to the particle, which can make convergence difficult. Values of w near 0 can make the search slower, yielding an unnecessary computational cost. An alternative is to update the value of w at each iteration, according to Eq. 3, where n_{ite} is the total number of iterations. At the beginning of the iterations, we have $w \approx 1$, increasing the exploratory characteristic of the algorithm. During iterations, we linearly decrease w, making the algorithm to implement a more refined search.

$$w(t+1) = w(t) - \frac{w(0)}{n_{ite}} \quad (3)$$

The size of the swarm and the number of iterations are other parameters of the PSO. The first one is the number of existing particles. A high number of particles allows for more parts of the search space to be verified at each iteration, which allows for better solutions to be found, if compared with solutions found in smaller swarms. However, this increases the computational cost, with the increase in execution time. The number of iterations depends on the problem. With few iterations, the algorithm could finish too early, whithou providing an acceptable solution. On the other hand, with a high number of iterations, the computational cost could be unnecessaryly high. Alg. 3 describes the PSO method.

Algorithm 3. PSO

Create and initialize a swarm with n particles;
repeat
    for $i = 1 \rightarrow n$ **do**
        Calculate the fitness of $particle_i$;
        if $Fitness_i \leq Pbest$ **then**
            Update $Pbest$ with the new position;
        end
        if $Pbest_i \leq Gbest_i$ **then**
            Update $Gbest_i$ with the new position;
        end
        Update the particle's velocity;
        Update the particle's position;
    end
until $Stop\ criteria = true$;

3.2 Communication between Processes

The parallel execution of the PSO method was done by allocating one instance of the algorithm to each processor of the network. The swarm was then equally divided among the processors. Each subswarm evolves independently and, periodically, *Gbest* is exchanged among the processors. This exchange of data was done based on three strategies: ring, neighbourhood and broadcast.

Fig. 7 describes the ring strategy, while Alg. 4 describes the PSO using this strategy for process communication. The neighbourhood strategy can be depicted by Fig. 8 and the PSO algorithm that implements this strategy is described by Alg. 5. Fig. 9 shows the broadcast strategy and Alg. 6 describes its use for process communication by the PSO algorithm.

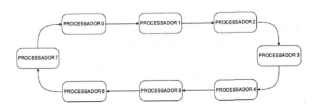

Fig. 7. Ring Strategy

3.3 Performance Figures

The PSO algorithm was used to minimize the Rosenbrock function, defined by Eq. 4 and whose curve is shown in Fig. 10. We used 1, 2, 4, 8, 16 and 32 processors

Algorithm 4. PSO with Ring Strategy

Create and initialize a swarm with n particles;
$id := processor identification$;
$tmpid := id - 1$;
$nproc := number of processors in the network$;
if $id \neq 0$ **then**
    $endprocess(id) := 0$;
end
$tmpid := id - 1$;
repeat
    for $j = 1 \rightarrow n$ **do**
        Calculate the fitness of $particle_i$;
        Update $Gbest(id)$ and $Pbest(id)$;
        Update the particle's velocity;
        Update the particle's position;
    end
    Copy $Gbest(id)$ to share the memory;
    Read $Gbest$ from $processor(tmpid)$;
    if $Gbest(tmpid) \leq Gbest(id)$ **then**
        $Gbest(id) := Gbest(tmpid)$;
    end
until $Stop\ criteria = true$;
if $id = 0$ **then**
    $Best := Gbest(id)$;
    $tmpid := id + 1$;
    for $k = 1 \rightarrow nproc - 1$ **do**
        Read $endprocess(tmpid)$;
        while $endprocess(tmpid) = 0$ **do**
            Read $endprocess(tmpid)$;
        end
        Read $Gbest$ from $processor(tmpid)$;
        if $Gbest(tmpid) \leq Best$ **then**
            $Best := Gbest(tmpid)$;
        end
        $tmpid := tmpid - 1$;
    end
else
    $endprocess(id) := 1$;
end

Algorithm 5. PSO with Neighbourhood Strategy

Create and initialize a swarm with n particles;
$id := processoridentification$;
$tmpid := id - 1$;
$nproc := numberofprocessorsinthenetwork$;
if $id \neq 0$ **then**
    $endprocess(id) := 0$;
end
repeat
    for $j = 1 \to n$ **do**
        Calculate the fitness of $particle_i$;
        Update $Gbest(id)$ and $Pbest(id)$;
        Update the particle's velocity;
        Update the particle's position;
    end
    Copy $Gbest(id)$ to share the memory;
    $tmpid := id + 1$;
    Read $Gbest$ from $processor(tmpid)$;
    if $Gbest(tmpid) \leq Gbest(id)$ **then**
        $Gbest(id) := Gbest(tmpid)$;
    end
    $tmpid := id - 1$;
    Read $Gbest$ from $processor(tmpid)$;
    if $Gbest(tmpid) \leq Gbest(id)$ **then**
        $Gbest(id) := Gbest(tmpid)$;
    end
until *Stop criteria* = *true*;
if $id = 0$ **then**
    $Best := Gbest(id)$;
    $tmpid := id + 1$;
    for $k = 1 \to nproc - 1$ **do**
        Read $endprocess(tmpid)$;
        while $endprocess(tmpid) = 0$ **do**
            Read $endprocess(tmpid)$;
        end
        Read $Gbest$ from $processor(tmpid)$;
        if $Gbest(tmpid) \leq Best$ **then**
            $Best := Gbest(tmpid)$;
        end
        $tmpid := tmpid + 1$;
    end
else
    $endprocess(id) := 1$;
end

Algorithm 6. PSO with Broadcast Strategy

Create and initialize a swarm with n particles;
$id := processor identification$;
$nproc := number of processors in the network$;
if $id \neq 0$ **then**
    $endprocess(id) := 0$;
end
repeat
    for $j = 1 \rightarrow n$ **do**
        Calculate the fitness of $particle_i$;
        Update $Gbest(id)$ and $Pbest(id)$;
        Update the particle's velocity;
        Update the particle's position;
    end
    Copy $Gbest(id)$ to share the memory;
    $tmpid := id + 1$;
    for $k = 1 \rightarrow nproc - 1$ **do**
        Read $Gbest$ from $processor(tmpid)$;
        if $tmpid = nproc - 1$ **then**
            $tmpid = 0$;
        else
            $tmpid := tmpid + 1$;
        end
    end
until $Stop\ criteria = true$;
if $id = 0$ **then**
    $Best := Gbest(id)$;
    $tmpid := id + 1$;
    for $k = 1 \rightarrow nproc - 1$ **do**
        Read $endprocess(tmpid)$;
        while $endprocess(tmpid) = 0$ **do**
            Read $endprocess(tmpid)$;
        end
        Read $Gbest$ from $processor(tmpid)$;
        if $Gbest(tmpid) \leq Best$ **then**
            $Best := Gbest(tmpid)$;
        end
        $tmpid := tmpid + 1$;
    end
else
    $endprocess(id) := 1$;
end

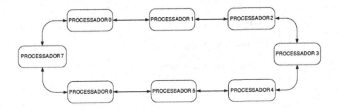

Fig. 8. Neighbourhood Strategy

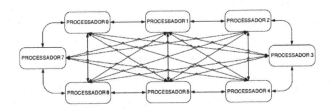

Fig. 9. Broadcast Strategy

for each simulation and considering each of the communication strategies, 64 particles, distributed among the processors, and the algorithm was run for 32 iterations. The speedup obtained is described by Fig. 11.

$$f(x,y) = 100(y - (x^2))^2 + (1 - x)^2 \tag{4}$$

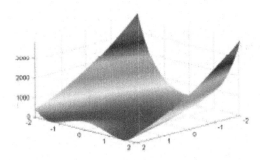

Fig. 10. Graphic of the Rosenbrock Function

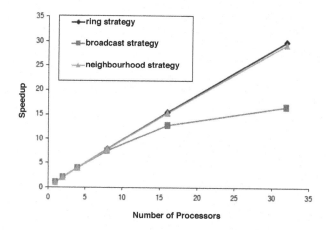

Fig. 11. Speedup obtained for the Execution of the Rosenbrock Function

4 Conclusions

In order to evaluate the performance offered by the proposed architecture, we executed the PSO method for the minimization of the Rosenbrock function, both sequentially and in parallel. The simulation was done for 1, 2, 4, 8, 16 and 32 processors, using a swarm of 64 particles and implementing 32 iterations. We exploited three communication strategies: ring, neighbourhood and broadcast. The speedup obtained demonstrated that the performance offered by the network increases with the number of processors. Another fact is that both ring and neighbourhood strategies have similar impact on the performance of the network, while the broadcast strategy decreases the performance. This decrease is due to the fact the the latter imposes much more interprocess communication than the former ones.

As for future work, we intend to explore other applications for parallelization, in order to analyse the impact of their behaviour specially concerning the interprocess communication; introduce cache memory, to improve performance; develop a microkernel, to implement task scheduling and explore multithread execution; explore other arbitration schemes; sintezise the architecture, in order to analyse the cost x performace relation.

Acknowledgments. We would like to thank the brazilian federal agency, CNPq, and the Rio de Janeiro state agency, FAPERJ, for their financial support.

References

1. Kongerita, P., et al.: Niagara: 32-way multithreaded Spark processor. IEEE MICRO 25(2), 21–29 (2005)
2. Patterson, D.A., Hennessy, J.L.: Computer Organization: the Hardware/Software Interface, 3rd edn. Morgan Kaufmann, San Francisco (2005)

3. Pande, P.T., Michele, G., et al.: Design, Synthesis, and Test of Networks on Chips. IEEE Design & Test of Computers (2005)
4. Matt, W.: Arbiters: Design Ideas and Coding Styles. Silicon Logic Engineering, Inc. (2001)
5. Duato, J., Yalamanchili, S., Ni, L.: Interconnection Networks: an Engineering Approach. Morgan Kaufmann, San Francisco (2003)
6. Ni, L.M.: Issues in Designing Truly Scalable Interconnection Networks. In: International Conference on Parallel Processing Workshop, pp. 74–83. IEEE Press, New York (1996)
7. OpenCores, http://www.opencores.org
8. Kennedy, J., Eberhart, R.: Particle swarm optimization. In: IEEE International Conference on Neural Networks, vol. 4, pp. 1942–1948. IEEE Press, New York (1995)
9. Engelbrecht, A.P.: Fundamentals of Computational Swarm Intelligence. John Wiley & Sons, Chichester (2006)
10. Tanenbaum, A.S.: Structured Computer Organization, 5th edn. PEARSON Prentice Hall, New Jersey (2006)

Modelling Higher Dimensional Data for GIS Using Generalised Maps

Ken Arroyo Ohori, Hugo Ledoux, and Jantien Stoter

Delft University of Technology

Abstract. Real-world phenomena have traditionally been modelled in 2D/3D GIS. However, powerful insights can be gained by integrating additional non-spatial dimensions, such as time and scale. While this integration to form higher-dimensional objects is theoretically sound, its implementation is problematic since the data models used in GIS are not appropriate. In this paper, we present our research on one possible data model/structure to represent higher-dimensional GIS datasets: generalised maps. It is formally defined, but is not directly applicable for the specific needs of GIS data, e.g. support for geometry, overlapping and disconnected regions, holes, complex handling of attributes, etc. We review the properties of generalised maps, discuss needs to be modified for higher-dimensional GIS, and describe the modifications and extensions that we have made to generalised maps. We conclude with where this research fits within our long term goal of a higher dimensional GIS, and present an outlook on future research.

1 Introduction

Spatial data modelling refers to the creation of abstract mathematical representations of real world objects embedded in space. This includes not only purely spatial aspects, such as the objects' geometry and topology, but also other characteristics required for their use, such as the ability to attach attributes (both for storage and for thematic aspects), mark visited objects, or to have efficient access to the objects within a region.

Spatial models have been developed largely independently in the disciplines that required information on spatial objects, including computer graphics, computer-aided design and manufacturing (CAD/CAM), geology, and geographic information systems (GIS) [1]. Because of their independent creation, they are a reflection of the idiosyncrasies of their domains and differ significantly in key issues. One consequence of this is that in many fields support for 3D data has been long widespread and the theoretical foundations for higher dimensions are well established. However, GIS still has limited support for 3D data, and higher-dimensional GIS, despite decades of frequent mentions in literature [2, 3, 4, 5], remains in most cases a theoretical discussion[1].

---

[1] Among the implementations that do exist, this term is most often used as a catch-phrase for any processing involving 3D space and time. However, time is usually treated as a mere attribute, and true 4D space is almost never used.

B. Murgante et al. (Eds.): ICCSA 2013, Part I, LNCS 7971, pp. 526–539, 2013.
© Springer-Verlag Berlin Heidelberg 2013

This slow progress in the GIS world is not due to a lack of applications in higher dimensions. While being limited to 3D space is acceptable to many users of geographic information, substantial work has been done regarding the integration of non spatial-dimensions [6], such as time [7, 8] and scale [9, 10], to spatial data models. This is done either by creating specific models for these non-spatial dimensions, or by treating them as additional spatial ones [5], yielding a *higher dimensional spatial model*. The latter case is more extensible and generic, allowing us to manipulate objects in a *dimension independent* manner [11]. It is also the focus of this paper, and therefore the notion of a higher dimensional spatial model is first explained in detail in Section 2.

Since there is both a need for higher-dimensional GIS, and an availability of such data models from other fields that are able to support higher-dimensional data, there is great potential in finding a suitable model and adapting it to the specific needs of real-world (GIS) data, such as: support for overlapping regions, holes, and complex handling of attributes and metadata; and providing the specific operations that are required for its use, such as good construction and querying operations, buffering and overlays [12]. A short summary of the most remarkable representations for higher dimensional objects developed in other fields and that could thus be adopted is given in Section 3.

Among these, we propose the use of *generalised maps*, which are explained in Section 4, a model capable of representing a wide class of objects in arbitrary dimensions. It has several advantageous properties, such as support for unbounded objects (useful for time and other unbounded dimensions [13]), avoiding problems with incompatible orientations (a common problem when objects are built independently), and providing a simple manner to attach attributes to the objects of every dimension (e.g. vertex, edge, facet, etc.). Practically, it also has the advantage of having been implemented in 3D (it is used in GOCAD[2] for geological modelling and in Moka[3] for geometric modelling).

However, generalised maps by themselves cannot support all the characteristics of real-world spatial data. To bring our ideas into practice, in Section 5 we therefore explain how we have modified and implemented generalised maps for this purpose, and how some specific challenging aspects of GIS data can be handled. We finalise with our conclusions, discussion and our plans for future work in Section 6.

2   Higher-Dimensional Spatial Information

The simplest technique to handle additional dimensions in spatial information, both in GIS and other fields, is using multiple independent representations. In practice, this means that these dimensions are considered as simple attributes which are attached to 2D or 3D objects. Such is the case in so-called 2.5D models for height, the 'snapshot' model for time [14] and most approaches to

---

[2] http://www.gocad.org/
[3] http://moka-modeller.sourceforge.net/

multi-scale data, including CityGML [15]. This approach is simple to understand and implement, but it also has important disadvantages:

- There is only a fixed (discrete) number of representations, which means that the objects being represented only have a known value at certain predefined points along the dimension. For example, a moving object's position is only known at certain moments.
- There is no link between the same object at different representations, which makes it difficult to maintain consistent representations after updates and precludes topological queries along this dimension. For instance, finding a moving object involves a brute force search, and checking if two objects at different scales are equivalent can only be inferred indirectly.
- The geometric and topological information is stored multiple times, which is wasteful in memory and can easily lead to changes being propagated incorrectly (or not at all), resulting in inconsistencies.

Many other approaches add some topology and additional information to these independent 2D or 3D representations. For instance, event-based models [16] connect successive moments in time with the changes that occurred in them, object-relationship models [17] add information to model the changes themselves, and the original $tGAP$ structure [18] links appropriate 2D objects at different levels of detail. These representations are lightweight and sufficient for many applications, but they do not solve any of the above mentioned problems in their entirety: there is still a fixed number of points along a dimension (e.g. levels of detail or moments in time), some topological relations are not possible to keep in an efficient manner (especially along the additional dimensions), and inconsistencies are easy to generate when combining data sources or manipulating objects without special care.

Because of this, others have proposed to treat all dimensions as spatial ones (see [19] for time and [20] for scale). This solution is more complex, but it means that objects have known geometry, topology and attributes at all possible values within a range. Alternatively, this can be seen as having access to all the topological relationships between the objects, down to the vertex-to-vertex level. This helps to avoid redundancies and inconsistencies in the data. What we mean by treating all dimensions as spatial ones is explained as follows.

For simplicity, let us first consider a case with 2D space, time as the third dimension, and only linear (flat) geometries. At any one point in time, an object would be represented as a polygon in 3D space, and it would be parallel to the 2D space plane and orthogonal to the time axis. Every object existing (and not moving or changing shape) during a time period would then be a prism, with its base and top parallel to the 2D space plane and the other facets orthogonal to it. An example of this situation is shown in Figure 1.

Extending this to a 4D representation of 3D space and time, every object at one point in time would be a polyhedron in 4D space, and an object that exists for a period of time would be a polychoron, i.e. the four-dimensional analogue of a polygon/polyhedron. If this object is not moving or changing shape, it would

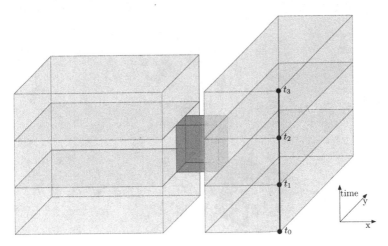

Fig. 1. A 2D space (x, y) + time (vertical axis) perspective view of the footprint of two separate buildings at time t_0, which were connected by a corridor (red) from time t_1 to time t_2 and then became disconnected again when the corridor was removed until time t_3. The moments in time are shown along the thick line representing the front right corner of the right building.

take the form of a prismatic polychoron, i.e. the four-dimensional analogue of a prism. A simple example of such an object, generated by successive extrusions of a 2D footprint from the GBKN[4] data set, is shown in Figure 2.

3   Higher-Dimensional Data Models

There are several data models that are able to support higher-dimensional objects. However, most of these are limited to point or raster data, which have trivial or no topology, and are thus much more straightforward to use and implement. Higher-dimensional point clouds are common in data mining [21], while higher-dimensional rasters are common in medical imaging [22], among other examples.

For vector data consisting of closed polytopes (i.e. the higher-dimensional analogue of a polygon/polyhedron), there are fewer options. A deceptively simple one involves the geometric subdivision of an n-dimensional polytope into n-dimensional simplices, i.e. an n-dimensional simplicial decomposition or n-dimensional triangulation, which can be easily represented and stored using an incidence model, as shown in Figure 3. In its simplest form, a data structure for this could be a list of vertex coordinates, and a list of simplices, each containing the $n + 1$ vertices that define it and the $n + 1$ simplices that are adjacent to it. This option has several advantages: the data structures are simple to use

---

[4] http://www.gbkn.nl, a Dutch large-scale topographic data set.

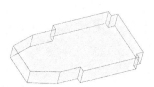

(a) The GBKN footprint of the Aula Congress Centre in Delft

(b) After extruding it to create a block shaped polyhedron (perspective projection of edges and facets only)

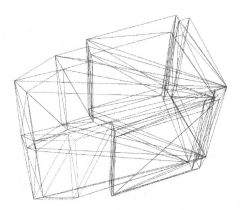

(c) After extruding it again (double perspective projection of the edges only)

Fig. 2. A 4D (3D+time) representation of the Aula Congress Centre in Delft

and implement, it can be compressed with relative ease [23, 24], and operations between simplices are much more straightforward than those between arbitrary polytopes (e.g. the intersection of two simplices of a certain dimension can yield only a limited number of different configurations).

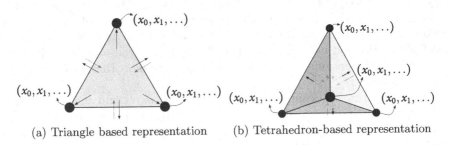

(a) Triangle based representation   (b) Tetrahedron-based representation

Fig. 3. n-simplex based data structures in 2D and 3D. Black arrows represent pointers from the element shown, while red arrows show the ones from other elements that point to this one.

However, doing this subdivision is extremely difficult in practice, since it requires the creation of an n-dimensional constrained triangulator, which has been described in theory [25, 26] for some cases, but has never been implemented. Doing so for the general case, especially in a robust manner, would prove very difficult.

Another option, and the one further discussed in this paper, is using *ordered topological models* [27], a type of boundary representations that are based on a single type of fundamental constructing element (e.g. a half-edge), on which a usually small number of pre-defined functions act. The more complex elements and connected components of such a model are only defined implicitly, e.g. as a set of fundamental elements. This allows objects to be represented as is, at least from a high level perspective, not needing to conform to a particular shape (unlike decomposition models like rasters).

Such data models also have the advantage of separating the topology of the objects (which is dealt with in the model directly), and their geometry (which is dealt with in an embedding model). This is a useful property, since it distinguishes the problems in geometric modelling from those in computational geometry [28]. Algorithms and methods from both fields can then be applied indistinctly to solve specific problems.

There are other possible models which are not discussed further in this paper but still represent interesting possibilities. Constructive models based on constructive solid geometry, alternate decompositions [29] or intersections of half-spaces have a strong theoretical background, but attaching attributes to individual elements is difficult, and realising an object involves complex geometric computations. Nef polyhedra [30] are very powerful, but require the construction of an n-dimensional hyperspherical projective kernel, which is also a very complex task in practice [31].

4   Generalised Maps

Generalised maps (sometimes shortened as G-maps) are an ordered topological model developed by Liendhardt [32] based on the concept of a combinatorial map, also known as a topological map, which was described by Edmonds [33]. They are roughly equivalent to the cell-tuple structure of Brisson [34], but were shown to be able to represent the topology of a wider class of objects, i.e. orientable or non-orientable cellular quasi-manifolds with or without boundary— manifolds partitioned into cells [35] that allow certain types of singularities, as long as every i-dimensional cell (i-cell) is incident to no more than two ($i+1$)-cells in a ($i + 2$)-cell.

A generalised map is composed of two elements: *darts* and *involutions* (α). The precise definition of a dart is complex [32], but since for our application we are only interested in representing linear (flat) geometries, a dart can be intuitively seen as a unique combination of a specific i-*cell* (a vertex being a 0-cell, an edge is a 1-cell, a facet is a 2-cell, and so on) in each dimension, all of which are incident to each other. Meanwhile, involutions are bijective operators connecting darts that are related along a certain dimension. In this manner, α_0 joins darts into edges, α_1 connects consecutive edges within a facet, α_2 connects adjacent facets within a volume, and so on. An example of a 3D generalised map (3-G-map) representation of two adjacent cubes is shown in Figure 4, where α_0 thus joins vertices to form edges, α_1 connects consecutive edges within a facet, α_2 connects adjacent facets within a volume, and so on.

One can traverse the combinatorial structure by the use of the *orbit* operator, which returns a set of darts that are reachable by following certain involutions only. To obtain the darts that are part of a certain i-cell only, one can start from any dart d belonging to the i-cell, following all involutions *except* for α_i. This is commonly denoted an $< \!\not\!\alpha_i \!>$ orbit of d [36]. Since α_i connects *adjacent* i-cells, not following it means staying within the same i-cell. For simple construction, the *sew* operation is used, connecting two objects of the same dimension along the common face, i.e. ($i - 1$)-cell, in their boundaries. Analogously, the *unsew* operation can be used to unset these involutions.

More formally, a n-dimensional generalised map is defined by a ($n + 2$)-tuple $G = (D, \alpha_0, \alpha_1, \ldots, \alpha_n)$, where D is a non-empty set of darts, and α_i is an involution (i.e. $\forall d \in D, \forall 0 \leq i \leq n, \alpha_i(\alpha_i(d)) = d$) that connects objects of dimension i, and $\forall 0 \leq i \leq n - 2, \forall i + 2 \leq j \leq n, \alpha_i(\alpha_j(d))$ is also an involution.

In order to traverse a G-map, the orbit operator $< A > (d) = < \alpha_{i1}, \alpha_{i2}, \ldots, \alpha_{in} > (d)$ obtains all the darts that can be reached from dart d by successive applications of the operators $\alpha_{i1}, \alpha_{i2}, \ldots, \alpha_{in} \in A$. For convenience, the operator $< \!\not\!\alpha_i \!> (d)$ is defined as well, which traverses all α involutions *except* for α_i [36], obtaining all the darts that are part of the same i-cell as d.

The construction of objects in its simplest form is based on the sewing operator, which joins two i-cells along the ($i - 1$)-cell that lies in their geometric common boundary. Thus, it takes two corresponding darts d_1 and d_2 ($\forall 0 \leq j \leq n, i \neq j \iff d_1$ and d_2 belong to the same j-cell) on opposite sides of the common ($i - 1$)-cell, computes their $< \alpha_0, \alpha_1, \ldots, \alpha_{i-1} > (d_1)$ and $< \alpha_0, \alpha_1, \ldots, \alpha_{i-1} >$

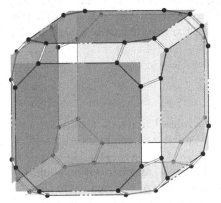

(a) A G-map representation of a cube.

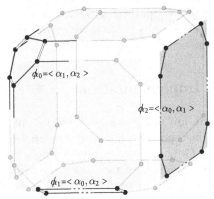

(b) The ϕ_i operator obtains all the darts belonging to a specific i-cell. Thus, ϕ_0 obtains the darts belonging to a vertex, ϕ_1 those belonging to an edge, and ϕ_2 those belonging to a facet.

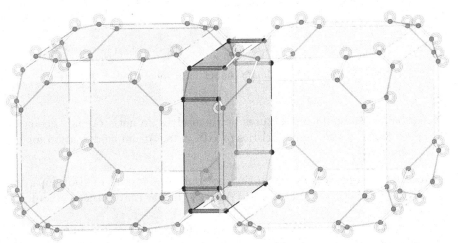

(c) A G-map representation of two cubes. Note how the individual cubes have identical involutions to those of (a), with the addition of an α_3 involution that connects the two cubes at their common face. In the other darts, this involution is not used.

Fig. 4. A 3D G-map representation of a pair of adjacent cubes, showing the α_0 (dashed red), α_1 (solid blue), α_2 (double green), α_3 (triple purple), and ϕ_i operators

(d_2) orbits, performs a parallel traversal of both, and connects them by adding α_i involutions that connect the corresponding darts from each orbit along the i-th dimension. Note that this implies the use of consistent ordering criteria in the orbit operator, such as always following the lowest possible α involution first. The unsew operator similarly uses a single $< \alpha_0, \alpha_1, \ldots, \alpha_{i-1} > (d)$ orbit (since the $(i-1)$-cells are now linked) to remove all the involutions between any darts in the orbit along the i-th dimension.

5   Implementation

There are many possible realisations of generalised maps as a data structure. For instance, a minimal data structure that stores the combinatorial aspect of an n-G-map could involve a single type of object, a `Dart` with $n+1$ pointers to other darts representing its involutions. However, another option could be to have a set of `Involutions` that store the identifiers of the two darts that each of them link. These two options are presented in Figure 5.

```
struct Dart {
    Dart *involutions [n+1];
};
```

```
struct Involution {
    id dart1, dart2;
};
```

(a) Based on darts   (b) Based on involutions

Fig. 5. A minimal G-maps implementation

Nevertheless, these data structures by themselves do not store any geometry or support many of the characteristics of GIS data. An implementation for use in GIS requires:

1. **Geometry and topology** Storing not only topological relationships, but also the geometry of the objects. At least linear (flat) objects should be supported.
2. **Attributes** Storing complex attributes of different types (e.g. numeric, text, an element of a discrete set of classes, etc.), possibly at every dimension (e.g. vertex, edge, face, etc.). Every i-cell can have a tuple of attributes of different types, but all the cells of a certain dimension generally have the same attribute types in their tuples.
3. **Construction** Constructing a model from both topological or non topological data. Topological data might need to be checked (in case the topological information does not match the actual geometry of the objects), while non topological construction should be performed in a consistent manner, generating valid topological information and ensuring that objects that are geometrically equivalent are only generated once.

4. **Queries** Answering geometric, topological and attribute based queries efficiently. In order to do this, all necessary links between the objects should be kept, and an external data structure for spatial indexing might be required as well.

5. **Holes** Storing and efficiently accessing void regions in possibly every dimension higher than 0. To ensure a consistent model, these holes should fit inside their containing object, which implies that they should be of the same dimensionality or lower.

6. **Disconnected and overlapping objects** Keeping track and traversing objects even when they form topologically disconnected groups. They might be disconnected by virtue of being geometrically disconnected, or also by being in a configuration that is not directly representable using generalised maps. This implies that a higher level structure that somehow maintains this information is required. This data structure can however have many possible forms.

The data structures presented previously are only sufficient to represent the combinatorial structure (**topology**) of a generalised map, equivalent to the topological relationships in a partition of space without holes. However, to represent the geometry and other characteristics of the model, some modifications and additional structures are needed. These are shown in Figure 6 and explained as follows.

To store **geometry**, *embedding* structures are used; each one of these containing the geometry of a specific *i*-cell. Since only linear geometries are required, only the 0-dimensional point embeddings are strictly necessary, which store the coordinates of each vertex. The geometry of the higher-dimensional embeddings can then be inferred from the points in their boundary. Since each dart represents a unique combination of an *i*-cell of each dimension, a dart can be linked to its corresponding embedding structure for each dimension. On the other direction, it is sufficient to link an embedding to any one dart representing part of its boundary.

Attributes and **holes** work in a similar manner. Since the tuples of attribute types of all *i*-cells (cells of equal dimension) are equal, one embedding data structure per dimension storing the attributes of that dimension, is sufficient. A list of holes present in that *i*-cell can be then kept as an additional attribute, its only practical requirement being that the dimensionality of the hole (represented as an embedded cell as well) should be equal or lower than that of the containing cell, and that its geometry should be fully inside the containing cell.

Meanwhile, **queries** and **disconnected objects** are handled through the use of a spatial index. For this purpose we have investigated several options, among which the most promising options are R-tree variants like the R*-tree, or a simple index using a single vertex per cell, such as the lexicographically smallest one. Usual R-tree implementations are not practical since they have problems when dealing with objects of heterogeneous dimension (e.g. an object with zero-length along a particular dimension has a volume[5] of zero in higher

---

[5] More precisely, a Lebesgue measure.

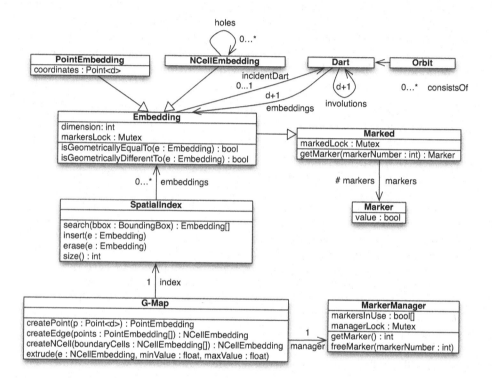

Fig. 6. Our implementation of generalised maps for real-world GIS data

dimensional space). The most important aspect of such a spatial index is that it allows us to maintain a connected graph.

Finally, we have developed two **construction** operators: a higher dimensional analogue of extrusion, which uses a d-dimensional object and a range along the $(d + 1)$-th dimension to create a $(d + 1)$-dimensional object; and an incremental construction methods that creates a $(d + 1)$-dimensional object described by the d-dimensional objects in its boundary. These are the focus of two upcoming articles and are not discussed here any further.

6   Conclusion and Future Work

This paper presents part of our ongoing research to implement a higher-dimensional GIS based on mathematical models that have been developed in other domains. This will enable the full integration of the separate dimensional aspects of GIS, such as 2D/3D space, time and scale [37]. We have modified and extended generalised maps into a data structure that is not much more complex than a basic implementation, but one that is able to support the real-world characteristics that are found in GIS data.

Our future work will cover: the visualisation of higher-dimensional data, efficient construction techniques, improved spatial indexing, keeping the consistency and validity of data, and improving the memory consumption of generalised maps.

Acknowledgements. This research is supported by the Dutch Technology Foundation STW, which is part of the Netherlands Organisation for Scientific Research (NWO), and which is partly funded by the Ministry of Economic Affairs (Project code: 11300).

References

[1] Frank, A.U.: Spatial concepts, geometric data models, and geometric data structures. Computers & Geosciences 18(4), 409–417 (1992)

[2] Hazelton, N., Leahy, F., Williamson, I.: On the design of temporally-referenced, 3-D geographical information systems: development of four-dimensional GIS. In: Proceedings of GIS/LIS 1990 (1990)

[3] Hansen, H.S.: A quasi-four dimensional database for the built environment. In: Westort, C.Y. (ed.) DEM 2001. LNCS, vol. 2181, pp. 48–59. Springer, Heidelberg (2001)

[4] O'Conaill, M.A., Bell, S.B.M., Mason, N.C.: Developing a prototype 4D GIS on a transputer array. ITC Journal (1), 47–54 (1992)

[5] Raper, J.: Multidimensional geographic information science. Taylor & Francis (2000)

[6] Worboys, M.F.: A unified model for spatial and temporal information. The Computer Journal 37(1), 26–34 (1994)

[7] Goodchild, M.F.: Geographical data modeling. Computers & Geosciences 18(4), 401–408 (1992)

[8] Peuquet, D.J.: Representations of Space and Time. Guilford Press (2002)

[9] Oosterom, P.v., Meijers, M.: Towards a true vario-scale structure supporting smooth-zoom. In: Proceedings of the 14th ICA/ISPRS Workshop on Generalisation and Multiple Representation, Paris (2011)

[10] Li, Z.: Reality in time-scale systems and cartographic representation. The Cartographic Journal 31(1), 50–51 (1994)

[11] Karimipour, F., Delavar, M.R., Frank, A.U.: A simplex-based approach to implement dimension independent spatial analyses. Computers & Geosciences 36(9), 1123–1134 (2010)

[12] Albrecht, J.: Universal Analytical GIS Operations. A Task-Oriented Systematization of Data Structure-Independent GIS Functionality Leading Towards a Geographic Modeling Language. PhD thesis, University of Vechta (1995)

[13] Thompson, R.J., van Oosterom, P.: Integrated representation of (potentially unbounded) 2D and 3D spatial objects for rigorously correct query and manipulation. In: Kolbe, T.H., König, G., Nagel, C. (eds.) Advances in 3D Geo-Information Sciences. Lecture Notes in Geoinformation and Cartography, pp. 179–196. Springer (2011)

[14] Basoglu, U., Morrison, J.: The efficient hierarchical data structure for the US historical boundary file. In: Dutton, G. (ed.) Harvard Papers on Geographic Information Systems, vol. 4, Addison-Wesley (1978)

[15] OGC: OpenGIS City Geography Markup Language (CityGML) Encoding Standard. Open Geospatial Consortium. 1.0.0 edn (August 2008)

[16] Peuquet, D.J., Duan, N.: An event-based spatiotemporal data model (ESTDM) for temporal analysis of geographical data. International Journal of Geographical Information Science 9(1), 7–24 (1995)

[17] Claramunt, C., Parent, C., Spaccapietra, S., Thériault, M.: Database modelling for environmental and land use changes. In: Stillwell, J.C.H., Geertman, S., Openshaw, S. (eds.) Geographical Information and Planning. Advances in Spatial Science, pp. 181–202. Springer, Heidelberg (1999)

[18] van Oosterom, P.: Variable-scale topological data structures suitable for progressive data transfer: The GAP-face tree and GAP-edge forest. Cartography and Geographic Information Science 32(4), 331–346 (2005)

[19] Tøssebro, E., Nygård, M.: Representing topological relationships for spatiotemporal objects. Geoinformatica 15, 633–661 (2011)

[20] van Oosterom, P., Meijers, M.: Vario-scale data structures supporting smooth zoom and progressive transfer of 2D and 3D data. In: Jaarverslag 2011. Nederlandse Commissie voor Geodesie (2012)

[21] Casali, A., Cicchetti, R., Lakhal, L.: Cube lattices: a framework for multidimensional data mining. In: Proceedings of the 3rd SIAM International Conference on Data Mining, pp. 304–308 (2003)

[22] McInerney, T., Terzopoulos, D.: A dynamic finite element surface model for segmentation and tracking in multidimensional medical images with application to cardiac 4D image analysis. Computerized Medical Imaging and Graphics 19(1), 69–83 (1995)

[23] Snoeyink, J., van Kreveld, M.: Good orders for incremental (re)construction. In: Proceedings of the 13th ACM Symposium on Computational Geometry, pp. 400–402 (1997)

[24] Blandford, D.K., Blelloch, G.E., Cardoze, D.E., Kadow, C.: Compact representations of simplicial meshes in two and three dimensions. International Journal of Computational Geometry and Applications 15(1), 3–24 (2005)

[25] Shewchuk, J.R.: Sweep algorithms for constructing higher-dimensional constrained Delaunay triangulations. In: Proceedings of the 16th Annual Symposium on Computational Geometry, pp. 350–359 (2000)

[26] Shewchuk, J.R.: General-dimensional constrained Delaunay and constrained regular triangulations, I: Combinatorial properties. Discrete & Computational Geometry 39(1003), 580–637 (2008)

[27] Lienhardt, P.: Topological models for boundary representation: a comparison with n-dimensional generalized maps. Computer-Aided Design 23(1), 59–82 (1991)

[28] Mäntylä, M.: An introduction to solid modeling. Computer Science Press, New York (1988)

[29] Bulbul, R., Frank, A.U.: AHD: The alternate simplicial decomposition of nonconvex polytopes (generalization of a convex polytope based spatial data model). In: Proceedings of the 17th International Conference on Geoinformatics, pp. 1–6 (2009)

[30] Bieri, H., Nef, W.: Elementary set operations with d-dimensional polyhedra. In: Noltemeier, H. (ed.) CG-WS 1988. LNCS, vol. 333, pp. 97–112. Springer, Heidelberg (1988)

[31] Granados, M., Hachenberger, P., Hert, S., Kettner, L., Mehlhorn, K., Seel, M.: Boolean operations on 3D selective nef complexes: Data structure, algorithms and implementation. In: Proceedings of the 11th Annual European Symposium on Algorithms, pp. 174–186 (September 2003)

[32] Lienhardt, P.: N-dimensional generalized combinatorial maps and cellular quasi-manifolds. International Journal of Computational Geometry and Applications 4(3), 275–324 (1994)

[33] Edmonds, J.: A combinatorial representation of polyhedral surfaces. Notices of the American Mathematical Society 7 (1960)

[34] Brisson, E.: Representing geometric structures in d dimensions: topology and order. In: Proceedings of the 5th Annual Symposium on Computational Geometry, pp. 218–227. ACM, New York (1989)

[35] Hatcher, A.: Algebraic Topology. Cambridge University Press (2002)

[36] Lévy, B., Mallet, J.L.: Cellular modeling in arbitrary dimension using generalized maps. Technical report, ISA-GOCAD (1999)

[37] van Oosterom, P., Stoter, J.: 5D data modelling: Full integration of 2D/3D space, time and scale dimensions. In: Fabrikant, S.I., Reichenbacher, T., van Kreveld, M., Schlieder, C. (eds.) GIScience 2010. LNCS, vol. 6292, pp. 310–324. Springer, Heidelberg (2010)

Constructing and Modeling Parcel Boundaries from a Set of Lines for Querying Adjacent Spatial Relationships

Nam Nguyen Vinh[1] and Bac Le[2]

[1] Vietnam Informatics and Mapping Corporation
nguyenvinhnam@vietbando.vn
[2] Faculty of Information Technology – University of Science
National University of Ho Chi Minh City, Vietnam
lhbac@fit.hcmus.edu.vn

Abstract. Geometrically, parcel boundaries are represented by polygons. Unfortunately, they are induced by a set of lines in some drawings (for example, CAD). In this article, we present an algorithm that polygonizes quickly and precisely a set of lines and stores the resulting polygons as well as adjacency relationships in a topological model. These pre-calculated topological relationships are efficient for basic functions, such as parcel division and parcel merge, in land parcel exploitation and management systems. Our algorithm is simple and easy to implement.

Keywords: Computational Geometry, Topology, CAD/CAM, GIS.

1 Introduction

In previous CAD systems, a few tools were available for parcel management. We could create geometry objects, such as lines and arcs, to represent the lot boundaries (see Fig.1, Fig.2) and then create a closed polyline to assist in determining the parcel area. With large drawings, polygonizing manually a set of lines is not efficient. Each parcel boundary is independent geometry object, so querying topological relationships, such as adjacency, connectivity and containment, are time-consuming and computationally costly, particularly where the data is very big.

Fig. 1. Parcel boundaries in a drawing   **Fig. 2.** Parcels boundaries are a set of lines

B. Murgante et al. (Eds.): ICCSA 2013, Part I, LNCS 7971, pp. 540–549, 2013.
© Springer-Verlag Berlin Heidelberg 2013

There are many polygon detection algorithms, however, they do not have a model that maintains spatial relationships for querying later [4, 6]. In general, pre-calculating the topological relationships is more efficient as the relationships are identified once and the resulting topology information then queries many times in parcel exploitation and management systems (e.g. what parcels is the parcel A adjacent to?). We follow this approach to design data structures for implementing our algorithm.

We used the VPF winged-edge topology data model [2] (see Fig.3) for storing spatial relationships, such as adjacency, and managing the integrity of coincident parcel boundaries. In this model, an edge knows its neighboring edges by *left edge* and *right edge* attributes. Each edge also knows its *left face* (left polygon or left parcel boundary) and *right face* (right polygon) attributes. Adjacency information for the topology model is updated incrementally at each step of the algorithm.

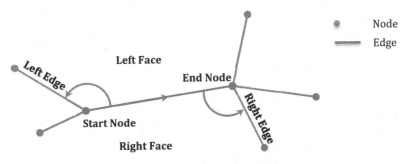

Fig. 3. Winged-edge components

To create automatically polygons (parcel boundaries) from a set of lines, we divide this task in three major steps. First, we decompose the set of lines into a collection of line segments and detect line segment intersections using plane sweep algorithm. The running time for finding intersections of a set S of N line segments in the plane is $O(NlogN + IlogN)$, where I is the number of intersection points of segments in S [1]. This process is known as line segmentation. Next, we build a planar graph (level-2 topology), in which no edges overlap [2]. *Left edge* and *right edge* attributes for each edge are updated at this step. Last, we update *left face* and *right face* attributes for edges, and then construct a set of polygons from full topology information of edges based on winged-edges algorithm [2].

The rest of the paper is organized as follows. Section 2 describes the steps of our algorithm. Section 3 presents the experimental results. Finally, we discuss conclusions and future work in section 4.

2   Algorithm

Let $S = \{s_1, s_2, s_3, \ldots, s_{n-1}, s_n\}$ be the set of line segments for which we want to polygonize and to pre-calculate adjacency relationships among resulting polygons (parcel boundaries). Fig. 4a is an example of a set of line segments and Fig. 4b illustrates a set of six resulting polygons that we expect. We construct level-3 winged-edge topology data model to solve this problem.

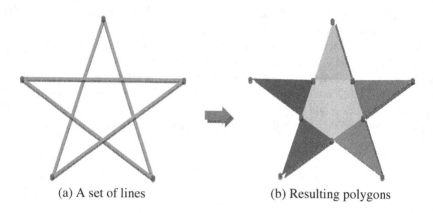

(a) A set of lines   (b) Resulting polygons

Fig. 4. Polygonize a set of lines

Our algorithm is depicted as follows:

Input: A set of line segments S and a threshold T
Output: A VPF level-3 topology model M_3 and a set of polygons P

$S_{new} \leftarrow LineSegmentation(S)$
$M_2 \leftarrow BuildLevel2Topology(S_{new}, T)$
$M_3 \leftarrow BuildLevel3Topology(M_2)$
$P \leftarrow ExtractPolygons(M_3)$

T is the distance range in which all vertices (points) are considered identical or coincident. Its value chosen depends on the precision level of input data. M_2 is the VPF model in level 2. M_3 is the VPF model in level 3. The $ExtractPolygons$ function uses information in M_3 to extract polygons (parcels).

2.1   Data Structures

Our algorithm relates to four primitives of VPF winged-edge topology data model [2]: node (connected node), edge, ring and face. These primitives have data structures as following:

- Node

```
struct TopoNode{
    int      id; // Node id
    double   x;  // coordinate X
    double   y;  // coordinate Y
    TopoEdge firstEdge; // FE
}
```

- Edge

```
struct TopoEdge{
    int       id; // Edge id
    TopoNode startNode;  // SN
    TopoNode endNode;    // EN
    TopoEdge leftEdge;   // LE
    TopoEdge rightEdge;  // RE
    TopoFace leftFace;   // LF
    TopoFace rightFace;  // RF
}
```

- Ring

```
struct TopoRing{
    int       id; // Ring Id
    TopoFace face;
    TopoEdge startEdge;
}
```

- Face

```
struct TopoFace{
    int       id; // Face Id
    TopoRing outerRing;
}
```

- TopologyModel

```
struct TopoModel{
    Hashset<TopoNode> nodes;
    Hashset<TopoEdge> edges;
    Hashset<TopoRing> rings;
    Hashset<TopoFace> faces;
}
```

2.2   Line Segmentation

A set of line segments S frequently exist many intersections between these line segments. To construct polygons, we have to compute all intersections and create a new set of line segments in which any pair of line segments shares at most one endpoint. Fig.5 shows some pairs of line segments and creates new line segments for each case.

We can simply take each pair of line segments, compute whether they intersect. This brute-force algorithm clearly requires $O(n^2)$. We want to void testing pairs of line segments that are far apart to speed up our algorithm, so we use plane sweep algorithm [1].

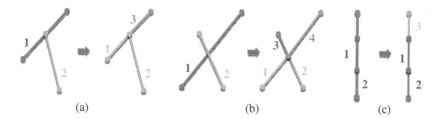

Fig. 5. Intersection cases in line segmentation

2.3   Build Level-2 Topology

At this step, our algorithm constructs a list of *TopoNode* nodes *Nodes* and a list of *TopoEdge* edges *Edges* from the new set of line segments that created at above step with the threshold T. For each node in *Nodes* we assign a value to *firstEdge* field. For each edge in *Edges* we assign values to *startNode, endNode, leftEdge* and *rightEdge* attributes. To compute these values, for each segment we create 2 segment endpoints which has the structure as following:

```
struct SegmentEndpoint{
    double  x; // x-coordinate
    double  y; // y-coordinate
    int     segmentId; // segment identification
    bool    isStart; // whether it is start point of segment
    double  angle; // resulted by this segment and x-axis at
                   //(x, y)
}
```

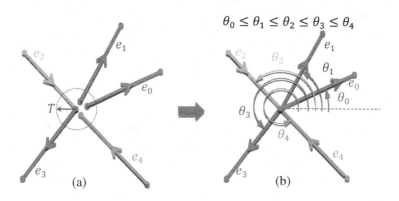

Fig. 6. Build the level-2 topology model

Our algorithm begins by sorting ascending the $2n$ segment endpoints by their x-coordinates, y-coordinates and angles, in $O(nlogn)$. Fig.6 describes how to build the level-2 topology model. Further details can be found in [3]. Topological relationships of nodes and edges in Fig.6 are shown in Table 1 and Table 2. Neighborhood vertices inside the threshold T are grouped to create a $TopoNode$ node, such as e_2^e. The edge e_0 has the smallest angle θ so e_0 is chosen as FE of e_2^e.

Table 1. Topological relationships for nodes in Fig.6b

Node	FE
e_2^e	e_0

Table 2. Topological relationships for edges in Fig.6b

Edge	SN	EN	LE	RE
e_0	e_2^e	e_0^e	e_1	
e_1	e_2^e	e_1^e	e_2	
e_2	e_2^s	e_2^e		e_3
e_3	e_2^e	e_3^e	e_4	
e_4	e_4^s	e_2^e		e_0

2.4   Build Level-3 Topology

This procedure creates a list of $TopoRing$ rings $Rings$ and a list of $TopoFace$ faces $Faces$, and updates values to $leftFace$ and $rightFace$ attributes for each edge in $Edges$. For each e_i in $Edges$, we consider   sequentially values $e_i.leftFace$ and $e_i.rightFace$. For example, we begin with $e_i.leftFace$. If $e_i.leftFace$ has not yet assigned we create a new $TopoFace$ object f_{new} and assign this object to it. e_i is the edge that belongs to f_{new}. From M_2 we know $e_i.leftEdge$ also is a part of f_{new}. We consider $e_i.leftEdge$. To determine f_{new} is assigned to $e_i.leftEdge.leftFace$ or $e_i.leftEdge.rightFace$, we find node that e_i and $e_i.leftEdge$ share. If sharing node is $e_i.leftEdge.startNode$ we assign f_{new} to $e_i.leftEdge.rightFace$, otherwise $e_i.leftEdge.leftFace$. If f_{new} is assigned to $e_i.leftEdge.rightFace$, the next edge for considering is $e_i.leftEdge.rightEdge$ otherwise $e_i.leftEdge.leftEdge$. We repeat this process until $e_i.leftFace$ is chosen to assign to.

The *BuildLevel3Topology* function is depicted as follows:

```
void BuildLevel3Topology(M){
   foreach e_i in M.Edges{
      endpoints.Add(e_i.startNode)
      endpoints.Add(e_i.endNode)
      foreach node in endpoints{
         nextEdge  ← Null
         nextNode  ← Null
         newFace   ← Null
```

```
if(eᵢ.startNode = node){
  if(eᵢ.rightFace = Null){
    newFace ← CreateNewFace()
    eᵢ.rightFace ← newFace
    nextNode ← eᵢ.endNode
    nextEdge ← eᵢ.rightEdge
  }
}
else{ // if(eᵢ.endNode = node)
    newFace ← CreateNewFace()
    eᵢ.leftFace ← newFace
    nextNode ← eᵢ.startNode
    nextEdge ← eᵢ.leftEdge
}
if(newFace ≠ Null){
  newRing ← CreateNewRing()
  newRing.face ← newFace
  newRing.startEdge ← eᵢ
  M.Rings.Add(newRing)
  newFace.outerRing ← newRing
  M.Faces.Add(newFace)
  while((nextEdge ≠ eᵢ)OR (nextNode ≠ node)){
    if(nextEdge.startNode = nextNode){
      nextEdge.rightFace ← newFace
      nextNode ← nextEdge.endNode
      nextEdge ← nextEdge.rightEdge
    }
    else{ // if(newEdge.endNode = nextNode){
      nextEdge.leftFace ← newFace
      nextNode ← nextEdge.startNode
      nextEdge ← nextEdge.leftEdge
    }
  }
}
}
}
}
}
```

The function *BuildLevel3Topology* only constructs faces without holes so number
of faces equals to number of rings in the resulting topology model. With dataset
shown in Fig.4, this function creates six polygons (faces) (see in Fig.4b) and a large
star-shaped polygon (face) containing the remain polygons. To remove unexpected
exterior polygons, we perform the following procedure:

```
void RemoveExteriorPolygons(M){
  foreach face in M.Faces{
```

```
  nodes ← GetNodesOfRing(face.outerRing)
  nodeTopLeft ← FindTopLeftNode(nodes)
  edge ← FindEdgeOfFaceShareNode(face, nodeTopLeft)
  if(edge.startNode = nodeTopLeft){
    if(edge.rightFace == face)
      Remove(face)
  }
  else{ // if(edge.endNode = nodeTopLeft)
    if(edge.leftFace == face)
      Remove(face)
  }
 }
}
```

With this procedure, we always find out the leftmost edge that contains *nodeTop-Left*. Fig.7 shows how to detect exterior polygons (faces).

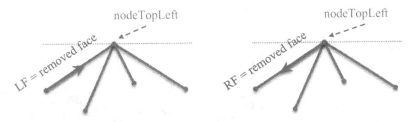

nodeTopLeft   nodeTopLeft

LF = removed face   RF = removed face

Fig. 7. Two cases that a face satisfies removal condition

2.5 Extract Polygons

With VPF level-3 topology model, it is easy to list all faces that stored in the model. All faces are defined by one ring, which is connected networks of edges that compose the face border. Each ring starts with a reference to a particular edge, and is defined by traveling in a consistent direction. Then the left and right edge on the edge are traversed, always keeping the face being defined on one side, until the ring returns to its starting edge [2].

```
void ExtractPolygons(M){
  P ← ∅
  foreach face in M.Faces{
    orderedNodes ← GetNodesOfRing(face.outerRing)
    p ← MakePolygon(orderedNodes)
    P.Add(p)
  }
}
```

3   Experiments

The experiments were done on a 3.40 GHz Intel® Core™ i7-3770 machine with 16GB main memory. The program was compiled by the Visual Studio C# 2010 compiler using level 3 optimization.

To evaluate our algorithm, we use three datasets: *VNAdmins* (a set of polygons), *HCMParcels* (a set of lines) and *USARoads* (a set of line segments, available from [5]). The size in line segments of these datasets are shown in Table 3. In addition, their shapes are shown in Fig. 8. Because of our experimental datasets in WGS84 coordinates, we choose the threshold $T = 0.0000001$.

Table 3. The dataset properties

Dataset Name	Number of segments
VNAdmins	1,574,832
HCMParcels	8,916,105
USARoads	29,166,672

(a) VNAdmins	(b) HCMParcels	(c) USARoads

Fig. 8. Visualize our datasets

Table 4 shows the number of nodes, edges, rings and faces stored in VPF level-3 topology model for each real dataset after applying our algorithm. The number of edges is less than the set of input line segments because our algorithm merges pairs of edges that share degree 2 nodes.

Table 5 shows spent CPU time at each step of the algorithm.

Table 4. Resulting topology model for each dataset

Dataset Name	Nodes	Edges	Rings	Faces
VNAdmins	1,559,963	1,574,824	15,517	15,517
HCMParcels	7,174,685	8,879,992	1,752,009	1,752,009
USARoads	16,712,376	21,619,341	4,889,241	4,889,241

Table 5. Execution time of our algorithm for each dataset

Dataset Name	Level-2 Topo	Level-3 Topo	Total Time
VNAdmins	6,622 ms	31 ms	6,653 ms
HCMParcels	53,067 ms	4,955 ms	58,022 ms
USARoads	507,680 ms	33,388 ms	541,068 ms

4   Conclusions and Future Work

This paper introduces a parcel detection algorithm based on VPF level-3 topology model. Adjacency relationships stored in the model are efficient for basic functions, such as parcel division and parcel merge, in land parcel exploitation and management systems. The proposed algorithm can work with large datasets. At present, our algorithm consumes much cost for removing coincident edges and only extracts polygons without holes. These drawbacks will research and solve in future.

References

1. de Berg, M., Cheong, O., van Kreveld, M., Overmars, M.: Computational Geometry - Algorithms and Applications, 3rd edn. Springer (2008)
2. Department of Defense Interface Standard for Vector Product Format, http://earth-info.nga.mil/publications/specs/printed/2407/2407_VPF.pdf
3. Nam, N.M., HoaiBac, L., Nam, N.V.: An Extreme Algorithm for Network-Topology Construction Based on Constrained Delaunay Triangulation. In: Proceedings of KSE 2009, pp. 179–184 (2009)
4. Ferreira, A., Fonseca, M.J., Jorge, J.A.: Polygon Detection from a Set of Lines. In: Proceedings of 12 o EncontroPortuguês de ComputaçãoGráfica (12th EPCG) (2003)
5. USARoads dataset, http://www.dis.uniroma1.it/challenge9/download.shtml
6. Krivograd, S., Trlep, M., Žalik, B.: A rapid algorithm for topology construction from a set of line segments. In: Proceedings of the 5th WSEAS International Conference on Telecommunications and Informatics, pp. 133–138 (2006)

Tiling 3D Terrain Models

Nuno Oliveira[1] and Jorge Gustavo Rocha[2]

[1] PT Inovação SA, Aveiro, Portugal
nuno-miguel-oliveira@ptinovacao.pt
[2] Universidade do Minho, Braga, Portugal
jgr@di.uminho.pt

Abstract. W3DS clients should be able to request 3D scenes, either using the GETSCENE or GETTILE operations, and directly display them without any additional manipulation or geographic positional correction.

In this paper we will discuss how tiles should be served to W3DS clients, since 3D tiles are more difficult to manage than 2D tiles. While 2D tiles just need to be placed side by side, 3D tiles also have to fit along the z axis.

The volume and complexity of 3D information requires more complicated logic on the client side. Clients can decide that geographic features far away from the viewer can have less detail than those near the observer. For that reason, 3D clients may want to join together tiles of different resolutions.

We will show an algorithm to perfectly slice 3D terrain models using the GDAL and CGAL libraries. It can be used in existing W3DS services to improve the quality of visualisations.

Keywords: Delaunay triangulation, Tessellation, Terrain models, Web 3D Service.

1 Introduction

3D information modelling has improved in recent years and Open Geospatial Consortium (OGC) standards like GML 3 and CityGML (which is a GML 3 profile) are becoming widely adopted. However, 3D visualisation lacks consensual approaches.

The evolution of the web, and specifically WebGL support in HTML5, has created a new opportunity to evaluate the Web 3D Service (W3DS) as an option for 3D visualisation. We developed an open source W3DS implementation on top of Geoserver [2], as a community module, based on the draft specification 0.4.0 [8].

The success of such a service will not only depend on the operations provided. It will depend on the availability of web based clients able to take advantage of WebGL support. The upstream work flow is also important. W3DS servers must be able to manage large amounts of 3D data to deliver scenes to be rendered by the client.

B. Murgante et al. (Eds.): ICCSA 2013, Part I, LNCS 7971, pp. 550–561, 2013.

In this paper we focus on the preparation of large 3D data sets, to be delivered as multi-resolution tiles. Tiles should be ready to be rendered, without any additional effort from the client.

But 3D data differs from 2D data. To create great 2D mosaics, we put the tiles side by side, according to their origin and resolution. Stitching 3D tiles is more difficult.

Side by side, 3D tiles must be well connected in the 3D space. The user should not notice where the tiles are joined together. This requirement is even more difficult to accomplish when we need to stitch 3D tiles of different resolutions.

In 2D stitching, in Google Maps, Bing Maps, OpenStreetMap, etc, we never use tiles of different scale or resolutions in the same map. When we zoom in, we zoom in the entire view equally. When we zoom out, all the tiles are replaced by others at a larger scale, but all requested tiles are at the same scale.

In 3D visualisation, where the amount of data can be quite large, it is desirable to keep low resolution data in places far from the user's point of view and have more detailed data where the user is focused.

In 3D, objects can become very far away even with a small rotation of the camera. In perspective views, objects are at different distances from the users. It makes sense to provide more detailed 3D data just for the objects closest to the user, and take advantage of the greater distance of some objects to provide less detailed models of them.

2   Background and Related Work

3D data is more difficult to model and to process. 3D navigation is complicated for the majority of users. But 3D models can be important in many application scenarios. It is not only used to visualise amazing 3D city models.

Spatial Data Infrastructures (SDI) are already adapting the procedures to incorporate 3D in their building blocks. A major initiative in the Netherlands, at the national level, is described in [14].

Incorporating 3D spatial data in national SDI will require major improvements at different levels. Map servers and clients for 3D are required and should be gradually integrated into the existing SDI.

2.1   Related OGC Activities

OGC working groups have already developed technologies and work flows to support spatial data infrastructures with a requirement for rapid visualisation of extremely large and complex 3D spatial data. From May to October 2011, OGC developed the first 3D Portrayal Interoperability Experiment (3DPIE).

Client Models. For the portrayal of spatial data, OGC employs a well-known four level visualisation pipeline, from the objects in the database to its representation on a display device. This pipeline can combine components residing in different servers. When developing client server applications for visualisation of geographic information, we can decide what components reside on the server side, and what components are delegated to the client side. As a consequence of this decision, clients can be more or less complex.

In Fig. 1, taken from the W3DS standard draft proposal, three client models are considered.

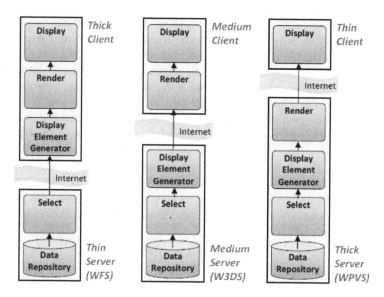

Fig. 1. Balancing schemes between client and server

Thick clients have direct access to features. They have more flexibility to manipulate the objects and produce sophisticated visualisations. We can consider globes as clients in this category, since globes usually provide the necessary logic to get and integrate features from different sources.

On the contrary, thin clients only access already rendered scenes. The client's logic can be very simple.

In between, medium clients have access to 3D scenes. They don't need to manipulate features. All features are already grouped in scenes. They have the flexibility to render the scene taking advantage of display graphic capabilities.

2.2 Virtual Globes

Virtual globes are good examples of thick clients. Google Earth and NASA World Wind, just to name two of the many virtual globes, are being used as a tool in geosciences. Globes are been used not only to visualise data, but to model complex processes [3]. To support such processes, globes can manipulate 3D geographic features, like terrains, buildings, etc.

With the WebGL support enabled in browsers, web based virtual globes are emerging. OpenWebGlobe [9] is able to provide much of the functionality of former globes, through the browser.

Virtual globes can play an important role in Spatial Data Infrastructures [10]. Globe evaluations are described in the literature, but these evaluations are mostly focused on the number of out-of-the-box features, and comparisons are always made against Google Earth [15].

Despite the number of virtual globes available, there is no standard API for globes. To model some specific geographical phenomena or to customise some specific interactions, developers must follow each virtual globe's API.

2.3 W3DS Clients

W3DS clients should be lighter than thick clients. According to the specification, "the Web 3D Service (W3DS) is a portrayal service for three-dimensional geodata such as landscape models, city models, textured building models, vegetation objects, and street furniture". W3DS clients should be flexible enough to provide interactive visualisations of 3D scenes. W3DS clients can perform much faster than thick clients, like virtual globes. No additional processing is required to display the scenes delivered by the server.

W3DS Operations. Information about the service, the supported operations, available layers and their properties, can be retrieved using the GETCAPABILITIES operation. Two additional operations that return information about features and their attributes are provided: GETFEATUREINFO and GETLAYERINFO.

Two operations are provided to return 3D data: GETSCENE and GETTILE. These two operations differ essentially in how the features are selected. GETSCENE allows the definition of an arbitrary rectangular box to spatially filter the features to compose the scene returned to the client. GETTILE returns a scene on-the-fly formed by features within a specific delimited cell, within a well-defined grid.

All five proposed operations, tagged as mandatory or optional are listed in Tab. 1.

Interactive scenes with terrain and relevant 3D geographic features will result from multiple GETSCENE and GETTILE requests to the service. These might be mostly called operations.

Table 1. W3DS operations

Operation	Use
GetCapabilities	mandatory
GetScene	mandatory
GetFeatureInfo	optional
GetLayerInfo	optional
GetTile	optional

W3DS Implementations. Until now, two implementations of the W3DS were known.

One is supporting the OpenStreetMap-3D Europe project, carried out by the GIScience group from Heidelberg University, Germany. The service is up and running, and can be explored with a java based client called XNavigator. The project not only contributed to the development of the W3DS service, but also became a major contribution to the OpenStreetMap project, since it opened a new opportunity to enhance the OSM with 3D building models [5].

The OpenStreetMap-3D also contributed with a new approach to enhancing 3D terrain models by integrating the roads directly into the triangulation system through the correction of the surface [12]. But due to limitations of the algorithm used, tiles in OSM-3D may not fit perfectly side by side, as shown in Fig. 2.

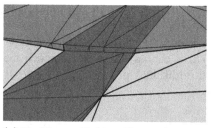

(a) Problems stitching tiles at the same resolution   (b) Problems stitching tiles at the different resolutions

Fig. 2. Tiles in OSM-3D may not fit perfectly at lower scales

Another W3DS implementation is CityServer3D [7] from the Fraunhofer Institute for Computer Graphics Research in Darmstadt, Germany. CityServer3D is more than a server, since it consists of several components to manage 3D city models.

2.4   Web View Service

The Web View Service (WVS) [6] follows previous efforts related with the Web Terrain Server (WTS) and Web Perspective View Service (WPVS). WVS delivers 3D spatial data as rendered images. Thin clients can access virtual 3D worlds through images. It can be considered the 3D equivalent of 2D map services.

3   3D Tiling

The usual way to access spatial data, in the Web Map Service (WMS) service for example, is to request one or more layers with the desired styles, within a certain bounding box. While this is the most flexible way to get the rendered maps, in some applications Web Map Tile Service (WMTS) are preferred over

WMS when performance is important. When data does not often change, this allows the offline preparation of the tiles, freeing the WMS from rendering the same area over and over again. Tiles are rendered once and served many times. WMTS trades the flexibility of custom rendered maps for fast delivery of fixed tiles.

Delivering fixed sets of tiles also enables simple scalability mechanisms. Tile caches can be distributed among different servers.

While the WMTS provides a complementary approach to the WMS for tiling maps, the W3DS includes both the flexibility of arbitrary scene creation though the GETSCENE operation and the fast delivery of tiles thought the GETTILE operation.

W3DS clients should be able to request tiles and display them directly without any additional 3D manipulation or geographic positional correction.

3.1 Basic Tiling Algorithm

Although we can create tiles from any source supported by Geospatial Data Abstraction Library (GDAL) [16], we will describe our algorithm starting from the simplest source, which is a list of points with elevation.

The first step is to create the Triangle Irregular Network (TIN) from the points, using the Delaunay triangulation. This triangulation is done using the Computational Geometry Algorithms Library (CGAL) library [1]. The Delaunay triangulation in CGAL is high configurable. For the TIN generation, we only need the 2.5 properties of the terrain, and the simple Delaunay triangulation applied to 2.5 data is adequate and the fastest approach.

The generated TIN can became quite large. In our approach, as we will show, we need to calculate the overall TIN. It is necessary to calculate this large TIN before dividing it into tiles.

To divide the TIN in tiles, we use 4 vertical planes to cut each tile. These planes start from the ground (elevation zero) and go to the highest possible elevation. Interception points are calculated. These are points on the edges that cross the vertical planes. This process is illustrated in Fig. 3.

Special care must be taken to create the four corners. These corner points are the ones in the vertical line where the vertical planes intercept. The corner is the point where that vertical line intercepts the TIN. It might be on a triangle that has no points inside the tile. The same corner can be shared by 4 different tiles.

All interceptions points will be used by both adjacent tiles divided by the same plane. If we keep these points in each tile, we guarantee that they will stitch perfectly.

Auxiliary Data Structure. The tiling algorithm described can scale quite well, maintaining a constant time per tile, if an adequate data structure is provided. We created a simple spatial index, called *InitialGrid*, to access all triangles that might be within one or more adjacent tiles. Using this index, for each tile, we

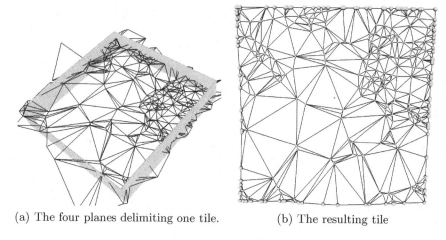

(a) The four planes delimiting one tile.   (b) The resulting tile

Fig. 3. Cutting tiles

only intercept the four planes with a small fraction of all triangles. The amount of time to cut each tile is thus constant, with a complexity of $O(1)$, in the big-O notation.

3.2 Joining Tiles Generated at different Levels

Tiles can be hierarchically organised. One tile can be split into four other tiles, occupying the same surface, but with higher accuracy (with more mass points or triangles within the same area). The notion of level is independent of the Level Of Detail (LOD) used in the GETSCENE operation.

Tiles can and should be generated at different levels. To do so, we start with the most accurate data. The described algorithm is used to provide the tiles for the higher level. Only afterwards, the next lower level of tiles is computed. The next lower level will occupy the area of four existing tiles. The interception points calculated in the previous levels around each four tiles are preserved. All other mass points within the four tiles are used to compute the lower level tile. So, the Delaunay triangulation is done to compute the lower level TIN, considering less mass points, but preserving all points of the border. For each lower level, the user can decide how many mass point are discarded (values like 1/4 or even 1/8 have been used, preserving the surface shape, while significantly reducing the number of triangles in each tile.

With such an algorithm, we preserve the points used in different tile levels. If the points are preserved, tile stitching will be perfect even when we put tiles from different levels side by side, as shown in Fig. 4.

Tile Storage. After being calculated, each tile is stored in a spatial database. Each tile is a row in the database and can be retrieved by its level, row and column number as keys. Alternatively, tiles can be stored as files, and served

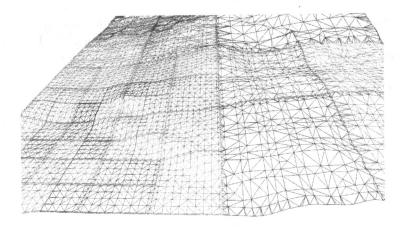

Fig. 4. Perfect composition of several tiles, at different resolutions

from the file system. The hierarchical organisation of the file system can use the three different keys (level, row and column) to organise the tiles in folders within folders.

4   Usage and Results

The GETTILE operation was implemented on top of GeoServer as a community module to take advantage of all the existing logic regarding the web service implementation. The module is open source and is available from the *github* repository.

A new service was created, called W3DS, with all operations defined in the draft proposal. The GETFEATUREINFO operation still has some limitations, though.

The GETCAPABILITIES operation provides the service's metadata to clients. Clients must know basic service and layer metadata, prior to any request.

Tiles can only be requested from layers tagged in the GETCAPABILITIES document with the key `<w3ds:Tiled>true</w3ds:Tiled>`. They must also have a `<TileSet>` definition associated, as illustrated in the following `<TileSet>` definition.

```
<w3ds:TileSet>
    <ows:Identifier>guimaraes</ows:Identifier>
    <w3ds:CRS>EPSG:27492</w3ds:CRS>
    <w3ds:TileSizes>4000 2000 1000 500 250</w3ds:TileSizes>
    <w3ds:LowerCorner>-17096.156 193503.057</w3ds:LowerCorner>
</w3ds:TileSet>
```

Basically, the `<TileSet>` provides the lower left corner of the grid, and all available tile sizes. Tile sizes are defined as an ordered list decreasing by tile size. The number of levels available is the length of the list. Levels are numbered from 0, starting at the largest tile size. Obviously, tiles are considered of equal size in both axes, according to the draft specification version 0.4.0. The version 0.4.1 introduced the possibility of supporting non-rectangular tiles.

4.1   W3DS Gettile Operation

The GETTILE parameters are listed in Tab. 2. These can be submitted either encoded as key-value pair in a GET request, or as an XML formatted document, in a POST request.

Table 2. GETTILE parameters

name	definition	use
crs	Coordinate Reference System of the returned tile	mandatory
layer	Identifier of the layer	mandatory
format	Tile encoding format	mandatory
tileLevel	Level of requested tile	mandatory
tileRow	Row index of requested tile	mandatory
tileCol	Column index of requested tile	mandatory
style	Identifiers of server styles to be applied	optional
exceptions	Format of exceptions	optional

Common GETTILE requests, with key and values pairs, are like this one:

```
http://localhost:9090/geoserver/w3ds?
    version=0.4&
    service=w3ds&
    request=GetTile&
    CRS=EPSG:27492&
    FORMAT=model/x3d+xml&
    LAYER=guimaraes&
    TILELEVEL=1&
    TILEROW=5&
    TILECOL=7
```

In our implementation, the tile encoding format can be X3D or HTML5. X3D is an XML encoded file format for representing 3D graphics. It is successor to the Virtual Reality Modeling Language (VRML). X3D is the ISO standard ISO/IEC 19775/19776/19777. It has been developed by the Web3D Consortium.

We also use the HTML5 format over X3D. With the HTML5 container, the GETTILE result can be viewed directly on a browser supporting WebGL, such as Mozilla Firefox or Google Chrome.

The integration of X3D within the HTML5 is provided by the open source X3DOM framework provided by the Fraunhofer Institute [4]. The HTML5 container will return a minimal HTML document with a header that includes:

```
<link rel="stylesheet"
    href="http://www.x3dom.org/x3dom/release/x3dom.css"/>
<script type="text/javascript"
    src="http://www.x3dom.org/x3dom/release/x3dom.js">
```

The HTML5 encoding comes in very handy for previewing data directly in the Geoserver administration interface. As soon as the administrator publishes the tiled layer it can be previewed from the administration interface.

Styling Tiles. The style optional parameter can specify one or more styles to be used. These must be chosen from the styles indicated in the GETCAPABILITIES response. If no style is provided in the request, the default style associated with the layer will be used. In the implementation of the GETTILE described here, we support image textures over the tiles, coming from static files (on the server) or from any WMS service. The exact boundaries of the tiles are appended to the GETMAP request against the remote WMS. For example, we can use for the same tile set one style that uses aerial imagery and another one with OpenStreetMap rendered images, as shown in Fig. 5.

(a) Aerial imagery over tiled X3D.   (b) OpenStreetMap rendered image over the same tile.

Fig. 5. The same X3D tile rendered with different styles

We are unable to support textures from other tile services (WMTS). This would require additional logic unless the tile set grids are exactly the same, i.e. the origin, the levels and each tile size are exactly the same.

5   Conclusions and Outlook

3D tile services enable visualisation of large 3D datasets. Additional logic on the client side is responsible for managing the tiles and their resolution levels, providing the user with a good experience while minimising the data transferred from the server.

While GETSCENE is important for providing selected 3D data on-the-fly, GETTILE is no less important in efficiently delivering less updated data, like terrain models, that can be pre-processed to improve the server's performance.

In this paper we discuss not only the GETTILE implementation, but also tile preparation. In tile preparation, special care must be paid to ensure the good stitching of the tiles. In 3D visualisation environments, tiles at different levels can be stitched, as opposed to 2D maps where only tiles of the same zoom level are displayed side by side.

The implementation described, either JAVA code contributions to GeoServer or C++ code using CGAL library, is released under GNU general public license.

5.1   Outlook

SDI will integrate more and more 3D spatial data and all related policies and technologies to manage 3D. New forms of displaying maps far beyond the historical 2D visualisation will be supported by web browsers. Since WebGL promises to have a major impact on the gaming industry, it will also have a major impact on GIS visualisation technologies.

But there are some doubts and challenges ahead. The HTML5 final specification will be only released in 2015. Until then, several parallel and complementary initiatives will exist, like X3D [4] and XML3D [13]. It is difficult to anticipate which technologies will be widely adopted.

The release of the described W3DS implementation as a GeoServer community module, available as open source, can play an important role towards W3DS consolidation. Open source implementations can attract a larger population of users. This larger community can inspect, validate, modify and improve the code. This process will ensure the reliability of the implementation. With trusted implementations, W3DS technology will start to appear at the SDI level.

The availability of an open source specification can also enable users to play with their own data. Different datasets and different use cases are very important for fully testing the implementation and the W3DS draft specification.

There are more features that could be included in the current W3DS draft specification, but it is wise to release version 1 and postpone other features for future versions. For example, a complementary extended SLD specification for 3D would improve the W3DS, as also suggested in [11]. Streaming support in W3DS could also enable effective fly-throughs, which is another feature of HTML5 that we can take advantage of.

References

[1] Cgal, Computational Geometry Algorithms Library, http://www.cgal.org
[2] Geoserver, open source Java-based map server, http://geoserver.org
[3] Bailey, J.E., Chen, A.: The role of Virtual Globes in geoscience. Computers & Geosciences 37(1), 1–2 (2011)
[4] Behr, J., Michael, Z.: X3DOM A DOM-based HTML5 / X3D Integration Model. Integration the Vlsi Journal 1(212), 127–136 (2009)
[5] Goetz, M., Zipf, A.: OpenStreetMap in 3D – Detailed Insights on the Current Situation in Germany. In: Gensel, J., Josselin, D., Vandenbroucke, D. (eds.) City, Proceedings of the AGILE 2012 International Conference on Geographic Information Science, Avignon, April 24-27, vol. (1), pp. 3–7 (2012)

[6] Hagedorn, B., et al.: Web View Service Discussion Paper. Technical report, Open Geospatial Consortium, Inc (2010)

[7] Haist, J., Coors, V.: The W3DS-Interface of Cityserver3D. In: Gröger, G., Kolbe, T. (eds.) Computer, vol. 49, pp. 63–67. EuroSDR Publication (2005)

[8] Kolbe, T.H., Schilling, A., Zipf, A., Neubauer, S., et al.: Draft for Candidate OpenGIS Web 3D Service Interface Standard. Technical report, Open Geospatial Consortium, Inc. (2010)

[9] Loesch, B., Christen, M., Nebiker, S.: OpenWebGlobe - an open source SDK for creating large-scale virtual globes on a WebGL basis. In: International Archives of the Photogrammetry Remote Sensing and Spatial Information Sciences XXII ISPRS Congress (2012)

[10] Nebiker, S., Bleisch, S., Gülch, E.: Virtual Globes. GIM International 24(7) (2010)

[11] Neubauer, S., Zipf, A.: Suggestions for Extending the OGC Styled Layer Descriptor (SLD) Specification into 3D – Towards Visualization Rules for 3D City Models. In: Applied Sciences, p. 133 (2007)

[12] Schilling, A., Lanig, S., Neis, P., Zipf, A.: Integrating Terrain Surface and Street Network for 3D Routing. In: Lee, J., Zlatanova, S. (eds.) 3D GeoInformation Sciences. Lecture Notes in Geoinformation and Cartography, pp. 109–126. Springer, Heidelberg (2009)

[13] Sons, K., Klein, F., Rubinstein, D., Byelozyorov, S., Slusallek, P.: XML3D: interactive 3D graphics for the web. In: Slusallek, P., Yoo, B., Polys, N. (eds.) Web3D 10 Proceedings of the 15th International Conference on Web 3D Technology, vol. 1, pp. 175–184. ACM (2010)

[14] Stoter, J., Vosselman, G., Goos, J., Zlatanova, S., Verbree, E., Klooster, R., Reuvers, M.: Towards a National 3D Spatial Data Infrastructure: Case of The Netherlands. Photogrammetrie Fernerkundung Geoinformation 2011(6), 405–420 (2011)

[15] Walker, M.: Comparison of Open Source Virtual Globes About Sourcepole. Source (2010)

[16] Warmerdam, F.: gdal, Geospatial Data Abstraction Library,
http://www.gdal.org

Mining Serial-Episode Rules
Using Minimal Occurrences with Gap Constraint

H.K. Dai and Z. Wang

Computer Science Department, Oklahoma State University
Stillwater, Oklahoma 74078, U.S.A.
{dai,wzhu}@cs.okstate.edu

Abstract. Data mining is a task of extracting useful patterns/episodes from large databases. Sequence data can be modeled using episodes. An episode is serial if the underlying temporal order is total. An episode rule of associating two episodes suggests a temporal implication of the antecedent episode to the consequent episode. We present two mining algorithms for finding frequent and confident serial-episode rules with their ideal occurrence/window widths, if exist, in event sequences based on the notion of minimal occurrences constrained by constant and mean maximum gap, respectively. A preliminary empirical study that illustrates the applicability of the episode-rule mining algorithms is performed with a set of earthquake data.

Keywords: data mining, episode-rule mining algorithms, minimal occurrences, gap constraint.

1 Preliminaries

Data mining is a task that extracts or "mines" knowledge from large amounts of data. Sequence data mining is one of the branches of data mining where the data can be viewed as a sequence of events and each event has an associated time of occurrence. Examples of such data are telecommunication network alarms, occurrences of recurrent illnesses, Web-site traversal actions, etc [1].

Sequence data can be modeled using episodes [6]. An episode is an acyclic directed graph representing a temporal order of a group of events. Episodes can be classified based on the topologies of their underlying acyclic directed graphs — according to the nature of temporal ordering among the events. An episode is parallel (serial) if the underlying temporal order is trivial (total, respectively), and complex episodes are composed via hierarchical structures of parallel and serial episodes.

The Apriori algorithm [2] [3] [7] can be used in mining frequent episodes. It employs a bottom-up strategy and utilizes the known/prior information to reduce the search space. There are two main steps in each iteration of the algorithm, candidate generation and frequent-episode recognition. The Apriori algorithm terminates when there is no more candidate generated.

B. Murgante et al. (Eds.): ICCSA 2013, Part I, LNCS 7971, pp. 562–572, 2013.

When working on sequence data mining, it is always desirable to maintain the relative sequence information in the output of date mining in terms of input data sequence. Serial episode is capable of keeping sequence information. However, the episode configuration is rigidly restricted and has to be totally ordered. The capability of expression of the serial episode therefore is limited to certain real-world scenarios. Parallel-episode mining does not take the event sequence information into account and is less useful in sequence data mining area. Complex episode is ideal for modeling the real world. However, mining complex episodes is not a trivial task, due to the topological complexity of episode generation and the computational complexity of episode recognition [4].

Denote by $\mathbb{N}$, $\mathbb{P}$, and $\mathbb{R}$ the sets of all nonnegative integers, all positive integers, and all reals, respectively.

Given a set E of event types, a (unit-time) event is an ordered pair $(A, t) \in E \times \mathbb{N}$ where A and t denote the event type and the occurrence (start) time of the event, respectively. We assume that all events last one time unit. An event sequence on E is a triple $(S, T_{(1)}, T_{(2)})$ such that: (1) for some $n \in \mathbb{P}$, $S = \{(A_i, t_i)\}_{i=1}^{n}$ with (A_i, t_i) representing the i-th event for each $i \in \{1, 2, \ldots, n\}$ and and $t_i \leq t_{i+1}$ for all $i \in \{1, 2, \ldots, n-1\}$, and (2) $T_{(1)}, T_{(2)} \in \mathbb{N}$ with $(\emptyset \neq)[T_{(1)}, T_{(2)}]$ denoting the time interval containing the occurrence times of all events of S, that is, $t_i \in [T_{(1)}, T_{(2)}]$ for all $i \in \{1, 2, \ldots, n\}$. A window on an event sequence $(S, T_{(1)}, T_{(2)})$ is an event sequence $(S', T'_{(1)}, T'_{(2)})$ such that $[T'_{(1)}, T'_{(2)}] \subseteq [T_{(1)}, T_{(2)}]$ and S' consists of all the events of S with occurrence times in $[T'_{(1)}, T'_{(2)}]$, that is, $S' = \{(A, t) \in S \mid t \in [T'_{(1)}, T'_{(2)}]\}$. The width of a window $(S, T_{(1)}, T_{(2)})$ is the time duration $T_{(2)} - T_{(1)}$ of the window.

An episode α in an event sequence $\mathcal{E}$ on an event type E is a triple $(V, \prec, \epsilon)$ such that $(V, \prec)$ is an acyclic directed graph with: (1) vertex set V: vertices associated with the event types of the underlying events of α via the function $\epsilon : V \to E$, and (2) edge set $\prec$: antireflexive, antisymmetric, and transitive temporal order on the underlying events of α preserved by ϵ, that is, for all vertices $v_1, v_2 \in V$, if $v_1 \prec v_2$ and $(\epsilon(v_1), t_1)$ and $(\epsilon(v_2), t_2)$ are two events of $\mathcal{E}$, then $t_1 < t_2$. The order of α, denoted by $|\alpha|$, is the order of the underlying acyclic directed graph, that is, $|\alpha| = |V|$.

For two episodes $\alpha_i = (V_i, \prec_i, \epsilon_i)$ for $i \in \{1, 2\}$ in an event sequence $\mathcal{E}$, α_1 is a subepisode of α_2 (equivalently, α_2 is a superepisode of α_1) if the underlying graph structure of α_1 is a subgraph of that of α_2, that is, there exists an injection $f : V_1 \to V_2$ such that $\epsilon_1 = \epsilon_2 \circ f$ and f preserves $\prec_1$ on V_1: for all $v, w \in V_1$, if $v \prec_1 w$, then $f(v) \prec_2 f(w)$.

An episode $\alpha = (V, \prec, \epsilon)$ is parallel (serial) if its temporal order $\prec$ on V is trivial/empty (total, respectively). Complex episodes are constructed via hierarchical combinations of parallel and serial (sub)episodes. For two episodes $\alpha_i = (V_i, \prec_i, \epsilon_i)$ for $i \in \{1, 2\}$, we may assume that $V_1 \cap V_2 = \emptyset$ via renaming. The parallel combination of α_1 and α_2 is the (super)episode $(V_1 \cup V_2, \prec_1 \cup \prec_2, \epsilon)$ where ϵ satisfies that $\epsilon|_{V_i} = \epsilon_i$ for $i \in \{1, 2\}$. The serial combination of α_1 and α_2, denoted by $\alpha_1 \diamond \alpha_2$, is the (super)episode $(V_1 \cup V_2, \prec, \epsilon)$ where $\prec = \prec_1 \cup \prec_2$

$\cup\{(v_1, v_2) \mid \text{outdegree}_{\alpha_1}(v_1) = 0 \text{ and indegree}_{\alpha_2}(v_2) = 0\}$ and $\epsilon|_{V_i} = \epsilon_i$ for $i = 1, 2$.

Our study is focused on serial episodes and their association rules, we provide all pertinent terminologies, notions, and notations as follows.

A serial episode $\alpha = (V, \prec, \epsilon)$ of order n with a totally $\prec$-ordered chain $v_1 \prec v_2 \prec \cdots \prec v_n$ is represented by the corresponding sequence of event types: $(\epsilon(v_1), \epsilon(v_2), \ldots, \epsilon(v_n))$. The prefix and suffix of an order-n serial episode $\alpha = (e_1, e_2, \ldots, e_n)$, denoted by $\text{pref}(\alpha)$ and $\text{suff}(\alpha)$ respectively, are the subepisodes of α: $(e_1, e_2, \ldots, e_{n-1})$ and (e_n), respectively.

For two serial episodes α_i for $i \in \{1, 2\}$, the episode (association) rule of α_1 and α_2, denoted by $\alpha_1 \Rightarrow \alpha_2$, suggests a temporal implication of the antecedent events of α_1 to the consequent events of α_2 — subject to a maximum-gap constraint and two measures applied to the episodes α_1 and $\alpha_1 \diamond \alpha_2$: frequency and confidence.

For an order-m serial episode $\alpha = (e_1, e_2, \ldots, e_m)$ in an event sequence $\mathcal{E} = (S = \{(A_i, t_i)\}_{i=1}^{n}, T_{(1)}, T_{(2)})$, a maximum-gap constraint imposed on α is a (user-prescribed) maximum-gap threshold $\text{MaxGap} (> 0) \in \mathbb{R}$ that asserts a temporal constraint (maximum elapsed time) between consecutive events of α.

In our study, we consider two types of maximum-gap constraint: type-1 of constant MaxGap and type-2 of mean MaxGap for our serial-episode-rule mining algorithms, and type-3 of ∞-MaxGap (without constraint) that facilitates the mining algorithms.

The serial episode α occurs in the event sequence $\mathcal{E}$ if there exists a subsequence $\{(A_{i_j}, t_{i_j})\}_{j=1}^{m}$ of S such that $\{(A_{i_j}, t_{i_j})\}_{j=1}^{m} = \{e_j\}_{j=1}^{m} (= \alpha)$ and, if $m \geq 2$,

type-1 (constant MaxGap): $(0 <) t_{i_{j+1}} - t_{i_j} \leq \text{MaxGap}$
    for all $j \in \{1, 2, \ldots, m-1\}$,

type-2 (mean MaxGap): $(0 <) \frac{t_{i_m} - t_{i_1}}{m-1} \leq \text{MaxGap}$, and

type-3 (∞-MaxGap): $(0 <) t_{i_{j+1}} - t_{i_j} < \infty$
    for all $j \in \{1, 2, \ldots, m-1\}$.

The three types of maximum-gap constraint can be ranked from strongest to weakest: type-1 < type-2 < type-3 and each of the two types with non-trivial maximum-gap constraint has its own implication for applications. We denote the context of type-τ maximum-gap constraint, where the type number $\tau \in \{1, 2, 3\}$, by the subscript τ.

Accordingly, for a type number $\tau \in \{1, 2, 3\}$, we denote by the time interval $[t_{i_1}, t_{i_m}]_{\tau}$ the type-τ occurrence of α in $\mathcal{E}$ — with occurrence width $t_{i_m} - t_{i_1}$. A type-τ minimal occurrence of α in $\mathcal{E}$ is a minimal time interval (with respect to containment) among all type-τ occurrences of α in $\mathcal{E}$. For occurrence width $w \in \mathbb{P}$, denote by $\text{mo}_{\tau}(\alpha; \mathcal{E}, w)$ and $\text{mo}_{\tau}(\alpha; \mathcal{E})$ the sets of all type-τ minimal occurrences with occurrence width w and all type-τ minimal occurrences, respectively, of α in $\mathcal{E}$.

With respect to the type of maximum-gap constraint, we define the two measures: frequency and confidence on episode rules. For a type number $\tau \in \{1, 2, 3\}$

and for a serial episode α in an event sequence $\mathcal{E}$ and an occurrence width $w \in \mathbb{P}$, the type-τ frequency of α with occurrence width at most w in $\mathcal{E}$, denoted by $\mathrm{freq}_\tau(\alpha; \mathcal{E}, w)$, is $\sum_{i=1}^{w} |\mathrm{mo}_\tau(\alpha; \mathcal{E}, i)|$. The type-$\tau$ (total) frequency of α in $\mathcal{E}$, denoted by $\mathrm{freq}_\tau(\alpha; \mathcal{E})$, is $\sum_{\mathrm{all}\ w} \mathrm{freq}_\tau(\alpha; \mathcal{E}, w)$.

For a type number $\tau \in \{1, 2, 3\}$ and for two serial episodes α_i for $i \in \{1, 2\}$ in an event sequence $\mathcal{E}$ and an occurrence width $w \in \mathbb{P}$, we define the type-τ frequency of the episode rule $\alpha_1 \Rightarrow \alpha_2$ for the occurrence width w, denoted by $\mathrm{freq}_\tau(\alpha_1 \Rightarrow \alpha_2; \mathcal{E}, w)$, to be $\mathrm{freq}_\tau(\alpha_1 \diamond \alpha_2; \mathcal{E}, w)$.

For a (user-prescribed) frequency threshold FreqThreshold $(> 0) \in \mathbb{R}$, a serial episode α in $\mathcal{E}$ (respectively, an episode rule $\alpha_1 \Rightarrow \alpha_2$ in $\mathcal{E}$) is type-τ frequent for an occurrence width $w \in \mathbb{P}$ if $\mathrm{freq}_\tau(\alpha; \mathcal{E}, w) \geq$ FreqThreshold (respectively, $\mathrm{freq}_\tau(\alpha_1 \Rightarrow \alpha_2; \mathcal{E}, w) (= \mathrm{freq}_\tau(\alpha_1 \diamond \alpha_2; \mathcal{E}, w)) \geq$ FreqThreshold).

Analogous to the notion of conditional probability, we define the type-τ confidence of an episode rule $\alpha_1 \Rightarrow \alpha_2$ in $\mathcal{E}$ for an occurrence width $w \in \mathbb{P}$, denoted by $\mathrm{conf}_\tau(\alpha_1 \Rightarrow \alpha_2; \mathcal{E}, w)$, to be the ratio of two frequencies of $\alpha_1 \Rightarrow \alpha_2$ and α_1 in proper contexts: "the frequency of type-τ minimal occurrence of $\alpha_1 \Rightarrow \alpha_2$ in $\mathcal{E}$ for occurrence width w" to "the frequency of minimal occurrence (without maximum-gap constraint) of α_1 in $\mathcal{E}$ for occurrence width w that can be augmented to type-τ minimal occurrences of $\alpha_1 \diamond \alpha_2$ for occurrence width w". Consequently, in type-1 context:

$$\mathrm{conf}_1(\alpha_1 \Rightarrow \alpha_2; \mathcal{E}, w) = \frac{\mathrm{freq}_1(\alpha_1 \diamond \alpha_2; \mathcal{E}, w)}{\mathrm{freq}_1(\alpha_1; \mathcal{E}, w)},$$

and in type-2 context:

$$\mathrm{conf}_2(\alpha_1 \Rightarrow \alpha_2; \mathcal{E}, w) = \frac{\mathrm{freq}_2(\alpha_1 \diamond \alpha_2; \mathcal{E}, w)}{\mathrm{freq}_3(\alpha_1; \mathcal{E}, w)}.$$

For a (user-prescribed) confidence threshold ConfThreshold $(\in (0, 1]) \in \mathbb{R}$, an episode rule $\alpha_1 \Rightarrow \alpha_2$ in $\mathcal{E}$ is type-τ confident for an occurrence width $w \in \mathbb{P}$ if $\mathrm{conf}_\tau(\alpha_1 \Rightarrow \alpha_2; \mathcal{E}, w) \geq$ ConfThreshold.

For our study, we present two mining algorithms for finding frequent and confident serial-episode rules with their ideal occurrence/window widths, if exist, in event sequences based on the notion of minimal occurrences constrained by constant and mean maximum gap (type-1 and type-2), respectively. A preliminary empirical study that illustrates the applicability of the episode-rule mining algorithms is performed with a set of earthquake data.

2   Mining Frequent Serial-Episode Rules with First Local-Maximum Confidence

The notion of first local-maximum confidence employed in constraint-based mining of frequent episode rules was introduced in [5], and has a few salient appeals:

1. Mathematical vigor:
   The coupling of (first) local-maximum confidence in $\mathrm{conf}_\tau(\alpha; \mathcal{E}, w)$ and ideal occurrence/window width w has a sound mathematical formulation — an instance of the integration of mathematical optimization and decision making in many theories and applications.
2. Filtering of randomness:
   Empirical studies in [5] suggest that first local-maximum confidence yielding ideal occurrence/window widths exists in real but not synthetic random data sets, which may reveal intrinsic dependencies of local-maximum confidence on the ideal occurrence/window widths.
3. Expert-knowledge integration:
   The requirement of the existence of first local-maximum confidence for episode rules results in a compact collection from a substantially larger pool of frequent and confident episode rules, which facilitates a viable integration of domain-expert knowledge and postmining of association rules.
4. Efficient implementation:
   With appropriate data structures, the algorithms for computing the functional dependency of first local-maximum confidence of episode rules corresponding to ideal occurrence/window widths, when exist, can be implemented efficiently.

For a type number $\tau \in \{1, 2\}$ and for two serial episodes α_1 and α_2 in an event sequence $\mathcal{E}$, the episode rule $\alpha_1 \Rightarrow \alpha_2$ has its type-τ first local-maximum confidence for an occurrence/window width $w \in \mathbb{P}$ if w exists as the least occurrence/window width satisfying the followings:

1. Frequent and confident:

$$\mathrm{freq}_\tau(\alpha_1 \Rightarrow \alpha_2; \mathcal{E}, w) \geq \mathrm{FreqThreshold}, \text{ and}$$
$$\mathrm{conf}_\tau(\alpha_1 \Rightarrow \alpha_2; \mathcal{E}, w) \geq \mathrm{ConfThreshold};$$

2. Strictly lower preceding confidence:
   For all occurrence/window widths $i \in \{1, 2, \ldots, w-1\}$,

$$\text{if } \mathrm{freq}_\tau(\alpha_1 \Rightarrow \alpha_2; \mathcal{E}, i) \geq \mathrm{FreqThreshold},$$
$$\text{then } \mathrm{conf}_\tau(\alpha_1 \Rightarrow \alpha_2; \mathcal{E}, i) < \mathrm{conf}_\tau(\alpha_1 \Rightarrow \alpha_2; \mathcal{E}, w);$$

   and
3. Lower succeeding confidence and significant decrement:
   For a (user-prescribed) decrement threshold $\mathrm{DecPercentage}\,(\in [0, 1)) \in \mathbb{R}$, there exists an occurrence/window width $w' \in \mathbb{P}$ such that:

$$w < w',$$
$$\text{for all occurrence/window widths } i \in (w, w'),$$
$$\mathrm{conf}_\tau(\alpha_1 \Rightarrow \alpha_2; \mathcal{E}, w) \geq \mathrm{conf}_\tau(\alpha_1 \Rightarrow \alpha_2; \mathcal{E}, i), \text{ and}$$
$$(1 - \mathrm{DecPercentage})\mathrm{conf}_\tau(\alpha_1 \Rightarrow \alpha_2; \mathcal{E}, w) \geq \mathrm{conf}_\tau(\alpha_1 \Rightarrow \alpha_2; \mathcal{E}, w').$$

A type-τ FLM-rule is an episode rule that attains its type-τ first local-maximum confidence at the ideal occurrence/window width.

We have designed and implemented two serial-episode-rule mining algorithms for finding all type-τ (for type number $\tau \in \{1, 2\}$) frequent FLM-rules together with their ideal occurrence/window widths. Note that the preliminary version of our study is limited to unit-order consequent episodes (that is, type-τ FLM-rules $\alpha_1 \Rightarrow \alpha_2$ with $|\alpha_2| = 1$). The data structures and algorithms employed can be readily generalized to higher-order consequent episodes.

The algorithmics for mining type-1 frequent FLM-rules follow the development in [5] with necessary optimization.

The notion of minimal prefix occurrence was introduced in [5] to overcome the type-1 maximum-gap constraint in computing mo_1. For a serial episode α in an event sequence $\mathcal{E}$, a type-1 occurrence $[t_1, t_2]_1$ of α in $\mathcal{E}$ is a minimal prefix occurrence in $\mathcal{E}$ provided that for every time interval $[t'_1, t'_2]_1 \in mo_1(\mathrm{pref}(\alpha), \mathcal{E})$, if $t_1 < t'_1$ then $t_2 \leq t'_2$.

Denote by $mpo(\alpha; \mathcal{E})$ the set of all minimal prefix occurrences of α in $\mathcal{E}$. For algorithmic efficiency, we organize the $mpo(\alpha; \mathcal{E})$, according to the common start time of the minimal prefix occurrences, into: (t_s, T_{t_s}) where T_{t_s} consists of the end times of all the minimal prefix occurrences of α in $\mathcal{E}$ with common start time t_s.

We unify all the supporting data structures and algorithmics for both type-1 and type-2 frequent FLM-rule mining algorithms except for the following type-dependent data structures: for a type number $\tau \in \{1, 2\}$, the type-τ "actual" minimal occurrence of a serial episode α in an event sequence $\mathcal{E}$ used in both algorithms is:

$$\text{a-mo}_\tau(\alpha; \mathcal{E}) = \begin{cases} mpo(\alpha; \mathcal{E}) & \text{if } \tau = 1, \\ mo_3(\alpha; \mathcal{E}) & \text{if } \tau = 2 \ (\text{due to the definition of conf}_2). \end{cases}$$

3 Supporting Algorithms

Table 1. Subalgorithms of two serial-episode-rule mining algorithms for finding all type-τ (for type number $\tau \in \{1, 2\}$) frequent FLM-rules together with their ideal occurrence/window widths.

Type-1 - constant MaxGap	Type-2 - mean MaxGap
unified algorithm WindowMiner-τ with type-dependent data structures a-mo$_\tau$	
unified algorithm ExploreNextLevel-τ with type-dependent data structures a-mo$_\tau$	
algorithm Join-1 with a-mo$_1$ = mpo	algorithm Join-2 with a-mo$_2$ = mo$_3$
unified algorithm FindFLM-τ with type-dependent data structures a-mo$_\tau$	

Algorithm. WindowMiner-τ (type number $\tau \in \{1, 2\}$)

Input: An event sequence $\mathcal{E}$ on an event type E, maximum-gap constraint: MaxGap, frequency and confidence thresholds: FreqThreshold and ConfThreshold, and decrement threshold: DecPercentage.

Output: The set of all type-τ FLM-rules (order-1 consequent episodes) with corresponding ideal occurrence/window widths.

1: $\{L_1$ consists of a-mo$_\tau(\alpha; \mathcal{E})$ of all frequent (serial) episodes α of order 1.$\}$
2: $L_1 := \emptyset$
3: **for all** $A \in E$ **do**
4:    a-mo$_\tau((A); \mathcal{E}) := \emptyset$
5: **end for**
6: $\{$Scan $\mathcal{E}$ to compute a-mo$_\tau$ for all (serial) episodes of order 1.$\}$
7: **for all** events $(A, t) \in \mathcal{E}$ **do**
8:    **if** $\tau = 1$ **then**
9:      a-mo$_\tau((A); \mathcal{E}) :=$ a-mo$_\tau((A); \mathcal{E}) \cup \{(t, \{t\})\}$
10:    **else** $\{\tau = 2\}$
11:      a-mo$_\tau((A); \mathcal{E}) :=$ a-mo$_\tau((A); \mathcal{E}) \cup \{[t, t]\}$
12:    **end if**
13: **end for**
14: **for all** $A \in E$ **do**
15:    **if** freq$_\tau((A); \mathcal{E}) \geq$ FreqThreshold **then**
16:      $L_1 := L_1 \cup$ a-mo$_\tau((A); \mathcal{E})$
17:    **end if**
18: **end for**
19: $\{$"Depth"-first search for all type-τ frequent FLM-rules.$\}$
20: **for all** a-mo$_\tau(\alpha; \mathcal{E}) \in L_1$ **do**
21:    Invoke ExploreNextLevel-τ(a-mo$_\tau(\alpha; \mathcal{E}), L_1$)
22: **end for**

Algorithm. ExploreNextLevel-τ (type number $\tau \in \{1, 2\}$)

Input: L_1: consists of a-mo$_\tau(\cdot; \mathcal{E})$ of all frequent (serial) episodes of order 1, and a-mo$_\tau(\alpha; \mathcal{E})$.

Output: The set of all type-τ FLM-rules with order-1 consequent episodes (with corresponding ideal occurrence/window widths).

1: **for all** a-mo$_\tau(\beta; \mathcal{E}) \in L_1$ **do**
2:    $\{$Decide if the serial-episode rule $\alpha \Rightarrow \beta$ is frequent.$\}$
3:    $\gamma := \alpha \diamond \beta$
4:    $\{$Temporal-join of a-mo$_\tau$s of antecedent and consequent episodes.$\}$
5:    a-mo$_\tau(\gamma; \mathcal{E}) :=$ Join-τ(a-mo$_\tau(\alpha; \mathcal{E})$, a-mo$_\tau(\beta; \mathcal{E})$)
6:    **if** freq$_\tau(\gamma; \mathcal{E}) \geq$ FreqThreshold; $\mathcal{E}$ **then**
7:      Invoke FindFLM-τ(a-mo$_\tau(\alpha; \mathcal{E})$, a-mo$_\tau(\gamma; \mathcal{E})$)
8:      Invoke ExploreNextLevel-τ(a-mo$_\tau(\gamma; \mathcal{E}), L_1$)
9:    **end if**
10: **end for**

Algorithm. Join-τ (type number $\tau = 1$)

Input: (Note: a-mo$_1(\cdot; \mathcal{E})$ is mpo$(\cdot; \mathcal{E})$ in this context.)
  a-mo$_1(\alpha; \mathcal{E}) = $ mpo$(\alpha; \mathcal{E})$ and a-mo$_1(\beta; \mathcal{E}) = $ mpo$(\beta; \mathcal{E})$.
Output: a-mo$_1(\gamma; \mathcal{E}) = $ mpo$(\gamma; \mathcal{E})$ where $\gamma = \alpha \diamond \beta$.
 1: $\gamma := \alpha \diamond \beta$;   mpo$(\gamma; \mathcal{E}) := \emptyset$
 2: {Temporal-join of mpos of antecedent and consequent episodes.}
 3: **for all** $(t_{\mathrm{s}}, T_{t_{\mathrm{s}}}) \in$ mpo$(\alpha; \mathcal{E})$ **do**
 4:    $L := \emptyset$
 5:    **for all** $t_{\mathrm{e}} \in T_{t_{\mathrm{s}}}$ **do**
 6:       $T' := \{t'_{\mathrm{s}} \mid (t'_{\mathrm{s}}, \{t'_{\mathrm{s}}\}) \in$ mpo$(\beta; \mathcal{E}), 0 < t'_{\mathrm{s}} - t_{\mathrm{e}} \leq$ MaxGap, and
                    for all $(t_1, T_{t_1}) \in$ mpo$(\alpha; \mathcal{E})$ if $t_{\mathrm{s}} < t_1$ then $t'_{\mathrm{s}} \leq \min\{t_2 \mid t_2 \in T_{t_1}\}\}$
 7:       $L := L \cup T'$
 8:    **end for**
 9:    **if** $L \neq \emptyset$ **then**
10:       mpo$(\gamma; \mathcal{E}) := $ mpo$(\gamma; \mathcal{E}) \cup \{(t_{\mathrm{s}}, L)\}$
11:    **end if**
12: **end for**

Algorithm. Join-τ (type number $\tau = 2$)

Input: (Note: a-mo$_2(\cdot; \mathcal{E})$ is mo$_3(\cdot; \mathcal{E})$ in this context.)
  a-mo$_2(\alpha; \mathcal{E}) = $ mo$_3(\alpha; \mathcal{E})$ and a-mo$_2(\beta; \mathcal{E}) = $ mo$_3(\beta; \mathcal{E})$.
Output: a-mo$_2(\gamma; \mathcal{E}) = $ mo$_3(\gamma; \mathcal{E})$ where $\gamma = \alpha \diamond \beta$.
 1: {Temporal-join of mo$_3$s of antecedent and consequent episodes.}
 2: $\gamma := \alpha \diamond \beta$;   mo$_3(\gamma; \mathcal{E}) := \emptyset$
 3: **for all** $[t_{\mathrm{s}}, t_{\mathrm{e}}] \in$ mo$_3(\alpha; \mathcal{E})$ **do**
 4:    $t_{\mathrm{e},\gamma} := \min\{t_{\mathrm{s}'} \mid [t_{\mathrm{s}'}, t_{\mathrm{s}'}] \in$ mo$_3(\beta; \mathcal{E}), t_{\mathrm{e}} < t_{\mathrm{s}'}$, and
                    for all $[t_1, t_2] \in$ mo$_3(\alpha; \mathcal{E})$ if $t_{\mathrm{s}} < t_1$ then $t'_{\mathrm{s}} \leq t_2\}$
 5:    **if** $t_{\mathrm{e},\gamma}$ exist **then**
 6:       mo$_3(\gamma; \mathcal{E}) := $ mo$_3(\gamma; \mathcal{E}) \cup \{[t_{\mathrm{s}}, t_{\mathrm{e},\gamma}]\}$
 7:    **end if**
 8: **end for**

Algorithm. FindFLM-τ (type number $\tau \in \{1, 2\}$)

Input: (Note: The serial episode $\gamma = \alpha \diamond \beta$ with $\alpha = \text{pref}(\gamma)$.)
   a-mo$_\tau(\alpha; \mathcal{E})$ and a-mo$_\tau(\gamma; \mathcal{E})$ of type-τ frequent serial episodes α and γ in $\mathcal{E}$, respectively.

Output: Decide if the serial-episode rule $\alpha \Rightarrow \text{suff}(\gamma)$ is a type-τ FLM-rule.

1: {Compute the set of all occurrence widths in the structure a-mo$_\tau(\alpha; \mathcal{E}) \cup$ a-mo$_\tau(\gamma; \mathcal{E})$ — in sorted order.}

2: $W := \emptyset$

3: **if** $\tau = 1$ **then**

4:    **for all** $(t_s, T_{t_s}) \in$ a-mo$_1(\alpha; \mathcal{E}) \cup$ a-mo$_1(\gamma; \mathcal{E})$ **do**

5:       $W := W \cup \{\min\{t_e \mid t_e \in T_{t_s}\} - t_s\}$

6:    **end for**

7: **else** {$\tau = 2$}

8:    **for all** $[t_s, t_e] \in$ a-mo$_2(\alpha; \mathcal{E}) \cup$ a-mo$_2(\gamma; \mathcal{E})$ **do**

9:       $W := W \cup \{t_e - t_s\}$

10:    **end for**

11: **end if**

12: Sort W in ascending order

13: {Decide the type-τ FLM-condition.}

14: maxConfidence := 0.0; $w_0 := 0$; $\neg$isFLM(γ)

15: **for all** $w \in W$ **do**

16:    confidence := conf$_\tau(\alpha \Rightarrow \text{suff}(\gamma); \mathcal{E}, w)$

17:    **if** maxConfidence < confidence **then**

18:       maxConfidence := confidence; $w_0 := w$

19:    **else if** freq$_\tau(\gamma; \mathcal{E}, w_0) \geq$ FreqThreshold $\wedge$ maxConfidence $\geq$ ConfThreshold
          $\wedge \, (1 - \text{DecPercentage})$maxConfidence $\geq$ confidence **then**

20:       isFLM(γ)

21:       Break for-loop

22:    **end if**

23: **end for**

---

4 Preliminary Empirical Study

The two serial-episode FLM-rule mining algorithms were studied in a small-scale experiment with a set of earthquake data. The earthquake database is available from Global Centroid-Moment-Tensor Project [8] at "http://www.globalcmt.org". The accessed data set consists of global seismicity captured during 1976 – 2010.

Earthquake activities are translated into an event sequence of events, in which the event type corresponds to the earthquake geographical location and the event occurrence time corresponds to the Unixtime of the earthquake reference time. Minimal preprocessing was performed on the data set. The studied event sequence of earthquakes consists of 33866 events on 672 event types.

Some obvious association rules are in the forms of series of earthquakes at near-by geographical locations lead to one in proximity region in relatively short time, including those with consequent episodes due to numerous aftershocks.

An empirical evaluation of the two suites of FLM-rule mining algorithms WindowMiner-τ for type number $\tau \in \{1, 2\}$ were performed on the same event sequence with: (1) the frequency and confidence thresholds: FreqThreshold $=$ 30 and ConfThreshold $= 0.90$ respectively, (2) the maximum-gap constraint MaxGap $\in \{12, 24, 36, 48\}$ (in hours), (3) the decrement threshold DecPercentage $\in \{0.10, 0.20, 0.30, 0.40\}$.

Since the maximum-gap constraint is more stringent for type-1 (constant MaxGap) than for type-2 (mean MaxGap), we expect that for each MaxGap $\in$ $\{12, 24, 36, 48\}$ (in hours) and each DecPercentage $\in \{0.10, 0.20, 0.30, 0.40\}$, the numbers of frequent episode rules, frequent and confident episode rules, and FLM-rules in type-1 are upper-bounded by those in type-2 respectively. These are evidenced in Table 2 for the case of DecPercentage $= 0.20$ and MaxGap $\in$ $\{12, 24, 36, 48\}$ (in hours).

Table 2. For DecPercentage $= 0.20$ and MaxGap $\in \{12, 24, 36, 48\}$ (in hours): the numbers of type-1 and type-2 frequent episode rules, frequent and confident episode rules, and FLM-rules

MaxGap	frequent rules	frequent and confident rules	FLM-rules
Type-1 - constant MaxGap:			
12	189	103	0
24	351	131	2
36	546	170	3
48	844	271	3
Type-2 - mean MaxGap:			
12	380	264	248
24	1007	674	356
36	14196	13561	653
48	92844	91566	1955

Furthermore, the FLM-condition seems to be much more demanding for type-1 than for type-2: only a few of type-1 frequent and confident episode rules can survive the "significant decrement"-requirement. Table 3 illustrates the statistics for the case of MaxGap $= 24$ hours and DecPercentage $\in \{0.10, 0.20, 0.30, 0.40\}$.

Among the discovered type-2 FLM-rules, most are saturated with identical event types — corresponding to continuing earthquake aftershock activities in the affected regions.

5   Conclusion

This work extends the existing mining algorithm for finding frequent and confident serial-episode rules using minimal occurrences constrained by constant

Table 3. For MaxGap = 24 hours and DecPercentage $\in \{0.10, 0.20, 0.30, 0.40\}$: the numbers of type-1 and type-2 FLM-rules

DecPercentage	type-1 FLM-rules	type-2 FLM-rules
0.10	39	356
0.20	2	356
0.30	0	356
0.40	0	354

maximum gap to one by mean maximum gap. Our work in progress includes an in-depth empirical study of earthquake data set based on current implementation. Our future work focuses on the topological and algorithmic aspects of episode and episode-rule mining for higher-order complex episode topologies.

References

1. Agrawal, R., Imielinski, T., Swami, A.: Mining Association Rules between Sets of Items in Large Databases. In: The 1993 Association for Computing Machinery Special Interest Group on Management of Data International Conference on Management of Data, pp. 207–216. Association for Computing Machinery (1993)
2. Agrawal, R., Srikant, R.: Fast Algorithms for Mining Association Rules. In: The 20th International Conference on Very Large Data Bases, pp. 487–499 (1994)
3. Agrawal, R., Srikant, R.: Mining Sequential Patterns. In: The Eleventh International Conference on Data Engineering, pp. 3–14 (1995)
4. Dai, H.K., Wang, G.: Mining a Class of Complex Episodes in Event Sequences. In: Megiddo, N., Xu, Y., Zhu, B. (eds.) AAIM 2005. LNCS, vol. 3521, pp. 460–471. Springer, Heidelberg (2005)
5. Méger, N., Rigotti, C.: Constraint-Based Mining of Episode Rules and Optimal Window Sizes. In: Boulicaut, J.-F., Esposito, F., Giannotti, F., Pedreschi, D. (eds.) PKDD 2004. LNCS (LNAI), vol. 3202, pp. 313–324. Springer, Heidelberg (2004)
6. Mannila, H., Toivonen, H., Verkamo, A.I.: Discovery of Frequent Episodes in Event Sequences. Data Mining and Knowledge Discovery 1(3), 259–289 (1997)
7. Pei, J., Han, J.: Can We Push More Constraints into Frequent Pattern Mining? In: The Sixth Association for Computing Machinery Special Interest Group on Knowledge Discovery and Data Mining International Conference on Knowledge Discovery and Data Mining, pp. 350–354. Association for Computing Machinery (2000)
8. Global Centroid-Moment-Tensor Project, http://www.globalcmt.org

Smoothing Kernel Estimator for the ROC Curve-Simulation Comparative Study

Maria Filipa Mourão[1], Ana C. Braga[2], and Pedro Nuno Oliveira[3]

[1] Basic Science and Computing, School of Technology and Management-IPVC,
4900-348 Viana do Castelo, Portugal
fmourao@estg.ipvc.pt
[2] Department of Production and Systems Engineering, University of Minho,
4710-057 Braga, Portugal
acb@dps.uminho.pt
[3] Biomedical Sciences Abel Salazar Institute, University of Porto,
4050-313 Porto, Portugal
pnoliveira@icbas.up.pt

Abstract. The kernel is a non-parametric estimation method of the probability density function of a random variable based on a finite sample of data. The estimated function is smooth and level of smoothness is defined by a parameter represented by h, called bandwidth or window. In this simulation work we compare, by the use of mean square error and bias, the performance of the normal kernel in smoothing the empirical ROC curve, using various amounts of bandwidth. In this sense, we intend to compare the performance of the normal kernel, for various values of bandwidth, in the smoothing of ROC curves generated from Normal distributions and evaluate the variation of the mean square error for these samples. Two methodologies were followed: replacing the distribution functions of positive cases (abnormal) and negative (normal), on the definition of the ROC curve, smoothed by nonparametric estimators obtained via the kernel estimator and the smoothing applied directly to the ROC curve. We conclude that the empirical ROC curve has higher standard error when compared with the smoothed curves, a small value for the bandwidth favors a higher standard error and a higher value of the bandwidth increasing bias estimation.

Keywords: ROC Curve, Kernel Estimator, bandwidth.

1 Introduction

In medical diagnosis, it is important to find the performance of a diagnostic test to discriminate between individuals "normal" (non-diseased) and "abnormal" (patients). When the test results are measured on a binary scale this performance is supplied through the same sensitivity and specificity. When the test results are of the continuous type, test performance is measured by plotting sensitivity versus (1-specificity) when we change the definition of a positive result of the test, called the ROC curve. Let $F(.)$ and $G(.)$ be the distribution functions

B. Murgante et al. (Eds.): ICCSA 2013, Part I, LNCS 7971, pp. 573–584, 2013.
© Springer-Verlag Berlin Heidelberg 2013

for "abnormal" and "normal", individuals, respectively. The ROC curve is a graphical representation of $(1 - F(t))$ vs $(1 - G(t))$, for $t \in (-\infty, +\infty)$, written as follows:

$$R(p) = 1 - F\left(G^{-1}\left(1 - p\right)\right), \text{for } p \in (0, 1) \tag{1}$$

where p is the fraction of false positives.

To estimate the ROC curve $R(p)$, parametric, semi-parametric and nonparametric methods have been developed. The first assume that F and G have an associated parametric distribution; seconds assume that there is a monotonic transformation on the results of continuous scale test such that the distribution for individuals "normal" and "abnormal" is Normal [4]. However, these methods can be sensitive to distributional assumptions and present only a narrow range of distribution. In this article, we use the empirical estimator of the ROC curve $R(p)$ for being the most widely used non-parametric methods. However, these methods can be sensitive to distributional assumptions and present only a narrow range of distribution. In this article, we use the empirical estimator of the ROC curve $R(p)$ for being the most widely used non-parametric method.

Let $\widehat{F}(.)$ and $\widehat{G}(.)$ be the empirical functions associated with $F(.)$ and $G(.)$, respectively. Let $x_1, x_2, ... x_{n_1}$ be the results of the diagnostic test for "abnormal" individuals and $y_1, y_2, ..., y_{n_0}$ be the results of the diagnostic test for "normal" individuals. Thus, empirical functions written in light of these results are as follows:

$$\widehat{F}(t) = \#\left\{\frac{x_i \leq t}{n_1}\right\} \text{ and } \widehat{G}(t) = \#\left\{\frac{y_i \leq t}{n_0}\right\} \tag{2}$$

The inverse function of $\widehat{G}(t)$ is defined by

$$\widehat{G}^{-1}(p) = inf\left\{t : \widehat{G}(t) \geq p\right\} \tag{3}$$

The empirical ROC curve can then be write as:

$$\widehat{R}(p) = 1 - \widehat{F}\left(\widehat{G}^{-1}\left(1 - p\right)\right) \tag{4}$$

This ROC curve has the advantage of being invariant with respect to monotonous changes in test results since it depends only of the range for observations of combined sample.

2   Kernel Estimator

In many situations arises the need for estimating the density function corresponding to a variable completely observed, using then an kernel estimator. Since the work of Rosenblatt [8], Whittle [11] and Parzen [12], the kernel estimator has become undoubtedly the most widely used method. It is also the most used in practice after the histogram upon which presents clear advantages, the most evident, a greater softness of the estimation obtained. References on the subject are for example Silverman [9] and Wand and Jones [10].

The Rosenblatt - Parzen estimator for a density function f at a point x is defined as

$$\widehat{f}_h(x) = \frac{1}{nh} K \left(\frac{x - X_i}{h} \right) \tag{5}$$

in which the kernel K satisfies the condition $\int k(x)dx = 1$ and h is called bandwidth or window. In practice, the kernel K is usually chosen to be a probability density function unimodal and symmetrical. In this case K satisfies the following conditions:

1. $\int K(y)dy = 1$
2. $\int yK(y)dy = 0$
3. $\int y^2 K(y)dy = \mu_2(K) > 0$

The role of the kernel is to control the shape of "highs" and the width of the window. As the kernel is not a decisive factor in the estimation, choosing an appropriate window is extremely important. Windows too small, give excessive variability while windows with high value cause biased estimates. There are many kernel mentioned in the literature such as normal, rectangular, Epanechnikov or triangular. The problem of the selection window appears on the definition of appropriate measures when estimating the error made by a density estimator $\widehat{f}$. The most used is the integrated mean squared error defined by

$$MISE(\widehat{f}) = E \left[\int \left(\widehat{f}(x) - f(x) \right)^2 \right] \tag{6}$$

In terms of the value of h, this requires always a compromise between the bias and variance of the estimator since the aim is to minimize the mean square error integrated in this case written as:

$$MISE(h) = \int \left\{ Var(f(x) + (bias(f(x)))^2 \right\} \tag{7}$$

Let $K(.)$ be a probability density function, continuous, with zero mean and finite variance σ_k^2. It is shown that

$$\int_{\Re} Var\left(f(x) \right) dx = \frac{R(K)}{nh} + O \left(\frac{1}{nh} \right) \tag{8}$$

$$\int_{\Re} bias(f(x))^2 dx = \frac{h^4 \sigma_k^4 R(f'')}{4} + O(h^4) \tag{9}$$

Thus, we can write

$$MISE(h) = AMISE(h) + O \left(\frac{1}{nh} + h^4 \right) \tag{10}$$

with

$$AMISE(h) = \frac{R(K)}{nh} + \frac{h^4 \sigma_k^4 R(f'')}{4} \qquad (11)$$

This way the optimal choice of h is given by the value that minimizes $AMISE$ which depends on the probability density function unknown, so the utility is not shown. For the choice of h, are known the following practical results:

- Silverman method: $h = \left(\frac{4}{3n}\right)^{1/5} \hat{\sigma}$
- Silverman modified method: $h = \left(\frac{4}{3n}\right)^{1/5} min\left(\hat{\sigma}; \frac{IQR}{1.35}\right)$
- Terrel method: $h = \left(\frac{R(K)}{35n}\right)^{1/5} \hat{\sigma}$

3   Smooth ROC Curve

The kernel methodology was extended to estimate the ROC curve, to obtain a smoothed and continuous curve regardless of the type of scale diagnostic in use. There are two main methods for obtaining the smoothed ROC curve, known in the literature by indirect method and direct method.

3.1   Smooth ROC Curve by the Indirect Method

In the indirect method the n_0 observations of the response variable for individuals classified as "normal", allow us to represent the histogram as an empirical estimator of the density function f_{n_0} associated with these theoretical results, as did the n_1 observations associated with the sample of individuals classified as "abnormal", g_{n_1}. The kernel method, applying smoothing techniques to the density functions associated with values of the diagnostic test of individuals classified as "abnormal" and "normal", produces a smooth and continuous curve. The work of Lloyd,[1], proposed to estimate the ROC curve from the kernel smoothing of distribution functions for the test results and presented formulas for the asymptotic bias and standard deviation of the estimator of the curve. The author, compare the performance of the asymptotic kernel estimator of the ROC curve with the performance of the empirical estimator of the ROC curve. Shows how the empirical estimator is deficient compared to the kernel estimator but this deficiency is masked when the sample size increases.

Providing two independent samples $X_{01}, ..., X_{0n_0}$ for the distribution F_0 associated with individuals classified as "normal" and $X_{11}, ..., X_{1n_1}$ for the distribution F_1 associated with individuals classified as "abnormal", in the case of not having information about these two distributions, their empirical nonparametric maximum likelihood estimates are given by the following expressions:

$$\hat{F}_0(c) = \frac{1}{n_0} \sum_{i=1}^{n_0} I\left(X_{0i} \leq c\right) \qquad (12)$$

$$\hat{F}_1(c) = \frac{1}{n_1} \sum_{j=1}^{n_1} I\left(X_{1j} \leq c\right) \tag{13}$$

The same author, proposed a smoothed estimator for the ROC curve based on the concept of kernel functions associated with the distribution of individuals classified as "'normal"' and "'abnormal"' described above. Let $k(x)$ be a continuous density function and $K(x) = \int_{-\infty}^{x} k(t)dt$.

The kernel estimators of F_0 and F_1 are given by

$$\tilde{F}_0(c) = \frac{1}{n_0} \sum_{i=1}^{n_0} K\left(\frac{c - X_{0i}}{h_0}\right) \tag{14}$$

$$\tilde{F}_1(c) = \frac{1}{n_1} \sum_{j=1}^{n_1} K\left(\frac{c - X_{1j}}{h_1}\right) \tag{15}$$

which require the choice of two bandwith one for each estimator.

In the work of Nadaraya, [7], the author suggest $h_i = O\left(n_i^{-1/3}\right)$ as an optimal bandwith, and in this case the kernel and the empirical estimators, have asymptotically the same mean and variance.

The smoothed estimator for the ROC curve, proposed by Lloyd [1], is based on these two estimators associated with distributions and can generally be written as

$$\tilde{R}_h(p) = 1 - \tilde{F}_1\left(\tilde{F}_0^{-1}(1 - p)\right), \text{ for } p \in (0, 1) \tag{16}$$

3.2   Smooth ROC Curve by the Direct Method

In his work, Peng [2], based on the idea that for obtain the kernel estimator of the distribution function, simply apply smoothing to the empirical function distribution, proposed a smoothed estimator for the ROC curve of the form:

$$\hat{R}_h(p) = 1 - \int \hat{F}_P\left(\hat{F}_N^{-1}(1 - p + hu)\, k(u)du\right) \tag{17}$$

in wish $\hat{F}_P$ and $\hat{F}_N$ are the empirical functions distributions of individuals classified as "positive" (abnormal) and "negative" (normal), respectively. Once the smoothing is applied directly to the ROC curve, only one "window" h is used, and not two as in the previous approach. The estimator of the ROC curve admits an explicit expression of the form

$$\hat{R}_h(p) = \frac{1}{n} \sum_{i=1}^{n} K\left(\frac{\hat{F}_N(Y_{Pi}) - 1 + p}{h}\right) \tag{18}$$

and the kernel estimator of the F function is written as

$$\hat{F}_h(x) = \frac{1}{n} \sum_{i=1}^{n} K\left(\frac{x - X_i}{h}\right) \tag{19}$$

4   Simulation Study

Based on the two estimation approaches described above, we carried out a simulation study in which we compared the performance of normal kernel, for various values of bandwidth, in the smoothing of ROC curves generated from normal distributions (for normal cases: $\mu_0 = 1.0, \sigma_0 = 0.75$ and for abnormal cases: $\mu_1 = 2.0, \sigma_1 = 0.75$ by varying the sample size to 10 , 50, 100. This process was repeated 50 times. The comparison is done based on the values of area under curves (AUC), the standard error (SE) and bias.

In the following figures, (Fig. 1 to Fig. 3) we present some empirical and smoothed ROC curves obtained by the directly and indirectly method, for $n = 10$, $n = 50$, $n = 100$ and varying the width of the window:

In the following tables (tab. 1 to tab. 4) , we present the results obtained in terms of areas under the ROC curve, standard error, mean square error and bias to the empirical and smoothed ROC curves:

Table 1. Mean of AUC to smooth ROC curves obtained by the direct method (bootstrap method)

h	$(10; 10)$	$(50; 50)$	$(100; 100)$
0.05	0.8381578	0.8283924	0.8241924
0.10	0.8357668	0.8269739	0.8230278
0.25	0.8247552	0.8171648	0.812564
0.50	0.7938438	0.7872557	0.7822923
1.00	0.7237794	0.7188639	0.714118
empirical	0.812044	0.8128005	0.8130962

Table 2. Standard Error (SE) to smooth ROC curves obtained by the direct method

h	$(10; 10)$	$(50; 50)$	$(100; 100)$
0.05	0.010829	0.004351	0.004013
0.10	0.010024	0.0042761	0.004171
0.25	0.010173	0.004162	0.003959
0.50	0.009653	0.003859	0.003547
1.00	0.007608	0.0032749	0.002921
empirical	0.010022	0.005203	0.005252

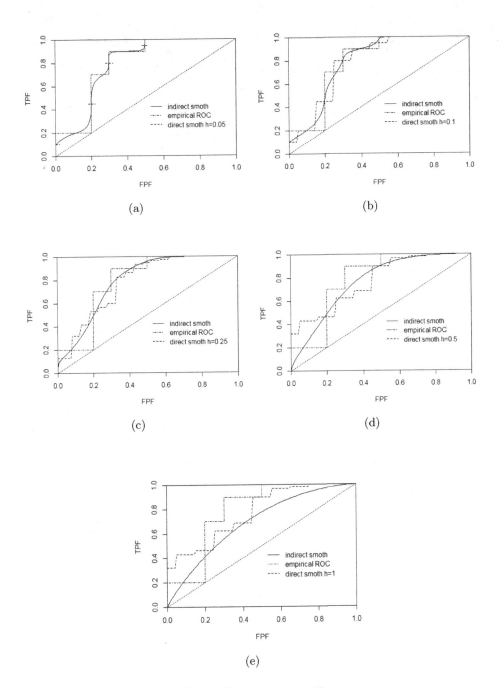

Fig. 1. ROC curves, n=10

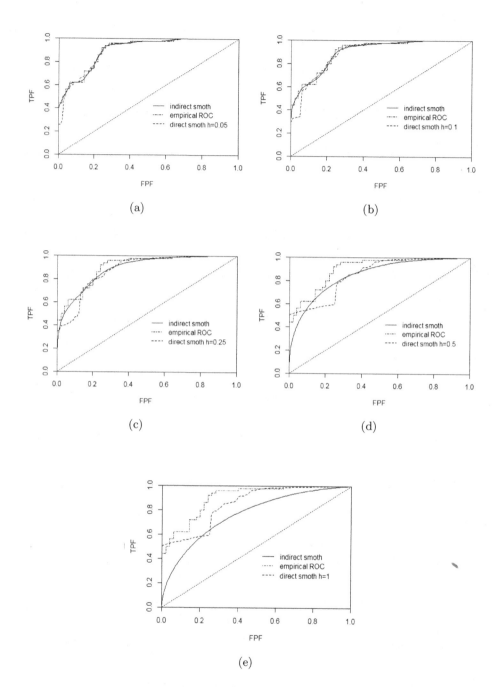

Fig. 2. ROC curves, n=50

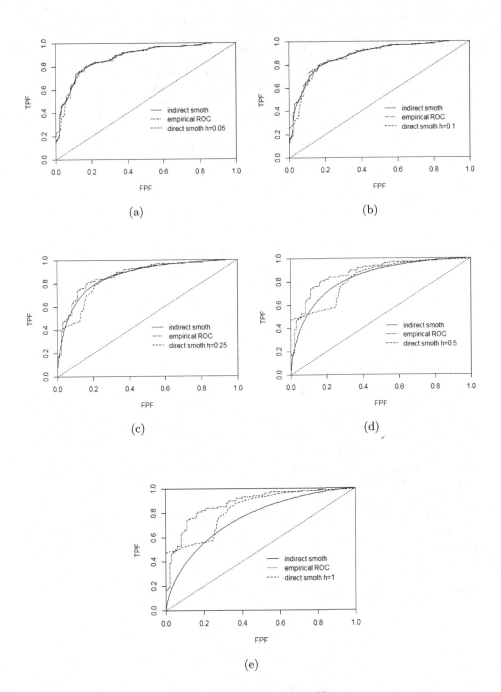

Fig. 3. ROC curves, n=100

Table 3. Bootstrap Bias of smoothed ROC curves by the direct method

h	$(10; 10)$	$(50; 50)$	$(100; 100)$
0.05	0.000172	0.0001429	-0.000118
0.10	-0.0001192	0.000102	0.000306
0.25	0.0001972	0.000141	0.000920
0.50	-0.0003742	0.000437	-0.000106
1.00	-0.000519	0.0008197	0.0007606
empirical	0.000844	-0.0003037	0.000763

Table 4. AUC of smoothed ROC curves by the indirect method

h	$(10; 10)$	$(50; 50)$	$(100; 100)$
0.05	– – –	0.8988	0.8773
0.10	0.7873	0.8932	0.8733
0.25	0.7800	0.8787	0.8582
0.50	0.7723	0.8534	0.8315
1.00	0.7723	0.8534	0.8315
empirical	0.7900	0.9040	0.8797

4.1 Example of Application

This section is intended to illustrate the methodology described through one practical example. The application includes a data set of 160 baby's (54.4% female, 45.6% male) of a Neonatology Intensive Care Unit of a Portuguese Hospital. The ratings given by the CRIB (Clinical Risk Index for Babies) scale regarding their clinical status, was 11.3% babies classified as "death" (38.9% female, 61.1% male) and 88.7% classified as "alive" (56.3% female, 43.7%male). The graph of Fig. 4 shows the empirical curve and the smoothed curves obtained by direct method, when applying a bandwidth of 0.05 and 0.10.

In table (tab. 5) we shows the values of AUCs for the generated curves and their standard errors (SE) obtained via bootstrap when considering 1500 replicates.

Table 5. AUC and Standard Error (SE) for empirical and smooth ROC curves obtained by the direct method

	AUC	SE
$h = 0.05$	0.926319	0.042179
$h = 0.10$	0.916483	0.029104
empirical	0.933881	0.028166

By the results presented in tab. 5 we can see that the AUC of the smoothed ROC curve decrease as the bandwidth increase, but the standard error tends to decrease. However, for this example, the empirical ROC curve performs better in either, the AUC and standard error.

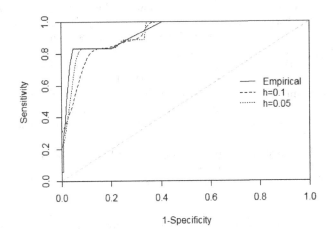

Fig. 4. Empirical and smoothed ROC curves

5 Conclusions

Analyzing tables of results presented, we found that, the AUC of the smoothed ROC curves, for the same sample size, decreases when increasing the value of the bandwdth. For the same value of h, increasing the sample sizes causes a decrease in the value of AUC of the smoothed ROC curve. In what concerns the empirical ROC curve, AUC increases when increasing the sample sizes. However, the empirical ROC curve, when analyzed in the same sample sizes, has a higher standard error as compared to the smoothed ROC curves. It is also observed that a small amount of bandwidth within the same sample size, favors a higher standard error for the AUC of the ROC curve. It was also observed that a greater value of the bandwidth favors, in most cases, a higher value for the bias. Analyzing also the figures submitted to the ROC curve, the choice of h lies with the value 0.1, which gives a smoothed ROC curve, by the direct method, with a more approximate shape of the shape of the empirical ROC curve. So, the results indicate that for the windows, and sample sizes chosen, the empirical ROC curve has higher standard error when compared with the smoothed ROC curves, that a small value for the bandwidth favors a higher standard error and a higher value of the bandwidth increasing bias estimation.

References

1. Lloyd, C.J., Yong, Z.: Kernel estimators of the ROC curve are better than empirical. Statistics & Probability Letters 44, 221–228
2. Peng, L., Zhou, X.-H.: Local linear smoothing of receiver operating characteristic (ROC) curves. Journal of Statistical Planning and Inference 118, 129–143
3. Zou, K.H., Hall, W.J., Shapiro, D.E.: Smooth nonparametric receiver operating characteristic (ROC) curves for continuous diagnostic tests. Statistics in Medicine 16, 2143–2156
4. Zou, K.H., Hall, W.J.: Two transformation models for estimating an ROC curve derived from continuous data. Journal of Applied Statistics 27, 621–631
5. Zhou, Y., Zhou, H., Ma, Y.: Smooth estimation of ROC curve in the presence of auxiliary information. J. Syst. Sci. Complex 24, 919–944
6. Zweig, M.H., Campbell, G.: Receiver operating characteristic (ROC) plots: a fundamental evaluation tool in clinical medicine. Clinical Chemistry 39, 561–577
7. Nadaraya, E.A.: Some new estimates for distribution functions. Theory Probab. Appl. 15, 497–500
8. Rosenblatt, M.: Remarks on Some Nonparametric Estimates of a Density Function. The Annals of Mathematical Statistics 27(3), 832
9. Silverman, B.W.: Density Estimation for Statistics and Data Analysis. Chapman and Hall, London
10. Wand, M.P., Jones, M.C.: Kernel Smoothing. Chapman and Hall, London
11. Whittle, P.: On the smoothing of probability density functions. Journal of the Royal Statistical Society. Series B 20, 334–343
12. Parzen, E.: On estimation of a density probability density function and mode. Ann. Math. Statist. 33, 1065–1076

Orthodontics Diagnostic Based
on Multinomial Logistic Regression Model

Ana Cristina Braga[1], Vanda Urzal[2], and A. Pinhão Ferreira[2]

[1] Department of Production and Systems, University of Minho,
4710-057 Braga, Portugal
acb@dps.uminho.pt
[2] Dental Medicine Faculty, Porto University,
4200-393 Porto, Portugal
vandaurzal@gmail.com, aferreira@fmd.up.pt

Abstract. The main objective of this study is evaluate the influence of several covariates in the occurrence of two types of vertical jaw dysplasia (open bite and deepbite) in orthodontics field.

The study of vertical jaw dysplasia is of great interest to the community of orthodontists, to ensure long-term stability of treatment, as this envolves complex etiological factors.

In this work we propose to build a multinomial logistic regression model that could assess the probability of an individual have open bite or deepbite taking into account some cephalometric measures of hyoid bone (HB) and some individual characteristics.

The study was conducted in a retrospective evaluation and consisted of 191 individuals random selected from a clinic in the Northern region of Portugal. We evaluated multiple factors in the construction of multinomial logistic regression model with 2 logit functions. The technique used to select variables to be included in the model, was the stepwise technique by choosing the smallest p value for the variable entering in the model.

Of the modeling process through multinomial logistic regression have resulted five position of the hyoid bone that have statistical significance and that can contribute as an auxiliary for the diagnosis of vertical jaw dysplasia.

Keywords: multinomial logistic regression, logit, stepwise, open bite, deepbite, hyoid bone (HB).

1 Introduction

The HB is unique because it has no bony articulation. However, it is an insertion element for muscles, ligaments, and fasciae attached to the mandible, clavicle, sternum, cranium, and cervical vertebrae [4]. It is an important part of the musculoskeletal apparatus of the craniofacial complex and factors affecting his system might have not only local but systemic effects as well [6].

B. Murgante et al. (Eds.): ICCSA 2013, Part I, LNCS 7971, pp. 585–595, 2013.
© Springer-Verlag Berlin Heidelberg 2013

It is a unique and important structure, differing from the other bones of the head and neck, has no bone joints, but provides support to muscles, ligaments and fascia of the pharynx, mandibles, skull and cervical spine. It is predominant in determining the physiological curvature of the spine and along with two muscle groups, supra-hyoid and infra-hyoid, are fundamental in the act of swallowing, preventing regurgitation of food and taking direct action in the control of mandibular dynamics [4]. There is a functional indivisible unit: the biomechanical relationship between skull, mandible, cervical spine, HB and airways [2]. The hyoid triangle is formed by the cephalometric points (RGN), hyoid (H) and the third cervical vertebra (C3), and allows the determination of the position of the HB in vertical plane [6].

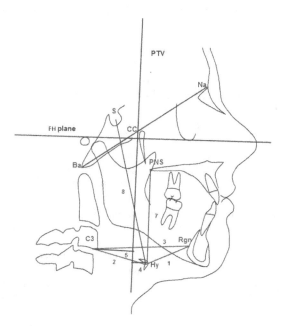

Fig. 1. Cephalometric landmarks used for evaluation of hyoid bone position (horizontal, vertical and angular measurements). 1. **Hy/Rgn**, linear distance from hyoidale to retrognation (Hy, the most anterior and superior point of the hyoid body; Rgn, the most posterior point of the mandible symphysis); 2. **Hy/C3**, linear distance from Hy to third cervical vertebra (C3, the point antero-inferior of the third cervical vertebra); 3. **C3/Rgn** linear distance from C3 to Rgn; 4, **C3/PTV**, linear distance between the C3 and Pterygoid (Pt, the most posterior point of the pterygomaxillary fissure) vertical reference, orthogonal to Frankfurt Horizontal; 5. **Hy/PTV**, linear distance between Hy to PTV; 6. **Hy/C3-Rgn**, linear distance between C3 and Rgn; 7. **Hy/PNS**, linear distance from Hy to posterior nasal spine (PNS); 8. **S/Hy**, linear distance from sella (S, midpoint of the pituitary fossa of sphenoid bone) to Hy. Cephalometric reference plane and line: Frankfurt horizontal plane (FH-plane): the horizontal plane that joins porion and orbital; Basal plane (BaNa): the plane joining nasion and basion; Vertical pterigoidea (PTV): a line perpendicular to FH-plane at the most posterior point of the pterigomaxillary fissure.

The spatial position of the HB can vary with the type of face, malocclusion, oral breathing, deglutition habits and adapted postural position of the head, being directly related to the mandibular movements, the tongue position and consequently influencing the deglutition, chewing and phonation functions [4].

Understanding better the HB position and the structures associated with it, may help the orthodontists to diagnose and treatment more accurately these malocclusions, thus obtaining successful stability.

In figures 1 and 2 are illustrated the cephalometric measures of hyoid bone (HB).

In statistics, logistic regression is a type of regression analysis used for predicting the outcome of a categorical dependent variable (a dependent variable that can take on a limited number of categories) based on one or more predictor variables. Multinomial logistic regression is defined for a response variable with three or more discrete outcomes. It is an extension of logistic regression based on the binomial distribution (i.e., where the response has only two outcomes). The multinomial model handles data analysis situations where a response variable is ordinal (the order of the response categories is important) or nominal (order of the response categories does not matter).

Like binary logistic regression, multinomial logistic regression uses maximum likelihood estimation to evaluate the probability of categorical membership.

The multinomial logistic regression does necessitate careful consideration of the sample size and examination for outlying cases.

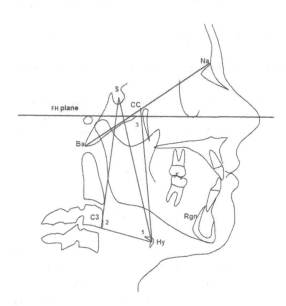

Fig. 2. Angular measurements: 1. **C3/Hy/S**, angle measured from C3, Hy and Sella; 2. **Hy/C3/S**, angle measured from Hy, C3 and Sella; 3. **NaBa/CC/Hy**, angle measured between Nasion Basion plane and Center of Cranium (intersection of facial axis (Pt-Gn) and NaBa) and Hy.

In this work, multinomial logistic regression is used to predict the probability of category membership on a dependent variable based on multiple independent variables. The independent variables can be either dichotomous (i.e., binary) or continuous (i.e., interval or ratio in scale).

2   Methodology

The study protocol, approved by the Ethical Committee of the Dental Medicine Faculty of Porto University (Portugal) and was outlined following the legal norms.

The three groups (control, open bite and deepbite) were retrospectively selected based on some clinical inclusion criteria, and the sample was composed by 191 subjects, with 94 males (49.2%) and 97 females (50.8%) aged between 12 and 55 years. The composition of the three groups based on the overbite (dental classification) were: 66 normal bite ($2.5 \pm 1mm$), 62 with open bite ($< 0mm$) and 63 with deepbite ($> 5mm$). The landmarks were done by the same investigator using the software Nemoceph, Nx06 from Nemotec. The cephalogram was rotating according to the natural position of the head, obtained by the lateral photo, through the true vertical line (TVL).

In this work we propose to build a multinomial logistic regression model that could assess the association of the vertical dysplasia (open bite or deepbite) with some cephalometric measures of HB and some individual characteristics (sex and age).

The binary logistic regression is the technique most often used to model the relationship between a binary outcome variable and a set of covariates. However, doing some modifications, can also be used when the dependent variable is polychotomous.

Considering r the number of categories in the dependent variable, Y, when $r = 2$, Y is dichotomous and we can model log of odds that an event occurs or does not occur. For binary logistic regression there is only 1 logit that we can form.

$$\text{logit}(\pi) = \log\left(\frac{\pi}{1 - \pi}\right)$$

When $r > 2$, we have a multi-category or polytomous dependent variable. There are $r(r-1)/2$ logits (odds) that we can form, but only $(r-1)$ are non-redundant. There are different ways to form a set of $(r-1)$ non-redundant logits, and these will lead to different polytomous logistic regression models.

For example, if the dependent variable Y has three categories 0 is the code for the reference category, 1 and 2 the remaining categories. For a model with three categories, we have two logit functions, one for $Y = 1$ vs $Y = 0$ and other for $Y = 2$ vs $Y = 0$. The other possible logit, $Y = 2$ vs $Y = 1$, is non-redundant and it results from the difference between the logit $Y = 2$ vs $Y = 0$ and $Y = 1$ vs $Y = 0$.

Let $\mathbf{x}$ be the vector of covariates of length $p+1$ with $x_0 = 1$ for entering with the constant term. We will designate the two logit functions by:

$$g_1(\mathbf{x}) = \ln\left(\frac{P(Y = 1|\mathbf{x})}{P(Y = 0|\mathbf{x})}\right)$$
$$= \beta_{10} + \beta_{11}x_1 + \beta_{12}x_2 + \cdots + \beta_{1p}x_p$$
$$= (1, \mathbf{x}')\boldsymbol{\beta}_1$$

$$g_2(\mathbf{x}) = \ln\left(\frac{P(Y = 2|\mathbf{x})}{P(Y = 0|x)}\right)$$
$$= \beta_{20} + \beta_{21}x_1 + \beta_{22}x_2 + \cdots + \beta_{2p}x_p$$
$$= (1, \mathbf{x}')\boldsymbol{\beta}_2$$

It follows that the three conditional probabilities for each category of dependent variable given the vector of covariates is:

$$P(Y = 0|\mathbf{x}) = \frac{1}{1 + \exp(g_1(\mathbf{x})) + \exp(g_2(\mathbf{x}))}$$

$$P(Y = 1|\mathbf{x}) = \frac{\exp(g_1(\mathbf{x}))}{1 + \exp(g_1(\mathbf{x})) + \exp(g_2(\mathbf{x}))}$$

$$P(Y = 2|\mathbf{x}) = \frac{\exp(g_2(\mathbf{x}))}{1 + \exp(g_1(\mathbf{x})) + \exp(g_2(\mathbf{x}))}$$

Following the convention given to the binary model, we can do $\pi_j(\mathbf{x}) = P(Y = j|\mathbf{x})$ for $j = 0, 1, 2$ each of which is a function of a vector with $2(p+1)$ parameters, $\boldsymbol{\beta}' = (\boldsymbol{\beta}_1', \boldsymbol{\beta}_2')$.

A general expression for the conditional probability for a model whose dependent variable has three categories, can be given by:

$$P(Y = j|\mathbf{x}) = \frac{\exp(g_j(\mathbf{x}))}{\sum_{k=0}^{2} \exp(g_k(\mathbf{x}))}$$

where the vector $\boldsymbol{\beta}_0 = \mathbf{0}$ and $g_0(\mathbf{x}) = 0$.

To construct the likelihood function, it should create three binary variables coded by "0" and "1" to indicate the group member of an observation. Should be noted that these variables are introduced merely to clarify the likelihood function. The variables are coded as follows:

- if $Y = 0$ than $Y_0 = 1$, $Y_1 = 0$ and $Y_2 = 0$;

- if $Y = 1$ than $Y_0 = 0$, $Y_1 = 1$ and $Y_2 = 0$;
- if $Y = 2$ than $Y_0 = 0$, $Y_1 = 0$ and $Y_2 = 1$.

We note so that independently of the value that Y takes, $\sum Y_i = 1$. The conditional likelihood function for an independent sample of n observations is:

$$l(\beta) = \prod_{i=1}^{n} [\pi_0(\mathbf{x}_i)^{y_{0i}} \pi_1(\mathbf{x}_i)^{y_{1i}} \pi_2(\mathbf{x}_i)^{y_{2i}}]$$

Making the logarithm and using the fact that the $\sum y_{ji} = 1$ for each i, the logarithm of the likelihood function is:

$$L(\beta) = \sum_{i=1}^{n} \{y_{1i} g_1(\mathbf{x}_i) + y_{2i} g_2(\mathbf{x}_i) - \ln[1 + e^{g_1(\mathbf{x}_i)} + e^{g_2(\mathbf{x}_i)}]\}$$

The likelihood equations are the result of first partial derivatives of $L(\beta)$ in order to the $2(p+1)$ unknown parameters. Using the notation $\pi_{ji} = \pi_j(\mathbf{x}_i)$, the form of these equations is:

$$\frac{\partial L(\beta)}{\partial \beta_{jk}} = \sum_{i=1}^{n} x_{ki}(y_{ji} - \pi_{ji}) \tag{1}$$

for $j = 1, 2$ e $k = 0, 1, 2, \ldots, p$; with $x_{0i} = 1$ for each case.

The maximum likelihood estimator (MLE), $\hat{\beta}$, is the result of the resolution of these set of equations equal to 0 in order to β. The method to solve this equation is iterative like in the binary logistic regression.

The interpretation of odds ratios in the multinomial outcome setting could be made generalizing the notation used in the binary outcome case to include the outcomes being compared as well as the values of the covariate [3]. Considering that the outcome labeled with $Y = 0$ is the reference outcome. The subscript on the odds ratio indicates which outcome is being compared to the reference outcome. The odds ratio of outcome $Y = j$ versus outcome $Y = 0$ for covariate values of $x = a$ versus $x = b$ is:

$$\Psi_j(a, b) = \frac{P(Y = j|x = a)/P(Y = 0|x = a)}{P(Y = j|x = b)/P(Y = 0|x = b)}$$

We could obtain the estimates values of odds ratio, labeled $\hat{\Psi}$, by exponentiating the estimated slope coefficients.

3   Results

The sample of 191 individuals consists of 50.8% females, with the remaining 49.2% of male. The distribution of individuals according to sex for each result of the dependent variable is the same, that is, for the three possible results, control, open bite and deepbite, the sample has almost the same distribution according to sex.

For the construction of logistic regression models were used initially 13 variables, 11 referring the HB position and 2 of characteristics of individual. According sample size guidelines for multinomial logistic regression indicate a minimum of 10 cases per independent variable [3]. In our work we have a ratio of 13/191 is greater than 1/10 which indicates that is possible the application of this technique.

Table 1 presents the designation of variables, his metric and categories.

Table 1. Variables identification

Variable	Designation	Scale	type	Code
Age (years)	Age	ratio	independent	
Sex	Sex	dichotomous	independent	1 = female
				2 = male
Hy/Rgn	position1_Hy	ratio	independent	
Hy/C3	position2_Hy	ratio	independent	
Hy/PNS	position3_Hy	ratio	independent	
C3/Rgn	position4_Hy	ratio	independent	
Hy/C3-Rgn	position5_Hy	ratio	independent	
S/Hy	position6_Hy	ratio	independent	
C3/Hy/S angle	position7_Hy	interval	independent	
Hy/C3/S angle	position8_Hy	interval	independent	
NaBa/CCHy angle	position9_Hy	ratio	independent	
C3 - PTV	position10_Hy	ratio	independent	
Hy - PTV	position11_Hy	ratio	independent	
Group	Group	categorical	dependent	0 = control
				1 = open bite
				2 = deepbite

To evaluate the significance of variables to dependent variable we start to run the multivariate analysis using the forward stepwise technique implemented in $IBM^{®}$ $SPSS^{®}$. The results are list in tables (Table 2 to Table 4).

Table 2. Model fitting information

	Model Fitting Criteria	Likelihood Ratio Tests		
Model	-2 Log Likelihood	Chi-Square	df	Sig.
Intercept Only	419.534			
Final	338.476	80.059	10	≈ 0

For model fitting information (Table 2) the probability of the model chi-square (80.059) was less than or equal to the level of significance of 0.05, so the null hypothesis that there was no difference between the model without independent variables and the model with independent variables was rejected. The existence of a relationship between the independent variables and the dependent variable was supported.

We could access the usefulness for this logistic model as suggested by Garson [1]. According Table 3 the classification accuracy rate was 59.2% which is greater than or equal to the proportional by chance accuracy criteria of 41.8% (1.25 x 33.4% = 41.8%).

Table 3. Classification table produced by SPSS

	Classification			
		Predicted		
Observed	Control	Open bite	Deepbite	Percent Correct
Control	30	18	18	45.5%
Open bite	12	43	7	69.4%
Deepbite	19	4	40	63.5%
Overall Percentage	31.9%	34.0%	34.0%	59.2%

In the construction of multinomial logistic regression model to variable selection technique was used the forward stepwise technique, in order to select a set of variables that could contribute to the outcome and that this contribution is reveal statistically significant. The objective is to achieve a model that contains all variables that are important with regard to pre-established criteria of p_E and p_R values (p values of entry and removal of the variable in the model by likelihood ratio test) chosen so that they become statistically significant. The values chosen for this analysis were: $p_R = 0.20$ and $p_E = 0.15$.

The result of parameters estimates was in Table 4.

Table 4. Parameter estimation for multinomial logistic regression

Group[a]		$\widehat{\beta}$	Std. Error	Wald	df	Sig.	$\widehat{\Psi}$	95% CI for Ψ	
								Lower Bound	Upper Bound
Openbite	Intercept	-21.061	5.065	17.288	1	.000			
	position8_Hy	.095	.037	6.710	1	.010	1.100	1.023	1.182
	position6_Hy	.028	.057	0.249	1	.618	1.029	.920	1.151
	position1_Hy	.061	.041	2.186	1	.139	1.063	.980	1.153
	position3_Hy	.138	.090	2.385	1	.122	1.148	.963	1.369
	position5_Hy	-.325	.089	13.290	1	.000	0.722	0.607	0.860
Deepbite	Intercept	7.288	4.979	2.143	1	.143			
	position8_Hy	-.089	.038	5.405	1	.020	0.915	0.848	0.986
	position6_Hy	.148	.057	6.676	1	.010	1.159	1.036	1.297
	position1_Hy	.084	.040	4.345	1	.037	1.087	1.005	1.177
	position3_Hy	-.329	.090	13.417	1	.000	0.720	0.603	0.858
	position5_Hy	.343	.094	13.210	1	.000	1.409	1.171	1.695

[a] The reference category is: Control.

In this analysis, two comparisons will be made, the open bite group will be compared to the control group and the deepbite group will be compared to the control group.

The contribution of each independent variable may be positive or negative depending on the sign of the estimated coefficient $(\widehat{\beta})$. Positive contribution means that when the variable in question is considered, increases the probability of being in the open bite group if we considered the logit 1 or in deepbite group if we considered the logit 2. In contrast, a negative contribution means that when the variable in question is verified, increases the probability of being in the control group.

After designing the model for each logit, we provide an interpretation of estimated coefficients aware of the Table 4, in terms of the odds ratio for each logits formed.

Thus we have for the logit 1, generally, the odds ratio is defined by:

$$\Psi_1 = \frac{P(Y = 1|X = a)/P(Y = 0|X = a)}{P(Y = 1|X = b)/P(Y = 0|X = b)}$$

- **Hy/C3/S angle:** $\widehat{\beta} = 0.095 \implies \widehat{\Psi}_1 = 1.100$. It means that the increase of one unit in the angle measurement Hy/C3/S contributes to the appearance of open bite. That is, for every increase of one unit in the angle Hy/C3/S, the possibility of belonging to the group of open bite is about 1.1 times greater when compared to the control group;
- **S/Hy :** $\widehat{\beta} = 0.028 \implies \widehat{\Psi}_1 = 1.029$. It means that the increase of one unit in the measurement S/Hy contributes to the appearance of open bite. That is, for every increase of one unit in S/Hy, the possibility of belonging to the group of open bite is about 1.03 times greater when compared to the control group;
- **Hy/Rgn :** $\widehat{\beta} = 0.061 \implies \widehat{\Psi}_1 = 1.063$. It means that the increase of one unit in the measurement Hy/Rgn contributes to the appearance of open bite. That is, for every increase of 5 units in Hy/Rgn, the possibility of belonging to the group of open bite is about 1.36 times greater when compared to the control group;
- **Hy/PNS :** $\widehat{\beta} = 0.138 \implies \widehat{\Psi}_1 = 1.143$. It means that the increase of one unit in the measurement Hy/PNS contributes to the appearance of open bite. That is, for every increase of 5 units in Hy/PNS, the possibility of belonging to the group of open bite is about 2 times greater when compared to the control group;
- **Hy/C3-Rgn:** $\widehat{\beta} = -0.325 \implies \widehat{\Psi}_1 = 0.722$. It means that the decrease of one unit in the measurement Hy/C3-Rgn contributes to the appearance of open bite. That is, for every decrease of one unit in Hy/C3-Rgn, the possibility of belonging to the group of open bite is about 1.38 times greater when compared to the control group.

Similarly for the logit 2, the odds ratio is defined by:

$$\Psi_2 = \frac{P(Y = 2|X = a)/P(Y = 0|X = a)}{P(Y = 2|X = b)/P(Y = 0|X = b)}$$

– **Hy/C3/S angle:** $\widehat{\beta} = -0.089 \implies \widehat{\Psi}_2 = 0.915$. It means that the decrease of one unit in the angle measurement Hy/C3/S contributes to the appearance of deepbite. That is, for every decrease of one unit in the angle Hy/C3/S, the possibility of belonging to the group of deepbite is about 1.1 times greater when compared to the control group;
– **S/Hy** : $\widehat{\beta} = 0.148 \implies \widehat{\Psi}_2 = 1.159$. It means that the increase of one unit in the measurement S/Hy contributes to the appearance of deepbite. That is, for every increase of one unit in S/Hy, the possibility of belonging to the group of open bite is about 1.16 times greater when compared to the control group;
– **Hy/Rgn** : $\widehat{\beta} = 0.084 \implies \widehat{\Psi}_2 = 1.087$. It means that the increase of one unit in the measurement Hy/Rgn contributes to the appearance of deepbite. That is, for every increase of 5 units in Hy/Rgn, the possibility of belonging to the group of deepbite is about 1.52 times greater when compared to the control group;
– **Hy/PNS** : $\widehat{\beta} = -0.329 \implies \widehat{\Psi}_2 = 0.720$. It means that the decrease of one unit in the measurement Hy/PNS contributes to the appearance of deepbite. That is, for every decrease of one unit in Hy/PNS, the possibility of belonging to the group of deepbite is about 1.4 times greater when compared to the control group;
– **Hy/C3-Rgn:** $\widehat{\beta} = 0.343 \implies \widehat{\Psi}_2 = 1.409$. It means that the increase of one unit in the measurement Hy/C3-Rgn contributes to the appearance of deepbite. That is, for every increase of one unit in Hy/C3-Rgn, the possibility of belonging to the group of open bite is about 1.4 times greater when compared to the control group.

According to these results, an individual with an angle Hy/C3/S = 98.6, S/Hy = 109.2 mm, Hy/Rgn = 43.1 mm, Hy/PNS = 65.8 mm and Hy/C3-Rgn = 8 mm, the probabilities of belonging to each group are:

$$P(Y = 0|\mathbf{x}) = 0.304561$$

$$P(Y = 1|\mathbf{x}) = 0.542462$$

$$P(Y = 2|\mathbf{x}) = 0.152976$$

So, this individual is more likely to belong to group 1, that is to belong a group with open bite.

4 Conclusions

According to the results obtained it was found that, to evaluate the probability of an individual be normal, have open bite or deepbite in the process of orthodontics judgment based on cephalometric measures of HB, variables that can influence the process are the measures of Hy/C3/S angle, S/Hy, Hy/Rgn, Hy/PNS and Hy/C3-Rgn.

This model contribute to understanding better the position of the hyoid bone and the structures associated with it, with one possible combination of these 5 variables. So we think that it will be a tool to predict more accurately the diagnose of vertical jaw dysplasia.

Acknowledgments. This work is funded by FEDER Funds through the Operational Programme Competitiveness Factors - COMPETE and National Funds through FCT - Foundation for Science and Technology under the Project: FCOMP-01-FEDER-0124-022674.

References

1. Garson, G.D.: Logistic Regression: Binomial and Multinomial. NC Statistical Associates Publishers, Asheboro (2012)
2. Haralabakis, N.B., Toutountzakis, N.M., Yiagtzis, S.C.: The hyoid bone position in adult individuals with open bite and normal occlusion. Eur. J. Orthod. 15, 265–271 (1993)
3. Hosmer, D., Lemeshow, S.: Applied Logistic Regression. John Wiley & Sons (2000)
4. Karacay, S., Gokce, S., Yildirima, E.: Evaluation of hyoid bone movements in subjects with open bite: a study with real-time balanced turbo field echo cine-magnetic resonance imaging. Korean J. Orthod. 42(6), 318–328 (2012)
5. Mickey, J., Greenland, S.: A study of confounder-selection criteria on effect estimation. American Journal of Epidemiology 129, 125–137 (1989)
6. Rocabado, M.: Analisis Biomecanico Craneo Cervical atraves de una Teleradiografia Lateral. Rev. Chil. de Ortodoncia, 5–15 (1984)

Vector-Projection Approach to Curve Framing for Extruded Surfaces

Abas Md. Said

Computer & Information Sciences Dept
Universiti Teknologi PETRONAS
Malaysia
abass@petronas.com.my

Abstract. Curve framing has numerous applications in computer graphics modeling, e.g., in the construction of extruded surfaces. Commonly used curve framing techniques such as the Frenet frame, parallel transport frame and 'up-vector' frame cannot handle all types of curves. Mismatch between the technique used and the curve being modeled may result in the extruded surfaces being twisted. We propose a simple end-to-end vector-projection approach to curve framing. Our results show that the technique yields less twists compared to those based on the Frenet frame.

Keywords: curve framing, Frenet frame, parallel transport frame.

1   Introduction

Curve framing is a procedure to associate coordinate frames to each point on a three-dimensional space curve. This can be depicted as in Fig. 1, where a number of local Cartesian coordinate frames are drawn on a curve. This is very useful when one needs to treat a coordinate locally, rather than globally. It has numerous applications in computer graphics, such as in the construction of extruded surfaces (Fig. 2), providing the orientation for flying objects and visualization during a fly-through.

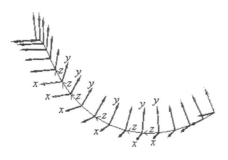

Fig. 1. Some local frames on the spine

B. Murgante et al. (Eds.): ICCSA 2013, Part I, LNCS 7971, pp. 596–607, 2013.
© Springer-Verlag Berlin Heidelberg 2013

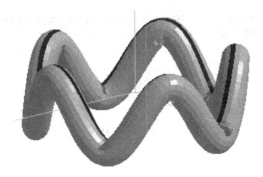

Fig. 2. Extruded surface of a curve

In these applications, frame orientations are essentially calculated based on the curve itself. While a few approaches commonly employed in these applications are the Frenet, parallel transport, 'up-vector' and rotation minimizing frames [1], [2], Cripps and Mullineux [3] make use of curvature and torsion profiles to obtain the frames. Having figured out the frames, the necessary points (or vectors) to construct the surface or orient an object can be easily obtained. While these approaches have been very useful, not all can successfully address all types of curves. At times, the inherent nature of a curve causes 'twists' to occur, giving distorted extruded surfaces or an object flying in a twist. In this paper we look at some of the problems with some of these approaches and propose an interactive technique to address the issue.

2   The Frenet Frame

A Frenet frame for a curve $C(t)$ is formed by a set of orthonormal vectors $T(t)$, $B(t)$ and $N(t)$ where

$$T(t) = \frac{C'(t)}{|C'(t)|} \tag{1}$$

is tangential to $C(t)$,

$$B(t) = \frac{C'(t) \times C''(t)}{|C'(t) \times C''(t)|} \tag{2}$$

is the binormal and

$$N(t) = B(t) \times T(t) \tag{3}$$

is the "inner" normal, i.e., on the osculating circle side of the curve [4], [5] (Fig. 3). A portion of the extruded surface of a curve is depicted in Fig. 3. Depending on the curve, it can be seen that at some point on the curve, i.e., near the point of inflection, the frame starts to twist (Fig. 4), where

$$C(t) = (\cos 5t, \sin 3t, t) \tag{4}$$

while on another (Fig. 5), where

$$C(t) = (\cos t, \sin t, 0) \tag{5}$$

there are no twists at all.

The twist in Eq. (4) occurs because the inner normal now starts to "change side".

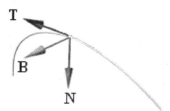

Fig. 3. A Frenet frame

Fig. 4. Twist

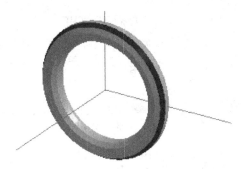

Fig. 5. No-twist

Another problem may also occur when the second derivative of $C(t)$ vanishes, i.e., on a momentary straight line [6]. This is clear from vector $B(t)$ as $C''(t)$ would be identically zero for a straight line.

3   Parallel Transport Frame

Parallel transport refers to a way of transporting geometrical data along smooth curves. Initial vectors u and v, not necessarily orthogonal, are transported along the original curve by some procedure to generate the other two parallel curves as in Fig. 5. Hanson and Ma [6] show that the method reduces twists significantly. In another similar application, Bergou et al. [7] successfully use parallel transport frames in modeling elastic discrete rods. Despite generating good extruded surfaces, the method may not meet requirements for a fly-through.

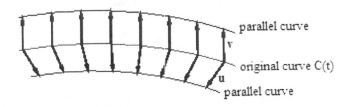

Fig. 6. Parallel curves

4   "Up-Vector" Frame

Another method that we can apply to avoid twists is to use the 'up-vector' [8]. In this approach, we view frames similar to Frenet's, with T_i's, B_i's and N_i's, moving along the curve. Without loss of generality, we can assume y-axis as the global vertical coordinate. We then find the unit 'radial' vector in the unit circle (spanned by the vectors B_i's and N_i's) that gives the highest y-coordinate value (Figure 7). The frame is then rotated about T_i until N_i coincides with the 'radial' vector

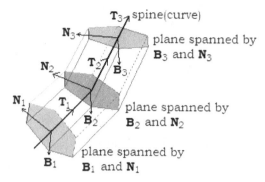

Fig. 7. An 'up-vector' construction

Fig. 8. An 'up-vector' result

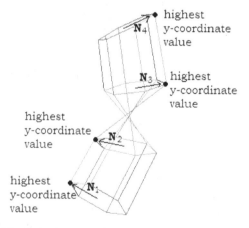

Fig. 9. An 'up-vector' construction and its problem

Fig. 10. An 'up-vector' result

While this approach works on many curves, e.g., as in Fig. 8, it fails when the curve is momentarily vertical or abruptly tilts the opposite 'side' (Figs. 9 and 10).

Applications-wise, this technique corrects the twists in extruded surfaces or a walk-through for many curves. However, it may not be realistic when simulating fast-moving objects, e.g., airplanes, due to issues such as inertia and momentum during flight [9].

5   Vector-Projection Approach

We propose a method to handle the twists by slowly decrementing or incrementing the angle of rotation for each segment of the extruded surface formed with respect to the previous segment.

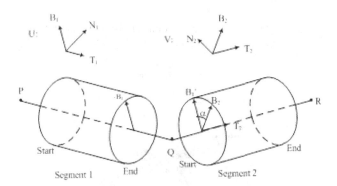

Fig. 11. Determining the angle of rotation

Let P, Q and R be three segment endpoints on a curve where an extruded surface is to be created (Fig. 11). Suppose that the two segments have local coordinate systems U and V respectively. Then the binormal vector B_1 at the end of Segment 1 can be projected onto the plane spanned by the binormal vector B_2 and normal vector N_2 (the start of Segment 2). In this projection, B_1 is mapped onto B_1', giving α, the angle by which the end of Segment 1 should be rotated to coincide with B_2. Alternatively, we can also take α as the angle between N_1' and N_2. Without loss of generality, the end of Segment 1 can be extended to the start of Segment 2, thus closing the gap at Q.

6   Results

We can visually compare the results of our approach with those of Frenet frames for a few curves. This is notably due to the wide-spread adoption of the Frenet frames to extrude surfaces. For a simple sine curve, Fig. 12 shows that the Frenet frame approach twists portions of the extruded surface periodically in the opposite direction whereas our approach produces no or negligible twists (Fig. 13).

Fig. 12. Using the Frenet frame for a sine curve

Fig. 13. Using the vector-projection approach for a sine curve

Fig. 14. Using the Frenet frame for a Lissajous curve

Fig. 15. Using the vector projection approach for a Lissajous curve

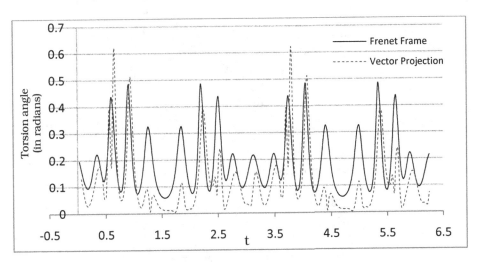

Fig. 16. The torsion angles between successive planes

Similar results are shown for a Lissajous curve (Figs. 14 and 15) with equation

$$C(t) = (2 \sin t, \sin 3t, \cos 5t) \tag{6}$$

We can also look at the performance of both approaches by comparing the angle of twist, or more precisely the torsion angle. The torsion angle in our case is defined as the angle between two successive planes, each plane spanned by B_i and T_i. Essentially, the angle α reduces to $\alpha = \cos^{-1}\left(\dfrac{N_i \cdot N_{i+1}}{\mid N_i \parallel N_{i+1} \mid}\right).$

Fig. 16 shows the angle between successive planes for the two approaches as t increases for the Lissajous curves. On the average, the vector projection approach clearly results in less twist (0.121817) compared to the Frenet frame (0.193118).

Further, we provide more results for three other curves (Figs. 17, 18 and 19). The figures show the images rendered and the average torsion angle between successive planes for the two approaches as t increases. On the average, the vector projection approach clearly results in less twist compared to the Frenet frame.

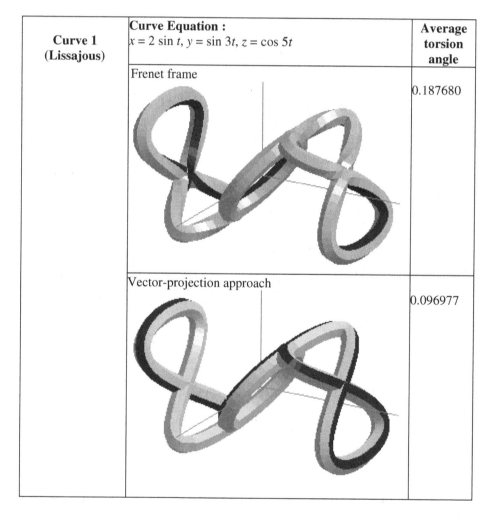

Curve 1 (Lissajous)	Curve Equation : $x = 2 \sin t$, $y = \sin 3t$, $z = \cos 5t$	Average torsion angle
	Frenet frame	0.187680
	Vector-projection approach	0.096977

Fig. 17. A Lissajous curve

Curve 2 (Knot)	Curve Equation : $x =$ $(3 - \cos (5t/3)) \cos t,$ $y =$ $- \sin (5t/3),$ $z =$ $(3 - \cos (5t/3)) \sin t$	Average torsion angle
	Frenet frame 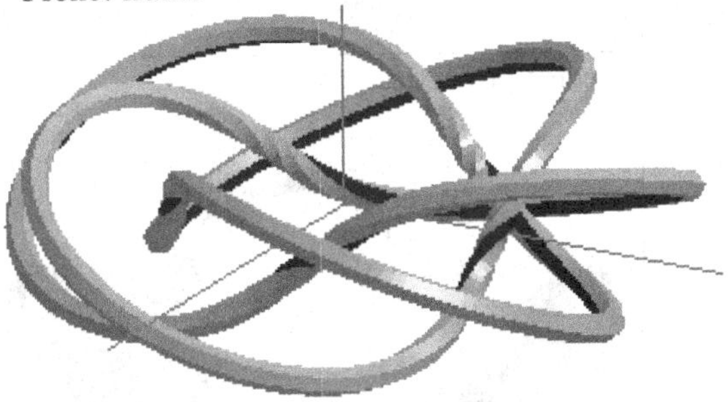	0.074063
	Vector-projection approach 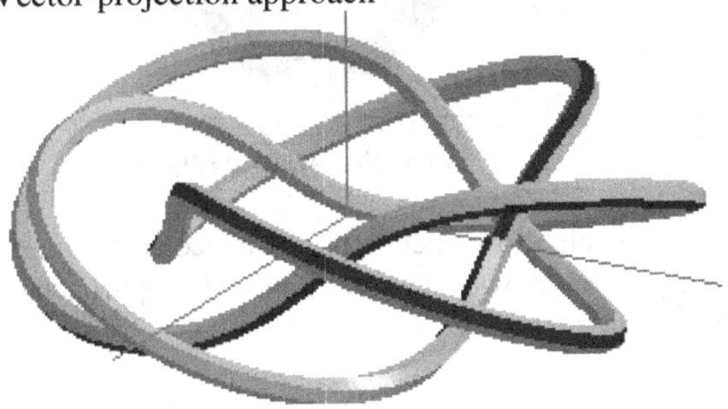	0.029289

Fig. 18. A knot

Curve 3 (Warped ring)	Curve Equation : $x = \cos t, y = \sin t, z = (\cos 3t)/3$	Average torsion angle
	Frenet frame 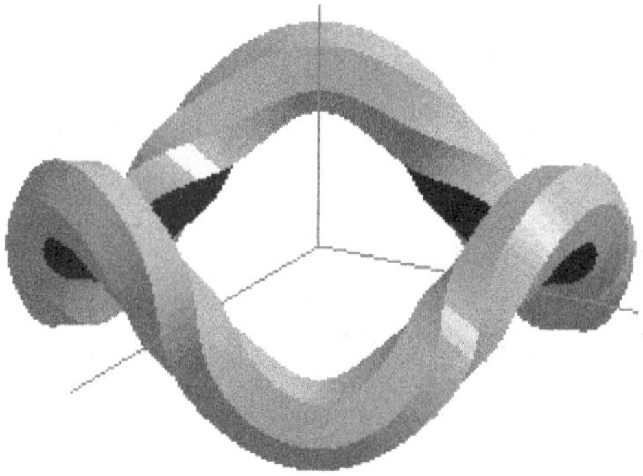 	0.097448
	Vector-projection approach 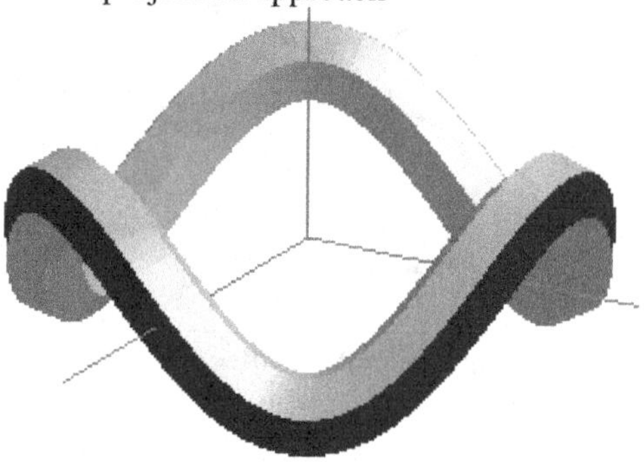 	0.051251

Fig. 19. A warped ring

Overall, our approach shows better performance in terms of the average torsion angle. Visual inspection, however, shows that the approach offers significant improvement in avoiding the twists.

7   Conclusions

Our approach to curve framing results in less average twist compared to the Frenet frame approach and guarantees that extreme twists are avoided. In this case, it is well-suited for extruded surface construction. However, it does not guarantee that the starting angle coincides with the end angle for a closed curve which may be required for some applications. Moreover, from our observation, this approach gives significant improvement when handling the twists occurring at the points of inflection of a curve.

References

1. Poston, T., Fang, S., Lawton, W.: Computing and approximating sweeping surfaces based on rotation minimizing frames. In: Proceedings of the 4th International Conference on CAD/CG, Wuhan, China (1995)
2. Cripps, R.J., Mullineux, G.: Constructing 3D motions from curvature and torsion profiles. Computer-Aided Design 44, 379–387 (2012)
3. Wang, W., Jüttler, B., Zheng, D., Liu, Y.: Computation of rotation minimizing frames. ACM Transactions on Graphics 27(1), 1–18 (2008)
4. Bishop, R.L.: There is more than one way to frame a curve. Amer. Math. Monthly 82(3), 246–251 (1975)
5. Hill, F.S.: Computer Graphics Using OpenGL. Prentice Hall, New Jersey (2001)
6. Hanson, A.J., Ma, H.: Parallel transport approach to curve framing. Technical Report 425, Indiana University Computer Science Department (1995)
7. Bergou, M., Wardetzky, M., Robinson, S., Audoly, S.B., Grinspun, E.: Discrete elastic rods. ACM Transactions on Graphics 27(3), Article 63 (2008)
8. Fujiki, J., Tomimatsu, K.: Design algorithm of the strings - string dance with sound. International Journal of Asia Digital Art and Design 2, 11–16 (2005)
9. Wu, J., Popovic, Z.: Realistic modeling of bird flight animations. In: International Conference on Computer Graphics and Interactive Techniques, ACM SIGGRAPH 2003 Papers, pp. 888–895 (2003)

Spatial Data Management
for Energy Efficient Envelope Retrofitting

Mattia Previtali[1], Luigi Barazzetti[1], and Fabio Roncoroni[2]

[1] Politecnico di Milano, Department of Architecture, Built Environment and Construction Engineering, Via Ponzio 31, 20133 Milano, Italy
mattia.previtali@mail.polimi.it, luigi.barazzetti@polimi.it
[2] Politecnico di Milano, Polo Territoriale di Lecco, Via Marco D'Oggiono 18/A, Lecco, Italy
fabio.roncoroni@polimi.it

Abstract. Retrofitting existing buildings is a key aspect for reaching the proposed energy consumption reduction targets in all countries. The availability of as-built building model is of primary importance for both thermal lack diagnosis and planning retrofitting works. In particular, integration between Building Information Models (BIMs) and Infrared Thermography (IRT) can be a powerful tool for evaluating building thermal efficiency. Indeed, thermography can be efficiently used to detect thermal bridges and heat losses. However, thermographic images does not allow geometrical measurements whereas precise localization of thermal defects is a main aspect for retrofitting intervention planning. For this reason building models can be efficiently integrated with thermal images and provide metric information to thermographic analysis. In this paper a methodology for combining thermographies, acquired by different platforms, and building models automatically derived from laser scanning technology is presented.

Keywords: Building, Energy retrofitting, Laser scanning, Image registration, Thermography.

1 Introduction

Increase in thermal efficiency of existing building is a major task to reach the energy saving targets fixed by different countries. For this reason the development of methodologies for thermal retrofitting has a primary importance. The thermal retrofit of a building is any improvement made to an existing structure which provides an increase in its overall energy efficiency. Although there are many factors influencing buildings' energy efficiency (e.g. the heating and cooling system, quality of the closing, etc.) one of the most important parameters is the composition of façade and cover because it may determine thermal bridges and heat losses due to low efficient insulation materials. Energy efficient envelop retrofitting interventions usually involves sealing building's thermal boundary, sizing and installing insulating panels, using special insulating paints, etc. One of the main problems in retrofitting existing buildings is the energy efficiency evaluation and the precise location of thermal defects.

B. Murgante et al. (Eds.): ICCSA 2013, Part I, LNCS 7971, pp. 608–621, 2013.

Integration of Terrestrial Laser Scanning (TLS) and Infrared Thermography (IRT) is a powerful tool in order to reach this task. Indeed, sensor integration and combination of different surveying techniques proved its usefulness in various fields [1], [2], [3].

Thermal evaluation of existing buildings is widely covered in literature. IRT analysis from the ground are successfully used for recording the irradiation of building façades [4], characterization of its thermal behavior, detecting thermal bridges and heat losses from the envelop [5]. IR images acquired by other platforms are used for different purposes. Satellite images are mainly used for fire detection [6], vegetation monitoring [7] and the analysis of urban heat islands [8] while airborne IRT-systems are applied for vehicle detection [9] and surveying malfunctioning in solar power plan. The development of micro Unmanned Aerial Vehicles (UAVs) in the last years opened new possibilities to IRT analysis [10], [11]. Indeed, UAVs equipped with thermal cameras give the possibility to explore areas inaccessible from the ground like higher floors of buildings and roofs. In addition, the reduction of camera-object distance enhances the ground sampling distance (pixel size on the object) of thermographies. However, the use of IRT data for metric purposes presents several limitations related to the reduced resolution of thermal sensors, the large distortion introduced by the thermographic lens systems, the impossibility to make precise geometrical measurements directly on the thermal images. For this reason, the combination of thermal information derived from thermographies and geometric information of building structure is of major importance.

In particular, 3D laser scanning and close range photogrammetry are the surveying techniques mainly used for producing 3D detailed models of buildings. These techniques have become useful as 3D modeling tools not only in heritage applications [12] where surfaces are complex and irregular, but also for modeling large structures. In particular, an increasing interest is paid to 3D modeling from TLS data. Indeed, TLS allows a rapid acquisition of point clouds describing the building surfaces. In addition, automation in scan registration reached in the last years significantly reduces time for data geo-referencing. Nowadays, great attention is paid to Building Information Models (BIMs) generation from laser scanning data. In contrast to traditional building design (which is largely based on two-dimensional drawings like plans, elevations, sections, etc.), the BIM concept not only extends design to three dimensional space but also augments the 3D representation of the building with time as the fourth dimension and cost as the fifth. Therefore, BIM models covers more than just building geometry [13], they are a combinations of objects, relations and attributes. The use of BIMs goes beyond the planning and design phase of the project, extending throughout the building life cycle. In particular, the BIM logic is based not only on a 3D building geometry but also on semantic and descriptive information that may efficiently support the decision-making process during the retrofitting design.

This paper presents a methodology for the combination of thermal images and building models. Integration is obtained by generating a thermographic textured as-built 3D BIM model of a building for supporting energy efficiency evaluation, diagnosis of thermal anomalies, retrofitting design and work control.

2   The Developed Methodology

The developed methodology for IRT images and geometrical data fusion is summarized in Fig.1 and it is based on mapping thermal data on a 3D façade model. In particular, the procedure can be divided into two main parts: the thermographic image processing and the point cloud analysis. These two surveys can be performed independently (even in different days) and also the data processing phase proceeds for many aspects in a parallel way.

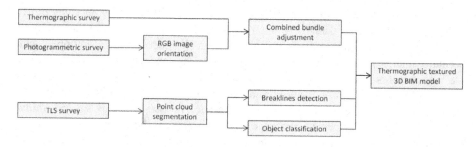

Fig. 1. Workflow of the proposed methodology for the generation of a thermographic textured 3D as-built BIM model

The IRT survey is based on the acquisition of a series of thermographic images of the building either from the ground and by using a UAV platform. In addition, some RGB images are also recorded. The co-registration of both datasets (thermogrphies and TLS point clouds) in the same reference system is a crucial step in order to perform obtain data integration. To this end, thermal images are accurately geo-referenced via standard Least Squares bundle adjustment techniques based on collinearity equations. Indeed, although UAV platforms are generally equipped with on-board GNSS and IMU sensors they are usually not accurate enough for direct geo-referencing of acquired images. However, bundle-adjustment of thermal images presents several challenges: (i) the low resolution of thermal cameras, (ii) the need of an adequate internal camera calibration, and (iii) the limited local contrast of thermal images. In order to partially overcome these problems a better estimation of thermal camera poses with a rigorous procedure was developed starting from thermal camera calibration up to photogrammetric image orientation of both thermal and RGB images in a combined bundle adjustment.

Façade modeling is performed in an automatic way starting from point clouds of building façades acquired with TLS. The developed procedure can be divided into three main steps: firstly the main elements constituting the façade are identified by means of a segmentation process, then breaklines are detected including some priors on architectural scenes, finally a further classification is performed to enrich data with semantics by means of a classification tree.

The last step is the integration of both data generating a thermal textured 3D model of the building in a BIM format. The developed methodology was tested on a building of Politecnico di Milano – Polo Territoriale di Lecco, that is the case study analyzed in this paper (Fig. 2).

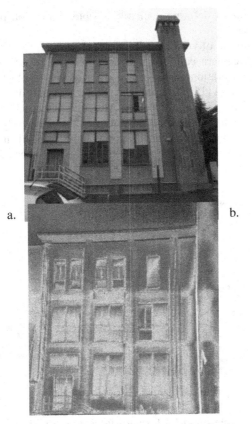

Fig. 2. The case study building: an RGB image (a) and a thermal one (b)

2.1   Thermographic Data Processing

Termographic data processing can be divided into three main steps: (i) thermal camera calibration, (ii) image acquisition and (iii) image orientation. All these phases are discussed in the following sections.

Thermal Camera Calibration

Geometric calibration of a thermal camera is necessary for photogrammetric purposes. In particular, for thermal cameras the *pinhole* model can be assumed for internal calibration. In other words, those cameras can be assumed as central projections with lens distortion and can be calibrated by using standard photogrammetric methods. However, IRT cameras were not designed for metric purposes and their calibration may not be an easy task for several reasons:

   i.   large lens distortion;
   ii.   larger pixel size of IRT with respect to RGB cameras due to longer wavelength of IR spectrum with respect to the visible one;
   iii.   low resolution of IRT sensors compared to RGB cameras;

    iv.   problems in defining a clearly visible calibration polygon for thermal images;

    v.   limited field of view of IRT cameras.

In our case thermal camera calibration was carried out by using the Brown model [14]. In particular, calibration is performed by acquiring a block of images of a so called calibration polygon [15]. The polygon (Fig. 3) consists on a wood frame, with a surface of about 2x4 m, with 40 iron nails. Although planar arrays could be used the nails form a more robust 3D calibration polygon. As previously said, an important aspect concerning the polygon is that its points have to be clearly visible in the IR images. Indeed, nails warm and cool faster than wood and become clearly visible in IR images.

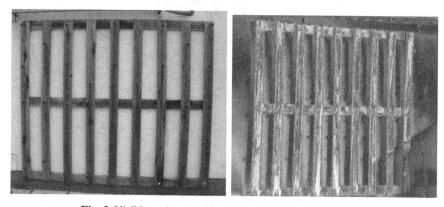

Fig. 3. Visible and thermal image of the calibration polygon

Each nail was also measured by forward intersection with a first order theodolite Leica TS30 (Fig.4.a). The geodetic measurements were included as Ground Control Points (GCPs) in the self-calibrating bundle adjustment with an average theoretical precision of ±2 mm.

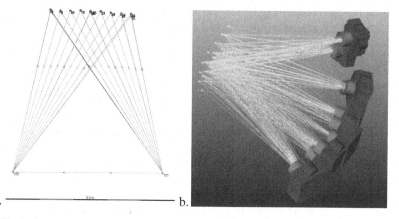

a.   b.

Fig. 4. Thermal camera calibration: the geodetic network for GCP measurement (a) and camera poses estimated with self-calibrating bundle adjustement (b)

Some aspects of the calibration procedure need a special remark. The calibrated camera (a FLIR Tau 640, see next section) showed a large distortion in particular at the image boundaries. For this reason in order to estimate the distortion coefficients in a reliable way, the points used for the bundle adjustment computation should be evenly distributed on the image, covering also regions close to the image borders. However, as distortion parameters are generally highly correlated, some rolled images were included in the adjustment and the radial distortion coefficient k_3 was removed from data processing in order to reduce correlation. A second problem concerns the precision of image coordinates measurements. Thermal images present a reduced resolution and a low local contrast. For this reason the precision of "thermal image points" is worse than that achievable with standard RGB cameras. Finally, the reduced field of view determines the difficulty in acquiring convergent images as requested for a robust calibration (Fig. 4.b).

Image Acquisition
The thermographic survey of the building façade was performed with a FLIR Tau 640 (Fig. 5 - sensor size of 640×480 pixels), equipped with a 19 mm lens system. The camera sensor is a vanadium oxide microbolometer sensitive to wavelengths in the range 7.5 - 13.5 μm. As previously mentioned the thermal images used for this works are acquired both with a UAV platform and from the ground. In particular, the employed UAV system was the Falcon 8 (Ascending Technologies).

Fig. 5. Thermal camera FLIR Tau 640 integrated in the actively stabilized Falcon 8 camera mount

During the thermographic survey the UAV was remotely controlled by a human operator while thermal images were acquired by using a laptop to record the video signal from the thermal camera (Fig. 6).

During the UAV survey some additional RGB images were acquired, by following a conventional motion (two vertical strips), with a customer grade camera (Sony NEX-5N). Those data were integrated from the ground with a photogrammetric block of RGB images (Nikon D700 equipped with a 20 mm lens) and some additional thermographic images.

Fig. 6. Thermal signal recording system (left) and UAV path in blue (right)

Image Orientation

The main issue for an accurate integration of thermographic images and laser scanning is the registration of scans and thermal images in the same reference system, i.e. the estimation of the spatial position of each image with respect to the defined datum. Registration of thermal images can be performed in different ways depending on to object geometry and acquisition scheme. In particular, *homographic* transformations are rigorous only in the case of planar façades. Otherwise, methods based on *collinearity equations* should be used. *Space-resection* techniques are mainly implemented in most commercial software. However, in this case images are processed independently, increasing the number of points to be measured on the images. In addition problems in overlapping zones between images may arise. Those problems can be partially solved by using a rigorous *bundle adjustment*. However, because of thermographic camera peculiarities it is difficult to obtain a block of thermal images suitable for a stable adjustment. Thermal image blocks generally present a low ratio between image baselines and camera-object distances. In addition, standard targets used for scan registration are not clearly visible in thermal images. Therefore interest points are directly measured on the scans in order to register the two datasets. However, this reduces the accuracy of measurements of a factor that is proportional to the scan sampling distance (in other words point density).

In this paper, in order to avoid instability problems and increase accuracy of camera poses estimation, thermal image orientation is performed by adjusting both thermal and RGB images in a combined way [16]. This solution allows a better control on the quality of results and reduces the number of points to be measured with respect to other traditional approaches.

As a first step only RGB images, acquired with a calibrated camera and fulfilling adequate requirements in terms of overlap and acquisition geometry, are oriented within a standard photogrammetric bundle adjustment. This first adjustment is based on a set of Tie Points (TPs) measured on the images, and some Ground Control Points (GCPs) that are used to register the project in the reference system of the laser scans. An important aspect deserves to be mentioned: TPs individuated in this first phase will be then adopted for registration of IRT images. This means that TPs should be preferably measured in correspondence of elements that are clearly visible in both RGB and IR images (e.g. window and door corners). Once RGB images have been registered, IR images can be added to the block and a final bundle adjustment including all data is then carried out to obtain the Exterior Orientation (EO) parameters of all images. This combined adjustment has several advantages:

i.   estimation of EO parameters better than those estimable with standard space resection techniques;

ii. GCPs are initially measured in RGB images instead of laser scans and therefore they provide a better identification;

iii. RGB and IR image acquisition may be carried out independently, even in different days;

iv. new data taken at different epochs can be added for evaluating changes in thermal behavior before and after the thermal retrofitting.

In this case 4 RGB and 3 thermal images were recorded from the UAV while 3 RGB and 1 additional thermal image were acquired from the ground. 14 checkerboard targets (measured with a theodolite Leica TS30) were employed for the registration of laser scans and were also used as GCPs during bundle adjustment for RGB images only (Fig. 7.a). In addition, 20 tie points in correspondence of window and door corners were measured on both RGB and thermal images to run the final bundle adjustment and obtain the EO parameters of all the images (Fig.7.b). Statistics of the combined bundle adjustment show a final sigma-naught of about ±0.9 pixels. Accuracy of check point is about 1 mm in the façade plane and 3 mm in the perpendicular direction.

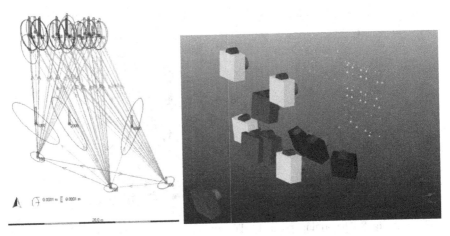

Fig. 7. Scheme of the geodetic network with error ellipses (left). Camera poses (right): thermal images (red), RGB acquired from the UAV (yellow) and RGB acquired from the ground (blue).

2.2   Scan Processing

Once scans are registered together the developed approach for façade modeling allows for automatic planar object extraction and 3D vector model generation [17]. The reader is referred to the cited paper for more information. In the next section only the main aspect of the procedure are described.

In particular, the reconstruction methodology used can be divided in three main steps: (i) point cloud segmentation, (ii) breaklines identification, and (iii) object classification.

Point Cloud Segmentation

As a first task the geo-referenced point cloud is segmented to identify the planar elements (Fig.8) constituting the building façade. In particular, a modified RANSAC algorithm is used in order to reduce bad-segmentation problems associated with massive point clouds. Those problems can be intended as:

- under-segmentation: several features are segmented as one;
- over-segmentation, one feature is segmented into several.

To avoid under-segmentation problem topology information between points is exploited. Indeed, points belonging to the same object should be sufficiently close whereas points belonging to different objects should have a gap area. For this reason, all points forming a connected component are determined by means of a raster bitmap.

Once all planar objects are detected the extracted planes are clustered together to reduce over-segmentation problem. Object clustering is performed by evaluating three parameters: (i) similarity of normal vectors, (ii) perpendicular distance between planes, (iii) intersection between clusters.

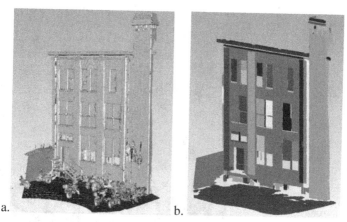

a.   b.

Fig. 8. The original point cloud (a) and its detected planar objects (b)

Breaklines Identification

Once all points are classified into different planes, edges in the point cloud are identified. Edges are extracted by identifying the boundary points of each detected planar element by using a binary image and a Delaunay triangulation. Then, dominant lines are identified for each boundary and a regularization criteria is imposed. Finally, breaklines are reconstructed as the intersection lines between different planes.

The detected lines define surfaces of the Building Information Model in the CityGML scheme.

Object Classification

Finally, object detected in the previous steps are further classified by using some recognition rules derived from knowledge of urban scenes and façade priors.

In particular, the following characteristics are evaluated:

i. object size: walls can be easily distinguished by non-wall elements because they have a larger area;

ii. object position and topology: for example, we assume that doors can be only at the bottom floor, roofs are always on the top of walls and the ground is the object at the lowest level;

iii. object orientation: walls are assumed vertical, while ground and roofs are generally horizontal or mildly sloped;

The above object characteristics and priors are summarized in a classification tree which is used to guide the classification phase in order to obtain a semantically reach model (Fig. 9).

Fig. 9. The CityGML façade model enriched with semantic information on a orthophoto

2.3   Textured As-Built BIM Model Generation

Once thermal images are oriented and building breaklines are extracted the thermographic textured 3D BIM model of the building façade is generated. In particular, the standard used in this work is the CityGML (Fig. 10). The extracted 3D breaklines are used as a constraints to generate the building surfaces by mean of a 2D constrained Delaunay triangulation. After that, triangles are reorganized into rectangular shapes and the following feature types are represented by using the results of the previously described classification step: *Building, RoofSurface, WallSurface, GroundSurface, Window, Door, BuildingInstallation, GroundSurface*. Additional information can be added to the file, e.g. the *Location* which indicate the global position of the building and its orientation. This can be used to evaluate façade exposition and sunlight or other descriptive data which are of major interest for energy efficiency evaluation like the insulation value (U-value) for windows and thermal resistance (R-value) for walls, floors and roofs.

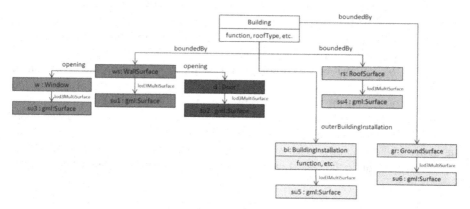

Fig. 10. CityGML feature structure as UML instance diagram

For each surface it is also possible to assign thermographic images as a texture. Indeed, once both data sets (thermal images and 3D façade model) are registered the pointwise correspondences between pixels in the image and points on the elementary surface to be textured is established by means of collinearity equations. Although some commercial software packages accomplish texture mapping, the process in many cases is manual, meaning that the measurement of correspond points (images ↔ model) is carried out interactively per each image. In the case the image orientations can be uploaded in the software (for a more automatic processing) a significant editing phase is needed to fix problems related to occlusions, variations in lighting and camera settings. The algorithm here adopted for texture mapping was implemented in order to minimize self-occlusions and texture-assignment problems [18]. Once texture is assigned, the thermography-textured 3D BIM model (Fig.11) of a building can be interactively browsed, opening in this way new possibilities for the investigators.

In addition, the derived model can be also converted into a .kml file and loaded in Google Earth®. This allows a global contextualization of the building, on-line exploration and sharing (Fig. 11). Starting from the 3D textured model also thermal orthophotos (Fig.12), can be derived and used in GIS environment for further exhaustive analysis.

3   Conclusion

This paper presented a procedure for integrating IRT analysis with BIM models automatically derived from TLS point clouds. The final product is a thermal textured 3D

Fig. 11. Generated 3D BIM model loaded in LandXplorer® (left) and in Google Earth® (right)

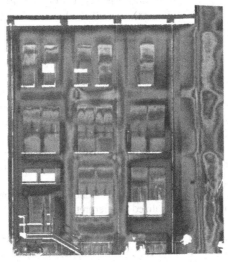

Fig. 12. IRT orthoimage derived by projecting and mosaicking acquired the IR images

BIM model of a building façade. Some relevant aspects of the whole process were covered. In particular, the registration of both sets of data is of major importance to obtain a correct data integration. For this reason, a rigorous combined bundle adjustment of both thermal and RGB images was set up in order to improve numerical stability and robustness of the adjustment.

A method to derive 3D BIM model of building façades from massive unstructured point clouds in an automatic way was also presented. The developed method is an extension of the RANSAC strategy that tries to avoid bad-segmentation results (reported in various papers in literature) by adding some topological information in the process. Starting from the detected planar clusters breaklines are derived by considering some constraints and regularization criteria. Finally, a semantic enrichment of the model is obtained by using a classification tree to recognize façade components. The oriented images and TLS data are then fused into a thermographic textured 3D BIM based on the CityGML standard. This standard can include both geometric and semantic information about the building. Moreover, further data concerning thermal and material properties can be added.

The derived BIM model can be used to identify thermal defects and heat losses in order to provide a support for energy retrofitting design. The possibility to integrate BIM functionalities allows one to perform a cost-benefit analysis maximizing the energy saving.

Finally, as occlusions and data lacks in the point clouds could determine some gaps in the automatically generated model, the derived models can be used as an alternative to manual modeling (in the case a simplified modeling is sufficient) or in conjunction with manual model (if a very accurate model is needed).

Acknowledgements. This work was supported by the EASEE project (Envelope Approach to improve Sustainability and Energy efficiency in Existing multi-storey multi-owner residential buildings). Call identifier FP7-2011-NMP EeB, EeB.NMP.2011-3 "Energy saving technologies for buildings envelope retrofitting".

References

1. Beraldin, J.-A.: Integration of laser scanning and close-range photogrammetry: the last decade and beyond. In: International Archives of Photogrammetry and Remote Sensing, XXth ISPRSCongress. ISPRS, Istanbul (2004)
2. Grubisic, I., Gjenero, L., Lipic, T., Sovic, I., Skala, T.: Active 3D scanning based 3D thermography system and medical applications. In: MIPRO, 2011 Proceedings of the 34th International Convention IEEE, 269 p (2011)
3. Brumana, R., Oreni, D., Cuca, B., Binda, L., Condoleo, P., Triggiani, M.: Strategy for integrated surveying techniques finalized to interpretive models in a byzantine church. International Journal of Architectural Heritage (2013), doi:10.1080/15583058.2011.605202
4. Klingert, M.: The usage of image processing methods for interpretation of thermography data. In: 17th International Conference on the Applications of Computer Science and Mathematics in Architecture and Civil Engineering, Weimar, Germany, July 12-14 (2006)
5. Lo, Y.T., Choi, K.T.W.: Building Defects Diagnosis by Infrared Thermography. Struct. Survey 22(5), 259–263 (2004)
6. Zhukov, B., Lorenz, E., Oertal, D., Wooster, M.J., Roberts, G.: Spaceborne detection and characterization of fires during the bi-spectral infrared detection (BIRD) experimental small satellite mission (2001–2004). Remote Sensing of Environment 100, 29–51 (2006), doi:10.1016/J.RSE.2005.09.019
7. Kerr, J.T., Ostrovsky, M.: From space to species: Ecological applications for remote sensing. Trends in Ecology & Evolution 18(6), 299–305 (2003)
8. Lo, C.P., Quattrochi, D.A.: Land-Use and Land-Cover Change, Urban Heat Island Phenomenon, and Health Implications: A Remote Sensing Approach. Photogrammetric Engineering & Remote Sensing 69(9), 1053–1063 (2003)
9. Hinz, S., Stilla, U.: Car detection in aerial thermal images by local and global evidence accumulation. Pattern Recognition Letter 27, 308–315 (2006)

10. Gaszczak, A., Breckon, T.P., Han, W.J.: Real-time people and vehicle detection from UAV imagery. In: Proc. SPIE Conference Intelligent Robots and Computer Vision XXVIII: Algorithms and Techniques, vol. 7878 (2011), doi:10.1117/12.876663
11. Dios, M.-D., Ollero, A.: Automatic detection of windows thermal heat losses in buildings using UAVs. In: World Automation Congress, WAC 2006, 6 p. IEEE (2006)
12. Oreni, D., Cuca, B., Brumana, R.: Three dimensional virtual models for better comprehension of architectural heritage construction techniques and its maintenance over time. In: Ioannides, M., Fritsch, D., Leissner, J., Davies, R., Remondino, F., Caffo, R. (eds.) EuroMed 2012. LNCS, vol. 7616, pp. 533–542. Springer, Heidelberg (2012)
13. Oreni, D., Brumana, R., Cuca, B.: Towards a Methodology for 3D Content Models. The Reconstruction of Ancient Vaults for Maintenance and Structural Behaviour in the logic of BIM management. In: Guidi, G., Addison, A.C. (eds.) Virtual Systems in the Information Society, pp. 475–482. IEEE, NJ (2012)
14. Brown, D.: Close-range camera calibration. Photogrammetric Engineering and Remote Sensing 37(8), 855–866 (1971)
15. Remondino, F., Fraser, C.: Digital camera calibration methods: considerations and comparisons. International Archives of Photogrammetry, Remote Sensing and Spatial Information Sciences 36(5), 266–272 (2006)
16. Scaioni, M., Rosina, E., Barazzetti, L., Previtali, M., Redaelli, V.: High-resolution texturing of building facades with thermal images. In: Proc. of SPIE Defence, Security and Sensing, Baltimore, USA, vol. 8354, pp. 23–27 (April 2012)
17. Previtali, M., Barazzetti, L., Brumana, R., Cuca, B., Oreni, D., Roncoroni, F., Scaioni, M.: Automatic façade segmentation for thermal retrofit. In: International Archives of the Photogrammetry, Remote Sensing and Spatial Information Sciences, 3D-ARCH 2013 - 3D Virtual Reconstruction and Visualization of Complex Architectures, Trento, Italy, February 25-26, vol. XL-5/W1 (2013)
18. Previtali, M., Scaioni, M., Barazzetti, L., Brumana, R., Oreni, D.: An algorithm for occlusion-free texture mapping from oriented images. In: SECETEC, Venice, Italy, March 28-29, 10 pages (2012)

Training on Laparoscopic Suturing
by Means of a Serious Game

Lucio Tommaso De Paolis

Department of Engineering for Innovation, University of Salento
Lecce, Italy
lucio.depaolis@unisalento.it

Abstract. Serious games usually refer to games that are used for training, advertising, simulation, or education and that provide a high fidelity simulation of particular environments and situations. Serious games are starting to be also employed in surgery-based training applications. This paper presents a serious game for training on suturing in laparoscopic surgery. In particular, it is focused on the physical modeling of the virtual environment and on the definition of a set of parameters used to assess the level of skills developed by the trainees. A pair of haptic devices has been utilized in order to simulate the manipulation of the surgical instruments.

Keywords: serious game, simulation, surgical training, laparoscopic suturing.

1 Introduction

The emerging use of Minimally Invasive Surgery (MIS), born with the aim to reduce the trauma and surgical incisions on patients, has led to increased demand for education and training of surgeons because the traditional approach to learning and teaching based on the use of animals, cadavers or dummies appears obsolete.

As compared to traditional surgical techniques in open surgery, in MIS they operate in a very small workspace, without having a direct view of the organs that are visualized by means of a camera. In this new context, some skills such as eye-hand coordination are fundamental for the success of the surgical operation.

The introduction of the simulation based on the Virtual Reality (VR) technology has completely changed the education and training tools and, since 1998, the virtual simulators are officially accepted as training tools for surgeons. Several studies have demonstrated how the virtual reality-based simulators, compared to training based on traditional methods, can significantly reduce the intra-operative errors [1].

The VR-based simulators allow reproducing different surgical scenarios without risks for the patient and have also the advantage of the repeatability of the training sessions in order to evaluate and study the mistakes; in addition, these simulators provide a objective measurement of the developed skills and it is also possible to reuse many times the simulator because the virtual environment does not undergo the degradation of the traditional toolbox used for training.

B. Murgante et al. (Eds.): ICCSA 2013, Part I, LNCS 7971, pp. 622–631, 2013.
© Springer-Verlag Berlin Heidelberg 2013

Nowadays the VR-based laparoscopic simulators have achieved a high level of maturity and they are by now training tools indispensable for the training of the surgeons in MIS. The use of haptic devices able to provide a force feedback to the user has contributed to the spread of the virtual simulators for surgery.

Among the most common tasks in minimally invasive surgery, the laparoscopic suturing is one of the most frequently performed. It is used for the closure of incisions made during the surgical procedure or the restoration of torn tissue. This task takes some experience in order to be carried out in the best way.

The new trend in the development of virtual surgical simulators is represented by the serious games that are games whose primary purpose is not the entertainment, but the teaching and learning. Although virtual simulations and serious games are conceptually similar and the same technology (hardware and software) can be used, the serious games introduce an element of entertainment and include some of the highlights of the videogame such as challenge, risk, reward and defeat.

The power of serious games in healthcare and surgery has the goal to educate and train people about treatments, medical and surgical procedures.

This paper presents a serious game for training on suturing in laparoscopic surgery. In particular, the system is focused on the physical modeling of the virtual environment and on the definition of a set of parameters used to assess the level of skills developed by the trainees. A pair of haptic devices has been utilized in order to simulate the manipulation of the surgical instruments and NVIDIA PhysX physics engine and Ogre3D graphics engine have been also used.

2   Virtual Surgical Training

The medical profession for centuries has used training based on an apprenticeship model, an approach to learning where a trainee observes a procedure and then practices it under supervision.

As technology has progressed, many different tools and techniques have been deployed to provide added value to the training process, such as using animals or cadavers or by practicing on mannequins. However, the interactions that occur in an animal's or cadaver's tissues differ from those of living humans due to varying anatomy or absence of physiological behaviour. This type of training also raises ethical issues. Mannequins that simulate part or all of a patient provide a limited range of anatomical variability.

An alternative approach that is becoming more and more accepted by the medical community is the use of virtual simulators that can train practitioners on a virtual patient and permit to have a feedback on the performed procedure. This feedback can then be used to refine the required skills until the operator reaches a target level of proficiency before doing the procedure on the real patient. In addition, virtual simulations can provide the user with the possibility to practice the surgical procedure on rare or difficult cases or on virtual models of patients with unconventional anatomy.

3D virtual models offer also the opportunity to be realized from patient medical data in order to replicate a real situation and produce a realistic simulation environment.

This kind of training is particularly important in laparoscopy that is a minimally invasive surgical procedure performed through small incisions using long thin tools [2]. The surgeon's view is provided by means of a camera inserted into the patient body, the manipulation of the surgical instruments is unintuitive and the distance between surgical tool and organ is difficult to be estimated. The practitioner needs long training using the commercial simulators available for this purpose before performing the operation on the real patient.

3   Serious Games

Serious games usually refer to games that are used for training, advertising, simulation, or education and are designed to run on personal computers or videogame consoles. Serious games provide a high fidelity simulation of particular environments and situations that focus on high level skills that are required in the field. They present situations in a complex interactive context coupled with interactive elements that are designed to engage the trainees.

The serious games inherit strengths and weaknesses from the videogames. Further benefits of serious games include improved self-monitoring, problem recognition and solving, improved short-and long-term memory, increased social skills and increased self-efficacy [3].

In contrast to traditional teaching environments whereby the teacher controls the learning (teacher centered), the serious games present a learner centered approach to education in which the trainee controls the learning through interactivity.

Such engagement may allow the trainee-player to learn via an active, critical learning approach. Game-based learning provides a methodology to integrate game design concepts with instructional design techniques to enhance the educational experience.

Virtual environments and videogames offer students the opportunity to practice their skills and abilities within a safe learning environment, leading to a higher level of self-efficacy when faced with real life situations where such skills and knowledge are required.

Serious games provide a balanced combination between challenge and learning. Playing the game must excite the user, while ensuring that the primary goal (acquiring knowledge or skills) is reached seemingly effortlessly, thus creating a 'stealth mode' of learning. Players are challenged to keep on playing to reach the game's objective.

Although game-based learning is becoming a new form of healthcare education, scientific research on this field is limited.

Wang et al. [4] present a physics-based thread simulator that enables realistic knot tying at haptic rendering rate. The virtual thread follows Newton's law and considers main mechanical properties of the real thread such as stretching, compressing, bending and twisting, as well as contact forces due to self-collision and interaction with the environment, and the effect of gravity.

Webster et al. [5] describe a new haptic simulation designed to teach basic suturing for simple wound closure. Needle holders are attached to the haptic device and the simulator incorporates several interesting components such as real-time modeling of deformable skin, tissue and suture material and real-time recording of state of activity during the task.

Le Duc et al. [6] present a suturing simulation using the mass-spring models. Various models for simulating a suture were studied, and a simple linear mass-spring model was determined to give good performance.

Choi et al. [7] explore the feasibility of using commodity physics engine to develop a suturing simulator prototype for manual skills training in the fields of nursing and medicine. Spring-connected boxes of finite dimension are used to simulate soft tissues, whereas needle and thread are modeled with chained segments. The needle insertion and thread advancement through the tissue is simulated and two haptic devices are used in order to provide a force feedback to the user.

Lenoir et al. [8] propose a surgical thread model for surgeons to practice a suturing task. We first model the thread as a spline animated by continuous mechanics. Moreover, to enhance realism, an adapted model of friction is proposed, which allows the thread to remain fixed at the piercing point or slides through it.

Shi et al. [9] present a physics-based haptic simulation designed to teach basic suturing techniques for simple skin or soft tissue wound closure. The objects are modeled using a modified mass-spring method.

Serious games are starting to be also employed in surgery-based training applications.

Dental Implant Training Simulation [10], developed for the Medical College of Georgia and funded through a grant by Nobel Biocare, is a groundbreaking project created to better teach and train dental school students and dental professionals on patient assessment and diagnosis protocol and to practice dental implant procedures in a realistic, 3D virtual environment. The game-based simulation has the aim to improve dental student learning outcomes in the area of diagnostics, decision-making and treatment protocols for enhanced patient therapy outcomes and risk management.

Total knee arthroplasty [11] is a commonly performed surgical procedure whereby knee joint surfaces are replaced with metal and polyethylene components that serve to function in the way that bone and cartilage previously had. The serious game has been designed to train orthopedic surgical residents on surgical procedures, and to gauge whether learning in an online serious gaming environment will enhance complex surgical skill acquisition.

Serious gaming can be used to enhance surgical skills. Ideally, these training instruments are used to measure certain parameters and to assess the trainees' performance. In these games, strict requirements should be met and the interpretation of the game metrics must be reliable and valid. Games used to train medical professionals need to be validated before they are integrated into teaching methods and applied to surgical training curricula [12].

4   Modeling of Virtual Objects

The developed serious game for laparoscopic suturing training was developed using OGRE 3D [13] graphics engine, NVIDIA PhysX for physical modeling of the objects and HAPI library for haptic interactions.

The NVIDIA PhysX [14] is a real-time physics simulation framework; a major advantage of this physics engine is the support of hardware acceleration when a compatible Graphics Processing Unit is installed. Developers can instruct the engine to utilize the processing power of the GPU in order to perform physics computation and to relieve the CPU for other tasks.

PhysX uses a position-based approach for the management of the body dynamics in order to reduce the instability problems that make unnatural the simulation [15].

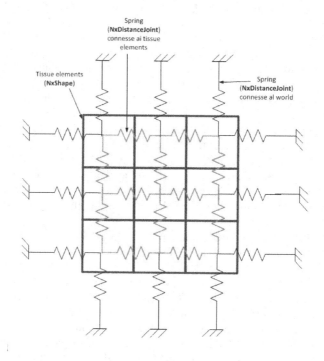

Fig. 1. Tissue modeling using the mass-spring method

The PhysX engine supports the simulation of both rigid bodies and soft objects, including cloth and fluids. It employs a scene graph to manage the objects in a virtual environment.

In order to provide a force feedback to the user, an haptic interface is used in our simulation; it is possible to use two SensAble Phantom Omni or two Novint Falcon haptic devices.

The choice of the preferred haptic devices is possible through the use of the multidevice HAPI library. In order to improve the usability of the simulator it is better to use two interfaces Phantom Omni devices that have a larger workspace and are provided of 6 DOFs [16].

Tissue Modelling

In the simulation, the virtual tissue has been modeled using the mass-spring method [17].

The tissue deformation is based on the dynamics of the masses, and the elasticity and damping of the springs that connect these masses. Additional springs have been used in order to fix the grid of springs in the virtual space, and these springs allow the tissue to resume its original shape when the effect of deformation is terminated.

In Fig. 1 is shown the tissue modeling using the mass-spring method.

Usually, in this kind of modeling, the masses have infinitesimal size, but PhysX does not allow the definition of dimensionless objects and, for this reason, a box of finite size has been used for the creation of a single tissue element.

The tissue rendering is obtained using Ogre3D [18].

In Fig. 2 is shown a wound in the tissue simulation and the fiducial points visualized on the tissue that represent the ideal insertion points of the needle.

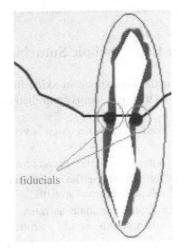

Fig. 2. Example of wound in the tissue simulation

Thread Modelling

The main technique used to model the dynamics of the surgical thread in virtual simulations is known in literature as "follow-the-leader" [19]. The thread is modeled by means of a chain of cylinders connected by joints that allow the bending of the thread.

At each step of the simulation, when an external force is applied to a cylinder, the new position of this cylinder is calculated, and, using the follow-the-leader approach, the new positions of all cylinders are computed; in addition also the collisions between elements of the thread are detected and the new configuration of the entire thread is displayed.

As for the tissue, also for the thread the PhysX features are exploited in order to manage the dynamics (calculation of position and collision detection).

In order to simulate the flexibility of the thread, two cylinders are connected through a spherical joint, which allows the rotation of the elements relative to each other. A spherical joint, shown in Fig. 3, is constrains two points located on two different rigid bodies to coincide in one point; a spherical joint has 3 free DOFs and 3 blocked DOFs. The visual model of the suturing thread is achieved through the rendering of all elements of the chain.

Fig. 3. The spherical joint used in order to simulate the flexibility of the thread

5   Serious Game for Laparoscopic Suturing Training

The serious game is used for the assessment of skills in performing a laparoscopic suture by means of a thread and the two clamps controlled by two haptic interfaces.

The aim of this procedure is:

- to acquire a good eye-hand coordination, that is very important in laparoscopic surgery;
- to improve the ability to manipulate the surgical instruments;
- to learn the techniques for performing the suture node.

The requirements of the developed serious game are:

- Simulation: the system should simulate as much as possible the appearance and the behavior of a real human tissue and a suture thread;
- Configuration: the size of the tissue, the number of elements of the thread, the number and position of the fiducial points on the tissue and the time duration of the test must be specified in an XML configuration file;
- Haptic device - surgical forceps interaction: the user must to be able to move the virtual surgical forceps by means of the haptic device;
- Skill evaluation: the game must implement algorithms for measuring the skill of the user during the execution of the task.

In the assessment of a suturing procedure it is possible to consider some parameters; these parameters are not extracted from the medical literature, which does not specify any quantitative metrics for evaluating the performance of this task.

Some numerical indicators can be achieved by a physically-based virtual simulation; the main parameters used for the assessment of a suturing procedure are:

- Duration time elapsed between the first contact of the needle with the tissue and the completion of the node; less time the surgeon spends and greater is the skill evaluated;
- Accuracy error that is the maximum distance between the ideal point (indicated by a marker) and the real point of entry of the needle into the tissue; smaller is the error and higher is the quality of the node;
- Force peak that is the value of the maximum force used during the simulation in order to piercing the tissue by means of the needle;

- Tissue damage that is the sum of the forces applied to the tissue over the threshold of breakage of the tissue;
- Angle of entry that is the difference between the normal to the surface and the tangent at the point of the needle entry;
- Overall score that is the average of all previous specified parameters;
- Needle total distance that is the total distance traveled from the needle in order to complete the task; shorter is this distance and greater is evaluated the skill of the surgeon. This parameter is not included in the calculation of the quality assessment of the suture task.

The software architecture of the serious game has been developed using the architectural pattern Model-View-Controller (MVC), whose use is not limited to the development of web applications, but also to virtual simulators [20].

The model manages the behavior and objects of the virtual environment, responds to requests for information about its state and responds to instructions to change state.

The view renders the model into a form suitable for interaction and is managed by OGRE 3D graphical engine.

The controller accepts input from the user and instructs the model to perform actions based on that input. There are two subcontrollers:

- The controller of the physical simulation that applies the laws of the physics to the elements of the model (by changing the position, velocity and acceleration of the objects) and handles the collision detection between objects in the scene. This controller is implemented by the PhysX library;
- The controller of the haptic simulation that allows the communication between the model and the haptic device. This controller returns the forces, but is not responsible for the calculation of these, that are computed by the controller of the physical simulation.

In Fig. 4 is shown the serious game using two Novint Falcon haptic devices.

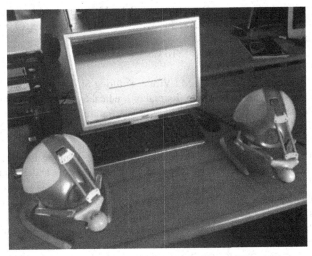

Fig. 4. The serious game using two Novint Falcon haptic devices

6   Conclusions and Future Work

This work has been focused on the design and development of the first prototype of a serious game for training on suturing in laparoscopic surgery.

The main phase of the development process has been the physical modeling of the objects involved in the serious game (the virtual tissue, the suture thread and needle) and the evaluating and measuring of the skills of the trainee. The surgical instruments manipulated by the surgeon are replaced in the serious game by two haptic interfaces.

A future work could be a more accurate implementation of the interaction between the tissue and suture thread and an improvement of the algorithms for the evaluation of skills could be provided of a system to store the results of the training sessions in order to evaluate the progresses in the execution of suturing task.

References

1. Seymour, N.E., Gallagher, A.G., Roman, S.A., O'Brien, M.K., Bansal, V.K., Andersen, D.K., Satava, R.M.: Virtual reality training improves operating room performance: Results of a randomized, double-blinded study. Annals of Surgery 236(4), 458–464 (2002)
2. Liu, A., Tendick, F., Cleary, K., Kaufmann, C.: A Survey of Surgical Simulation: Applications, Technology, and Education. Presence: Teleoperators and Virtual Environments 12(6), 599–614 (2003)
3. Susi, T., Johannesson, M., Backlund, P.: Serious Games – An Overview. Technical Report HS- IKI -TR-07-001, School of Humanities and Informatics, University of Skövde, Sweden
4. Wang, F., Burdet, E., Vuillemin, R., Bleuler, H.: Knot-tying with Visual and Force Feedback for VR Laparoscopic Training. In: Proc. of 27th Annual International Conference of the IEEE Engineering in Medicine and Biology Society (IEEE-EMBS), China (2005)
5. Webster, R.W., Zimmerman, D.I., Mohler, B.J., Melkonian, M.G., Haluck, R.S.: A Prototype Haptic Suturing Simulator. In: Westwood, J.D., et al. (eds.) Medicine Meets Virtual Reality 2001. IOS Press (2001)
6. LeDuc, M., Payandeh, S., Dill, J.: Toward Modeling of a Suturing Task. Graphics Interface, 273–279 (2003)
7. Choi, K.-S., Chan, S.-H., Pang, W.-M.: Virtual Suturing Simulation Based on Commodity Physics Engine for Medical Learning. Journal of Medical Systems, 1–13 (2010)
8. Lenoir, J., Meseure, P., Grisoni, L., Chaillou, C.: A Suture Model for Surgical Simulation. In: Cotin, S., Metaxas, D. (eds.) ISMS 2004. LNCS, vol. 3078, pp. 105–113. Springer, Heidelberg (2004)
9. Shi, H.F., Payandeh, S.: Suturing Simulation in Surgical Training Environment. In: The 2009 IEEE/RSJ International Conference on Intelligent Robots and Systems, St. Louis, USA, October 11-15 (2009)
10. Dental Implant Training Simulation, http://www.breakawaygames.com
11. Sabri, H., Cowan, B., Kapralos, B., Porte, M., Backstein, D., Dubrowskie, A.: Serious games for knee replacement surgery procedure education and training. Procedia - Social and Behavioral Sciences 2(2), 3483–3488 (2010)
12. NVIDIA Corporation, PhysX SDK 2.8 Reference (2008)

13. OGRE 3d, http://www.ogre3d.org
14. Graafland, M., Schraagen, J.M., Schijven, M.P.: Systematic review of serious games for medical education and surgical skills training. British Journal of Surgery 99, 1322–1330 (2012)
15. Müller, M., Heidelberger, B., Hennix, M., Ratcliff, J.: Position based dynamics. J. Vis. Comun. Image Represent. 18, 109–118 (2007)
16. Chan, L.S.-H., Choi, K.-S.: Integrating Physx and Openhaptics: Efficient Force Feedback Generation Using Physics Engine and Haptic Devices. In: Joint Conferences on Pervasive Computing (JCPC), Tamsui, Taipei, December 3-5, pp. 853–858 (2009)
17. Cotin, S., Delingette, H., Ayache, N.: Real-time elastic deformations of soft tissues for surgery simulation. IEEE TVCG 5(1) (1999)
18. Junker, G.: Pro OGRE 3D Programming. Apress (2006)
19. Brown, J., Latombe, J.-C., Montgomery, K.: Real-time knot tying simulation. The Visual Computer 20(2-3), 165–179
20. Maciel, A., Sankaranarayanan, G., Halic, T., Arikatla, V.S., Lu, Z., De, S.: Surgical model-view-controller simulation software framework for local and collaborative applications. International Journal of Computer Assisted Radiology and Surgery 5 (2010)

Walking in a Virtual Town to Understand and Learning About the Life in the Middle Ages

Lucio Tommaso De Paolis

Department of Engineering for Innovation, University of Salento
Lecce, Italy
lucio.depaolis@unisalento.it

Abstract. Edutainment refers to any form of entertainment aimed at an educational role; it enhances the learning environment and makes it much more engaging and fun-filled. The videogame is one of the most exciting and immediate tools of the edutainment applications since the game enables a type of multisensory and immersive relationship of the user through its interactive interface; Virtual Reality technology makes possible to create applications for edutainment purposes and to integrate different learning approaches. One of the important applications of edutainment is the reconstruction of 3D environments aimed at the study of cultural heritage. This paper presents some results of the MediaEvo Project that has led the researchers to use the reconstruction of a town in the Middle Ages in order to develop a multi-channel and multi-sensory platform for the edutainment in Cultural Heritage. MediaEvo project has permitted a didactic experimentation whereby simulation is considered as a precious teaching support tool. The educational MediaEvo game has prompted students to participate in and experience in a simulated and immersive environment of a town in the Middle Ages in order to connect the recreational actions, and to critically discover roles, functions and actions referring to Medieval life.

Keywords: Virtual Reality, edutainment, virtual cultural heritage.

1 Introduction

Virtual Reality (VR) technology enables to create edutainment applications also for the general public. Edutainment is a field that combines education with entertainment aspects; thus, enhancing the learning environment, making it much more engaging and fun-filled.

Virtual Cultural Heritage (VCH) is the use of electronic media to recreate or interpret culture and cultural artefacts as they are today or as they might have been in the past. Virtual Heritage applications employ different kinds of 3 dimensional representations created using Virtual Reality technology.

Building 3-dimensional renderings is an efficient way of storing information, a means of communicating a large amount of visual information, as well as a tool for constructing collaborative worlds where there is a combination of different media and

B. Murgante et al. (Eds.): ICCSA 2013, Part I, LNCS 7971, pp. 632–645, 2013.

methods for an integrated learning approach. By recreating or simulating an aspect of an ancient culture, virtual heritage applications are a bridge between the people of the ancient culture and the modern user.

One of the best uses of the virtual model is that of creating a mental tool to help students to learn about things and to explore ancient cultures and places that no longer exist or that might be too dangerous or too expensive to visit. In addition, it allows students to interact in a novel and very effective way.

Edutainment, a neologism created from the combination of the words education and entertainment, refers to any form of entertainment aimed at an educational role; thus, it enhances the learning environment and makes it much more engaging and fun-filled.

The videogame is one of the most exciting and immediate tools of the edutainment applications since the game enables a type of multisensory and immersive relationship of the user through its interactive interface; moreover, cyberspace of the videogame is a privileged sharing and socializing perspective for players. There are few investments related to the teaching usage of such technologies, since they are still restricted to the entertainment context.

One of the most important applications of VR in edutainment is the reconstruction of 3D environments aimed at the study of cultural heritage; in this field the use of Virtual Reality makes possible to examine the three-dimensional high-resolution environments reconstructed by using information retrieved from the archaeological and historical studies and to navigate in these in order to test new methodologies or to practically evaluate the assessment. The building of three-dimensional renderings is an efficient way of storing information, a means to communicate a large amount of visual information and a tool for constructing collaborative worlds with a combination of different media and methods. By recreating or simulating something concerning an ancient culture, virtual heritage applications are a bridge between people of the ancient culture and modern users.

A very effective way to use VR to teach students is to make them enter the virtual environment as a shared social space and allow them to play as members of that virtual society. The virtual models help students to learn and explore ancient places and different cultures that no longer exist or that might be too dangerous or too expensive to visit.

Several VR applications in Cultural Heritage have been developed, but only very few of these with an edutainment aim.

Song et al. [1] present the historical and cultural content of the reconstructed 3D virtual environment to the general public in a pedagogical and entertaining way; they incorporate interactive storytelling techniques into a Digital Heritage application. Because they believe interactive storytelling techniques can enrich the process of exploring the VE since each visitor can walk away with a different virtual experience.

Kiefer et al. [2] describe a subclass of location-based games, Geogames, which are characterized by a specific spatial-temporal structuring of the game events and assert that spatial-temporal structuring makes it easy to integrate educational content into the game process.

Cutrì et al. [3] study the use of mobile technologies equipped with global positioning systems as an information aid for archaeological visits. They sum up that the use of this kind of technology is an effective tool to promote the archeo-geographical value of the site.

Luyten et al. [4] present Archie, a mobile guide system that uses a social-constructionist approach to enhance the learning experience for museum visitors. They created a collaborative game for youngsters that is built on top of a generic mobile guide framework. The framework offers a set of services such as a rich interactive presentation, communication facilities among the visitors, and the possibility to customize the interface according to the user group.

This paper describes the results of the MediaEvo Project, a digital heritage application aimed to serve as an edutainment tool. Virtual players of the MediaEvo game could explore a reconstructed town and learn about the medieval world through the virtual experience. The game is an immersive environment where the user benefits from multisensory experiences and multimodal learning resources, which enable the learner to use the "visual grammar", typical of a contemporary media system, yet little exploited in educational contexts.

2   Edutainment in Cultural Heritage

The recreational aspects of the game, together with the realistic reconstruction of the scenario components, favor significant learning can be active, constructive, collaborative, intentional, conversational, contextualized and reflexive, at a time.

In the game students carry out didactic activities in which they are responsible for exploration, research and interaction with the objects on the screen, establishing a synergy between the physical and the digital environments, thus actively participating, being aware of reaching mutual goals. On screen activities, moreover, reflect the behaviors and the actions of a real world of the past, also via the reconstruction of local details and the peculiarities of the territory as it was in the past.

The involvement of the users in a sharing and exploration activity in a virtual environment is the assumption for the development of a "culture of participation." This new type of culture demands new abilities, which the contemporary communication scenario and the digital environments contribute to develop.

The educational videogame prompts students to participate in and experience (experiencing phase) - in a simulated and immersive way - the reality of the past and the daily life and material culture of the period. Multimedia resources integrating the videogame, connected to the recreational actions, enable to delve into the theoretical links and to discover - critically - roles, functions and actions referring to medieval life (conceptualizing and analyzing). Throughout the didactic process, the student may easily reconstruct, independently and fully, the "reality" of the time, and apply the concepts acquired to a more familiar scenario, closer to his own experience (applying phase).

The implementation of such knowledge phases is favored by placing the MediaEvo game into a more complex type of educational narration that includes the various tools and supports to learn medieval history.

Evolution in research methodology corresponds to a general debate on communication and education closely linked to the characteristics of a changing perception of teaching, oscillating between experimental impulses and conservative attitudes.

The improvement in technological capabilities enriches the possibilities for research and protection and enhances the value of cultural heritage, thus halting their demise. Firstly, the increased speed of communication and data exchange within the research community offers the dimension of real time interconnectivity. Secondly, the overall amount of information originating from both qualitative and quantitative exploration with the support of technologically advanced equipment, compared with that of a few decades ago, leads to the possibility of an extremely detailed description of reality.

Simulation enables the construction of a platform that adds the definition of game rules and plots to interaction and immersion. The final goal of the definition of the historical landscape is a cybernetic world that will create infinite possible simulations, not necessarily bound to physical reality, based on the algorithms that encode the understanding of ancient situations [5], [6].

At present, many experiences of interactive reconstruction have been presented; these primarily concern the elaboration of algorithmic models in order to better comprehend and reconstruct the sites, technological applications for Virtual Reality on cultural heritage in order to facilitate access to and reading of the cultural heritage.

The reconstruction of the site of Faragola (Foggia) by the University of Foggia, undertaken as part of the Itinera Time Machine Project, fits within the trend of an experiential relationship within an archaeological context (http://www.itinera.puglia.it).

The Appia Antica Project, a digital archive of the monuments of the park, employing different technologies for a 3D representation of the landscape and integrating instruments for topographic surveys and methodologies of surveying on site (http://www.appia.itabc.cnr.it).

Virtual Rome is an Open Source web VR project, based on geospecific data, 3d models and multimedia contents for the interpretation, reconstruction and 3d exploration of the archaeological and potential landscape of ancient Rome (http://www.virtualrome.itabc.cnr.it). The purpose is the creation of a three-dimensional on line 3D environment, embedded into a web-browser.

Rome Reborn [7] is an international initiative whose goal is the creation of 3D digital models illustrating the urban development of ancient Rome from the first settlement in the late Bronze Age (ca. 1000 B.C.) to the depopulation of the city in the early Middle Ages (ca. A.D. 550). The leaders of the project decided that A.D. 320 was the best moment in time to begin the work of modeling; at that time, Rome had reached the peak of its population, and major Christian churches were just beginning to be built. Having started with A.D. 320, the Rome Reborn team intends to move both backwards and forwards in time until the entire span of time foreseen by our mission has been covered. The project has led to the building of the "Ancient Rome 3D" layer in Google Earth in 2008.

Medieval Dublin (http://www.medievaldublin.ie) is an interactive DVD box set that illustrates life in Dublin between 800 and 1540 AD. A 3D model and an interactive timeline display 700 years of history covering daily life, historical events, interesting stories and famous characters.

On a strongly interactive level, related specifically to multichannel edutainment, examples of applications utilizing Virtual Collaborative Environments are:

- City Cluster (http://www.fabricat.com).
- The Quest Atlantis Project (http://atlantis.crlt.indiana.edu).
- Integrated Technologies of Robotics and Virtual Environment in Archaeology Project (http://www.vhlab.itabc.cnr.it).

3   The MediaEvo Project

The MediaEvo Project aims to develop a multi-channel and multi-sensory platform in Cultural Heritage and to test new data processing technologies for the realization of a digital didactic game oriented to the knowledge of Medieval history and society [8], [9].

In addition, the MediaEvo Project investigates the use of digital entertainment and performance media to enhance the communication of cultural heritage and history in order to increase our knowledge of such a relevant part of our history. The idea of the MediaEvo Project is the development of an innovative game engine that has to be specialized for an educational application with multichannel and distributed fruition with an edutainment purpose [10].

MediaEvo project has conducted didactic experimentation and research activities whereby simulation is considered as a precious teaching support tool. Simulation involved in the didactic game enables:

- to create a significant learning environment.
- to trigger the culture of participation.
- to develop experiential learning processes in remote spatial-temporal contexts, other than those of students.
- to increase motivation and involvement of students.

The game is meant to be a tool to experience a loyal representation of the possible scenarios of the everyday life and activities of the people during the reign of Frederick II (XIII century), describing them in simple language but with the utmost fidelity to history.

The framework will have features of strategy games, in which the decision capabilities of a user have a big impact on the result, which in our case is the achievement of a learning target. Nevertheless, the strategy and tactics are in general opposed by unforeseeable factors (provided by the game), connected with the edutainment modules, in order to provide a higher level of participation, which is expressed in terms of the effortlessness with which it is learnt. The idea is to trigger competition between the players, during their learning.

The system, on the basis of a well-defined learning target and, eventually, based on the knowledge of the user, will continuously propose a learning path (learning path composed by a sequence of learning objects), in order to enable the achievement of particular learning results.

The use of digital entertainment and performance media, then, can enhance the communication of cultural heritage and history in order to increase our knowledge of such a relevant part of our history.

In the MediaEvo Project, other application fields have been tested on the game: new peripherals for motion and interaction, virtual treasure hunts, Augmented Reality and evaluation of the scheme for territorial marketing and touristic promotion.

The scheme of the MediaEvo Project work organization is shown in Figure 1.

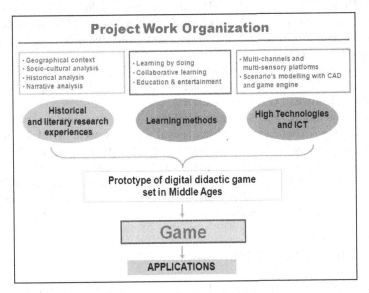

Fig. 1. Scheme of project work organization for MediaEvo Project

We chose Otranto as an example town; Otranto is located in the easternmost tip of the Italian peninsula, in Puglia, in what is dubbed Italy's heel. Due to its geographical position, Otranto was like a bridge between East and West and played an important connective role in the Middle Ages, from a historical and cultural point of view. The town of Otranto was identified as a unique and eloquent historical setting for the project.

Through its art, spatial relationships and landscape, Otranto provides evidence of the close contact between Mediterranean cultures, particularly those of Western Roman Catholicism, Byzantium and Islam. The year considered emblematic for the medieval reconstruction is 1227, the year in which Emperor Frederick II of Swabia and his court entered the city for the first time to embark for the Sixth Crusade [11], [12].

From the analysis of monuments and documents, numerous useful points that facilitate the multicultural experience appear to enhance the educative platform for immediate reference.

The cultural melting pot of the town has produced a particular blend of knowledge and traditions, still recognizable in some of the customs, handicrafts, and figurative art and in the expression of space in Otranto.

Given the knowledge we have of the town in the Middle Ages, it may be almost impossible to carry out a full reconstruction; what could be experienced in the game is the immersion in a virtual environment that can easily enhance the communication of historical research and understanding of the birth and life of cultural heritage.

During its definition, the platform, that has been planned for educational purposes, has proved to be useful for testing researchers' hypotheses about the ancient town and its everyday life.

For the reconstruction on the virtual town, the general information we first collected Digital Terrain Models (DTM), thematic, technical, hydro-geological, nautical charts. Locally, surveyor operations produced maps of street organization, urban limits and fortifications, monuments and materials, referenced to absolute coordinates (mapscape).

Unfortunately, during the last centuries, there has been a substantial loss of historical documents and the ones that survived are not enough to describe resourcefully the town of Otranto in Middle Ages.

The numerous gaps regarding, above all, the urban structure and placement of notable buildings, monumental and functional contexts were filled in part by a historical-urban and architectonic analysis in order to establish the spatial hierarchy, the urban poles, the lot sizes and the typological distribution.

The material elements were compared with analogous situations relating to surrounding areas or cities and modulation grids on a typological-functional basis were used for the built environment, the objects, the clothes and activities.

4   Building of the Virtual Town

Architectural contents were modelled using the Torque Constructor editor of the Torque 3D engine.

Torque Constructor has proved to be an efficient tool for the direct implementation of basic 3D graphics models. In particular, it has many geometrical tools for the graphic processing of the reality context and different controls to select the top of the structure or individual brush model.

We found some difficulties in modelling complex buildings because of the lack of many useful features implemented in professional 3D software. For this reason, the reconstruction of big monuments has been carried out using a CAM in order to obtain a more accurate definition of the architectural structures. Subsequently, we imported these models into the Torque 3D engine.

A stable algorithm was implemented to import CAD objects into the Torque Game Engine platform and to ensure navigation into each graphic structure. This technique, together with an efficient system for exporting textures and paintings, will be used to perform graphic complex environments for the 2D/3D reconstruction in cultural heritage. All the models have been imported into the Torque 3D engine.

St. Peter's Church was found to be useful for testing the importing system both for its characteristic modularity and for its historical relevance as a single byzantine building located in a medieval context. Figure 2 shows the reconstruction of St. Peter's Church; in particular, 2(a) depicts the scheme of the reconstructed church with (in black) a chapel that existed in the Middle Ages and was afterwards destroyed [13].

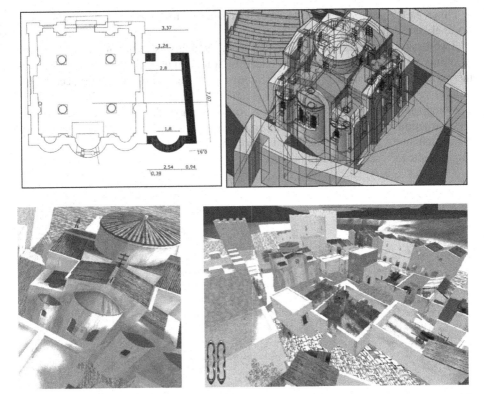

Fig. 2. The reconstruction of St. Peter's Church: (a) scheme of the reconstructed church with the later removed chapel in black; (b) virtual model made using a CAD software; (c) the reconstructed church; (d) the church in its surroundings

Figure 3 shows the reconstruction of Otranto Cathedral and the mosaic of the internal floor of the church.

Fig. 3. The reconstruction of the Otranto Cathedral

Other parts, walls and external structures such as towers or gates, complete the original landscape for the game.

Two basic houses have been deducted from the analysis of urban tissue. The first dwelling consists of a unit cell surrounded by a rectangular court; this is considered the initial settlement model for all ancient towns; the second one is the terraced house unit.

Composing and varying those units on the particular scheme leads to the reconstruction of the urban medieval space in the game. The modular elementary residential units have been designed according to the local medieval unit system. Figure 4 shows a set of residential units.

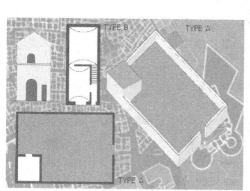 

Fig. 4. A set of residential units

5   Interaction Modalities in the Virtual Town

For the context of edutainment for cultural heritage, in this issue various virtual interactions into the Torque Game Engine platform have been produced. These interactions have been placed into some checkpoints of the Torque virtual environment; these checkpoints make it possible to trigger particular audio or/and video events during the navigation of the game player. Each event manages a particular action in the game mission; hence, events can be used for controlling the collisions with game objects placed in the virtual environment [14].

When the game starts, the player sees a multimedia presentation (a videoclip developed by experts in medieval history/art) with a short introduction to the history of Otranto. Then, the player enters the virtual world and s/he could choose between navigation with a guide or free surfing without the guide.

The guide is a facilitator and gives the player suggestions to follow a specific navigation path. The player can start multiple paths and select among four different Interest Points (IPs): the Cathedral (IP1), St. Peter's Church (IP2), the Castle (IP3) and the Town Walls (IP4). Figure 5 shows the Interest Points of the game.

The player can ask the guide to be tele-transported to some points of interest. This option is a learning strategy to turn the player's attention to specific educational objectives.

There may also be a case whereby the player, without the help of a guide, can freely surf the virtual environment; in this case the only helping tools are some "road signs" located at the crossroads in order to direct the player to the Points of Interest.

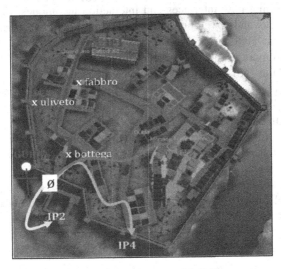

Fig. 5. The Interest Points in the MediaEvo game

There are also some "Intermediate Interest Points" (IIPs), such as the workshop, the blacksmith's shop, the olive tree grove, etc. These IIPs will provide other multimedia educational contents (videos, texts, audio files, images) for the player. In fact, it is also possible to insert audio elements and to run video clips when the player reaches some specific checkpoints (yellow cubes) distributed in the virtual environment.

Figure 6 shows the market with the checkpoint in yellow.

Fig. 6. The market with the checkpoint

6   Walking in a Virtual Town

The Human-Computer Interaction (HCI) technology provides methodologies and methods for designing new interfaces and interaction techniques, for evaluating and comparing interfaces and developing descriptive and predictive models and theories of interaction. The HCIs improve interactions between users and computers by making computers more usable and receptive to the user's needs.

Researches in the HCI field focus on developing new design methodologies and new hardware devices, as well as on exploring new interaction paradigms and theories. The end point in the interface design would then lead to a paradigm in which the interaction with computers becomes similar to the one between human beings.

The techniques for navigation within virtual environments have covered a broad variety of approaches ranging from directly manipulating the environment with gestures of the hands, to indirectly navigating using hand-held widgets, to identifying body gestures or recognizing speech commands. Perhaps the most popular style of navigation control for virtual environments is directly manipulating the environment with gestures or movements of part of the user's body.

In the last few years, systems based on locomotion interfaces and on control navigation by walking in place for the navigation in a virtual environment have also been developed.

In the MediaEvo Project we present an application of navigation and interaction in a virtual environment using the Wiimote (word obtained as a combination of "Wii" and "Remote") and the Balance Board of Nintendo (http://www.nintendo.com).

The aim is to make interaction easier for users without any experience in navigation in a virtual world and more efficient for trained users. Hence, the need to use some intuitive input devices oriented to its purpose and that can increase the sense of immersion.

Wii is the last console produced by Nintendo; the reasons for the success of this device can be undoubtedly found in the new approach that the gaming console gives the user in terms of interaction that effectively makes it usable and enjoyable by a large part of users. The secret of this usability is the innovative interaction system; the Wiimote replaces the traditional gamepad controller type (with cross directional stick and several buttons) with a common object: the remote control.

The Wiimote is provided with an infrared camera that can sense the infrared LED of a special bar (Sensor Bar) and it can interpret, by means of a built-in accelerometer, the movements of translation, rotation and tilt.

The Wiimote has been equipped with a series of accessories that increase its potential, such as the Balance Board, that, by means of four pressure sensors at each corner, is able to interpret the movements of the body in order to control the actions of the user in a videogame.

Because we walk on our feet, controlling walking in Virtual Reality could be felt as more natural when done with the feet rather than with other modes of input. For this reason we used the Nintendo Balance Board as an input device for navigation that offers a new and accessible way to gain input. It is a low-cost interface that transmits the sensor data via Bluetooth to the computer and enables the calculation of the direction the user is leaning to. Figure 7 shows the interaction modalities of Wiimote and Balance Board.

In addition, in order to implement the control of different views and to change the point of view of the user, in our application we use the Nintendo Wiimote and the interaction by means of this device has the aim to simulate the use of the mouse.

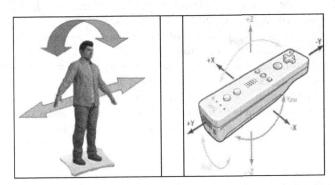

Fig. 7. Interaction modalities of Wiimote and Balance Board

Figure 8 shows the use of Wiimote and Balance Board in the MediaEvo game. Since the frequency of communication between the Wii console and the Wiimote/Balance Board is that of the standard Bluetooth, these devices can be used as tools to interact with any computer equipped with the same technology. Appropriate libraries have been implemented to enable interfacing between these devices and a computer.

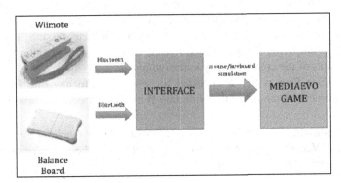

Fig. 8. Use of Wiimote and Balance Board in the MediaEvo game

A software layer that allows the Balance Board and the Wiimote to be used as input devices for any application that runs on a computer has been performed. The aim is to make it possible to receive signals and instructions from the Wiimote and the Balance Board and to translate these into orders for the computer to emulate the keyboard and the mouse.

The application, created to provide a new system of interaction in the virtual world of the MediaEvo Project, can be coupled to any application of navigation in a virtual world.

The modalities of interaction provided by the application involve the use of the Wiimote and Balance Board simultaneously. In particular, the user is able to move the avatar in the virtual environment by tipping the scales in the direction in which he/she wants to move; an imbalance in forward or reverse leads to a movement forward or backward of the virtual character, while the lateral imbalance corresponds to the so-called "strafe" in video games, where the movement is made on the horizontal axis while maintaining a fixed pointing direction of the gaze.

Regarding the interaction by means of the Wiimote, the aim is to simulate the mouse using two modalities of interaction.

The developed software uses two open-source libraries in C# and the WiimoteLib, a library for interfacing the Nintendo Wiimote and other devices (such as the Balance Board) in an environment .NET (http://www.brianpeek.com). The purpose of this library within the application is to simulate the use of a mouse and a keyboard starting from the properly interpreted and translated inputs received from the Wiimote and Balance Board.

7   Conclusions

The aim of the MediaEvo Project is the development of a multi-channel and multi-sensory platform for the edutainment in Cultural Heritage.

This paper presents the some results of the project that has led researchers to use the reconstruction of the city of Otranto in the Middle Ages in order to determine the conditions for testing more elements of interaction in a virtual environment and a multisensory mediation in which merge objects, subjects and experiential context.

A series of properties have been defined to give the game platform an effective educational value; the immersive environment appears to be particularly innovative at the level of design, as it has welcomed interdisciplinary issues and accurate methodological and critical suggestions, related to the didactic potentials of the new media, as well as of the contemporary narrative forms.

The MediaEvo Project has also tested the possibility to navigate in a complex virtual environment by means of the Nintendo Wiimote and Balance Board.

Acknowledgements. The author wishes to thank Maria Grazia Celentano, Luigi Oliva, Pietro Vecchio and Massimo Limoncelli for their contribution in the development of the MediaEvo platform.

References

1. Song, M., Elias, T., Martinovic, I., Mueller-Wittig, W., Chan, T.K.Y.: Digital heritage application as an edutainment tool. In: Proceedings of the 2004 ACM SIGGRAPH International Conference on Virtual Reality Continuum and Its Applications in Industry, Singapore, pp. 163–167 (2004)
2. Kiefer, P., Matyas, S., Schlieder, C.: Learning about Cultural Heritage by playing geogames. In: Harper, R., Rauterberg, M., Combetto, M. (eds.) ICEC 2006. LNCS, vol. 4161, pp. 217–228. Springer, Heidelberg (2006)

3. Cutrí, G., Naccarato, G., Pantano, E.: Mobile Cultural Heritage: the case study of Locri. In: Pan, Z., Zhang, X., El Rhalibi, A., Woo, W., Li, Y. (eds.) Edutainment 2008. LNCS, vol. 5093, pp. 410–420. Springer, Heidelberg (2008)
4. Luyten, K., Schroyen, J., Robert, K., Gabriëls, K., Teunkens, D., Coninx, K., Flerackers, E., Manshoven, E.: Collaborative gaming in the Gallo-Roman museum to increase attractiveness of learning cultural heritage for youngsters. In: Second International Conference on Fun and Games 2008, Eindhoven, The Netherlands, October 20-21 (2008)
5. Pescarin, S.: Reconstructing Ancient Landscape, Archaeolingua, Budapest, pp. 21–23 (2009)
6. Pescarin, S., Valentini, L.: Databases and Virtual Environments: a Good Match for Communicating Complex Cultural Sites. In: ACM SIGGRAPH 2004, Los Angeles (2004)
7. Wells, S., Frischer, B., Ross, D., Keller, C.: Rome Reborn in Google Earth. In: 37th Proceedings of the CAA Conference, CAA 2009. Making History Interactive, Williamsburg, Virginia, March 22-26, pp. 373–379. Archaeopress, Oxford (2010)
8. De Paolis, L.T., Celentano, M.G., Vecchio, P., Oliva, L., Aloisio, G.: Otranto in the Middle Ages: a Virtual Cultural Heritage Application. In: 10th VAST International Symposium on Virtual Reality, Archaeology and Cultural Heritage (VAST 2009), Malta, September 22-25 (2009)
9. De Paolis, L.T., Celentano, M.G., Vecchio, P., Oliva, L., Aloisio, G.: A Multi-Channel and Multi-Sensorial Platform for the Edutainment in Cultural Heritage. In: IADIS International Conference Web Virtual Reality and Three-Dimensional Worlds 2010, Freiburg, Germany, July 27-29 (2010)
10. Oliva, L., Aloisio, G., De Paolis, L.T.: Otranto nelMedioevo. Ricerca e Ricostruzione Urbana per l'Edutainment. REM 1(2), 199–212 (2009)
11. Houben, H.: Otranto nelMedioevo: traBisanzio e l'Occidente, Congedo, Galatina, Lecce (2007)
12. Houben, H.: La conquistaturca di Otranto (1480) trastoria e mito, Congedo, Galatina, Lecce (2008)
13. Safran, L.: Pietro at Otranto: Byzantine Art in South Italy, Rari Nantes, Roma (1992)
14. Finney, K.C.: 3D Game Programming All in One. Thomson Course Technology, Boston (2004)

Interactive Mesh Generation with Local Deformations in Multiresolution

Renan Dembogurski[1,*], Bruno Dembogurski[2],
Rodrigo Luis de Souza da Silva[1], and Marcelo Bernardes Vieira[1]

[1] Department of Computer Science, Universidade Federal de Juiz de Fora (UFJF)
DCC/ICE, R. Lourenço Kelmer, 36036-330, Juiz de Fora, MG, Brazil
[2] Computer Institute - IC, Universidade Federal Fluminense (UFF) - IC/UFF
Rua Passo da Pátria 156 - Bloco E, 24210-240, Niterói, RJ, Brazil
{renan.augusto,rodrigoluis,marcelo.bernardes}@ice.ufjf.br
bdembogurski@ic.uff.br

Abstract. This work presents a method to model a spherical mesh by modifying its heightmap in an augmented reality environment. Our contribution is the use of the hierarchical structure of semiregular A4-8 meshes to represent a dynamic deformable mesh suitable for modeling. It defines only a fraction of the overall terrain that is subjected to local deformations. The modeling of spherical terrains is achieved with proper subdivision constraints at the singularities of the parametric space. An error metric dependent on the observer and on the geometry of the topography was used to provide fast visualization and editing. The results demonstrate that the use of the A4-8 mesh combined with the tangible augmented reality system is flexible to shape spherical terrains and can be easily modified to deal with other topologies, such as the torus and the cylinder.

Keywords: Mesh Deformation, 4-8 Semi-regular Adaptive Meshes, Augmented Reality.

1 Introduction

Mesh deformation is an important resource for the object modeling area, allowing the modification of a surface to suit a particular purpose. Due to its importance, several deformation techniques have been developed, as seen in [1], [2] and [3], where some of these take into account the decomposition of highly detailed surfaces in hierarchical levels [4], [5]. Representing a surface by multiple levels of detail allows changes to be made at any of these levels, resulting in a fine control of the mesh.

A hierarchical structure that has favorable properties for representation and modeling of a terrain at various levels of detail is a semi-regular 4-8 adaptive mesh (A4-8 mesh [6]). Through this structure, it is possible to represent terrains

---

* Authors thank to Fundação de Amparo à Pesquisa do Estado de Minas Gerais and CAPES for financial support.

B. Murgante et al. (Eds.): ICCSA 2013, Part I, LNCS 7971, pp. 646–661, 2013.

with different resolutions, allowing punctual control over the generated mesh. The A4-8 can also define a parametric space to calculate coordinates on a $\mathbb{R}^3$ Euclidean space.

One of the possible ways to interact with 3D models is through fiducial markers in an augmented reality environment. In this scenario, real and virtual elements are combined to create the impression that they coexist in the same scene. To interact with virtual models an interface called Tangible Augmented Reality (TAR) [7] can be used, where each virtual object is associated with a physical object and the user interacts with the virtual objects by manipulating tangible objects.

This work proposes a method to deform spheric terrains through the combination of an A4-8 structure and tangible markers. Local deformations are applied to a fraction of the overall terrain, resulting in an application that allows both visualization and modification of a terrain in real time. The A4-8 hierarchical structure provides the method with a precise control over the surface to be modified.

2   Related Work

There are not many works that deal with the problem of deforming meshes through augmented reality. Also, the focus of most studies involving object modeling and augmented reality is in the deformation models in general, or the development of applications for remote areas such as surgical operations simulation [8], clothes modeling [9] and accidents modeling [10].

A work that allows interactive modification of multiresolution polygon meshes through a 3D interface can be found in [11]. The authors propose an interface called *inTouch*, which contains a projected 3D scene where the model is modified and a 2D menu for operations. The interface has a multiresolution meshes editing subsystem, which takes as input the position of the haptic device probe projected onto the scene and the direction of the applied force.

When the mesh is modified at a particular resolution, a set of vertices is moved. The change is then propagated to the higher detailed levels through mesh subdivision and to the less detailed levels through smoothing. The mesh editing subsystem receives a triangle, a contact point and a motion vector as input. In our method, the control point is a vertex and the deformation is propagated only to the resolution levels which are higher than that point.

An example of work that uses augmented reality and a tangible user interface for modeling 3D objects, can be found in [12], where a system called *3DARModeler* is presented. With this system it is possible to create a 3D model through one or several primitive geometries, apply textures, add animations, estimate real light sources and cast shadows on it. The difference between this system and the one developed in this work, with respect to modeling, is that the 3D models used by the *3DARModeler* are static. In this paper, the proposal is to dynamically change the geometry of a surface at multiple levels of detail, providing greater user interaction with the model and, also, overall control over the mesh.

3   A4-8 Meshes

Among the various categories of regular meshes, it is necessary to choose one that not only represent a terrain, but also can support simplification and refinement methods. The semi-regular meshes of type 4-8 (vertices with valence equal to 4 or 8, except at the edges) fit the desired profile to solve the problem addressed in this work. The semi-regular 4-8 mesh is a hierarchical structure for subdivision surfaces, or a complex cellular homeomorphic to a $[4, 8^2]$ Laves tiling [6].

The semi-regular 4-8 meshes can be divided into adaptive and non-adaptive. In the case of non-adaptive, there is only one hierarchy in the mesh, meaning that the modifications made in a region of the mesh affect the resolution of the entire grid. In the case of adaptive ones, there is a family of hierarchies, so that there is no interference between local modifications at each level.

This work uses adaptive meshes in order to deform a terrain at different levels of detail, independent of each other, and view it in different resolutions through simplifications or refinements. An example of this structure can be seen in Figure 1, where it is shown a basic block of the structure and two steps of consecutive refinement over it.

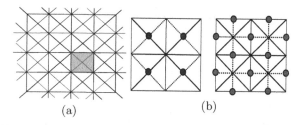

(a)   (b)

Fig. 1. Structure of a 4-8 mesh, (a) shows the basic construction block in purple and (b) shows two steps of refinement, the first generates the vertices in blue and the second, the vertices in red

An A4-8 mesh can be constructed using two approaches: top-down and bottom-up. The top-down approach, also known as refinement method, characterizes itself for creating a mesh in the lowest possible resolution and, progressively, increasing the resolution through the creation and addition of new triangles to the mesh. The bottom-up approach, or simplification method, creates a mesh in the highest possible resolution and, in a similar way, reduces the resolution by uniting the mesh triangles.

In this work, the mesh is constructed using a strictly top-down approach in the tessellation process. This way, it is possible for the final application to maintain its efficiency even with modifications being done to the mesh in real time.

3.1   Triangle *Bintree*

This work uses the binary triangle tree structure, or triangle *bintree* [13], to represent a terrain. Consider an A4-8 mesh M with a set of vertices $\mathbf{V} \subset \mathbb{R}^2$

given by $\mathbf{V} := \{\vec{v_i} := (u_i, v_i) \in \mathbb{R}^2, i = 1, ..., n\}$, where n is the number of vertices. A triangle bintree is a binary tree where, geometrically, every node of the tree represents a triangle.

The root triangle, $T = (\vec{v_a}, \vec{v_0}, \vec{v_1})$ where $\vec{v_a}, \vec{v_0}, \vec{v_1} \in \mathbf{V}$, is define as a right isosceles triangle in the coarser level, ($l = 0$). In the next level, $l = 1$, the children of the root are defined through the insertion of an edge defined between the apex vertex $\vec{v_a}$ and the middle point $\vec{v_c} \in \mathbf{V}$ of the base edge $(\vec{v_0}, \vec{v_1})$. The left children is $T_0 = (\vec{v_c}, \vec{v_a}, \vec{v_0})$ and the right children is $T_1 = (\vec{v_c}, \vec{v_1}, \vec{v_a})$. Repeating this process recursively, the rest of the tree is obtained. Figure 2 shows a few levels of a triangle bintree.

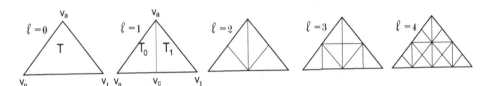

Fig. 2. Levels $0-4$ of a triangle bintree

It is important to note that the described structure needs to be adapted in a way that the A4-8 mesh can represent a spherical terrain. When subdividing a triangle at the first or last meridian, a middle point vertex must be created for both meridians. These vertices are "connected", meaning that any modification done to one of them must be applied to the other in order to prevent cracks on the mesh. This procedure ensures that the plane is conformed according to the cylinder topology.

Another important factor to be mentioned is that, for each vertex $\vec{v} \in \mathbf{V}$, two values must be stored, w e ε. The first value, w, represents the height of a vertex defined as $w = g(f(u, v)) = g(x, y, z)$, where f is a mapping function $f : \mathbb{R}^2 \to \mathbb{R}^3$ and g is a function that uses the mapped vertex coordinates, x, y and z, to calculate a height value. The resulting scalar is stored in a scalar matrix (heightmap M_a) at the vertex $\vec{v}$ coordinates u and v.

The second value, ε, represents the error value of each vertex $\vec{v} \in \mathbf{V}$. This value will be detailed in Section 5.1. In a resumed way, the vertex error is defined as $\varepsilon = \varepsilon_T$, where ε_T is the triangle error that has $\vec{v}$ as apex vertex. In practice, a triangle error value ε_T, for $T = (\vec{v_a}, \vec{v_0}, \vec{v_1})$, is stored in a scalar matrix (error map M_e) at the triangle T apex vertex $\vec{v_a}$ coordinates, u and v.

4   Parametrization

A parametric surface in the Euclidean space $\mathbb{R}^3$ is defined by a parametric equation with two parameters, named in this work as u and v. In particular, we want a mapping $f : \Omega \to \mathbf{S}$, where f is a mapping function, Ω is an open subset $\Omega \subset \mathbb{R}^2$ and $\mathbf{S}$ is a surface $\mathbf{S} \in \mathbb{R}^3$.

Orthographic projection was used as a function f to map Ω in S. More specifically, a plane, represented by A4-8 mesh, in a sphere. Consider, the vertex set $\mathbf{V}$ of a A4-8 mesh is Ω and all vertices $\vec{v} \in \mathbf{V}$ are associated with coordinates u and v in the plane. Consider also, that $0 \leq u \leq m_u$ and $0 \leq v \leq m_v$.

To define a function f in terms of latitude and longitude of a vertex $\vec{v} \in \mathbf{V}$, it is necessary to calculate the latitude of $\vec{v}$ using Equation 1 and its longitude using Equation 2.

$$latitude = \frac{(v * \pi)}{m_v} - \frac{\pi}{2}. \tag{1}$$

$$longitude = \frac{(2u * \pi)}{m_u} - \pi. \tag{2}$$

Now consider that the radius of the sphere defined by $\mathbf{S}$ is r. The domain surface V and S can be written using Equations 3 and 4, respectively.

$$\mathbf{V} = \{(u, v) \in \mathbb{R}^2 : 0 \leq u \leq m_u, 0 \leq v \leq m_v\}. \tag{3}$$

$$\mathbf{S} = \{(x, y, z) \in \mathbb{R}^3 : x^2 + y^2 + z^2 = r^2\}. \tag{4}$$

A plane to sphere mapping function is defined in Equation 5.

$$f(u, v) = \begin{pmatrix} r * \cos(longitude) * \cos(latitude), \\ r * \sin(longitude) * \cos(latitude), \\ r * \sin(latitude) \end{pmatrix}. \tag{5}$$

From this point on in this work, $\vec{v}$ will be used as a notation for describing a vertex that belongs to $\mathbf{V}$, which is a vertex of the mesh in discrete domain. To differentiate, $\vec{\mathbf{v}}$ will be used when the vertex is in continuous domain, in other words, the vertex mapped by $f(u, v)$ which has coordinates x, y and z.

When the set of vertices to be deformed is defined, the values of height are calculated in the continuous domain for each one of these vertices. The calculation is done through a function g and the values obtained are stored in the heightmap. This way, the terrain deformation is the modification made to the values of height w of the vertices inside a deformation area, in the discrete domain.

4.1   Parameter Space Partitioning

To obtain a partition $\mathbf{N} \subset \mathbf{V}$ endowed with all vertices $\vec{v}$ that when mapped to $\mathbb{R}^3$ are within a certain distance of a control vertex $\vec{\mathbf{v_c}} \in \mathbb{R}^3$, it is necessary to discretize the continuous object that represents the area of deformation in $\mathbb{R}^3$. More specifically, it is desired to obtain a partition of $\mathbf{V}$ having all the vertices $\vec{v}$ that, when mapped by $f(u, v)$, reside within a certain distance from a control vertex $\vec{\mathbf{v_c}} \in \mathbb{R}^3$. In other words, a partition that has all vertices $\vec{v} \in \mathbf{V}$ that reside in a deformation area defined in Euclidean $\mathbb{R}^3$ space.

To obtain $\mathbf{N}$, the Euclidean distance d in $\mathbb{R}^3$ between the vertices $\vec{v}^T = [x, y, z]^T$ and the control vertex $\vec{v_c}^T = [x_c, y_c, z_c]^T$ is initially calculated for all vertices $\vec{v} \in \mathbf{V}$. Then, a function is defined to classify the vertices $\vec{v_i}$ as

$$I(u, v) = \begin{cases} 1, \text{if } d \leq r_d \\ 0, \text{otherwise} \end{cases}, \tag{6}$$

where $I(u, v)$ is a function that uses the coordinates u and v of a vertex $\vec{v}$ to indicate if it is inside or outside of a deformation area. The value r_d is defined dynamically by the user.

This classification of the vertex according to the I function will be called property $I(u, v)$. If the function returns one, the vertex has the property, otherwise, it has not. If the vertex has the property $I(u, v)$, it is placed in the set of vertices that are inside the deformation area. This set of vertices will be called $\mathbf{V}_d$.

It's Important to note that the set $\mathbf{V}_d$ is defined in the discrete domain, but the distance is calculated in the continuous domain. This happens because, if the distance was calculated in the discrete domain, the resulting deformation area would be distorted by the mapping according to its location on the sphere.

5   Error Metric

This work uses the error metric presented by [13] in their algorithm named Real-time Optimally Adapting Meshes (ROAM). To decide if a triangle must be refined or simplified, the ROAM algorithm uses the concept of nested spaces. Being $\mathbf{C}$ the descendants of a vertex $\vec{v_i} \in \mathbf{V}$, the error $\varepsilon_{\vec{v_i}} \geq e_{\vec{v_j}}, \forall j \in \mathbf{C}$. This means that the error metric guarantees the monotonicity of the error values, in other words, the "parents" in a coarser level always have a higher error than their children, that reside in finer levels of the terrain. To use this error metric, the geometric error and the view-dependent error must be calculated for each triangle of the mesh.

5.1   Geometric Error

The ROAM algorithm uses the concept of a *wedgie* to define the geometric error. Consider that T is a triangle of the triangle bintree and the same notation of neighborhood from Section 3.1 will be used. A *wedgie* is defined as the world volume that has the vertices (u, v, w), in a way that $(u, v) \in T$ and $|w - w_T(\vec{v})| \leq e_T$, for a certain *wedgie* thickness $\varepsilon_T \geq 0$. The line segment from $(u, v, w - e_T)$ to $(u, v, w + e_T)$ is called thickness of a segment for a vertex $\vec{v}$.

The nested errors of the *wedgies* are calculated in a bottom-up manner. It is assumed that $\varepsilon_T = 0$ for all triangles in the finest possible level. The *wedgie* thickness of a triangle T is calculated based on the *wedgie* thickness of its children, ε_{T_0} and ε_{T_1}. Consider a triangle T with a left neighbor T_0 and right neighbor T_1. The error ε_T can be calculated as

$$\varepsilon_T = max(\varepsilon_{T_0}, \varepsilon_{T_1}) + |w(\vec{v_c}) - w_T(\vec{v_c})|,$$

where $\vec{v_c}$ is the vertex obtained from the bisection of the edge opposite to the apex vertex in the triangle T and $w_T(\vec{v_c})$ is given by

$$w_T(\vec{v_c}) = \frac{(w(\vec{v_0}) + w(\vec{v_1}))}{2}.$$

5.2 View-Dependent Error

The view-dependent error, called geometric screen distortion, is simply the calculation of the geometric distortion. This calculation represents the distance between the position where each vertex of the surface **S** should be in the screen space and where the triangulation touches the screen.

Be $s(\vec{v})$ the correct position in screen space of a vertex $\vec{v}$ given by $f(u_{\vec{v}}, v_{\vec{v}})$ and $s_T(\vec{v})$ its approximation by the triangulation T. The error in this point is defined as:

$$dist(\vec{v}) = \|s(\vec{v}) - s_T(\vec{v})\|$$

In the entire image, the maximum error is given by $dist_{max} = max_{\vec{v} \in \mathbf{V}_f}(dist(\vec{v}))$, where $\mathbf{V}_f$ is the set of vertices $\vec{v}$ of the mesh that, when mapped to the sphere, reside in the view frustum (region in space that appear on the screen).

In practice, a superior limit is calculated for the maximum distortion. For each triangle T in the triangulation, a local limit is calculated, projecting the *wedgie* in the screen space. This limit is defined as the higher thickness size ε_T of all vertices $\vec{v} \in T$ projected to the screen.

6   Proposed Method

In order to deform a terrain through the modification of its height map, a method is proposed with the following steps: Definition of the control point, definition of a deformation area and propagation of heights and errors. The first stage defines where on the terrain the deformation is made. The second stage uses the control point to define a dynamic area of deformation. The last step is the most important, where error and height values are propagated to the vertices of the terrain contained within the area of deformation.

6.1   Control Point Definition

This work uses one control vertex $\vec{v_c}$, which defines the center of the propagation area of height and error values, where the proposed method starts. It is obtained during the mesh tessellation, representing the vertex of the mesh $\vec{v} \in \mathbf{V}$ that, when mapped to the sphere, is the closest one to the deformation object (pre-defined fiducial marker).

If the camera position approaches the area pointed by the user, it increases the model scale and the mesh goes through a refinement process that can modify the control vertex $\vec{v_c}$. Likewise, if the user moves the deformation object, $\vec{v_c}$ changes dynamically.

6.2   Defining the Deformation Area

The area of deformation is a set of vertices $\mathbf{V}_d \subset V$ that, when mapped to the sphere through $f(u, v)$, reside at a distance less than or equal to a radius in relation to a control point $\vec{v_c} \in \mathbb{R}^3$ (Section 4.1). This value is dynamically defined by the user through the user interface and will be called, henceforth, *radius of deformation*, or r_d. The area of deformation is constructed according to a set of constraints described below.

Let $\mathbf{V}$ be the set of vertices of the A4-8 mesh. Subregions $\mathbf{V}_i \in \mathbf{V}$, representing partitions of $\mathbf{V}$ endowed with some property $I(\mathbf{V}_i)$, can be constructed by the following restrictions:

1. $\bigcup_{i=1}^{n} \mathbf{V}_i = \mathbf{V}$;
2. $\mathbf{V}_i$ is a connected region, $i = 1, 2, ..., n$;
3. $\mathbf{V}_i \bigcap \mathbf{V}_j = \emptyset \mid \forall\, i, j$ where $i \neq j$;
4. $I(\mathbf{V}_i) = 1$ for $i = 1, 2, ..., n$;
5. $I(\mathbf{V}_i \bigcup \mathbf{V}_j) = 0$ for $i \neq j$, where $\mathbf{V}_i$ and $\mathbf{V}_j$ are adjacent.

The first restriction means that the partition should be complete, in other words, all vertices of $\mathbf{V}_i$ must be contained in $\mathbf{V}$. The second requires that all vertices of the sub-region are "connected". The third constraint indicates that different subregions should be disconnected. The fourth requires that all vertices $\vec{v_i} \in \mathbf{V}_i$ must satisfy the properties $I(\mathbf{V}_i)$. The last restriction means that distinct sub-regions have different properties.

The restrictions shown above can be reinterpreted according to the problem addressed here. A property called $I(u, v)$ is defined for each vertex of the mesh according to Equation 6. This way, it is desired to partition the terrain into two disconnected sets, a set $\mathbf{V}_d$ that has vertices with the property $I(u, v)$ and another complementary to it $\bar{\mathbf{V}}_d$. Henceforth in this paper, the vertices of $\mathbf{V}_d$ will be called internal and vertices of $\bar{\mathbf{V}}_d$ will be called external.

The formulation proposed above follows the same principle of the image segmentation procedures, where usually one wants to separate the background and a region of interest. This choice is due to the fact that the almost completely regular structure of the semi-regular meshes allows the use of traditional algorithms of image processing.

6.3   Region Growing

The region growing technique was chosen as the implementation of the expansion function to define the set $\mathbf{V}_d$ ([14]). The starting point is the control vertex $\vec{v_c}$ and the expansion occurs in all of its neighbors directions while there are vertices with the property $I(u, v)$. Henceforth, the set of neighbors of $\vec{v_c}$ that have the property $I(u, v)$ will be called $\mathbf{V}_{d_c}$.

Initially, the algorithm tests the neighbors of $\vec{v_c}$ and the vertices with the property $I(u, v)$ are put in a set of vertices called $\mathbf{V}_{d_c}$. This set is passed to a recursive expansion function so the set $\mathbf{V}_d$ can be obtained. This function expands $\mathbf{V}_{d_c}$ from $\vec{v_c}$ toward the direction of each neighboring vertices $\vec{v_i} \in \mathbf{V}_{d_c}$. The neighborhood is covered in the A4-8 mesh, but the distance test is done in Euclidean space $\mathbb{R}^3$. An example of the test of a vertex $\vec{v_i}$ neighbor of $\vec{v_c}$, through the distance $\mathbf{d_{ic}}$ between them, can be seen in Figure 3(a). The final set of vertices $\mathbf{V}_{d_c}$ can be seen in Figure 3(b).

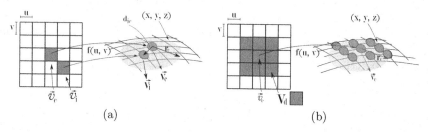

(a)   (b)

Fig. 3. (a) Beginning of the expansion process. The control vertex is $\vec{v_c}$ in the discrete domain and $\vec{\mathbf{v_c}}$ in continuous domain. The radius that defines the area of deformation is r_d and $\mathbf{d_{ic}}$ is the distance in $\mathbb{R}^3$ between the control vertex $\vec{v_c}$ and of its neighbors $\vec{v_i}$. (b) The set $\mathbf{V}_{d_c}$, in green, represents the vertices that have the property $I(u, v)$.

The recursive function is responsible for "walking" across the mesh in the direction of each of the neighbors of $\vec{v_c}$. It is possible to separate the eight neighborhoods of a vertex in two cases, diagonal direction and vertical/horizontal direction. A neighbor is in a diagonal direction if both coordinates of the original vertex in the parameter space, u and v, change in one step of the algorithm in that direction. For the vertical/horizontal direction, one step of the algorithm modifies only one coordinate, u or v, of the original vertex.

The combination of these two cases mentioned above allows the algorithm to go through the entire deformation area over the mesh. If the direction is diagonal, the recursion just repeats the direction of expansion of the previous step (Figure 4a). In the case of the vertical/horizontal direction, the recursion is called for the expansion direction of the previous step and its two adjacent diagonal directions (Figure 4b). The situation after one step of the algorithm for all internal vertices neighboring $\vec{v_c}$ is shown in Figure 4c.

Border Treatment. When a vertex $\vec{v_i} \in \mathbf{V}_{d_c}$ is at the border of the mesh, meaning it has coordinates $u = \{0 \text{ or } m_u\}$ and/or $v = \{0 \text{ or } m_v\}$, the method needs to modify its routine to promote expansion. If the vertex coordinate reaches $u = 0$ the next step of the algorithm will map this coordinate to $u = m_u - 1$ and, similarly, if the coordinate is $u = m_u$ the next step of the algorithm will map this coordinate to $u = 1$. In the case of the v coordinate (latitude), if a vertex reaches a border coordinate, the algorithm simply stops propagating.

The leap of one coordinate for the longitude case is deliberate, reflecting that for coordinates $u = 0$ and $u = m_u$ there is an overlap. If a vertex has a border

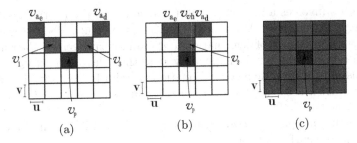

Fig. 4. Diagonal expansion example (a) and vertical/horizontal example (b). A single algorithm pass, after all neighbors of $\vec{v_c}$ are calculated, is presented in (c).

coordinate u on the terrain, any change made in the heightmap and error maps for this vertex must be repeated at the other end of the heightmap to prevent cracks on the terrain. An example of border treatment can be seen in Figure 5.

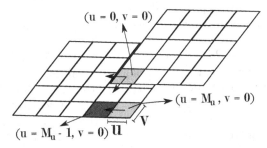

Fig. 5. Border treatment in the case of vertices trying to expand beyond the dimensions of the terrain. For the latitude, if a vertex $\vec{v}$ goes beyond the dimensions of the terrain, the expansion is interrupted in that direction. In the case of longitude, the coordinate of the vertex u is modified so that it can continue expanding.

6.4 Error and Height Propagation

Determined the area of deformation, or set $\mathbf{V}_d \in \mathbf{V}$, it is necessary to propagate error and height values for all vertices in the set $\mathbf{V}_d$. The propagation algorithm can be defined as an extension of the area determination algorithm, since both run through all vertices that have the property $I(u, v)$.

Height Definition and Propagation. The function used in this paper to calculate the height values is the Gaussian distribution function $g(x)$ described by Equation 7. The function has a maximum at the control vertex and a minimum at the border of the area of deformation. In this function, the standard deviation is σ and the average is μ.

$$g(x) = \frac{1}{\sigma\sqrt{2\pi}} e^{-\frac{(x-\mu)^2}{2\sigma^2}}. \tag{7}$$

The process involves a mapping from the discrete to the continuous domain and vice versa. A vertex $\vec{v} \in \mathbf{V}$ is mapped on the sphere and tested to see whether it is or not within the area of deformation. If it is an internal vertex, the Euclidean distance in $\mathbb{R}^3$ of the vertex $\vec{v}$, mapped by f, to the control vertex $\vec{v_c}$, is passed to the Gaussian function defined by Equation 7 and a height value is obtained. This value is then associated with a scalar w and stored in the heightmap at its coordinates (u, v). The new value of w is accessed on the next tessellation step, where the mapping function will transmit the modification made on the heightmap to the surface of the terrain.

Metric and Error Propagation. In the same way as the propagation of height values, it is desired to cover all vertices $\vec{v} \in \mathbf{V}_d$ for the error propagation. One restriction to be recalled is that the error metric must preserve the monotonicity, meaning that the parents have higher error value than their children.

To propagate error in the correct way, it is necessary to use an expansion function that traverse the mesh in a hierarchical manner. This is possible using the concept of alternate step. Consider a number of iterations $i = 0, 1, 2, ..., n$ and a step size p_a. Knowing that every leaf node in the triangle bintree has error $\varepsilon_T = 0$ and that the error grows as we reduce the resolution level until the root node is reached, it is possible to notice in the tree that the position of the nodes is spaced in a homogeneous way for similar error values.

Consider, for instance, Figure 6a that shows the displacement of the vertices that have error $\varepsilon_T = 0$ in a triangle in the highest level of resolution. It is possible to notice that one can walk on the structure covering all vertices in this resolution, using steps p_l of odd steps in the vertical direction, followed by "walks" in both diagonal directions of the vertex reached by an initial step.

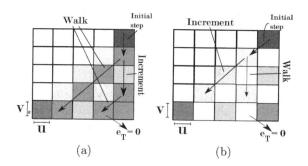

Fig. 6. Hierarchical expansion example. In (a) the first case occurs, vertical step and diagonal walk. In (b) the second case happens, diagonal step and vertical walk, both on the zero value iteration.

In a more specific way it is possible to state in this case, that the initial step of each iteration i is given by Equation 8 and the step increment i_c is given by Equation 9. In this case where the resolution is maximum, the iteration is $i = 0$, the initial step $p_l = 1$, the step increment $i_c = 2$ in the vertical direction and the step for the diagonal walk $p_d = p_l$.

$$p_l = 2^i. \tag{8}$$

$$i_c = 2 * p_l. \tag{9}$$

To the level immediately below the maximum, the model modifies itself (Fig. 6b). The initial step and the increment are obtained in the same way as in the previous case, following Equations 8 and 9. The iteration is again $i = 0$, $p_l = 1$ and $i_c = 2$. In this case, although, the initial step is done in the diagonal direction and the walk through the structure is done in the vertical direction (the inverse of the previous case). The recursion walks vertically with a step $p_a = i_c$, being $p_a = 2$ for this resolution.

The combination of these cases allows that all vertices are traversed. An algorithm can be easily extracted from these two cases, representing a hierarchical manner to cover the A4-8 mesh. The monotonicity is guaranteed this way, because the error can still be calculated with a bottom-up approach.

7   Results

All experiments were done on a computer with an AMD FX(tm)-6100 six cores processor, 8GB RAM and GeForce GTX 550 Ti with 2GB of RAM. The Microsoft LifeCam camera was used as video capture device. A resolution of 640 × 480 *pixels* was defined for all visualization windows. In order to validate the method, meshes with maximum resolution of 2049 × 2049 and 8193 × 8193 vertices were chosen. The maximum number of triangles generated by the tesselation process was defined as 70000.

As the user modifies the distance from the terrain to the camera, the view-dependent error changes, and the mesh adapts itself through refinement or simplification. Figure 7 shows an example of mesh change due to terrain scale modification. The initial position of the terrain can be seen in Figure 7b, mesh modifications when the camera approaches the terrain is seen in Figure 7c and when it moves away in Figure 7a.

The deformation of a spherical terrain with the proposed method can be seen in Figure 8. The image shows a sequence of frames, from left to right, where the user performs a movement with the marker through the terrain surface causing deformation.

Here, the maximum image error $dist_{max}$ tested varied from 0.1 to 6.0. Experiments shown that the ideal value, that promote balance between visualization quality and number of triangles, is $dist_{max} = 4.0$.

The radius of deformation r_d used during testing ranged from 150km to 600km, being the radius of the spherical terrain equal to 6372.79km. In all experiments, the area of deformation change in the $\mathbb{R}^3$ space for three different radius of deformation defined as: 150km, 300km and 600km.

(a)   (b)   (c)

Fig. 7. Mesh modifications related to the distance between the camera and the terrain. Terrain initial position (b), camera close to the terrain (c) and moved away from it (a).

Fig. 8. A sequence of frames spaced in time showing the mesh being deformed using the proposed method

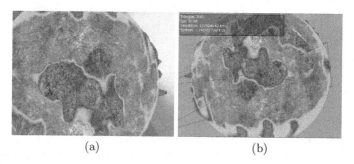

(a)   (b)

Fig. 9. Final terrain visualization generated with the proposed method. Global scale (a) and local scale (b).

The average time that the application takes to render one frame with the terrain being deformed was calculated for different resolutions. The camera was fixed in a single position for all values contained in Tables 1 and 2. The chosen position was kept at a distance of three times the radius of the sphere to its center, being approximately 19,000km.

In the case of deformation, the worst case scenario was chosen to evaluate the method. In other words, the longitude was set at 180°, where the vertices have coordinates $u = 0$ and $u = m_u$ for variables values of latitude. Latitude values begin at 0° and increase 30° until they reach 180°, in other words, from one pole to the other. The average time to render a frame without deformation being applied to the terrain was 0.033 seconds.

The values in Table 1 represent the average time to render a frame with the terrain being deformed for a maximum resolution of 2049 × 2049 vertices. It is possible to affirm that, for a radius of deformation $r_d = 150$km, any region of the planet may be deformed without the user noticing drops in the number of frames per second. For a radius of $r_d = 600$km, the application takes three times longer to render a frame with deformation occurring at the pole than at the equator, but the number of vertices deformed at the pole is about 90 times greater when compared to the equator region.

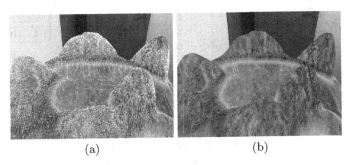

(a)   (b)

Fig. 10. A complete deformation scenario. Terrain with edges drawn (a) and without (b).

Table 2 lists values for the higher terrain resolution tested, 8193×8193 vertices. For a deformation radius of $r_d = 600$km, one realizes that the average time start to rise for all tested latitudes. This means that, as the resolution increases, it is necessary to reduce the deformation radius in order to exchange application performance and display quality.

A complete deformation scenario can be seen in Figure 10. A planet generated with the proposed method is shown in Figure 9. The implementation of the proposed method can cope with deformations in real time even if the speed of rotation of the planet is set to a high value. It is important to note that adaptive meshes are usually used only for visualization, here the final application can both visualize and modify a terrain in real time.

Table 1. Average rendering time of the terrain in seconds when a number of vertices are modified at a certain latitude. Maximum resolution of 2049 × 2049 vertices.

Latitude	Average time(s)			Latitude	Modified Vertices		
0	0.064	0.096	0.117	0	32764	63471	126700
30	0.041	0.047	0.051	30	698	2789	11467
60	0.041	0.047	0.044	60	397	1593	6650
90	0.041	0.046	0.043	90	360	1431	5737
120	0.038	0.044	0.050	120	405	1629	6560
150	0.045	0.047	0.050	150	704	2916	11543
180	0.063	0.086	0.116	180	32759	63463	126888
	$r_d = 150$km	$r_d = 300$km	$r_d = 600$km				

Table 2. Average rendering time of the terrain in seconds when a number of vertices are modified at a certain latitude. Maximum Resolution 8193 × 8193 vertices.

Latitude	Average time(s)			Latitude	Modified vertices		
0	0.356	0.711	1.328	0	507744	1006861	2013967
30	0.052	0.053	0.114	30	11302	45437	180019
60	0.045	0.052	0.098	60	6532	25752	106635
90	0.035	0.047	0.082	90	5765	23056	89330
120	0.046	0.051	0.083	120	6511	26024	101521
150	0.045	0.059	0.107	150	11329	45625	179929
180	0.350	0.720	1.333	180	507805	1007216	2013721
	$r_d = 150$km	$r_d = 300$km	$r_d = 600$km				

8   Conclusion

Interactive systems providing real 3D modeling freedom are rare. In this work, we propose an interactive method for 3D modeling of spherical terrains in real time through augmented reality. The entire process is easy to be controlled by the user. Nevertheless, the generated terrain can be quite complex, natural-looking, even if defined by a small number of user actions.

The proposed method proved to be efficient in dealing with different resolutions. The increase of the average rendering time is lower than the increase in the number of modified vertices for all tests. The hierarchical propagation allows that all vertices in an area of deformation are visited, complying with the monotonicity constraint of the error metric.

The use of a physical marker to deform a mesh, provides an intuitive interface to create shapes and patterns across the terrain. This interface allows easy surface modeling and manipulation, without requiring prior knowledge in 3D modeling softwares such as *Maia* and *3D Studio*.

The combination of known techniques with the A4-8 mesh structure produced the desired results, it is possible to not only generate a terrain with a few movements, but also add detail at any resolution. Although the visualization and modeling steps are done sequentially, the application runs in real time for various resolutions. This means that a large but highly detailed terrain can be generated with the proposed method.

References

1. Coquillart, S.: Extended free-form deformation: a sculpturing tool for 3d geometric modeling. In: Proceedings of the 17th Annual Conference on Computer Graphics and Interactive Techniques, SIGGRAPH 1990, pp. 187–196. ACM, New York (1990)
2. Singh, K., Fiume, E.: Wires: A geometric deformation technique (1998)
3. Sumner, R.W., Schmid, J., Pauly, M.: Embedded deformation for shape manipulation. ACM Trans. Graph. 26, 80 (2007)
4. Kobbelt, L.P., Bareuther, T.: peter Seidel, H.: Multiresolution shape deformations for meshes with dynamic vertex connectivity (2000)
5. Sha, C., Liu, B., Ma, Z., Zhang, H.: Multi-resolution meshes deformation based on pyramid coordinates. In: CGIV, pp. 200–204. IEEE Computer Society (2007)
6. Velho, L., Gomes, J.: Variable resolution 4-k meshes: Concepts and applications. Comput. Graph. Forum 19(4), 195–212 (2000)
7. Kato, H., Billinghurst, M.: Marker tracking and hmd calibration for a video-based augmented reality conferencing system. In: Proceedings of 2nd IEEE and ACM International Workshop on Augmented Reality (IWAR 1999), pp. 85–94 (1999)
8. Nealen, A., Mueller, M., Keiser, R., Boxerman, E., Carlson, M.: Physically Based Deformable Models in Computer Graphics. Computer Graphics Forum 25(4), 809–836 (2006)
9. Baraff, D., Witkin, A.: Large steps in cloth simulation. In: Proceedings of the 25th Annual Conference on Computer Graphics and Interactive Techniques, SIGGRAPH 1998, pp. 43–54. ACM, New York (1998)
10. O'Brien, J.F., Hodgins, J.K.: Graphical modeling and animation of brittle fracture. In: Proceedings of the 26th Annual Conference on Computer Graphics and Interactive Techniques, SIGGRAPH 1999, pp. 137–146. ACM Press/Addison-Wesley Publishing Co., New York, NY (1999)
11. Gregory, A.D., Ehmann, S.A., Lin, M.C.: intouch: Interactive multiresolution modeling and 3d painting with a haptic interface. In: Proc. of IEEE VR Conference, pp. 45–52 (1999)
12. Do, T.V., Lee, J.-w.: 3darmodeler: a 3d modeling system in augmented reality environment. Systems Engineering 4, 2 (2010)
13. Duchaineau, M.A., Wolinsky, M., Sigeti, D.E., Miller, M.C., Aldrich, C., Mineev-Weinstein, M.B.: Roaming terrain: real-time optimally adapting meshes. In: IEEE Visualization, pp. 81–88 (1997)
14. Stockman, G., Shapiro, L.G.: Computer Vision, 1st edn. Prentice Hall PTR, Upper Saddle River (2001)

Online Analysis and Visualization of Agent Based Models

Arnaud Grignard[1], Alexis Drogoul[1,2], and Jean-Daniel Zucker[1,2]

[1] UMMISCO/UPMC, Paris, France
[2] MSI-UMMISCO/IRD, Hanoi, Vietnam

Abstract. Agent-based modeling is used to study many kind of complex systems in different fields such as biology, ecology, or sociology. Visualization of the execution of a such complex systems is crucial in the capacity to apprehend its dynamics. The ever increasing complexification of requirements asked by the modeller has highlighted the need for more powerful tools than the existing ones to represent, visualize and interact with a simulation and extract data online to discover imperceptible dynamics at different spatio-temporal scales. In this article we present our research in advanced visualization and online data analysis developed in GAMA an agent-based, spatially explicit, modeling and simulation platform.

Keywords: Agent Based Modeling, online analysis, visualization, interaction, 3D, complex systems.

1 Introduction

Agent-based modeling has undergone an incredible, albeit slow, evolution over the last 30 years. From simple to apprehend abstractions that merely served to underline intriguing dynamics [15] under the KISS (Keep It Simple, Stupid) paradigm, it has progressively become the technology of choice for designing large, complex models that follow the KIDS (Keep It Descriptive, Stupid) approach advocated by [7]. In this approach, a model of a real system is built to provide modelers or end-users an experimental playground that allows altering any component of the model and explore the dynamics that result from these modifications. Instead of providing an analytic vision of the system, agent-based modeling becomes, under this paradigm, a heuristic method to understand, by simulation, the possible evolutions of the described system. This evolution has made a number of new requirements emerge, among which: (a) the possibility to set up a complete experimental environments around models, based on rich visual feedback and interactivity; (b) the possibility to extract interesting dynamics in the simulations thanks to the use of analysis tools, data-mining and high-performance computing techniques [4].

As a matter of fact, it is now widely recognized that achieving a comprehensive understanding of a complex Agent Based Model (ABM) through an experimental approach has to rely on: (1) dynamic, realistic and intelligible visualization of

B. Murgante et al. (Eds.): ICCSA 2013, Part I, LNCS 7971, pp. 662–672, 2013.

its evolution under different setups, (2) the ability to easily (re)define this visualization, ideally in a an interactive way, at different spatial and temporal scales or using different points of view, (3) the ability to extract abstract properties and knowledge from the dynamics generated by the simulation and to visualize them as well in real-time, (4) the ability to interact with such visualizations in a natural way in order to modify the model itself in an interactive design approach.

Nowadays, these different requirements are hot research topics [1], but have led, so far, to few usable implementations. Some graphical languages dedicated to the visualization of agent-based models have been proposed in different ABM platforms, but they are often too simple or too complicated to be used by non computer scientists. There is still no possibility, in any of these platforms, to have different points of view on a given model, to confront them and make them evolve during simulations. In addition, statistical or data-mining tools are mostly available as independent tools that are used to analyze the outputs of simulations after and not during their executions. Building abstractions online and using them to facilitate the understanding of the dynamics of a model through advanced visualization techniques, is still out of reach in existing ABM tools.

After giving an overview of the current practices and related work in agent-based model visualization we propose an on-going research on a language for model visualization and online tools to analyze and abstract data (based on graph representation) online. Most of the tools proposed here have already been implemented and tested in GAMA version 1.5 [17], others are at the stage of design and are implemented in the soon to be released in version 1.6.

2   Related Work

In already existing popular platforms (NetLogo, Repast, Mason, Swarm, Anylogic), languages for visualization do exist [14] but most of them lack tools for building, observing and interacting with models. In effect, Most of them offer basic tools such as 2D animation window representing agents with dynamic color, inspector, model parameter display and chart output. On top of it, standardization emerges and help the user to build a meaningful model representation that follows design technique guidelines which are often based on a non hierarchical categorization of ABM models such as conventional ABM visualization (natural phenomena or mathematical representations), structured ABM visualization (aggregation of many agents creating an emergent pattern), unstructured ABM visualization (irregular distribution of agents) [6]. Unfortunately, this description only stay focused on the agent representation and do not propose any abstraction and multi-level representation or only by using ad-hoc solutions and post simulation treatment that are not yet standardized.

In existing platforms outputing data in a file and data recording can easily be scheduled like any other action but the analysis is done offline and most of the time when the simulation is finished. Only few platform propose built-in analysis tools that run online and serve as feedback or indicator during the simulation by representing macro-behaviour and using graph analysis or data clustering to build groups of agents [11]. However, recent work on the multi-level approach [9] propose solution to consider entities at several levels of scales through an analytical framework and an automated description of group dynamics by the use of statistical based tools [3].

To conclude visualization is often seen as a cosmetic aspect of the model, and rarely considered as an essential part of the modelers abstraction and comprehension work. Visualization is not used as an experimental approach but rather as a final result. We present in this paper a new way to represent model and simulation as well as tools to extract abstract data online.

3   New Way to Visualize Agent-Based Model

Building a visualization is like building a visualization model on the model itself and the construction of this "visual" model requires its own language and need t o be independent of the execution of the model to define different representations of a given model without modifying the model itself. A dedicated language for visualization allows to describe graphical primitive (from simple geometry to more complex objects such as graph, dendrogram or 3d object) to construct a sensitive representation of the model, allowing feedback on the model and its dynamic from the visualization, and offering modellers a simple way to declare it. In addition, a complete understanding of a model cannot be dissociated from (1) basic visualization concepts, such as: overview, zoom and filter, details on demand [16]. (2) basic interaction concepts to help the user to adjust the model during the simulation through the use of probes, dynamic filtering, highlighted details and different level of details representations. Above the model representation, a model is included in a scene with at least one point of view , a description of phenomenon such as lighting, shading, transparency with a given resolution both in time and space [12].

Starting from geometric representation of 2D and 3D to graph visualization, GIS data integration and multi-level representation, the techniques described below are implemented in the platform Gama, whose default display, based on the Java2D API, has shown some limits when dealing with large-scale models and realistic rendering. The limits address by JAVA2D was first the impossibility to define and represent 3D geometry but it also showde limits in term of real time rendering when dealing with large models. To address this limitation OpenGL display has been integrated in GAMA 1.5. This not only helps the user to build aesthetic visualization by offering most of the common metaphors for visualizing complex systems but also offers much more in term of graphical performance.

3.1   From Data to Form

There are many ways to visualize and distinguish data elements: size, value, texture, color, orientation and shape. For each agent the value of one of its attributes can be represented in different ways such as a text, a color, a 2D or 3D geometry. In many cases the shape of the agent can be used to represent one or several values of its attributes. Gama handles in a seamless way vector data that are integrated as the simulation environment as an input but also as an output to store resulting experiment [17]. Graphic operator applied on vector data are used to compute distance, surface, neighbourhood or even more complicated operation such as shortest-path computation, geometry intersection, union, difference and many other spatial algorithms. Those graphical operators are also used for the agent aspect definition through pre-built graphical primitive described in figure 1.

	Point	Line	Circle	Square	Polygon	Bitmap	GIS
2D	.	/	●	■	⬟	🐜	🌐

	Sphere	Plan	Cylinder	Cube	Polyhedron	3D Object	3D GIS
3D	●	◣	▮	◆	●	🫖	🌐3D

Fig. 1. Common graphical primitives used in ABM visualization

It is straightforward to define several aspects for a given agent. Figure 2 shows different displays applied on the same 6x6 Grid where each agent (a cell) has a specific value represented by 4 different aspects (text, square, circle, cylinder).

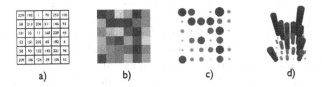

a)   b)   c)   d)

Fig. 2. Different aspects of the same model (6x6 Grid). a) value is displayed as a text. b) value is displayed as a colored square. c) value is displayed as a circle with a variable radius. d) value is displayed as an cylinder with a variable elevation.

Once agent aspect have been defined, a 3D scene displays all the agents present in the model with their respective aspect on different layer. By using intuitive 3D navigation any location in the model can be reach to observe the model from different point of view. In addition, any agent can be selected, inspected and modified through the user interface. The layer control is a smart way to make visible layer or not in the 3 dimensions or to superpose them for a better readability. Finally when a user wants to share his work or record a given step of the simulation with a given point of view the snapshot tool is here to capture the model at a given spatio-temporal step. All the features described in figure 3 has been implemented in GAMA.

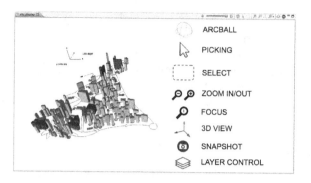

Fig. 3. Interaction tools

3.2 Integration with GIS Data

A large scale of geographical vector datasets are now available and used to face problem that integrate spatial dimension. Using this type of data is now often required to make the simulations closer to the field situation and it allows to use tools, like spatial analysis, coming from Geographic Information Systems (GIS) to manage these data. GIS technology first aim is to represent map and it cannot provide adequate performance with very large datasets on a huge amount of iteration when applied in dynamic simulation where spatial analysis is necessary [5]. To face this issue, in Gama, any geographical object is transformed in an agent with its own internal state and behavior and it can go further by easily converting any agent to a geographical object by spatializing it. Some of the already existing platform support geographical data [5] but GAMA offers much more in term of GIS services and in operations on geographical vector data [17]. We use functions from GeoTools to import and export data and Java Topology Suite (JTS) for data manipulation. As the integration of GIS data is seamless and straightforward it is very easy to develop model mixing data coming from GIS and built-in agent in a 3D environment. Figure 4 is an illustration of the mixing of GIS data with the integration of building agent.

3.3 Multi-level Modeling

Complex system is by definition mixing different entities at different levels of organization and the visualization of those different level is crucial for the good understanding. Most of the already existing approaches are called emergentist approach where only the lower model is present in the model and where the higher level is only an observation of the model but is not reified in the model itself as it emerges during the simulation. Recently new research has been produced in considering entities at different levels [9]. This approach focus on how to describe the articulation and influence of different levels. In this kind of approach, a powerful visualization toolkit is, among other statistic tools, one of the more intuitive way to understand, describe and perceive the multi-level paradigm.

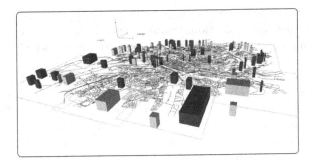

Fig. 4. GIS integration in ABM

Gama meta-model has been design to take in account simultaneously several levels in the same model [18]. Thus, we can easily represent multi-level model using multi-layer rendering where layers can be then placed on different z value to represent different level of organisations as shown in figure 5.

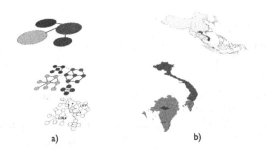

Fig. 5. Multi-Level 3D representation. a) Simple graph, clustered graph and macro graph. b) district, country and world level.

4 MODAVI - Multi-Level Online Data Analysis and Visualization

Agent-Based Simulations generate massive loads of data, which are usually analysed at the end of simulations. In large scale model it's not always easy to have a clear idea of all the interactions occurring between agent. Interaction analysis is part of the different frameworks to explore a complex systems where macro-level dynamics and structure are caused by the interactions of a relatively high number of agents. MODAVI is an exploratory tool that addresses the problem of online analysis of an ABM simulation using the new available graphic feature developed in GAMA such as graph agentification, multi-level modeling and 3D visualization, to handle online data abstraction. It proposes to build an online network abstraction to represent interaction that can appear between entities (i.e gene regulatory network, infectious contact, betweenness individual in a population, etc). This process is related to data stream mining which is the process of extracting knowledge structures from continuous, rapid data records. The MODAVI approach supports representing the interdependence between various interacting entities to be observed as abstract agent. Such an online mining tool may be used to explore any kind of interaction between agents [13]. The visualization of the results and their interpretation plays a key role in the understanding of a simulation and it answers to questions in a manner that is not possible with empirical observation and experimentation. MODAVI is at the boundary between visualization and analysis and can be used to describe and represent indicator to both assess and explore a model. It goes beyond model visualization as the information displayed is the result of mining algorithms such as clustering that are performed on the stream of data from the ABM simulation. This tool provides new exploration capabilities and strategies to generate, analyze and visualize a large amount of alternative simulation experiments. MODAVI is based on two concepts developed in Gama, the proxy graph pattern that creates an interaction graph from a population of agent and a macro graph generator that creates a macroscopic description of the interaction graph.

4.1   Proxy Agent Pattern

From a given population of agent, the proxy graph pattern creates an interaction graph. In a proxy graph, a proxy node is an agent that mirrors a given agent. By creating proxy node we insure that the target population is not affected by the creation of the graph. The link between the agent and the proxy node is dynamic insuring the mapping between the proxy node and its target node at any time of the simulation. Two proxy nodes are connected together if their distance is smaller that a given threshold as shown in figure 6. This proxy agent provides more in term of representation in the sense that it enables to manipulate and thus altering the representation of the proxy agent without modifying the target agent.

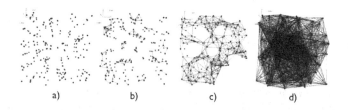

Fig. 6. From a population of n agents in red, a proxy graph is created where proxy node in blue are connected with other proxy node if them distance between them is smaller that a given threshold. a) threshold=5 - b) threshold = 10 - c) threshold= 20 - d) threshold= 50

4.2 Macro Graph Generation

From an already existing graph at a micro level, a macro graph that clusters groups of agents and summarizes the number of links between each class of agents is created online see figure 7. A macro node represents all the aggregated agents of a given population at each iteration. A macro node is represented by a sphere whose radius evolves according to the number of aggregated nodes. A macro edge linked a macro node representing (aggregating) all the agents A and a macro node representing(aggregating) all the agents B. A macro edge is the representation of the total number of edges between agent A and agent B. The macro edge is created thanks to a linkage function that measures the distance between two groups. We define here four different linkage function:

- Single linkage: the similarity of two clusters is the similarity of their most similar members.

$$D(X,Y) = \min_{x \in X, y \in Y} d(x,y)$$

- Complete linkage: the similarity of two clusters is the similarity of their most dissimilar members.

$$D(X,Y) = \max_{x \in X, y \in Y} d(x,y)$$

- Average linkage: the similarity of two clusters is computed as the average distance between objects from the first cluster and objects from the second cluster.

$$D(X,Y) = \frac{1}{N_X N_Y} \sum_{i=1}^{N_X} \sum_{j=1}^{N_Y} d(x_i, y_i)_{x_i \in X, y_i \in Y}$$

- Average group linkage: the similarity of two clusters is computed as the distance between the average values (the mean vectors or centroids) of the two clusters.

$$D(X,Y) = \rho(\bar{x}, \bar{y}); \bar{x} = \frac{1}{N_X} \sum_{i=1}^{N_X} \bar{y} = \frac{1}{N_Y} \sum_{i=1}^{N_Y}$$

5 Usability and Results

GAMA is being used as a decision-support tool for natural resource management in the MAELIA [8] project. This project aims at studying the social, economic and ecological impact of water management in the Adour-Garonne Basin (France) by the integration a huge amount of geographical data and models (from ecological models to human decision-making models). The model produces thus data that cannot be understood without advanced visualization features. In particular, to visualize various views of the same model with georeferenced or

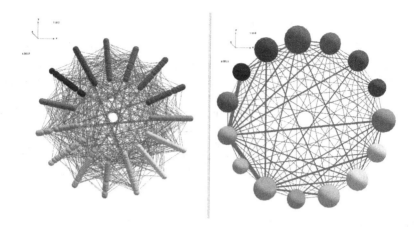

Fig. 7. Micro Graph and the corresponding Macro Graph using an average linkage function. In the macro graph a node is the aggregation of all the agent in the micro graph with the same color, and an edge between macronode A and macronode B is the aggregation of all the edges between micronode a and micronode b.

aggregated data and to display at the same time the areas with water lacks, the agricultural activities and the plant growth states with a spatial representation and time series of water flows aggregated at the level of the whole basin. All these information need separated displays to be understandable.

GAMA is also used as a decision-support tool for studying daily urban dynamic in the MIRO project [2]. The Miro model addresses the issue of sustainable cities. Therefore, improving urban accessibility merely results in increasing the traffic and its negative externalities, while reducing the accessibility of people to the city. Given real data the simulator is used to realise scenarios determined by geographers for quantifying service accessibility and identifying cities management strategies. Such simulation produce huge amount of data especially geolocalized ones. To sum up them we imagine 3D visualisation associating data, space and time such as proposed in time-geography domain. In spite of their importance spatio-temporal prism is a visualisation proposed by time-geography that have not any response in simulation domain. Proposing this kind of visualisation in the simulation will open new perspective of the geography domain.

6   Conclusion and Perspectives

As modeling complex systems is becoming more and more pervasive there is a growing need for online visualization tools that support making sense of simulations. To that extent, this paper focus on: (1) the setting of a graphical experimental environment based on high-level graphics and dedicated graphical language, (2) MODAVI, an approach that abstracts dynamics and provide interactive online analysis during the simulation. Most of the concepts described in this article have been implemented in GAMA while some others are still in under development. Besides those researches we also work on rendering optimization to achieve high performance rendering and to deal with real time application. Future work will includes a graphical language to modify the model – in a more intuitive way than with the code – will explore new way (e.g such as ones offered by web technology) to share simulation where a model could be played at runtime (online) or in a replay mode (offline) and where any agent could be inspected at any spatio-temporal scale during the simulation on different devices such as computers, or embedded devices. The next version of GAMA will (V. 1.6) also support exporting simulation in an open standard xml called COLADA. It will support exchanging digital assets and especially simulation and replay them on different architecture later on and share knowledge about model. Thus the whole dynamics is captured and one can replay the simulation with the ability to move in the scene and inspect any agent at any time step. In term of spatio-temporal scale we are working on spatio-temporal geo-visualization where different techniques have evolved over the years such as 2D Maps, Animation and SpaceTime Cubes where the time dimension is orthogonal to the surface of the map [10]. Finally, high-level Graphics library mixed with web technology open new way to develop participatory simulation where multiple humans control or design individual agents in the simulation or in new field such as crowdsourcing where a groups of humans, working together in parallel, produce a work impossible to achieve alone or serially.

References

1. Allan, R.: Survey of agent based modelling and simulation tools. Science & Technology Facilities Council (2010)
2. Banos, A., Marilleau, N.: Improving individual accessibility to the city: an agent-based modelling approach. In: ECCS (2012)
3. Caillou, P., Gil-Quijano, J.: Simanalyzer: automated description of groups dynamics in agent-based simulations. In: AAMAS 2012, International Foundation for Autonomous Agents and Multiagent Systems, Richland, SC, pp. 1353–1354 (2012)
4. Chuffart, F., Dumoulin, N., Faure, T., Deffuant, G.: Simexplorer: Programming experimental designs on models and managing quality of modelling process. IJAEIS (2010)
5. Crooks, A.T., Castle, C.J.E.: Agent-Based Models of Geographical Systems. Springer, Netherlands (2012)
6. Daniel Kornhauser, U.W., Rand, W.: Design Guidelines for Agent Based Model Visualization. Journal of Artificial Societies and Social Simulation 12(2) (2009)
7. Edmonds, B., Moss, S.: From kiss to kids–an anti-simplistic modelling approach. Multi-Agent and Multi-Agent-Based Simulation, 130–144 (2005)
8. Gaudou, B., et al.: The maelia multi-agent platform for integrated assessment of low-water management issues. In: MABS, Multi-Agent-Based Simulation XIV - International Workshop (to appear, 2013)
9. Gil-Quijano, J., Louail, T., Hutzler, G.: From biological to urban cells: lessons from three multilevel agent-based models. In: Principles and Practice of Multi-Agent Systems, pp. 620–635 (2012)

10. Kjellin, A., Pettersson, L.W., Seipel, S., Lind, M.: Evaluating 2d and 3d visualizations of spatiotemporal information. ACM Transactions on Applied Perception (TAP) 7(3), 19 (2010)
11. Lamarche-perrin, R., Demazeau, Y., Vincent, J., Demazeau, Y., Vincent, J.M.: How to Build the Best Macroscopic Description of your Multi-agent System?, 1–18 (2013)
12. Navarro, L., Flacher, F., Corruble, V.: Dynamic Level of Detail for Large Scale. In: AAMAS 2011: The 10th International Conference on Autonomous Agents and Multiagent Systems, vol. 2, pp. 701–708 (2011)
13. Prifti, E., Zucker, J.D., Clement, K., Henegar, C.: FunNet: an integrative tool for exploring transcriptional interactions. Bioinformatics 24(22), 2636–2638 (2008)
14. Railsback, S.F., Lytinen, S.L., Jackson, S.K.: Agent-based Simulation Platforms: Review and Development Recommendations. Simulation, 609–623 (2006)
15. Schelling, T.C.: A process of residential segregation: neighborhood tipping. Racial Discrimination in Economic Life 157, 174 (1972)
16. Shneiderman, B.: The eyes have it: a task by data type taxonomy for information visualizations. In: IEEE Symposium on Visual Languages (1996)
17. Taillandier, P., Vo, D.-A., Amouroux, E., Drogoul, A.: GAMA: A simulation platform that integrates geographical information data, agent-based modeling and multi-scale control. In: Desai, N., Liu, A., Winikoff, M. (eds.) PRIMA 2010. LNCS, vol. 7057, pp. 242–258. Springer, Heidelberg (2012)
18. Vo, D.A., Drogoul, A., Zucker, J.D.: An operational meta-model for handling multiple scales in agent-based simulations. In: 2012 IEEE RIVF International Conference on Computing and Communication Technologies, Research, Innovation, and Vision for the Future (RIVF), pp. 1–6. IEEE (2012)

Real-Time Fall Detection and Activity Recognition Using Low-Cost Wearable Sensors

Cuong Pham[1,2] and Tu Minh Phuong[1]

[1] Department of Computer Science,
Posts & Telecommunications Institute of Technology,
Ha Noi, Vietnam
cuongpham.ptit@gmail.com, phuongtm@ptit.edu.vn
[2] Affiliation: Culture Lab, Newcastle University, Newcastle upon Tyne, NE17RU, UK

Abstract. We present a real-time fall detection and activity recognition system (FDAR) that can be easily deployed using Wii Remotes worn on human body. Features extracted from continuous accelerometer data streams are used for training pattern recognition models, then the models are used for detecting falls and recognizing 14 fine grained activities including unknown activities in realtime. An experiment on 12 subjects was conducted to rigorously evaluate the system performance. With the recognition rates as high as 91% precision and recall for 10-fold cross validation and as high as 82% precision and recall for leave one subject out evaluations, the results demonstrated that the development of real-time fall detection and activity recognition systems using low-cost sensors is feasible.

1   Introduction

Falls are one of most high-risk problems for old people. A study conducted by the Centers for Disease Control and Prevention [1] shows that up to 33% adult people aged over 65 falls at least once a year. Of which about 30% cases cause medium to severe injuries that can lead to death. This is an obstacle on the elderly's living independent at homes. Although there are a number of previous studies on fall detection and yielded significant results (i.e. accuracy of 80-90%), in fact falls might occur while the elderly performs daily life activities such as walking, jumping, going-up stair, going-down stair, running etc. The information of such activities can provide warning the elderly for preventing the falls (i.e. jumping might highly cause a fall). Moreover, a study low-level daily activity information can not only be a fundamental element for situated services to support people [2, 23], but also be useful for health & energy expenditure monitoring [3]. The development of system integrated both activity recognition and fall detection with low-cost technologies at people's homes has real potential to provide age-related impaired people with a more autonomous lifestyle, while at the same time reducing the financial burden on the state and these people and their families.

Majority of existing activity recognition systems and fall detection systems require specific hardware and software design which often cost hundreds of US dollars or

B. Murgante et al. (Eds.): ICCSA 2013, Part I, LNCS 7971, pp. 673–682, 2013.
© Springer-Verlag Berlin Heidelberg 2013

more. Moreover, pervasive sensing deployment often requires many sensors in the environmental surroundings [3], which might exceed the budgets of the poor and middle-class people. This is especially true in Vietnam where up to 29% of Vietnamese population is classified as poor or below (according to the UNDP standard [4]). Therefore proposing a low-cost sensing and easy deployment technology for prototyping the FDAR system is a key component of this study to make the pervasive computing technologies support people's lives and help the elderly to live more independent at their homes.

Our contribution is twofold:

First, we prototype a fall detection and activity recognition system that can automatically detect human falls and can recognize 13 activities in real-time. Our work is distinct from other works on activity recognitions [7][11][14][25] as we address a set of low-level activities (rather than high-level activity set) and utilize the use of easily deployable and low-cost wearable sensors.

Second, we evaluate the system on an open-dataset (i.e. including unknown activities) collected from 12 subjects who performed 12 activities and 144 falls at their homes under 10-fold cross validation and leave one-subject out protocols.

2   Related Work

Activity recognition

Two common approaches to activity recognition are computer vision based and sensors based. In the first approach, computer vision technology is used to analyze the video streams from digital cameras installed in the environments (i.e. [11][12]) to infer human activities. In the second approach, activity recognition is performed by analyzing the sensing data from sensor streams. Ubiquitous sensors can be embedded into objects [2][13], environments [14][15], or worn on different parts of human body [3][5][6][7]. In this study, we focus on the wearable sensing as embedding sensors into objects or environments often requires many sensors while cameras invade people's privacy. Among sensors available on the market we choose Wii Remotes as they are cheap and easy deployment. Previously, Wii Remotes were also used for recognizing food preparation activities [13].

Fall detection

Majority of fall detection applications are implemented on smart-phone [16][17] which utilizes the acceleration data from accelerometer integrated inside the phone. Although those studies have significant results (~90% accuracy), it is noticed that falls often occur while people are doing other activities such as jumping, running, going-up stairs, going-down-stair etc. Some of them might lead to the elderly fall. Previously, a small number of research addressed fall detection and activity recognition problem (i.e. [18]), but using cameras that can invade people's privacy and without real-time implementation. Moreover, computer vision approach's performance is significant affected by light conditions in the environment (i.e. how it can work at night?). To our knowledge, no existing studies explored the recognition a set of low-level activities and detection of falls using low-cost sensing technologies with real-time implementation.

3   Hardware

In contrast to most of accelerometers often used in research labs (i.e. [4] [5] [6][7]) or on the market but quite expensive (i.e. [8][9][10]), or relatively complex deployment (i.e. requires base-station for communication to the computer), Wii Remotes are relatively cheap, available on the market, and simple deployment as Wii communicates with the computer via a Bluetooth dongle (also very cheap).

The Wii Remote [19] is a consumer off-the-shelf wireless sensing system and games controller which supports two functionalities of relevance to our application: (i) input detection through an embedded accelerometer; and (ii) data communications through Bluetooth. A Wii Remote comprises a printed circuit board (which is encapsulated by a white case) and uses an AXDL 330 accelerometer [3] and a Broardcom BCM2042 chip that integrates the entire profile, application, and Bluetooth protocol stack. Based on Micro Electro Mechanical System (MEMS) technology the AXDL 330 accelerometer is a small, low power, 3-axis accelerometer with signal conditioned voltage outputs. The AXDL 330 accelerometer can sense acceleration in three axes with a minimum full-scale range of ±3g. While the static acceleration of gravity can be used to implement tilt-sensing in applications, dynamic acceleration measurement can be detected through the quantifiers of motion, shock or vibration.

Fig. 1. Wii Remote worn on wrist (left) and Wii Remote worn on hip (right)

The Broadcom BCM2042 board is a system-on-chip which integrates an on-board 8051 microprocessor, random access memory/read only memory, human interface device profile (HID), application, and Bluetooth protocol stack. Furthermore, multiple peripherals and an expansion port for external add-ons are embedded on the board. The integration of these components and the technology's adoption in a mass market consumer games console has significantly reduced the cost of BCM2042. The Wii

Remote's input capabilities include buttons, an infrared sensor and an accelerometer. The infrared sensor is embedded in a camera which detects IR light coming from an external sensor bar. The accelerations are measured in X, Y, and Z axes (relative to the accelerometer) and the three directions of the movement (X, Y, Z) can be computed through tilt angles. Wii Remote inputs and sensor values are communicated to a Bluetooth host through the standard Bluetooth HID protocol. Values for acceleration are transmitted with a sampling frequency of 100Hz (100 samples per second).

| *Cleaning* | *stretching* | *falling* |

Fig. 2. Examples of accelerometer signals for three human activities and falls

In this study, subjects were asked to worn 2 Wii Remotes: one worn on hip and the other worn on right-hand's wrist. While the sensor worn on hip can provide good features for the detection of falls, running, walking, going-up stairs, the sensor worn on wrist might be useful for recognizing activities performed with hand such as cleaning, typing, and brushing. We will use the combination of both sensing data streams from 2 sensors for the detection of falls and recognition of activities.

4   Real-Time Fall Detection and Activity Recognition

The real-time FDAR algorithm works in 4 steps:

- Signal processing: sensing data is filtered for removing the noise or resampling the lost data.
- Segmentation: a sliding window/frame is used to segment the signal stream into frames of fixed length.
- Feature extraction: from each frame, different features are extracted.
- Classification: the system uses the features extracted from the previous step as input for a HMM based classifier.
— In the following sections, we describe each of these steps in detail.

4.1   Signal Processing

Sensing data from sensors are often noisy and ambiguity. Ideally, at a sampling frequency of 100Hz each second contains 100 samples of X, Y, Z acceleration triplets (i.e. one sample per 10 miliseconds). In practice, real-world factors mean that some samples are lost or dropped (e.g. metallic items placed between the sensors and the receiver). Furthermore, the sensors themselves can yield noisy readings (e.g. too large or small). In such cases, a filter is applied to remove noise and to fill out lost samples. In this step, the data filter performs both a low-pass filtering (removing abnormally low sample values) and a high-pass filtering (removing abnormally high sample values). After that, samples are grouped into sliding windows or frames. If a frame contains less that 75% of its full complement, it is discarded on the grounds that there is insufficient information to classify activities. Otherwise, it is resampled using a cubic spline interpolation method [20] to fill out the lost samples.

Along with acceleration X,Y,Z, we compute pitch, roll for each triplet:

$$Pitch = 2\ arctan\left(\frac{y}{\sqrt{x^2+z^2}}\right) \tag{1}$$

$$Roll = 2\ arctan\left(\frac{x}{\sqrt{y^2+z^2}}\right) \tag{2}$$

where x, y, z are acceleration values of the three axis

4.2   Segmentation

Previous studies showed that the length of sliding window has significantly impact on the performance of the pattern recognition algorithms [5][13]. In this study, we did a pilot study on the subset of collected dataset for selecting a reasonable length for sliding window. We varied the window length 1 second, 1.2 second, 1.5 seconds, 1.8 seconds, 2 seconds and 2.5 seconds and we stick on the window length of 1.8 seconds. The reason for the choosing window length of 1.8 second is that this length allows avoiding delay from continuously real-time processing while providing a reasonable recognition rate.

4.3   Feature Extraction

For each frame of size n where n is number of time points, the following features are extracted:

$$Mean(x) = \frac{\sum_{i=1}^{n} x_i}{n} \tag{3}$$

$$Standard\ deviation(x):\ \delta_x = \sqrt{\frac{1}{n}\sum_{i=1}^{n}(x_i^2) - [Mean(x)]^2} \tag{4}$$

$$Energy(x) = \frac{\sum_{i=1}^{n} x_i^2}{n} \tag{5}$$

$$Entropy(x) = -\sum_{i=1}^{n} p(x_i) log\ (p(x_i)) \tag{6}$$

where x_i is an acceleration value, $p(x_i)$, a probability distribution of x_i within the sliding window, can be estimated as the number of x_i in the window divided by n.

$$Correlation(X,Y)= \frac{cov(x,\ y)}{\delta_x \delta_y} \tag{7}$$

In which cov(x,y) is covariance and δ_x , δ_y are standard deviations of x and y.

Peak/bottom acceleration: for each sliding window, we also extracted 3 peak values and 3 bottom values of acceleration.

These features are combined into a 58-dimentional feature vector, composed of *Mean X, Standard deviation X, Energy X, Entropy X, Mean Y, Standard deviation Y, Energy Y, Entropy Y, Mean Z, Standard deviation Z, Energy Z, Entropy Z, Mean Pitch, Standard deviation Pitch, Energy Pitch, Entropy Pitch, Mean Roll, Standard deviation Roll, Energy Roll, Entropy Roll, Correlation XY, Correlation YZ, Correlation ZX, peak value of X, peak of Y, peak of Z, bottom of X, bottom of Y,* and *bottom value of Z.* These feature vectors are then used for training the pattern recognition algorithms. In the next section, we present the hidden Markov models [24] that we employed for real-time fall detection and activity recognition.

4.4   Hidden Markov Model-Based Classifier

In brief, a hidden Markov model [24] is a stochastic model that can be used to characterize statistical properties of a sensing signal. An HMM is associated with a stochastic process not directly observable (i.e. hidden), but it can be observed indirectly through a set of other outputs. The key problem is to determine the hidden parameters given a model and observedparameters. Then, the extracted model parameters can be used to perform further analysis in future learning. An HMM is based on the Markov assumption that the current state depends only on the precedingstate.

In our domain, we use HMMs (one HMM for one activity) with a mixture of Gaussians of for state's observation distribution. The number of hidden statesis manually tailored for each model. In the training phase, the parameters of each model (i.e. initial state probabilities, state transition probabilities, observation probability distribution) were estimated using the Baum-Welch algorithm (implemented in Murphy's HMM ToolKit [25]). After that, we use the trained models for classifying the falls and activities. The Viterbi algorithm is implementedand is used for the computation of the log likelihood probability of each observation sequence O (i.e. a feature vector computed from a continuous sliding window) given the trained model. The classifier will choose the model that produces the maximum log likelihood given feature vectors computed from test data.

5   Experimental Evaluation

5.1   Data Collection and Annotation

Twelve students from our institute (Post &Telecommunication Institute of Technology) were recruited, each wore 2 Wii Remotes, one on wrist and the other on hip. Each student was asked to perform 12 activities including walking, jumping, going-up stair, going-down stair, running, stretching, cleaning, typing, standing-to-sit, sitting-to-stand, brushing teeth, vacuuming and 12 (intentional) falls with various postures. No order of performing activities is required and no time constraint to each activity performed by the subject. Each activity is required to be performed as naturally as possible. In addition to sensing data, we recorded videos of subjects performing the activities.

To synchronize between the collected videos and the accelerometer data from Wii remotes, at the beginning time of each session the subject was asked to shake the body and the hand 3 times to make distinct signals. The subject was apparently shown on the videos. In addition, along with each sample, a timestamp was written to the acceleration data log files.

The subjects were given a list of 12 labels of activities to annotate the collected videos using ELAN Multimedia Annotator Tool [22]. Movements that are not one of 12 activities and falls were automatically labeled with "unknown".

5.2   Performance Metrics

Recognition results are reported as frame-wise precision, recall and F-measure values. The *precision* for an activity was calculated by dividing the number of correctly classified frames by the total number of frames classified as being a particular activity (i.e. *true positives/(true positives + false positives)*). *Recall* was calculated accordingly as the ratio of the number of correctly classified frames to the total number of frames of an activity (i.e. *true positives/total number of frames of an activity*). And, F-measure is the harmonic mean of precision and recall.

5.3   Results for 10-Fold Cross Validation

Under 10-fold cross validation procedure, the dataset was randomly partitioned into 10 parts of equal size. Nineof them are used for training and the remaining one is used for testing. Then, the process is repeated for all 10 parts and the results are averaged. The results are shown on the Table 1.   Note that both training and test sets may contain data from the same subject.

Table 1. -fold cross validationresults (numbers are in percent)

Activity	Precision	Recall	F-measure
brushing teeth	94.64	90.56	92.56
cleaning	89.98	82.43	86.04
falling	93.06	91.67	92.36
going-down stairs	96.76	93.06	94.87
going-up stairs	98.72	95.3	96.98
jumping	98.48	97.98	98.23
running	97.39	96.52	96.95
sitting-to-stand	87.43	89.22	88.32
standing-to-sit	96.75	94.16	95.44
stretching	76.71	69.41	72.88
typing	96.4	95.89	96.14
vacuuming	97.92	97.51	97.71
walking	96.77	95.35	96.05
unknown	86.17	86.63	86.4
Average	**92.58**	**90.54**	**91.55**

Overall, precision, recall, and F-measure are over 90% for 10-fold cross validation(i.e. subject dependent) analysis. Majority number of activities including falls has precision and recall as high as over 90% except for stretching activity which is often misclassified as cleaning. It is noticed that the recognition rate of falling is 93% precision and 91.6% recall.

5.4   Results for Leave-One-Subject-Out Evaluation

In addition to 10-fold cross validation which is often used for systems for personal use or adaptation. We envisage to evaluate the system under the *leave-one-subject-out* protocol. In which, we used 11 subjects for training and left the remaining one for testing. The process was repeated for all 12 subjects, and the results were averaged. Table 2 shows the results. It is noticed that the tested subject was not included in the training data.

Table 2. Leave-one-subject-out results (numbers are in percent)

Activity	Precision	Recall	F-measure
brushing teeth	85.71	86.99	86.35
cleaning	81.2	77.5	79.31
falling	84.03	82.64	83.33
going-down stairs	91.67	84.72	88.06
going-up stairs	94.44	90.17	92.26
jumping	97.47	96.46	96.96
running	96.23	93.91	95.06
sitting-to-stand	73.65	74.85	74.25
standing-to-sit	78.57	77.92	78.24
stretching	64.38	63.93	64.15
typing	95.37	95.12	95.24
vacuuming	96.05	85.45	90.44
walking	95.15	88.89	91.91
unknown	72.25	70.51	71.37
Average	**85.2**	**82.16**	**83.65**

The overall recognition rates for leave one-subject out evaluation are 85% precision, 82% recall, and 83.6% F-measure, which are lower than thoseof 10-fold cross-validation. Note that, leave-one-subject-out is more difficult then cross-validation because the system has to recognize activities for unseen subjects. This setting is also more similar to practical conditions where a system trained on a set of subjects is used to make recognitions for new subjects, data about which are unknown at training time. Again,stretching proved to be the most difficult activity to recognize with the F-measure as low as 63% while for running, jumping, and typing the system achieved F-measures higher than 90%.This is consistent with 10-fold cross validation results.

6   Conclusion

We have presented a solution for detecting and recognizing falls and 12 other human activities. Our method uses inexpensive sensing devices such as Wii Remotes worn on human body as sources of signals, thus provides a low-cost solution. From accelerometer signals, the system extracts several types of features that summarize different aspects of movements. A hidden Markov model is used to map these features into hidden states, which corresponds to different activities. An empirical study with data collected from 12 subjects demonstrates the effectiveness of the proposed method. The system achieved average F-measures of 91.55% for 10-fold cross-validation and 83.6% for leave-one-subject-out settings respectively. With relatively high accuracy while being simple and inexpensive, the proposed solution can be used for practical applications requiring the recognition of human activities.

References

1. Falls Among Older Adults: An Overview, http://www.cdc.gov/HomeandRecreationalSafety/Falls/adultfalls.html (accessed on January 28, 2013)
2. Pham, C., Hooper, C., Lindsay, S., Jackson, D., Shearer, J., Wagner, J., Ladha, C., Ladha, K., Plötz, T., Olivier, P.: The Ambient Kitchen: A Pervasive Sensing Environment for Situated Services. In: Demonstration in the Designing Interactive Systems Conference (DIS 2012). ACM, Newcastle Upon Tyne (2012)
3. Albinali, F., Intille, S., Haskell, W., Rosenberger, M.: Using Wearable Activity Type Detection to Improve Physical Activity Energy Expenditure Estimation. In: Proceedings of the 12th ACM International Conference on Ubiquitous Computing (Ubicomp 2010), pp. 311–320. ACM, New York (2010)
4. General office for population family planning, http://www.gopfp.gov.vn (accessed on January 28, 2013)
5. Bao, L., Intille, S.: Activity Recognition from User-Annotated Acceleration Data. In: Ferscha, A., Mattern, F. (eds.) PERVASIVE 2004. LNCS, vol. 3001, pp. 1–17. Springer, Heidelberg (2004)
6. Ravi, N., Dandekar, N., Mysore, P., Littman, M.L.: Activity Recognition from Accelerometer Data. In: Porter, B. (ed.) Proceedings of the 17th Conference on Innovative Applications of Artificial Intelligence (IAAI 2005), vol. 3, pp. 1541–1546. AAAI Press (2005)
7. Huỳnh, T., Blanke, U., Schiele, B.: Scalable Recognition of Daily Activities with Wearable Sensors. In: Hightower, J., Schiele, B., Strang, T. (eds.) LoCA 2007. LNCS, vol. 4718, pp. 50–67. Springer, Heidelberg (2007)
8. SUNSPOT, http://www.sunspotworld.com/products/ (accessed on January 28, 2013)
9. WAX3, http://axivity.com/v2/ (accessed on January 28, 2013)
10. Microstrain, http://www.microstrain.com/parameters/acceleration (accessed on January 28, 2013)
11. Wu, J., Osuntogun, A., Choudhury, T., Philipose, M., James, M.R.: A Scalable Approach to Activity Recognition based on Object Use. In: Proceedings of the 11th International Conference on Computer Vision (ICCV 2007), pp. 1–8. IEEE Press, Rio de Janeiro (2007)

12. Duong, T.V., Bui, H.H., Dinh, Q.P., Venkatesh, S.: Activity Recognition and Abnormality Detection with the Switching Hidden Semi-Markov Model. In: Proceedings of the 2005 IEEE Computer Society Conference on Computer Vision and Pattern Recognition (CVPR 2005), vol. 1, pp. 838–845. IEEE Computer Society, Washington, DC (2005)

13. Pham, C., Olivier, P.: Slice&Dice: Recognizing Food Preparation Activities using Embedded Accelerometers. In: Proc. Europ. Conf. Ambient Intell., pp. 34–43

14. Buettner, M., Prasad, R., Philipose, M., Wetherall, D.: Recognizing Daily Activities with RFID-based Sensors. In: 11th International Conference on Ubiquitous Computing (Ubicomp 2009), pp. 51–60. ACM, New York (2009)

15. Tapia, E.M., Intille, S.S., Larson, K.: Activity Recognition in the Home Using Simple and Ubiquitous Sensors. In: Ferscha, A., Mattern, F. (eds.) PERVASIVE 2004. LNCS, vol. 3001, pp. 158–175. Springer, Heidelberg (2004)

16. Abbate, S., Avvenuti, M., Bonatesta, F., Cola, G., Corsini, P.V.: A Smartphone-based Fall Detection System. Pervasive Mob. Comput. 8(6), 883–899 (2012)

17. Sposaro, F., Tyson, G.: iFall: An Android Application for Fall Monitoring and Response.In: In: Proceedings of 31st Annual International Conference of the IEEE EMBS, Minneapolis, Minnesota, USA (2009)

18. Olivieri, D.N., Conde, I.G., Sobrino, X.A.V.: Eigenspace-based Fall Detection and Activity Recognition from Motion Templates and Machine Learning. Expert Syst. Appl. 39(5), 5935–5945 (2012)

19. Wii Remote, http://en.wikipedia.org/wiki/Wii_Remote (accessed on January 28, 2013)

20. Spline Interpolation, http://en.wikipedia.org/wiki/Spline_interpolation (accessed on January 29, 2013)

21. Ward, A.J., Lukowicz, P., Gellersen, W.H.: Performance Metrics for Activity Recognition. ACM Trans. Intell. Syst. Technol. 2(1), Article 6, 23 pages (2011)

22. ELAN, http://tla.mpi.nl/tools/tla-tools/elan/ (accessed on February 19, 2013)

23. Wherton, P.J., Monk, F.A.: Technological Opportunities for Supporting People with Dementia Who Are Living at Home. Int. J. Hum.-Comput. Stud. 66(8), 571–586 (2008)

24. Rabiner, L.: A Tutorial on Hidden Markov Models and Selected Applications in Speech Recognition. In: Waibel, A., Lee, K.-F. (eds.) Readings in Speech Recognition, pp. 267–296. Morgan Kaufmann Publishers Inc., San Francisco

25. Kasteren, T., Noulas, V., Englebienne, A., Kröse, G.: Accurate Activity Recognition in a Home Setting. In: Proceedings of the 10th International Conference on Ubiquitous Computing (UbiComp 2008), pp. 1–9. ACM, New York (2008)

26. HMM ToolKit, http://www.cs.ubc.ca/~murphyk/Software/HMM/hmm.html (accessed on March 07, 2013)

A New Approach for Animating 3D Signing Avatars

Nour Ben Yahia and Mohamed Jemni

University of Tunis
Research Laboratory on Technologies of Information
and Communication & Electrical Engineering (Latice), Tunisia
benyahia.nour@gmail.com, mohamed.jemni@fst.rnu.tn
http://www.latice.rnu.tn

Abstract. In this paper, we present the description of an animation solver that allows the automatic moving of the different parts of a signing avatar body. The objective is to obtain a realistic animation of the virtual character that enables deaf person to visualize realistic gestures. Our approach is based on a thorough study of sign language and gestures classification. We identified the main constraints required for automatic generation of postures in sign language, and we developed a graphical interface in order to facilitate editing. This interface allows, in particular, the selection of the sign components and the rendering of the choice made on the avatar postures. The challenge of this project is to find a good compromise between computational time and realistic representation that should be closer to real-time generation signs.

Keywords: Animation, Inverse kinematics, sign language.

1 Introduction

To improve dialogue between human and machine [1] new methods are required. These methods facilitate access to information for everyone. In response to this context, the development of virtual persons can improve this interaction. We note that the new technologies of information and communication invest more and more our daily space. We observe the emergence of new services that tend to facilitate the generation, the propagation and the consultation of information [2]. However, such technical progress is not accessible to everyone. An individual may encounter difficulties in accessing information for varieties of reasons that may be economic, cultural, linguistic or physical. We seek to enable the machine to generate information into sign language. Nowadays more and more software are available to facilitate access to information for deaf through visualization tools of their native language. These applications concern in particular the real-time remote communication, the teaching assisted computer and signed and oral broadcasts. Further new forms of communication incorporate virtual agents that improve the ability to interact with users [3]. These agents can play the role of assistants in interactive multimedia applications. In this context, our

B. Murgante et al. (Eds.): ICCSA 2013, Part I, LNCS 7971, pp. 683–696, 2013.
© Springer-Verlag Berlin Heidelberg 2013

work consists in developing an animation solver that allow automatical anima-tion of the different body parts of an avatar in order to improve the fluidity of the avatar movement. This approach is based on a depth study of sign language and gestures classification [4]. We identified the parameters to be used for auto-matic generation of sign language postures. Detection and avoidance of collisions between body parts of the avatar must be mastered to allow the contacts and to prevent unacceptable movements. To further facilitate editing, we developed a human machine interface to create signs [5] [6]. The challenge of this project is to find the tradeoff between computational time and realistic representation that must be closer to the real-time generation signs.

2   Sign Language

In many minds, the sign language (SL) is what the Braille is for the blind, i.e. to ensure habitual communication for this community. This language has been designed recently, as a substitute of the ear canal in favor to a visual channel. It avoids the isolation in which the deaf could be immersed if they have no means of communication. However, this vision is wrong perceived and does not reflect the sociolinguistic reality of SL because a sign doesn't have one representation and closely relation with its environment. SL is a mean of expression used by the community of deaf and hearing person to communicate [7]. It's a real language with a lexicon and syntax [4] and it's considered as the most advanced form of gestural communication. In these kind of languages the message is transmitted through the gesture channel rather than oral sign [8]. The expression of SL sen-tences cannot be reduced to only gestures produced by the two hands but to the whole body that is often involved. Indeed, the signer uses the space around him to make signs and to place different elements of speech and refer to it. To better describe signs, it is essential to integrate this concept in automated systems that analyze these languages. Nowadays, most of signs capture systems are studied in a context of machine translation and are based on specific features of motion capture [9] [10]. The goal of this work is to access to this information through the design of a system that is able to generate gesture for SL automatically. Even if the SLs are considered as full languages, a minority of hearing people use them, there are often ignored in favor of the official dominant surrounding languages. It is the same for minority spoken languages like regional languages all over the world. That is problematic for deaf person especially when SL is not part of the official languages. The use of written language is difficult, in fact 80% of deaf people in the world are illiterate, according to the World Federation of Deaf in 2003. The result is an increase in need and demand of the deaf population, especially in education, information dissemination and dialogue. SL is used by deaf people to ensure all functions performed by the other natural languages (so-called "spoken languages" or "voice"). They are visual-gestural for the deaf and the only really appropriate linguistic mode, which allows them a cognitive and psychological development in an equivalent way to what is an oral language for a hearing. We need to underline that the SL is not a universal way of com-munication, the transcription system requires an apprenticeship [11]. Moreover, there is no one universal SL but many like French Sign Language (FSL), the American Sign Language (ASL or American Sign Language) [12] [13], English sign language (British Sign Language or BSL). And, like spoken language, each one has its own history, meaningful units and lexicon. Ethnologue.com site lists 121 different sign languages and Wittmann provides a classification [14]. The development of SL depends on the vivacity of the community, as a language for voice. But despite the differences between SLs all over the world, understanding and communication is possible between two people mastering different sign languages due to the proximity of syntactic structures and the existence of very close iconic structures [15]. Iconicity is the basis of SL. It expresses the illustration of an object or a person. This mechanism is based on formatting gestures to convey information. Iconicity is the process by which the speaker will make iconic context. This process provides access to illustrated target that the speaker aims to rebuild in an imagery manner. This target is due to cognitive mechanisms that select the experience that can or should be iconized and restore the language in the form of statements. We can keep in mind that there are two ways to say in SL: "say by showing" (referred iconicity) and "say by no showing" (standard lexicon).

3   Sign Language Transcription

Three mains approaches exist in the literature: the first one is related to writing or drawing symbols, the second one is based on video and the third one is based on 3D sequences and the animation of a virtual person according to a standard [16]. Drawing was the first transcription manner of SL and the way generally used to replace writing. Later on, several transcription systems appeared such as HamNoSys (Hamburg Notation System) [17] and SignWriting in spite of difficulty to encode them in a linear way. The video based systems consist of a video sequence record of human SL interpreter. Much more sophisticated tools exist nowadays on the market based on debit and quality. In the new technological context, the modeling of a virtual character can be achieved either according to a segmented model, gotten by a hierarchical graph of the anatomical 3D segments. In both cases, the surfaces are represented either with a polygonal stitch, or by a mathematical analysis, otherwise according to implicit functions of the skeleton. The creation of a virtual character could be achieved either by a modeler of geometric primitives, or with the help of a 3D scanner. The approach of segmented virtual character exists in the H-Anim specifications, as well as in MPEG-4 FBA (Face & Body Animation), whereas the representation by virtual character is processed in MPEG-4 BBA (Bone-Based Animation) [18] [5]. The most comfortable solutions for the deaf are to involve interpreters who translate in real time the statements in oral language. Research in signing avatars on SL machine translation systems led to various prototypes of systems to translate statements comply with the use of SL for the deaf communication. Generation of statements in SL involves humanoid able to reproduce gestural sequences. From a user interface, research on animated virtual character assume that human users are able to interact more efficiently and nicer with the machine that mean that the mode of interaction is close to the natural mode of communication equivalent to a human-human dialogue.

4   Animation Solver

Many statistics have confirmed that many deaf are enabled to access to written information. As a solution, computer applications [19] designed for deaf person, in particular who are illiterate, have been created [20]. Therefore, to facilitate access to information, new methods improving the dialogue between human and machine are required. The development of virtual characters (avatar) able to generate postures in sign language, could be a response to this request [21]. This development is based on an automatic gestures generation. Generally, within the set of admissible postures, the user is free to manipulate the avatar rather than specifying the value of each individual degree of freedom [22] [23] [24]. The Inverse Kinematics (IK) method automatically computes these values in order to satisfy a given task usually expressed in cartesian coordinates [25]. This technique requires the resolution of nonlinear complex equations and is usually expressed as a constraint-satisfaction problem. However, this is a laborious task because of the high number of degrees of freedom present in the model i.e. fifty for a human model without considering the fingers. The control of a human model by means of IK requires a simultaneous multitasks application. For example, we may consider tasks that control the position of left hand and another of right hand. The balance of human model will be controlled by another task which provides some information concerning forces and mass distribution. In this context, this work consists in developing an animation solver. This system permits to animate automatically different avatar body parts in order to improve realistic animation of virtual character gesture [26] and allow deaf person to visualize realistic gestures [27]. The approach is based on a thorough study of sign language and gestures classification. We have identified the constraints that need to be used for automatic generation postures sign language i.e. symmetry relations, that facilitate the specification of movement of the non-dominant hand knowledge of those dominant articulator and various types of repetition in sign. In addition, sign language requires contacts between different joints. Detection and avoidance of collisions between body parts of the avatar must be mastered to allow the contacts and to prevent unacceptable movement. To further facilitate editing, a graphical interface has been developed. It allows the selection of the sign components with a direct view of the choice made on the avatar postures. The challenge of this project is to find the tradeoff between computational time and realistic representation. Indeed it must be closer to the maximum generation of real-time signs.

4.1   Animation Solver Description

The goal is to develop an animation engine to animate virtual characters, and to generate precise SL postures. For the automatic creation of signs, our Web-Sign project [5] can generate these postures, in a simple and intuitive way. This application uses an interface that allows non-IT users to create their own descriptions of signs and participate in the development of their dictionary communities. This interface uses the technique of key frames to select the important moments assumed (the key moment) for training movement. But this technique raises difficulties especially in terms of time spent in the creation of postures and had lack of precision in the generation of gestures. The main problem induced by the use of direct kinematics is to achieve the desired position by the avatar. Therefore, the first step is to rotate the shoulder of the virtual character then the elbow and the wrist. This causes a multitude of solutions to generate this posture. So the user must intervene to choose the right solution (the best meets their needs). This operation is costly in terms of time and presents several challenges to create a fluid motion understandable by the deaf.

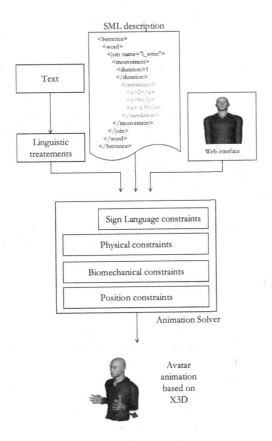

Fig. 1. Architecture of animation engine

The system uses the architecture illustrated in figure 1. It relies on several inputs. We can from a web interface create the sign by selecting the desired joint and move along different axis (x, y, z) to the target position. The system, based on IK, calculate automatically in a real-time the angles of rotation of each joint that are required to generate the posture. From a biomechanical study, we identify the degrees of freedom of each joint. This study has allowed us to generate real time and natural postures of the virtual character. We can from a SML description [3] [28] [29], containing the position of the desired joint as shown (Figure 1), generate the animation of the avatar from a text and perform linguistic process in order to extract some information for the description of a sign. Thus, we can animate the virtual characters in real-time.

4.2 Sign Language Decomposition

Complex movements are infrequent in sign language. They can take the form of zigzags or sinusoids, but in most cases, cannot be simply described by a prede-fined type of trajectory [30] [31]. In this context we decompose the movement in several parts: the description of the initial hand posture, the dynamic step and the final posture, manual specifications including the parameters that are related to the hand configuration, position and orientation, and contact point [32]. These parameters can vary or remain constant during the movement.In case where the hand comes into contact with a body part (chest, face, hand or other arm), the wrist position is no longer relevant, which is the manual contact point that must be included in the manual specification. The arm movement is already specified by the shape of the wrist path. Many grammatical processes affect its amplitude, so the movement modifier must inform it the latter in order to be amplified, shortened, or stopped. This problem appears in some local hand signs, as the vibration of the fingers. More generally, we denote by secondary movement such repeated movement that occur in the fingers, wrist, or even the forearm (but never affects the position of the wrist). It can be superimposed on the main movement or just be the only dynamic character of the sign.

5   Inverse Kinematics

The IK is a method for calculating the postures, in order to satisfy a given task by the user, based on degree of freedom of each joint. This method plays an important role in the animation and simulation of virtual characters to improve and facilitate the use of interfaces. The IK finds its application in various fields such as robotics [33], medicine and the gaming industry. These methods have been implemented in order to facilitate or to control various virtual characters [34]. The IK problem has been widely studied in the recent decades. It was first used in the field of robotics, particularly for computing the robots poses. Since, it widely used in the world of computer assisted animation, and the game industry to compute the position of articulated figures but the production of realistic and plausible movement still an open challenge in robotics and animation commu-nities [25]. Several models have been implemented to try to solve this problem. Solving the IK task coincides with the problem of finding a local minimum of a set of non-linear equations that defines the constraints of a cartesian space.

5.1   Background

To solve the IK problem, the most popular approach use the numerical Jacobian matrix to find a linear approximation [35]. The Jacobian solutions model the movement of end effectors with respect to changes of the system represented by links and joints angles [36]. Several methods have been presented for the approximation or calculation like the inverse Jacobian, Jacobian transposition, damped least squares (DLS), damped least squares with Singular Value Decomposition (SVD-DLS), selectively damped least squares (LDI) and many other extensions [37] [38]. Solutions based on the inverse Jacobian produce realistic postures, but most of these approaches suffer from high computational cost given that the matrix calculations are very complex and singularity problems serve to cause further calculation [39] [40]. An alternative approach is proposed by Peshev [41] where the problem of IK is solved without using the inverse of the matrix. The IK second family solver is based on Newton methods [42]. These algorithms look for target configurations that solutions arise from a minimization problem. The best known methods are the Broyden method [43] and The Powell method [44]. However, Newton methods are complex, difficult to implement and have a high computational cost [45] [46]. Recently, the sequential Monte Carlo (MCMS) approach has been proposed based on particle filtration. This method gives good results but they are costly and suffer from singularity [47] [48]. A very popular IK method is the Cyclic Coordinate Descent (CCD) algorithm [49], which was first introduced by Wang and Chen [40] and then the biomechanical constrained by Welman [25]. CCD has been extensively used in the computer games industry and has recently been adapted for protein structure prediction. CCD is a heuristic iterative method with low computational cost for each joint per iteration, which can solve the IK problem without matrix manipulations; consequently it formulates a solution very quickly. The cyclic coordinate descent is a greedy approach to the problem of IK. It browse chain by optimizing the elements sequentially one by one, seeking position as close as possible to each effector target. The process is iterated as many times as necessary until the system stabilize. This method is particularly suitable for solving problems with only one effector but offers poor performance in the case of a multi-resolution. The fact that this method is simple makes its attractive. However as, any iterative optimization method, the speed of each optimization step must be reduced to avoid instability near solutions that slow down animation engine has to be. Another method, called the triangulation algorithm, did not use an iterative approach is presented in [50]. This method use the cosine law to calculate each joint angle from the root of the kinematic chain to the end effector. It guarantees to find a solution when used without joints constrain and when the target is in achievable range. The triangulation algorithm incurs a lower computational cost of than CCD algorithm because it needs just one iteration to achieve the target. However, the results obtained are not realistic. The triangulation method can be applied to solve problems with only one target that can be manipulated by the user; kinematic chains with multiple end effectors can not be resolved then this method can not be used such complex characters. We also tested several approaches like

Newton methods and the family of Jacobian matrix. But, we noticed that these alogorithms don't obey in real-time and this logical due to the matrix size used and the operations performed on it.

5.2   Inverse Kinematic Method

The used IK method here, involve the previously calculated positions of the joints to find the solution to achieve the target. This method minimize the error of the system by adjusting each joint angle. The proposed method starts from the last joint of the IK chain, iterate and adjusting each joint along the chain. Thereafter, it iterate in the reverse way, in order to complete the adjustment. This method, instead of calculating directly the angle rotations, try to find the joint locations. Hence, the time dedicated to compute and resolve the constraints can be saved. So, we can generate realistic and human animation in real time.

Algorithm 1. Inverse Kinematic algorithm

Input: The joint positions p_i for $i = 1, ..., n$, the target position t and the distances between each joint $di = |p_{i+1} - p_i|$, for $i = 1, ..., n - 1$.
Output: The new joint positions p_i for $i = 1, ..., n$.
  $dist \leftarrow |p_1 - t|$
  if $dist > d_1 + d_2 + ... + d_{n-1}$ **then**
    for $i = 1$ to $n - 1$ **do**
      $r_i \leftarrow |t - p_i|$
      $k_i \leftarrow d_i/r_i$
      $p_{i+1} \leftarrow (1 - k_i)p_i + k_it$
    end for
  else
    $temp \leftarrow p_1$
    $dif \leftarrow |p_n - t|$
    while $dif > tolerance$ **do**
      $p_n \leftarrow t$
      for $i = n - 1$ to 1 **do**
        $r_i \leftarrow |p_{i+1} - p_i|$
        $k_i \leftarrow d_i/r_i$
        $p_i \leftarrow (1 - k_i)p_{i+1} + k_ip_i$
      end for
      $p_1 \leftarrow temp$
      for $i = 1$ to $n - 1$ **do**
        $r_i \leftarrow |p_{i+1} - p_i|$
        $k_i \leftarrow d_i/r_i$
        $p_i \leftarrow (1 - k_i)p_i + k_ip_{i+1}$
      end for
      $dif \leftarrow |p_n - t|$
    end while
  end if

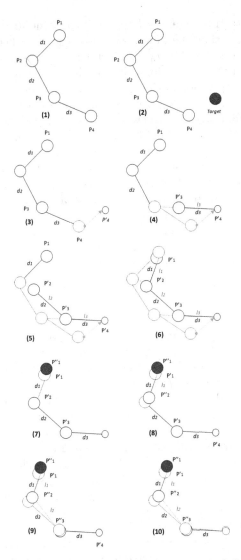

Fig. 2. Inverse kinematics method. (1) The initial position of the manipulator with 4 joints. (2) The initial position of the manipulator and the target. (3) move the end effector p_4 to the target. (4) find the joint p'_3 which lies on the line l_3 that passes through the points p'_4 and p_3, and has distance d_3 from the joint p'_4. (5) (6) continue the algorithm for the rest of the joints. (7) move the root joint p'_1 to its initial position. (8) (9) (10) repeat the same procedure but this time start from the base and move outwards to the end effector. The algorithm is repeated until the position of the end effector reaches the target or gets sufficiently close.

If we consider (Figure 2, Algorithm 1) $p_1, ..., p_n$ the joints of the avatar. Note that p_1 is the root joint and p_n is the end effector. The target is symbolized by a red point. The method used is illustrated with a single target and 4 joints

2. First we calculate the distances between each joint $di = |p_{i+1}-p_i|$, for $i = 1, ..., n-1$. Assuming that the new position of the end effector be the target position, $p'_n = t$, we find the line, l_{n-1}, which passes through the joint p_{n-1} and p'_n. The new position of the p'_{n-1} joint, lies on that line with distance d_{n-1} from p'_n. Similarly, the new position of the p'_{n-2} joint, can be calculated using the line l_{n-2}, which passes through the p_{n-2} and p'_{n-1}, and has distance d_{n-2} from p'_{n-1}. The algorithm continues until all new joint positions are calculated. The new position of the root joint, p'_1, should not be different from its initial position. Another iteration is completed when the same procedure is repeated but this time starting from the root joint and moving outwards to the end effector. Thus, let the new position p''_n, then find the line l_1 that passes through the points p''_1 and p'_2, we define the new position of the joint p''_2 as the point on that line with distance d_1 from p''_n. This procedure is repeated for all the joints, including the end effector. After a complete iteration, the solution may not achieve the desired target. The procedure is then repeated, as many time as needed, until the end effector is the same or achieve to the desired target.

5.3 Results

The IK method converges to any given goal positions, seeing that the target is reachable (Table 1). However, if the target is not within the reachable area, there is a termination condition that compares the previous and the current position of the end effector, if this distance is less than an indicated tolerance. The method gives very good results and converges quickly to the desired position. This method is based only on the position computing and avoids the tedious rotations computing. The animation engine is integrated into the project Websign to facilitate the task of creating signs (Figure 3). In reality, the signing avatar model is comprised of several kinematic chains, and each chain generally has different end effector. Therefore, it is essential for an IK solver to be able to solve problems with multiple end effectors and targets. The animation engine can be extended to process models with multiple end effectors and achieve the different targets positions. We tested the animation engine by adding constraints of sign language as the notion of symmetry either total or anti-symmetry and of course along different axis x, y and z. A video demonstration is available on the official site of the research laboratory [51].

The figure 3 shows the use of this notion by considering several kinematic chains and several end effector (left wrist and right wrist) by applying the notion

Table 1. Results for a single kinematic chain with 10 joints with reachable target

Method	Number of iterations	Execution time (sec)
our IK method	15	0.014
CDD	26	0.13

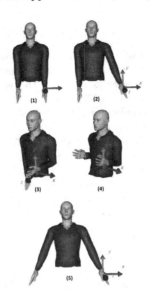

Fig. 3. Signing avatar using Inverse kinematics method. (1) initial posture of the avatar. (2) movement of left wrist along the x axis. (3) movement of left wrist along the x,y and z axis. (4) activate symmetry for different axis. (5) activate antisymmetry for x axis.

of symmetry that uses two kinematic chains, the first correspond to the avatar left side and the second to the right side.

6   Conclusion

Generation of Signs is based on different parameters such as manual configuration, orientation of hands, the location where the sign is made, the movement made by hand and the facial expression accompanying the realization of the sign. We take in account all these parameters and this system deals with figures which have many degrees of freedom. We use the technique of inverse kinematics to ensure precise movements, control and maintain the balance of characters. This method applied in user interface facilitates the manipulation of the avatar. Inverse Kinematics is defined as the problem of determining a set of appropriate joint configurations for which the end effectors move to desired positions rapidly as possible. However, many of the currently available methods suffer from high computational cost and generation of unrealistic poses. Through this study, we succeed in solving this complex system in real-time. We integrate this solver in the Websign project for automatic translation of text to sign language by the means of avatars. The perspective of this work is to extend the animation engine by adding methods to avoid collision between the various articulations of virtual character.

References

1. Jemni, M., Ghoul, O., Ben Yahia, N., Boulares, M.: Sign language mms to make cell phones accessible to the deaf and hard-of-hearing community. In: Conference and Workshop on Assistive Technology for People with Vision and Hearing Impairments (CVHI), Granada, Spain, August 28-31 (2007)
2. Jemni, M., Elghoul, O., Makhlouf, S.: A web-based tool to create online courses for deaf pupils. In: International Conference on Interactive Mobile and Computer Aided Learning, Amman, Jordan, April 18-20 (2007)
3. Jemni, M., Elghoul, O.: A system to make signs using collaborative approach. In: Miesenberger, K., Klaus, J., Zagler, W.L., Karshmer, A.I. (eds.) ICCHP 2008. LNCS, vol. 5105, pp. 670–677. Springer, Heidelberg (2008)
4. Liddell, S.K., Johnson, R.E.: American sign language: The phonological base. Sign Language Studies 64, 195–278 (1989)
5. Jemni, M., Elghoul, O.: An avatar based approach for automatic interpretation of text to sign language. In: 9th European Conference for the Advancement of the Assistive Technologies in Europe, San Sebastin, Spain, October 3-5, pp. 260–270 (2007)
6. Boulares, M., Jemni, M.: Mobile sign language translation system for deaf community. In: Proceedings of the International Cross-Disciplinary Conference on Web Accessibility, Lyon, France, April 16-17, p. 37. ACM (2012)
7. Klima, E.S., Bellugi, U.: The signs of language. Harvard University Press (1979)
8. Messing, L.: Bimodal communication, signing skill, and tenseness. Sign Language Studies 84, 209–220 (1994)
9. Jaballah, K., Jemni, M.: Toward automatic sign language recognition from web3d based scenes. In: Miesenberger, K., Klaus, J., Zagler, W., Karshmer, A. (eds.) ICCHP 2010, Part II. LNCS, vol. 6180, pp. 205–212. Springer, Heidelberg (2010)
10. Othman, A., Jemni, M.: Statistical sign language machine translation: from english written text to american sign language gloss. International Journal of Computer Science Issues 8(3) (September 2011)
11. Shantz, M., Poizner, H.: A computer program to synthesize american sign language. Behavior Research Methods 14(5), 467–474 (1982)
12. Poizner, H.: Visual and phonetic coding of movement: Evidence from american sign language. Science 212(4495), 691–693 (1981)
13. Battison, R.: Phonological deletion in american sign language. In: Lexical Borrowing in American Sign Language, vol. 5, pp. 1–14. Linstok Press, Silver spring (1974)
14. Wittmann, H.: Classification linguistique des langues signées non vocalement. Revue Québécoise de Linguistique Théorique et Appliquée 10(1), 215–288 (1991)
15. Cuxac, C.: Iconicité et mouvement des signes en langue des signes française. Le Mouvement: des Boucles Sensorimotrices aux Représentations Langagières, 205–218 (1997)
16. Gavrila, D., Davis, L.: 3-d model-based tracking of humans in action: a multi-view approach. In: Proceedings of the 1996 IEEE Computer Society Conference on Computer Vision and Pattern Recognition, CVPR 1996, College Park, MD, USA, pp. 73–80. IEEE Institute of Electrical and Electronics Engineers (1996)
17. Prillwitz, S.: HamNoSys Version 2.0: Hamburg notation system for sign languages: An introductory guide. Signum (1989)
18. Pandzic, I.S., Forchheimer, R.: The origins of the mpeg-4 facial animation standard. MPEG-4 Facial Animation: The Standard, Implementation and Applications, pp. 3–13 (Janury 2003)

19. Lebourque, T., Gibet, S.: Synthesis of hand-arm gestures. In: Proceedings of Gesture Workshop on Progress in Gestural Interaction, London, UK, March 19, pp. 217–225. Springer (1996)

20. Frishberg, N., Corazza, S., Day, L., Wilcox, S., Schulmeister, R.: Sign language interfaces. In: Proceedings of the INTERACT 1993 and CHI 1993 Conference on Human Factors in Computing Systems, Amsterdam, Netherlands, April 24-29, pp. 194–197. ACM the Association for Computing Machinery (1993)

21. Badler, N., Chi, D., Chopra, S.: Virtual human animation based on movement observation and cognitive behavior models. In: Proceedings of the Computer Animation, Geneva, Switzerland, pp. 128–137. IEEE Institute of Electrical and Electronics Engineers (May 1999)

22. Chang, L., Pollard, N.: Method for determining kinematic parameters of the in vivo thumb carpometacarpal joint. IEEE Transactions on Biomedical Engineering 55(7), 1897–1906 (2008)

23. Hollister, A., Buford, W., Myers, L., Giurintano, D., Novick, A.: The axes of rotation of the thumb carpometacarpal joint. Journal of Orthopaedic Research 10(3), 454–460 (1992)

24. Klopčar, N., Tomšič, M., Lenarčič, J.: A kinematic model of the shoulder complex to evaluate the arm-reachable workspace. Journal of Biomechanics 40(1), 86–91 (2007)

25. Welman, C.: Inverse kinematics and geometric constraints for articulated figure manipulation. PhD thesis, Simon Fraser University (1993)

26. Liddell, S.: Structures for representing handshape and local movement at the phonemic level. In: Theoretical Issues in Sign Language Research, Chicago, IL, USA, vol. 1, pp. 37–66. University of Chicago Press (December 1990)

27. Kendon, A.: How gestures can become like words. Cross-cultural Perspectives in Nonverbal Communication, 131–141 (1988)

28. Ben Yahia, N., Jemni, M.: Animating signing avatar using descriptive sign language. In: International Conference on Electrical Engineering and Software Applications (ICEESA), Hammamet, Tunisia, March 21-32. IEEE (2013)

29. Ben Yahia, N., Jemni, M.: Toward a descriptive language for signs. In: Conference and Workshop on Assistive Technologies for People with Vision and Hearing Impairments (CVHI), Worclaw, Poland, April 20-23 (2009)

30. McNeill, D.: Hand and mind: What gestures reveal about thought. University of Chicago Press, Chicago (1996)

31. Stokoe, W., Casterline, D., Croneberg, C.: A dictionary of American Sign Language on linguistic principles. Linstok Press, Silver Spring (1976)

32. Filhol, M.: Une approche géometrique pour la modélisation des lexiques en langues signées. In: Verbum ex machina: actes de la 13e Conférence sur le Traitement Automatique des Langues Naturelles, Leuven, Belgium, vol. 2, p. 736. Presses univ. de Louvain (April 2006)

33. Craig, J.: Introduction to robotics: mechanics and control. Prentice Hall (August 2004)

34. Tolani, D., Goswami, A., Badler, N.: Real-time inverse kinematics techniques for anthropomorphic limbs. Graphical Models and Image Processing 62(5), 353–388 (2000)

35. Wolovich, W., Elliott, H.: A computational technique for inverse kinematics. In: The 23rd IEEE Conference Decision and Control, Las Vegas, USA, vol. 23, pp. 1359–1363. IEEE Institute of Electrical and Electronics Engineers (December 1984)

36. Meredith, M., Maddock, S.: Real-time inverse kinematics: The return of the jacobian. Technical report, Technical Report No. CS-04-06, Department of Computer Science, University of Sheffield (2004)
37. Buss, S.R.: Introduction to inverse kinematics with jacobian transpose, pseudoinverse and damped least squares methods. IEEE Journal of Robotics and Automation 17 (April 2004)
38. Albert, A.: Regression and the Moore-Penrose pseudoinverse, vol. 94. Academic Press (November 1972)
39. Der, K., Sumner, R., Popović, J.: Inverse kinematics for reduced deformable models. In: ACM Transactions on Graphics (TOG) - Proceedings of ACM SIGGRAPH, vol. 25, pp. 1174–1179. ACM the Association for Computing Machinery, New York (2006)
40. Wang, L., Chen, C.: A combined optimization method for solving the inverse kinematics problems of mechanical manipulators. IEEE Transactions on Robotics and Automation 7(4), 489–499 (1991)
41. Pechev, A.: Inverse kinematics without matrix inversion. In: IEEE International Conference on Robotics and Automation, Pasadena, USA, IEEE Institute of Electrical and Electronics Engineers (April 2012)
42. Luenberger, D., Ye, Y.: Linear and nonlinear programming, vol. 116. Springer (2008)
43. Broyden, C.: Quasi-newton methods and their application to function minimization. Math. Comput. 21(99), 368–381 (1967)
44. Powell, M.: Restart procedures for the conjugate gradient method. Mathematical Programming 12(1), 241–254 (1977)
45. Unzueta, L., Peinado, M., Boulic, R., Suescun, Á.: Full-body performance animation with sequential inverse kinematics. Graphical Models 70(5), 87–104 (2008)
46. Boulic, R., Varona, J., Unzueta, L., Peinado, M., Suescun, A., Perales, F.: Evaluation of on-line analytic and numeric inverse kinematics approaches driven by partial vision input. Virtual Reality 10(1), 48–61 (2006)
47. Grochow, K., Martin, S., Hertzmann, A., Popović, Z.: Style-based inverse kinematics. In: ACM Transactions on Graphics (TOG), vol. 23, pp. 522–531. ACM the Association for Computing Machinery (2004)
48. Sumner, R., Zwicker, M., Gotsman, C., Popović, J.: Mesh-based inverse kinematics. In: ACM Transactions on Graphics (TOG) - Proceedings of ACM SIGGRAPH, vol. 24, pp. 488–495. ACM the Association for Computing Machinery, New York (2005)
49. Canutescu, A., Dunbrack Jr., R.: Cyclic coordinate descent: A robotics algorithm for protein loop closure. Protein Science 12(5), 963–972 (2003)
50. Muller-Cajar, R., Mukundan, R.: Triangulation: a new algorithm for inverse kinematics. In: Proceedings of Image and Vision Computing, New Zealand, University of Canterbury. Computer Science and Software Engineering, pp. 181–186 (December 2007)
51. WebSign, T.: WebSign web site (2013), http://www.latice.rnu.tn/websign (online; accessed 2013)

A Real-Time Rendering Technique for View-Dependent Stereoscopy Based on Face Tracking

Anh Nguyen Hoang[1], Viet Tran Hoang[1], and Dongho Kim[2]

[1] Soongsil University
{anhnguyen,vietcusc}@magiclab.kr
[2] 511 Sangdo-dong Dongjak-gu, South Korea
cg@su.ac.kr

Abstract. In our research, we propose and implement a virtual reality system with the common and widely used devices such as 3D screen and digital webcam. Our approach involves the combination of 3D stereoscopic rendering and face tracking technique in order to render a stereo scene based on the position of the viewer. Our approach is to calculate the offset values of face position to assign to the virtual camera position relatively. We employ a technique to change the typical symmetric frustum into asymmetric to achieve the head-coupled perspective. With our system, the rendered scene observed by human eyes remains realistic and the viewport can be seen as the physical window in the real environment. Therefore, the right perspective can be maintained regardless of viewer position.

Keywords: View-dependent, asymmetric frustum, stereoscopy, face tracking.

1   Introduction

Virtual reality came into existence during the early 1960s. It has been developed and applied widely in recent years because of the development of computer in both hardware and software. A virtual reality system at least consists of these kinds of devices: a computer, input devices for position or gesture tracking, and output devices for displaying information to the user. For a simple virtual reality system, it is typical to implement using a common desktop computer with a digital webcam to track the viewer's face and adjust the 3D rendered scene appearing in the system window. Virtual reality is often used to deliver the visual contents in various application forms which are mostly associated with 3D stereoscopic environment. Therefore, using 3D displays such as interlaced screens or anaglyph with red-cyan glasses would be a simple and appropriated approach for our system.

Stereoscopic rendering has been researched and implemented comprehensively recently. The techniques for stereo display have been applied in common usages such as the mass production of 3D screen with polarized glasses or the 3D Vision

B. Murgante et al. (Eds.): ICCSA 2013, Part I, LNCS 7971, pp. 697–707, 2013.

Fig. 1. Main interface of system. Stereoscopic rendering and face tracking.

equipment from NVIDIA. Using those kinds of technique as well as the 3D stereoscopic rendering fundamentally well-known theories, we are able to produce and transfer the 3D stereoscopic rendered scene to the viewer in real-time. It is required to keep our system as simple and applicable as it should be, we are not employing the expensive and high-technology devices, the cheap and accessible devices are employed instead. This is one of main principles in our system. Using the common devices for representing an interactive system between the viewer pose and 3D stereoscopic contents involves the high challenges in the technique to improve the performance of the face tracking as well as stereoscopic rendering.

View-dependent rendering is the terminology describing the changes of rendered scene regarding to the view of observer. In our system, view-dependent rendering is used to adjust the virtual camera position to display a scene dynamically with the viewer's face. View-dependent rendering was used mostly in terms of perspective change when the viewer is looking through the window. The virtual scene will be changed and displayed interactively. In many researches, the basic and simple techniques are being used to track the head position and by controlling the virtual camera position relatively to the viewer, the scenes are changed but they do not actually reflect the right human perspective in real environment. In these cases, the frustum defined by the graphic API will retain the normal symmetric shape and it therefore cannot present a right perspective to human perception.

In our system, we also present a tracking technique based on optical flow to track the viewer's face. In computer vision, face tracking is always an important and interesting research area. Face tracking in our approach is understood as the ability to retrieve the position of a human head relatively to the computer screen. Although it seems to be the challenges in computer vision, the recent works of researches have been introducing variety of face tracking processes with the high accuracy in real-time. We do not make a novel technique. Instead, we employ a high accuracy technique with a modification to enhance the tracking speed. More particularly, the face tracking technique will make the interactive recognition and response to the human perspective changes. The Fig. 1 gives a basic illustration for our system. As the result, when the human head is tracked and the frustum is in the arbitrary form, the rendered scene will appear with the right perspective to human view and the observed scene will look at most realistic.

2   Related Works

There are a number of previous researches which have the same purpose in implementation of system using view-dependent rendering. P. Slotbo [1] introduced fundamental theories of rendering a 3D stereoscopic scene. He also implemented system for 3D interactive and view-dependent rendering. The proposed concept in his research was implemented using two inexpensive off-the-shelf web-cameras and a low-end desktop computer. He used the information from cameras as the input to track viewer's position for setting up a view-dependent rendering. There were some makers used to detect the interactions of the user. The developed system showed the abilities to render the scene in real-time but non-stereo mode. With stereoscopy enabled, the system ran slowly in only 12 fps. It also tracked the viewer position but indirectly.

In another related study, Jens Garstka and Gabriel Peters [7] calculated a view-dependent 3D projection of a scene which was projected on flat surfaces. They implemented a system that enabled the single viewer to explore the scene while walking around. They introduced a novel approach for head tracking using depth-images retrieved from a Microsoft KINECT 3D sensor. Such device had done most of hard works in detection head's orientation. The delays in adaptation of projected scene remained as a limitation. Their system is the first right step of expandability to 3D stereoscopic view-dependent rendering.

Sebastien [4] presented a non-intrusive system that gave the illusion of a 3D depth and interactive environment with 2D projectors. His research involved the head-coupled perspective to give the user a 3D scene which was projected onto a flat surface. The projected image was such that the perspective was consistent with the viewpoint of the user. Kenvin [2] discussed characteristics of head-coupled stereo display; he also dealt with the issues involved in implementing head-coupled perspective correctly. His research gave a first and fundamental theory on head-coupled stereo display. Buchanan [8] proposed a view-dependent rendering set-up for home computer use. In his research, view-dependent rendering used a parallax effect to give the illusion of depth. His face tracking method was based on the Lucas-Kanade algorithm.

3D stereoscopic rendering technique has been known and implemented recently on many researches. Paul Baker [6] introduced a right method to stereoscopic rendering with an off axis setting for two offset cameras. His studies have provided straightforward and fundamental theories to 3D stereoscopic rendering. In 2011, Samuel Gateau from NVIDIA [3] presented that of technique with an overview but concise and comprehensive information. The concrete implementation for such system can be found in the course BYO3D [12] by MIT media laboratory. We utilize this work in our system as a part of handling the 3D stereoscopic rendering.

3   Technique Background

3.1   3D Stereoscopic Rendering

Stereoscopic rendering has been well studied and implemented widely by many researchers in computer graphics field. With the development of display technology and the power of Graphics Card, implementing a 3D stereoscopic render system in real-time is currently not a challenge. The systems involved that of technique always aim to create the correct illusion of 3D depth by presenting a left and right image to the human eyes. Human perception of depth will in turn proceed the visual information taken from two eyes. Therefore, if the left and right image are correctly sent to human eyes, the strong sense of depth can be presented to the viewer as seeing the real world.

There are many methods for viewing a 3D stereoscopic rendered frame. In our system, we use two methods to present the stereo images to the viewer. We employ the common and traditional red and cyan glasses for a rapid 3D display which is known as anaglyph, we also use another method called row-interlaced stereoscopy using 3D display screen with glasses. Anaglyph is an old approach but it is used to illustrate the basic principle of transmitting stereo images. Anaglyph does not represent the colours faithfully. The interlaced display can preserve the image colours but it has the restriction in view in position. The viewer must look the screen in a straight direction and only moves the head horizontally to observe the scene, otherwise the 3D stereo effects will no longer be available.

In the standard graphics API, the frame is rendered and viewed by projecting a virtual 3D scene onto the computer screen as a 2D image. A perspective projection performs perspective division to shorten and shrink objects. View frustum defines those parts of the scene that can be seen from a particular position which is camera position. In our approach with stereoscopy, we consider two cameras presenting two eyes and a single projection plane. The object will be projected differently based on the position of camera and the visibility of object will be considered as well. It can be thought of as a window through which we can see the world. It is obvious that when the viewer moves, the scene will change in order that there are some parts of world being seen from one position but not from other position, and the objects in scene are observed from different angles.

We want the system to render the scene view-dependently. It means that the scene will always have the right perspective to the viewer position. When the viewer moves, the 3D scene should be rendered in such a way that the viewer is observing the real objects. The rendered scenes will change in real-time but the scene perceived by the human eyes must remain stable. Using perspective projection, the scene will be scaled based on the depth, this projection will then be able provide a depth cue. Therefore, such kinds of projection are called on-axis because the points in frustum are chosen to be the points on the near plane along the z-axis which is perpendicular to the screen plane. That projection is implemented by most of graphics API. However, using it will result in various distortions of virtual scene

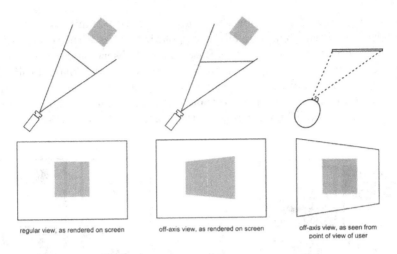

regular view, as rendered on screen   off-axis view, as rendered on screen   off-axis view, as seen from
                                                                                       point of view of user

Fig. 2. The concept of off-axis view

because the rendered frames are observed monocularly from an incorrect view-point in case they should be seen in a straight pose towards the screen.

In order to implement that aspect, we use head-coupled perspective [2] which means that we directly control the camera position by attaching it relatively to the human face position. Furthermore, we also want to achieve the most correct perspective of a 3D stereoscopic rendered scene, we employ an off-axis view which is illustrated by an asymmetric frustum. By using this frustum in the system, the distortion of object arising when the viewer looks at the screen from different angles will be reduced at most. The Fig. 2 illustrates the distortion appeared with different viewing angles [10].

3.2 Face Tracking

When applying a head-coupled perspective into the real-time rendering system, there is always a tracking technique involved because the system must know where the user's eyes are located. The user eyes will be considered as two virtual cameras and their positions will be assigned to the virtual camera's positions respectively. Those techniques are used with many researches on view-dependent rendering because they are straightforward and easy to implement.

In our approach, we employ the stereo display methods which rely on the glasses. The viewers must wear glasses so that they can perceive the 3D illusion of the virtual scene. Wearable glasses will affect the accuracy of eyes detection and tracking process because the features used to detect eyes are no longer true and distinctive enough. Thus, the result is not trust-able for tracking eyes positions and assign to the virtual cameras. Addressing this problem, we choose to track a human face position using digital camera. The eyes position can be computed correctly because it is true that the eye's position is always remained relative to the face position. In addition to that issue, the algorithm used in detection the human face is robust and adaptable with many changes. For instance, the viewer can wear a anaglyph or polarized glass without any negative effects on the first face detection step. Based on that observation, we use a robust face detection framework implemented by OpenCV called Viola-Johns[5]. This framework describes a approach using machine learning to detect visual objects. It combines four concepts which are Haar features, an integral image for rapid feature detection, a machine-learning method and a cascaded classifiers[5].

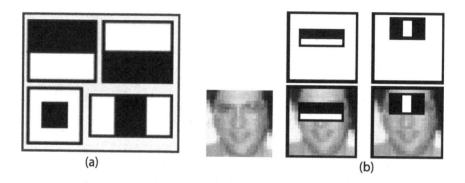

Fig. 3. The Fig. (a) shows the examples of Haar features in OpenCV. The Fig. (b) shows the first two Haar features in the Viola-Jones cascade.

In their detection framework [5], the detection features, which are the combination of sequence rectangles, are based on Haar wavelets. The sequent rectangles form the better method to visual detection tasks. Because of the difference from original Haar wavelet, the features they use are called Haar-like features. The Fig. 3 shows some examples of the Haar features used in OpenCV and the first two Haar features which are used in the original Viola-Jones cascades. To check whether the Haar-like features are being presented in the image, they simply use subtraction between the average dark-region and light-region pixel value. The differences will be compared with a threshold achieved during a machine learning process to decide the presence of those features in image. They use the technique called Integral image. The integral value is calculated for each pixel. This value is recursively computed by adding the sum of all the pixels from its top left pixels. The process starts from the top left pixel of the image and traverse to the right down [11]. Thus, the features are detected efficiently in every location with some scales.

They use a machine-learning method called AdaBoost to decide the Haar-like features and specify the threshold value. AdaBoost method selects a small number of critical visual features from a larger set and makes an extremely efficient classifiers. AdaBoost combines many classifiers which are assigned the weight to create one effective classifier. They use a training face set as the input for AdaBoost learning method, the threshold at each filter level is set to a value at most that face examples in the training set can pass. This method is illustrated

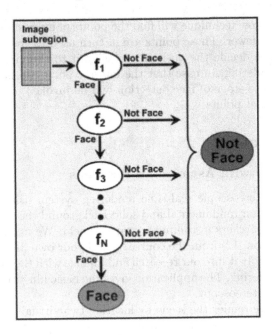

Fig. 4. The classifier cascade

in the Fig. 4 [11]. The image sub-region will pass through the filters, if that region can pass all the filters in the chain, it will be classified as a human face.

After the first stage of face detection, we use another technique to track the face. There are many tracking techniques based on the movement or the colour histograms from the object we want to track. Using the techniques based on the colour of object is fast, effective and rather widely implemented. However, in our system when the viewer wears a glass, especially the red-cyan glass, the incorrect tracking may occur. Tracking by using histogram can potentially miss the face in the frame captured by camera due to the presence of some regions having the same colours with the face such as the neck or some regions in environment around the viewer. The inaccuracy as well as the failure of face tracking breaks our system stability. Therefore, we involve a tracking technique based on the movement of object, that technique is called Lucas-Kanade (LK) optical flow [9].

The LK optical flow technique is typically used to track the optical flow of a video. In our research, this technique will track the location of some specific points in the face detected across multiple frames which the digital camera is capturing. The basic idea of the LK algorithm rests on three assumptions. A pixel from the image of an object in the scene will retain its appearance when moving from frame to frame; this is known as brightness constancy. The motion of a surface region changes slowly in time relatively to the scale of motion in the image. The point belonging to the same surface will have the similar motion with the neighbouring points in a scene.

The LK optical flow technique will find the points of interest which are located on the face of the viewer. Those points are in turn used to track the motion the face so that we can decide the position to assign it to the virtual camera used in stereoscopic rendering. In case that the interest points do not track the face feature precisely, the stage of face detection will be invoked to relocate the face and compute interest points.

4   Implementation

4.1   Stereoscopy with Asymmetric Frustums

We implement a stereoscopic real-time rendering system using OpenGL. It is developed based upon fundamental and solid background theory on stereoscopy made by many researchers in computer graphics field. We utilize a stereoscopic rendering application [12] from the course "Build your own 3D display" in SIG-GRAPH 2011 to apply it into our research and finalize with the best performance based on GPU rendering. The application gives the basic illustration of rendering a scene off axis in stereoscopy.

In the system, we render the scene by locating two virtual cameras relatively to the eye positions and the distance between two virtual cameras are computed by eyes separation value. The pseudo code in below illustrates the steps in order to create asymmetric frustum to each virtual camera in OpenGL. Two virtual cameras positions will be computed by shifting main camera a half eyes separation distance to the left and right alternatively. The width and height of the window viewport are also computed in centimetre. The *screen.pitch* will return the size of a pixel in centimetre. We continue to compute other values to form the parameters to construct the asymmetric frustum.

OpenGL initializing asymmetric frustum pseudo code

```
compute x value of left/right camera position alternatively;
depthRatio = camera.near/camera.z;
halfWidth = window_width * screen.pitch/2;
halfHeight = window_height * screen.pitch/2;
left = -depthRatio * (x + halfWidth);
right = -depthRatio * (x - halfWidth);
bottom = -depthRatio * (camera.y + halfHeight);
top = -depthRatio * (camera.y - halfHeight);
set glFrustum() based on six standard frustum values;
```

In anaglyph stereo rendering, we use a fragment shader to specify three colour modes of anaglyph which are full-colour, half-colour and optimized anaglyph. The fragment shader will compute the output colour according to the user-selected mode. We compute the anaglyph's RGB values (r_a, g_a, b_a) from the RGB values of the left (r_1, g_1, b_1) and right images (r_2, g_2, b_2)[13]. We also use a fragment shader for row-interlaced rendering mode. For each view of camera, the shader evaluates the texture coordinate and extract the view mask value. The fragment colour will then be assigned for the interlaced image.

4.2   Face Tracking with OpenCV

In our system, we set up a digital webcam by mounting it into the top of the screen where it lies exactly in the middle point of the top screen edge. The viewer position is restricted to the distance from 40 centimetres to 100 centimetres in order that the camera can capture a face with the best quality for detection task.

The frame as individual image extracted from the webcam will be used as the input for the face detection. The viewer face will be detected as soon as the frames are outputted from the camera. However, there is a restriction in enabling the tracking step. We delay tracking until the viewer's face moves into the centre position of the sequence frames taken by camera in both vertical and horizontal direction. Otherwise, we must adjust the camera vertically so that the face appears in centre. Using such restriction, the relationship between face and virtual camera's position can be proportional to each other because we assume that the starting position of face is to give a look towards the screen in the straight orientation. In addition to this issue, we are able to calculate the centre point of two eyes because it also has the relative position to the face. Finally, we can retrieve the spatial changes from centre of eyes and assign it to the virtual cameras.

We use the implementation in OpenCV for those theories which have been discussed in section 3 - Technique background. OpenCV already implemented the Haar Cascade classifier in order to use not only for face detection but also for object detection generally. The face detector will examine each image location and classify whether that location contains a face. The classification uses a scale in size 50 x 50 pixels to scan faces and it runs through the image several times to search for faces within a range of scales. OpenCV provides an XML file name `haarcascade_frontalface_default.xml` which is the output after the learning process on faces data. The classifier uses that file for decision on which location is a face in the image. The face detector task is under form of an OpenCV method called `cvHaarDetectObjects`. This method will use the classifier as a parameter. The classifier is represented by `CvHaarClassifierCascade` class in OpenCV.

As discussed in the previous section, we use the algorithm called Lucas-Kanade optical flow to track the face which is already detected in the detection step. The first thing to do with this algorithm is getting the feature we want to keep track in frame by frame. In our case, that is the face. We invoke the OpenCV method `cvGoodFeaturesToTrack` to get the interest points in the face region we have from the detecting phase. We then use those points as the good features for tracking. After that, we continue to invoke the method `cvCalcOpticalFlowPyrLK`. This optical flow function makes use of the good tracking features and indicates whether the tracking is proceeding well [9]. By using the combination of these two functions, we can obtain the effective tracking. Our system keeps running in real-time for both face tracking and stereoscopic rendering.

5   Discussion and Result

Our paper presents a technique for real-time rendering a stereoscopic scene which involves the view-dependent aspect. The view-dependent rendering of a normal 3D scene was studied and implemented by many researchers. However, the studies on that of stereoscopy are still interesting research field. Some implementations have been introduced while the challenges on building the real-time system still remain attractive because the computational cost for both stereoscopy and face tracking is extremely high.

We also present an off axis frustum or asymmetric frustum by which the perspective will be reflected as much realistic as the scene should appear in a real environment. These techniques would be effective in case of implementing a virtual reality system using the popular and simple devices. The user will need only display devices for 3D stereoscopy and they will be able to look at the real scene by only sitting front of the working place.

We develop and run our system on Windows 7 Ultimate version with service pack 1. Our test platform is a CPU 3.7GHz Intel Core i7, NVIDIA GTX 480 graphics card as well as 16GB of RAM. For the stereo display, we use a 27-inch 3D screen with the polarized glass. Our system runs fast and smoothly with an average of 30 FPS. We perform many changes in viewer position. The 3D object appears stable to the viewer perception as a real object is floating out of the screen or located behind the screen.

In the future, we will improve the performance of the system. With the 3D stereoscopic scene changed by human view, the system would interest the learner in case it is applied for education. We also would like to involve the hand gesture tracking system in order to enable the interaction between the observer and the virtual 3D objects rendered by our system.

6   Conclusion

In our research, we present and implement a 3D stereoscopic rendering system in real-time based on the viewer's face position. We make the combination of 3D stereoscopic rendering and face tracking techniques. By improving them, we achieve a system having a good performance. By using our system, the user can perceive the realistic scene in 3D and the observed scene therefore can be considered as the scene in reality. The combination of those techniques helps us implement a widely applicable system.

Acknowledgments. This research was supported by Next-Generation Information Computing Development Program through the National Research Foundation of Korea(NRF) funded by the Ministry of Education, Science and Technology (No. 2012M3C4A7032182).

References

1. Slotsbo, P.: 3D Interactive and View Dependent Stereo Rendering. M.Sc. thesis at Technical University of Denmark (2004)
2. Kevin, A.: 3D task performance using head-coupled stereo displays. M.Sc thesis at The university of Bristish Columbia, pp. 14–24 (1993)
3. Samuel, G.: 3D Vision Technology Develop, Design, Play in 3D Stereo. NVIDA (2011)
4. Sbastien, P., Vincent, P., Antoine, L., Marc, D.: An Interactive and Immersive System that dynamically adapts 2D projections to the location of a users eyes. In: Proceedings of International Conference on 3D Imaging (2012)
5. Paul, V., Michael, J.: Rapid Object Detection using a Boosted Cascade of Simple Features. In: Proceedings of Conference on Computer Vision and Recognition (2001)
6. Paul, B., Peter, M.: Stereoscopy: Theory and Practice. In: Workshop at VSMM. Queensland University of Technology (2007)
7. Jens, G., Gabriele, P.: View-dependent 3D Projection using Depth-Image-based Head Tracking. In: 8th IEEE International Workshop on Projector Camera Systems (2011)
8. Buchanan, P., Green, R.: Creating a view dependent rendering system for mainstream use. In: 23rd International Conference on Image and Vision Computing New Zealand, IVCNZ 2008, pp. 1–6 (2008)
9. Gary, B., Adrian, K.: Learning OpenCV. O'Reilly Media publisher (2009)
10. Opera development article, http://dev.opera.com/articles/view/head-tracking-with-webrtc/ (visited on February 20, 2013)
11. Cognotics article, http://www.cognotics.com/opencv/servo_2007_series/index.html (visited on February 22, 2013)
12. Build your own 3D displays course, http://web.media.mit.edu/~mhirsch/byo3d/ (visited on January 20, 2013)
13. Anaglyph Methods Comparison, http://3dtv.at/Knowhow/AnaglyphComparison_en.aspx (visited on January 15, 2013)

Virtual Exhibitions on the Web: From a 2D Map to the Virtual World

Osvaldo Gervasi, Luca Caprini, and Gabriele Maccherani

Department of Mathematics and Computer Science, University of Perugia
Via Vanvitelli, 1 - 06123 Perugia, Italy
osvaldo@unipg.it,{luca.caprini,m4cch3}@gmail.com

Abstract. The optimization of the process of implementing virtual exhibitions on the Web represents an interesting research area devoted to simplify the design and the production of virtual exhibitions, particularly for people with a scarce technical skill and strong competencies in the area to which the exhibition is related to.

In the present paper we propose a web environment facilitating the production of virtual exhibition through a series of operations: from the virtual representation of the physical structure, starting from the image of a 2D map that is processed to optimize the walls identification, to the management of the artworks shown in the exhibition, including the selection of the artworks and their positioning in the physical structure. Finally the virtual exhibition may be released to the visitors, making the virtual world available for the download on the client side.

Several technologies have been adopted in order to reach a user friendly backoffice environment, suitable for being used by naive (from the technology point of view) users, concentrated on the subject related to the virtual exhibition: the OpenCV library which make possible to produce the physical scenario in which the exhibition will take place evaluating the measures from 2D maps, the SVG graphic format per representing the map after having detected the boundaries of the provided 2D map, Ajax, X3D and X3DOM which enable the virtual environment representation.

As a testbed we present the virtual museum of Villa Fidelia in Spello, PG, Italy.

Keywords: Virtual exhibition, Web3D, OpenCV, X3D, X3DOM.

1 Introduction

In the last years the Virtual Reality is spreading in many areas: from video-games to medicine and biology, from architecture to art history and cultural heritage. Even at the business level, many companies are producing virtual reality services based on the navigation and the exploration of interesting buildings, museums and exhibitions that the final user can enjoy through virtual tours on the web.

In this paper we present a web environment which enables the production of a virtual exhibitions, including the rendering of the physical environment, in a

B. Murgante et al. (Eds.): ICCSA 2013, Part I, LNCS 7971, pp. 708–722, 2013.
© Springer-Verlag Berlin Heidelberg 2013

intuitive and easy way, particularly oriented to people unskilled in technology but designated of arranging artworks in a exhibition environment.

Our work has been carried out in order to deploy two components: firstly we implemented a tool which, extensively using the OpenCV (Open Source Computer Vision) library [1,2], produces the virtual world of the venue of the virtual exhibition starting from a 2D map. The virtual world related to the exhibition area is then made using the Ajax [3] technology and producing an output file in X3D language [4].

The procedure uses an algorithm which, as a function of the type of map used (image or photo), performs a series of image manipulation in order to identify the contours related to the physical structures (walls, corridors, etc) of the exhibition area. After having produced the digital map of the exhibition environment, having detected all contours in the provided 2D map image, the user may customize the scene, in a friendly manner on a suitable web environment, based on the X3DOM [5,6] technology, with textures and other shape properties. In our research group X3DOM has been successfully used in various context (see for example [7]) and we proved the real advantages of such technology that avoids the adoption of external software plugins and allows to deploy virtual reality applications entirely on the web.

Secondly, we implemented a back-office environment for the deployment of the virtual exhibition. The artworks are managed by a MySQL database management system, which makes possible the artworks classification and retrieval. Once defined the virtual world related to the exhibition area, the user can upload to the database new artworks and/or select from the set of objects stored on it, the artworks which are part of the virtual exhibition. The user arranges each artwork in the exhibition area, using some intuitive tools available on the web page. Once all artworks are properly located, the user can produce the archive file containing the virtual exhibition, including all the necessary files.

The virtual exhibition can be advertised in a website, and the final user, after having downloaded the files, may visit the virtual exhibition using the X3D player Instant Reality [8].

The innovative aspects of our approach are mainly related to the algorithm we used to build the physical structure from a bi-dimensional map, combining several image processing techniques in order to detect the walls properly, and to the tools implemented for managing and positioning the artworks in the environment for deploying the virtual exhibition. Moreover, the web environment results easy to use by non expert people for customizing the exhibition area. Last, but not least, we emphasize the portability of the application among various platforms.

The paper is structured as follows: in section 2 the existing works related to the implementation of virtual exhibition are discussed; in section 3 the algorithm implemented for building the exhibition area from 2D maps is illustrated; in section 4 the environment and tools available for customizing the exhibition area are presented; in section 5 the artworks management and the deployment of the virtual exhibition are discussed; finally in section 6 some conclusions and the future work are presented.

2   Related Works

Among the various tools and applications developed in the last years to deploy virtual reality environments, the virtual exhibition is one of the most relevant subjects. Google Art Project is one of the most famous and appealing tools, that allow to visit museums and visualize artifacts at high resolution, being based on the Street View technology, where the visitor is enabled to move inside the scene. However, we can notice that there are no real interactive tools, the scene is essentially static. Such approach works well on open spaces, but it is not easy to use inside buildings and in small spaces and it may induce confusion and disorientation in the user. Furthermore, Street View technology requires the definition of the Viewpoints that need to be defined and stored in the database. This may result in a difficult and complex work to be made by experts, often unfeasible for medium and small museums.

A very nice application is VEX-CMS [9–11], a software which enables unskilled people to deploy virtual exhibitions. The environment requires in input a 3D model of the building where the exhibition will take place, implemented using an authoring software. The software provides easy to use tools for setting up the environment and the virtual exhibition and for defining the viewpoints. During the set up phase the manager can add pictures, statues and artworks in general and add textual information, images or web URLs. Moreover, the manager can redefine size and position of each artwork, using the anchor points to facilitate movements. In the final phase VEX-CMS allows to define a series of viewpoints and a virtual tour, that the final user will experience through the VEX-CMS Viewer. VEX-CMS is really a nice and good tool, despite the fact that the components have to be implemented using 3D modeling software. VEX-CMS is available for Windows operating systems.

3   Building the Structure from a 2D Map

One of the innovative aspects of our work is the possibility to easily create the physical environment and the various rooms where the virtual exhibition will take place from 2D maps, processing the related image by means of the OpenCV routines, that are a powerful set of image processing routines released under the BSD Open Source license. The library is composed by more than 500 routines covering the areas of Image Filtering, Image Transformation, Histograms, Object and Feature Detection, and Shape Description.

In Figure 1 the adopted algorithm is shown. In the first step the user specifies if the map is represented by a photo or an image. In the first case, the photo is processed to optimize the subsequent phases, in particular the identification of contours. The photo related to the map is processed by the following functions, in order to magnify the lines, reducing interferences and noise:

GaussianBlur: blurs the image using a Gaussian filter reducing the graininess and flaws;

adaptiveThreshold: optimize the image analysis when the image has variable brightness and noise;

dilate: connects objects that should belong to a single form, making the overall picture clearer;

erode: reduces the area of the objects making the image darker.

If the map is represented by a high quality image, as that produced by a specific application (however not a photo), the image is processed identifying the pixel with the lower value (closest to black) in order to emphasize the lines and contours from the image background. Such value is used when calling the *threshold* function, in *inverse binary* mode, in order to remove the intermediate gray areas.

1. **Input** the type of file associated with the 2D map (photo or image)
2. **if not photo:**
   a. Find the lowest value pixel
   b. call *threshold* in *inverse binary* model
   else:
   call *GaussianBlur*
   call *variableThreshold*
   call *dilate*
   call *erode*
3. call *findContours*
4. contour approximation
5. write the X3D file

Fig. 1. Algorithm implemented to identify the contours of a 2D map

The main phases of the process of building the exhibition environment from a 2D map are the following:

- The image is processed by the OpenCV *thresholding* function, that uses some filters to reduce the noise and magnify the lines of the map
- The function *findContours* returns all vertex and lines represented in the map to identify the walls of the structure
- By means of the function *approxPolyDP*, contours are approximated to obtain wall corners and the structure edges, removing some vagueness introduced by the previous functions or by a non-optimal image quality;
- Once identified the walls present in the map, the height of the ceiling is added and the virtual world is implemented in a web page making use of X3DOM.

3.1   Thresholding

Such function transforms the pixel intensity of the image, expressed by the value of m ranging in the interval $[0, 255]$, in a binary image, made of black and white if $m <= t$ or $m > t$, respectively, being t a selected thresholding value. The threshold t may be constant or variable as a function of the image type. The result of the thresholding function is the separation of a region of interest from the image background. In general the global thresholding approach is suitable for images having a single object of interest on a uniform background, like those obtained from CAD or graphic programs or obtained scanning professional maps.

The thresholding function is expressed by the prototype shown in Listing 1.1.

```
double cvThreshold(const CvArr * src , CvArr * dst , double
    threshold , double max_value , int threshold_type);
```

Listing 1.1. Declaration of *cvThreshold* function

In addition to the parameters specifying the source and destination filenames, (src and dst, respectively) our attention is drawn to the following parameters: max_value, the maximum value to be assigned to the considered pixel, threshold, the threshold value, and threshold_type, the type of thresholding function invoked (global or adaptive). We set such values to 120, 255 and *THRESH_BINARY_INV*, since for contours identification we need a binary image with the pixel values set to 0 (black) or 255 (white). This may be expressed from the rules shown in eq. 1:

$$threshold\_type = CV\_THRESH\_BINARY\_INV$$

$$dst(x, y) = \begin{cases} 0 & \text{if } src(x, y) > threshold \\ max\_value & \text{otherwise} \end{cases} \tag{1}$$

The images of Figures 2(a) and 2(b) respectively show the considered map image and the associated pixel values graph before applying the thresholding operation. The blue line represents the adopted threshold value.

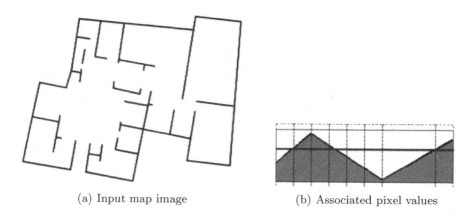

(a) Input map image   (b) Associated pixel values

Fig. 2. Map image before thresholding

The images of Figures 3(a) and 3(b), show the output map image and the associated pixel values graph. It is interesting to notice that, thanks to the applied type of threshold function, pixel values above threshold have been set to white (255), while the remaining are set to black (0).

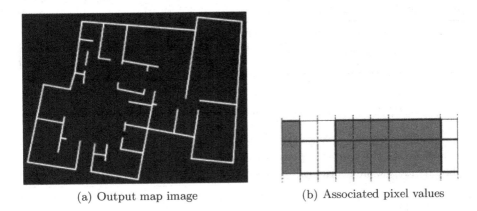

(a) Output map image   (b) Associated pixel values

Fig. 3. Map image after thresholding

If the map image presents a high variation of luminous intensity, we may apply the adaptive threshold function, that determines the right threshold, evaluating the average luminous intensity in the neighborhood of the considered pixel.

3.2 FindContours

The identification of boundaries is one of the fundamental techniques of image processing. The function returns a sequence of coordinates that once connected represent the boundaries. The algorithm analyzes the image as a matrix, starting from the upper left pixel. The rows are incremented from the top to the bottom, while columns are incremented from left to right.

The algorithm has been implemented in two variants: in the first one the image is analyzed and all contours are extracted using a tree structure, that describes the degree of relationship (external or internal boundary), while the second one ignores the internal components returning the list more quickly.

The process of identifying the boundaries is described in detail as follows:

1. Starting from the top-left element, the pixel values of the image are analyzed. Whenever the value of the two contiguous pixels are $(0, 1)$ for an external boundary, or $(\geq 1, 0)$ for an internal boundary, a new search of a boundary is started.
2. The search process considers the eight neighboring pixels and proceeds finding the contour, setting the pixel value to 2, until there are not anymore pixels with value 1 or the original pixel is reached.

3. The algorithm continues the search, according to the described method, until the end of the image is reached.
4. The boundaries (pixel with value 2), are extracted and converted as output values.

The algorithm is implemented by the function *findContours* whose prototype is presented in Listing 1.2.

```
void findContours(const Mat& image, vector<vector<Point>>&
    contours, vector<Vec4i>& hierarchy, int mode, int method,
    Point offset=Point())
```

Listing 1.2. Declaring *findContours* function

The first parameter represents the matrix associated to the input image, while the second one is the vector containing the output values, `hierarchy` describes the hierarchy of the contours, `mode` the recognition mode, and `method` the approximation method. In this particular case we considered only external boundaries.

The listing 1.3 shows the call to the function *findContours*.

```
findContours(src, contours, hierarchy, CV_RETR_EXTERNAL,
    CV_CHAIN_APPROX_SIMPLE);
```

Listing 1.3. Calling *findContours* function

3.3   ApproxPolyDP

ApproxPolyDP is a function based on the algorithm Ramer-Douglas-Peucker that, starting from a curve composed by a set of segments, allows to derive a second curve similar to the first but composed of a small set of segments. The algorithm receives in input the set of points defining the curve and a value ϵ that defines a distance. The ending points of the input curve are set as the ending points of the new curve. The algorithm will select two points of the curve recursively together with the most distant point from the segment which connects the two points: if the distance is less than ϵ, then the point is discarded and will not be part of the new curve. On the other hand, if the distance is greater than ϵ, then the point will be selected and will be part of the new curve. Recursively the procedure will be applied to the new two segments created by the new point: the one made by the initial point and the selected point and the other one made by the selected point and the final point. Once completed, the function will return the curve made only by the selected points. In Listing 1.4 the call of the function *approxPolyDP* is shown, where `Mat(contours[i])` is the input matrix, `approx` is the output curve, 1.5 is the selected value for ϵ and the last value indicates if the final curve has to be closed or not. Figure 4(a) shows a contour before applying ApproxPolyDP function while Figure 4(b) shows the result after having applyied the named function.

```
approxPolyDP(Mat(contours[i]), approx, 1.5, true);
```

Listing 1.4. Calling *approxPolyDP* function

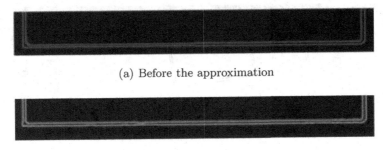

(a) Before the approximation

(b) After the approximation

Fig. 4. Contours approximation

After having applied the *approxPolyDP* routine, the contours are interpreted and used for building the X3D virtual world related to the exhibition area.

4   Customization of the Exhibition Area

In this section we are presenting the back-office website that produces the virtual exhibition. Once uploaded the file containing the image of the 2D map, the map is processed to obtain the coordinates of the boundaries detected, according to the method described in section 3. The result is shown in Figure 5: the map is created on the client side using the Javascript library *Raphael*, that simplify the process of making vector graphics on the web. The library uses the SVG graphic standard format [12], displaying images in the Document Object Module (DOM); furthermore the library manages the events associated to the user interaction. From such website the user is enabled to define the logical rooms (reserved to the exhibition area), customizing the appearance of the floor and the walls. The website is based also on X3DOM technology, to display the X3D scene as part of the DOM, avoiding the need of adopting an external software plugin for such purpose.

4.1   The SVG Map

The points of the contours that have been identified processing the 2D map provided by the user are stored in a file. Each contour is defined in a row, which contain the sequence of 2D coordinates identified in the map provided by the user. Each point is separated from the subsequent by a pipe ("|") character.

A PHP program parses the file and displays the line identified by two subsequent 2D coordinates, using the Raphael routines. In Listing 1.5 is shown a sample call to the **paper.path** API to draw a line from point $(100, 100)$ to point $(150, 150)$ with a specific set of attributes.

```
var line1 = paper.path('M100 100 L150 150').attr({
        fill: 'none', stroke: '#000', 'stroke-width': 3});
```

Listing 1.5. Sample call to the Raphael library to draw a line to build the map

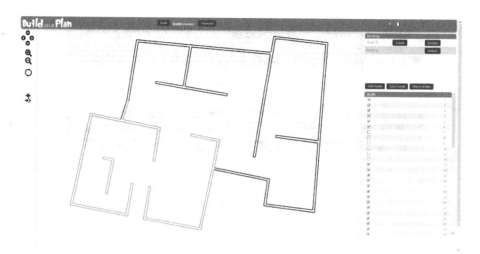

Fig. 5. The map rendered in the web page, made using the boundaries detected processing the 2D map. It may be customized by the back-office user, defining the logical rooms, the wall and floor aspect.

4.2   Functionalities

The Figure 6 shows the web interface, which is composed by two distinct parts: on the left hand side the X3D rendering of the virtual environment related to the 2D map provided in input by the user is presented, while on the right hand side there are several controls and a map, which enable the back-office user to customize the environment.

The main functionalities may be summarized as follows:

- the movements on the 2D map on left hand side, are reflected into the virtual world on the left panels;
- the user may select the entrance point of the exhibition area;
- it is possible to select one or more walls on the map;
- the selected walls may be used to define rooms;
- the wall may be customized with colors or textures;
- the virtual world may be exported in an archive.

5   Deploying a Virtual Exhibition

In this section we will describe the backoffice of the virtual museum and its functionalities. The backoffice plays a main role in the entire system, allowing administrator to virtually manage paintings and exhibitions in a straightforward and intuitive way.

Once logged the administrator access a complete summary of all the data stored in the system: paintings, packages, rooms and exhibitions. It is very important to distinguish between data that are uploaded by the administrator and

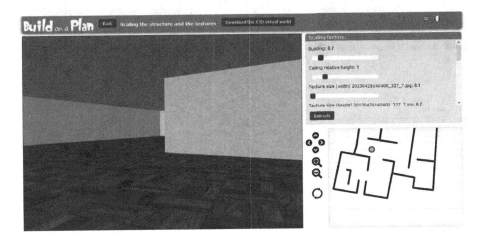

Fig. 6. The Exhibition area, customized by the back-office user, is ready to host the artworks the user may select from the items in the database, or adding new objects and artworks

those that already exist in the system since the first utilization; therefore, it is necessary to consider the hierarchy with which the data are stored. A package can be viewed as an entire museum (or one of its floors) containing different rooms, each one of that can be setup in a different way, using the paintings stored in the system.

The backoffice consists in two different parts sharing the same information, principally stored in XML files. The first part is entirely implemented using PHP pages and functions, linked to opportune Javascript functions, amply using jQuery library; this part has the task to create and modify the XML files related to paintings and exhibitions and to manage the upload of multimedia files. The second part has the aim of supplying the backoffice administrator for all the necessary tools to setup the exhibition and makes available all the files to the final user, by means of Javascript functions and the X3DOM exceptional potentialities.

5.1   Paintings

The interface (Figure 7) used for the paintings management is divided into two sections: on the left side there is a list of all the paintings stored by the system; in the right one all the information about the selected painting are available.

The procedure for the creation of a new painting is very rapid and consists in few simple steps. Once clicked the "New" button, a new painting will be automatically added to the right side list and all the customizable fields will be available on the right. In this way the user can add the information filling out the related fields, specifying, for example, the title and the size of the painting. Then,

Fig. 7. The page containing the list of the created paintings

it is possible to upload a picture of the painting and, if it is the case, multimedia audio and video files. Clicking on the "Save" button the information are sent to the server that will immediately manage them, letting the user visualize the new painting in the list.

The user can modify a painting simply selecting it from the list on the left and clicking on "Modify" button; the procedure is pretty similar to the painting creation. It is possible to update any information in the fields, including the title, the size and the associated picture. Only once the "Save" button is clicked, the information will be sent to the server. Moreover, the user has the opportunity to delete one or more paintings, simply selecting them in the list using the related checkbox and clicking o the "Discard" button. Obviously, the cancellation of a painting implies the withdraw of all multimedia attributes related to it.

5.2   Exhibitions

As the previous one, the interface used for the exhibition management (Figure 8) is also divided into two sections: a list on the left side and all the information about the selected exhibition on the right side.

By means of "New" button, a new exhibition is added to the list and it becomes possible to define the related information using the fields on the right side. Once inserted the name and selected the room to setup, the administrator can add a brief description that could be useful to the final user to understand, for example, what the exhibition is about; in fact, this description will appear in the download section of the web site together with the exhibition name, but only if the "Visible" box has been checked. In this way, the backoffice administrator can properly setup an exhibition and define all details before it will become visible to the public.

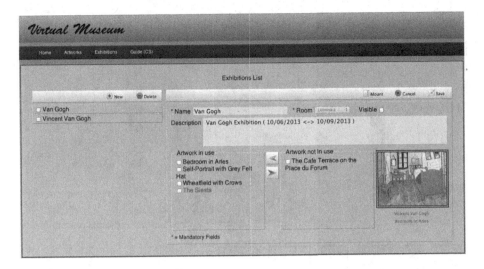

Fig. 8. The interface containing the list of the created exhibitions

To add paintings to the exhibition it is sufficient to select the selected artwork from the list of the available artworks on the right hand side panel. Clicking on the green arrow button, the artworks will be visualized on the left hand side; the removal of an artwork can be obtained in a similar way, selecting the paintings to remove in the left side panel and clicking on the red arrow button. If a painting moves from a list to another, its color becomes red, in order to let the change be more evident. Clicking on the name of a painting, a little overview of the painting appears sideways and, together with the title and the author, facilitates the definition of the exhibition.

As already seen for paintings, the procedures of change and removal are realized by selecting the paintings on the left list and clicking on the "Modify" or "Delete" buttons, respectively. The changes become effective only after having saved the modifications.

5.3 Virtual Exhibition Setup

Once the exhibition and its paintings have been defined, clicking on the "Setup" button it is possible to start to arrange paintings in the specified room.

The new interface (see Figure 9) contains in the central part the X3DOM scene representing the room where the virtual exhibition has to be setup; on the right side a panel allows to move and modify the paintings into the scene; finally, at the bottom of the page the collection of paintings related to the exhibition is shown.

Adding a painting to the exhibition is an easy and intuitive task. It is necessary to move along the room using the X3DOM typical controls and once positioned in front of the wall where to place the painting, choosing the painting from the library and then clicking on the wall, it will be added into the scene.

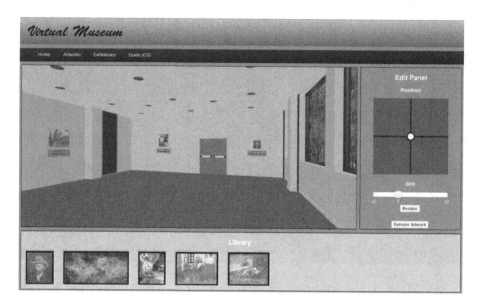

Fig. 9. The interface for the exhibition setup

After having added the painting, it is possible to select it with a click (the selected object will be highlighted) and using the blue panel on the right, shown in Figure 9 one can move it along the named wall. In fact, moving the white circle, the picture will be moved accordingly along the wall on the scene.

It is important to note that the default size of the paintings are proportional to the exhibition area, but it may be appropriated to scale the original size of a painting, to increase the magnificence of the scene. We may do so, since we are arranging a virtual exhibition and the real measures may be modified without particular problems. For this reason, after a painting has been selected, its size can be proportionally changed thanks to a slider; however, the administrator can restore the original size of the painting using the "Restore" button anytime.

The removal procedure can be realized selecting the painting into the scene and then clicking on the "Remove" button; this operation deletes the painting from the current exhibition, but not from the library, allowing its inclusion in the exhibition at a later stage, through the procedure previously described.

All changes will be applied only after the backoffice user performs the saving operation. In fact, using the "Setup" menu on the top, it is possible to cancel all changes performed till then, reloading the last version saved. The "Setup" menu also offers the option of saving, that produces the X3D file containing the virtual exhibition shown on the X3DOM panel, including all multimedia files related to it. All files are compressed in a ZIP archive, that is placed in a predetermined folder, that will be scanned during the download requests from the visitors.

The download section of the site allows to download a ZIP archive containing the virtual world related to the physical structure, without any exhibitions. Such package may have a size of many hundreds of megabytes in the case of very detailed textures or elaborate virtual reproductions. For this reason, the exhibitions and the virtual world representing the physical structure can be separately downloaded.

6 Conclusions

Virtual Exhibitions made available through the web may facilitate the communication in several fields, enhancing the possibility of institutions and company to disseminate concepts, to advertise relevant artworks and sites and to promote the diffusion of culture and beauty.

Our work was targeted on implementing an application able to simplify the creation process of virtual exhibitions, starting from the creation of the exhibition environment (walls, corridors, exhibition rooms, etc) and then providing an environment in which the artworks may be inserted into a database, classified using metadata, enriched with multimedia information and managed, in order to be included in a given exhibition. Once created a virtual exhibition may be saved for future reference and use.

The approach described in this paper has been successfully applied to the virtual exhibition of the "Sala della Limonaia" of the magnificent *Villa Fidelia*, in Spello, Perugia, Italy. However the same approach may be extended not only in the cultural heritage area as a whole, but also in other disciplines, in particular it may facilitate the production of virtual laboratories, very useful in e-learning applications [13–15].

In the digital era, we think such an approach and systems like the one we implemented may facilitate the preservation of memory in several areas. Unfortunately one drawback is the dependency on the software used for rendering the deployed virtual exhibition on the final user client device. In fact, we choose to let the user to download the virtual world related to the exhibition and to navigate in it using the Instant Player software, instead of navigate on the web page using the X3DOM approach. One of the future works will be the implementation in a X3DOM page of the deployed virtual exhibition. We have not yet implemented all phases in such way mainly because we still are facing with some current limitations of the X3DOM implementation.

Acknowledgments. We acknowledge the "Provincia di Perugia", Perugia, Italy for the collaboration and for supporting this research.

References

1. Bradski, G., Kaehler, A.: Learning OpenCV: Computer Vision with the OpenCV Library. O'Reilly, Cambridge (2008)
2. Laganière, R.: OpenCV 2 Computer Vision Application Programming Cookbook. Packt Publishing (May 2011)
3. Holdener III, A.T.: Ajax: The Definitive Guide. O'Reilly Media (2008)

4. Brutzman, D., Daly, L.: Extensible 3D Graphics for web authors. Morgan Kaufmann (2007)
5. Behr, J., Eschler, P., Jung, Y., Zöllner, M.: X3dom: a dom-based html5/x3d integration model. In: Proceedings of the 14th International Conference on 3D Web Technology, Web3D 2009, pp. 127–135. ACM, New York (2009)
6. Behr, J., Jung, Y., Keil, J., Drevensek, T., Zoellner, M., Eschler, P., Fellner, D.: A scalable architecture for the html5/x3d integration model x3dom. In: Proceedings of the 15th International Conference on Web 3D Technology, Web3D 2010, pp. 185–194. ACM, New York (2010)
7. Zollo, F., Caprini, L., Gervasi, O., Costantini, A.: X3dmms: an x3dom tool for molecular and material sciences. In: Proceedings of the 16th International Conference on 3D Web Technology, Web3D 2011, pp. 129–136. ACM, New York (2011)
8. http://www.instantreality.org
9. Chittaro, L., Ieronutti, L., Ranon, R., Siotto, E., Visintini, D.: A high-level tool for curators of 3d virtual visits and ist application to a virtual exhibition of renaissance frescoes. In: Proceedings of the 11th International Conference on Virtual Reality, Archaeology and Cultural Heritage, VAST 2010, Aire-la-Ville, Switzerland, Switzerland, pp. 147–154. Eurographics Association (2010)
10. Chittaro, L., Ieronutti, L., Ranon, R.: VEX-CMS: A tool to design virtual exhibitions and walkthroughs that integrates automatic camera control capabilities. In: Taylor, R., Boulanger, P., Krüger, A., Olivier, P. (eds.) Smart Graphics. LNCS, vol. 6133, pp. 103–114. Springer, Heidelberg (2010)
11. Chittaro, L., Ranon, R., Ieronutti, L.: 3d object arrangement for novice users: the effectiveness of combining a first-person and a map view. In: Proceedings of the 16th ACM Symposium on Virtual Reality Software and Technology, VRST 2009, pp. 171–178. ACM, New York (2009)
12. Eisenberg, D.J.: SVG Essentials. O'Reilly Media (2002)
13. Gervasi, O., Riganelli, A., Pacifici, L., Laganà, A.: Vmslab-g: a virtual laboratory prototype for molecular science on the grid. Future Generation Computer Systems 20(5), 717–726 (2004); Computational Chemistry and Molecular Dynamics
14. Gervasi, O., Tasso, S., Laganá, A.: Immersive Molecular Virtual Reality Based on X3D and Web Services. In: Gavrilova, M.L., Gervasi, O., Kumar, V., Tan, C.J.K., Taniar, D., Laganá, A., Mun, Y., Choo, H. (eds.) ICCSA 2006. LNCS, vol. 3980, pp. 212–221. Springer, Heidelberg (2006)
15. Andrew Davies, R.: James S. Maskery, and Nigel W. John. Chemical education using feelable molecules. In: Proceedings of the 14th International Conference on 3D Web Technology, Web3D 2009, pp. 7–14. ACM, New York (2009)

Author Index